THOMPSON & THOMPSON
GENETICS IN MEDICINE

THOMPSON & THOMPSON
GENETICS IN MEDICINE

EIGHTH EDITION

Robert L. Nussbaum, MD, FACP, FACMG
Holly Smith Chair of Medicine and Science
Professor of Medicine, Neurology, Pediatrics and Pathology
Department of Medicine and Institute for Human Genetics
University of California San Francisco
San Francisco, California

Roderick R. McInnes, CM, MD, PhD, FRS(C), FCAHS, FCCMG
Alva Chair in Human Genetics
Canada Research Chair in Neurogenetics
Professor of Human Genetics and Biochemistry
Director, Lady Davis Institute
Jewish General Hospital
McGill University
Montreal, Quebec, Canada

Huntington F. Willard, PhD
President and Director
The Marine Biological Laboratory
Woods Hole, Massachusetts
and
Professor of Human Genetics
University of Chicago
Chicago, Illinois

With Clinical Case Studies updated by:

Ada Hamosh, MD, MPH
Professor of Pediatrics
McKusick-Nathans Institute of Genetic Medicine
Scientific Director, OMIM
Johns Hopkins University School of Medicine
Baltimore, Maryland

ELSEVIER

ELSEVIER

1600 John F. Kennedy Blvd.
Ste 1800
Philadelphia, PA 19103-2899

THOMPSON & THOMPSON GENETICS IN MEDICINE,
EIGHTH EDITION

ISBN: 978-1-4377-0696-3

Notices

Knowledge and best practice in this field are constantly changing. As new research and experience broaden our understanding, changes in research methods, professional practices, or medical treatment may become necessary.

Practitioners and researchers must always rely on their own experience and knowledge in evaluating and using any information, methods, compounds, or experiments described herein. In using such information or methods they should be mindful of their own safety and the safety of others, including parties for whom they have a professional responsibility.

With respect to any drug or pharmaceutical products identified, readers are advised to check the most current information provided (i) on procedures featured or (ii) by the manufacturer of each product to be administered, to verify the recommended dose or formula, the method and duration of administration, and contraindications. It is the responsibility of practitioners, relying on their own experience and knowledge of their patients, to make diagnoses, to determine dosages and the best treatment for each individual patient, and to take all appropriate safety precautions. To the fullest extent of the law, neither the Publisher nor the authors, contributors, or editors, assume any liability for any injury and/or damage to persons or property as a matter of products liability, negligence or otherwise, or from any use or operation of any methods, products, instructions, or ideas contained in the material herein.

Previous editions copyrighted 2007, 2004, 2001, 1991, 1986, 1980, 1973, 1966.

Library of Congress Cataloging-in-Publication Data

Nussbaum, Robert L., 1950- , author.
  Thompson & Thompson genetics in medicine / Robert L. Nussbaum, Roderick R. McInnes,
Huntington F. Willard.—Eighth edition.
      p. ; cm.
  Genetics in medicine
  Thompson and Thompson genetics in medicine
  Includes bibliographical references and index.
  ISBN 978-1-4377-0696-3 (alk. paper)
  I. McInnes, Roderick R., author. II. Willard, Huntington F., author. III. Title.
IV. Title: Genetics in medicine. V. Title: Thompson and Thompson genetics in medicine.
  [DNLM: 1. Genetics, Medical. QZ 50]
  RB155
  616'.042—dc23
                                                                              2015009828

Content Strategist: Meghan Ziegler
Senior Content Development Specialist: Joan Ryan
Publishing Services Manager: Jeff Patterson
Senior Project Manager: Mary Pohlman
Design Direction: Xiaopei Chen

Printed in Canada.

Last digit is the print number: 9 8 7 6 5 4 3

Preface

In their preface to the first edition of *Genetics in Medicine,* published nearly 50 years ago, James and Margaret Thompson wrote:

Genetics is fundamental to the basic sciences of preclinical medical education and has important applications to clinical medicine, public health and medical research. ... This book has been written to introduce the medical student to the principles of genetics as they apply to medicine, and to give him (her) a background for his own reading of the extensive and rapidly growing literature in the field. If his (her) senior colleagues also find it useful, we shall be doubly satisfied.

What was true then is even more so now as our knowledge of genetics and of the human genome is rapidly becoming an integral part of public health and the practice of medicine. This new edition of *Genetics in Medicine,* the eighth, seeks to fulfill the goals of the previous seven by providing an accurate exposition of the fundamental principles of human and medical genetics and genomics. Using illustrative examples drawn from medicine, we continue to emphasize the genes and mechanisms operating in human diseases.

Much has changed, however, since the last edition of this book. The rapid pace of progress stemming from the Human Genome Project provides us with a refined catalogue of all human genes, their sequence, and an extensive, and still growing, database of human variation around the globe and its relationship to disease. Genomic information has stimulated the creation of powerful new tools that are changing human genetics research and medical genetics practice. Throughout, we have continued to expand the scope of the book to incorporate the concepts of personalized health care

and precision medicine into *Genetics in Medicine* by providing more examples of how genomics is being used to identify the contributions made by genetic variation to disease susceptibility and treatment outcomes.

The book is not intended to be a compendium of genetic diseases nor is it an encyclopedic treatise on human genetics and genomics in general. Rather, the authors hope that the eighth edition of *Genetics in Medicine* will provide students with a framework for understanding the field of medical genetics and genomics while giving them a basis on which to establish a program of continuing education in this area. The Clinical Cases—first introduced in the sixth edition to demonstrate and reinforce general principles of disease inheritance, pathogenesis, diagnosis, management, and counseling—continue to be an important feature of the book. We have expanded the set of cases to add more common complex disorders to the set of cases. To enhance further the teaching value of the Clinical Cases, we continue to provide a case number (highlighted in green) throughout the text to direct readers to the case in the Clinical Case Studies section that is relevant to the concepts being discussed at that point in the text.

Any medical or genetic counseling student, advanced undergraduate, graduate student in genetics or genomics, resident in any field of clinical medicine, practicing physician, or allied medical professional in nursing or physical therapy should find this book to be a thorough but not exhaustive (or exhausting!) presentation of the fundamentals of human genetics and genomics as applied to health and disease.

Robert L. Nussbaum, MD
Roderick R. McInnes, MD, PhD
Huntington F. Willard, PhD

Acknowledgments

The authors wish to express their appreciation and gratitude to their many colleagues who, through their ideas, suggestions, and criticisms, improved the eighth edition of *Genetics in Medicine*. In particular, we are grateful to Anthony Wynshaw-Boris for sharing his knowledge and experience in molecular dysmorphology and developmental genetics in the writing of Chapter 14 and to Ada Hamosh for her continuing dedication to and stewardship of the Clinical Case Studies.

We also thank Mark Blostein, Isabelle Carrier, Eduardo Diez, Voula Giannopoulos, Kostas Pantopoulos, and Prem Ponka of the Lady Davis Institute, McGill University; Katie Bungartz; Peter Byers of the University of Washington; Philippe Campeau of the Ste Justine University Hospital Research Center; Ronald Cohn, Chris Pearson, Peter Ray, Johanna Rommens, and Stephen Scherer of the Hospital for Sick Children, Toronto; Gary Cutting and Ada Hamosh of Johns Hopkins School of Medicine; Beverly Davidson of the Children's Hospital of Philadelphia; Harold C. Dietz of the Howard Hughes Medical Institute and Johns Hopkins School of Medicine; Evan Eichler of the Howard Hughes Medical Institute and the University of Washington; Geoffrey Ginsburg of Duke University Medical Center; Douglas R. Higgs and William G. Wood of the Weatherall Institute of Molecular Medicine, Oxford University; Katherine A. High of the Howard Hughes Medical Institute and the Children's Hospital of Philadelphia; Ruth Macpherson of the University of Ottawa Heart Institute; Mary Norton at the University of California San Francisco; Crista Lese Martin of the Geisinger Health System; M. Katharine Rudd and Lora Bean of Emory University School of Medicine; Eric Shoubridge of McGill University; Peter St. George-Hyslop of the University of Toronto and the Cambridge Institute for Medical Research; Paula Waters of the University of British Columbia; Robin Williamson; Daynna Wolff of the Medical University of South Carolina; and Huda Zoghbi of the Howard Hughes Medical Institute and Baylor College of Medicine.

We extend deep thanks to our ever persistent, determined, and supportive editors at Elsevier, Joan Ryan, Mary Pohlman, and Meghan Ziegler. Most importantly, we once again thank our families for their patience and understanding for the many hours we spent creating this, the eighth edition of *Genetics in Medicine*.

And, lastly and most profoundly, we express our deepest gratitude to Dr. Margaret Thompson for providing us the opportunity to carry on the textbook she created nearly 50 years ago with her late husband, James S. Thompson. Peggy passed away at the age of 94 shortly after we completed this latest revision of her book. The book, known widely and simply as "Thompson and Thompson", lives on as a legacy to their careers and to their passion for genetics in medicine.

Contents

CHAPTER 19
Ethical and Social Issues
in Genetics and Genomics *383*

CASES
Clinical Case Studies
Illustrating Genetic Principles *391*

CHAPTER 1

Introduction

THE BIRTH AND DEVELOPMENT OF GENETICS AND GENOMICS

Few areas of science and medicine are seeing advances at the pace we are experiencing in the related fields of genetics and genomics. It may appear surprising to many students today, then, to learn that an appreciation of the role of genetics in medicine dates back well over a century, to the recognition by the British physician Archibald Garrod and others that Mendel's laws of inheritance could explain the recurrence of certain clinical disorders in families. During the ensuing years, with developments in cellular and molecular biology, the field of **medical genetics** grew from a small clinical subspecialty concerned with a few rare hereditary disorders to a recognized medical specialty whose concepts and approaches are important components of the diagnosis and management of many disorders, both common and rare.

At the beginning of the 21st century, the **Human Genome Project** provided a virtually complete sequence of human DNA—our **genome** (the suffix -*ome* coming from the Greek for "all" or "complete")—which now serves as the foundation of efforts to catalogue all human genes, understand their structure and regulation, determine the extent of variation in these genes in different populations, and uncover how genetic variation contributes to disease. The human genome of any individual can now be studied in its entirety, rather than one gene at a time. These developments are making possible the field of **genomic medicine**, which seeks to apply a large-scale analysis of the human genome and its products, including the control of gene expression, human gene variation, and interactions between genes and the environment, to medical care.

GENETICS AND GENOMICS IN MEDICINE
The Practice of Genetics

The medical geneticist is usually a physician who works as part of a team of health care providers, including many other physicians, nurses, and genetic counselors, to evaluate patients for possible hereditary diseases. They characterize the patient's illness through careful history taking and physical examination, assess possible modes of inheritance, arrange for diagnostic testing, develop treatment and surveillance plans, and participate in outreach to other family members at risk for the disorder.

However, genetic principles and approaches are not restricted to any one medical specialty or subspecialty; they permeate many, and perhaps all, areas of medicine. Here are just a few examples of how genetics and genomics are applied to medicine today:

- A pediatrician evaluates a child with multiple congenital malformations and orders a high-resolution genomic test for submicroscopic chromosomal deletions or duplications that are below the level of resolution of routine chromosome analysis (Case 32).
- A genetic counselor specializing in hereditary breast cancer offers education, testing, interpretation, and support to a young woman with a family history of hereditary breast and ovarian cancer (Case 7).
- An obstetrician sends a chorionic villus sample taken from a 38-year-old pregnant woman to a cytogenetics laboratory for confirmation of abnormalities in the number or structure of the fetal chromosomes, following a positive screening result from a noninvasive prenatal blood test (see Chapter 17).
- A hematologist combines family and medical history with gene testing of a young adult with deep venous thrombosis to assess the benefits and risks of initiating and maintaining anticoagulant therapy (Case 46).
- A surgeon uses gene expression array analysis of a lung tumor sample to determine prognosis and to guide therapeutic decision making (see Chapter 15).
- A pediatric oncologist tests her patients for genetic variations that can predict a good response or an adverse reaction to a chemotherapeutic agent (Case 45).
- A neurologist and genetic counselor provide *APOE* gene testing for Alzheimer disease susceptibility for a woman with a strong family history of the disease so she can make appropriate long-term financial plans (Case 4).
- A forensic pathologist uses databases of genetic polymorphisms in his analysis of DNA samples obtained from victims' personal items and surviving relatives to identify remains from an airline crash.
- A gastroenterologist orders genome sequence analysis for a child with a multiyear history of life-threatening and intractable inflammatory bowel disease. Sequencing reveals a mutation in a previously unsuspected

gene, clarifying the clinical diagnosis and altering treatment for the patient (see Chapter 16).

• Scientists in the pharmaceutical industry sequence cancer cell DNA to identify specific changes in oncogenic signaling pathways inappropriately activated by a somatic mutation, leading to the development of specific inhibitors that reliably induce remissions of the cancers in patients (Case 10).

Categories of Genetic Disease

Virtually any disease is the result of the combined action of genes and environment, but the relative role of the genetic component may be large or small. Among disorders caused wholly or partly by genetic factors, three main types are recognized: chromosome disorders, single-gene disorders, and multifactorial disorders.

In **chromosome disorders**, the defect is due not to a single mistake in the genetic blueprint but to an excess or a deficiency of the genes located on entire chromosomes or chromosome segments. For example, the presence of an extra copy of one chromosome, chromosome 21, underlies a specific disorder, Down syndrome, even though no individual gene on that chromosome is abnormal. Duplication or deletion of smaller segments of chromosomes, ranging in size from only a single gene up to a few percent of a chromosome's length, can cause complex birth defects like DiGeorge syndrome or even isolated autism without any obvious physical abnormalities. As a group, chromosome disorders are common, affecting approximately 7 per 1000 liveborn infants and accounting for approximately half of all spontaneous abortions occurring in the first trimester of pregnancy. These types of disorders are discussed in Chapter 6.

Single-gene defects are caused by pathogenic mutations in individual genes. The mutation may be present on both chromosomes of a pair (one of paternal origin and one of maternal origin) or on only one chromosome of a pair (matched with a normal copy of that gene on the other copy of that chromosome). Single-gene defects often cause diseases that follow one of the classic inheritance patterns in families (autosomal recessive, autosomal dominant, or X-linked). In a few cases, the mutation is in the mitochondrial rather than in the nuclear genome. In any case, the cause is a critical error in the genetic information carried by a single gene. Single-gene disorders such as cystic fibrosis (Case 12), sickle cell anemia (Case 42), and Marfan syndrome (Case 30) usually exhibit obvious and characteristic pedigree patterns. Most such defects are rare, with a frequency that may be as high as 1 in 500 to 1000 individuals but is usually much less. Although individually rare, single-gene disorders as a group are responsible for a significant proportion of disease and death. Overall, the incidence of serious single-gene disorders in the pediatric population has been estimated to be approximately 1 per 300 liveborn infants; over an entire lifetime, the prevalence of single-gene disorders is 1 in 50. These disorders are discussed in Chapter 7.

Multifactorial disease with complex inheritance describes the majority of diseases in which there is a genetic contribution, as evidenced by increased risk for disease (compared to the general public) in identical twins or close relatives of affected individuals, and yet the family history does not fit the inheritance patterns seen typically in single-gene defects. Multifactorial diseases include congenital malformations such as Hirschsprung disease (Case 22), cleft lip and palate, and congenital heart defects, as well as many common disorders of adult life, such as Alzheimer disease (Case 4), diabetes, and coronary artery disease. There appears to be no single error in the genetic information in many of these conditions. Rather, the disease is the result of the combined impact of variant forms of many different genes; each variant may cause, protect from, or predispose to a serious defect, often in concert with or triggered by environmental factors. Estimates of the impact of multifactorial disease range from 5% in the pediatric population to more than 60% in the entire population. These disorders are the subject of Chapter 8.

ONWARD

During the 50-year professional life of today's professional and graduate students, extensive changes are likely to take place in the discovery, development, and use of genetic and genomic knowledge and tools in medicine. Judging from the quickening pace of discovery within only the past decade, it is virtually certain that we are just at the beginning of a revolution in integrating knowledge of genetics and the genome into public health and the practice of medicine. An introduction to the language and concepts of human and medical genetics and an appreciation of the genetic and genomic perspective on health and disease will form a framework for lifelong learning that is part of every health professional's career.

GENERAL REFERENCES

Feero WG, Guttmacher AE, Collins FS: Genomic medicine—an updated primer, *N Engl J Med* 362:2001–2011, 2010.
Ginsburg G, Willard HF, editors: *Genomic and personalized medicine* (vols 1 & 2), ed 2, New York, 2012, Elsevier.

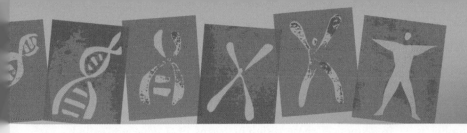

CHAPTER 2

Introduction to the Human Genome

Understanding the organization, variation, and transmission of the **human genome** is central to appreciating the role of genetics in medicine, as well as the emerging principles of genomic and personalized medicine. With the availability of the sequence of the human genome and a growing awareness of the role of genome variation in disease, it is now possible to begin to exploit the impact of that variation on human health on a broad scale. The comparison of individual genomes underscores the first major take-home lesson of this book—*every individual has his or her own unique constitution of gene products, produced in response to the combined inputs of the genome sequence and one's particular set of environmental exposures and experiences.* As pointed out in the previous chapter, this realization reflects what Garrod termed *chemical individuality* over a century ago and provides a conceptual foundation for the practice of genomic and personalized medicine.

Advances in genome technology and the resulting explosion in knowledge and information stemming from the **Human Genome Project** are thus playing an increasingly transformational role in integrating and applying concepts and discoveries in genetics to the practice of medicine.

THE HUMAN GENOME AND THE CHROMOSOMAL BASIS OF HEREDITY

Appreciation of the importance of genetics to medicine requires an understanding of the nature of the hereditary material, how it is packaged into the human genome, and how it is transmitted from cell to cell during cell division and from generation to generation during reproduction. The human genome consists of large amounts of the chemical deoxyribonucleic acid (**DNA**) that contains within its structure the genetic information needed to specify all aspects of embryogenesis, development, growth, metabolism, and reproduction—essentially all aspects of what makes a human being a functional organism. Every nucleated cell in the body carries its own copy of the human genome, which contains, depending on how one defines the term, approximately 20,000 to 50,000 **genes** (see Box later). Genes,

CHROMOSOME AND GENOME ANALYSIS IN CLINICAL MEDICINE

Chromosome and genome analysis has become an important diagnostic procedure in clinical medicine. As described more fully in subsequent chapters, these applications include the following:
- *Clinical diagnosis.* Numerous medical conditions, including some that are common, are associated with changes in chromosome number or structure and require chromosome or genome analysis for diagnosis and genetic counseling (see Chapters 5 and 6).
- *Gene identification.* A major goal of medical genetics and genomics today is the identification of specific genes and elucidating their roles in health and disease. This topic is referred to repeatedly but is discussed in detail in Chapter 10.
- *Cancer genomics.* Genomic and chromosomal changes in somatic cells are involved in the initiation and progression of many types of cancer (see Chapter 15).
- *Disease treatment.* Evaluation of the integrity, composition, and differentiation state of the genome is critical for the development of patient-specific pluripotent stem cells for therapeutic use (see Chapter 13).
- *Prenatal diagnosis.* Chromosome and genome analysis is an essential procedure in prenatal diagnosis (see Chapter 17).

which at this point we consider simply and most broadly as functional units of genetic information, are encoded in the DNA of the genome, organized into a number of rod-shaped organelles called **chromosomes** in the nucleus of each cell. The influence of genes and genetics on states of health and disease is profound, and its roots are found in the information encoded in the DNA that makes up the human genome.

Each species has a characteristic chromosome complement (**karyotype**) in terms of the number, morphology, and content of the chromosomes that make up its genome. The genes are in linear order along the chromosomes, each gene having a precise position or **locus**. A **gene map** is the map of the genomic location of the genes and is characteristic of each species and the individuals within a species.

The study of chromosomes, their structure, and their inheritance is called **cytogenetics**. The science of human cytogenetics dates from 1956, when it was first established that the normal human chromosome number is 46. Since that time, much has been learned about human chromosomes, their normal structure and composition, and the identity of the genes that they contain, as well as their numerous and varied abnormalities.

With the exception of cells that develop into gametes (the **germline**), all cells that contribute to one's body are called **somatic cells** (*soma*, body). The genome contained in the nucleus of human somatic cells consists of 46 chromosomes, made up of 24 different types and arranged in 23 pairs (Fig. 2-1). Of those 23 pairs, 22 are alike in males and females and are called **autosomes**, originally numbered in order of their apparent size from the largest to the smallest. The remaining pair comprises the two different types of **sex chromosomes:** an X and a Y chromosome in males and two X chromosomes in females. Central to the concept of the human genome, each chromosome carries a different subset of genes that are arranged linearly along its DNA. Members of a pair of chromosomes (referred to as **homologous chromosomes** or **homologues**) carry matching genetic information; that is, they typically have the same genes in the same order. At any specific locus, however, the homologues either may be identical or may vary slightly in sequence; these different forms of a gene are called **alleles**. One member of each pair of chromosomes is inherited from the father, the other from the mother. Normally, the members of a pair of autosomes are microscopically indistinguishable from each other. In females, the sex chromosomes, the two **X chromosomes,** are likewise largely indistinguishable. In males, however, the sex chromosomes differ. One is an X, identical to the Xs of the female, inherited by a male from his mother and transmitted to his daughters; the other, the **Y chromosome,** is inherited from his father and transmitted to his sons. In Chapter 6, as we explore the chromosomal and genomic basis of disease, we will look at some exceptions to the simple and almost universal rule that human females are XX and human males are XY.

In addition to the nuclear genome, a small but important part of the human genome resides in mitochondria in the cytoplasm (see Fig. 2-1). The mitochondrial chromosome, to be described later in this chapter, has a number of unusual features that distinguish it from the rest of the human genome.

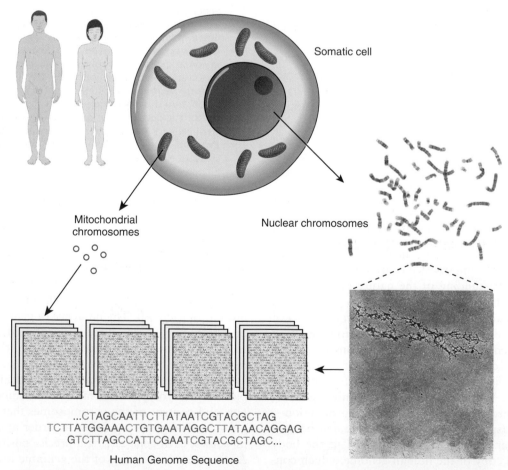

Somatic cell

Mitochondrial chromosomes

Nuclear chromosomes

...CTAGCAATTCTTATAATCGTACGCTAG
TCTTATGGAAACTGTGAATAGGCTTATAACAGGAG
GTCTTAGCCATTCGAATCGTACGCTAGC...

Human Genome Sequence

Figure 2-1 The human genome, encoded on both nuclear and mitochondrial chromosomes. *See Sources & Acknowledgments.*

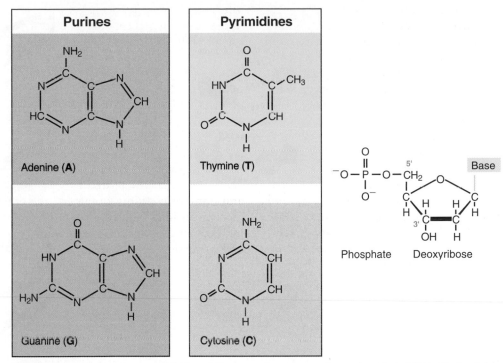

Figure 2-2 The four bases of DNA and the general structure of a nucleotide in DNA. Each of the four bases bonds with deoxyribose (through the nitrogen shown in *magenta*) and a phosphate group to form the corresponding nucleotides.

GENES IN THE HUMAN GENOME

What is a gene? And how many genes do we have? These questions are more difficult to answer than it might seem.

The word *gene*, first introduced in 1908, has been used in many different contexts since the essential features of heritable "unit characters" were first outlined by Mendel over 150 years ago. To physicians (and indeed to Mendel and other early geneticists), a gene can be defined by its observable impact on an organism and on its statistically determined transmission from generation to generation. To medical geneticists, a gene is recognized clinically in the context of an observable variant that leads to a characteristic clinical disorder, and today we recognize approximately 5000 such conditions (see Chapter 7).

The Human Genome Project provided a more systematic basis for delineating human genes, relying on DNA sequence analysis rather than clinical acumen and family studies alone; indeed, this was one of the most compelling rationales for initiating the project in the late 1980s. However, even with the finished sequence product in 2003, it was apparent that our ability to recognize features of the sequence that point to the existence or identity of a gene was sorely lacking. Interpreting the human genome sequence and relating its variation to human biology in both health and disease is thus an ongoing challenge for biomedical research.

Although the ultimate catalogue of human genes remains an elusive target, we recognize two general types of gene, those whose product is a protein and those whose product is a functional RNA.

- The number of **protein-coding genes**—recognized by features in the genome that will be discussed in Chapter 3—is estimated to be somewhere between 20,000 and 25,000. In this book, we typically use approximately 20,000 as the number, and the reader should recognize that this is both imprecise and perhaps an underestimate.

- In addition, however, it has been clear for several decades that the ultimate product of some genes is not a protein at all but rather an RNA transcribed from the DNA sequence. There are many different types of such RNA genes (typically called **noncoding genes** to distinguish them from protein-coding genes), and it is currently estimated that there are at least another 20,000 to 25,000 noncoding RNA genes around the human genome.

Thus overall—and depending on what one means by the term—the total number of genes in the human genome is of the order of approximately 20,000 to 50,000. However, the reader will appreciate that this remains a moving target, subject to evolving definitions, increases in technological capabilities and analytical precision, advances in informatics and digital medicine, and more complete genome annotation.

DNA Structure: A Brief Review

Before the organization of the human genome and its chromosomes is considered in detail, it is necessary to review the nature of the DNA that makes up the genome. DNA is a polymeric nucleic acid macromolecule composed of three types of units: a five-carbon sugar, deoxyribose; a nitrogen-containing base; and a phosphate group (Fig. 2-2). The bases are of two types, **purines** and **pyrimidines**. In DNA, there are two purine bases, **adenine** (A) and **guanine** (G), and two pyrimidine

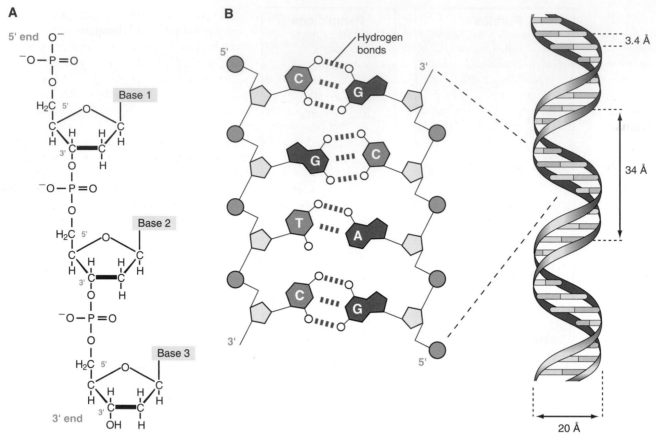

Figure 2-3 **The structure of DNA. A,** A portion of a DNA polynucleotide chain, showing the 3′-5′ phosphodiester bonds that link adjacent nucleotides. **B,** The double-helix model of DNA, as proposed by Watson and Crick. The horizontal "rungs" represent the paired bases. The helix is said to be right-handed because the strand going from lower left to upper right crosses over the opposite strand. The detailed portion of the figure illustrates the two complementary strands of DNA, showing the AT and GC base pairs. Note that the orientation of the two strands is antiparallel. *See Sources & Acknowledgments.*

bases, **thymine** (T) and **cytosine** (C). Nucleotides, each composed of a base, a phosphate, and a sugar moiety, polymerize into long polynucleotide chains held together by 5′-3′ phosphodiester bonds formed between adjacent deoxyribose units (Fig. 2-3A). In the human genome, these polynucleotide chains exist in the form of a double helix (Fig. 2-3B) that can be hundreds of millions of nucleotides long in the case of the largest human chromosomes.

The anatomical structure of DNA carries the chemical information that allows the exact transmission of genetic information from one cell to its daughter cells and from one generation to the next. At the same time, the primary structure of DNA specifies the amino acid sequences of the polypeptide chains of proteins, as described in the next chapter. DNA has elegant features that give it these properties. The native state of DNA, as elucidated by James Watson and Francis Crick in 1953, is a double helix (see Fig. 2-3B). The helical structure resembles a right-handed spiral staircase in which its two polynucleotide chains run in opposite directions, held together by hydrogen bonds between pairs of bases: T of one chain paired with A of the other, and G with

C. The specific nature of the genetic information encoded in the human genome lies in the sequence of C's, A's, G's, and T's on the two strands of the double helix along each of the chromosomes, both in the nucleus and in mitochondria (see Fig. 2-1). Because of the complementary nature of the two strands of DNA, knowledge of the sequence of nucleotide bases on one strand automatically allows one to determine the sequence of bases on the other strand. The double-stranded structure of DNA molecules allows them to replicate precisely by separation of the two strands, followed by synthesis of two new complementary strands, in accordance with the sequence of the original template strands (Fig. 2-4). Similarly, when necessary, the base complementarity allows efficient and correct repair of damaged DNA molecules.

Structure of Human Chromosomes

The composition of genes in the human genome, as well as the determinants of their expression, is specified in the DNA of the 46 human chromosomes in the nucleus plus the mitochondrial chromosome. *Each human chromosome consists of a single, continuous DNA double helix;* that is, each chromosome is one long, double-stranded DNA molecule, and the nuclear genome consists, therefore, of 46 linear DNA molecules, totaling more than 6 billion nucleotide pairs (see Fig. 2-1).

Chromosomes are not naked DNA double helices, however. Within each cell, the genome is packaged as **chromatin,** in which genomic DNA is complexed with

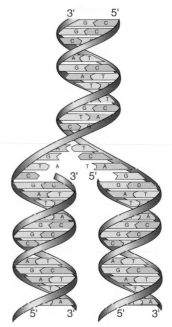

Figure 2-4 Replication of a DNA double helix, resulting in two identical daughter molecules, each composed of one parental strand and one newly synthesized strand.

several classes of specialized proteins. Except during cell division, chromatin is distributed throughout the nucleus and is relatively homogeneous in appearance under the microscope. When a cell divides, however, its genome condenses to appear as microscopically visible chromosomes. Chromosomes are thus visible as discrete structures only in dividing cells, although they retain their integrity between cell divisions.

The DNA molecule of a chromosome exists in chromatin as a complex with a family of basic chromosomal proteins called *histones*. This fundamental unit interacts with a heterogeneous group of nonhistone proteins, which are involved in establishing a proper spatial and functional environment to ensure normal chromosome behavior and appropriate gene expression.

Five major types of histones play a critical role in the proper packaging of chromatin. Two copies each of the four core histones H2A, H2B, H3, and H4 constitute an octamer, around which a segment of DNA double helix winds, like thread around a spool (Fig. 2-5). Approximately 140 base pairs (bp) of DNA are associated with each histone core, making just under two turns around the octamer. After a short (20- to 60-bp) "spacer" segment of DNA, the next core DNA complex forms, and so on, giving chromatin the appearance of beads on a string. Each complex of DNA with core histones is called a **nucleosome** (see Fig. 2-5), which is the basic structural unit of chromatin, and each of the 46 human chromosomes contains several hundred thousand to well over a million nucleosomes. A fifth histone, H1, appears to bind to DNA at the edge of each nucleosome, in the internucleosomal spacer region. The amount of

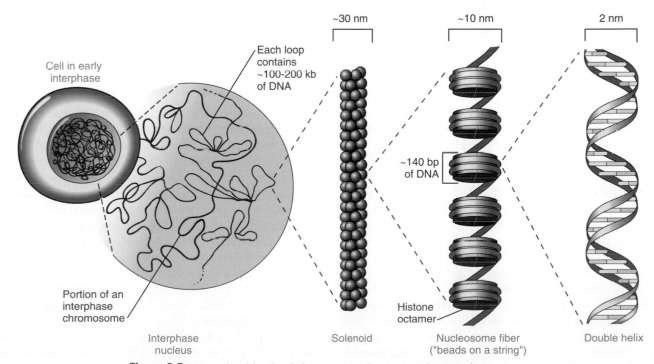

Figure 2-5 Hierarchical levels of chromatin packaging in a human chromosome.

DNA associated with a core nucleosome, together with the spacer region, is approximately 200 bp.

In addition to the major histone types, a number of specialized histones can substitute for H3 or H2A and confer specific characteristics on the genomic DNA at that location. Histones can also be modified by chemical changes, and these modifications can change the properties of nucleosomes that contain them. As discussed further in Chapter 3, the pattern of major and specialized histone types and their modifications can vary from cell type to cell type and is thought to specify how DNA is packaged and how accessible it is to regulatory molecules that determine gene expression or other genome functions.

During the cell cycle, as we will see later in this chapter, chromosomes pass through orderly stages of condensation and decondensation. However, even when chromosomes are in their most decondensed state, in a stage of the cell cycle called **interphase,** DNA packaged in chromatin is substantially more condensed than it would be as a native, protein-free, double helix. Further, the long strings of nucleosomes are themselves compacted into a secondary helical structure, a cylindrical "solenoid" fiber (from the Greek *solenoeides*, pipe-shaped) that appears to be the fundamental unit of chromatin organization (see Fig. 2-5). The solenoids are themselves packed into **loops** or domains attached at intervals of approximately 100,000 bp (equivalent to 100 kilobase pairs [kb], because 1 kb = 1000 bp) to a protein **scaffold** within the nucleus. It has been speculated that these loops are the functional units of the genome and that the attachment points of each loop are specified along the chromosomal DNA. As we shall see, one level of control of gene expression depends on how DNA and genes are packaged into chromosomes and on their association with chromatin proteins in the packaging process.

The enormous amount of genomic DNA packaged into a chromosome can be appreciated when chromosomes are treated to release the DNA from the underlying protein scaffold (see Fig. 2-1). When DNA is released in this manner, long loops of DNA can be visualized, and the residual scaffolding can be seen to reproduce the outline of a typical chromosome.

The Mitochondrial Chromosome

As mentioned earlier, a small but important subset of genes encoded in the human genome resides in the cytoplasm in the mitochondria (see Fig. 2-1). Mitochondrial genes exhibit exclusively maternal inheritance (see Chapter 7). Human cells can have hundreds to thousands of mitochondria, each containing a number of copies of a small circular molecule, the mitochondrial chromosome. The mitochondrial DNA molecule is only 16 kb in length (just a tiny fraction of the length of even the smallest nuclear chromosome) and encodes only 37 genes. The products of these genes function in

mitochondria, although the vast majority of proteins within the mitochondria are, in fact, the products of nuclear genes. Mutations in mitochondrial genes have been demonstrated in several maternally inherited as well as sporadic disorders (Case 33) (see Chapters 7 and 12).

The Human Genome Sequence

With a general understanding of the structure and clinical importance of chromosomes and the genes they carry, scientists turned attention to the identification of specific genes and their location in the human genome. From this broad effort emerged the **Human Genome Project,** an international consortium of hundreds of laboratories around the world, formed to determine and assemble the sequence of the 3.3 billion base pairs of DNA located among the 24 types of human chromosome.

Over the course of a decade and a half, powered by major developments in DNA-sequencing technology, large sequencing centers collaborated to assemble sequences of each chromosome. The genomes actually being sequenced came from several different individuals, and the consensus sequence that resulted at the conclusion of the Human Genome Project was reported in 2003 as a "reference" sequence assembly, to be used as a basis for later comparison with sequences of individual genomes. This reference sequence is maintained in publicly accessible databases to facilitate scientific discovery and its translation into useful advances for medicine. Genome sequences are typically presented in a 5′ to 3′ direction on just one of the two strands of the double helix, because—owing to the complementary nature of DNA structure described earlier—if one knows the sequence of one strand, one can infer the sequence of the other strand (Fig. 2-6).

Organization of the Human Genome

Chromosomes are not just a random collection of different types of genes and other DNA sequences. Regions of the genome with similar characteristics tend to be clustered together, and the functional organization of the genome reflects its structural organization and sequence. Some chromosome regions, or even whole chromosomes, are high in gene content ("gene rich"), whereas others are low ("gene poor") (Fig. 2-7). The clinical consequences of abnormalities of genome structure reflect the specific nature of the genes and sequences involved. Thus abnormalities of gene-rich chromosomes or chromosomal regions tend to be much more severe clinically than similar-sized defects involving gene-poor parts of the genome.

As a result of knowledge gained from the Human Genome Project, it is apparent that the organization of DNA in the human genome is both more varied and

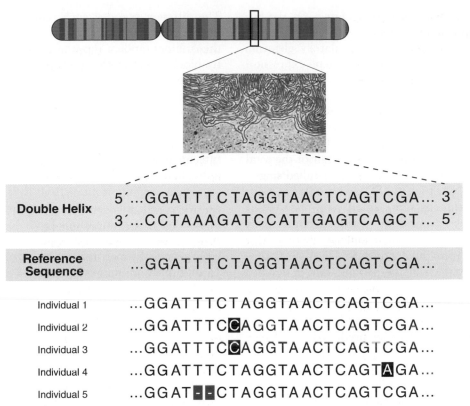

Figure 2-6 **A portion of the reference human genome sequence.** By convention, sequences are presented from one strand of DNA only, because the sequence of the complementary strand can be inferred from the double-stranded nature of DNA (shown above the reference sequence). The sequence of DNA from a group of individuals is similar but not identical to the reference, with single nucleotide changes in some individuals and a small deletion of two bases in another.

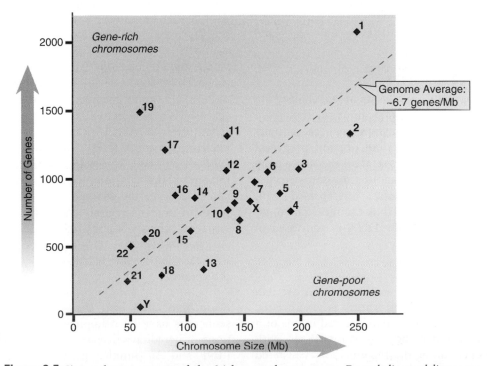

Figure 2-7 **Size and gene content of the 24 human chromosomes.** *Dotted diagonal line* corresponds to the average density of genes in the genome, approximately 6.7 protein-coding genes per megabase (Mb). Chromosomes that are relatively gene rich are above the diagonal and trend to the upper left. Chromosomes that are relatively gene poor are below the diagonal and trend to the lower right. *See Sources & Acknowledgments.*

more complex than was once appreciated. Of the billions of base pairs of DNA in any genome, less than 1.5% actually encodes proteins. Regulatory elements that influence or determine patterns of gene expression during development or in tissues were believed to account for only approximately 5% of additional sequence, although more recent analyses of chromatin characteristics suggest that a much higher proportion of the genome may provide signals that are relevant to genome functions. Only approximately half of the total linear length of the genome consists of so-called **single-copy** or **unique DNA**, that is, DNA whose linear order of specific nucleotides is represented only once (or at most a few times) around the entire genome. This concept may appear surprising to some, given that there are only four different nucleotides in DNA. But, consider even a tiny stretch of the genome that is only 10 bases long; with four types of bases, there are over a million possible sequences. And, although the order of bases in the genome is not entirely random, any particular 16-base sequence would be predicted by chance alone to appear only once in any given genome.

The rest of the genome consists of several classes of **repetitive DNA** and includes DNA whose nucleotide sequence is repeated, either perfectly or with some variation, hundreds to millions of times in the genome. Whereas most (but not all) of the estimated 20,000 protein-coding genes in the genome (see Box earlier in this chapter) are represented in single-copy DNA, sequences in the repetitive DNA fraction contribute to maintaining chromosome structure and are an important source of variation between different individuals; some of this variation can predispose to pathological events in the genome, as we will see in Chapters 5 and 6.

Single-Copy DNA Sequences

Although single-copy DNA makes up at least half of the DNA in the genome, much of its function remains a mystery because, as mentioned, sequences actually encoding proteins (i.e., the coding portion of genes) constitute only a small proportion of all the single-copy DNA. Most single-copy DNA is found in short stretches (several kilobase pairs or less), interspersed with members of various repetitive DNA families. The organization of genes in single-copy DNA is addressed in depth in Chapter 3.

Repetitive DNA Sequences

Several different categories of repetitive DNA are recognized. A useful distinguishing feature is whether the repeated sequences ("repeats") are clustered in one or a few locations or whether they are interspersed with single-copy sequences along the chromosome. Clustered repeated sequences constitute an estimated 10% to 15% of the genome and consist of arrays of various short repeats organized in tandem in a head-to-tail fashion.

The different types of such tandem repeats are collectively called **satellite DNAs,** so named because many of the original tandem repeat families could be separated by biochemical methods from the bulk of the genome as distinct ("satellite") fractions of DNA.

Tandem repeat families vary with regard to their location in the genome and the nature of sequences that make up the array. In general, such arrays can stretch several million base pairs or more in length and constitute up to several percent of the DNA content of an individual human chromosome. Some tandem repeat sequences are important as tools that are useful in clinical cytogenetic analysis (see Chapter 5). Long arrays of repeats based on repetitions (with some variation) of a short sequence such as a pentanucleotide are found in large genetically inert regions on chromosomes 1, 9, and 16 and make up more than half of the Y chromosome (see Chapter 6). Other tandem repeat families are based on somewhat longer basic repeats. For example, the α-satellite family of DNA is composed of tandem arrays of an approximately 171-bp unit, found at the **centromere** of each human chromosome, which is critical for attachment of chromosomes to microtubules of the spindle apparatus during cell division.

In addition to tandem repeat DNAs, another major class of repetitive DNA in the genome consists of related sequences that are dispersed throughout the genome rather than clustered in one or a few locations. Although many DNA families meet this general description, two in particular warrant discussion because together they make up a significant proportion of the genome and because they have been implicated in genetic diseases. Among the best-studied dispersed repetitive elements are those belonging to the so-called *Alu* **family**. The members of this family are approximately 300 bp in length and are related to each other although not identical in DNA sequence. In total, there are more than a million *Alu* family members in the genome, making up at least 10% of human DNA. A second major dispersed repetitive DNA family is called the long interspersed nuclear element (**LINE**, sometimes called L1) family. LINEs are up to 6 kb in length and are found in approximately 850,000 copies per genome, accounting for nearly 20% of the genome. Both of these families are plentiful in some regions of the genome but relatively sparse in others—regions rich in GC content tend to be enriched in *Alu* elements but depleted of LINE sequences, whereas the opposite is true of more AT-rich regions of the genome.

Repetitive DNA and Disease. Both *Alu* and LINE sequences have been implicated as the cause of mutations in hereditary disease. At least a few copies of the LINE and *Alu* families generate copies of themselves that can integrate elsewhere in the genome, occasionally causing insertional inactivation of a medically important gene. The frequency of such events causing genetic disease in humans is unknown, but they may account for as many as 1 in 500 mutations. In addition, aberrant recombination events between different LINE repeats or *Alu* repeats can also be a cause of mutation in some genetic diseases (see Chapter 12).

An important additional type of repetitive DNA found in many different locations around the genome includes sequences that are duplicated, often with extraordinarily high sequence conservation. Duplications involving substantial segments of a chromosome, called **segmental duplications**, can span hundreds of kilobase pairs and account for at least 5% of the genome. When the duplicated regions contain genes, genomic rearrangements involving the duplicated sequences can result in the deletion of the region (and the genes) between the copies and thus give rise to disease (see Chapters 5 and 6).

VARIATION IN THE HUMAN GENOME

With completion of the reference human genome sequence, much attention has turned to the discovery and cataloguing of variation in sequence among different individuals (including both healthy individuals and those with various diseases) and among different populations around the globe. As we will explore in much more detail in Chapter 4, there are many tens of millions of common sequence variants that are seen at significant frequency in one or more populations; any given individual carries at least 5 million of these sequence variants. In addition, there are countless very rare variants, many of which probably exist in only a single or a few individuals. In fact, given the number of individuals in our species, *essentially each and every base pair in the human genome is expected to vary in someone somewhere around the globe.* It is for this reason that the original human genome sequence is considered a "reference" sequence for our species, but one that is actually identical to no individual's genome.

Early estimates were that any two randomly selected individuals would have sequences that are 99.9% identical or, put another way, that an individual genome would carry two *different* versions (**alleles**) of the human genome sequence at some 3 to 5 million positions, with different bases (e.g., a T or a G) at the maternally and paternally inherited copies of that particular sequence position (see Fig. 2-6). Although many of these allelic differences involve simply one nucleotide, much of the variation consists of insertions or deletions of (usually) short sequence stretches, variation in the number of copies of repeated elements (including genes), or inversions in the order of sequences at a particular position (**locus**) in the genome (see Chapter 4).

The total amount of the genome involved in such variation is now known to be substantially more than originally estimated and approaches 0.5% between any two randomly selected individuals. As will be addressed in future chapters, any and all of these types of variation can influence biological function and thus must be accounted for in any attempt to understand the contribution of genetics to human health.

TRANSMISSION OF THE GENOME

The chromosomal basis of heredity lies in the copying of the genome and its transmission from a cell to its progeny during typical cell division and from one generation to the next during reproduction, when single copies of the genome from each parent come together in a new embryo.

To achieve these related but distinct forms of genome inheritance, there are two kinds of cell division, mitosis and meiosis. **Mitosis** is ordinary somatic cell division by which the body grows, differentiates, and effects tissue regeneration. Mitotic division normally results in two daughter cells, each with chromosomes and genes identical to those of the parent cell. There may be dozens or even hundreds of successive mitoses in a lineage of somatic cells. In contrast, **meiosis** occurs only in cells of the germline. Meiosis results in the formation of reproductive cells (**gametes**), each of which has only 23 chromosomes—one of each kind of autosome and either an X or a Y. Thus, whereas somatic cells have the **diploid** (*diploos*, double) or the 2n chromosome complement (i.e., 46 chromosomes), gametes have the **haploid** (*haploos*, single) or the n complement (i.e., 23 chromosomes). Abnormalities of chromosome number or structure, which are usually clinically significant, can arise either in somatic cells or in cells of the germline by errors in cell division.

The Cell Cycle

A human being begins life as a fertilized ovum (**zygote**), a diploid cell from which all the cells of the body (estimated to be approximately 100 trillion in number) are derived by a series of dozens or even hundreds of mitoses. Mitosis is obviously crucial for growth and differentiation, but it takes up only a small part of the life cycle of a cell. The period between two successive mitoses is called **interphase,** the state in which most of the life of a cell is spent.

Immediately after mitosis, the cell enters a phase, called G_1, in which there is no DNA synthesis (Fig. 2-8). Some cells pass through this stage in hours; others spend a long time, days or years, in G_1. In fact, some cell types, such as neurons and red blood cells, do not divide at all once they are fully differentiated; rather, they are permanently arrested in a distinct phase known as G_0 ("G zero"). Other cells, such as liver cells, may enter G_0 but, after organ damage, return to G_1 and continue through the cell cycle.

The cell cycle is governed by a series of **checkpoints** that determine the timing of each step in mitosis. In

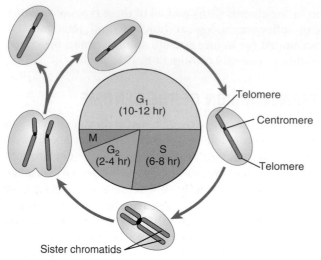

Figure 2-8 A typical mitotic cell cycle, described in the text. The telomeres, the centromere, and sister chromatids are indicated.

addition, checkpoints monitor and control the accuracy of DNA synthesis as well as the assembly and attachment of an elaborate network of microtubules that facilitate chromosome movement. If damage to the genome is detected, these mitotic checkpoints halt cell cycle progression until repairs are made or, if the damage is excessive, until the cell is instructed to die by programmed cell death (a process called **apoptosis**).

During G_1, each cell contains one diploid copy of the genome. As the process of cell division begins, the cell enters **S phase,** the stage of programmed DNA synthesis, ultimately leading to the precise replication of each chromosome's DNA. During this stage, each chromosome, which in G_1 has been a single DNA molecule, is duplicated and consists of two **sister chromatids** (see Fig. 2-8), each of which contains an identical copy of the original linear DNA double helix. The two sister chromatids are held together physically at the **centromere**, a region of DNA that associates with a number of specific proteins to form the **kinetochore**. This complex structure serves to attach each chromosome to the microtubules of the **mitotic spindle** and to govern chromosome movement during mitosis. DNA synthesis during S phase is not synchronous throughout all chromosomes or even within a single chromosome; rather, along each chromosome, it begins at hundreds to thousands of sites, called **origins of DNA replication.** Individual chromosome segments have their own characteristic time of replication during the 6- to 8-hour S phase. The ends of each chromosome (or chromatid) are marked by **telomeres,** which consist of specialized repetitive DNA sequences that ensure the integrity of the chromosome during cell division. Correct maintenance of the ends of chromosomes requires a special enzyme called **telomerase,** which ensures that the very ends of each chromosome are replicated.

The essential nature of these structural elements of chromosomes and their role in ensuring genome integrity is illustrated by a range of clinical conditions that result from defects in elements of the telomere or kinetochore or cell cycle machinery or from inaccurate replication of even small portions of the genome (see Box). Some of these conditions will be presented in greater detail in subsequent chapters.

CLINICAL CONSEQUENCES OF ABNORMALITIES AND VARIATION IN CHROMOSOME STRUCTURE AND MECHANICS

Medically relevant conditions arising from abnormal structure or function of chromosomal elements during cell division include the following:
- A broad spectrum of congenital abnormalities in children with inherited defects in genes encoding key components of the mitotic spindle checkpoint at the kinetochore
- A range of **birth defects and developmental disorders** due to anomalous segregation of chromosomes with multiple or missing centromeres (see Chapter 6)
- A variety of cancers associated with overreplication (amplification) or altered timing of replication of specific regions of the genome in S phase (see Chapter 15)
- **Roberts syndrome** of growth retardation, limb shortening, and microcephaly in children with abnormalities of a gene required for proper sister chromatid alignment and cohesion in S phase
- **Premature ovarian failure** as a major cause of female infertility due to mutation in a meiosis-specific gene required for correct sister chromatid cohesion
- The so-called **telomere syndromes,** a number of degenerative disorders presenting from childhood to adulthood in patients with abnormal telomere shortening due to defects in components of telomerase
- And, at the other end of the spectrum, common gene variants that correlate with the number of copies of the repeats at telomeres and with life expectancy and **longevity**

By the end of S phase, the DNA content of the cell has doubled, and each cell now contains two copies of the diploid genome. After S phase, the cell enters a brief stage called G_2. Throughout the whole cell cycle, the cell gradually enlarges, eventually doubling its total mass before the next mitosis. G_2 is ended by mitosis, which begins when individual chromosomes begin to condense and become visible under the microscope as thin, extended threads, a process that is considered in greater detail in the following section.

The G_1, S, and G_2 phases together constitute interphase. In typical dividing human cells, the three phases take a total of 16 to 24 hours, whereas mitosis lasts only 1 to 2 hours (see Fig. 2-8). There is great variation, however, in the length of the cell cycle, which ranges from a few hours in rapidly dividing cells, such as those of the dermis of the skin or the intestinal mucosa, to months in other cell types.

Mitosis

During the mitotic phase of the cell cycle, an elaborate apparatus ensures that each of the two daughter cells receives a complete set of genetic information. This result is achieved by a mechanism that distributes one chromatid of each chromosome to each daughter cell (Fig. 2-9). The process of distributing a copy of each chromosome to each daughter cell is called **chromosome segregation**. The importance of this process for normal cell growth is illustrated by the observation that many tumors are invariably characterized by a state of genetic imbalance resulting from mitotic errors in the distribution of chromosomes to daughter cells.

The process of mitosis is continuous, but five stages, illustrated in Figure 2-9, are distinguished: prophase, prometaphase, metaphase, anaphase, and telophase.

- *Prophase.* This stage is marked by gradual condensation of the chromosomes, formation of the mitotic spindle, and formation of a pair of **centrosomes,** from which microtubules radiate and eventually take up positions at the poles of the cell.
- *Prometaphase.* Here, the nuclear membrane dissolves, allowing the chromosomes to disperse within the cell and to attach, by their kinetochores, to microtubules of the mitotic spindle.
- *Metaphase.* At this stage, the chromosomes are maximally condensed and line up at the equatorial plane of the cell.
- *Anaphase.* The chromosomes separate at the centromere, and the sister chromatids of each chromosome now become independent **daughter chromosomes,** which move to opposite poles of the cell.
- *Telophase.* Now, the chromosomes begin to decondense from their highly contracted state, and a nuclear membrane begins to re-form around each of the two daughter nuclei, which resume their interphase appearance. To complete the process of cell division, the cytoplasm cleaves by a process known as **cytokinesis.**

There is an important difference between a cell entering mitosis and one that has just completed the process. A cell in G_2 has a fully replicated genome (i.e., a 4n complement of DNA), and each chromosome consists of a pair of sister chromatids. In contrast, after mitosis, the chromosomes of each daughter cell have only one copy of the genome. This copy will not be duplicated until a daughter cell in its turn reaches the S phase of the next cell cycle (see Fig. 2-8). The entire process of mitosis thus ensures the orderly duplication and distribution of the genome through successive cell divisions.

The Human Karyotype

The condensed chromosomes of a dividing human cell are most readily analyzed at metaphase or prometaphase. At these stages, the chromosomes are visible under the microscope as a so-called **chromosome spread;** each chromosome consists of its sister chromatids, although in most chromosome preparations, the two chromatids are held together so tightly that they are rarely visible as separate entities.

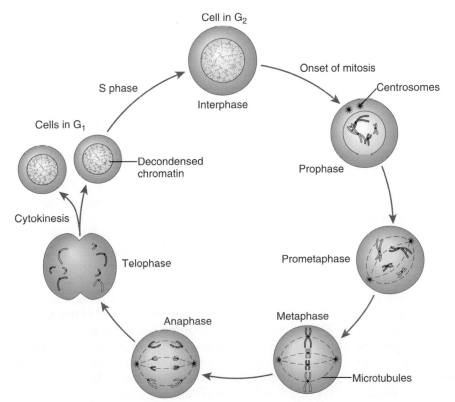

Figure 2-9 Mitosis. Only two chromosome pairs are shown. For details, see text.

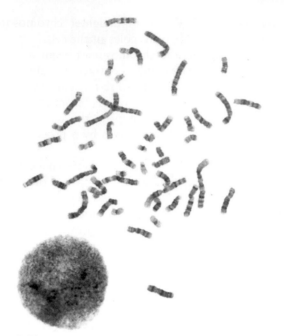

Figure 2-10 A chromosome spread prepared from a lymphocyte culture that has been stained by the Giemsa-banding (G-banding) technique. The darkly stained nucleus adjacent to the chromosomes is from a different cell in interphase, when chromosomal material is diffuse throughout the nucleus. *See Sources & Acknowledgments.*

As stated earlier, there are 24 different types of human chromosome, each of which can be distinguished cytologically by a combination of overall length, location of the centromere, and sequence content, the latter reflected by various staining methods. The centromere is apparent as a **primary constriction,** a narrowing or pinching-in of the sister chromatids due to formation of the kinetochore. This is a recognizable cytogenetic landmark, dividing the chromosome into two **arms,** a short arm designated **p** (for *petit*) and a long arm designated **q.**

Figure 2-10 shows a prometaphase cell in which the chromosomes have been stained by the Giemsa-staining (**G-banding**) method (also see Chapter 5). Each chromosome pair stains in a characteristic pattern of alternating light and dark bands (G bands) that correlates roughly with features of the underlying DNA sequence, such as base composition (i.e., the percentage of base pairs that are GC or AT) and the distribution of repetitive DNA elements. With such banding techniques, all of the chromosomes can be individually distinguished, and the nature of many structural or numerical abnormalities can be determined, as we examine in greater detail in Chapters 5 and 6.

Although experts can often analyze metaphase chromosomes directly under the microscope, a common procedure is to cut out the chromosomes from a digital image or photomicrograph and arrange them in pairs in a standard classification (Fig. 2-11). The completed picture is called a **karyotype.** The word *karyotype* is also used to refer to the standard chromosome set of an individual ("a normal male karyotype") or of a species

("the human karyotype") and, as a verb, to the process of preparing such a standard figure ("to karyotype").

Unlike the chromosomes seen in stained preparations under the microscope or in photographs, the chromosomes of living cells are fluid and dynamic structures. During mitosis, the chromatin of each interphase chromosome condenses substantially (Fig. 2-12). When maximally condensed at metaphase, DNA in chromosomes is approximately 1/10,000 of its fully extended state. When chromosomes are prepared to reveal bands (as in Figs. 2-10 and 2-11), as many as 1000 or more bands can be recognized in stained preparations of all the chromosomes. Each cytogenetic band therefore contains as many as 50 or more genes, although the density of genes in the genome, as mentioned previously, is variable.

Meiosis

Meiosis, the process by which diploid cells give rise to haploid gametes, involves a type of cell division that is unique to germ cells. In contrast to mitosis, meiosis consists of one round of DNA replication followed by *two* rounds of chromosome segregation and cell division (see meiosis I and meiosis II in Fig. 2-13). As outlined here and illustrated in Figure 2-14, the overall sequence of events in male and female meiosis is the same; however, the timing of gametogenesis is very different in the two sexes, as we will describe more fully later in this chapter.

Meiosis I is also known as the **reduction division** because it is the division in which the chromosome number is reduced by half through the pairing of homologues in prophase and by their segregation to different cells at anaphase of meiosis I. Meiosis I is also notable because it is the stage at which genetic **recombination** (also called **meiotic crossing over**) occurs. In this process, as shown for one pair of chromosomes in Figure 2-14, homologous segments of DNA are exchanged between nonsister chromatids of each pair of homologous chromosomes, thus ensuring that none of the gametes produced by meiosis will be identical to another. The conceptual and practical consequences of recombination for many aspects of human genetics and genomics are substantial and are outlined in the Box at the end of this section.

Prophase of meiosis I differs in a number of ways from mitotic prophase, with important genetic consequences, because homologous chromosomes need to pair and exchange genetic information. The most critical early stage is called **zygotene,** when homologous chromosomes begin to align along their entire length. The process of meiotic pairing—called **synapsis**—is normally precise, bringing corresponding DNA sequences into alignment along the length of the entire chromosome pair. The paired homologues—now called **bivalents**—are held together by a ribbon-like proteinaceous structure

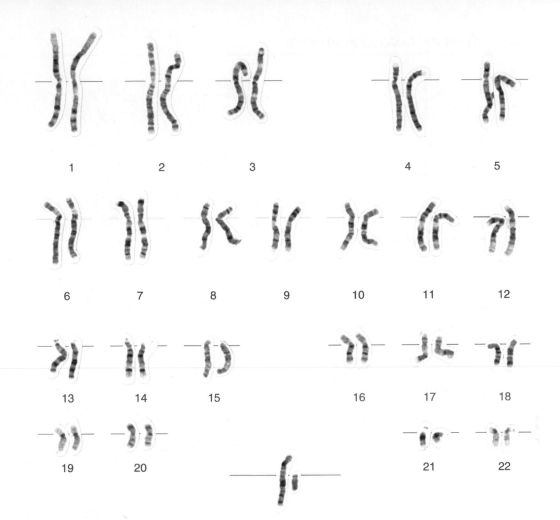

SEX CHROMOSOMES

Figure 2-11 A human male karyotype with Giemsa banding (G banding). The chromosomes are at the prometaphase stage of mitosis and are arranged in a standard classification, numbered 1 to 22 in order of length, with the X and Y chromosomes shown separately. *See Sources & Acknowledgments.*

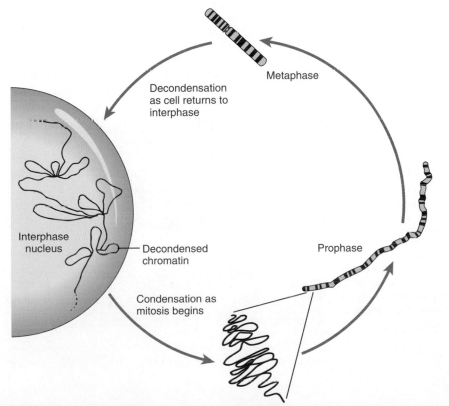

Metaphase

Decondensation as cell returns to interphase

Interphase nucleus

Decondensed chromatin

Prophase

Condensation as mitosis begins

Figure 2-12 Cycle of condensation and decondensation as a chromosome proceeds through the cell cycle.

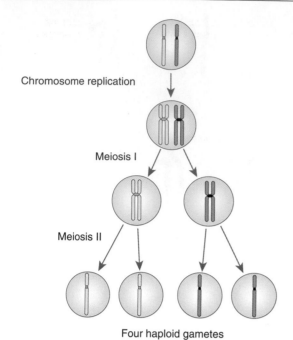

Chromosome replication

Meiosis I

Meiosis II

Four haploid gametes

Figure 2-13 A simplified representation of the essential steps in meiosis, consisting of one round of DNA replication followed by two rounds of chromosome segregation, meiosis I and meiosis II.

called the **synaptonemal complex,** which is essential to the process of recombination. After synapsis is complete, meiotic crossing over takes place during **pachytene,** after which the synaptonemal complex breaks down.

Metaphase I begins, as in mitosis, when the nuclear membrane disappears. A spindle forms, and the paired chromosomes align themselves on the equatorial plane with their centromeres oriented toward different poles (see Fig. 2-14).

Anaphase of meiosis I again differs substantially from the corresponding stage of mitosis. Here, it is the two members of each bivalent that move apart, not the sister chromatids (contrast Fig. 2-14 with Fig. 2-9). The homologous centromeres (with their attached sister

Figure 2-14 **Meiosis and its consequences.** A single chromosome pair and a single crossover are shown, leading to formation of four distinct gametes. The chromosomes replicate during interphase and begin to condense as the cell enters prophase of meiosis I. In meiosis I, the chromosomes synapse and recombine. A crossover is visible as the homologues align at metaphase I, with the centromeres oriented toward opposite poles. In anaphase I, the exchange of DNA between the homologues is apparent as the chromosomes are pulled to opposite poles. After completion of meiosis I and cytokinesis, meiosis II proceeds with a mitosis-like division. The sister kinetochores separate and move to opposite poles in anaphase II, yielding four haploid products.

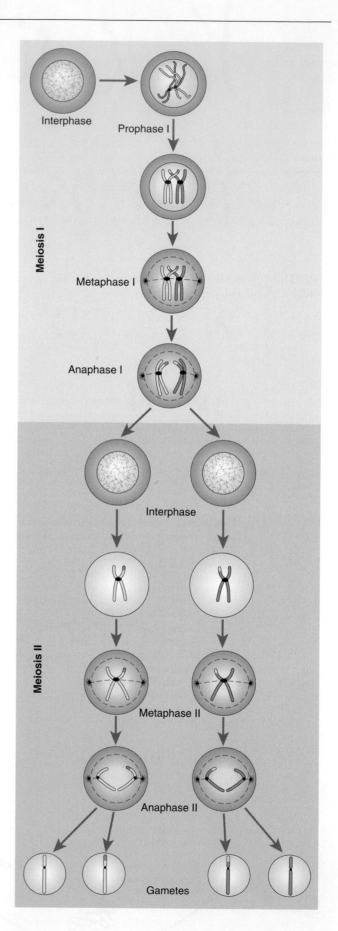

Interphase Prophase I

Meiosis I

Metaphase I

Anaphase I

Interphase

Meiosis II

Metaphase II

Anaphase II

Gametes

chromatids) are drawn to opposite poles of the cell, a process termed **disjunction**. Thus the chromosome number is halved, and each cellular product of meiosis I has the haploid chromosome number. The 23 pairs of homologous chromosomes assort independently of one another, and as a result, the original paternal and maternal chromosome sets are sorted into random combinations. The possible number of combinations of the 23 chromosome pairs that can be present in the gametes is 2^{23} (more than 8 million). Owing to the process of crossing over, however, the variation in the genetic material that is transmitted from parent to child is actually much

GENETIC CONSEQUENCES AND MEDICAL RELEVANCE OF HOMOLOGOUS RECOMBINATION

The take-home lesson of this portion of the chapter is a simple one: **the genetic content of each gamete is unique,** because of random assortment of the parental chromosomes to shuffle the combination of sequence variants *between* chromosomes and because of homologous recombination to shuffle the combination of sequence variants *within* each and every chromosome. This has significant consequences for patterns of genomic variation among and between different populations around the globe and for diagnosis and counseling of many common conditions with complex patterns of inheritance (see Chapters 8 and 10).

The **amounts and patterns of meiotic recombination** are determined by sequence variants in specific genes and at specific "hot spots" and differ between individuals, between the sexes, between families, and between populations (see Chapter 10).

Because recombination involves the physical intertwining of the two homologues until the appropriate point during meiosis I, it is also critical for ensuring proper chromosome segregation during meiosis. Failure to recombine properly can lead to **chromosome missegregation (nondisjunction)** in meiosis I and is a frequent cause of pregnancy loss and of chromosome abnormalities like Down syndrome (see Chapters 5 and 6).

Major ongoing efforts to **identify genes and their variants responsible for various medical conditions** rely on tracking the inheritance of millions of sequence differences within families or the sharing of variants within groups of even unrelated individuals affected with a particular condition. The utility of this approach, which has uncovered thousands of gene-disease associations to date, depends on patterns of homologous recombination in meiosis (see Chapter 10).

Although homologous recombination is normally precise, areas of repetitive DNA in the genome and genes of variable copy number in the population are prone to occasional **unequal crossing over** during meiosis, leading to variations in clinically relevant traits such as drug response, to common disorders such as the thalassemias or autism, or to abnormalities of sexual differentiation (see Chapters 6, 8, and 11).

Although homologous recombination is a normal and essential part of meiosis, it also occurs, albeit more rarely, in somatic cells. Anomalies in **somatic recombination** are one of the causes of **genome instability in cancer** (see Chapter 15).

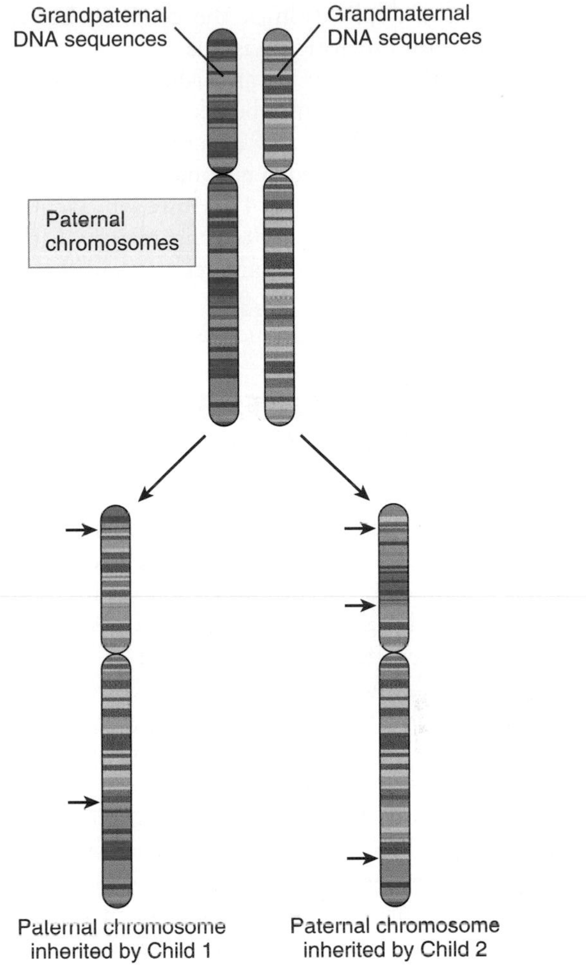

Figure 2-15 The effect of homologous recombination in meiosis. In this example, representing the inheritance of sequences on a typical large chromosome, an individual has distinctive homologues, one containing sequences inherited from his father (*blue*) and one containing homologous sequences from his mother (*purple*). After meiosis in spermatogenesis, he transmits a single complete copy of that chromosome to his two offspring. However, as a result of crossing over (*arrows*), the copy he transmits to each child consists of alternating segments of the two grandparental sequences. Child 1 inherits a copy after two crossovers, whereas child 2 inherits a copy with three crossovers.

greater than this. As a result, each chromatid typically contains segments derived from each member of the original parental chromosome pair, as illustrated schematically in Figure 2-14. For example, at this stage, a typical large human chromosome would be composed of three to five segments, alternately paternal and maternal in origin, as inferred from DNA sequence variants that distinguish the respective parental genomes (Fig. 2-15).

After telophase of meiosis I, the two haploid daughter cells enter meiotic interphase. In contrast to mitosis, this interphase is brief, and meiosis II begins. The notable point that distinguishes meiotic and mitotic interphase is that there is no S phase (i.e., no DNA synthesis and duplication of the genome) between the first and second meiotic divisions.

Meiosis II is similar to an ordinary mitosis, except that the chromosome number is 23 instead of 46; the chromatids of each of the 23 chromosomes separate, and one chromatid of each chromosome passes to each daughter cell (see Fig. 2-14). However, as mentioned earlier, because of crossing over in meiosis I, the chromosomes of the resulting gametes are not identical (see Fig. 2-15).

HUMAN GAMETOGENESIS AND FERTILIZATION

The cells in the germline that undergo meiosis, primary spermatocytes or primary oocytes, are derived from the zygote by a long series of mitoses before the onset of meiosis. Male and female gametes have different histories, marked by different patterns of gene expression that reflect their developmental origin as an XY or XX embryo. The human primordial germ cells are recognizable by the fourth week of development outside the embryo proper, in the endoderm of the yolk sac. From there, they migrate during the sixth week to the genital ridges and associate with somatic cells to form the primitive gonads, which soon differentiate into testes or ovaries, depending on the cells' sex chromosome constitution (XY or XX), as we examine in greater detail in Chapter 6. Both spermatogenesis and oogenesis require meiosis but have important differences in detail and timing that may have clinical and genetic consequences for the offspring. Female meiosis is initiated once, early during fetal life, in a limited number of cells. In contrast, male meiosis is initiated continuously in many cells from a dividing cell population throughout the adult life of a male.

In the female, successive stages of meiosis take place over several decades—in the fetal ovary before the female in question is even born, in the oocyte near the time of ovulation in the sexually mature female, and after fertilization of the egg that can become that female's offspring. Although postfertilization stages can be studied in vitro, access to the earlier stages is limited. Testicular material for the study of male meiosis is less difficult to obtain, inasmuch as testicular biopsy is included in the assessment of many men attending infertility clinics. Much remains to be learned about the cytogenetic, biochemical, and molecular mechanisms involved in normal meiosis and about the causes and consequences of meiotic irregularities.

Spermatogenesis

The stages of spermatogenesis are shown in Figure 2-16. The seminiferous tubules of the testes are lined with **spermatogonia**, which develop from the primordial

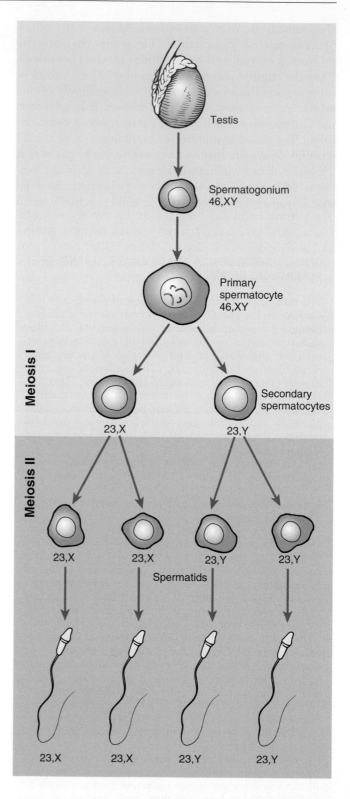

Figure 2-16 Human spermatogenesis in relation to the two meiotic divisions. The sequence of events begins at puberty and takes approximately 64 days to be completed. The chromosome number (46 or 23) and the sex chromosome constitution (X or Y) of each cell are shown. *See Sources & Acknowledgments.*

germ cells by a long series of mitoses and which are in different stages of differentiation. **Sperm** (spermatozoa) are formed only after sexual maturity is reached. The last cell type in the developmental sequence is the **primary spermatocyte,** a diploid germ cell that undergoes meiosis I to form two haploid **secondary spermatocytes.** Secondary spermatocytes rapidly enter meiosis II, each forming two **spermatids,** which differentiate without further division into sperm. In humans, the entire process takes approximately 64 days. The enormous number of sperm produced, typically approximately 200 million per ejaculate and an estimated 10^{12} in a lifetime, requires several hundred successive mitoses.

As discussed earlier, normal meiosis requires pairing of homologous chromosomes followed by recombination. The autosomes and the X chromosomes in females present no unusual difficulties in this regard; but what of the X and Y chromosomes during spermatogenesis? Although the X and Y chromosomes are different and are not homologues in a strict sense, they do have relatively short identical segments at the ends of their respective short arms (Xp and Yp) and long arms (Xq and Yq) (see Chapter 6). Pairing and crossing over occurs in both regions during meiosis I. These homologous segments are called **pseudoautosomal** to reflect their autosome-like pairing and recombination behavior, despite being on different sex chromosomes.

Oogenesis

Whereas spermatogenesis is initiated only at the time of puberty, oogenesis begins during a female's development as a fetus (Fig. 2-17). The **ova** develop from **oogonia,** cells in the ovarian cortex that have descended from the primordial germ cells by a series of approximately 20 mitoses. Each oogonium is the central cell in a developing follicle. By approximately the third month of fetal development, the oogonia of the embryo have begun to develop into **primary oocytes,** most of which have already entered prophase of meiosis I. The process of oogenesis is not synchronized, and both early and late stages coexist in the fetal ovary. Although there are several million oocytes at the time of birth, most of these degenerate; the others remain arrested in prophase I (see Fig. 2-14) for decades. Only approximately 400 eventually mature and are ovulated as part of a woman's menstrual cycle.

After a woman reaches sexual maturity, individual follicles begin to grow and mature, and a few (on average one per month) are ovulated. Just before ovulation, the oocyte rapidly completes meiosis I, dividing in such a way that one cell becomes the secondary oocyte (an egg or **ovum**), containing most of the cytoplasm with its organelles; the other cell becomes the first polar body (see Fig. 2-17). Meiosis II begins promptly and proceeds to the metaphase stage during ovulation, where it halts again, only to be completed if fertilization occurs.

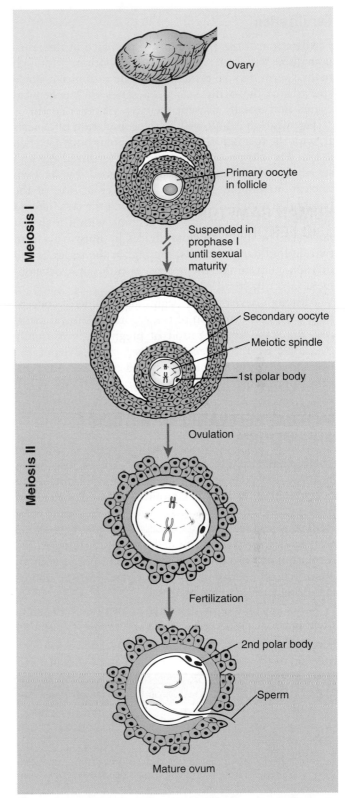

Figure 2-17 Human oogenesis and fertilization in relation to the two meiotic divisions. The primary oocytes are formed prenatally and remain suspended in prophase of meiosis I for years until the onset of puberty. An oocyte completes meiosis I as its follicle matures, resulting in a secondary oocyte and the first polar body. After ovulation, each oocyte continues to metaphase of meiosis II. Meiosis II is completed only if fertilization occurs, resulting in a fertilized mature ovum and the second polar body.

Fertilization

Fertilization of the egg usually takes place in the fallopian tube within a day or so of ovulation. Although many sperm may be present, the penetration of a single sperm into the ovum sets up a series of biochemical events that usually prevent the entry of other sperm.

Fertilization is followed by the completion of meiosis II, with the formation of the second polar body (see Fig. 2-17). The chromosomes of the now-fertilized egg and sperm form **pronuclei,** each surrounded by its own nuclear membrane. It is only upon replication of the parental genomes after fertilization that the two haploid genomes become one diploid genome within a shared nucleus. The diploid **zygote** divides by mitosis to form two diploid daughter cells, the first in the series of cell divisions that initiate the process of embryonic development (see Chapter 14).

Although development begins at the time of conception, with the formation of the zygote, in clinical medicine the stage and duration of pregnancy are usually measured as the "menstrual age," dating from the beginning of the mother's last menstrual period, typically approximately 14 days before conception.

MEDICAL RELEVANCE OF MITOSIS AND MEIOSIS

The biological significance of mitosis and meiosis lies in ensuring the constancy of chromosome number—and thus the integrity of the genome—from one cell to its progeny and from one generation to the next. The medical relevance of these processes lies in errors of one or the other mechanism of cell division, leading to the formation of an individual or of a cell lineage with an abnormal number of chromosomes and thus an abnormal dosage of genomic material.

As we see in detail in Chapter 5, meiotic nondisjunction, particularly in oogenesis, is the most common mutational mechanism in our species, responsible for chromosomally abnormal fetuses in at least several percent of all recognized pregnancies. Among pregnancies that survive to term, chromosome abnormalities are a leading cause of developmental defects, failure to thrive in the newborn period, and intellectual disability.

Mitotic nondisjunction in somatic cells also contributes to genetic disease. Nondisjunction soon after fertilization, either in the developing embryo or in extraembryonic tissues like the placenta, leads to chromosomal mosaicism that can underlie some medical conditions, such as a proportion of patients with Down syndrome. Further, abnormal chromosome segregation in rapidly dividing tissues, such as in cells of the colon, is frequently a step in the development of chromosomally abnormal tumors, and thus evaluation of chromosome and genome balance is an important diagnostic and prognostic test in many cancers.

GENERAL REFERENCES

Green ED, Guyer MS, National Human Genome Research Institute: Charting a course for genomic medicine from base pairs to bedside, *Nature* 470:204–213, 2011.

Lander ES: Initial impact of the sequencing of the human genome, *Nature* 470:187–197, 2011.

Moore KL, Presaud TVN, Torchia MG: *The developing human: clinically oriented embryology,* ed 9, Philadelphia, 2013, WB Saunders.

REFERENCES FOR SPECIFIC TOPICS

Deininger P: Alu elements: know the SINES, *Genome Biol* 12:236, 2011.

Frazer KA: Decoding the human genome, *Genome Res* 22:1599–1601, 2012.

International Human Genome Sequencing Consortium: Initial sequencing and analysis of the human genome, *Nature* 409:860–921, 2001.

International Human Genome Sequencing Consortium: Finishing the euchromatic sequence of the human genome, *Nature* 431:931–945, 2004.

Venter J, Adams M, Myers E, et al: The sequence of the human genome, *Science* 291:1304–1351, 2001.

PROBLEMS

1. At a certain locus, a person has two alleles, *A* and *a*.
   a. What alleles will be present in this person's gametes?
   b. When do *A* and *a* segregate (1) if there is no crossing over between the locus and the centromere of the chromosome? (2) if there is a single crossover between the locus and the centromere?

2. What is the main cause of numerical chromosome abnormalities in humans?

3. Disregarding crossing over, which increases the amount of genetic variability, estimate the probability that all your chromosomes have come to you from your father's mother and your mother's mother. Would you be male or female?

4. A chromosome entering meiosis is composed of two sister chromatids, each of which is a single DNA molecule.
   a. In our species, at the end of meiosis I, how many chromosomes are there per cell? How many chromatids?
   b. At the end of meiosis II, how many chromosomes are there per cell? How many chromatids?
   c. When is the diploid chromosome number restored? When is the two-chromatid structure of a typical metaphase chromosome restored?

5. From Figure 2-7, estimate the number of genes per million base pairs on chromosomes 1, 13, 18, 19, 21, and 22. Would a chromosome abnormality of equal size on chromosome 18 or 19 be expected to have greater clinical impact? On chromosome 21 or 22?

CHAPTER 3

The Human Genome: Gene Structure and Function

Over the past three decades, remarkable progress has been made in our understanding of the structure and function of genes and chromosomes. These advances have been aided by the applications of molecular genetics and genomics to many clinical problems, thereby providing the tools for a distinctive new approach to medical genetics. In this chapter, we present an overview of gene structure and function and the aspects of molecular genetics required for an understanding of the genetic and genomic approach to medicine. To supplement the information discussed here and in subsequent chapters, we provide additional material online to detail many of the experimental approaches of modern genetics and genomics that are becoming critical to the practice and understanding of human and medical genetics.

The increased knowledge of genes and of their organization in the genome has had an enormous impact on medicine and on our perception of human physiology. As 1980 Nobel laureate Paul Berg stated presciently at the dawn of this new era:

> Just as our present knowledge and practice of medicine relies on a sophisticated knowledge of human anatomy, physiology, and biochemistry, so will dealing with disease in the future demand a detailed understanding of the molecular anatomy, physiology, and biochemistry of the human genome.... We shall need a more detailed knowledge of how human genes are organized and how they function and are regulated. We shall also have to have physicians who are as conversant with the molecular anatomy and physiology of chromosomes and genes as the cardiac surgeon is with the structure and workings of the heart.

INFORMATION CONTENT OF THE HUMAN GENOME

How does the 3-billion-letter digital code of the human genome guide the intricacies of human anatomy, physiology, and biochemistry to which Berg referred? The answer lies in the enormous amplification and integration of information content that occurs as one moves from genes in the genome to their products in the cell and to the observable expression of that genetic information as cellular, morphological, clinical, or biochemical traits—what is termed the **phenotype** of the individual. This hierarchical expansion of information from the genome to phenotype includes a wide range of structural and regulatory RNA products, as well as protein products that orchestrate the many functions of cells, organs, and the entire organism, in addition to their interactions with the environment. Even with the essentially complete sequence of the human genome in hand, we still do not know the precise number of genes in the genome. Current estimates are that the genome contains approximately 20,000 **protein-coding genes** (see Box in Chapter 2), but this figure only begins to hint at the levels of complexity that emerge from the decoding of this digital information (Fig. 3-1).

As introduced briefly in Chapter 2, the product of protein-coding genes is a protein whose structure ultimately determines its particular functions in the cell. But if there were a simple one-to-one correspondence between genes and proteins, we could have at most approximately 20,000 different proteins. This number seems insufficient to account for the vast array of functions that occur in human cells over the life span. The answer to this dilemma is found in two features of gene structure and function. First, many genes are capable of generating multiple different products, not just one (see Fig. 3-1). This process, discussed later in this chapter, is accomplished through the use of alternative coding segments in genes and through the subsequent biochemical modification of the encoded protein; these two features of complex genomes result in a substantial amplification of information content. Indeed, it has been estimated that in this way, these 20,000 human genes can encode many hundreds of thousands of different proteins, collectively referred to as the **proteome**. Second, individual proteins do not function by themselves. They form elaborate networks, involving many different proteins and regulatory RNAs that respond in a coordinated and integrated fashion to many different genetic, developmental, or environmental signals. The combinatorial nature of protein networks results in an even greater diversity of possible cellular functions.

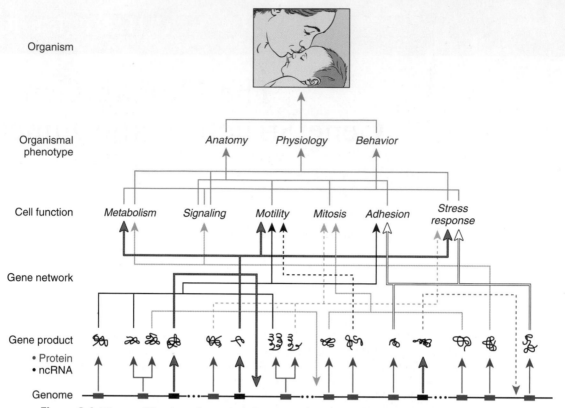

Figure 3-1 The amplification of genetic information from genome to gene products to gene networks and ultimately to cellular function and phenotype. The genome contains both protein-coding genes (*blue*) and noncoding RNA (ncRNA) genes (*red*). Many genes in the genome use alternative coding information to generate multiple different products. Both small and large ncRNAs participate in gene regulation. Many proteins participate in multigene networks that respond to cellular signals in a coordinated and combinatorial manner, thus further expanding the range of cellular functions that underlie organismal phenotypes.

Genes are located throughout the genome but tend to cluster in particular regions on particular chromosomes and to be relatively sparse in other regions or on other chromosomes. For example, chromosome 11, an approximately 135 million-bp (megabase pairs [Mb]) chromosome, is relatively gene-rich with approximately 1300 protein-coding genes (see Fig. 2-7). These genes are not distributed randomly along the chromosome, and their localization is particularly enriched in two chromosomal regions with gene density as high as one gene every 10 kb (Fig. 3-2). Some of the genes belong to families of related genes, as we will describe more fully later in this chapter. Other regions are gene-poor, and there are several so-called gene deserts of a million base pairs or more without any known protein-coding genes. Two caveats here: first, the process of gene identification and genome annotation remains very much an ongoing challenge; despite the apparent robustness of recent estimates, it is virtually certain that there are some genes, including clinically relevant genes, that are currently undetected or that display characteristics that we do not currently recognize as being associated with genes. And second, as mentioned in Chapter 2, many genes are not protein-coding; their products

are functional RNA molecules (**noncoding RNAs** or ncRNAs; see Fig. 3-1) that play a variety of roles in the cell, many of which are only just being uncovered.

For genes located on the autosomes, there are two copies of each gene, one on the chromosome inherited from the mother and one on the chromosome inherited from the father. For most autosomal genes, both copies are expressed and generate a product. There are, however, a growing number of genes in the genome that are exceptions to this general rule and are expressed at characteristically different levels from the two copies, including some that, at the extreme, are expressed from only one of the two homologues. These examples of **allelic imbalance** are discussed in greater detail later in this chapter, as well as in Chapters 6 and 7.

THE CENTRAL DOGMA: DNA → RNA → PROTEIN

How does the genome specify the functional complexity and diversity evident in Figure 3-1? As we saw in the previous chapter, genetic information is contained in DNA in the chromosomes within the cell nucleus. However, protein synthesis, the process through which

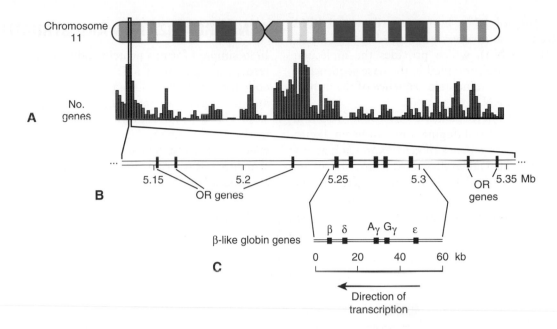

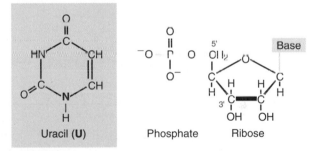

Figure 3-2 Gene content on chromosome 11, which consists of 135 Mb of DNA. **A,** The distribution of genes is indicated along the chromosome and is high in two regions of the chromosome and low in other regions. **B,** An expanded region from 5.15 to 5.35 Mb (measured from the short-arm telomere), which contains 10 known protein-coding genes, five belonging to the olfactory receptor (OR) gene family and five belonging to the globin gene family. **C,** The five β-like globin genes expanded further. *See Sources & Acknowledgments.*

information encoded in the genome is actually used to specify cellular functions, takes place in the cytoplasm. This compartmentalization reflects the fact that the human organism is a **eukaryote**. This means that human cells have a nucleus containing the genome, which is separated by a nuclear membrane from the cytoplasm. In contrast, in prokaryotes like the intestinal bacterium *Escherichia coli*, DNA is not enclosed within a nucleus. Because of the compartmentalization of eukaryotic cells, information transfer from the nucleus to the cytoplasm is a complex process that has been a focus of much attention among molecular and cellular biologists.

The molecular link between these two related types of information—the DNA code of genes and the amino acid code of protein—is **ribonucleic acid (RNA).** The chemical structure of RNA is similar to that of DNA, except that each nucleotide in RNA has a ribose sugar component instead of a deoxyribose; in addition, uracil (U) replaces thymine as one of the pyrimidine bases of RNA (Fig. 3-3). An additional difference between RNA and DNA is that RNA in most organisms exists as a single-stranded molecule, whereas DNA, as we saw in Chapter 2, exists as a double helix.

The informational relationships among DNA, RNA, and protein are intertwined: genomic DNA directs the synthesis and sequence of RNA, RNA directs the synthesis and sequence of polypeptides, and specific proteins are involved in the synthesis and metabolism of DNA and RNA. This flow of information is referred to as the **central dogma** of molecular biology.

Figure 3-3 The pyrimidine uracil and the structure of a nucleotide in RNA. Note that the sugar ribose replaces the sugar deoxyribose of DNA. Compare with Figure 2-2.

Genetic information is stored in the DNA of the genome by means of a code (the **genetic code,** discussed later) in which the sequence of adjacent bases ultimately determines the sequence of amino acids in the encoded polypeptide. First, RNA is synthesized from the DNA template through a process known as **transcription.** The RNA, carrying the coded information in a form called **messenger RNA (mRNA),** is then transported from the nucleus to the cytoplasm, where the RNA sequence is decoded, or translated, to determine the sequence of amino acids in the protein being synthesized. The process of **translation** occurs on **ribosomes,** which are cytoplasmic organelles with binding sites for all of the interacting molecules, including the mRNA, involved in protein synthesis. Ribosomes are themselves made up of many different structural proteins in association with specialized types of RNA known as **ribosomal RNA (rRNA)**. Translation involves yet a third type of RNA, **transfer RNA (tRNA)**, which provides the molecular link between the code contained in the base sequence of each mRNA and the amino acid sequence of the protein encoded by that mRNA.

Because of the interdependent flow of information represented by the central dogma, one can begin discussion of the molecular genetics of gene expression at any of its three informational levels: DNA, RNA, or protein. We begin by examining the structure of genes in the genome as a foundation for discussion of the genetic code, transcription, and translation.

GENE ORGANIZATION AND STRUCTURE

In its simplest form, a protein-coding gene can be visualized as a segment of a DNA molecule containing the code for the amino acid sequence of a polypeptide chain and the regulatory sequences necessary for its expression. This description, however, is inadequate for genes in the human genome (and indeed in most eukaryotic genomes) because few genes exist as continuous coding sequences. Rather, in the majority of genes, the coding sequences are interrupted by one or more noncoding regions (Fig. 3-4). These intervening sequences, called **introns**, are initially transcribed into RNA in the

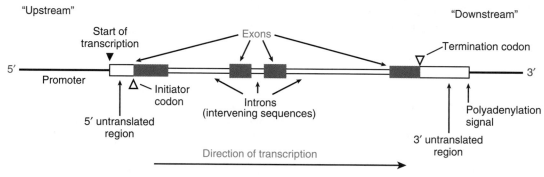

A

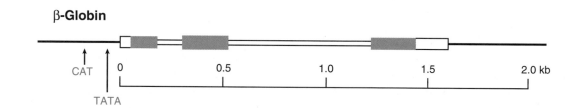

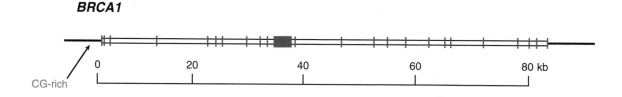

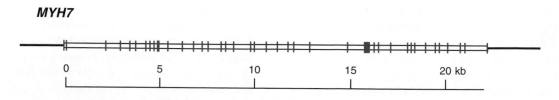

B

Figure 3-4 A, General structure of a typical human gene. Individual labeled features are discussed in the text. **B,** Examples of three medically important human genes. Different mutations in the β-globin gene, with three exons, cause a variety of important disorders of hemoglobin (Cases 42 and 44). Mutations in the *BRCA1* gene (24 exons) are responsible for many cases of inherited breast or breast and ovarian cancer (Case 7). Mutations in the β-myosin heavy chain (*MYH7*) gene (40 exons) lead to inherited hypertrophic cardiomyopathy.

nucleus but are not present in the mature mRNA in the cytoplasm, because they are removed ("spliced out") by a process we will discuss later. Thus information from the intronic sequences is not normally represented in the final protein product. Introns alternate with **exons,** the segments of genes that ultimately determine the amino acid sequence of the protein. In addition, the collection of coding exons in any particular gene is flanked by additional sequences that are transcribed but untranslated, called the 5′ and 3′ untranslated regions (see Fig. 3-4). Although a few genes in the human genome have no introns, most genes contain at least one and usually several introns. In many genes, the cumulative length of the introns makes up a far greater proportion of a gene's total length than do the exons. Whereas some genes are only a few kilobase pairs in length, others stretch on for hundreds of kilobase pairs. Also, few genes are exceptionally large; for example, the dystrophin gene on the X chromosome (mutations in which lead to Duchenne muscular dystrophy [Case 14]) spans more than 2 Mb, of which, remarkably, less than 1% consists of coding exons.

Structural Features of a Typical Human Gene

A range of features characterize human genes (see Fig. 3-4). In Chapters 1 and 2, we briefly defined *gene* in general terms. At this point, we can provide a molecular definition of a gene as *a sequence of DNA that specifies production of a functional product,* be it a polypeptide or a functional RNA molecule. A gene includes not only the actual coding sequences but also adjacent nucleotide sequences required for the proper expression of the gene—that is, for the production of normal mRNA or other RNA molecules in the correct amount, in the correct place, and at the correct time during development or during the cell cycle.

The adjacent nucleotide sequences provide the molecular "start" and "stop" signals for the synthesis of mRNA transcribed from the gene. Because the primary RNA transcript is synthesized in a 5′ to 3′ direction, the transcriptional start is referred to as the 5′ end of the transcribed portion of a gene (see Fig. 3-4). By convention, the genomic DNA that precedes the transcriptional start site in the 5′ direction is referred to as the "upstream" sequence, whereas DNA sequence located in the 3′ direction past the end of a gene is referred to as the "downstream" sequence. At the 5′ end of each gene lies a **promoter** region that includes sequences responsible for the proper initiation of transcription. Within this region are several DNA elements whose sequence is often conserved among many different genes; this conservation, together with functional studies of gene expression, indicates that these particular sequences play an important role in gene regulation. Only a subset of genes in the genome is expressed in any given tissue or at any given time during development. Several

different types of promoter are found in the human genome, with different regulatory properties that specify the patterns as well as the levels of expression of a particular gene in different tissues and cell types, both during development and throughout the life span. Some of these properties are encoded in the genome, whereas others are specified by features of chromatin associated with those sequences, as discussed later in this chapter. Both promoters and other **regulatory elements** (located either 5′ or 3′ of a gene or in its introns) can be sites of mutation in genetic disease that can interfere with the normal expression of a gene. These regulatory elements, including **enhancers, insulators,** and **locus control regions,** are discussed more fully later in this chapter. Some of these elements lie a significant distance away from the coding portion of a gene, thus reinforcing the concept that the genomic environment in which a gene resides is an important feature of its evolution and regulation.

The 3′ untranslated region contains a signal for the addition of a sequence of adenosine residues (the so-called polyA tail) to the end of the mature RNA. Although it is generally accepted that such closely neighboring regulatory sequences are part of what is called a gene, the precise dimensions of any particular gene will remain somewhat uncertain until the potential functions of more distant sequences are fully characterized.

Gene Families

Many genes belong to gene families, which share closely related DNA sequences and encode polypeptides with closely related amino acid sequences.

Members of two such gene families are located within a small region on chromosome 11 (see Fig. 3-2) and illustrate a number of features that characterize gene families in general. One small and medically important gene family is composed of genes that encode the protein chains found in hemoglobins. The β-globin gene cluster on chromosome 11 and the related α-globin gene cluster on chromosome 16 are believed to have arisen by duplication of a primitive precursor gene approximately 500 million years ago. These two clusters contain multiple genes coding for closely related globin chains expressed at different developmental stages, from embryo to adult. Each cluster is believed to have evolved by a series of sequential gene duplication events within the past 100 million years. The exon-intron patterns of the functional globin genes have been remarkably conserved during evolution; each of the functional globin genes has two introns at similar locations (see the β-globin gene in Fig. 3-4), although the sequences contained within the introns have accumulated far more nucleotide base changes over time than have the coding sequences of each gene. The control of expression of the various globin genes, in the normal state as well as in the many inherited disorders of hemoglobin, is considered in more detail both later in this chapter and in Chapter 11.

The second gene family shown in Figure 3-2 is the family of olfactory receptor (OR) genes. There are estimated to be as many as 1000 OR genes in the genome. ORs are responsible for our acute sense of smell that can recognize and distinguish thousands of structurally diverse chemicals. OR genes are found throughout the genome on nearly every chromosome, although more than half are found on chromosome 11, including a number of family members near the β-globin cluster.

Pseudogenes

Within both the β-globin and OR gene families are sequences that are related to the functional globin and OR genes but that do not produce any functional RNA or protein product. DNA sequences that closely resemble known genes but are nonfunctional are called **pseudogenes,** and there are tens of thousands of pseudogenes related to many different genes and gene families located all around the genome. Pseudogenes are of two general types, processed and nonprocessed. **Nonprocessed pseudogenes** are thought to be byproducts of evolution, representing "dead" genes that were once functional but are now vestigial, having been inactivated by mutations in critical coding or regulatory sequences. In contrast to nonprocessed pseudogenes, **processed pseudogenes** are pseudogenes that have been formed, not by mutation, but by a process called **retrotransposition,** which involves transcription, generation of a DNA copy of the mRNA (a so-called **cDNA**) by reverse transcription, and finally integration of such DNA copies back into the genome at a location usually quite distant from the original gene. Because such pseudogenes are created by retrotransposition of a DNA copy of processed mRNA, they lack introns and are not necessarily or usually on the same chromosome (or chromosomal region) as their progenitor gene. In many gene families, there are as many or even more pseudogenes as there are functional gene members.

Noncoding RNA Genes

As just discussed, many genes are protein coding and are transcribed into mRNAs that are ultimately translated into their respective proteins; their products comprise the enzymes, structural proteins, receptors, and regulatory proteins that are found in various human tissues and cell types. However, as introduced briefly in Chapter 2, there are additional genes whose functional product appears to be the RNA itself (see Fig. 3-1). These so-called **noncoding RNAs (ncRNAs)** have a range of functions in the cell, although many do not as yet have any identified function. By current estimates, there are some 20,000 to 25,000 ncRNA genes in addition to the approximately 20,000 protein-coding genes

that we introduced earlier. Thus the collection of ncRNAs represents approximately half of all identified human genes. Chromosome 11, for example, in addition to its 1300 protein-coding genes, has an estimated 1000 ncRNA genes.

Some of the types of ncRNA play largely generic roles in cellular infrastructure, including the tRNAs and rRNAs involved in translation of mRNAs on ribosomes, other RNAs involved in control of RNA splicing, and small nucleolar RNAs (snoRNAs) involved in modifying rRNAs. Additional ncRNAs can be quite long (thus sometimes called long ncRNAs, or **lncRNAs**) and play roles in gene regulation, gene silencing, and human disease, as we explore in more detail later in this chapter.

A particular class of small RNAs of growing importance are the **microRNAs (miRNAs),** ncRNAs of only approximately 22 bases in length that suppress translation of target genes by binding to their respective mRNAs and regulating protein production from the target transcript(s). Well over 1000 miRNA genes have been identified in the human genome; some are evolutionarily conserved, whereas others appear to be of quite recent origin during evolution. Some miRNAs have been shown to down-regulate hundreds of mRNAs each, with different combinations of target RNAs in

NONCODING RNAS AND DISEASE

The importance of various types of ncRNAs for medicine is underscored by their roles in a range of human diseases, from early developmental syndromes to adult-onset disorders.

- Deletion of a cluster of miRNA genes on chromosome 13 leads to a form of **Feingold syndrome,** a developmental syndrome of skeletal and growth defects, including microcephaly, short stature, and digital anomalies.
- Mutations in the miRNA gene *MIR96*, in the region of the gene critical for the specificity of recognition of its target mRNA(s), can result in **progressive hearing loss** in adults.
- Aberrant levels of certain classes of miRNAs have been reported in a wide variety of cancers, central nervous system disorders, and cardiovascular disease (see Chapter 15).
- Deletion of clusters of snoRNA genes on chromosome 15 results in **Prader-Willi syndrome,** a disorder characterized by obesity, hypogonadism, and cognitive impairment (see Chapter 6).
- Abnormal expression of a specific lncRNA on chromosome 12 has been reported in patients with a pregnancy-associated disease called **HELLP syndrome.**
- Deletion, abnormal expression, and/or structural abnormalities in different lncRNAs with roles in long-range regulation of gene expression and genome function underlie a variety of disorders involving telomere length maintenance, monoallelic expression of genes in specific regions of the genome, and X chromosome dosage (see Chapter 6).

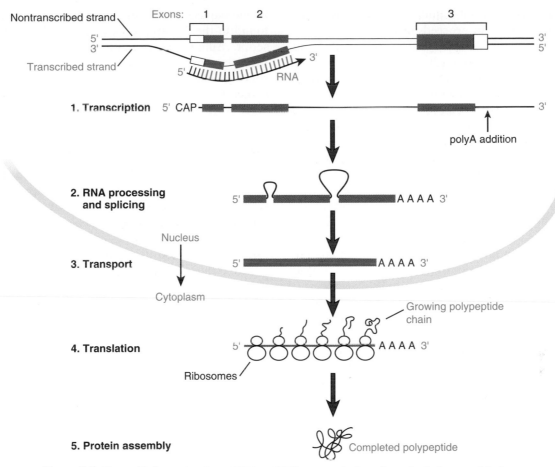

Figure 3-5 Flow of information from DNA to RNA to protein for a hypothetical gene with three exons and two introns. Within the exons, *purple* indicates the coding sequences. Steps include transcription, RNA processing and splicing, RNA transport from the nucleus to the cytoplasm, and translation.

different tissues; combined, the miRNAs are thus predicted to control the activity of as many as 30% of all protein-coding genes in the genome.

Although this is a fast-moving area of genome biology, mutations in several ncRNA genes have already been implicated in human diseases, including cancer, developmental disorders, and various diseases of both early and adult onset (see Box).

FUNDAMENTALS OF GENE EXPRESSION

For genes that encode proteins, the flow of information from gene to polypeptide involves several steps (Fig. 3-5). Initiation of transcription of a gene is under the influence of promoters and other regulatory elements, as well as specific proteins known as **transcription factors,** which interact with specific sequences within these regions and determine the spatial and temporal pattern of expression of a gene. Transcription of a gene is initiated at the transcriptional "start" site on chromosomal DNA at the beginning of a 5′ transcribed but *un*translated *r*egion (called the 5′ UTR), just upstream from the coding sequences, and continues along the

chromosome for anywhere from several hundred base pairs to more than a million base pairs, through both introns and exons and past the end of the coding sequences. After modification at both the 5′ and 3′ ends of the primary RNA transcript, the portions corresponding to introns are removed, and the segments corresponding to exons are spliced together, a process called **RNA splicing.** After splicing, the resulting mRNA (containing a central segment that is now colinear with the coding portions of the gene) is transported from the nucleus to the cytoplasm, where the mRNA is finally translated into the amino acid sequence of the encoded polypeptide. Each of the steps in this complex pathway is subject to error, and mutations that interfere with the individual steps have been implicated in a number of inherited disorders (see Chapters 11 and 12).

Transcription

Transcription of protein-coding genes by RNA polymerase II (one of several classes of RNA polymerases) is initiated at the transcriptional start site, the point in the 5′ UTR that corresponds to the 5′ end of the final

RNA product (see Figs. 3-4 and 3-5). Synthesis of the primary RNA transcript proceeds in a 5′ to 3′ direction, whereas the strand of the gene that is transcribed and that serves as the template for RNA synthesis is actually read in a 3′ to 5′ direction with respect to the direction of the deoxyribose phosphodiester backbone (see Fig. 2-3). Because the RNA synthesized corresponds both in polarity and in base sequence (substituting U for T) to the 5′ to 3′ strand of DNA, this 5′ to 3′ strand of non-transcribed DNA is sometimes called the *coding*, or **sense**, DNA strand. The 3′ to 5′ strand of DNA that is used as a template for transcription is then referred to as the *noncoding*, or **antisense**, strand. Transcription continues through both intronic and exonic portions of the gene, beyond the position on the chromosome that eventually corresponds to the 3′ end of the mature mRNA. Whether transcription ends at a predetermined 3′ termination point is unknown.

The primary RNA transcript is processed by addition of a chemical "cap" structure to the 5′ end of the RNA and cleavage of the 3′ end at a specific point downstream from the end of the coding information. This cleavage is followed by addition of a polyA tail to the 3′ end of the RNA; the polyA tail appears to increase the stability of the resulting polyadenylated RNA. The location of the polyadenylation point is specified in part by the sequence AAUAAA (or a variant of this), usually found in the 3′ untranslated portion of the RNA transcript. All of these post-transcriptional modifications take place in the nucleus, as does the process of RNA splicing. The fully processed RNA, now called mRNA, is then transported to the cytoplasm, where translation takes place (see Fig. 3-5).

Translation and the Genetic Code

In the cytoplasm, mRNA is translated into protein by the action of a variety of short RNA adaptor molecules, the tRNAs, each specific for a particular amino acid. These remarkable molecules, each only 70 to 100 nucleotides long, have the job of bringing the correct amino acids into position along the mRNA template, to be added to the growing polypeptide chain. Protein synthesis occurs on ribosomes, macromolecular complexes made up of rRNA (encoded by the 18S and 28S rRNA genes), and several dozen ribosomal proteins (see Fig. 3-5).

The key to translation is a code that relates specific amino acids to combinations of three adjacent bases along the mRNA. Each set of three bases constitutes a **codon**, specific for a particular amino acid (Table 3-1). In theory, almost infinite variations are possible in the

TABLE 3-1 The Genetic Code

First Base		Second Base							Third Base
		U		C		A		G	
U	UUU	phe	UCU	ser	UAU	tyr	UGU	cys	U
	UUC	phe	UCC	ser	UAC	tyr	UGC	cys	C
	UUA	leu	UCA	ser	UAA	stop	UGA	stop	A
	UUG	leu	UCG	ser	UAG	stop	UGG	trp	G
C	CUU	leu	CCU	pro	CAU	his	CGU	arg	U
	CUC	leu	CCC	pro	CAC	his	CGC	arg	C
	CUA	leu	CCA	pro	CAA	gln	CGA	arg	A
	CUG	leu	CCG	pro	CAG	gln	CGG	arg	G
A	AUU	ile	ACU	thr	AAU	asn	AGU	ser	U
	AUC	ile	ACC	thr	AAC	asn	AGC	ser	C
	AUA	ile	ACA	thr	AAA	lys	AGA	arg	A
	AUG	met	ACG	thr	AAG	lys	AGG	arg	G
G	GUU	val	GCU	ala	GAU	asp	GGU	gly	U
	GUC	val	GCC	ala	GAC	asp	GGC	gly	C
	GUA	val	GCA	ala	GAA	glu	GGA	gly	A
	GUG	val	GCG	ala	GAG	glu	GGG	gly	G

Abbreviations for Amino Acids

ala (A)	alanine	leu (L)	leucine
arg (R)	arginine	lys (K)	lysine
asn (N)	asparagine	met (M)	methionine
asp (D)	aspartic acid	phe (F)	phenylalanine
cys (C)	cysteine	pro (P)	proline
gln (Q)	glutamine	ser (S)	serine
glu (E)	glutamic acid	thr (T)	threonine
gly (G)	glycine	trp (W)	tryptophan
his (H)	histidine	tyr (Y)	tyrosine
ile (I)	isoleucine	val (V)	valine

Stop, Termination codon.
Codons are shown in terms of mRNA, which are complementary to the corresponding DNA codons.

arrangement of the bases along a polynucleotide chain. At any one position, there are four possibilities (A, T, C, or G); thus, for three bases, there are 4^3, or 64, possible triplet combinations. These 64 codons constitute the **genetic code**.

Because there are only 20 amino acids and 64 possible codons, most amino acids are specified by more than one codon; hence the code is said to be **degenerate**. For instance, the base in the third position of the triplet can often be either purine (A or G) or either pyrimidine (T or C) or, in some cases, any one of the four bases, without altering the coded message (see Table 3-1). Leucine and arginine are each specified by six codons. Only methionine and tryptophan are each specified by a single, unique codon. Three of the codons are called **stop** (or **nonsense**) **codons** because they designate termination of translation of the mRNA at that point.

Translation of a processed mRNA is always initiated at a codon specifying methionine. Methionine is therefore the first encoded (amino-terminal) amino acid of each polypeptide chain, although it is usually removed before protein synthesis is completed. The codon for methionine (the **initiator codon**, AUG) establishes the **reading frame** of the mRNA; each subsequent codon is read in turn to predict the amino acid sequence of the protein.

The molecular links between codons and amino acids are the specific tRNA molecules. A particular site on each tRNA forms a three-base **anticodon** that is complementary to a specific codon on the mRNA. Bonding between the codon and anticodon brings the appropriate amino acid into the next position on the ribosome for attachment, by formation of a peptide bond, to the carboxyl end of the growing polypeptide chain. The ribosome then slides along the mRNA exactly three bases, bringing the next codon into line for recognition by another tRNA with the next amino acid. Thus proteins are synthesized from the amino terminus to the carboxyl terminus, which corresponds to translation of the mRNA in a 5′ to 3′ direction.

As mentioned earlier, translation ends when a stop codon (UGA, UAA, or UAG) is encountered in the same reading frame as the initiator codon. (Stop codons in either of the other unused reading frames are not read, and therefore have no effect on translation.) The completed polypeptide is then released from the ribosome, which becomes available to begin synthesis of another protein.

Transcription of the Mitochondrial Genome

The previous sections described fundamentals of gene expression for genes contained in the nuclear genome. The mitochondrial genome has its own transcription and protein-synthesis system. A specialized RNA polymerase, encoded in the nuclear genome, is used to transcribe the 16-kb mitochondrial genome, which contains

INCREASING FUNCTIONAL DIVERSITY OF PROTEINS

Many proteins undergo extensive post-translational packaging and processing as they adopt their final functional state (see Chapter 12). The polypeptide chain that is the primary translation product folds on itself and forms intramolecular bonds to create a specific three-dimensional structure that is determined by the amino acid sequence itself. Two or more polypeptide chains, products of the same gene or of different genes, may combine to form a single multiprotein complex. For example, two α-globin chains and two β-globin chains associate noncovalently to form a tetrameric hemoglobin molecule (see Chapter 11). The protein products may also be modified chemically by, for example, addition of methyl groups, phosphates, or carbohydrates at specific sites. These modifications can have significant influence on the function or abundance of the modified protein. Other modifications may involve cleavage of the protein, either to remove specific amino-terminal sequences after they have functioned to direct a protein to its correct location within the cell (e.g., proteins that function within mitochondria) or to split the molecule into smaller polypeptide chains. For example, the two chains that make up mature insulin, one 21 and the other 30 amino acids long, are originally part of an 82–amino acid primary translation product called proinsulin.

two related promoter sequences, one for each strand of the circular genome. Each strand is transcribed in its entirety, and the mitochondrial transcripts are then processed to generate the various individual mitochondrial mRNAs, tRNAs, and rRNAs.

GENE EXPRESSION IN ACTION

The flow of information outlined in the preceding sections can best be appreciated by reference to a particular well-studied gene, the β-globin gene. The β-globin chain is a 146–amino acid polypeptide, encoded by a gene that occupies approximately 1.6 kb on the short arm of chromosome 11. The gene has three exons and two introns (see Fig. 3-4). The β-globin gene, as well as the other genes in the β-globin cluster (see Fig. 3-2), is transcribed in a centromere-to-telomere direction. The orientation, however, is different for different genes in the genome and depends on which strand of the chromosomal double helix is the coding strand for a particular gene.

DNA sequences required for accurate initiation of transcription of the β-globin gene are located in the promoter within approximately 200 bp upstream from the transcription start site. The double-stranded DNA sequence of this region of the β-globin gene, the corresponding RNA sequence, and the translated sequence of the first 10 amino acids are depicted in Figure 3-6 to illustrate the relationships among these three

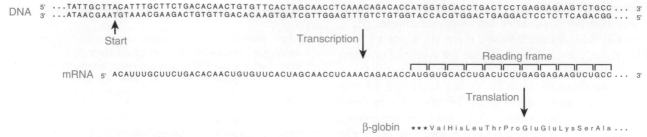

Figure 3-6 Structure and nucleotide sequence of the 5′ end of the human β-globin gene on the short arm of chromosome 11. Transcription of the 3′ to 5′ (*lower*) strand begins at the indicated start site to produce β-globin messenger RNA (mRNA). The translational reading frame is determined by the AUG initiator codon (∗∗∗); subsequent codons specifying amino acids are indicated in *blue*. The other two potential frames are not used.

information levels. As mentioned previously, it is the 3′ to 5′ strand of the DNA that serves as the template and is actually transcribed, but it is the 5′ to 3′ strand of DNA that directly corresponds to the 5′ to 3′ sequence of the mRNA (and, in fact, is identical to it except that U is substituted for T). Because of this correspondence, the 5′ to 3′ DNA strand of a gene (i.e., the strand that is *not* transcribed) is the strand generally reported in the scientific literature or in databases.

In accordance with this convention, the complete sequence of approximately 2.0 kb of chromosome 11 that includes the β-globin gene is shown in Figure 3-7. (It is sobering to reflect that a printout of the entire human genome at this scale would require over 300 books the size of this textbook!) Within these 2.0 kb are contained most, but not all, of the sequence elements required to encode and regulate the expression of this gene. Indicated in Figure 3-7 are many of the important structural features of the β-globin gene, including conserved promoter sequence elements, intron and exon boundaries, 5′ and 3′ UTRs, RNA splice sites, the initiator and termination codons, and the polyadenylation signal, all of which are known to be mutated in various inherited defects of the β-globin gene (see Chapter 11).

Initiation of Transcription

The β-globin promoter, like many other gene promoters, consists of a series of relatively short functional elements that interact with specific regulatory proteins (generically called **transcription factors**) that control transcription, including, in the case of the globin genes, those proteins that restrict expression of these genes to erythroid cells, the cells in which hemoglobin is produced. There are well over a thousand sequence-specific, DNA-binding transcription factors in the genome, some of which are ubiquitous in their expression, whereas others are cell type– or tissue-specific.

One important promoter sequence found in many, but not all, genes is the **TATA box**, a conserved region rich in adenines and thymines that is approximately 25

to 30 bp upstream of the start site of transcription (see Figs. 3-4 and 3-7). The TATA box appears to be important for determining the position of the start of transcription, which in the β-globin gene is approximately 50 bp upstream from the translation initiation site (see Fig. 3-6). Thus in this gene, there are approximately 50 bp of sequence at the 5′ end that are transcribed but are not translated; in other genes, the 5′ UTR can be much longer and can even be interrupted by one or more introns. A second conserved region, the so-called CAT box (actually CCAAT), is a few dozen base pairs farther upstream (see Fig. 3-7). Both experimentally induced and naturally occurring mutations in either of these sequence elements, as well as in other regulatory sequences even farther upstream, lead to a sharp reduction in the level of transcription, thereby demonstrating the importance of these elements for normal gene expression. Many mutations in these regulatory elements have been identified in patients with the hemoglobin disorder β-thalassemia (see Chapter 11).

Not all gene promoters contain the two specific elements just described. In particular, genes that are constitutively expressed in most or all tissues (so-called housekeeping genes) often lack the CAT and TATA boxes, which are more typical of tissue-specific genes. Promoters of many housekeeping genes contain a high proportion of cytosines and guanines in relation to the surrounding DNA (see the promoter of the *BRCA1* breast cancer gene in Fig. 3-4). Such CG-rich promoters are often located in regions of the genome called **CpG islands,** so named because of the unusually high concentration of the dinucleotide 5′-CpG-3′ (the *p* representing the phosphate group between adjacent bases; see Fig. 2-3) that stands out from the more general AT-rich genomic landscape. Some of the CG-rich sequence elements found in these promoters are thought to serve as binding sites for specific transcription factors. CpG islands are also important because they are targets for **DNA methylation.** Extensive DNA methylation at CpG islands is usually associated with repression of gene transcription, as we will discuss further later in the

5′....agccacaccctagggtttgg`ccaat`ctactcccaggagcagggagggcaggagccagggctgggc`ataaaa`

                                                                                                       * * *

gtcagggcagagccatctctattgctt`ACATTTGCTTCTGACACAACTGTGTTCACTAGCAACCTCAAACAGACACCATG`

Exon 1
```
ValHisLeuThrProGluGluLysSerAlaValThrAlaLeuTrpGlyLysValAsnValAspGluValGlyGlyGlu
GTGCACCTGACTCCTGAGGAGAAGTCTGCCGTTACTGCCCTGTGGGGCAAGGTGAACGTGGATGAAGTTGGTGGTGAG
AlaLeuGlyAr-
GCCCTGGGCAGgttggtatcaaggttacaagacaggtttaaggagaccaatagaaactgggcatgtggagacagagaag
```

Intron 1
```
                                                           -gLeuLeuValValTyr
actcttgggtttctgataggcactgactctctctgcctattggtctattttcccacccttagGCTGCTGGTGGTCTAC
```

Exon 2
```
ProTrpThrGlnArgPhePheGluSerPheGlyAspLeuSerThrProAspAlaValMetGlyAsnProLysValLys
CCTTGGACCCAGAGGTTCTTTGAGTCCTTTGGGGATCTGTCCACTCCTGATGCTGTTATGGGCAACCCTAAGGTGAAG
AlaHisGlyLysLysValLeuGlyAlaPheSerAspGlyLeuAlaHisLeuAspAsnLeuLysGlyThrPheAlaThr
GCTCATGGCAAGAAAGTGCTCGGTGCCTTTAGTGATGGCCTGGCTCACCTGGACAACCTCAAGGGCACCTTTGCCACA
LeuSerGluLeuHisCysAspLysLeuHisValAspProGluAsnPheArg
CTGAGTGAGCTGCACTGTGACAAGCTGCACGTGGATCCTGAGAACTTCAGGgtgagtctatgggacccttgatgtttt
```

Intron 2
```
ctttccccttctttctatggttaagttcatgtcataggaaggggagaagtaacagggtacagtttagaatgggaaac
agacgaatgattgcatcagtgtggaagtctcaggatcgttttagtttcttttatttgctgttcataacaattgttttc
ttttgtttaattcttgctttctttttttttcttctccgcaattttactattatacttaatgccttaacattgtgtat
aacaaaaggaaatatctctgagatacattaagtaacttaaaaaaaaaactttacacagtctgcctagtacattactatt
tggaatatatgtgtgcttatttgcatattcataatgtccctactttatttttcttttatttttaattgatacataatca
ttatacatatttatgggttaaagtgtaatgttttaatatgtgtacacatattgaccaaatcagggtaattttgcatt
tgtaattttaaaaaatgctttcttcttttaatatacttttttgtttatcttatttctaatactttccctaatctcttt
ctttcagggcaataatgatacaatgtatcatgcctctttgcaccattctaaagaataacagtgataatttctgggtta
aggcaatagcaatatttctgcatataaatatttctgcatataaattgtaactgatgtaagagggtttcatattgctaa
tagcagctacaatccagctaccattctgcttttattttatggttgggataaggctggattattctgagtccaagctag
```

Exon 3
```
                                   LeuLeuGlyAsnValLeuValCysValLeuAla
gcccttttgctaatcatgttcatacctcttatcttcctcccacagCTCCTGGGCAACGTGCTGGTCTGTGTGCTGGCC
HisHisPheGlyLysGluPheThrProProValGlnAlaAlaTryGlnLysValValAlaGlyValAlaAsnAlaLeu
CATCACTTTGGCAAAGAATTCACCCCACCAGTGCAGGCTGCCTATCAGAAAGTGGTGGCTGGTGTGGCTAATGCCCTG
AlaHisLysTyrHisTer
GCCCACAAGTATCACTAAGCTCGCTTTCTTGCTGTCCAATTTCTATTAAAGGTTCCTTTGTTCCCTAAGTCCAACTAC
TAAACTGGGGGATATTATGAAGGGCCTTGAGCATCTGGATTCTGCCTAATAAAAAACATTTATTTTCATTGCaatgat
```

gtatttaaattatttctgaatattttactaaaaagggaatgtgggaggtcagtgcatttaaaacataaagaaatgatg
agctgttcaaaccttgggaaaatacactatatcttaaactccatgaaagaaggtgaggctgcaaccagctaatgcaca
ttggcaacagcccctgatgcctatgccttattcatccctcagaaaaggattcttgtagaggcttga....   3′

Figure 3-7 Nucleotide sequence of the complete human β-globin gene. The sequence of the 5′ to 3′ strand of the gene is shown. *Tan* areas with capital letters represent exonic sequences corresponding to mature mRNA. Lowercase letters indicate introns and flanking sequences. The CAT and TATA box sequences in the 5′ flanking region are indicated in *brown*. The GT and AG dinucleotides important for RNA splicing at the intron-exon junctions and the AATAAA signal important for addition of a polyA tail also are highlighted. The ATG initiator codon (AUG in mRNA) and the TAA stop codon (UAA in mRNA) are shown in *red* letters. The amino acid sequence of β-globin is shown above the coding sequence; the three-letter abbreviations in Table 3-1 are used here. *See Sources & Acknowledgments.*

context of chromatin and its role in the control of gene expression.

Transcription by RNA polymerase II (RNA pol II) is subject to regulation at multiple levels, including binding to the promoter, initiation of transcription, unwinding of the DNA double helix to expose the template strand, and elongation as RNA pol II moves along the DNA. Although some silenced genes are devoid of RNA pol II binding altogether, consistent with their inability to be transcribed in a given cell type, others have RNA pol II poised bidirectionally at the transcriptional start site, perhaps as a means of fine-tuning transcription in response to particular cellular signals.

In addition to the sequences that constitute a promoter itself, there are other sequence elements that can markedly alter the efficiency of transcription. The best characterized of these "activating" sequences are called **enhancers**. Enhancers are sequence elements that can act at a distance from a gene (often several or even hundreds of kilobases away) to stimulate transcription. Unlike promoters, enhancers are both position and orientation independent and can be located either 5′ or 3′ of the transcription start site. Specific enhancer elements function only in certain cell types and thus appear to be involved in establishing the tissue specificity or level of expression of many genes, in concert with one or more transcription factors. In the case of the β-globin gene, several tissue-specific enhancers are present both within the gene itself and in its flanking regions. The interaction of enhancers with specific regulatory proteins leads to increased levels of transcription.

Normal expression of the β-globin gene during development also requires more distant sequences called the **locus control region (LCR)**, located upstream of the ε-globin gene (see Fig. 3-2), which is required for establishing the proper chromatin context needed for appropriate high-level expression. As expected, mutations that disrupt or delete either enhancer or LCR sequences interfere with or prevent β-globin gene expression (see Chapter 11).

RNA Splicing

The primary RNA transcript of the β-globin gene contains two introns, approximately 100 and 850 bp in length, that need to be removed and the remaining RNA segments joined together to form the mature mRNA. The process of **RNA splicing**, described generally earlier, is typically an exact and highly efficient one; 95% of β-globin transcripts are thought to be accurately spliced to yield functional globin mRNA. The splicing reactions are guided by specific sequences in the primary RNA transcript at both the 5′ and the 3′ ends of introns. The 5′ sequence consists of nine nucleotides, of which two (the dinucleotide GT [GU in the RNA transcript] located in the intron immediately adjacent to the splice site) are virtually invariant among splice sites in different genes (see Fig. 3-7). The 3′ sequence consists of approximately a dozen nucleotides, of which, again, two—the AG located immediately 5′ to the intron-exon boundary—are obligatory for normal splicing. The splice sites themselves are unrelated to the reading frame of the particular mRNA. In some instances, as in the case of intron 1 of the β-globin gene, the intron actually splits a specific codon (see Fig. 3-7).

The medical significance of RNA splicing is illustrated by the fact that mutations within the conserved sequences at the intron-exon boundaries commonly impair RNA splicing, with a concomitant reduction in the amount of normal, mature β-globin mRNA; mutations in the GT or AG dinucleotides mentioned earlier invariably eliminate normal splicing of the intron containing the mutation. Representative splice site mutations identified in patients with β-thalassemia are discussed in detail in Chapter 11.

Alternative Splicing

As just discussed, when introns are removed from the primary RNA transcript by RNA splicing, the remaining exons are spliced together to generate the final, mature mRNA. However, for most genes, the primary transcript can follow multiple alternative splicing pathways, leading to the synthesis of multiple related but different mRNAs, each of which can be subsequently translated to generate different protein products (see Fig. 3-1). Some of these alternative events are highly tissue- or cell type–specific, and, to the extent that such events are determined by primary sequence, they are subject to allelic variation between different individuals. Nearly all human genes undergo alternative splicing to some degree, and it has been estimated that there are an average of two or three alternative transcripts per gene in the human genome, thus greatly expanding the information content of the human genome beyond the approximately 20,000 protein-coding genes. The regulation of alternative splicing appears to play a particularly impressive role during neuronal development, where it may contribute to generating the high levels of functional diversity needed in the nervous system. Consistent with this, susceptibility to a number of neuropsychiatric conditions has been associated with shifts or disruption of alternative splicing patterns.

Polyadenylation

The mature β-globin mRNA contains approximately 130 bp of 3′ untranslated material (the 3′ UTR) between the stop codon and the location of the polyA tail (see Fig. 3-7). As in other genes, cleavage of the 3′ end of the mRNA and addition of the polyA tail is controlled, at least in part, by an AAUAAA sequence approximately 20 bp before the polyadenylation site. Mutations in this polyadenylation signal in patients with β-thalassemia document the importance of this signal for proper 3′ cleavage and polyadenylation (see Chapter 11). The 3′ UTR of some genes can be up to several kb in length. Other genes have a number of alternative polyadenylation sites, selection among which may influence the stability of the resulting mRNA and thus the steady-state level of each mRNA.

RNA Editing and RNA-DNA Sequence Differences

Recent findings suggest that the conceptual principle underlying the central dogma—that RNA and protein sequences reflect the underlying genomic sequence—may not always hold true. RNA editing to change the nucleotide sequence of the mRNA has been demonstrated in a number of organisms, including humans. This process involves deamination of adenosine at particular sites, converting an A in the DNA sequence to an inosine in the resulting RNA; this is then read by the translational machinery as a G, leading to changes in gene expression and protein function, especially in the nervous system. More widespread RNA-DNA differences involving other bases (with corresponding changes in the encoded amino acid sequence) have also been reported, at levels that vary among individuals. Although the mechanism(s) and clinical relevance of these events remain controversial, they illustrate the existence of a range of processes capable of increasing transcript and proteome diversity.

EPIGENETIC AND EPIGENOMIC ASPECTS OF GENE EXPRESSION

Given the range of functions and fates that different cells in any organism must adopt over its lifetime, it is apparent that not all genes in the genome can be actively expressed in every cell at all times. As important as completion of the Human Genome Project has been for contributing to our understanding of human biology and disease, identifying the genomic sequences and features that direct developmental, spatial, and temporal aspects of gene expression remains a formidable challenge. Several decades of work in molecular biology have defined critical regulatory elements for many individual genes, as we saw in the previous section, and more recent attention has been directed toward performing such studies on a genome-wide scale.

In Chapter 2, we introduced general aspects of chromatin that package the genome and its genes in all cells. Here, we explore the specific characteristics of chromatin that are associated with active or repressed genes as a step toward identifying the regulatory code for expression of the human genome. Such studies focus on reversible changes in the chromatin landscape as determinants of gene function rather than on changes to the genome sequence itself and are thus called *epi*genetic or, when considered in the context of the entire genome, *epig*enomic (Greek *epi-*, over or upon).

The field of **epigenetics** is growing rapidly and is the study of heritable changes in cellular function or gene expression that can be transmitted from cell to cell (and even generation to generation) as a result of chromatin-based molecular signals (Fig. 3-8). Complex epigenetic states can be established, maintained, and transmitted by a variety of mechanisms: modifications to the DNA, such as **DNA methylation;** numerous **histone modifications** that alter chromatin packaging or access; and substitution of specialized **histone variants** that mark chromatin associated with particular sequences or regions in the genome. These chromatin changes can be highly dynamic and transient, capable of responding rapidly and sensitively to changing needs in the cell, or they can be long lasting, capable of being transmitted through multiple cell divisions or even to subsequent generations. In either instance, the key concept is that epigenetic mechanisms do *not* alter the underlying DNA sequence, and this distinguishes them from genetic mechanisms, which are sequence based. Together, the epigenetic marks and the DNA sequence make up the set of signals that guide the genome to express its genes at the right time, in the right place, and in the right amounts.

Increasing evidence points to a role for epigenetic changes in human disease in response to environmental or lifestyle influences. The dynamic and reversible nature of epigenetic changes permits a level of adaptability or plasticity that greatly exceeds the capacity of DNA sequence alone and thus is relevant both to the origins and potential treatment of disease. A number of large-scale epigenomics projects (akin to the original Human Genome Project) have been initiated to catalogue DNA methylation sites genome-wide (the so-called methylome), to evaluate CpG landscapes across the genome, to discover new histone variants and modification patterns in various tissues, and to document positioning of nucleosomes around the genome in different cell types, and in samples from both asymptomatic individuals and those with cancer or other diseases. These analyses are part of a broad effort (called the **ENCODE Project,** for *Encyclopedia of DNA Elements*) to explore epigenetic patterns in chromatin genome-wide in order to better understand control of gene expression in different tissues or disease states.

DNA Methylation

DNA methylation involves the modification of cytosine bases by methylation of the carbon at the fifth position in the pyrimidine ring (Fig. 3-9). Extensive DNA methylation is a mark of repressed genes and is a widespread mechanism associated with the establishment of specific programs of gene expression during cell differentiation and development. Typically, DNA methylation occurs on the C of CpG dinucleotides (see Fig. 3-8) and inhibits gene expression by recruitment of specific methyl-CpG–binding proteins that, in turn, recruit chromatin-modifying enzymes to silence transcription. The presence of 5-methylcytosine (5-mC) is considered to be a stable epigenetic mark that can be faithfully transmitted through cell division; however, altered methylation states are frequently observed in cancer, with hypomethylation of large genomic segments or with regional hypermethylation (particularly at CpG islands) in others (see Chapter 15).

Extensive demethylation occurs during germ cell development and in the early stages of embryonic development,

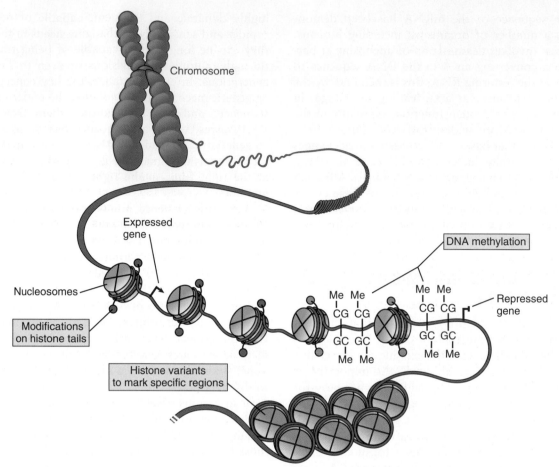

Figure 3-8 Schematic representation of chromatin and three major epigenetic mechanisms: DNA methylation at CpG dinucleotides, associated with gene repression; various modifications (indicated by different colors) on histone tails, associated with either gene expression or repression; and various histone variants that mark specific regions of the genome, associated with specific functions required for chromosome stability or genome integrity. Not to scale.

consistent with the need to "re-set" the chromatin environment and restore totipotency or pluripotency of the zygote and of various stem cell populations. Although the details are still incompletely understood, these reprogramming steps appear to involve the enzymatic conversion of 5-mC to 5-hydroxymethylcytosine (5-hmC; see Fig. 3-9), as a likely intermediate in the demethylation of DNA. Overall, 5-mC levels are stable across adult tissues (approximately 5% of all cytosines), whereas 5-hmC levels are much lower and much more variable (0.1% to 1% of all cytosines). Interestingly, although 5-hmC is widespread in the genome, its highest levels are found in known regulatory regions, suggesting a possible role in the regulation of specific promoters and enhancers.

Histone Modifications

A second class of epigenetic signals consists of an extensive inventory of modifications to any of the core histone types, H2A, H2B, H3, and H4 (see Chapter 2). Such modifications include histone methylation,

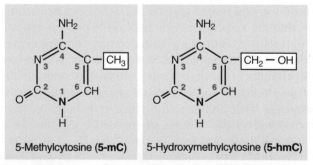

Figure 3-9 The modified DNA bases, 5-methylcytosine and 5-hydroxymethylcytosine. Compare to the structure of cytosine in Figure 2-2. The added methyl and hydroxymethyl groups are boxed in *purple*. The atoms in the pyrimidine rings are numbered 1 to 6 to indicate the 5-carbon.

phosphorylation, acetylation, and others at specific amino acid residues, mostly located on the N-terminal "tails" of histones that extend out from the core nucleosome itself (see Fig. 3-8). These epigenetic modifications are believed to influence gene expression by affecting chromatin compaction or accessibility and by signaling protein complexes that—depending on the nature of the signal—activate or silence gene expression at that site. There are dozens of modified sites that can be experimentally queried genome-wide by using antibodies that recognize specifically modified sites—for example, histone H3 methylated at lysine position 9 (H3K9 methylation, using the one-letter abbreviation K for lysine; see Table 3-1) or histone H3 acetylated at lysine position 27 (H3K27 acetylation). The former is a repressive mark associated with silent regions of the genome, whereas the latter is a mark for activating regulatory regions.

Specific patterns of different histone modifications are associated with promoters, enhancers, or the body of genes in different tissues and cell types. The ENCODE Project, introduced earlier, examined 12 of the most common modifications in nearly 50 different cell types and integrated the individual chromatin profiles to assign putative functional attributes to well over half of the human genome. This finding implies that much more of the genome plays a role, directly or indirectly, in determining the varied patterns of gene expression that distinguish cell types than previously inferred from the fact that less than 2% of the genome is "coding" in a traditional sense.

Histone Variants

The histone modifications just discussed involve modification of the core histones themselves, which are all encoded by multigene clusters in a few locations in the genome. In contrast, the many dozens of histone variants are products of entirely different genes located elsewhere in the genome, and their amino acid sequences are distinct from, although related to, those of the canonical histones.

Different histone variants are associated with different functions, and they replace—all or in part—the related member of the core histones found in typical nucleosomes to generate specialized chromatin structures (see Fig. 3-8). Some variants mark specific regions or loci in the genome with highly specialized functions; for example, the CENP-A histone is a histone H3-related variant that is found exclusively at functional centromeres in the genome and contributes to essential features of centromeric chromatin that mark the location of kinetochores along the chromosome fiber. Other variants are more transient and mark regions of the genome with particular attributes; for example, H2A.X is a histone H2A variant involved in the response to DNA damage to mark regions of the genome that require DNA repair.

Chromatin Architecture

In contrast to the impression one gets from viewing the genome as a linear string of sequence (see Fig. 3-7), the genome adopts a highly ordered and dynamic arrangement within the space of the nucleus, correlated with and likely guided by the epigenetic and epigenomic signals just discussed. This three-dimensional landscape is highly predictive of the map of all expressed sequences in any given cell type (the **transcriptome**) and reflects dynamic changes in chromatin architecture at different levels (Fig. 3-10). First, large chromosomal domains (up to millions of base pairs in size) can exhibit coordinated patterns of gene expression at the chromosome level, involving dynamic interactions between different intrachromosomal and interchromosomal points of contact within the nucleus. At a finer level, technical advances to map and sequence points of contact around the genome in the context of three-dimensional space have pointed to ordered loops of chromatin that position and orient genes precisely, exposing or blocking critical regulatory regions for access by RNA pol II, transcription factors, and other regulators. Lastly, specific and dynamic patterns of nucleosome positioning differ among cell types and tissues in the face of changing environmental and developmental cues (see Fig. 3-10). The biophysical, epigenomic, and/or genomic properties that facilitate or specify the orderly and dynamic packaging of each chromosome during each cell cycle, without reducing the genome to a disordered tangle within the nucleus, remain a marvel of landscape engineering.

GENE EXPRESSION AS THE INTEGRATION OF GENOMIC AND EPIGENOMIC SIGNALS

The gene expression program of a cell encompasses the specific subset of the approximately 20,000 protein-coding genes in the genome that are actively transcribed and translated into their respective functional products, the subset of the estimated 20,000 to 25,000 ncRNA genes that are transcribed, the amount of products produced, and the particular sequence (alleles) of those products. The gene expression profile of any particular cell or cell type in a given individual at a given time (whether in the context of the cell cycle, early development, or one's entire life span) and under a given set of circumstances (as influenced by environment, lifestyle, or disease) is thus the integrated sum of several different but interrelated effects, including the following:

- The primary sequence of genes, their allelic variants, and their encoded products
- Regulatory sequences and their epigenetic positioning in chromatin
- Interactions with the thousands of transcriptional factors, ncRNAs, and other proteins involved in the control of transcription, splicing, translation, and post-translational modification
- Organization of the genome into subchromosomal domains

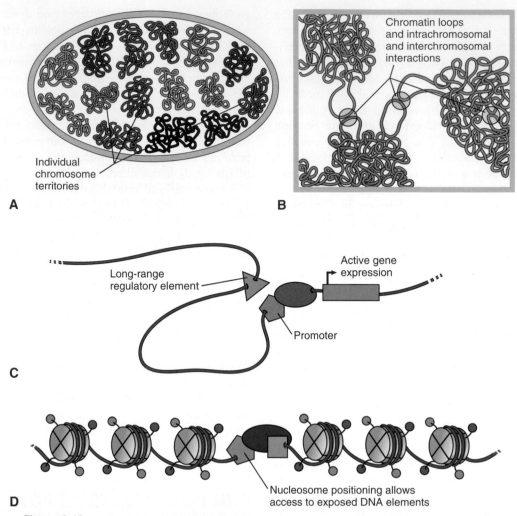

Figure 3-10 Three-dimensional architecture and dynamic packaging of the genome, viewed at increasing levels of resolution. **A,** Within interphase nuclei, each chromosome occupies a particular territory, represented by the different colors. **B,** Chromatin is organized into large subchromosomal domains within each territory, with loops that bring certain sequences and genes into proximity with each other, with detectable intrachromosomal and interchromosomal interactions. **C,** Loops bring long-range regulatory elements (e.g., enhancers or locus-control regions) into association with promoters, leading to active transcription and gene expression. **D,** Positioning of nucleosomes along the chromatin fiber provides access to specific DNA sequences for binding by transcription factors and other regulatory proteins.

- Programmed interactions between different parts of the genome
- Dynamic three-dimensional chromatin packaging in the nucleus

All of these orchestrate in an efficient, hierarchical, and highly programmed fashion. Disruption of any one—due to genetic variation, to epigenetic changes, and/or to disease-related processes—would be expected to alter the overall cellular program and its functional output (see Box).

ALLELIC IMBALANCE IN GENE EXPRESSION

It was once assumed that genes present in two copies in the genome would be expressed from both homologues at comparable levels. However, it has become increasingly evident that there can be extensive imbalance between alleles, reflecting both the amount of sequence variation in the genome and the interplay between genome sequence and epigenetic patterns that were just discussed.

THE EPIGENETIC LANDSCAPE OF THE GENOME AND MEDICINE

- Different chromosomes and chromosomal regions occupy characteristic territories within the nucleus. The probability of physical proximity influences the incidence of specific chromosome abnormalities (see Chapters 5 and 6).
- The genome is organized into megabase-sized domains with locally shared characteristics of base pair composition (i.e., GC rich or AT rich), gene density, timing of replication in the S phase, and presence of particular histone modifications (see Chapter 5).
- Modules of coexpressed genes correspond to distinct anatomical or developmental stages in, for example, the human brain or the hematopoietic lineage. Such coexpression networks are revealed by shared regulatory networks and epigenetic signals, by clustering within genomic domains, and by overlapping patterns of altered gene expression in various disease states.
- Although monozygotic twins share virtually identical genomes, they can be quite discordant for certain traits, including susceptibility to common diseases. Significant changes in DNA methylation occur during the lifetime of such twins, implicating epigenetic regulation of gene expression as a source of diversity.
- The epigenetic landscape can integrate genomic and environmental contributions to disease. For example, differential DNA methylation levels correlate with underlying sequence variation at specific loci in the genome and thereby modulate genetic risk for rheumatoid arthritis.

In Chapter 2, we introduced the general finding that any individual genome carries two different alleles at a minimum of 3 to 5 million positions around the genome, thus distinguishing by sequence the maternally and paternally inherited copies of that sequence position (see Fig. 2-6). Here, we explore ways in which those sequence differences reveal allelic imbalance in gene expression, both at autosomal loci and at X chromosome loci in females.

By determining the sequences of all the RNA products—the transcriptome—in a population of cells, one can quantify the relative level of transcription of all the genes (both protein-coding and noncoding) that are transcriptionally active in those cells. Consider, for example, the collection of protein-coding genes. Although an average cell might contain approximately 300,000 copies of mRNA in total, the abundance of specific mRNAs can differ over many orders of magnitude; among genes that are active, most are expressed at low levels (estimated to be < 10 copies of that gene's mRNA per cell), whereas others are expressed at much higher levels (several hundred to a few thousand copies of that mRNA per cell). Only in highly specialized cell types are particular genes expressed at very high levels (many tens of thousands of copies) that account for a significant proportion of all mRNA in those cells.

Now consider an expressed gene with a sequence variant that allows one to distinguish between the RNA products (whether mRNA or ncRNA) transcribed from each of two alleles, one allele with a T that is transcribed to yield RNA with an A and the other allele with a C that is transcribed to yield RNA with a G (Fig. 3-11). By sequencing individual RNA molecules and comparing the number of sequences generated that contain an A or G at that position, one can infer the ratio of transcripts from the two alleles in that sample. Although most genes show essentially equivalent levels of biallelic expression, recent analyses of this type have demonstrated widespread unequal allelic expression for 5% to 20% of autosomal genes in the genome (Table 3-2). For most of these genes, the extent of imbalance is twofold or less, although up to tenfold differences have been observed for some genes. This allelic imbalance may reflect interactions between genome sequence and gene regulation; for example, sequence changes can alter the relative binding of various transcription factors or other transcriptional regulators to the two alleles or the extent of DNA methylation observed at the two alleles (see Table 3-2).

Monoallelic Gene Expression

Some genes, however, show a much more complete form of allelic imbalance, resulting in monoallelic gene expression (see Fig. 3-11). Several different mechanisms have been shown to account for allelic imbalance of this type for particular subsets of genes in the genome: DNA rearrangement, random monoallelic expression, parent-of-origin imprinting, and, for genes on the X chromosome in females, X chromosome inactivation. Their distinguishing characteristics are summarized in Table 3-2.

Somatic Rearrangement

A highly specialized form of monoallelic gene expression is observed in the genes encoding **immunoglobulins** and **T-cell receptors**, expressed in B cells and T cells, respectively, as part of the immune response. Antibodies are encoded in the germline by a relatively small number of genes that, during B-cell development, undergo a unique process of somatic rearrangement that involves the cutting and pasting of DNA sequences in lymphocyte precursor cells (but *not* in any other cell lineages) to rearrange genes in somatic cells to generate enormous antibody diversity. The highly orchestrated DNA rearrangements occur across many hundreds of kilobases but involve only one of the two alleles, which is chosen randomly in any given B cell (see Table 3-2). Thus expression of mature mRNAs for the immunoglobulin heavy or light chain subunits is exclusively monoallelic.

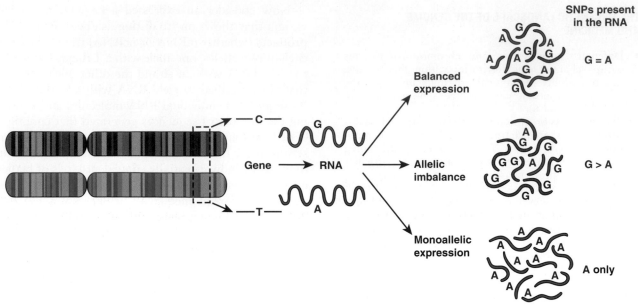

Figure 3-11 Allelic expression patterns for a gene sequence with a transcribed DNA variant (here, a C or a T) to distinguish the alleles. As described in the text, the relative abundance of RNA transcripts from the two alleles (here, carrying a G or an A) demonstrates whether the gene shows balanced expression *(top)*, allelic imbalance *(center)*, or exclusively monoallelic expression *(bottom)*. Different underlying mechanisms for allelic imbalance are compared in Table 3-2. SNP, Single nucleotide polymorphism.

TABLE 3-2 Allelic Imbalance in Gene Expression

Type	Characteristics	Genes Affected	Basis	Developmental Origin
Unbalanced expression	Unequal RNA abundance from two alleles due to DNA variants and associated epigenetic changes; usually < twofold difference in expression	5%-20% of autosomal genes	Sequence variants cause different levels of expression at the two alleles	Early embryogenesis
Monoallelic expression				
• Somatic rearrangement	Changes in DNA organization to produce functional gene at one allele, but not other	Immunoglobulin genes, T-cell receptor genes	Random choice of one allele	B- and T-cell lineages
• Random allelic silencing or activation	Expression from only one allele at a locus, due to differential epigenetic packaging at locus	Olfactory receptor genes in sensory neurons; other chemosensory or immune system genes; up to 10% of all genes in other cell types	Random choice of one allele	Specific cell types
• Genomic imprinting	Epigenetic silencing of allele(s) in imprinted region	>100 genes with functions in development	Imprinted region marked epigenetically according to parent of origin	Parental germline
• X chromosome inactivation	Epigenetic silencing of alleles on one X chromosome in females	Most X-linked genes in females	Random choice of one X chromosome	Early embryogenesis

This mechanism of somatic rearrangement and random monoallelic gene expression is also observed at the T-cell receptor genes in the T-cell lineage. However, such behavior is unique to these gene families and cell lineages; the rest of the genome remains highly stable throughout development and differentiation.

Random Monoallelic Expression

In contrast to this highly specialized form of DNA rearrangement, monoallelic expression typically results from differential epigenetic regulation of the two alleles. One well-studied example of random monoallelic expression involves the OR gene family described earlier (see Fig. 3-2). In this case, only a single allele of one OR gene is expressed in each olfactory sensory neuron; the many hundred other copies of the OR family remain repressed in that cell. Other genes with chemosensory or immune system functions also show random monoallelic expression, suggesting that this mechanism may be a general one for increasing the diversity of responses for cells that interact with the outside world. However, this mechanism is apparently not restricted to the immune and sensory systems, because a substantial subset of all human genes (5% to 10% in different cell types) has been shown to undergo random allelic silencing; these genes are broadly distributed on all autosomes, have a wide range of functions, and vary in terms of the cell types and tissues in which monoallelic expression is observed.

Parent-of-Origin Imprinting

For the examples just described, the choice of which allele is expressed is not dependent on parental origin; either the maternal or paternal copy can be expressed in different cells and their clonal descendants. This distinguishes *random* forms of monoallelic expression from **genomic imprinting,** in which the choice of the allele to be expressed is *nonrandom* and is determined solely by parental origin. Imprinting is a normal process involving the introduction of epigenetic marks (see Fig. 3-8) in the germline of one parent, but not the other, at specific locations in the genome. These lead to monoallelic expression of a gene or, in some cases, of multiple genes within the imprinted region.

Imprinting takes place during gametogenesis, before fertilization, and marks certain genes as having come from the mother or father (Fig. 3-12). After conception, the parent-of-origin imprint is maintained in some or all of the somatic tissues of the embryo and silences gene expression on allele(s) within the imprinted region; whereas some imprinted genes show monoallelic expression throughout the embryo, others show tissue-specific imprinting, especially in the placenta, with biallelic expression in other tissues. The imprinted state persists postnatally into adulthood through hundreds of cell divisions so that only the maternal or paternal copy of the gene is expressed. Yet, imprinting must be reversible: a paternally derived allele, when it is inherited by a female, must be converted in her germline so that she can then pass it on with a maternal imprint to her offspring. Likewise, an imprinted maternally derived allele, when it is inherited by a male, must be converted in his germline so that he can pass it on as a paternally imprinted allele to his offspring (see Fig. 3-12). Control over this conversion process appears to be governed by specific DNA elements called **imprinting control regions** or **imprinting centers** that are located within imprinted regions throughout the genome; although their precise mechanism of action is not known, many appear to involve ncRNAs that initiate the epigenetic change in chromatin, which then spreads outward along the chromosome over the imprinted region. Notably, although the imprinted region can encompass more than a single gene, this form of monoallelic expression is confined to a delimited genomic segment, typically a few hundred kilobase pairs to a few megabases in overall size; this distinguishes genomic imprinting both from the more general form of random monoallelic expression described earlier (which appears to involve individual genes under locus-specific control) and from X chromosome inactivation, described in the next section (which involves genes along the entire chromosome).

To date, approximately 100 imprinted genes have been identified on many different autosomes. The involvement of these genes in various chromosomal disorders is described more fully in Chapter 6. For clinical conditions due to a single imprinted gene, such as **Prader-Willi syndrome** (Case 38) and **Beckwith-Wiedemann syndrome** (Case 6), the effect of genomic imprinting on inheritance patterns in pedigrees is discussed in Chapter 7.

X Chromosome Inactivation

The chromosomal basis for sex determination, introduced in Chapter 2 and discussed in more detail in Chapter 6, results in a dosage difference between typical males and females with respect to genes on the X chromosome. Here we discuss the chromosomal and molecular mechanisms of X chromosome inactivation, the most extensive example of random monoallelic expression in the genome and a mechanism of **dosage compensation** that results in the epigenetic silencing of most genes on one of the two X chromosomes in females.

In normal female cells, the choice of which X chromosome is to be inactivated is a random one that is then maintained in each clonal lineage. Thus females are mosaic with respect to X-linked gene expression; some cells express alleles on the paternally inherited X but not the maternally inherited X, whereas other cells do the opposite (Fig. 3-13). This mosaic pattern of gene expression distinguishes most X-linked genes from imprinted genes, whose expression, as we just noted, is determined strictly by parental origin.

Although the inactive X chromosome was first identified cytologically by the presence of a heterochromatic mass (called the **Barr body**) in interphase cells, many epigenetic features distinguish the active and inactive X chromosomes, including DNA methylation, histone modifications, and a specific histone variant, macroH2A, that is particularly enriched in chromatin on the inactive X. As well as providing insights into the mechanisms of X inactivation, these features can be useful diagnostically for identifying inactive X chromosomes in clinical material, as we will see in Chapter 6.

Although X inactivation is clearly a chromosomal phenomenon, not all genes on the X chromosome show monoallelic expression in female cells. Extensive

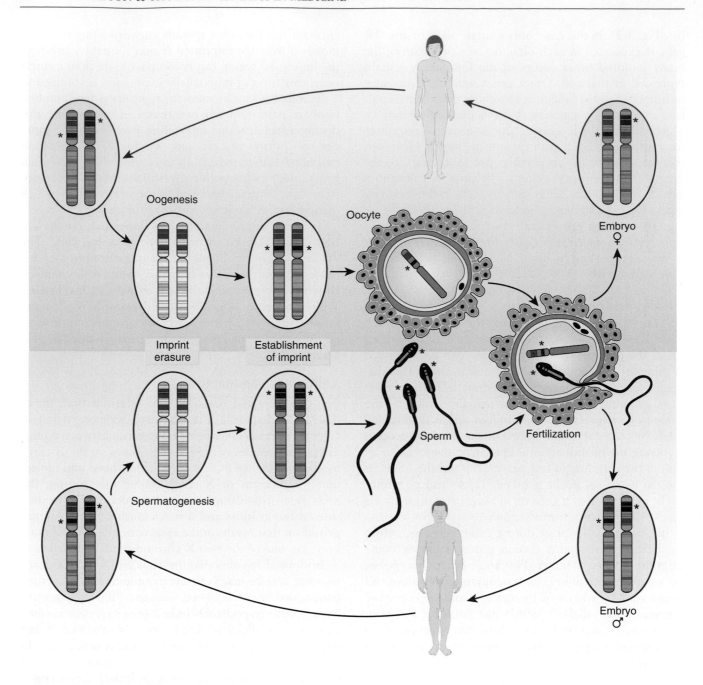

Figure 3-12 **Genomic imprinting and conversion of maternal and paternal imprints during passage through male or female gametogenesis.** Within a hypothetical imprinted region on an pair of homologous autosomes, paternally imprinted genes are indicated in *blue*, whereas a maternally imprinted gene is indicated in *red*. After fertilization, both male and female embryos have one copy of the chromosome carrying a paternal imprint and one copy carrying a maternal imprint. During oogenesis *(top)* and spermatogenesis *(bottom)*, the imprints are erased by removal of epigenetic marks, and new imprints determined by the sex of the parent are established within the imprinted region. Gametes thus carry a monoallelic imprint appropriate to the parent of origin, whereas somatic cells in both sexes carry one chromosome of each imprinted type.

analysis of expression of nearly all X-linked genes has demonstrated that at least 15% of the genes show biallelic expression and are expressed from both active and inactive X chromosomes, at least to some extent; a proportion of these show significantly higher levels of mRNA production in female cells compared to male cells and are interesting candidates for a role in explaining sexually dimorphic traits.

A special subset of genes is located in the pseudo-autosomal segments, which are essentially identical on

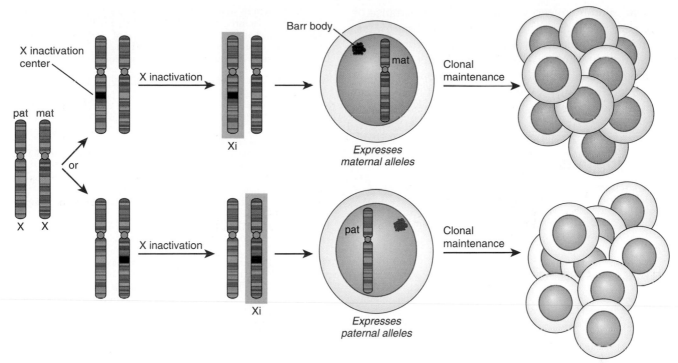

Figure 3-13 Random X chromosome inactivation early in female development. Shortly after conception of a female embryo, both the paternally and maternally inherited X chromosomes (pat and mat, respectively) are active. Within the first week of embryogenesis, one or the other X is chosen at random to become the future inactive X, through a series of events involving the X inactivation center *(black box)*. That X then becomes the inactive X (Xi, indicated by the *shading*) in that cell and its progeny and forms the Barr body in interphase nuclei. The resulting female embryo is thus a clonal mosaic of two epigenetically determined cell types: one expresses alleles from the maternal X (*pink* cells), whereas the other expresses alleles from the paternal X (*blue* cells). The ratio of the two cell types is determined randomly but varies among normal females and among females who are carriers of X-linked disease alleles (see Chapters 6 and 7)

the X and Y chromosomes and undergo recombination during spermatogenesis (see Chapter 2). These genes have two copies in both females (two X-linked copies) and males (one X-linked and one Y-linked copy) and thus do not undergo X inactivation; as expected, these genes show balanced biallelic expression, as one sees for most autosomal genes.

The X Inactivation Center and the XIST Gene. X inactivation occurs very early in female embryonic development, and determination of which X will be designated the inactive X in any given cell in the embryo is a random choice under the control of a complex locus called the **X inactivation center**. This region contains an unusual ncRNA gene, *XIST,* that appears to be a key master regulatory locus for X inactivation. *XIST (*an acronym for inactive X [*Xi*]–specific *t*ranscripts) has the novel feature that it is expressed only from the allele on the inactive X; it is transcriptionally silent on the active X in both male and female cells. Although the exact mode of action of *XIST* is unknown, X inactivation cannot occur in its absence. The product of *XIST* is a long ncRNA that stays in the nucleus in close association with the inactive X chromosome.

Additional aspects and consequences of X chromosome inactivation will be discussed in Chapter 6, in the context of individuals with structurally abnormal X chromosomes or an abnormal number of X chromosomes, and in Chapter 7, in the case of females carrying deleterious mutant alleles for X-linked disease.

VARIATION IN GENE EXPRESSION AND ITS RELEVANCE TO MEDICINE

The regulated expression of genes in the human genome involves a set of complex interrelationships among different levels of control, including proper gene dosage (controlled by mechanisms of chromosome replication and segregation), gene structure, chromatin packaging and epigenetic regulation, transcription, RNA splicing, and, for protein-coding loci, mRNA stability, translation, protein processing, and protein degradation. For some genes, fluctuations in the level of functional gene product, due either to inherited variation in the structure of a particular gene or to changes induced by nongenetic factors such as diet or the environment, are of relatively little importance. For other genes, even relatively minor changes in the level of expression can have dire clinical consequences, reflecting the importance of those gene products in particular biological pathways. The nature of inherited variation in the structure and function of chromosomes, genes, and the genome, combined with the influence of this variation on the expression of specific traits, is the very essence of medical and molecular genetics and is dealt with in subsequent chapters.

GENERAL REFERENCES

Brown TA: *Genomes*, ed 3, New York, 2007, Garland Science.

Lodish H, Berk A, Kaiser CA, et al: *Molecular cell biology*, ed 7, New York, 2012, WH Freeman.

Strachan T, Read A: *Human molecular genetics*, ed 4, New York, 2010, Garland Science.

REFERENCES FOR SPECIFIC TOPICS

Bartolomei MS, Ferguson-Smith AC: Mammalian genomic imprinting, *Cold Spring Harbor Perspect Biol* 3:1002592, 2011.

Beck CR, Garcia-Perez JL, Badge RM, et al: LINE-1 elements in structural variation and disease, *Annu Rev Genomics Hum Genet* 12:187–215, 2011.

Berg P: Dissections and reconstructions of genes and chromosomes (Nobel Prize lecture), *Science* 213:296–303, 1981.

Chess A: Mechanisms and consequences of widespread random monoallelic expression, *Nat Rev Genet* 13:421–428, 2012.

Dekker J: Gene regulation in the third dimension, *Science* 319:1793–1794, 2008.

Djebali S, Davis CA, Merkel A, et al: Landscape of transcription in human cells, *Nature* 489:101–108, 2012.

ENCODE Project Consortium: An integrated encyclopedia of DNA elements in the human genome, *Nature* 489:57–74, 2012.

Gerstein MB, Bruce C, Rozowsky JS, et al: What is a gene, post-ENCODE? *Genome Res* 17:669–681, 2007.

Guil S, Esteller M: Cis-acting noncoding RNAs: friends and foes, *Nat Struct Mol Biol* 19:1068–1074, 2012.

Heyn H, Esteller M: DNA methylation profiling in the clinic: applications and challenges, *Nature Rev Genet* 13:679–692, 2012.

Hubner MR, Spector DL: Chromatin dynamics, *Annu Rev Biophys* 39:471–489, 2010.

Li M, Wang IX, Li Y, et al: Widespread RNA and DNA sequence differences in the human transcriptome, *Science* 333:53–58, 2011.

Nagano T, Fraser P: No-nonsense functions for long noncoding RNAs, *Cell* 145:178–181, 2011.

Willard HF: The human genome: a window on human genetics, biology and medicine. In Ginsburg GS, Willard HF, editors: *Genomic and personalized medicine*, ed 2, New York, 2013, Elsevier.

Zhou VW, Goren A, Bernstein BE: Charting histone modifications and the functional organization of mammalian genomes, *Nat Rev Genet* 12:7–18, 2012.

PROBLEMS

1. The following amino acid sequence represents part of a protein. The normal sequence and four mutant forms are shown. By consulting Table 3-1, determine the double-stranded sequence of the corresponding section of the normal gene. Which strand is the strand that RNA polymerase "reads"? What would the sequence of the resulting mRNA be? What kind of mutation is each mutant protein most likely to represent?

Normal	-lys-arg-his-his-tyr-leu-
Mutant 1	-lys-arg-his-his-cys-leu-
Mutant 2	-lys-arg-ile-ile-ile-
Mutant 3	-lys-glu-thr-ser-leu-ser-
Mutant 4	-asn-tyr-leu-

2. The following items are related to each other in a hierarchical fashion: chromosome, base pair, nucleosome, kilobase pair, intron, gene, exon, chromatin, codon, nucleotide, promoter. What are these relationships?

3. Describe how mutation in each of the following might be expected to alter or interfere with normal gene function and thus cause human disease: promoter, initiator codon, splice sites at intron-exon junctions, one base pair deletion in the coding sequence, stop codon.

4. Most of the human genome consists of sequences that are not transcribed and do not directly encode gene products. For each of the following, consider ways in which these genome elements might contribute to human disease: introns, *Alu* or LINE repetitive sequences, locus control regions, pseudogenes.

5. Contrast the mechanisms and consequences of RNA splicing and somatic rearrangement.

6. Consider different ways in which mutations or variation in the following might lead to human disease: epigenetic modifications, DNA methylation, miRNA genes, lncRNA genes.

7. Contrast the mechanisms and consequences of genomic imprinting and X chromosome inactivation.

CHAPTER 4

Human Genetic Diversity: Mutation and Polymorphism

The study of genetic and genomic variation is the conceptual cornerstone for genetics in medicine and for the broader field of human genetics. During the course of evolution, the steady influx of new nucleotide variation has ensured a high degree of genetic diversity and individuality, and this theme extends through all fields in human and medical genetics. Genetic diversity may manifest as differences in the organization of the genome, as nucleotide changes in the genome sequence, as variation in the copy number of large segments of genomic DNA, as alterations in the structure or amount of proteins found in various tissues, or as any of these in the context of clinical disease.

This chapter is one of several in which we explore the nature of genetically determined differences among individuals. The sequence of nuclear DNA is approximately 99.5% identical between any two unrelated humans. Yet it is precisely the small fraction of DNA sequence difference among individuals that is responsible for the genetically determined variability that is evident both in one's daily existence and in clinical medicine. Many DNA sequence differences have little or no effect on outward appearance, whereas other differences are directly responsible for causing disease. Between these two extremes is the variation responsible for genetically determined variability in anatomy, physiology, dietary intolerances, susceptibility to infection, predisposition to cancer, therapeutic responses or adverse reactions to medications, and perhaps even variability in various personality traits, athletic aptitude, and artistic talent.

One of the important concepts of human and medical genetics is that diseases with a clearly inherited component are only the most obvious and often the most extreme manifestation of genetic differences, one end of a continuum of variation that extends from rare deleterious variants that cause illness, through more common variants that can increase susceptibility to disease, to the most common variation in the population that is of uncertain relevance with respect to disease.

THE NATURE OF GENETIC VARIATION

As described in Chapter 2, a segment of DNA occupying a particular position or location on a chromosome is a locus (plural loci). A locus may be large, such as a segment of DNA that contains many genes, such as the major histocompatibility complex locus involved in the response of the immune system to foreign substances; it may be a single gene, such as the β-globin locus we introduced in Chapter 3; or it may even be just a single base in the genome, as in the case of a single nucleotide variant (see Fig. 2-6 and later in this chapter). Alternative versions of the DNA sequence at a locus are called **alleles**. For many genes, there is a single prevailing allele, usually present in more than half of the individuals in a population, that geneticists call the **wild-type** or common allele. (In lay parlance, this is sometimes referred to as the "normal" allele. However, because genetic variation is itself very much "normal," the existence of different alleles in "normal" individuals is commonplace. Thus one should avoid using "normal" to designate the most common allele.) The other versions of the gene are **variant** (or **mutant**) alleles that differ from the wild-type allele because of the presence of a **mutation**, a permanent change in the nucleotide sequence or arrangement of DNA. Note that the terms *mutation* and *mutant* refer to DNA, but not to the human beings who carry mutant alleles. The terms denote a change in sequence but otherwise do not carry any connotation with respect to the function or fitness of that change.

The frequency of different variants can vary widely in different populations around the globe, as we will explore in depth in Chapter 9. If there are two or more relatively common alleles (defined by convention as having an allele frequency > 1%) at a locus in a population, that locus is said to exhibit **polymorphism** (literally "many forms") in that population. Most variant alleles, however, are not frequent enough in a population to be considered polymorphisms; some are so rare as to be found in only a single family and are known as **"private" alleles**.

The Concept of Mutation

In this chapter, we begin by exploring the nature of **mutation**, ranging from the change of a single nucleotide to alterations of an entire chromosome. To recognize a change means that there has to be a "gold standard," compared to which the variant shows a difference. As we saw in Chapter 2, there is no single individual whose genome sequence could serve as such a standard for the human species, and thus one arbitrarily designates the most common sequence or arrangement in a population at any one position in the genome as the so-called reference sequence (see Fig. 2-6). As more and more genomes from individuals around the globe are sampled (and thus as more and more variation is detected among the currently 7 billion genomes that make up our species), this reference genome is subject to constant evaluation and change. Indeed, a number of international collaborations share and update data on the nature and frequency of DNA variation in different populations in the context of the reference human genome sequence and make the data available through publicly accessible databases that serve as essential resources for scientists, physicians, and other health care professionals (Table 4-1).

Mutations are sometimes classified by the size of the altered DNA sequence and, at other times, by the functional effect of the mutation on gene expression. Although classification by size is somewhat arbitrary, it can be helpful conceptually to distinguish among mutations at three different levels:

- Mutations that leave chromosomes intact but change the number of chromosomes in a cell (**chromosome mutations**)
- Mutations that change only a portion of a chromosome and might involve a change in the copy number of a subchromosomal segment or a structural rearrangement involving parts of one or more chromosomes (**regional** or **subchromosomal mutations**)
- Alterations of the sequence of DNA, involving the substitution, deletion, or insertion of DNA, ranging from a single nucleotide up to an arbitrarily set limit of approximately 100 kb (**gene** or **DNA mutations**) The basis for and consequences of this third type of mutation are the principal focus of this chapter, whereas both chromosome and regional mutations will be presented at length in Chapters 5 and 6.

The functional consequences of DNA mutations, even those that change a single base pair, run the gamut from being completely innocuous to causing serious illness, all depending on the precise location, nature, and size of the mutation. For example, even a mutation within a coding exon of a gene may have no effect on how a gene is expressed if the change does not alter the primary amino acid sequence of the polypeptide product; even if it does, the resulting change in the encoded amino acid sequence may not alter the functional properties of the protein. Not all mutations, therefore, are manifest in an individual.

The Concept of Genetic Polymorphism

The DNA sequence of a given region of the genome is remarkably similar among chromosomes carried by many different individuals from around the world. In fact, any randomly chosen segment of human DNA approximately 1000 bp in length contains, on average, only one base pair that is different between the two homologous chromosomes inherited from that individual's parents (assuming the parents are unrelated). However, across all human populations, many tens of millions of single nucleotide differences and over a million more complex variants have been identified and catalogued. Because of limited sampling, these figures are likely to underestimate the true extent of genetic diversity in our species. Many populations around

TABLE 4-1 Useful Databases of Information on Human Genetic Diversity

Description	URL
The **Human Genome Project**, completed in 2003, was an international collaboration to sequence and map the genome of our species. The draft sequence of the genome was released in 2001, and the "essentially complete" reference genome assembly was published in 2004.	http://www.genome.gov/10001772 http://genome.ucsc.edu/cgi-bin/hgGateway http://www.ensembl.org/Homo_sapiens/Info/Index
The **Single Nucleotide Polymorphism Database** (**dbSNP**) and the **Structural Variation Database** (**dbVar**) are databases of small-scale and large-scale variations, including single nucleotide variants, microsatellites, indels, and CNVs.	http://www.ncbi.nlm.nih.gov/snp/ http://www.ncbi.nlm.nih.gov/dbvar/
The **1000 Genomes Project** is sequencing the genomes of a large number of individuals to provide a comprehensive resource on genetic variation in our species. All data are publicly available.	www.1000genomes.org
The **Human Gene Mutation Database** is a comprehensive collection of germline mutations associated with or causing human inherited disease (currently including over 120,000 mutations in 4400 genes).	www.hgmd.org
The **Database of Genomic Variants** is a curated catalogue of structural variation in the human genome. As of 2012, the database contains over 400,000 entries, including over 200,000 CNVs, 1000 inversions, and 34,000 indels.	http://dgv.tcag.ca
The Japanese Single Nucleotide Polymorphisms Database (**JSNP Database**) reports SNPs discovered as part of the Millennium Genome Project.	http://snp.ims.u-tokyo.ac.jp/

CNV, Copy number variant; SNP, single nucleotide polymorphism.

Updated from Willard HF: The human genome: a window on human genetics, biology and medicine. In Ginsburg GS, Willard HF, editors: *Genomic and personalized medicine*, ed 2, New York, 2013, Elsevier.

the globe have yet to be studied, and, even in the populations that have been studied, the number of individuals examined is too small to reveal most variants with minor allele frequencies below 1% to 2%. Thus, as more people are included in variant discovery projects, additional (and rarer) variants will certainly be uncovered.

Whether a variant is formally considered a polymorphism or not depends entirely on whether its frequency in a population exceeds 1% of the alleles in that population, and not on what kind of mutation caused it, how large a segment of the genome is involved, or whether it has a demonstrable effect on the individual. The location of a variant with respect to a gene also does not determine whether the variant is a polymorphism. Although most sequence polymorphisms are located between genes or within introns and are inconsequential to the functioning of any gene, others may be located in the coding sequence of genes themselves and result in different protein variants that may lead in turn to distinctive differences in human populations. Still others are in regulatory regions and may also have important effects on transcription or RNA stability.

One might expect that deleterious mutations that cause rare monogenic diseases are likely to be too rare to achieve the frequency necessary to be considered a polymorphism. Although it is true that the alleles responsible for most clearly inherited clinical conditions are rare, some alleles that have a profound effect on health—such as alleles of genes encoding enzymes that metabolize drugs (for example, sensitivity to abacavir in some individuals infected with human immunodeficiency virus [HIV]) (Case 1), or the sickle cell mutation in African and African American populations (see Chapter 11) (Case 42)—are relatively common. Nonetheless, these are exceptions, and, as more and more genetic variation is discovered and catalogued, it is clear that the vast majority of variants in the genome, whether common or rare, reflect differences in DNA sequence that have no known significance to health.

Polymorphisms are key elements for the study of human and medical genetics. The ability to distinguish different inherited forms of a gene or different segments of the genome provides critical tools for a wide array of applications, both in research and in clinical practice (see Box).

INHERITED VARIATION AND POLYMORPHISM IN DNA

The original Human Genome Project and the subsequent study of now many thousands of individuals worldwide have provided a vast amount of DNA sequence information. With this information in hand, one can begin to characterize the types and frequencies of polymorphic variation found in the human genome and to generate catalogues of human DNA sequence

---

POLYMORPHISMS AND INHERITED VARIATION IN HUMAN AND MEDICAL GENETICS

Allelic variants can be used as "markers" for tracking the inheritance of the corresponding segment of the genome in families and in populations. Such variants can be used as follows:

- As powerful research tools for mapping a gene to a particular region of a chromosome by linkage analysis or by allelic association (see Chapter 10)
- For prenatal diagnosis of genetic disease and for detection of carriers of deleterious alleles (see Chapter 17), as well as in blood banking and tissue typing for transfusions and organ transplantation
- In forensic applications such as identity testing for determining paternity, identifying remains of crime victims, or matching a suspect's DNA to that of the perpetrator (this chapter)
- In the ongoing efforts to provide genomic-based personalized medicine (see Chapter 18) in which one tailors an individual's medical care to whether or not he or she carries variants that increase or decrease the risk for common adult disorders (such as coronary heart disease, cancer, and diabetes; see Chapter 8) or that influence the efficacy or safety of particular medications

---

diversity around the globe. DNA polymorphisms can be classified according to how the DNA sequence varies between the different alleles (Table 4-2 and Figs. 4-1 and 4-2).

Single Nucleotide Polymorphisms

The simplest and most common of all polymorphisms are **single nucleotide polymorphisms (SNPs)**. A locus characterized by a SNP usually has only two alleles, corresponding to the two different bases occupying that particular location in the genome (see Fig. 4-1). As mentioned previously, SNPs are common and are observed on average once every 1000 bp in the genome. However, the distribution of SNPs is uneven around the genome; many more SNPs are found in noncoding parts of the genome, in introns and in sequences that are some distance from known genes. Nonetheless, there is still a significant number of SNPs that do occur in genes and other known functional elements in the genome. For the set of protein-coding genes, over 100,000 exonic SNPs have been documented to date. Approximately half of these do not alter the predicted amino acid sequence of the encoded protein and are thus termed **synonymous**, whereas the other half do alter the amino acid sequence and are said to be **nonsynonymous**. Other SNPs introduce or change a stop codon (see Table 3-1), and yet others alter a known splice site; such SNPs are candidates to have significant functional consequences.

The significance for health of the vast majority of SNPs is unknown and is the subject of ongoing research.

TABLE 4-2 Common Variation in the Human Genome

Type of Variation	Size Range (approx.)	Basis for the Polymorphism	Number of Alleles
Single nucleotide polymorphisms	1 bp	Substitution of one or another base pair at a particular location in the genome	Usually 2
Insertion/deletions (indels)	1 bp to > 100 bp	*Simple*: Presence or absence of a short segment of DNA 100-1000 bp in length *Microsatellites*: Generally, a 2-, 3-, or 4-nucleotide unit repeated in tandem 5-25 times	*Simple*: 2 *Microsatellites*: typically 5 or more
Copy number variants	10 kb to > 1 Mb	Typically the presence or absence of 200-bp to 1.5-Mb segments of DNA, although tandem duplication of 2, 3, 4, or more copies can also occur	2 or more
Inversions	Few bp to > 1 Mb	A DNA segment present in either of two orientations with respect to the surrounding DNA	2

bp, Base pair; kb, kilobase pair; Mb, megabase pair.

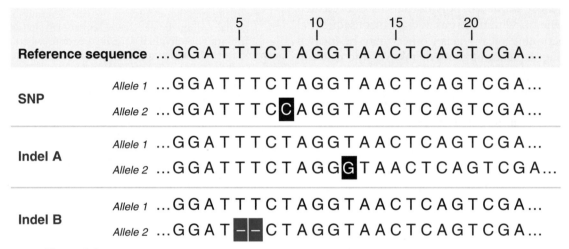

Figure 4-1 Three polymorphisms in genomic DNA from the segment of the human genome reference assembly shown at the top (see also Fig. 2-6). The single nucleotide polymorphism (SNP) at position 8 has two alleles, one with a T (corresponding to the reference sequence) and one with a C. There are two indels in this region. At indel A, allele 2 has an insertion of a G between positions 11 and 12 in the reference sequence (allele 1). At indel B, allele 2 has a 2-bp deletion of positions 5 and 6 in the reference sequence.

The fact that SNPs are common does not mean that they are without effect on health or longevity. What it does mean is that any effect of common SNPs is likely to involve a relatively subtle altering of disease susceptibility rather than a direct cause of serious illness.

Insertion-Deletion Polymorphisms

A second class of polymorphism is the result of variations caused by **insertion** or **deletion** (**in/dels** or simply **indels**) of anywhere from a single base pair up to approximately 1000 bp, although larger indels have been documented as well. Over a million indels have been described, numbering in the hundreds of thousands in any one individual's genome. Approximately half of all indels are referred to as "simple" because they have only two alleles—that is, the presence or absence of the inserted or deleted segment (see Fig. 4-1).

Microsatellite Polymorphisms

Other indels, however, are multiallelic due to variable numbers of the segment of DNA that is inserted in tandem at a particular location, thereby constituting what is referred to as a **microsatellite**. They consist of stretches of DNA composed of units of two, three, or four nucleotides, such as TGTGTG, CAACAACAA, or AAATAAATAAAT, repeated between one and a few dozen times at a particular site in the genome (see Fig. 4-2). The different alleles in a microsatellite polymorphism are the result of differing numbers of repeated nucleotide units contained within any one microsatellite and are therefore sometimes also referred to as **short tandem repeat (STR) polymorphisms**. A microsatellite locus often has many alleles (repeat lengths) that can be rapidly evaluated by standard laboratory procedures to distinguish different individuals and to infer familial

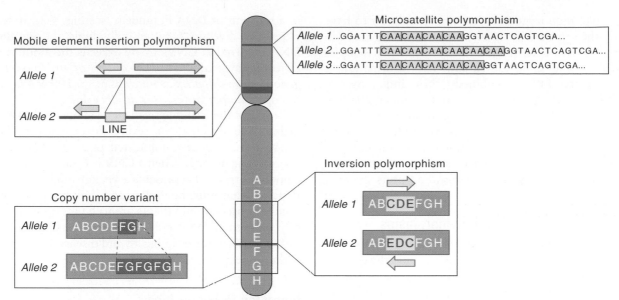

Figure 4-2 **Examples of polymorphism in the human genome larger than SNPs.** *Clockwise from upper right*: The microsatellite locus has three alleles, with four, five, or six copies of a CAA trinucleotide repeat. The inversion polymorphism has two alleles corresponding to the two orientations (indicated by the *arrows*) of the genomic segment shown in *green*; such inversions can involve regions up to many megabases of DNA. Copy number variants involve deletion or duplication of hundreds of kilobase pairs to over a megabase of genomic DNA. In the example shown, allele 1 contains a single copy, whereas allele 2 contains three copies of the chromosomal segment containing the F and G genes; other possible alleles with zero, two, four, or more copies of F and G are not shown. The mobile element insertion polymorphism has two alleles, one with and one without insertion of an approximately 6 kb LINE repeated retroelement; the insertion of the mobile element changes the spacing between the two genes and may alter gene expression in the region.

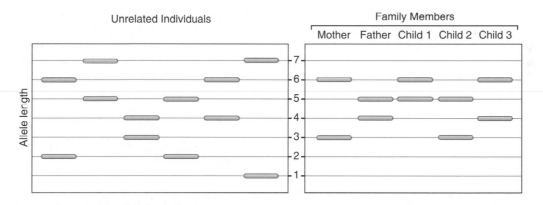

Figure 4-3 **A schematic of a hypothetical microsatellite marker in human DNA.** The different-sized alleles (numbered 1 to 7) correspond to fragments of genomic DNA containing different numbers of copies of a microsatellite repeat, and their relative lengths are determined by separating them by gel electrophoresis. The shortest allele (allele 1) migrates toward the bottom of the gel, whereas the longest allele (allele 7) remains closest to the top. *Left*, For this multiallelic microsatellite, each of the six unrelated individuals has two different alleles. *Right*, Within a family, the inheritance of alleles can be followed from each parent to each of the three children.

relationships (Fig. 4-3). Many tens of thousands of microsatellite polymorphic loci are known throughout the human genome.

Microsatellites are a particularly useful group of indels. Determining the alleles at multiple microsatellite loci is currently the method of choice for **DNA fingerprinting** used for identity testing. For example, the Federal Bureau of Investigation (FBI) in the United States currently uses the collection of alleles at 13 such loci for its DNA fingerprinting panel. Two individuals (other than monozygotic twins) are so unlikely to have exactly the same alleles at all 13 loci that the panel will allow definitive determination of whether two samples came from the same individual. The information is stored in the FBI's Combined DNA Index System (CODIS), which has grown as of December 2014 to include over 11,548,700 offender profiles, 1,300,000 arrestee profiles, and 601,600 forensic profiles (material obtained at crime scenes). Many states and the U.S. Department of Defense have similar databases of DNA fingerprints, as do corresponding units in other countries.

Mobile Element Insertion Polymorphisms

Nearly half of the human genome consists of families of repetitive elements that are dispersed around the genome (see Chapter 2). Although most of the copies of these repeats are stationary, some of them are mobile and contribute to human genetic diversity through the process of **retrotransposition,** a process that involves transcription into an RNA, reverse transcription into a DNA sequence, and insertion (i.e., transposition) into another site in the genome, as we introduced in Chapter 3 in the context of processed pseudogenes. The two most common mobile element families are the *Alu* and LINE families of repeats, and nearly 10,000 mobile element insertion polymorphisms have been described in different populations. Each polymorphic locus consists of two alleles, one with and one without the inserted mobile element (see Fig. 4-2). Mobile element polymorphisms are found on all human chromosomes; although most are found in nongenic regions of the genome, a small proportion of them are found within genes. At least 5000 of these polymorphic loci have an insertion frequency of greater than 10% in various populations.

Copy Number Variants

Another important type of human polymorphism includes **copy number variants (CNVs).** CNVs are conceptually related to indels and microsatellites but consist of variation in the number of copies of larger segments of the genome, ranging in size from 1000 bp to many hundreds of kilobase pairs. Variants larger than 500 kb are found in 5% to 10% of individuals in the general population, whereas variants encompassing more than 1 Mb are found in 1% to 2%. The largest CNVs are sometimes found in regions of the genome characterized by repeated blocks of homologous sequences called **segmental duplications** (or **segdups**). Their importance in mediating duplication and deletion of the corresponding segments is discussed further in Chapter 6 in the context of various chromosomal syndromes.

Smaller CNVs in particular may have only two alleles (i.e., the presence or absence of a segment), similar to indels in that regard. Larger CNVs tend to have multiple alleles due to the presence of different numbers of copies of a segment of DNA in tandem (see Fig. 4-2). In terms of genome diversity between individuals, the amount of DNA involved in CNVs vastly exceeds the amount that differs because of SNPs. *The content of any two human genomes can differ by as much as 50 to 100 Mb because of copy number differences at CNV loci.*

Notably, the variable segment at many CNV loci can include one to as many as several dozen genes, and thus CNVs are frequently implicated in traits that involve altered gene dosage. When a CNV is frequent enough to be polymorphic, it represents a background of common variation that must be understood if alterations in copy number observed in patients are to be interpreted properly. As with all DNA polymorphism, the significance of different CNV alleles in health and disease susceptibility is the subject of intensive investigation.

Inversion Polymorphisms

A final group of polymorphisms to be discussed is inversions, which differ in size from a few base pairs to large regions of the genome (up to several megabase pairs) that can be present in either of two orientations in the genomes of different individuals (see Fig. 4-2). Most inversions are characterized by regions of sequence homology at the edges of the inverted segment, implicating a process of homologous recombination in the origin of the inversions. In their balanced form, inversions, regardless of orientation, do not involve a gain or loss of DNA, and the inversion polymorphisms (with two alleles corresponding to the two orientations) can achieve substantial frequencies in the general population. However, anomalous recombination can result in the duplication or deletion of DNA located between the regions of homology, associated with clinical disorders that we will explore further in Chapters 5 and 6.

THE ORIGIN AND FREQUENCY OF DIFFERENT TYPES OF MUTATIONS

Along the spectrum of diversity from rare variants to more common polymorphisms, the different kinds of mutations arise in the context of such fundamental processes of cell division as DNA replication, DNA repair, DNA recombination, and chromosome segregation in mitosis or meiosis. The **frequency of mutations per locus per cell division** is a basic measure of how error prone these processes are, which is of fundamental importance for genome biology and evolution. However, of greatest importance to medical geneticists is the **frequency of mutations per disease locus per generation**, rather than the overall mutation rate across the genome per cell division. Measuring disease-causing mutation rates can be difficult, however, because many mutations cause early embryonic lethality before the mutation can be recognized in a fetus or newborn, or because some people with a disease-causing mutation may manifest the condition only late in life or may never show signs of the disease. Despite these limitations, we have made great progress is determining the overall frequency—sometimes referred to as the **genetic load**—of all mutations affecting the human species.

The major types of mutation briefly introduced earlier occur at appreciable frequencies in many different cells in the body. In the practice of genetics, we are principally concerned with inherited genome variation; however, all such variation had to originate as a new *(de novo)* change occurring in germ cells. At that point, such a variant would be quite rare in the population (occurring just once), and its ultimate frequency in the population over time depends on chance and on the principles of inheritance and population genetics (see Chapters 7 and 9). Although the original mutation would have occurred only in the DNA of cells in the **germline,** anyone who inherits that mutation would then carry it as a constitutional mutation in all the cells of the body.

In contrast, **somatic mutations** occur throughout the body but cannot be transmitted to the next generation. Given the rate of mutation (see later in this section), one would predict that, in fact, every cell in an individual has a slightly different version of his or her genome, depending on the number of cell divisions that have occurred since conception to the time of sample acquisition. In highly proliferative tissues, such as intestinal epithelial cells or hematopoietic cells, such genomic heterogeneity is particularly likely to be apparent. However, most such mutations are not typically detected, because, in clinical testing, one usually sequences DNA from collections of many millions of cells; in such a collection, the most prevalent base at any position in the genome will be the one present at conception, and rare somatic mutations will be largely invisible and unascertained. Such mutations can be of clinical importance, however, in disorders caused by mutation in only a subset of cells in certain tissues, leading to somatic mosaicism (see Chapter 7).

The major exception to the expectation that somatic mutations will be typically undetected within any multicell DNA sample is in cancer, in which the mutational basis for the origins of cancer and the clonal nature of tumor evolution drives certain somatic changes to be present in essentially all the cells of a tumor. Indeed, 1000 to 10,000 somatic mutations (and sometimes many more) are readily found in the genomes of most adult cancers, with mutation frequencies and patterns specific to different cancer types (see Chapter 15).

Chromosome Mutations

Mutations that produce a change in chromosome number because of chromosome missegregation are among the most common mutations seen in humans, with a rate of one mutation per 25 to 50 meiotic cell divisions. This estimate is clearly a minimal one because the developmental consequences of many such events are likely so severe that the resulting fetuses are aborted spontaneously shortly after conception without being detected (see Chapters 5 and 6).

Regional Mutations

Mutations affecting the structure or regional organization of chromosomes can arise in a number of different ways. Duplications, deletions, and inversions of a segment of a single chromosome are predominantly the result of homologous recombination between DNA segments with high sequence homology located at more than one site in a region of a chromosome. Not all structural mutations are the result of homologous recombination, however. Others, such as chromosome translocations and some inversions, can occur at the sites of spontaneous double-stranded DNA breaks. Once breakage occurs at two places anywhere in the genome, the two broken ends can be joined together even without any obvious homology in the sequence between the two ends (a process termed *nonhomologous end-joining repair*). Examples of such mutations will be discussed in depth in Chapter 6.

Gene Mutations

Gene or DNA mutations, including base pair substitutions, insertions, and deletions (Fig. 4-4), can originate by either of two basic mechanisms: errors introduced during DNA replication or mutations arising from a failure to properly repair DNA after damage. Many such mutations are spontaneous, arising during the normal (but imperfect) processes of DNA replication and repair, whereas others are induced by physical or chemical agents called **mutagens.**

DNA Replication Errors

The process of DNA replication (see Fig. 2-4) is typically highly accurate; the majority of replication errors (i.e., inserting a base other than the complementary base that would restore the base pair at that position in the double helix) are rapidly removed from the DNA and corrected by a series of DNA repair enzymes that first recognize which strand in the newly synthesized double helix contains the incorrect base and then replace it with the proper complementary base, a process termed **DNA proofreading.** DNA replication needs to be a remarkably accurate process; otherwise, the burden of mutation on the organism and the species would be intolerable. The enzyme DNA polymerase faithfully duplicates the two strands of the double helix based on strict base-pairing rules (A pairs with T, C with G) but introduces one error every 10 million bp. Additional proofreading then corrects more than 99.9% of these

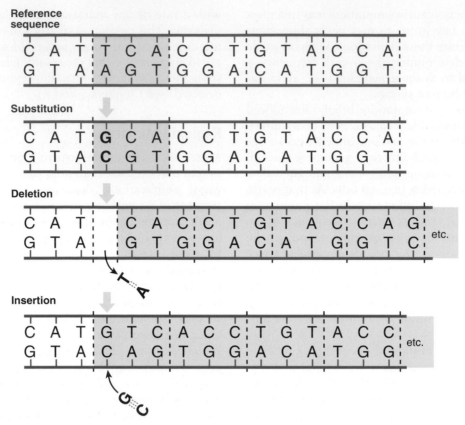

Figure 4-4 Examples of mutations in a portion of a hypothetical gene with five codons shown (delimited by the *dotted lines*). The first base pair of the second codon in the reference sequence (shaded in *blue*) is mutated by a base substitution, deletion, or insertion. The base substitution of a G for the T at this position leads to a codon change (shaded in *green*) and, assuming that the upper strand is the sense or coding strand, a predicted nonsynonymous change from a serine to an alanine in the encoded protein (see genetic code in Table 3-1); all other codons remain unchanged. Both the single base pair deletion and insertion lead to a frameshift mutation in which the translational reading frame is altered for all subsequent codons (shaded in *green*), until a termination codon is reached.

errors of DNA replication. Thus the overall mutation rate per base as a result of replication errors is a remarkably low 1×10^{-10} per cell division—*fewer than one mutation per genome per cell division.*

Repair of DNA Damage

It is estimated that, in addition to replication errors, between 10,000 and 1 million nucleotides are damaged per human cell per day by spontaneous chemical processes such as depurination, demethylation, or deamination; by reaction with chemical mutagens (natural or otherwise) in the environment; and by exposure to ultraviolet or ionizing radiation. Some but not all of this damage is repaired. Even if the damage is recognized and excised, the repair machinery may create mutations by introducing incorrect bases. Thus, in contrast to replication-related DNA changes, which are usually corrected through proofreading mechanisms, nucleotide changes introduced by DNA damage and repair often result in permanent mutations.

A particularly common spontaneous mutation is the substitution of T for C (or A for G on the other strand).

The explanation for this observation comes from considering the major form of epigenetic modification in the human genome, DNA methylation, introduced in Chapter 3. Spontaneous deamination of 5-methylcytosine to thymidine (compare the structures of cytosine and thymine in Fig. 2-2) in the CpG doublet gives rise to C to T or G to A mutations (depending on which strand the 5-methylcytosine is deaminated). Such spontaneous mutations may not be recognized by the DNA repair machinery and thus become established in the genome after the next round of DNA replication. More than 30% of all single nucleotide substitutions are of this type, and they occur at a rate 25 times greater than those of any other single nucleotide mutations. Thus the CpG doublet represents a true "hot spot" for mutation in the human genome.

Overall Rate of DNA Mutations

Although the rate of DNA mutations at specific loci has been estimated using a variety of approaches over the past 50 years, the overall impact of replication and repair errors on the occurrence of new mutations throughout the genome can now be determined directly by **whole-genome sequencing** of trios consisting of a child and both parents, looking for new mutations in the child that are not present in the genome sequence of either parent. The overall rate of new mutations averaged between maternal and paternal gametes is approximately 1.2×10^{-8} mutations per base pair per generation. *Thus every person is likely to receive approximately 75 new mutations in his or her genome from one or the other parent.* This rate, however, varies from gene to gene around the genome and perhaps from population to population or even individual to individual. Overall, this rate, combined with considerations of population growth and dynamics, predicts that there must be an enormous number of relatively new (and thus very rare) mutations in the current worldwide population of 7 billion individuals.

As might be predicted, the vast majority of these mutations will be single nucleotide changes in noncoding portions of the genome and will probably have little or no functional significance. Nonetheless, at the level of populations, the potential collective impact of these new mutations on genes of medical importance should not be overlooked. In the United States, for example, with over 4 million live births each year, approximately 6 million new mutations will occur in coding sequences; thus, even for a *single* protein-coding gene of average size, we can anticipate several hundred newborns each year with a new mutation in the coding sequence of that gene.

Conceptually similar studies have determined the rate of mutations in CNVs, where the generation of a new length variant depends on recombination, rather than on errors in DNA synthesis to generate a new base pair. The measured rate of formation of new CNVs ($\approx 1.2 \times 10^{-2}$ per locus per generation) is orders of magnitude higher than that of base substitutions.

Rate of Disease-Causing Gene Mutations

The most direct way of estimating the rate of disease-causing mutations per locus per generation is to measure the incidence of new cases of a genetic disease that is not present in either parent and is caused by a single mutation that causes a condition that is clearly recognizable in all neonates who carry that mutation. **Achondroplasia,** a condition of reduced bone growth leading to short stature (Case 2), is a condition that meets these requirements. In one study, seven achondroplastic children were born in a series of 242,257 consecutive births. All seven were born to parents of normal stature, and, because achondroplasia always manifests when a mutation is present, all were considered to represent new mutations. The new mutation rate at this locus can be calculated to be seven new mutations in a total of $2 \times 242,257$ copies of the relevant gene, or approximately 1.4×10^{-5} disease-causing mutations per locus per generation. This high mutation rate is particularly striking because it has been found that virtually all cases of achondroplasia are due to the *identical* mutation, a G to A mutation that changes a glycine codon to an arginine in the encoded protein.

The rate of gene mutations that cause disease has been estimated for a number of other disorders in which the occurrence of a new mutation was determined by the appearance of a detectable disease (Table 4-3). The measured rates for these and other disorders vary over a 1000-fold range, from 10^{-4} to 10^{-7} mutations per locus per generation. The basis for these differences may be related to some or all of the following: the size of different genes; the fraction of all mutations in that gene that will lead to the disease; the age and sex of the parent in whom the mutation occurred; the mutational mechanism; and the presence or absence of mutational hot spots in the gene. Indeed, the high rate of the particular site-specific mutation in achondroplasia may be partially explained by the fact that the mutation on the other strand is a C to T change in a position that undergoes CpG methylation and is a hot spot for mutation by deamination, as discussed earlier.

Notwithstanding this range of rates among different genes, the median gene mutation rate is approximately 1×10^{-6}. Given that there are at least 5000 genes in the human genome in which mutations are currently known to cause a discernible disease or other trait (see Chapter 7), *approximately 1 in 200 persons is likely to receive*

TABLE 4-3 Estimates of Mutation Rates for Selected Human Disease Genes

Disease	Locus (Protein)	Mutation Rate*
Achondroplasia (Case 2)	*FGFR3* (fibroblast growth factor receptor 3)	1.4×10^{-5}
Aniridia	*PAX6* (Pax6)	$2.9-5 \times 10^{-6}$
Duchenne muscular dystrophy (Case 14)	*DMD* (dystrophin)	$3.5-10.5 \times 10^{-5}$
Hemophilia A (Case 21)	*F8* (factor VIII)	$3.2-5.7 \times 10^{-5}$
Hemophilia B (Case 21)	*F9* (factor IX)	$2-3 \times 10^{-6}$
Neurofibromatosis, type 1 (Case 34)	*NF1* (neurofibromin)	$4-10 \times 10^{-5}$
Polycystic kidney disease, type 1 (Case 37)	*PKD1* (polycystin)	$6.5-12 \times 10^{-5}$
Retinoblastoma (Case 39)	*RB1* (Rb1)	$5-12 \times 10^{-6}$

*Expressed as mutations per locus per generation.
Based on data in Vogel F, Motulsky AG: *Human genetics,* ed 3, Berlin, 1997, Springer-Verlag.

a new mutation in a known disease-associated gene from one or the other parent.

Sex Differences and Age Effects on Mutation Rates

Because the DNA in sperm has undergone far more replication cycles than has the DNA in ova (see Chapter 2), there is greater opportunity for errors to occur; one might predict, then, that many mutations will be more often paternal rather than maternal in origin. Indeed, where this has been explored, new mutations responsible for certain conditions (e.g., achondroplasia, as we just discussed) are usually missense mutations that arise nearly always in the paternal germline. Furthermore, the older a man is, the more rounds of replication have preceded the meiotic divisions, and thus the frequency of paternal new mutations might be expected to increase with the age of the father. In fact, correlations of the increasing age of the father have been observed with the incidence of gene mutations for a number of disorders (including achondroplasia) and with the incidence of regional mutations involving CNVs in **autism spectrum disorders** (Case 5). In other diseases, however, the parent-of-origin and age effects on mutational spectra are, for unknown reasons, not as striking.

TYPES OF MUTATIONS AND THEIR CONSEQUENCES

In this section, we consider the nature of different mutations and their effect on the genes involved. Each type of mutation discussed here is illustrated by one or more disease examples. Notably, the specific mutation found in almost all cases of achondroplasia is the exception rather than the rule, and the mutations that underlie a single genetic disease are more typically heterogeneous among a group of affected individuals. Different cases of a particular disorder will therefore usually be caused by different underlying mutations (Table 4-4). In Chapters 11 and 12, we will turn to the ways in which mutations in specific disease genes cause these diseases.

Nucleotide Substitutions
Missense Mutations

A single nucleotide substitution (or **point mutation**) in a gene sequence, such as that observed in the example of achondroplasia just described, can alter the code in a triplet of bases and cause the nonsynonymous replacement of one amino acid by another in the gene product (see the genetic code in Table 3-1 and the example in Fig. 4-4). Such mutations are called **missense mutations** because they alter the coding (or "sense") strand of the gene to specify a different amino acid. Although not all missense mutations lead to an observable change in the function of the protein, the resulting protein may fail to

TABLE 4-4 Types of Mutation in Human Genetic Disease

Type of Mutation	Percentage of Disease-Causing Mutations
Nucleotide Substitutions	
• Missense mutations (amino acid substitutions)	50%
• Nonsense mutations (premature stop codons)	10%
• RNA processing mutations (destroy consensus splice sites, cap sites, and polyadenylation sites or create cryptic sites)	10%
• Splice-site mutations leading to frameshift mutations and premature stop codons	10%
• Long-range regulatory mutations	Rare
Deletions and Insertions	
• Addition or deletions of a small number of bases	25%
• Larger gene deletions, inversions, fusions, and duplications (may be mediated by DNA sequence homology either within or between DNA strands)	5%
• Insertion of a LINE or *Alu* element (disrupting transcription or interrupting the coding sequence)	Rare
• Dynamic mutations (expansion of trinucleotide or tetranucleotide repeat sequences)	Rare

work properly, may be unstable and rapidly degraded, or may fail to localize in its proper intracellular position. In many disorders, such as **β-thalassemia** (Case 44), most of the mutations detected in different patients are missense mutations (see Chapter 11).

Nonsense Mutations

Point mutations in a DNA sequence that cause the replacement of the normal codon for an amino acid by one of the three termination (or "stop") codons are called **nonsense mutations**. Because translation of messenger RNA (mRNA) ceases when a termination codon is reached (see Chapter 3), a mutation that converts a coding exon into a termination codon causes translation to stop partway through the coding sequence of the mRNA. The consequences of premature termination mutations are twofold. First, the mRNA carrying a premature mutation is often targeted for rapid degradation (through a cellular process known as **nonsense-mediated mRNA decay**), and no translation is possible. And second, even if the mRNA is stable enough to be translated, the truncated protein is usually so unstable that it is rapidly degraded within the cell (see Chapter 12 for examples).

Whereas some point mutations create a premature termination codon, others may destroy the normal termination codon and thus permit translation to continue until another termination codon in the mRNA is reached further downstream. Such a mutation will lead to an abnormal protein product with additional amino acids at its carboxyl terminus, and may also disrupt regulatory functions normally provided by the 3′ untranslated region downstream from the normal stop codon.

Mutations Affecting RNA Transcription, Processing, and Translation

The normal mechanism by which initial RNA transcripts are made and then converted into mature mRNAs (or final versions of noncoding RNAs) requires a series of modifications, including transcription factor binding, 5′ capping, polyadenylation, and splicing (see Chapter 3). All of these steps in RNA maturation depend on specific sequences within the RNA. In the case of splicing, two general classes of splicing mutations have been described. For introns to be excised from unprocessed RNA and the exons spliced together to form a mature RNA requires particular nucleotide sequences located at or near the exon-intron (5′ donor site) or the intron-exon (3′ acceptor site) junctions. Mutations that affect these required bases at either the splice donor or acceptor site interfere with (and in some cases abolish) normal RNA splicing at that site. A second class of splicing mutations involves base substitutions that do not affect the donor or acceptor site sequences themselves but instead create alternative donor or acceptor sites that compete with the normal sites during RNA processing. Thus at least a proportion of the mature mRNA or noncoding RNA in such cases may contain improperly spliced intron sequences. Examples of both types of mutation are presented in Chapter 11.

For protein-coding genes, even if the mRNA is made and is stable, point mutations in the 5′ and 3′-untranslated regions can also contribute to disease by changing mRNA stability or translation efficiency, thereby reducing the amount of protein product that is made.

Deletions, Insertions, and Rearrangements

Mutations can also be caused by the insertion, deletion, or rearrangement of DNA sequences. Some deletions and insertions involve only a few nucleotides and are generally most easily detected by direct sequencing of that part of the genome. In other cases, a substantial segment of a gene or an entire gene is deleted, duplicated, inverted, or translocated to create a novel arrangement of gene sequences. Depending on the exact nature of the deletion, insertion, or rearrangement, a variety of different laboratory approaches can be used to detect the genomic alteration.

Some deletions and insertions affect only a small number of base pairs. When such a mutation occurs in a coding sequence and the number of bases involved is not a multiple of three (i.e., is not an integral number of codons), the reading frame will be altered beginning at the point of the insertion or deletion. The resulting mutations are called **frameshift mutations** (see Fig. 4-4). From the point of the insertion or deletion, a different sequence of codons is thereby generated that encodes incorrect amino acids followed by a termination codon in the shifted frame, typically leading to a functionally altered protein product. In contrast, if the number of base pairs inserted or deleted *is* a multiple of three, then no frameshift occurs and there will be a simple insertion or deletion of the corresponding amino acids in the otherwise normally translated gene product. Larger insertions or deletions, ranging from approximately 100 to more than 1000 bp, are typically referred to as "indels," as we saw in the case of polymorphisms earlier. They can affect multiple exons of a gene and cause major disruptions of the coding sequence.

One type of insertion mutation involves insertion of a mobile element, such as those belonging to the LINE family of repetitive DNA. It is estimated that, in any individual, approximately 100 copies of a particular subclass of the LINE family in the genome are capable of movement by **retrotransposition,** introduced earlier. Such movement not only generates genetic diversity in our species (see Fig. 4-2) but can also cause disease by insertional mutagenesis. For example, in some patients with the severe bleeding disorder **hemophilia A** (Case 21), LINE sequences several kilobase pairs long are found to be inserted into an exon in the factor VIII gene, interrupting the coding sequence and inactivating the gene. LINE insertions throughout the genome are also common in colon cancer, reflecting retrotransposition in somatic cells (see Chapter 15).

As we discussed in the context of polymorphisms earlier in this chapter, duplications, deletions, and inversions of a larger segment of a single chromosome are predominantly the result of homologous recombination between DNA segments with high sequence homology (Fig. 4-5). Disorders arising as a result of such exchanges can be due to a change in the dosage of otherwise wild-type gene products when the homologous segments lie outside the genes themselves (see Chapter 6). Alternatively, such mutations can lead to a change in the nature of the encoded protein itself when recombination occurs between different genes within a gene family (see Chapter 11) or between genes on different chromosomes (see Chapter 15). Abnormal pairing and recombination between two similar sequences in opposite orientation on a single strand of DNA leads to inversion. For example, nearly half of all cases of hemophilia A are due to recombination that inverts a number of exons, thereby disrupting gene structure and rendering the gene incapable of encoding a normal gene product (see Fig. 4-5).

Dynamic Mutations

The mutations in some disorders involve amplification of a simple nucleotide repeat sequence. For example, simple repeats such as $(CCG)_n$, $(CAG)_n$, or $(CCTG)_n$

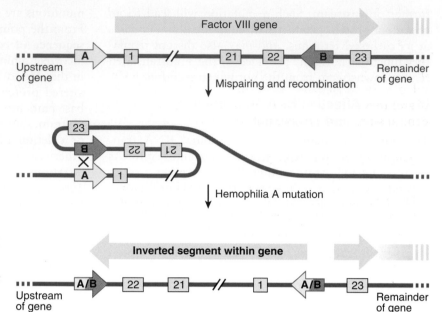

Figure 4-5 Inverted homologous sequences, labeled *A* and *B*, located 500 kb apart on the X chromosome, one upstream of the factor VIII gene, the other in an intron between exons 22 and 23 of the gene. Intrachromosomal mispairing and recombination results in inversion of exons 1 through 22 of the gene, thereby disrupting the gene and causing severe hemophilia.

located in the coding portion of an exon, in an untranslated region of an exon, or even in an intron may expand during gametogenesis, in what is referred to as a **dynamic mutation,** and interfere with normal gene expression or protein function. An expanded repeat in the coding region will generate an abnormal protein product, whereas repeat expansion in the untranslated regions or introns of a gene may interfere with transcription, mRNA processing, or translation. How dynamic mutations occur is not completely understood; they are conceptually similar to microsatellite polymorphisms but expand at a rate much higher than typically seen for microsatellite loci.

The involvement of simple nucleotide repeat expansions in disease is discussed further in Chapters 7 and 12. In disorders caused by dynamic mutations, marked parent-of-origin effects are well known and appear characteristic of the specific disease and/or the particular simple nucleotide repeat involved (see Chapter 12). Such differences may be due to fundamental biological differences between oogenesis and spermatogenesis but may also result from selection against gametes carrying certain repeat expansions.

VARIATION IN INDIVIDUAL GENOMES

The most extensive current inventory of the amount and type of variation to be expected in any given genome comes from the direct analysis of individual diploid human genomes. The first of such genome sequences, that of a male individual, was reported in 2007. Now, tens of thousands of individual genomes have been sequenced, some as part of large international research consortia exploring human genetic diversity in health and disease, and others in the context of clinical

sequencing to determine the underlying basis of a disorder in particular patients.

What degree of genome variation does one detect in such studies? Individual human genomes typically carry 5 to 10 million SNPs, of which—depending in part on the population—as many as a quarter to a third are novel (see Box). This suggests that the number of SNPs described for our species is still incomplete, although presumably the fraction of such novel SNPs will decrease as more and more genomes from more and more populations are sequenced.

Within this variation lie variants with known, likely, or suspected clinical impact. Based on studies to date, each genome carries 50 to 100 variants that have previously been implicated in known inherited conditions. In addition, each genome carries thousands of nonsynonymous SNPs in protein-coding genes around the genome, some of which would be predicted to alter protein function. Each genome also carries approximately 200 to 300 likely loss-of-function mutations, some of which are present at both alleles of genes in that individual. Within the clinical setting, this realization has important implications for the interpretation of genome sequence data from patients, particularly when trying to predict the impact of mutations in genes of currently unknown function (see Chapter 16).

An interesting and unanticipated aspect of individual genome sequencing is that the reference human genome assembly still lacks considerable amounts of undocumented and unannotated DNA that are discovered in literally every individual genome being sequenced. These "new" sequences are revealed only as additional genomes are sequenced. Thus the complete collection of all human genome sequences to be found in our current population of 7 billion individuals, estimated to be 20 to 40 Mb larger than the extant reference assembly, still remains to be fully elucidated.

As impressive as the current inventory of human genetic diversity is, it is clear that we are still in a mode of discovery; no doubt millions of additional SNPs and other variants remain to be uncovered, as does the degree to which any of them might affect an individual's clinical status in the context of wellness and health care.

VARIATION DETECTED IN A TYPICAL HUMAN GENOME

Individuals vary greatly in a wide range of biological functions, determined in part by variation among their genomes. Any individual genome will contain the following:

- ≈5-10 million SNPs (varies by population)
- 25,000-50,000 rare variants (private mutations or seen previously in < 0.5% of individuals tested)
- ≈75 new base pair mutations not detected in parental genomes
- 3-7 new CNVs involving ≈500 kb of DNA
- 200,000-500,000 indels (1-50 bp) (varies by population)
- 500-1000 deletions 1-45 kb, overlapping ≈200 genes
- ≈150 in-frame indels
- ≈200-250 shifts in reading frame
- 10,000-12,000 synonymous SNPs
- 8,000-11,000 nonsynonymous SNPs in 4,000-5,000 genes
- 175-500 rare nonsynonymous variants
- 1 new nonsynonymous mutation
- ≈100 premature stop codons
- 40-50 splice site-disrupting variants
- 250-300 genes with likely loss-of-function variants
- ≈25 genes predicted to be completely inactivated

Clinical Sequencing Studies

In the context of genomic medicine, a key question is to what extent variation in the sequence and/or expression of one's genome influences the likelihood of disease onset, determines or signals the natural history of disease, and/or provides clues relevant to the management of disease. As just discussed, variation in one's constitutional genome can have a number of different direct or indirect effects on gene function.

Sequencing of entire genomes (so-called **whole-genome sequencing**) or of the subset of genomes that include all of the known coding exons (so-called **whole-exome sequencing**) has been introduced in a number of clinical settings, as will be discussed in greater detail in Chapter 16. Both whole-exome and whole-genome sequencing have been used to detect *de novo* mutations (both point mutations and CNVs) in a variety of conditions of complex and/or unknown etiology, including, for example, various neurodevelopmental or neuropsychiatric conditions, such as autism, schizophrenia, epilepsy, or intellectual disability and developmental delay.

Clinical sequencing studies can target either germline or somatic variants. In cancer, especially, various strategies have been used to search for somatic mutations in tumor tissue to identify genes potentially relevant to cancer progression (see Chapter 15).

PERSONAL GENOMICS AND THE ROLE OF THE CONSUMER

The increasing ability to sequence individual genomes is not only enabling research and clinical laboratories, but also spawning a social and information revolution among consumers in the context of **direct-to-consumer (DTC) genomics**, in which testing of polymorphisms genome-wide and even sequencing of entire genomes is offered directly to potential customers, bypassing health professionals.

It is still largely unclear what degree of genome surveillance will be most useful for routine clinical practice, and this is likely to evolve rapidly in the case of specific conditions, as our knowledge increases, as professional practice guidelines are adopted, and as insurance companies react. Some groups have raised substantial concerns about privacy and about the need to regulate the industry. At the same time, however, other individuals are willing to make genome sequence data (and even medical information) available more or less publicly.

Attitudes in this area vary widely among professionals and the general public alike, depending on whether one views knowing the sequence of one's genome to be a fundamentally medical or personal activity. Critics of DTC testing and policymakers, in both the health industry and government, focus on issues of clinical utility, regulatory standards, medical oversight, availability of genetic counseling, and privacy. Proponents of DTC testing and even consumers themselves, on the other hand, focus more on freedom of information, individual rights, social and personal awareness, public education, and consumer empowerment.

The availability of individual genome information is increasingly a commercial commodity and a personal reality. In that sense, and notwithstanding or minimizing the significant scientific, ethical, and clinical issues that lie ahead, it is certain that individual genome sequences will be an active part of medical practice for today's students.

IMPACT OF MUTATION AND POLYMORPHISM

Although it will be self-evident to students of human genetics that new deleterious mutations or rare variants in the population may have clinical consequences, it may appear less obvious that *common* polymorphic variants can be medically relevant. For the proportion of polymorphic variation that occurs in the genes themselves, such loci can be studied by examining variation in the proteins encoded by the different alleles. It has long been estimated that any one individual is likely to carry two distinct alleles determining structurally differing polypeptides at approximately 20% of all protein-coding loci; when individuals from different geographic or ethnic groups are compared, an even greater fraction of proteins has been found to exhibit detectable polymorphism. In addition, even when the gene product is identical, the levels of expression of that product may be very different among different individuals, determined by a combination of genetic and epigenetic variation, as we saw in Chapter 3.

Thus a striking degree of biochemical individuality exists within the human species in its makeup of enzymes and other gene products. Furthermore, because the products of many of the encoded biochemical and regulatory pathways interact in functional and physiological networks, one may plausibly conclude that each individual, regardless of his or her state of health, has a unique, genetically determined chemical makeup and thus responds in a unique manner to environmental, dietary, and pharmacological influences. This concept of **chemical individuality**, first put forward over a century ago by Garrod, the remarkably prescient British physician introduced in Chapter 1, remains true today. The broad question of what is normal—an essential concept in human biology and in clinical medicine—remains very much an open one when it comes to the human genome.

The following chapters will explore this concept in detail, first in the context of genome and chromosome mutations (Chapters 5 and 6) and then in terms of gene mutations and polymorphisms that determine the inheritance of genetic disease (Chapter 7) and influence its likelihood in families and populations (Chapters 8 and 9).

GENERAL REFERENCES

Olson MV: Human genetic individuality, *Ann Rev Genomics Hum Genet* 13:1–27, 2012.

Strachan T, Read A: *Human molecular genetics*, ed 4, New York, 2010, Garland Science.

The 1000 Genomes Project Consortium: An integrated map of genetic variation from 1,092 human genomes, *Nature* 491:56–65, 2012.

Willard HF: The human genome: a window on human genetics, biology and medicine. In Ginsburg GS, Willard HF, editors: *Genomic and personalized medicine*, ed 2, New York, 2013, Elsevier.

REFERENCES FOR SPECIFIC TOPICS

Alkan C, Coe BP, Eichler EE: Genome structural variation discovery and genotyping, *Nature Rev Genet* 12:363–376, 2011.

Bagnall RD, Waseem N, Green PM, Giannelli F: Recurrent inversion breaking intron 1 of the factor VIII gene is a frequent cause of severe hemophilia A, *Blood* 99:168–174, 2002.

Crow JF: The origins, patterns and implications of human spontaneous mutation, *Nature Rev Genet* 1:40–47, 2000.

Gardner RJ: A new estimate of the achondroplasia mutation rate, *Clin Genet* 11:31–38, 1977.

Kong A, Frigge ML, Masson G, et al: Rate of *de novo* mutations and the importance of father's age to disease risk, *Nature* 488:471–475, 2012.

Lappalainen T, Sammeth M, Friedlander MR, et al: Transcriptome and genome sequencing uncovers functional variation in humans, *Nature* 501:506–511, 2013.

MacArthur DG, Balasubramanian S, Rrankish A, et al: A systematic survey of loss-of-function variants in human protein-coding genes, *Science* 335:823–828, 2012.

McBride CM, Wade CH, Kaphingst KA: Consumers' view of direct-to-consumer genetic information, *Ann Rev Genomics Hum Genet* 11:427–446, 2010.

Stewart C, Kural D, Stromberg MP, et al: A comprehensive map of mobile element insertion polymorphisms in humans, *PLoS Genet* 7:e1002236, 2011.

Sun JX, Helgason A, Masson G, et al: A direct characterization of human mutation based on microsatellites, *Nature Genet* 44:1161–1165, 2012.

PROBLEMS

1. Polymorphism can arise from a variety of mechanisms, with different consequences. Describe and contrast the types of polymorphism that can have the following effects:
   a. A change in dosage of a gene or genes
   b. A change in the sequence of multiple amino acids in the product of a protein-coding gene
   c. A change in the final structure of an RNA produced from a gene
   d. A change in the order of genes in a region of a chromosome
   e. No obvious effect

2. Aniridia is an eye disorder characterized by the complete or partial absence of the iris and is always present when a mutation occurs in the responsible gene. In one population, 41 children diagnosed with aniridia were born to parents of normal vision among 4.5 million births during a period of 40 years. Assuming that these cases were due to new mutations, what is the estimated mutation rate at the aniridia locus? On what assumptions is this estimate based, and why might this estimate be either too high or too low?

3. Which of the following types of polymorphism would be most effective for distinguishing two individuals from the general population: a SNP, a simple indel, or a microsatellite? Explain your reasoning.

4. Consider two cell lineages that differ from one another by a series of 100 cell divisions. Given the rate of mutation for different types of variation, how different would the genomes of those lineages be?

5. Compare the likely impact of each of the following on the overall rate of mutation detected in any given genome: age of the parents, hot spots of mutation, intrachromosomal homologous recombination, genetic variation in the parental genomes.

CHAPTER 5

Principles of Clinical Cytogenetics and Genome Analysis

Clinical cytogenetics is the study of chromosomes, their structure, and their inheritance, as applied to the practice of medicine. It has been apparent for over 50 years that chromosome abnormalities—microscopically visible changes in the number or structure of chromosomes—could account for a number of clinical conditions that are thus referred to as **chromosome disorders**. With their focus on the complete set of genetic material, cytogeneticists were the first to bring a genome-wide perspective to the practice of medicine. Today, chromosome analysis—with increasing resolution and precision at both the cytological and genomic levels—is an important diagnostic procedure in numerous areas of clinical medicine. Current genome analyses that use approaches to be explored in this chapter, including **chromosomal microarrays** and **whole-genome sequencing,** represent impressive improvements in capacity and resolution, but ones that are conceptually similar to microscopic methods focusing on chromosomes (Fig. 5-1).

Chromosome disorders form a major category of genetic disease. They account for a large proportion of all reproductive wastage, congenital malformations, and intellectual disability and play an important role in the pathogenesis of cancer. Specific cytogenetic disorders are responsible for hundreds of distinct syndromes that collectively are more common than all the single-gene diseases together. Cytogenetic abnormalities are present in nearly 1% of live births, in approximately 2% of pregnancies in women older than 35 years who undergo prenatal diagnosis, and in fully half of all spontaneous, first-trimester abortions.

The spectrum of analysis from microscopically visible changes in chromosome number and structure to anomalies of genome structure and sequence detectable at the level of whole-genome sequencing encompasses literally the entire field of medical genetics (see Fig. 5-1). In this chapter, we present the general principles of chromosome and genome analysis and focus on the **chromosome mutations** and **regional mutations** introduced in the previous chapter. We restrict our discussion to disorders due to genomic imbalance—either for the hundreds to thousands of genes found on individual chromosomes or for smaller numbers of genes located within a particular chromosome region. Application of these principles to some of the most common and best-known chromosomal and genomic disorders will then be presented in Chapter 6.

INTRODUCTION TO CYTOGENETICS AND GENOME ANALYSIS

The general morphology and organization of human chromosomes, as well as their molecular and genomic composition, were introduced in Chapters 2 and 3. To be examined by chromosome analysis for clinical purposes, cells must be capable of proliferation in culture. The most accessible cells that meet this requirement are white blood cells, specifically T lymphocytes. To prepare a short-term culture that is suitable for cytogenetic analysis of these cells, a sample of peripheral blood is obtained, and the white blood cells are collected, placed in tissue culture medium, and stimulated to divide. After a few days, the dividing cells are arrested in **metaphase** with chemicals that inhibit the mitotic spindle. Cells are treated with a hypotonic solution to release the chromosomes, which are then fixed, spread on slides, and stained by one of several techniques, depending on the particular diagnostic procedure being performed. They are then ready for analysis.

Although ideal for rapid clinical analysis, cell cultures prepared from peripheral blood have the disadvantage of being short-lived (3 to 4 days). Long-term cultures suitable for permanent storage or further studies can be derived from a variety of other tissues. Skin biopsy, a minor surgical procedure, can provide samples of tissue that in culture produce **fibroblasts,** which can be used for a variety of biochemical and molecular studies as well as for chromosome and genome analysis. White blood cells can also be transformed in culture to form **lymphoblastoid** cell lines that are potentially immortal. **Bone marrow** has the advantage of containing a high proportion of dividing cells, so that little if any culturing

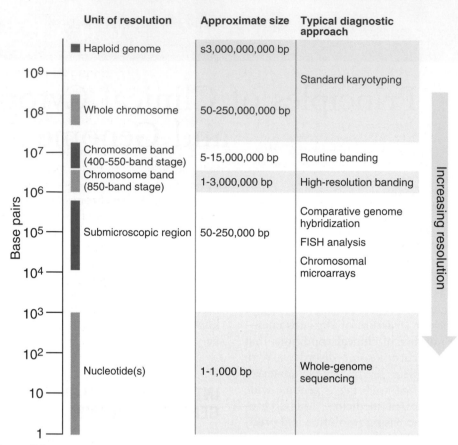

Figure 5-1 **Spectrum of resolution in chromosome and genome analysis.** The typical resolution and range of effectiveness are given for various diagnostic approaches used routinely in chromosome and genome analysis. See text for details and specific examples. FISH, Fluorescence in situ hybridization.

is required; however, it can be obtained only by the relatively invasive procedure of marrow biopsy. Its main use is in the diagnosis of suspected hematological malignancies. **Fetal cells** derived from amniotic fluid (amniocytes) or obtained by chorionic villus biopsy can also be cultured successfully for cytogenetic, genomic, biochemical, or molecular analysis. Chorionic villus cells can also be analyzed directly after biopsy, without the need for culturing. Remarkably, small amounts of **cell-free fetal DNA** are found in the maternal plasma and can be tested by whole-genome sequencing (see Chapter 17 for further discussion).

Molecular analysis of the genome, including whole-genome sequencing, can be carried out on any appropriate clinical material, provided that good-quality DNA can be obtained. Cells need not be dividing for this purpose, and thus it is possible to study DNA from tissue and tumor samples, for example, as well as from peripheral blood. Which approach is most appropriate for a particular diagnostic or research purpose is a rapidly evolving area as the resolution, sensitivity, and ease of chromosome and genome analysis increase (see Box).

Chromosome Identification

The 24 types of chromosome found in the human genome can be readily identified at the cytological level by specific staining procedures. The most common of these, Giemsa banding (**G banding**), was developed in the early 1970s and was the first widely used whole-genome analytical tool for research and clinical diagnosis (see Figs. 2-1 and 2-10). It has been the gold standard for the detection and characterization of structural and numerical genomic abnormalities in clinical diagnostic settings for both constitutional (postnatal or prenatal) and acquired (cancer) disorders.

G-banding and other staining procedures can be used to describe individual chromosomes and their variants or abnormalities, using an internationally accepted system of chromosome classification. Figure 5-2 is an ideogram of the banding pattern of a set of normal human chromosomes at metaphase, illustrating the alternating pattern of light and dark bands used for chromosome identification. The pattern of bands on each chromosome is numbered on each arm from the centromere to the telomere, as shown in detail in Figure 5-3 for several chromosomes. The identity of any particular

CLINICAL INDICATIONS FOR CHROMOSOME AND GENOME ANALYSIS

Chromosome analysis is indicated as a routine diagnostic procedure for a number of specific conditions encountered in clinical medicine. Some general clinical situations indicate a need for cytogenetic and genome analysis:

- *Problems of early growth and development.* Failure to thrive, developmental delay, dysmorphic facies, multiple malformations, short stature, ambiguous genitalia, and intellectual disability are frequent findings in children with chromosome abnormalities. Unless there is a definite nonchromosomal diagnosis, chromosome and genome analysis should be performed for patients presenting with any combination of such problems.

- *Stillbirth and neonatal death.* The incidence of chromosome abnormalities is much higher among stillbirths (up to approximately 10%) than among live births (approximately 0.7%). It is also elevated among infants who die in the neonatal period (approximately 10%). Chromosome analysis should be performed for all stillbirths and neonatal deaths that that do not have a clear basis to rule out a chromosome abnormality. In such cases, karyotyping (or other comprehensive ways of scanning the genome) is essential for accurate genetic counseling. These analyses may provide important information for prenatal diagnosis in future pregnancies.

- *Fertility problems.* Chromosome studies are indicated for women presenting with amenorrhea and for couples with a history of infertility or recurrent miscarriage. A chromosome abnormality is seen in one or the other parent in 3% to 6% of cases in which there is infertility or two or more miscarriages.

- *Family history.* A known or suspected chromosome or genome abnormality in a first degree relative is an indication for chromosome and genome analysis.

- *Neoplasia.* Virtually all cancers are associated with one or more chromosome abnormalities (see Chapter 15). Chromosome and genome evaluation in the tumor itself, or in bone marrow in the case of hematological malignant neoplasms, can offer diagnostic or prognostic information.

- *Pregnancy.* There is a higher risk for chromosome abnormality in fetuses conceived by women of increased age, typically defined as older than 35 years (see Chapter 17). Fetal chromosome and genome analysis should be offered as a routine part of prenatal care in such pregnancies. As a screening approach for the most common chromosome disorders, noninvasive prenatal testing using whole-genome sequencing is now available to pregnant women of all ages.

band (and thus the DNA sequences and genes within it) can be described precisely and unambiguously by use of this regionally based and hierarchical numbering system.

Human chromosomes are often classified into three types that can be easily distinguished at metaphase by the position of the **centromere**, the primary constriction visible at metaphase (see Fig. 5-2): **metacentric** chromosomes, with a more or less central centromere and arms of approximately equal length; **submetacentric**

chromosomes, with an off-center centromere and arms of clearly different lengths; and **acrocentric** chromosomes, with the centromere near one end. A potential fourth type of chromosome, **telocentric,** with the centromere at one end and only a single arm, does not occur in the normal human karyotype, but it is occasionally observed in chromosome rearrangements. The human acrocentric chromosomes (chromosomes 13, 14, 15, 21, and 22) have small, distinctive masses of chromatin known as **satellites** attached to their short arms by narrow stalks (called secondary constrictions). The stalks of these five chromosome pairs contain hundreds of copies of genes for ribosomal RNA (the major component of ribosomes; see Chapter 3) as well as a variety of repetitive sequences.

In addition to changes in banding pattern, nonstaining gaps—called **fragile sites**—are occasionally observed at particular sites on several chromosomes that are prone to regional genomic instability. Over 80 common fragile sites are known, many of which are heritable variants. A small proportion of fragile sites are associated with specific clinical disorders; the fragile site most clearly shown to be clinically significant is seen near the end of the long arm of the X chromosome in males with a specific and common form of X-linked intellectual disability, **fragile X syndrome** (Case 17), as well as in some female carriers of the same genetic defect.

High-Resolution Chromosome Analysis

The standard G-banded karyotype at a 400- to 550-band stage of resolution, as seen in a typical metaphase preparation, allows detection of deletions and duplications of greater than approximately 5 to 10 Mb anywhere in the genome (see Fig. 5-1). However, the sensitivity of G-banding at this resolution may be lower in regions of the genome in which the banding patterns are less specific.

To increase the sensitivity of chromosome analysis, high-resolution banding (also called **prometaphase banding**) can be achieved by staining chromosomes that have been obtained at an early stage of mitosis (prophase or prometaphase), when they are still in a relatively uncondensed state (see Chapter 2). High-resolution banding is especially useful when a subtle structural abnormality of a chromosome is suspected. Staining of prometaphase chromosomes can reveal up to 850 bands or even more in a haploid set, although this method is frequently replaced now by microarray analysis (see later). A comparison of the banding patterns at three different stages of resolution is shown for one chromosome in Figure 5-4, demonstrating the increase in diagnostic precision that one obtains with these longer chromosomes. Development of high-resolution chromosome analysis in the early 1980s allowed the discovery of a number of new so-called **microdeletion syndromes**

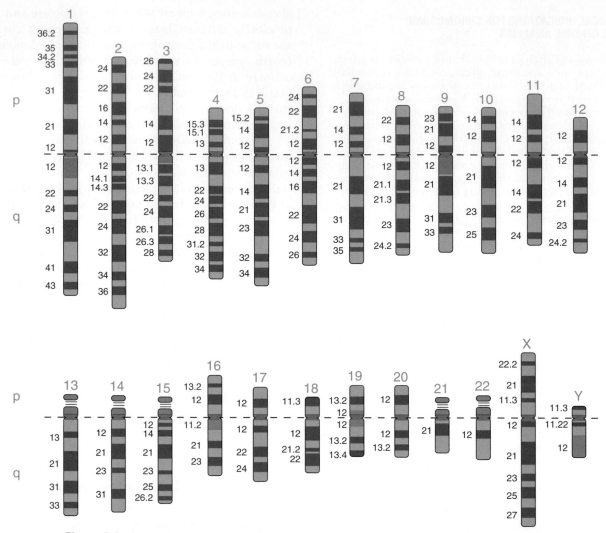

Figure 5-2 Ideogram showing G-banding patterns for human chromosomes at metaphase, with approximately 400 bands per haploid karyotype. As drawn, chromosomes are typically represented with the sister chromatids so closely aligned that they are not recognized as distinct entities. Centromeres are indicated by the primary constriction and narrow *dark gray* regions separating the p and q arms. For convenience and clarity, only the G-dark bands are numbered. For examples of full numbering scheme, see Figure 5-3. *See Sources & Acknowledgments.*

caused by smaller genomic deletions or duplications in the 2- to 3-Mb size range (see Fig. 5-1). However, the time-consuming and technically difficult nature of this method precludes its routine use for whole-genome analysis.

Fluorescence In Situ Hybridization

Targeted high-resolution chromosome banding was largely replaced in the early 1990s by **fluorescence in situ hybridization (FISH)**, a method for detecting the presence or absence of a particular DNA sequence or for evaluating the number or organization of a chromosome or chromosomal region in situ (literally, "in place") in the cell. This convergence of genomic and cytogenetic approaches—variously termed *molecular cytogenetics, cytogenomics,* or *chromonomics*—dramatically

expanded both the scope and precision of chromosome analysis in routine clinical practice.

FISH technology takes advantage of the availability of ordered collections of recombinant DNA clones containing DNA from around the entire genome, generated originally as part of the Human Genome Project. Clones containing specific human DNA sequences can be used as probes to detect the corresponding region of the genome in chromosome preparations or in interphase nuclei for a variety of research and diagnostic purposes, as illustrated in Figure 5-5:

- DNA probes specific for individual chromosomes, chromosomal regions, or genes can be labeled with different fluorochromes and used to identify particular chromosomal rearrangements or to rapidly diagnose the existence of an abnormal chromosome number in clinical material.

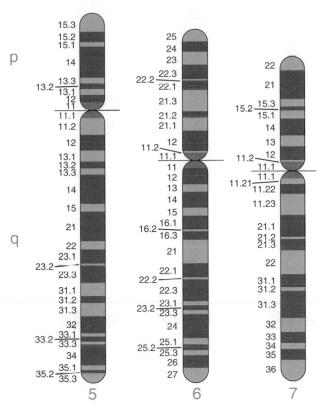

Figure 5-3 Examples of G-banding patterns for chromosomes 5, 6, and 7 at the 550-band stage of condensation. Band numbers permit unambiguous identification of each G-dark or G-light band, for example, chromosome 5p15.2 or chromosome 7q21.2. *See Sources & Acknowledgments.*

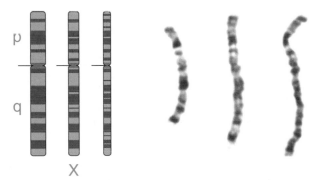

Figure 5-4 The X chromosome: ideograms and photomicrographs at metaphase, prometaphase, and prophase (*left to right*). *See Sources & Acknowledgments.*

- Repetitive DNA probes allow detection of satellite DNA or other repeated DNA elements localized to specific chromosomal regions. Satellite DNA probes, especially those belonging to the α-satellite family of centromere repeats (see Chapter 2), are widely used for determining the number of copies of a particular chromosome.

Although FISH technology provides much higher resolution and specificity than G-banded chromosome analysis, it does not allow for efficient analysis of the entire genome, and thus its use is limited by the need to

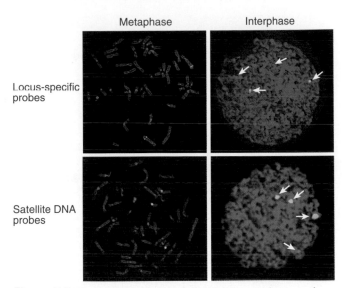

Figure 5-5 Fluorescence in situ hybridization to human chromosomes at metaphase and interphase, with different types of DNA probe. *Top,* Single-copy DNA probes specific for sequences within bands 4q12 (*red* fluorescence) and 4q31.1 (*green* fluorescence). *Bottom,* Repetitive α-satellite DNA probes specific for the centromeres of chromosomes 18 *(aqua),* X *(green),* and Y *(red).* *See Sources & Acknowledgments.*

target a specific genomic region based on a clinical diagnosis or suspicion.

Genome Analysis Using Microarrays

Although the G-banded karyotype remains the front-line diagnostic test for most clinical applications, it has been complemented or even replaced by genome-wide approaches for detecting copy number imbalances at higher resolution (see Fig. 5-1), extending the concept of targeted FISH analysis to test the entire genome. Instead of examining cells and chromosomes in situ one probe at a time, chromosomal microarray techniques simultaneously query the whole genome represented as an ordered array of genomic segments on a microscope slide containing overlapping or regularly spaced DNA segments that represent the entire genome. In one approach based on **comparative genome hybridization (CGH)**, one detects relative copy number gains and losses in a genome-wide manner by hybridizing two samples—one a control genome and one from a patient—to such microarrays. An excess of sequences from one or the other genome indicates an overrepresentation or underrepresentation of those sequences in the patient genome relative to the control (Fig. 5-6). An alternative approach uses "single nucleotide polymorphism (SNP) arrays" that contain versions of sequences corresponding to the two alleles of various SNPs around the genome (as introduced in Chapter 4). In this case, the relative representation and intensity of alleles in different regions of the genome indicate if a chromosome or chromosomal region is present at the appropriate dosage (see Fig. 5-6).

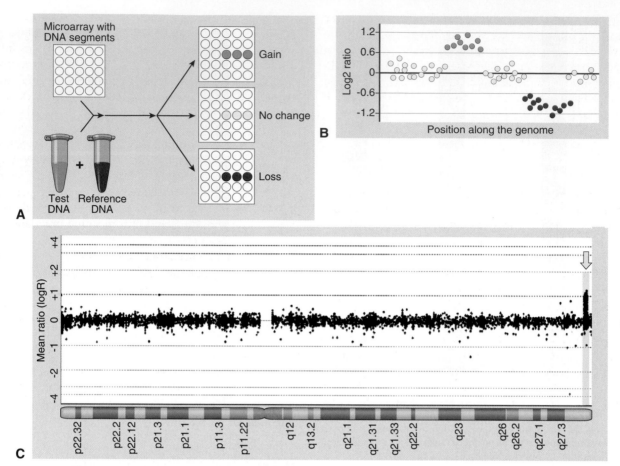

Figure 5-6 Chromosomal microarray to detect chromosome and genomic dosage. A, Schematic of an array assay based on comparative genome hybridization (CGH), where a patient's genome (denoted in *green*) is cohybridized to the array with a control reference genome (denoted in *red*). The probes are mixed and allowed to hybridize to their complementary sequences on the array. Relative intensities of hybridization of the two probes are measured, indicating equivalent dosage between the two genomes *(yellow)* or a relative gain *(green)* or loss *(red)* in the patient sample. **B,** A typical output plots the logarithm of the fluorescence ratios as a function of the position along the genome. **C,** Array CGH result for a patient with Rett syndrome **(Case 40)**, indicating a duplication of approximately 800 kb in band Xq28 containing the *MECP2* gene. LogR of fluorescence ratios are plotted along the length of the X chromosome. Each dot represents the ratio for an individual sequence on the array. Sequences corresponding to the *MECP2* gene and its surrounding region are duplicated in the patient's genome, leading to an increased ratio, indicated by the *green arrow* and *shaded box* in that region of the chromosome. *See Sources & Acknowledgments.*

For routine clinical testing of suspected chromosome disorders, probe spacing on the array provides a resolution as high as 250 kb over the entire unique portion of the human genome. A higher density of probes can be used to achieve even higher resolution (<25-50 kb) over regions of particular clinical interest, such as those associated with known developmental disorders or congenital anomalies (see Fig. 5-6; for other examples, see Chapter 6). This approach, which is being used in an increasing number of clinical laboratories, complements conventional karyotyping and provides a much more sensitive, high-resolution assessment of the genome. Microarrays have been used successfully to identify chromosome and genome abnormalities in children with unexplained developmental delay, intellectual disability, or birth defects, revealing a number of pathogenic genomic alterations that were not detectable by conventional G banding. Based on this significantly increased yield, genome-wide arrays are replacing the G-banded karyotype as the routine frontline test for certain patient populations.

Two important limitations of this technology bear mentioning, however. First, array-based methods measure only the relative copy number of DNA sequences but not whether they have been translocated or rearranged from their normal position(s) in the genome. Thus confirmation of suspected chromosome or genome abnormalities by karyotyping or FISH is important to determine the nature of an abnormality and thus its risk for recurrence, either for the individual or for other family members. And second, high-resolution genome analysis can reveal variants, in particular small differences in copy number, that are of uncertain clinical significance. An increasing number of such variants are being documented and catalogued even within the general population. As we saw in Chapter 4, many are likely to be benign **copy number variants**. Their existence underscores the unique nature of each individual's genome and emphasizes the diagnostic challenge of assessing what is considered a "normal" karyotype and what is likely to be pathogenic.

Genome Analysis by Whole-Genome Sequencing

At the extreme end but on the same spectrum as cytogenetic analysis and microarray analysis, the ultimate resolution for clinical tests to detect chromosomal and genomic disorders would be to sequence patient genomes in their entirety. Indeed, as the efficiency of whole-genome sequencing has increased and its costs have fallen, it is becoming increasingly practical to consider sequencing patient samples in a clinical setting (see Fig. 5-1).

The principles underlying such an approach are straightforward, because the number and composition of any particular segment of an individual's genome will be reflected in the DNA sequences generated from that genome. Although the sequences routinely obtained with today's technology are generally short (approximately 50 to 500 bp) compared to the size of a chromosome

or even a single gene, a genome with an abnormally low or high representation of those sequences from a particular chromosome or segment of a chromosome is likely to have a numerical or structural abnormality of that chromosome. To detect numerical abnormalities of an entire chromosome, it is generally not necessary to sequence a genome to completion; even a limited number of sequences that align to a particular chromosome of interest should reveal whether those sequences are found in the expected number (e.g., equivalent to two copies per diploid genome for an autosome) or whether they are significantly overrepresented or underrepresented (Fig. 5-7). This concept is now being applied to the prenatal diagnosis of fetal chromosome imbalance (see Chapter 17).

To detect balanced rearrangements of the genome, however, in which no DNA in the genome is either gained or lost, a more complete genome sequence is required. Here, instead of sequences that align perfectly to the reference human genome sequence, one finds rare sequences that align to two *different* and *normally noncontiguous* regions in the reference sequence (whether on the same chromosome or on different chromosomes) (see Fig. 5-7). This approach has been used to identify the specific genes involved in some cancers, and in children with various congenital defects due to translocations, involving the juxtaposition of sequences that are

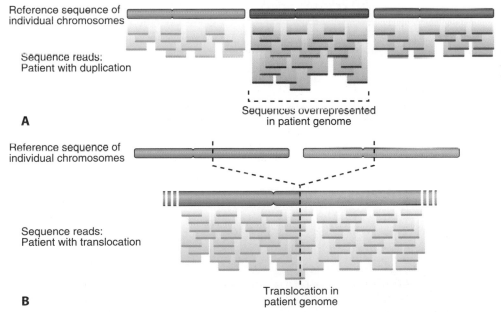

Figure 5-7 Strategies for detection of numerical and structural chromosome abnormalities by whole-genome sequence analysis. Although only a small number of reads are illustrated schematically here, in practice many millions of sequence reads are analyzed and aligned to the reference genome to obtain statistically significant support for a diagnosis of aneuploidy or a structural chromosome abnormality. **A,** Alignment of sequence reads from a patient's genome to the reference sequence of three individual chromosomes. Overrepresentation of sequences from the *red* chromosome indicates that the patient is aneuploid for this chromosome. **B,** Alignment of sequence reads from a patient's genome to the reference sequence of two chromosomes reveals a number of reads that contain contiguous sequences from *both* chromosomes. This indicates a translocation in the patient's genome involving the *blue* and *orange* chromosomes at the positions designated by the *dotted lines.*

normally located on different chromosomes (see Chapters 6 and 15).

CHROMOSOME ABNORMALITIES

Abnormalities of chromosomes may be either numerical or structural and may involve one or more autosomes, sex chromosomes, or both simultaneously. The overall incidence of chromosome abnormalities is approximately 1 in 154 live births (Fig. 5-8), and their impact is therefore substantial, both in clinical medicine and for society. By far the most common type of clinically significant chromosome abnormality is **aneuploidy,** an abnormal chromosome number due to an extra or missing chromosome. An aneuploid karyotype is always associated with physical or mental abnormalities or both. **Structural abnormalities** (rearrangements involving one or more chromosomes) are also relatively common (see Fig. 5-8). Depending on whether or not a structural rearrangement leads to an imbalance of genomic content, these may or may not have a phenotypic effect. However, as explained later in this chapter, even balanced chromosome abnormalities may be at an increased risk for abnormal offspring in the subsequent generation.

Chromosome abnormalities are described by a standard set of abbreviations and nomenclature that indicate the nature of the abnormality and (in the case of analyses performed by FISH or microarrays) the technology used. Some of the more common abbreviations and examples of abnormal karyotypes and abnormalities are listed in Table 5-1.

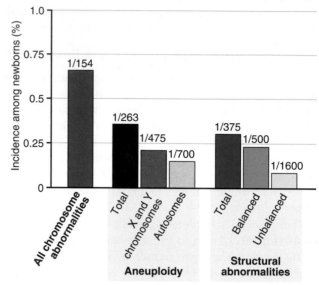

Figure 5-8 Incidence of chromosome abnormalities in newborn surveys, based on chromosome analysis of over 68,000 newborns. *See Sources & Acknowledgments.*

TABLE 5-1 Some Abbreviations Used for Description of Chromosomes and Their Abnormalities, with Representative Examples

Abbreviation	Meaning	Example	Condition
		46,XX	Normal female karyotype
		46,XY	Normal male karyotype
cen	Centromere		
del	Deletion	46,XX,del(5)(q13)	Female with terminal deletion of one chromosome 5 distal to band 5q13
der	Derivative chromosome	der(1)	Translocation chromosome derived from chromosome 1 and containing the centromere of chromosome 1
dic	Dicentric chromosome	dic(X;Y)	Translocation chromosome containing the centromeres of both the X and Y chromosomes
dup	Duplication		
inv	Inversion	inv(3)(p25q21)	Pericentric inversion of chromosome 3
mar	Marker chromosome	47,XX,+mar	Female with an extra, unidentified chromosome
mat	Maternal origin	47,XY,+der(1)mat	Male with an extra der(1) chromosome inherited from his mother
p	Short arm of chromosome		
pat	Paternal origin		
q	Long arm of chromosome		
r	Ring chromosome	46,X,r(X)	Female with ring X chromosome
rob	Robertsonian translocation	rob(14;21)(q10;q10)	Breakage and reunion have occurred at band 14q10 and band 21q10 in the centromeric regions of chromosomes 14 and 21
t	Translocation	46,XX,t(2;8)(q22;p21)	Female with balanced translocation between chromosomes 2 and 8, with breaks in bands 2q22 and 8p21
+	Gain of	47,XX,+21	Female with trisomy 21
−	Loss of	45,XY,−22	Male with monosomy 22
/	Mosaicism	46,XX/47,XX,+21	Female with two populations of cells, one with a normal karyotype and one with trisomy 21

Abbreviations from Shaffer LG, McGowan-Jordan J, Schmid M, editors: *ISCN 2013: an international system for human cytogenetic nomenclature*, Basel, 2013, Karger.

Gene Dosage, Balance and Imbalance

For chromosome and genomic disorders, it is the *quantitative* aspects of gene expression that underlie disease, in contrast to single-gene disorders, in which pathogenesis often reflects *qualitative* aspects of a gene's function. The clinical consequences of any particular chromosome abnormality will depend on the resulting imbalance of parts of the genome, the specific genes contained in or affected by the abnormality, and the likelihood of its transmission to the next generation.

The central concept for thinking about chromosome and genomic disorders is that of **gene dosage** and its balance or imbalance. As we shall see in later chapters, this same concept applies generally to considering some single-gene disorders and their underlying mutational basis (see Chapters 7, 11, and 12); however, it takes on uniform importance for chromosome abnormalities, where we are generally more concerned with the dosage of genes within the relevant chromosomal region than with the actual normal or abnormal sequence of those genes. Here, the sequence of the genes is typically quite unremarkable and would not lead to any clinical condition except for the fact that their dosage is incorrect.

Most genes in the human genome are present in two doses and are expressed from both copies. Some genes, however, are expressed from only a single copy (e.g., imprinted genes and X linked genes subject to X inactivation; see Chapter 3). Extensive analysis of clinical cases has demonstrated that the relative dosage of these genes is critical for normal development. One or three doses instead of two is generally not conducive to normal function for a gene or set of genes that are typically expressed from two copies. Similarly, abnormalities of genomic imprinting or X inactivation that cause the anomalous expression of two copies of a gene or set of genes instead of one invariably lead to clinical disorders.

Predicting clinical outcomes for chromosomal and genomic disorders can be an enormous challenge for genetic counseling, particularly in the prenatal setting. Many such diagnostic dilemmas will be presented throughout this section and in Chapters 6 and 17, but there are a number of general principles that should be kept in mind as we explore specific types of chromosome abnormality in the sections that follow (see Box).

Abnormalities of Chromosome Number

A chromosome complement with any chromosome number other than 46 is said to be **heteroploid**. An exact multiple of the haploid chromosome number (n) is called **euploid**, and any other chromosome number is **aneuploid**.

UNBALANCED KARYOTYPES AND GENOMES IN LIVEBORNS: GENERAL GUIDELINES FOR COUNSELING

- *Monosomies are more deleterious than trisomies.* Complete monosomies are generally not viable, except for monosomy for the X chromosome. Complete trisomies are viable for chromosomes 13, 18, 21, X, and Y.
- *The phenotype in partial aneuploidy depends on a number of factors,* including the size of the unbalanced segment, which regions of the genome are affected and which genes are involved, and whether the imbalance is monosomic or trisomic.
- *Risk in cases of inversions depends on the location of the inversion with respect to the centromere and on the size of the inverted segment.* For inversions that do not involve the centromere (paracentric inversions), there is a very low risk for an abnormal phenotype in the next generation. But, for inversions that do involve the centromere (pericentric inversions), the risk for birth defects in offspring may be significant and increases with the size of the inverted segment.
- *For a mosaic karyotype involving any chromosome abnormality, all bets are off!* Counseling is particularly challenging because the degree of mosaicism in relevant tissues or relevant stages of development is generally unknown. Thus there is uncertainty about the severity of the phenotype.

Triploidy and Tetraploidy

In addition to the diploid (2n) number characteristic of normal somatic cells, two other euploid chromosome complements, **triploid** (3n) and **tetraploid** (4n), are occasionally observed in clinical material. Both triploidy and tetraploidy have been seen in fetuses. Triploidy is observed in 1% to 3% of recognized conceptions; triploid infants can be liveborn, although they do not survive long. Among the few that survive at least to the end of the first trimester of pregnancy, most result from fertilization of an egg by two sperm (dispermy). Other cases result from failure of one of the meiotic divisions in either sex, resulting in a diploid egg or sperm. The phenotypic manifestation of a triploid karyotype depends on the source of the extra chromosome set; triploids with an extra set of maternal chromosomes are typically aborted spontaneously early in pregnancy, whereas those with an extra set of paternal chromosomes typically have an abnormal degenerative placenta (resulting in a so-called **partial hydatidiform mole**), with a small fetus. Tetraploids are always 92,XXXX or 92,XXYY and likely result from failure of completion of an early cleavage division of the zygote.

Aneuploidy

Aneuploidy is the most common and clinically significant type of human chromosome disorder, occurring in at least 5% of all clinically recognized pregnancies. Most aneuploid patients have either **trisomy** (three instead of the normal pair of a particular chromosome) or, less often, **monosomy** (only one representative of a particular chromosome). Either trisomy or monosomy can have severe phenotypic consequences.

Trisomy can exist for any part of the genome, but trisomy for a whole chromosome is only occasionally compatible with life. By far the most common type of trisomy in liveborn infants is **trisomy 21,** the chromosome constitution seen in 95% of patients with **Down syndrome** (karyotype 47,XX,+21 or 47,XY,+21) (Fig. 5-9). Other trisomies observed in liveborns include trisomy 18 and trisomy 13. It is notable that these autosomes (13, 18, and 21) are the three with the lowest number of genes located on them (see Fig. 2-7); presumably, trisomy for autosomes with a greater number of genes is lethal in most instances. Monosomy for an entire chromosome is almost always lethal; an important exception is monosomy for the X chromosome, as seen in **Turner syndrome** (Case 47). These conditions are considered in greater detail in Chapter 6.

Although the causes of aneuploidy are not fully understood, the most common chromosomal mechanism is **meiotic nondisjunction.** This refers to the failure of a pair of chromosomes to disjoin properly during one of the two meiotic divisions, usually during meiosis I. The genomic consequences of nondisjunction during meiosis I and meiosis II are different (Fig. 5-10). If the error occurs during meiosis I, the gamete with 24 chromosomes contains both the paternal and the maternal members of the pair. If it occurs during meiosis II, the gamete with the extra chromosome contains both copies of either the paternal or the maternal chromosome. (Strictly speaking, these statements refer only to the paternal or maternal centromere, because recombination between homologous chromosomes has usually taken place in the preceding meiosis I, resulting in some genetic differences between the chromatids and thus between the corresponding daughter chromosomes; see Chapter 2.)

Proper disjunction of a pair of homologous chromosomes in meiosis I appears relatively straightforward (see Fig. 5-10). In reality, however, it involves a feat of complex engineering that requires precise temporal and spatial control over alignment of the two homologues, their tight connections to each other (synapsis), their interactions with the meiotic spindle, and, finally, their release and subsequent movement to opposite poles and to different daughter cells. The propensity of a chromosome pair to nondisjoin has been strongly associated with aberrations in the frequency or placement, or both, of recombination events in meiosis I, which are critical for maintaining proper synapsis. A chromosome pair with too few (or even no) recombinations, or with recombination too close to the centromere or telomere, may be more susceptible to nondisjunction than a chromosome pair with a more typical number and distribution of recombination events.

In some cases, aneuploidy can also result from premature separation of sister chromatids in meiosis I instead of meiosis II. If this happens, the separated chromatids may by chance segregate to the oocyte or to the polar body, leading to an unbalanced gamete.

Nondisjunction can also occur in a mitotic division after formation of the zygote. If this happens at an early cleavage division, clinically significant **mosaicism** may result (see later section). In some malignant cell lines and some cell cultures, mitotic nondisjunction can lead to highly abnormal karyotypes.

Abnormalities of Chromosome Structure

Structural rearrangements result from chromosome breakage, recombination, or exchange, followed by reconstitution in an abnormal combination. Whereas rearrangements can take place in many ways, they are together less common than aneuploidy; overall, structural abnormalities are present in approximately 1 in 375 newborns (see Fig. 5-8). Like numerical abnormalities, structural rearrangements may be present in all cells of a person or in mosaic form.

Structural rearrangements are classified as **balanced,** if the genome has the normal complement of chromosomal material, or **unbalanced,** if there is additional or missing material. Clearly these designations depend on the resolution of the method(s) used to analyze a particular rearrangement (see Fig. 5-1); some that appear balanced at the level of high-resolution banding, for example, may be unbalanced when studied with chromosomal microarrays or by DNA sequence analysis. Some rearrangements are stable, capable of passing through mitotic and meiotic cell divisions unaltered, whereas others are unstable. Some of the more common types of structural rearrangements observed in human chromosomes are illustrated schematically in Figure 5-11.

Unbalanced Rearrangements

Unbalanced rearrangements are detected in approximately 1 in 1600 live births (see Fig. 5-8); the phenotype is likely to be abnormal because of deletion or duplication of multiple genes, or (in some cases) both. Duplication of part of a chromosome leads to **partial trisomy** for the genes within that segment; deletion leads to **partial monosomy.** As a general concept, any change that disturbs normal gene dosage balance can result in abnormal development; a broad range of phenotypes can result, depending on the nature of the specific genes whose dosage is altered in a particular case.

Large structural rearrangements involving imbalance of at least a few megabases can be detected at the level of routine chromosome banding, including high-resolution karyotyping. Detection of smaller changes, however, generally requires higher resolution analysis, involving FISH or chromosomal microarray analysis.

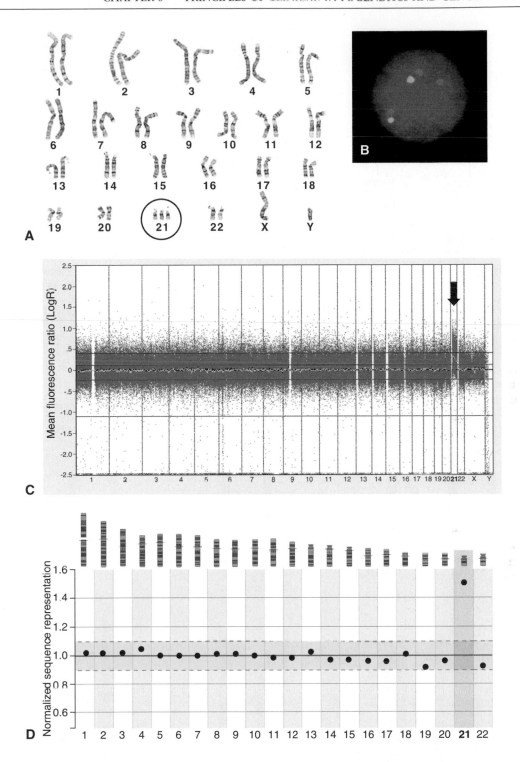

Figure 5-9 Chromosomal and genomic approaches to the diagnosis of trisomy 21. **A,** Karyotype from a male patient with Down syndrome, showing three copies of chromosome 21. **B,** Interphase fluorescence in situ hybridization analysis using locus-specific probes from chromosome 21 (*red,* three spots) and from a control autosome (*green,* two spots). **C,** Detection of trisomy 21 in a female patient by whole-genome chromosomal microarray. Increase in the fluorescence ratio for sequences from chromosome 21 are indicated by the *red arrow.* **D,** Detection of trisomy 21 by whole-genome sequencing and overrepresentation of sequences from chromosome 21. Normalized sequence representation for individual chromosomes (± SD) in chromosomally normal samples is indicated by the *gray shaded region.* A normalized ratio of approximately 1.5 indicates three copies of chromosome 21 sequences instead of two, consistent with trisomy 21. *See Sources & Acknowledgments.*

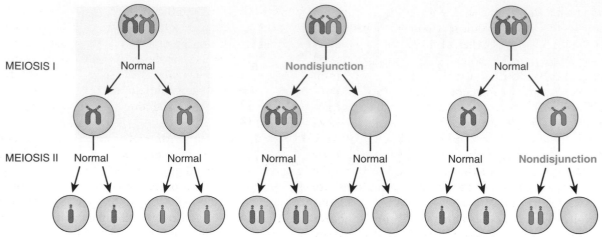

Figure 5-10 The different consequences of nondisjunction at meiosis I (*center*) and meiosis II (*right*), compared with normal disjunction (*left*). If the error occurs at meiosis I, the gametes either contain a representative of both members of the chromosome 21 pair or lack a chromosome 21 altogether. If nondisjunction occurs at meiosis II, the abnormal gametes contain two copies of one parental chromosome 21 (and no copy of the other) or lack a chromosome 21.

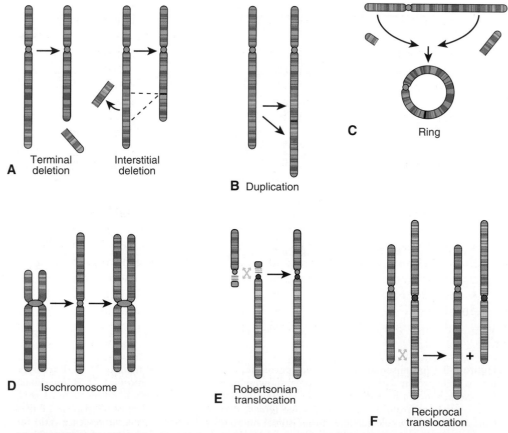

Figure 5-11 **Structural rearrangements of chromosomes, described in the text. A,** Terminal and interstitial deletions, each generating an acentric fragment that is typically lost. **B,** Duplication of a chromosomal segment, leading to partial trisomy. **C,** Ring chromosome with two acentric fragments. **D,** Generation of an isochromosome for the long arm of a chromosome. **E,** Robertsonian translocation between two acrocentric chromosomes, frequently leading to a pseudodicentric chromosome. Robertsonian translocations are nonreciprocal, and the short arms of the acrocentrics are lost. **F,** Translocation between two chromosomes, with reciprocal exchange of the translocated segments.

Deletions and Duplications. Deletions involve loss of a chromosome segment, resulting in chromosome imbalance (see Fig. 5-11). A carrier of a chromosomal deletion (with one normal homologue and one deleted homologue) is monosomic for the genetic information on the corresponding segment of the normal homologue. The clinical consequences generally reflect **haplo-insufficiency** (literally, the inability of a single copy of the genetic material to carry out the functions normally performed by two copies), and, where examined, their severity reflects the size of the deleted segment and the number and function of the specific genes that are deleted. Cytogenetically visible autosomal deletions have an incidence of approximately 1 in 7000 live births. Smaller, submicroscopic deletions detected by microarray analysis are much more common, but as mentioned earlier, the clinical significance of many such variants has yet to be fully determined.

A deletion may occur at the end of a chromosome (**terminal**) or along a chromosome arm (**interstitial**). Deletions may originate simply by chromosome breakage and loss of the acentric segment. Numerous deletions have been identified in the course of prenatal diagnosis or in the investigation of dysmorphic patients or patients with intellectual disability; specific examples of such cases will be discussed in Chapter 6.

In general, duplication appears to be less harmful than deletion. However, because duplication in a gamete results in chromosomal imbalance (i.e., partial trisomy), and because the chromosome breaks that generate it may disrupt genes, duplication often leads to some phenotypic abnormality.

Marker and Ring Chromosomes. Very small, unidentified chromosomes, called **marker chromosomes**, are occasionally seen in chromosome preparations, frequently in a mosaic state. They are usually in addition to the normal chromosome complement and are thus also referred to as **supernumerary chromosomes** or **extra structurally abnormal chromosomes**. The prenatal frequency of de novo supernumerary marker chromosomes has been estimated to be approximately 1 in 2500. Because of their small and indistinctive size, higher resolution genome analysis is usually required for precise identification.

Larger marker chromosomes contain genomic material from one or both chromosome arms, creating an imbalance for whatever genes are present. Depending on the origin of the marker chromosome, the risk for a fetal abnormality can range from very low to 100%. For reasons not fully understood, a relatively high proportion of such markers derive from chromosome 15 and from the sex chromosomes.

Many marker chromosomes lack telomeres and are **ring chromosomes** that are formed when a chromosome undergoes two breaks and the broken ends of the chromosome reunite in a ring structure (see Fig. 5-11). Some rings experience difficulties at mitosis, when the two sister chromatids of the ring chromosome become tangled in their attempt to disjoin at anaphase. There may be breakage of the ring followed by fusion, and larger and smaller rings may thus be generated. Because of this mitotic instability, it is not uncommon for ring chromosomes to be found in only a proportion of cells.

Isochromosomes. An isochromosome is a chromosome in which one arm is missing and the other duplicated in a mirror-image fashion (see Fig. 5-11). A person with 46 chromosomes carrying an isochromosome therefore has a single copy of the genetic material of one arm (partial monosomy) and three copies of the genetic material of the other arm (partial trisomy). Although isochromosomes for a number of autosomes have been described, the most common isochromosome involves the long arm of the X chromosome—designated i(X)(q10)—in a proportion of individuals with Turner syndrome (see Chapter 6). Isochromosomes are also frequently seen in karyotypes of both solid tumors and hematological malignant neoplasms (see Chapter 15).

Dicentric Chromosomes. A dicentric chromosome is a rare type of abnormal chromosome in which two chromosome segments, each with a centromere, fuse end to end. Dicentric chromosomes, despite their two centromeres, can be mitotically stable if one of the two centromeres is inactivated epigenetically or if the two centromeres always coordinate their movement to one or the other pole during anaphase. Such chromosomes are formally called **pseudodicentric**. The most common pseudodicentrics involve the sex chromosomes or the acrocentric chromosomes (so-called Robertsonian translocations; see later).

Balanced Rearrangements

Balanced chromosomal rearrangements are found in as many as 1 in 500 individuals (see Fig. 5-8) and do not usually lead to a phenotypic effect because all the genomic material is present, even though it is arranged differently (see Fig. 5-11). As noted earlier, it is important to distinguish here between *truly* balanced rearrangements and those that *appear* balanced cytogenetically but are really unbalanced at the molecular level. Because of the high frequency of copy number polymorphisms around the genome (see Chapter 4), collectively adding up to differences of many megabases between genomes of unrelated individuals, the concept of what is balanced or unbalanced is subject to ongoing investigation and continual refinement.

Even when structural rearrangements are truly balanced, they can pose a threat to the subsequent generation because carriers are likely to produce a significant frequency of unbalanced gametes and therefore have an increased risk for having abnormal offspring with unbalanced karyotypes; depending on the specific rearrangement, that risk can range from 1% to as high as 20%. There is also a possibility that one of the chromosome breaks will disrupt a gene, leading to mutation. Especially with the use of whole-genome sequencing to examine the nature of apparently balanced rearrangements in patients who present with significant phenotypes, this is an increasingly well-documented cause of disorders in carriers of balanced translocations (see Chapter 6); such translocations can be a useful clue to the identification of the gene responsible for a particular genetic disorder.

Translocations. Translocation involves the exchange of chromosome segments between two chromosomes. There are two main types: reciprocal and nonreciprocal.

Reciprocal Translocations. This type of rearrangement results from breakage or recombination involving nonhomologous chromosomes, with reciprocal exchange of the broken-off or recombined segments (see Fig. 5-11). Usually only two chromosomes are involved, and because the exchange is reciprocal, the total chromosome number is unchanged. Such translocations are usually without phenotypic effect; however, like other balanced structural rearrangements, they are associated with a high risk for unbalanced gametes and abnormal progeny. They come to attention either during prenatal diagnosis or when the parents of a clinically abnormal child with an unbalanced translocation are karyotyped. Balanced translocations are more commonly found in couples who have had two or more spontaneous abortions and in infertile males than in the general population.

The existence of translocations presents challenges for the process of chromosome pairing and homologous recombination during meiosis (see Chapter 2). When the chromosomes of a carrier of a balanced reciprocal translocation pair at meiosis, as shown in Figure 5-12, they must form a **quadrivalent** to ensure proper alignment of homologous sequences (rather than the typical bivalents seen with normal chromosomes). In typical segregation, two of the four chromosomes in the quadrivalent go to each pole at anaphase; however, the chromosomes can segregate from this configuration in several ways, depending on which chromosomes go to

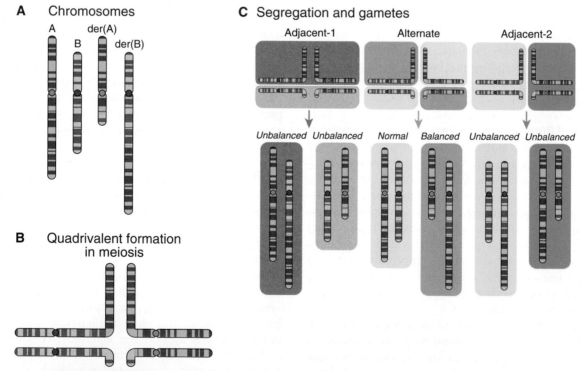

Figure 5-12 A, Diagram illustrating a balanced translocation between two chromosomes, involving a reciprocal exchange between the distal long arms of chromosomes A and B. **B,** Formation of a quadrivalent in meiosis is necessary to align the homologous segments of the two derivative chromosomes and their normal homologues. **C,** Patterns of segregation in a carrier of the translocation, leading to either balanced or unbalanced gametes, shown at the bottom. Adjacent-1 segregation (in *red,* top chromosomes to one gamete, bottom chromosomes to the other) leads only to unbalanced gametes. Adjacent-2 segregation (in *green,* left chromosomes to one gamete, right chromosomes to the other) also leads only to unbalanced gametes. Only alternate segregation (in *gray,* upper left/lower right chromosomes to one gamete, lower left/upper right to the other) can lead to balanced gametes.

which pole. **Alternate segregation,** the usual type of meiotic segregation, produces balanced gametes that have either a normal chromosome complement or contain the two reciprocal chromosomes. Other segregation patterns, however, always yield unbalanced gametes (see Fig. 5-12).

Robertsonian Translocations. Robertsonian translocations are the most common type of chromosome rearrangement observed in our species and involve two acrocentric chromosomes that fuse near the centromere region with loss of the short arms (see Fig. 5-11). Such translocations are nonreciprocal, and the resulting karyotype has only 45 chromosomes, including the translocation chromosome, which in effect is made up of the long arms of two acrocentric chromosomes. Because, as noted earlier, the short arms of all five pairs of acrocentric chromosomes consist largely of various classes of satellite DNA, as well as hundreds of copies of ribosomal RNA genes, loss of the short arms of two acrocentric chromosomes is not deleterious; thus, the karyotype is considered to be balanced, despite having only 45 chromosomes. Robertsonian translocations are typically, although not always, pseudodicentric (see Fig. 5-11), reflecting the location of the breakpoint on each acrocentric chromosome.

Although Robertsonian translocations can involve all combinations of the acrocentric chromosomes, two—designated rob(13;14)(q10;q10) and rob(14;21) (q10;q10)—are relatively common. The translocation involving 13q and 14q is found in approximately 1 person in 1300 and is thus by far the single most common chromosome rearrangement in our species. Rare individuals with two copies of the same type of Robertsonian translocation have been described; these phenotypically normal individuals have only 44 chromosomes and lack any normal copies of the involved acrocentrics, replaced by two copies of the translocation.

Although a carrier of a Robertsonian translocation is phenotypically normal, there is a risk for unbalanced gametes and therefore for unbalanced offspring. The risk for unbalanced offspring varies according to the particular Robertsonian translocation and the sex of the carrier parent; carrier females in general have a higher risk for transmitting the translocation to an affected child. The chief clinical importance of this type of translocation is that carriers of a Robertsonian translocation involving chromosome 21 are at risk for producing a child with translocation Down syndrome, as will be explored further in Chapter 6.

Insertions. An insertion is another type of nonreciprocal translocation that occurs when a segment removed from one chromosome is inserted into a different chromosome, either in its usual orientation with respect to the centromere or inverted. Because they require three chromosome breaks, insertions are relatively rare. Abnormal segregation in an insertion carrier can produce offspring with duplication or deletion of the inserted segment, as well as normal offspring and balanced carriers. The average risk for producing an abnormal child can be up to 50%, and prenatal diagnosis is therefore indicated.

Inversions. An inversion occurs when a single chromosome undergoes two breaks and is reconstituted with the segment between the breaks inverted. Inversions are of two types (Fig. 5-13): **paracentric,** in which both breaks occur in one arm (Greek *para,* beside the centromere); and **pericentric,** in which there is a break in each arm (Greek *peri,* around the centromere). Pericentric inversions can be easier to identify cytogenetically when they change the proportion of the chromosome arms as well as the banding pattern.

An inversion does not usually cause an abnormal phenotype in carriers because it is a balanced rearrangement. Its medical significance is for the progeny; a carrier of either type of inversion is at risk for producing abnormal gametes that may lead to unbalanced offspring because, when an inversion is present, a loop needs to form to allow alignment and pairing of homologous segments of the normal and inverted chromosomes in meiosis I (see Fig. 5-13). When recombination occurs within the loop, it can lead to the production of unbalanced gametes: gametes with balanced chromosome complements (either normal or possessing the inversion) and gametes with unbalanced complements are formed, depending on the location of recombination events. When the inversion is paracentric, the unbalanced recombinant chromosomes are acentric or dicentric and typically do not lead to viable offspring (see Fig. 5-13); thus, the risk that a carrier of a paracentric inversion will have a liveborn child with an abnormal karyotype is very low indeed.

A pericentric inversion, on the other hand, can lead to the production of unbalanced gametes with both **duplication** and **deficiency** of chromosome segments (see Fig. 5-13). The duplicated and deficient segments are the segments that are distal to the inversion. Overall, the risk for a carrier of a pericentric inversion leading to a child with an unbalanced karyotype is estimated to be 5% to 10%. Each pericentric inversion, however, is associated with a particular risk, typically reflecting the size and content of the duplicated and deficient segments.

Mosaicism for Chromosome Abnormalities

When a person has a chromosome abnormality, whether numerical or structural, the abnormality is usually present in all of his or her cells. Sometimes, however, two or more different chromosome complements are present in an individual; this situation is called **mosaicism.** Mosaicism is typically detected by conventional karyotyping but can also be suspected on the

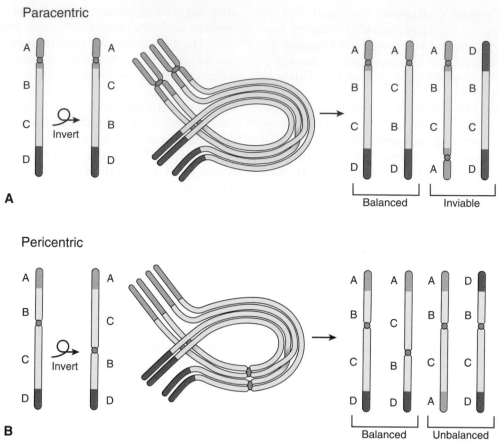

Figure 5-13 Crossing over within inversion loops formed at meiosis I in carriers of a chromosome with segment B-C inverted (order A-C-B-D, instead of the normal order A-B-C-D). **A,** Paracentric inversion. Gametes formed after the second meiosis usually contain either a normal (A-B-C-D) or a balanced (A-C-B-D) copy of the chromosome because the acentric and dicentric products of the crossover are inviable. **B,** Pericentric inversion. Gametes formed after the second meiosis may be balanced (normal or inverted) or unbalanced. Unbalanced gametes contain a copy of the chromosome with a duplication or a deficiency of the material flanking the inverted segment (A-B-C-A or D-B-C-D).

basis of interphase FISH analysis or chromosomal microarrays.

A common cause of mosaicism is nondisjunction in an early postzygotic mitotic division. For example, a zygote with an additional chromosome 21 might lose the extra chromosome in a mitotic division and continue to develop as a 46/47,+21 mosaic. The effects of mosaicism on development vary with the timing of the nondisjunction event, the nature of the chromosome abnormality, the proportions of the different chromosome complements present, and the tissues affected. It is often believed that individuals who are mosaic for a given trisomy, such as mosaic Down syndrome or mosaic Turner syndrome, are less severely affected than nonmosaic individuals.

When detected in lymphocytes, in cultured cell lines or in prenatal samples, it can be difficult to assess the significance of mosaicism, especially if it is identified prenatally. The proportions of the different chromosome complements seen in the tissue being analyzed (e.g., cultured amniocytes or lymphocytes) may not necessarily reflect the proportions present in other tissues or in the embryo during its early developmental stages. Mosaicism can also arise in cells in culture *after* they were taken from the individual; thus, cytogeneticists attempt to differentiate between **true mosaicism,** present in the individual, and **pseudomosaicism,** which has occurred in the laboratory. The distinction between these types is not always easy or certain and can lead to major interpretive difficulties in prenatal diagnosis (see Box earlier and Chapter 17).

Incidence of Chromosome Anomalies

The incidence of different types of chromosomal aberration has been measured in a number of large population surveys and was summarized earlier in Figure 5-8. The major numerical disorders of chromosomes observed in liveborns are three autosomal trisomies (trisomy 21, trisomy 18, and trisomy 13) and four types of sex chromosomal aneuploidy: Turner syndrome (usually 45,X), Klinefelter syndrome (47,XXY), 47,XYY,

TABLE 5-2 Outcome of 10,000 Pregnancies*

Outcome	Pregnancies	Spontaneous Abortions (%)	Live Births
Total	10,000	1500 (15)	8500
Normal chromosomes	9,200	750 (8)	8450
Abnormal chromosomes	800	750 (94)	50
Specific Abnormalities			
Triploid or tetraploid	170	170 (100)	0
45,X	140	139 (99)	1
Trisomy 16	112	112 (100)	0
Trisomy 18	20	19 (95)	1
Trisomy 21	45	35 (78)	10
Trisomy, other	209	208 (99.5)	1
47,XXY, 47,XXX, 47,XYY	19	4 (21)	15
Unbalanced rearrangements	27	23 (85)	4
Balanced rearrangements	19	3 (16)	16
Other	39	37 (95)	2

*These estimates are based on observed frequencies of chromosome abnormalities in spontaneous abortuses and in liveborn infants. It is likely that the frequency of chromosome abnormalities in all conceptuses is much higher than this, because many spontaneously abort before they are recognized clinically.

and 47,XXX (see Chapter 6). Triploidy and tetraploidy account for only a small percentage of cases, typically in spontaneous abortions. The classification and incidence of chromosomal defects measured in these surveys can be used to consider the fate of 10,000 conceptuses, as presented in Table 5-2.

Live Births

As mentioned earlier, the overall incidence of chromosome abnormalities in newborns has been found to be approximately 1 in 154 births (0.65%) (see Fig. 5-8). Most of the autosomal abnormalities can be diagnosed at birth, but most sex chromosome abnormalities, with the exception of Turner syndrome, are not recognized clinically until puberty (see Chapter 6). Unbalanced rearrangements are likely to come to clinical attention because of abnormal appearance and delayed physical and mental development in the chromosomally abnormal individual. In contrast, balanced rearrangements are rarely identified clinically unless a carrier of a rearrangement gives birth to a child with an unbalanced chromosome complement and family studies are initiated.

Spontaneous Abortions

The overall frequency of chromosome abnormalities in spontaneous abortions is at least 40% to 50%, and the kinds of abnormalities differ in a number of ways from those seen in liveborns. Somewhat surprisingly, the single most common abnormality in abortuses is 45,X (the same abnormality found in Turner syndrome), which accounts for nearly 20% of chromosomally abnormal spontaneous abortuses but less than 1% of chromosomally abnormal live births (see Table 5-2). Another difference is the distribution of kinds of trisomy; for example, trisomy 16 is not seen at all in live births but accounts for approximately one third of trisomies in abortuses.

CHROMOSOME AND GENOME ANALYSIS IN CANCER

We have focused in this chapter on constitutional chromosome abnormalities that are seen in most or all of the cells in the body and derive from chromosome or regional mutations that have been transmitted from a parent (either inherited or occurring de novo in the germline of a parent) or that have occurred in the zygote in early mitotic divisions.

However, such mutations also occur in somatic cells throughout life and are a hallmark of cancer, both in hematological neoplasias (e.g., leukemias and lymphomas) and in the context of solid tumor progression. An important area in cancer research is the delineation of chromosomal and genomic changes in specific forms of cancer and the relation of the breakpoints of the various structural rearrangements to the process of oncogenesis. The chromosome and genomic changes seen in cancer cells are numerous and diverse. The association of cytogenetic and genome analysis with tumor type and with the effectiveness of therapy is already an important part of the management of patients with cancer; these are discussed further in Chapter 15.

GENERAL REFERENCES

Gardner RJM, Sutherland GR, Shaffer LG: *Chromosome abnormalities and genetic counseling*, ed 4, Oxford, England, 2012, Oxford University Press.

Shaffer LG, McGowan-Jordan J, Schmid M, editors: *ISCN 2013: an international system for human cytogenetic nomenclature*, Basel, 2013, Karger.

Trask B: Human cytogenetics: 46 chromosomes, 46 years and counting, *Nature Rev Genet* 3:769–778, 2002.

REFERENCES FOR SPECIFIC TOPICS

Baldwin EK, May LF, Justice AN, et al: Mechanisms and consequences of small supernumerary marker chromosomes, *Am J Hum Genet* 82:398–410, 2008.

Coulter ME, Miller DT, Harris DJ, et al: Chromosomal microarray testing influences medical management, *Genet Med* 13:770–776, 2011.

Dan S, Chen F, Choy KW, et al: Prenatal detection of aneuploidy and imbalanced chromosomal arrangements by massively parallel sequencing, *PLoS ONE* 7:e27835, 2012.

Debatisse M, Le Tallec B, Letessier A, et al: Common fragile sites: mechanisms of instability revisited, *Trends Genet* 28:22–32, 2012.

Fantes JA, Boland E, Ramsay J, et al: FISH mapping of de novo apparently balanced chromosome rearrangements identifies characteristics associated with phenotypic abnormality, *Am J Hum Genet* 82:916–926, 2008.

Firth HV, Richards SM, Bevan AP, et al: DECIPHER: database of chromosomal imbalance and phenotype in humans using Ensembl resources, *Am J Hum Genet* 84:524–533, 2009.

Green RC, Rehm HL, Kohane IS: Clinical genome sequencing. In Ginsburg GS, Willard HF, editors: *Genomic and personalized medicine*, ed 2, New York, 2013, Elsevier, pp 102–122.

Higgins AW, Alkuraya FS, Bosco AF, et al: Characterization of apparently balanced chromosomal rearrangements from the Developmental Genome Anatomy Project, *Am J Hum Genet* 82:712–722, 2008.

Kearney HM, South ST, Wolff DJ, et al: American College of Medical Genetics recommendations for the design and performance expectations for clinical genomic copy number microarrays intended for use in the postnatal setting for detection of constitutional abnormalities, *Genet Med* 13:676–679, 2011.

Ledbetter DH, Riggs ER, Martin CL: Clinical applications of whole-genome chromosomal microarray analysis. In Ginsburg GS, Willard

HF, editors: *Genomic and personalized medicine*, ed 2, New York, 2013, Elsevier, pp 133–144.

Lee C: Structural genomic variation in the human genome. In Ginsburg GS, Willard HF, editors: *Genomic and personalized medicine*, ed 2, New York, 2013, Elsevier, pp 123–132.

Miller DT, Adam MP, Aradhya S, et al: Consensus statement: chromosomal microarray is a first-tier clinical diagnostic test for individuals with developmental disabilities or congenital anomalies, *Am J Hum Genet* 86:749–764, 2010.

Nagaoka SI, Hassold TJ, Hunt PA: Human aneuploidy: mechanisms and new insights into an age-old problem, *Nat Rev Genet* 13:493–504, 2012.

Reddy UM, Page GP, Saade GR, et al: Karyotype versus microarray testing for genetic abnormalities after stillbirth, *N Engl J Med* 367:2185–2193, 2012.

Talkowski ME, Ernst C, Heilbut A, et al: Next-generation sequencing strategies enable routine detection of balanced chromosome rearrangements for clinical diagnostics and genetic research, *Am J Hum Genet* 88:469–481, 2011.

PROBLEMS

1. You send a blood sample from a dysmorphic infant to the chromosome laboratory for analysis. The laboratory's report states that the child's karyotype is 46,XY,del(18)(q12).
   a. What does this karyotype mean?
   b. The laboratory asks for blood samples from the clinically normal parents for analysis. Why?
   c. The laboratory reports the mother's karyotype as 46,XX and the father's karyotype as 46,XY,t(7;18)(q35;q12). What does the latter karyotype mean? Referring to the normal chromosome ideograms in Figure 5-2, sketch the translocation chromosome or chromosomes in the father and in his son. Sketch these chromosomes in meiosis in the father. What kinds of gametes can he produce?
   d. In light of this new information, what does the child's karyotype mean now? What regions are monosomic? trisomic? Given information from Chapters 2 and 3, estimate the number of genes present in the trisomic or monosomic regions.

2. A spontaneously aborted fetus is found to have trisomy 18.
   a. What proportion of fetuses with trisomy 18 are lost by spontaneous abortion?
   b. What is the risk that the parents will have a liveborn child with trisomy 18 in a future pregnancy?

3. A newborn child with Down syndrome, when karyotyped, is found to have two cell lines: 70% of her cells have the typical 47,XX,+21 karyotype, and 30% are normal 46,XX. When did the nondisjunctional event probably occur? What is the prognosis for this child?

4. Which of the following persons is or is expected to be phenotypically normal?
   a. A female with 47 chromosomes, including a small supernumerary chromosome derived from the centromeric region of chromosome 15
   b. A female with the karyotype 47,XX,+13
   c. A male with deletion of a band on chromosome 4
   d. A person with a balanced reciprocal translocation

   e. A person with a pericentric inversion of chromosome 6

   What kinds of gametes can each of these individuals produce? What kinds of offspring might result, assuming that the other parent is chromosomally normal?

5. For each of the following, state whether chromosome or genome analysis is indicated or not. For which family members, if any? For what kind of chromosome abnormality might the family in each case be at risk?
   a. A pregnant 29-year-old woman and her 41-year-old husband, with no history of genetic defects
   b. A pregnant 41-year-old woman and her 29-year-old husband, with no history of genetic defects
   c. A couple whose only child has Down syndrome
   d. A couple whose only child has cystic fibrosis
   e. A couple who has two boys with severe intellectual disability

6. Explain the nature of the chromosome abnormality and the method of detection indicated by the following nomenclature.
   a. inv(X)(q21q26)
   b. 46,XX,del(1)(1qter → p36.2:)
   c. 46,XX.ish del(15)(q11.2q11.2)(SNRPN–,D15S10–)
   d. 46,XX,del(15)(q11q13).ishdel(15)(q11.2q11.2)(SNRPN–,D15S10–)
   e. 46,XX.arrcgh1p36.3(RP11-319A11,RP11-58A11,RP11-92O17) × 1
   f. 47,XY,+mar.ish r(8)(D8Z1+)
   g. 46,XX,rob(13;21)(q10;q10),+21
   h. 45,XY,rob(13;21)(q10;q10)

7. Using the nomenclature system in Table 5-1, describe the "molecular karyotypes" that correspond to the microarray data in Figures 5-6C and 5-9C.
   a. Referring to Figure 5-6C, is the individual whose array result is shown a male or a female? How do you know?
   b. Referring to Figure 5-9C, is the individual whose array result is shown a male or a female? How do you know?

The Chromosomal and Genomic Basis of Disease: Disorders of the Autosomes and Sex Chromosomes

In this chapter, we present several of the most common and best understood chromosomal and genomic disorders encountered in clinical practice, building on the general principles of clinical cytogenetics and genome analysis introduced in the previous chapter. Each of the disorders presented here illustrates the principles of **dosage balance and imbalance** at the level of chromosomes and subchromosomal regions of the genome. Because a wide range of phenotypes seen in clinical medicine involve chromosome and subchromosomal mutations, we include in this chapter the spectrum of disorders that are characterized by intellectual disability or by abnormal or ambiguous sexual development. Although many such disorders can be determined by single genes, the clinical approach to evaluation of such phenotypes frequently includes detailed chromosome and genome analysis.

MECHANISMS OF ABNORMALITIES

In this section, we consider abnormalities that illustrate the major chromosomal and genomic mechanisms that underlie genetic imbalance of entire chromosomes or chromosomal regions. Overall, we distinguish five different categories of such abnormalities, each of which can lead to disorders of clinical significance:

- Disorders due to **abnormal chromosome segregation** (nondisjunction)
- Disorders due to **recurrent chromosomal syndromes,** involving deletions or duplications at genomic hot spots
- Disorders due to **idiopathic chromosomal abnormalities,** typically de novo
- Disorders due to **unbalanced familial chromosomal abnormalities**
- Disorders due to chromosomal and genomic events that reveal regions of **genomic imprinting**

The distinguishing features of the underlying mechanisms are summarized in Table 6-1. Although the categories of defects that result from these mechanisms can involve any chromosomes, we introduce them here in the context of autosomal abnormalities.

ANEUPLOIDY

The most common mutation in our species involves errors in chromosome segregation, typically leading to production of an abnormal gamete that has two copies or no copies of the chromosome involved in the nondisjunction event. Notwithstanding the high frequency of such errors in meiosis and, to a lesser extent, in mitosis, there are only three well-defined nonmosaic chromosome disorders compatible with postnatal survival in which there is an abnormal dose of an entire autosome: **trisomy 21** (Down syndrome), **trisomy 18,** and **trisomy 13.** It is surely no coincidence that these chromosomes are the ones with the smallest number of genes among all autosomes (see Fig. 2-7). Imbalance for more gene-rich chromosomes is presumably incompatible with long-term survival, and aneuploidy for some of these is frequently associated with pregnancy loss (see Table 5-2).

Each of these autosomal trisomies is associated with growth retardation, intellectual disability, and multiple congenital anomalies (Table 6-2). Nevertheless, each has a fairly distinctive phenotype that is immediately recognizable to an astute clinician in the newborn nursery. Trisomy 18 and trisomy 13 are both less common than trisomy 21; survival beyond the first year is rare, in contrast to Down syndrome, in which average life expectancy is over 50 years of age.

The developmental abnormalities characteristic of any one trisomic state must be determined by the extra dosage of the particular genes on the additional chromosome. Knowledge of the specific relationship between the extra chromosome and the consequent developmental abnormality has been limited to date. Current research, however, is beginning to localize specific genes on the extra chromosome that are responsible for specific aspects of the abnormal phenotype, through direct or indirect modulation of patterning events during early development (see Chapter 14). The principles of gene dosage and the likely role of imbalance for individual genes that underlie specific developmental aspects of the phenotype apply to all aneuploid conditions; these are

TABLE 6-1 Mechanisms of Chromosome Abnormalities and Genomic Imbalance

Category	Underlying Mechanism	Consequences/Examples
Abnormal chromosome segregation	Nondisjunction	Aneuploidy (Down syndrome, Klinefelter syndrome) Uniparental disomy
Recurrent chromosomal syndromes	Recombination at segmental duplications	Duplication/deletion syndromes Copy number variation
Idiopathic chromosome abnormalities	Sporadic, variable breakpoints	Deletion syndromes (cri du chat syndrome, 1p36 deletion syndrome)
	De novo balanced translocations	Gene disruption
Unbalanced familial abnormalities	Unbalanced segregation	Offspring of balanced translocations Offspring of pericentric inversions
Syndromes involving genomic imprinting	Any event that reveals imprinted gene(s)	Prader-Willi/Angelman syndromes

TABLE 6-2 Features of Autosomal Trisomies Compatible with Postnatal Survival

Feature	Trisomy 21	Trisomy 18	Trisomy 13
Incidence (live births)	1 in 850	1 in 6,000-8,000	1 in 12,000-20,000
Clinical presentation	Hypotonia, short stature, loose skin on nape, palmar crease, clinodactyly	Hypertonia, prenatal growth deficiency, characteristic fist clench, rocker-bottom feet	Microcephaly, sloping forehead, characteristic fist clench, rocker-bottom feet, polydactyly
Dysmorphic facial features	Flat occiput, epicanthal folds, Brushfield spots	Receding jaw, low-set ears	Ocular abnormalities, cleft lip and palate
Intellectual disability	Moderate to mild	Severe	Severe
Other common features	Congenital heart disease Duodenal atresia Risk for leukemia Risk for premature dementia	Severe heart malformations Feeding difficulties	Severe CNS malformations Congenital heart defects
Life expectancy	55 yr	Typically less than a few months; almost all <1 yr	50% die within first month, >90% within first year

CNS, Central nervous system.

illustrated here in the context of Down syndrome, whereas the other conditions are summarized in Table 6-2.

Down Syndrome

Down syndrome is by far the most common and best known of the chromosome disorders and is the single most common genetic cause of moderate intellectual disability. Approximately 1 child in 850 is born with Down syndrome (see Table 5-2), and among liveborn children or fetuses of mothers 35 years of age or older, the incidence of trisomy 21 is far higher (Fig. 6-1).

Down syndrome can usually be diagnosed at birth or shortly thereafter by its dysmorphic features, which vary among patients but nevertheless produce a distinctive phenotype (Fig. 6-2). Hypotonia may be the first abnormality noticed in the newborn. In addition to characteristic dysmorphic facial features (see Fig. 6-2), the patients are short in stature and have brachycephaly with a flat occiput. The neck is short, with loose skin on the nape. The hands are short and broad, often with a single transverse palmar crease ("simian crease") and incurved fifth digits (termed clinodactyly).

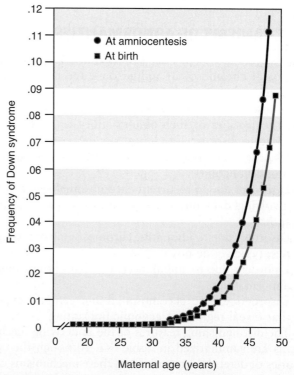

Figure 6-1 Maternal age dependence on incidence of trisomy 21 at birth and at time of amniocentesis. *See Sources & Acknowledgments.*

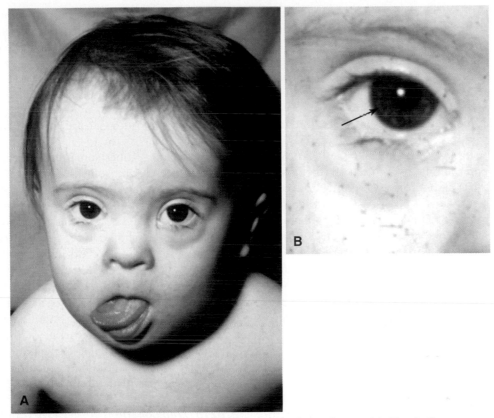

Figure 6-2 Phenotype of Down syndrome. **A,** Young infant. The nasal bridge is flat; ears are low-set and have a characteristic folded appearance; the eyes show characteristic epicanthal folds and upslanting palpebral fissures; the mouth is open, showing a protruding tongue. **B,** Brushfield spots around the margin of the iris *(arrow)*. *See Sources & Acknowledgments.*

A major cause for concern in Down syndrome is intellectual disability. Even though in early infancy the child may not seem delayed in development, the delay is usually obvious by the end of the first year. Although the extent of intellectual disability varies among patients from moderate to mild, many children with Down syndrome develop into interactive and even self-reliant persons, and many attend local schools.

There is a high degree of variability in the phenotype of Down syndrome individuals; specific abnormalities are detected in almost all patients, but others are seen only in a subset of cases. Congenital heart disease is present in at least one third of all liveborn Down syndrome infants. Certain malformations, such as duodenal atresia and tracheoesophageal fistula, are much more common in Down syndrome than in other disorders.

Only approximately 20% to 25% of trisomy 21 conceptuses survive to birth (see Table 5-2). Among Down syndrome conceptuses, those least likely to survive are those with congenital heart disease; approximately one fourth of the liveborn infants with heart defects die before their first birthday. There is a fifteen fold increase in the risk for leukemia among Down syndrome patients who survive the neonatal period. Premature dementia,

associated with the neuropathological findings characteristic of Alzheimer disease (cortical atrophy, ventricular dilatation, and neurofibrillar tangles), affects nearly all Down syndrome patients, several decades earlier than the typical age at onset of Alzheimer disease in the general population.

As a general principle, it is important to think of this constellation of clinical findings, their variation, and likely outcomes in terms of gene imbalance—the relative overabundance of specific gene products; their impact on various critical pathways in particular tissues and cell types, both early in development and throughout life; and the particular alleles present in a particular patient's genome, both for genes on the trisomic chromosome and for the many other genes inherited from his or her parents.

The Chromosomes in Down Syndrome

The clinical diagnosis of Down syndrome usually presents no particular difficulty. Nevertheless, karyotyping is necessary for confirmation and to provide a basis for genetic counseling. Although the specific abnormal karyotype responsible for Down syndrome usually has little effect on the phenotype of the patient, it is essential for determining the recurrence risk.

Trisomy 21. In at least 95% of all patients, the Down syndrome karyotype has 47 chromosomes, with an extra copy of chromosome 21 (see Fig. 5-9). This trisomy results from meiotic nondisjunction of the chromosome 21 pair. As noted earlier, the risk for having a child with trisomy 21 increases with maternal age, especially after the age of 30 years (see Fig. 6-1). The meiotic error responsible for the trisomy usually occurs during maternal meiosis (approximately 90% of cases), predominantly in meiosis I, but approximately 10% of cases occur in paternal meiosis, often in meiosis II. Typical trisomy 21 is a sporadic event, and thus recurrences are infrequent, as will be discussed further later.

Approximately 2% of Down syndrome patients are mosaic for two cell populations—one with a normal karyotype and one with a trisomy 21 karyotype. The phenotype may be milder than that of typical trisomy 21, but there is wide variability in phenotypes among mosaic patients, presumably reflecting the variable proportion of trisomy 21 cells in the embryo during early development.

Robertsonian Translocation. Approximately 4% of Down syndrome patients have 46 chromosomes, one of which is a Robertsonian translocation between chromosome 21q and the long arm of one of the other acrocentric chromosomes (usually chromosome 14 or 22) (see Fig. 5-11). The translocation chromosome replaces one of the normal acrocentric chromosomes, and the karyotype of a Down syndrome patient with a Robertsonian translocation between chromosomes 14 and 21 is therefore 46,XX or XY,rob(14;21)(q10;q10),+21 (see Table 5-1 for nomenclature). Despite having 46 chromosomes, patients with a Robertsonian translocation involving chromosome 21 are trisomic for genes on the entirety of 21q.

A carrier of a Robertsonian translocation, involving, for example, chromosomes 14 and 21, has only 45 chromosomes; one chromosome 14 and one chromosome 21 are missing and are replaced by the translocation chromosome. The gametes that can be formed by such a carrier are shown in Figure 6-3, and such carriers are at risk for having a child with translocation Down syndrome.

Unlike standard trisomy 21, translocation Down syndrome shows no relation to maternal age but has a relatively high recurrence risk in families when a parent, especially the mother, is a carrier of the translocation. For this reason, karyotyping of the parents and possibly other relatives is essential before accurate genetic counseling can be provided.

21q21q Translocation. A 21q21q translocation chromosome is seen in a few percent of Down syndrome

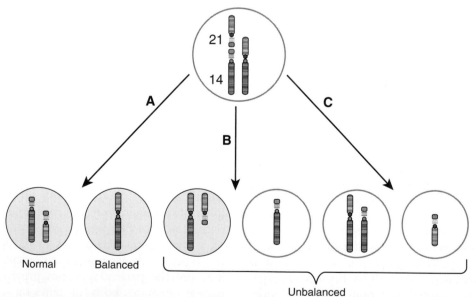

Figure 6-3 Chromosomes of gametes that theoretically can be produced by a carrier of a Robertsonian translocation, rob(14;21). **A,** Normal and balanced complements. **B,** Unbalanced, one product with both the translocation chromosome and the normal chromosome 21, and the reciprocal product with chromosome 14 only. **C,** Unbalanced, one product with both the translocation chromosome and chromosome 14, and the reciprocal product with chromosome 21 only. Theoretically, there are six possible types of gamete, but three of these appear unable to lead to viable offspring. Only the three *shaded* gametes at the left can lead to viable offspring. Theoretically, the three types of gametes will be produced in equal numbers, and thus the theoretical risk for a Down syndrome child should be 1 in 3. However, extensive population studies have shown that unbalanced chromosome complements appear in only approximately 10% to 15% of the progeny of carrier mothers and in only a few percent of the progeny of carrier fathers who have translocations involving chromosome 21.

patients and is thought to originate as an isochromosome. It is particularly important to evaluate if a parent is a carrier because all gametes of a carrier of such a chromosome must either contain the 21q21q chromosome, with its double dose of chromosome 21 genetic material, or lack it and have no chromosome 21 representative at all. The potential progeny, therefore, inevitably have either Down syndrome or monosomy 21, which is rarely viable. Mosaic carriers are at an increased risk for recurrence, and thus prenatal diagnosis should be considered in any subsequent pregnancy.

Partial Trisomy 21. Very rarely, Down syndrome is diagnosed in a patient in whom only a part of the long arm of chromosome 21 is present in triplicate. These patients are of particular significance because they can show what region of chromosome 21 is likely to be responsible for specific components of the Down syndrome phenotype and what regions can be triplicated *without* causing that aspect of the phenotype. The most notable success has been identification of a less than 2-Mb region that is critical for the heart defects seen in approximately 40% of Down syndrome patients. Sorting out the specific genes crucial to the expression of the Down syndrome phenotype from those that merely happen to be syntenic with them on chromosome 21 is critical for determining the pathogenesis of the various clinical findings.

Risk for Down Syndrome

A frequent problem in genetic counseling is to assess the risk for the birth of a Down syndrome child. The risk depends chiefly on the mother's age but also on both parents' karyotypes, as discussed previously. Down syndrome can be detected prenatally by karyotyping, by chromosomal microarray analysis, or by genome-wide sequencing of chorionic villus or amniotic fluid cells (see Fig. 5-9). Screening for Down syndrome is also possible now by **noninvasive prenatal screening** (**NIPS**) of cell-free fetal DNA in maternal plasma. As will be discussed in more detail in Chapter 17, although all pregnancies should be offered prenatal diagnosis, a decision to undergo invasive methods of prenatal testing balances the risk that a fetus has Down syndrome and the risk that the procedure of amniocentesis or chorionic villus sampling used to obtain fetal tissue for chromosome analysis will lead to fetal loss. However, with NIPS emerging as a screening test for Down syndrome and other relatively common aneuploid conditions, this paradigm and counseling considerations are likely to change in the years ahead (see Chapter 17).

The population incidence of Down syndrome in live births is currently estimated to be approximately 1 in 850, reflecting the maternal age distribution for all births and the proportion of older mothers who make use of prenatal diagnosis and selective termination. At approximately the age of 30 years, the risk begins to rise sharply, approaching 1 in 10 births in the oldest maternal age-group (see Fig. 6-1). Even though younger mothers have a much lower risk, their birth rate is much higher, and therefore more than half of the mothers of all Down syndrome babies are younger than 35 years. The risk for Down syndrome due to translocation or partial trisomy is unrelated to maternal age. The paternal age appears to have no influence on the risk.

Recurrence Risk

The recurrence risk for trisomy 21 or any other autosomal trisomy, after one such child has been born in a family, is approximately 1% overall. The risk is approximately 1.4% for mothers younger than 30 years, and it is the same as the age-related risk for older mothers; that is, there is a slight but significant increase in risk for the younger mothers but not for the older mothers, whose risk is already elevated. The reason for the increased risk for the younger mothers is not known. A history of trisomy 21 elsewhere in the family, although often a cause of maternal anxiety, does not appear to significantly increase the risk for having a Down syndrome child.

The recurrence risk for Down syndrome due to a translocation is much higher, as described previously.

UNIPARENTAL DISOMY

Chromosome nondisjunction most commonly results in trisomy or monosomy for the particular chromosome involved in the segregation error. However, less commonly, it can also lead to a disomic state in which *both* copies of a chromosome derive from the *same* parent, rather than one copy being inherited from the mother and the other from the father. This situation, called **uniparental disomy**, is defined as the presence of a disomic cell line containing two chromosomes, or portions thereof, that are inherited from only one parent (see Table 6-1). If the two chromosomes are derived from identical sister chromatids, the situation is described as **isodisomy**; if both homologues from one parent are present, the situation is **heterodisomy**.

The most common explanation for uniparental disomy is trisomy "rescue" due to chromosome nondisjunction in cells of a trisomic conceptus to restore a disomic state. The cause of the originating trisomy is typical meiotic nondisjunction in one of the parental germlines; the rescue results from a *second* nondisjunction event, this one occurring mitotically at an early postzygotic stage, thus "rescuing" a fetus that otherwise would most likely be aborted spontaneously (the most common fate for any trisomic fetus; see Table 5-2). Depending on the stage and parent of the original nondisjunction event (i.e., maternal or paternal meiosis I or II), the location of meiotic recombination events, and which chromosome is subsequently lost in the postzygotic mitotic nondisjunction event, the resulting fetus or liveborn can have complete or partial isodisomy or heterodisomy for the relevant chromosome.

Although it is not known how common uniparental disomy is overall, it has been documented for most chromosomes in the karyotype by demonstrating uniparental inheritance of polymorphisms in a family. Clinical abnormalities, however, have been demonstrated for only some of these, typically in cases when an imprinted region is present in two copies from one parent (see the section on genomic imprinting later in this chapter) or when a typically recessive condition (which would ordinarily imply that both parents are obligate carriers; see Chapter 7) is observed in a patient who has only one documented carrier parent. It is important to stress that, although such conditions frequently come to clinical attention because of mutations in individual genes or in imprinted regions, the underlying pathogenomic mechanism in cases of uniparental disomy is abnormal chromosome segregation.

GENOMIC DISORDERS: MICRODELETION AND DUPLICATION SYNDROMES

Dozens of syndromes characterized by developmental delay, intellectual disability, and a specific constellation of dysmorphic features and birth defects are known to be associated with recurrent subchromosomal or regional abnormalities (see Table 6-1). These small but sometimes cytogenetically visible deletions and/or duplications lead to a form of genetic imbalance referred to as **segmental aneusomy**. These deletions (and, in some cases, their reciprocal duplications) are typically detected by chromosomal microarrays. The term **contiguous gene syndrome** has been applied to many of these conditions, because the phenotype is often attributable to extra or deficient copies of multiple, contiguous genes within the deleted or duplicated region. For other such disorders, however, the phenotype is apparently due to deletion or duplication of only a single gene within the region, despite being associated typically with a chromosomal abnormality that encompasses several genes.

For many of these syndromes, although the clinical phenotype in different patients can be quite variable, the nature of the underlying genomic abnormality is highly similar. Indeed, for the syndromes listed in Table 6-3, high-resolution genomic studies have demonstrated that the centromeric and telomeric breakpoints cluster among different patients, suggesting the existence of genomic sequences that predispose to the rearrangements. Fine mapping in a number of these disorders has shown that the breakpoints localize to low-copy repeated sequences in the genome termed **segmental duplications** (see Chapter 4). Aberrant recombination between nearby copies of the repeats causes the deletions and/or duplications, which typically span several hundred to several thousand kilobase pairs. Extensive analysis of over 30,000 patients worldwide has now implicated this general sequence-dependent mechanism in 50 to 100 syndromes involving contiguous gene rearrangements, which collectively are sometimes referred to as **genomic disorders**.

It is this mechanistic association with segmental duplications that distinguishes this subgroup of deletion and duplication syndromes from others whose breakpoints are highly variable and are not associated with any identifiable genomic feature(s), and whose mechanistic basis appears idiopathic (see Table 6-1). Here we focus on syndromes involving chromosome 22 to illustrate underlying genomic features of this class of disorders.

Deletions and Duplications Involving Chromosome 22q11.2

Several deletions and duplications mediated by unequal recombination between segmental duplications have been documented within the proximal long arm of chromosome 22 and illustrate the general concept of genomic disorders (Fig. 6-4). A particularly common microdeletion involves chromosome 22q11.2 and is associated with diagnoses of **DiGeorge syndrome, velocardiofacial syndrome,** and **conotruncal anomaly face syndrome.** All

---

TABLE 6-3 Examples of Genomic Disorders Involving Recombination between Segmental Duplications

Disorder	Location	Genomic Rearrangement	
		Type	Size (Mb)
1q21.1 deletion/duplication syndrome	1q21.1	Deletion/duplication	≈0.8
Williams syndrome	7q11.23	Deletion	≈1.6
Prader-Willi/Angelman syndrome	15q11-q13	Deletion	≈3.5
16p11.2 deletion/duplication syndrome	16p11.2	Deletion/duplication	≈0.6
Smith-Magenis syndrome	17p11.2	Deletion	≈3.7
dup(17)(p11.2p11.2)		Duplication	
DiGeorge syndrome/velocardiofacial syndrome	22q11.2	Deletion	≈3.0, 1.5
Cat eye syndrome/22q11.2 duplication syndrome		Duplication	
Azoospermia (AZFc)	Yq11.2	Deletion	≈3.5

Based on Lupski JR, Stankiewicz P: *Genomic disorders: the genomic basis of disease*, Totowa, NJ, 2006, Humana Press; Cooper GM, Coe BP, Girirajan S, et al: A copy number variation morbidity map of developmental delay. *Nat Genet* 43:838-846, 2011; and Weischenfeldt J, Symmns O, Spitz F, et al: Phenotypic impact of genomic structural variation: insights from and for human disease. *Nat Rev Genet* 14:125-138, 2013.

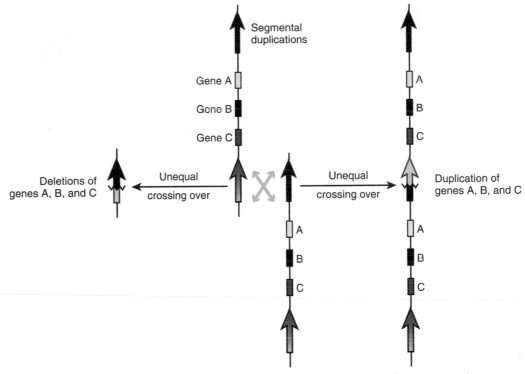

Figure 6-4 **Model of rearrangements underlying genomic disorders.** Unequal crossing over between misaligned sister chromatids or homologous chromosomes containing highly homologous copies of segmentally duplicated sequences can lead to deletion or duplication products, which differ in the number of copies of genes normally located between the repeats. The copy number of any gene or genes (e.g., A, B, and C) that lie between the copies of the repeat will change as a result of these genome rearrangements. For examples of genomic disorders, segmental duplications, and the size of the deleted or duplicated region, see Table 6-3.

three clinical syndromes are autosomal conditions with variable clinical expression, caused by a deletion of approximately 3 Mb within 22q11.2 on one copy of chromosome 22. The microdeletion and other rearrangements of this region shown in Figure 6-5 are each mediated by homologous recombination between segmental duplications in the region. The deletions are detected in approximately 1 in 4000 live births, making this one of the most common genomic rearrangements associated with important clinical phenotypes.

Patients show characteristic craniofacial anomalies, intellectual disability, immunodeficiency, and heart defects, likely reflecting haploinsufficiency for one or more of the several dozen genes that are normally found in this region. Deletion of the *TBX1* gene in **22q11.2 deletion syndrome** is thought to play a role in as many as 5% of all congenital heart defects and is a particularly frequent cause of left-sided outflow tract abnormalities.

Compared to the relatively common deletion of 22q11.2, the reciprocal duplication of 22q11.2 is much rarer and leads to a series of distinct dysmorphic malformations and birth defects called the **22q11.2 duplication syndrome** (see Fig. 6-5). Diagnosis of this duplication generally requires analysis by fluorescence in situ

LESSONS FROM GENOMIC DISORDERS

Genomic disorders collectively illustrate a number of concepts of general importance for considering the causes and consequences of chromosomal or genomic imbalance.

- First, with few exceptions, altered gene dosage for any extensive chromosomal or genomic region is likely to result in a clinical abnormality, the phenotype of which will, in principle, reflect **haploinsufficiency** for or **overexpression** of one or more genes encoded within the region. In some cases, the clinical presentation appears to be accounted for by dosage imbalance for just a single gene; in other syndromes, however, the phenotype appears to reflect imbalance for multiple genes across the region.
- Second, the distribution of these duplication/deletion disorders around the genome appears not to be random, because the location of families of **segmental duplications**, especially in pericentromeric and subtelomeric regions, predisposes particular regions to the unequal recombination events that underlie these syndromes.
- And third, even patients carrying what appears to be the same chromosomal deletion or duplication can present with a **range of variable phenotypes**. Although the precise basis for this variability is unknown, it could be due to nongenetic causes, to underlying genetic variation in the region on the non-deleted chromosome, or to differences elsewhere in the genome among unrelated individuals.

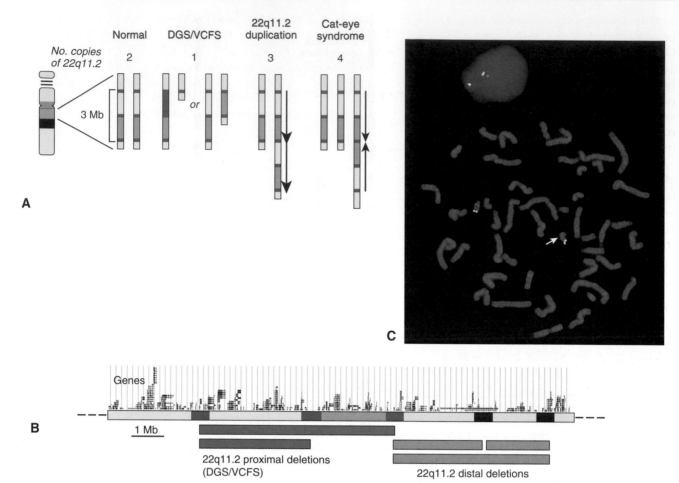

Figure 6-5 Chromosomal deletions, duplications, and rearrangements in 22q11.2 mediated by homologous recombination between segmental duplications. **A,** Normal karyotypes show two copies of 22q11.2, each containing multiple copies of a family of related segmental duplications within the region (*dark blue*). In DiGeorge syndrome (DGS) or velocardiofacial syndrome (VCFS), a 3-Mb region is deleted from one homologue, removing approximately 30 genes; in approximately 10% of patients, a smaller 1.5-Mb deletion (nested within the larger segment) is deleted. The reciprocal duplication is seen in patients with dup(22)(q11.2q11.2). Tetrasomy for 22q11.2 is seen in patients with cat eye syndrome. Note that the duplicated region in the cat eye syndrome chromosome is in an inverted orientation relative to the duplication seen in dup(22) patients, indicating a more complex genomic rearrangement involving these segmental duplications. **B,** Expanded view of the 22q11.2 genomic region, indicating the common DGS/VCFS deletions (*red*) and more distal deletions (also mediated by recombination involving segmental duplications) that are seen in patients with other phenotypes (*orange*). Genes in the region (from www.genome.ucsc.edu browser) are indicated above the region. **C,** Two-color fluorescence in situ hybridization analysis of proband with DGS, demonstrating deletion of 22q11.2 on one homologue. *Green* signal is hybridization to a control region in distal 22q. *Red* signal shows hybridization to a region in proximal 22q that is present on one copy of the chromosome but deleted from the other *(arrow)*. See *Sources & Acknowledgments.*

hybridization (FISH) on interphase cells or by chromosomal microarray.

The general concepts illustrated for disorders associated with 22q11.2 also apply to many other chromosomal and genomic disorders, some of the most common or more significant of which are summarized in Table 6-3. Together, these recurrent syndromes emphasize several important principles in human and medical genetics (see Box).

IDIOPATHIC CHROMOSOME ABNORMALITIES

Whereas the abnormalities just described are mediated by the landscape of specific genomic features in particular chromosomal regions, many other chromosome abnormalities are due to deletions or rearrangements that have no definitive mechanistic basis (see Table 6-1). There are many reports of cytogenetically detectable abnormalities in dysmorphic patients involving events such as deletions, duplications, or translocations of one or more chromosomes in the karyotype (see Fig. 5-11). Overall, cytogenetically visible autosomal deletions occur with an estimated incidence of 1 in 7000 live births. Most of these have been seen in only a few patients and are not associated with recognized clinical syndromes. Others, however, are sufficiently common to allow delineation of clearly recognizable syndromes in which a series of patients have similar abnormalities.

The defining mechanistic feature of this class of abnormalities is that the underlying chromosomal event is idiopathic (see Table 6-1); most of them occur de novo and have highly variable breakpoints in the particular chromosomal region, thus distinguishing them as a class from those discussed in the previous section.

Autosomal Deletion Syndromes

One long-recognized syndrome is the **cri du chat syndrome,** in which there is either a terminal or interstitial deletion of part of the short arm of chromosome 5. This deletion syndrome was given its common name because crying infants with this disorder sound like a mewing cat. The facial appearance, shown in Figure 6-6, is distinctive and includes microcephaly, hypertelorism, epicanthal folds, low-set ears, sometimes with preauricular tags, and micrognathia. The overall incidence of the deletion is estimated to be as high as 1 in 15,000 live births.

Most cases of cri du chat syndrome are sporadic; only 10% to 15% of the patients are the offspring of translocation carriers. The breakpoints and extent of the deleted segment of chromosome 5p is highly variable among different patients, but the critical region missing in all patients with the phenotype has been identified as band 5p15. Many of the clinical findings have been attributed to haploinsufficiency for a gene or genes within specific regions; the degree of intellectual impairment usually correlates with the size of the deletion, although genomic studies suggest that haploinsufficiency for particular regions within 5p14-p15 may contribute disproportionately to severe intellectual disability (see Fig. 6-6).

Although many large deletions can be appreciated by routine karyotyping, detection of other idiopathic deletions requires more detailed analysis by microarrays; this is particularly true for abnormalities involving subtelomeric bands of many chromosomes, which can be difficult to visualize well by routine chromosome banding. For example, one of the most common idiopathic abnormalities, the **chromosome 1p36 deletion syndrome,** has a population incidence of 1 in 5000 and involves a wide range of different breakpoints, all within the terminal 10 Mb of chromosome 1p. Approximately 95% of cases are de novo, and many (e.g., the case illustrated in Fig. 6-6) are not detectable by routine chromosome analysis.

Detailed genomic analysis of various autosomal deletion syndromes underscores the idiopathic nature of these abnormalities. Typically, and in contrast to the genomic disorders presented in Table 6-3, the breakpoints are highly variable and reflect a range of different mechanisms, including terminal deletion of the chromosome arm with telomere healing, interstitial deletion of a subtelomeric segment, or recombination between copies of repetitive elements, such as *Alu* or L1 elements (see Chapter 2).

Balanced Translocations with Developmental Phenotypes

Reciprocal translocations are relatively common (see Chapter 5). Most are balanced and involve the precise exchange of chromosomal material between nonhomologous chromosomes; as such, they usually do not have an obvious phenotypic effect. However, among the approximately 1 in 2000 newborns who has a de novo balanced translocation, the risk for a congenital abnormality is empirically elevated several-fold, leading to the suggestion that some balanced translocations involve direct disruption of a gene or genes by one or both of the translocation breakpoints.

Detailed analysis of a number of such cases by FISH, microarrays, and targeted or whole-genome sequencing has identified defects in protein-coding or noncoding RNA genes in patients with various phenotypes, ranging from developmental delay to congenital heart defects to autism spectrum disorders. Although the clinical abnormalities in these cases can be ascribed to mutations in individual genes located at the site of the translocations, the underlying mechanism in each case is the chromosomal rearrangement itself (see Table 6-1).

SEGREGATION OF FAMILIAL ABNORMALITIES

Although most of the idiopathic abnormalities just described are sporadic, other clinical presentations can occur because of unbalanced segregation of familial chromosome abnormalities. In these cases, the underlying mechanism for the clinical phenotype is not the chromosomal abnormality itself, but rather its transmission in an unbalanced state from a parent who is a balanced carrier to the subsequent generation (see Table 6-1).

The mechanism of pathogenesis here is distinguished from the mechanism of nondisjunction described earlier in this chapter. In contrast to aneuploidy or uniparental disomy, it is not the process of segregation that is abnormal in these cases; rather, it is the random nature of events during segregation that leads to unbalanced karyotypes and thus to offspring with abnormal phenotypes.

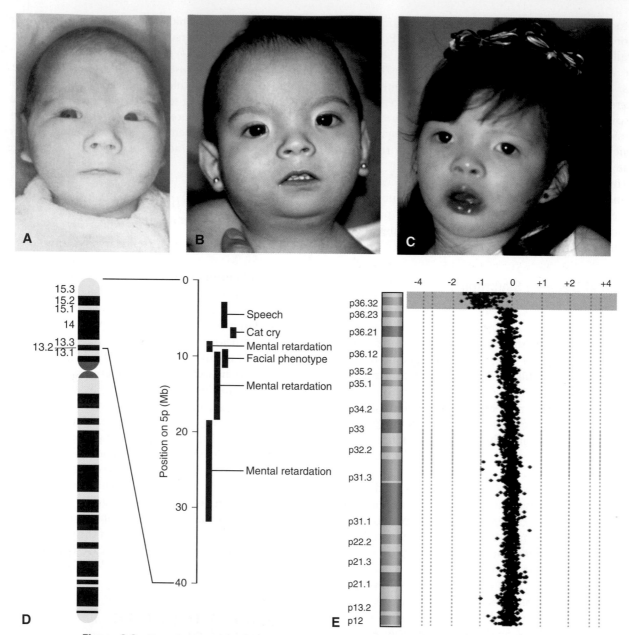

Figure 6-6 **Idiopathic deletion syndromes. A-C,** Three different children with cri du chat syndrome, which results from deletion of part of chromosome 5p. Note, even among unrelated individuals, the characteristic facies with hypertelorism, epicanthus, and retrognathia. **D,** Phenotype-karyotype map of chromosome 5p, based on chromosomal microarray analysis of a series of del(5p) patients. **E,** Chromosomal microarray analysis of approximately 5-Mb deletion in band 1p36.3 *(red),* which is undetectable by conventional karyotyping. *See Sources & Acknowledgments.*

In the case of balanced translocations, for example, because the chromosomes involved form a quadrivalent in meiosis, the particular combination of chromosomes transmitted to a given gamete can lead to genomic imbalance (see Fig. 5-12), even though the segregation is itself normal.

Another type of familial structural abnormality that illustrates this mechanism involves inversion chromosomes. In this case, segregation of the inverted

chromosome and its normal homologue during meiosis is typically uneventful; however, unbalanced gametes can be produced as a result of the process of recombination occurring within the inverted segment, in particular for pericentric inversions (see Fig. 5-13). Different inversion chromosomes carry different risks for abnormal offspring, presumably reflecting both the likelihood that a recombination event will occur within the inverted segment and the likelihood that an unbalanced gamete can lead to viable offspring. This overall risk must be determined empirically for use in genetic counseling. Several well-described inversions illustrate this point.

A pericentric inversion of chromosome 3 is one of the few for which sufficient data have been obtained to allow an estimate of the transmission of the inversion chromosome to the offspring of carriers. The inv(3)(p25q21) originated in a couple from Newfoundland in the early 1800s and has since been reported in a number of families whose ancestors can be traced to the Atlantic provinces of Canada. Carriers of the inv(3) chromosome are normal, but some of their offspring have a characteristic abnormal phenotype associated with the presence of a recombinant chromosome 3, in which there is duplication of the segment distal to 3q21 and deficiency of the segment distal to 3p25. The other predicted unbalanced gamete, with a duplication of distal 3p and deficiency of distal 3q, does not lead to viable offspring. The empirical risk for an abnormal pregnancy outcome in inv(3) carriers is greater than 40% and indicates the importance of family chromosome studies to identify carriers and to offer genetic counseling and prenatal diagnosis.

Not all pericentric inversions have a risk for abnormal offspring, however. One of the most common inversions seen in human chromosomes is a small pericentric inversion of chromosome 9, which is present in up to 1% of all individuals. The inv(9)(p11q12) has no known deleterious effect on carriers and does not appear to be associated with a significant risk for miscarriage or unbalanced offspring; the empirical risk is not different from that of the population at large, and it is therefore generally considered a normal variant.

DISORDERS ASSOCIATED WITH GENOMIC IMPRINTING

For some disorders, the expression of the disease phenotype depends on whether the mutant allele or abnormal chromosome has been inherited from the father or from the mother. As we introduced in Chapter 3, such parent-of-origin effects are the result of genomic imprinting.

The effect of genomic imprinting on inheritance patterns in pedigrees will be discussed in Chapter 7. Here, we focus on the relevance of imprinting to clinical cytogenetics, as many imprinting effects come to light because of chromosome abnormalities. Evidence of genomic imprinting has been obtained for a number of chromosomes or chromosomal regions throughout the genome, as revealed by comparing phenotypes of individuals carrying the same cytogenetic abnormality affecting either the maternal or paternal homologue. Although estimates vary, it is likely that as many as several hundred genes in the human genome show imprinting effects. Some regions contain a single imprinted gene; others contain clusters of multiple imprinted genes, spanning in some cases well over 1 Mb along a chromosome.

The hallmark of imprinted genes that distinguishes them from other autosomal loci is that only one allele, either maternal or paternal, is expressed in the relevant tissue. The effect of such mechanisms on the clinical phenotype will necessarily depend on whether a mutational event occurred on the maternal or paternal homologue. Among the best-studied examples of the role of genomic imprinting in human disease are **Prader-Willi syndrome** (Case 38) and **Angelman syndrome,** and we discuss these next to illustrate the genetic and genomic features of imprinting conditions. An additional example, **Beckwith-Wiedemann syndrome,** is presented in Case 6.

Prader-Willi and Angelman Syndromes

Prader-Willi syndrome is a relatively common dysmorphic syndrome characterized by neonatal hypotonia followed by obesity, excessive and indiscriminate eating habits, small hands and feet, short stature, hypogonadism, and intellectual disability (Fig. 6-7). Prader-Willi syndrome results from the absence of a paternally expressed imprinted gene or genes. In approximately 70% of cases of the syndrome, there is a cytogenetic deletion of the proximal long arm of chromosome 15 (15q11.2-q13); the deletion is mediated by recombination involving segmental duplications that flank a region of approximately 5 to 6 Mb and in that sense is mechanistically similar to other genomic disorders described earlier (see Table 6-3). However, within this region lies a smaller interval that contains a number of monoallelically expressed genes, some of which are normally expressed only from the paternal copy and others of which are expressed only from the maternal copy (see Fig. 6-7). In Prader-Willi syndrome, the deletion is found only on the chromosome 15 inherited from the patient's father (Table 6-4). Thus the genomes of these patients have genomic information in 15q11.2-q13 that derives

TABLE 6-4 Genomic Mechanisms Causing Prader-Willi and Angelman Syndromes

Mechanism	Prader-Willi Syndrome	Angelman Syndrome
15q11.2-q13 deletion	≈70% (paternal)	≈70% (maternal)
Uniparental disomy	≈20-30% (maternal)	≈7% (paternal)
Imprinting center mutation	≈2.5%	≈3%
Gene mutations	Rare (small deletions within snoRNA gene cluster)	~10% (*UBE3A* mutations)
Unidentified	<1%	≈10%

snoRNA, Small nucleolar RNA.
Data from Cassidy SB, Schwartz S, Miller JL, et al: Prader-Willi syndrome. *Genet Med* 14:10-26, 2012; Dagli AI, Williams CA: Angelman syndrome. In Pagon RA, Adam MP, Bird TD, et al, editors: GeneReviews [Internet], Seattle, 1993-2013, University of Washington, Seattle, http://www.ncbi.nlm.nih.gov/books/NBK1144/.

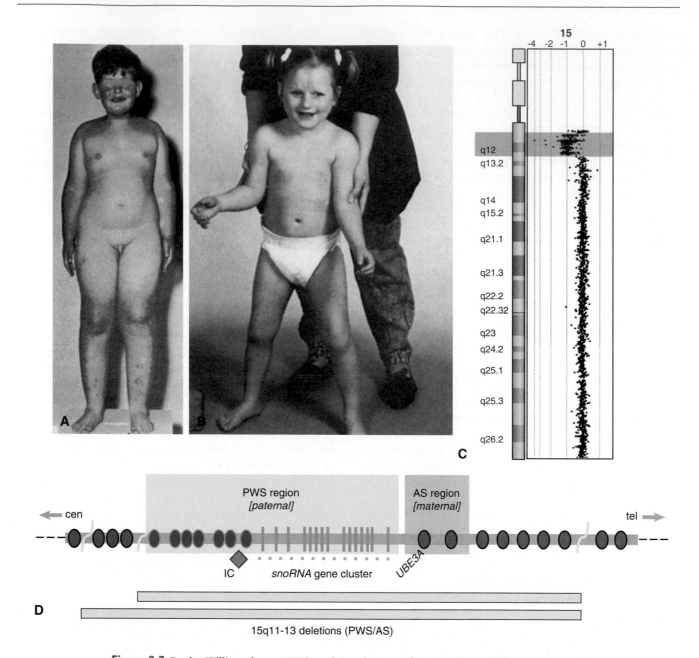

Figure 6-7 Prader-Willi syndrome (PWS) and Angelman syndrome (AS). **A,** PWS in a 9½-year-old boy with obesity, hypogonadism, and small hands and feet who also has short stature and developmental delay. **B,** Angelman syndrome in a 4-year-old girl. Note wide stance and position of arms. **C,** Chromosomal microarray detection of approximately 5-Mb deletion in 15q11.2-q13.1 (*red*). **D,** Schematic of the 15q11.2-q13 region. The PWS region (shaded in *blue*) contains a series of imprinted genes (*blue*) that are expressed only from the paternal copy. The AS region (shaded in *pink*) contains two imprinted genes that are expressed only from the maternal copy, including the *UBE3A* gene, which is imprinted in the central nervous system and mutations in which can cause AS. The region is flanked by nonimprinted genes (*purple*) that are expressed from both maternal and paternal copies. Common deletions of the PWS/AS region, caused by recombination between pairs of segmental duplications, are shown in *green* at the bottom. Smaller deletions of the imprinting center (IC; *orange*) and of a subset of genes in the small nucleolar RNA (snoRNA) gene cluster can also lead to PWS. cen, Centromere; tel, telomere. *See Sources & Acknowledgments.*

only from their mothers, and the syndrome results from the loss of expression of one or more of the normally paternally expressed genes in the region.

Notably, the low-copy repeats that flank the Prader-Willi and Angelman syndrome regions have also been implicated in other disorders, including duplication or triplication of the region or inverted duplication of chromosome 15. This underscores that although imprinting is responsible for the inheritance and specific clinical findings in Prader-Willi and Angelman syndromes, the underlying pathogenomic mechanism of all these disorders involves unequal recombination of the segmental duplications in the region.

In contrast, in most patients with the rare Angelman syndrome, which is characterized by unusual facial appearance, short stature, severe intellectual disability, spasticity, and seizures (see Fig. 6-7), there is a deletion of the same chromosomal region, but, now on the chromosome 15 inherited from the mother. Patients with Angelman syndrome therefore have genetic information in 15q11.2-q13 derived *only from their fathers*. This unusual circumstance demonstrates strikingly that the parental origin of genetic material (in this case, in a segment of chromosome 15) can have a profound effect on the clinical expression of a defect.

Some patients with Prader-Willi syndrome do not have cytogenetically detectable deletions; instead, they have two cytogenetically normal chromosome 15s, both of which were inherited from the mother (see Table 6-4). This situation illustrates **uniparental disomy**, introduced previously in this chapter in the section on abnormal chromosome segregation. A smaller percentage of patients with Angelman syndrome also have uniparental disomy, but in their case with two intact chromosome 15s of paternal origin (see Table 6-4). These patients add further emphasis that, although genomic imprinting is responsible for bringing such cases to clinical attention, the underlying defect in a proportion of cases is one of chromosome segregation, not one of imprinting per se, which is completely normal in these cases.

Primary defects in the imprinting process are seen, however, in a few patients with Prader-Willi syndrome and Angelman syndrome, who have abnormalities in the **imprinting center** itself (see Fig. 6-7). As a result, the switch from female to male imprinting during spermatogenesis or from male to female imprinting during oogenesis (see Fig. 3-12) fails to occur. Fertilization by a sperm carrying an abnormally persistent female imprint would produce a child with Prader-Willi syndrome; fertilization of an egg that bears an inappropriately persistent male imprint would result in Angelman syndrome (see Table 6-4).

Finally, there is evidence that the major features of the Prader-Willi and Angelman syndrome phenotypes can be accounted for by defects at particular genes within the imprinted region. Mutations in the maternal copy of a single gene, the ubiquitin-protein ligase E3A gene (*UBE3A*), have been found to cause Angelman syndrome (see Table 6-4). The *UBE3A* gene is located within the 15q11.2-q13 imprinted region and is normally expressed only from the maternal allele in the central nervous system. Maternally inherited single-gene mutations in *UBE3A* account for approximately 10% of Angelman syndrome cases. In Prader-Willi syndrome, several patients have been described with deletions of a much smaller region on the paternally inherited chromosome 15, specifically implicating the noncoding small nucleolar RNA (snoRNA)116 gene cluster in the etiology of the syndrome (see Fig. 6-7).

Other Disorders due to Uniparental Disomy of Imprinted Regions

Although it is unclear how common uniparental disomy is, it may provide an explanation for a disease when an imprinted region is present in two copies from one parent. Thus physicians and genetic counselors must keep imprinting in mind as a possible cause of genetic disorders.

For example, a few patients with cystic fibrosis and short stature have been described with two identical copies of most or the entirety of their maternal chromosome 7. In these cases, the mother happened to be a carrier for **cystic fibrosis (Case 12)**, and because the child received two maternal copies of the mutant cystic fibrosis gene and no paternal copy of the normal allele at this locus, the child developed the disease. The growth failure was unexplained but might be related to loss of unidentified paternally imprinted genes on chromosome 7.

THE SEX CHROMOSOMES AND THEIR ABNORMALITIES

The X and Y chromosomes have long attracted interest because they differ between the sexes, because they have their own specific patterns of inheritance, and because they are involved in primary sex determination. They are structurally distinct and subject to different forms of genetic regulation, yet they pair in male meiosis. For all these reasons, they require special attention. In this section, we review the common sex chromosome abnormalities and their clinical consequences, the current state of knowledge concerning the control of sex determination, and abnormalities of sex development.

The Chromosomal Basis of Sex Determination

The different sex chromosome constitution of normal human male and female cells has been appreciated for more than 50 years. Soon after cytogenetic analysis became feasible, the fundamental basis of the XX/XY system of sex determination became apparent. Males with Klinefelter syndrome have 47 chromosomes with two X chromosomes as well as a Y chromosome (karyotype 47,XXY), whereas most Turner syndrome females have only 45 chromosomes with a single X chromosome (karyotype 45,X). These findings unambiguously establish the crucial role of the Y chromosome in normal male development. Furthermore, compared with the dramatic consequences of autosomal aneuploidy, these karyotypes underscore the relatively modest effect of varying the number of X chromosomes in either males or females. The basis for both observations can be explained in terms of the unique biology of the Y and X chromosomes.

The process of sex determination can be thought of as occurring in distinct but interrelated steps (Fig. 6-8):

- Establishment of **chromosomal sex** (i.e., XY or XX) at the time of fertilization
- Initiation of alternate pathways to differentiation of one or the other **gonadal sex,** as determined normally by the presence or absence of the testis-determining gene on the Y chromosome
- Continuation of **sex-specific differentiation** of internal and external sexual organs
- Especially after puberty, development of distinctive secondary sexual characteristics to create the corresponding **phenotypic sex,** as a male or female

Whereas the sex chromosomes play a determining role in specifying chromosomal and gonadal sex, a number of genes located on both the sex chromosomes and the autosomes are involved in sex determination and subsequent sexual differentiation. In most instances, the role of these genes has come to light as a result of patients with various conditions known as **disorders of sex development,** and many of these are discussed later in this chapter.

The Y Chromosome

The structure of the Y chromosome and its role in sex development has been determined at both the molecular and genomic levels (Fig. 6-9). In male meiosis, the X and Y chromosomes normally pair by segments at the ends of their short arms (see Chapter 2) and undergo recombination in that region. The pairing segment includes the **pseudoautosomal region** of the X and Y chromosomes, so called because the X- and Y-linked copies of this region are essentially identical to one another and undergo homologous recombination in meiosis I, like pairs of autosomes. (A second, smaller pseudoautosomal segment is located at the distal ends of Xq and Yq.) By comparison with autosomes and the X chromosome, the Y chromosome is relatively gene poor (see Fig. 2-7) and contains fewer than 100 genes (some of which belong to multigene families), specifying only approximately two dozen distinct proteins. Notably, the

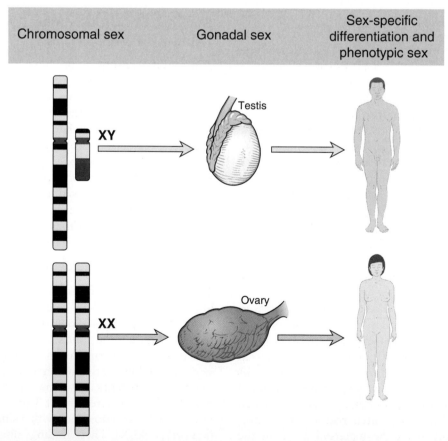

Figure 6-8 The process of sex determination and development: establishment of chromosomal sex at fertilization; commitment to the male or female pathway of gonadal differentiation; sex-specific differentiation of internal and external genitalia and development of secondary sexual characteristics (phenotypic sex). Whereas the sex chromosomes play a determining role in specifying chromosomal sex, many genes located on both the sex chromosomes and the autosomes are involved in sex determination and subsequent sexual differentiation (see Table 6-8).

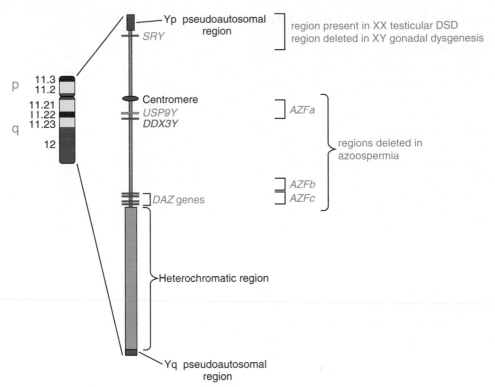

Figure 6-9 The Y chromosome in sex determination and in disorders of sex development (DSDs). Individual genes and regions implicated in sex determination, DSDs, and defects of spermatogenesis are indicated, as discussed in the text.

functions of a high proportion of these genes are restricted to gonadal and genital development.

Embryology of the Reproductive System

The effect of the Y chromosome on the embryological development of the male and female reproductive systems is summarized in Figure 6-10. By the sixth week of development in both sexes, the primordial germ cells have migrated from their earlier extraembryonic location to the paired genital ridges, where they are surrounded by the sex cords to form a pair of primitive gonads. Up to this time, the developing gonad is ambipotent, regardless of whether it is chromosomally XX or XY.

Development into an ovary or a testis is determined by the coordinated action of a sequence of genes in finely balanced pathways that lead to ovarian development when no Y chromosome is present but tip to the side of testicular development when a Y is present. Under normal circumstances, the ovarian pathway is followed unless a particular Y-linked gene, originally designated testis-determining factor *(TDF)*, diverts development into the male pathway.

If no Y chromosome is present, the gonad begins to differentiate to form an ovary, beginning as early as the eighth week of gestation and continuing for several

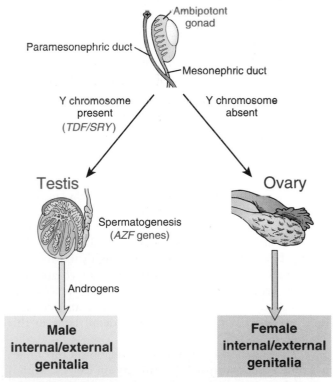

Figure 6-10 Scheme of developmental events in sex determination and differentiation of the male and female gonads from the ambipotent gonad. See text for discussion.

weeks; the cortex develops, the medulla regresses, and oogonia begin to develop within follicles (see Fig. 6-10). Beginning at approximately the third month, the oogonia enter meiosis I, but (as described in Chapter 2) this process is arrested at dictyotene until ovulation occurs many years later.

In the presence of a normal Y chromosome (with the *TDF* gene), however, the medullary tissue forms typical testes with seminiferous tubules and Leydig cells that, under the stimulation of chorionic gonadotropin from the placenta, become capable of androgen secretion (see Fig. 6-10). Spermatogonia, derived from the primordial germ cells by successive mitoses, line the walls of the seminiferous tubules, where they reside together with supporting Sertoli cells, awaiting the onset of puberty to begin spermatogenesis.

While the primordial germ cells are migrating to the genital ridges, thickenings in the ridges indicate the developing genital ducts, the **mesonephric** (also called **wolffian**) and **paramesonephric** (also called **müllerian**) ducts, under the influence of hormones produced by specific cell types in the developing gonad. Duct formation is usually completed by the third month of gestation.

In the early embryo, the external genitalia consist of a genital tubercle, paired labioscrotal swellings, and paired urethral folds. From this undifferentiated state, male external genitalia develop under the influence of androgens, beginning at around 12 weeks of gestation. In the absence of a testis (or, more specifically, in the absence of androgens), female external genitalia are formed regardless of whether an ovary is present.

SRY is the Major Testis-Determining Gene

The earliest cytogenetic studies established the male-determining function of the Y chromosome. In the ensuing three decades, chromosomal and genomic analysis of individuals with different submicroscopic abnormalities of the Y chromosome and well-studied disorders of sex development allowed identification of the primary testis-determining region on Yp.

Whereas the X and Y chromosomes normally exchange in meiosis I within the Xp/Yp pseudoautosomal region, in rare instances, genetic recombination occurs outside of the pseudoautosomal region (Fig. 6-11). This leads to two rare but highly informative abnormalities—males with a 46,XX karyotype and females with a 46,XY karyotype—that involve an inconsistency between chromosomal sex and gonadal sex, as we will explore in greater detail later in this chapter.

The *SRY* gene (*sex-determining region on the Y*) lies near the pseudoautosomal boundary on the Y chromosome. It is present in many males with an otherwise normal 46,XX karyotype (Case 41) and is deleted or mutated in a proportion of females with an otherwise normal 46,XY karyotype, thus strongly implicating *SRY*

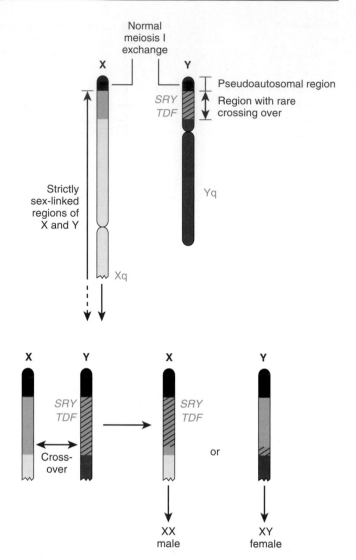

Figure 6-11 **Etiological factors of phenotypic males with a 46,XX karyotype or phenotypic females with a 46,XY karyotype by aberrant exchange between X- and Y-linked sequences.** X and Y chromosomes normally recombine within the Xp/Yp pseudoautosomal segment in male meiosis. If recombination occurs below the pseudoautosomal boundary, between the X-specific and Y-specific portions of the chromosomes, sequences responsible for male gonadal sex determination (including the *SRY* gene) may be translocated from the Y to the X. Fertilization by a sperm containing such an X chromosome leads to a phenotypic male with XX testicular DSD. In contrast, fertilization by a sperm containing a Y chromosome that has lost *SRY* will lead to a phenotypic female with XY complete gonadal dysgenesis.

in normal male sex determination (see Fig. 6-11). *SRY* is expressed only briefly early in development in cells of the germinal ridge just before differentiation of the testis. *SRY* encodes a DNA-binding protein that is likely to be a transcription factor, which up-regulates a key autosomal gene, *SOX9*, in the ambipotent gonad, leading ultimately to testes differentiation. Thus, by all available genetic and developmental criteria, *SRY* is equivalent to the *TDF* gene on the Y chromosome. If *SRY* is absent or not functioning properly, then the female sex differentiation pathway ensues (see Fig. 6-10).

Although there is clear evidence demonstrating the critical role of *SRY* in normal male sexual development, the presence or absence of *SRY/TDF* does not explain all cases of abnormal sex determination. Other genes are involved in the sex determination pathway and are discussed later in this chapter.

Y-Linked Genes in Spermatogenesis

The prevalence of Y chromosome deletions and microdeletions in the general male population is reported to be approximately 1 in 2000 to 3000 males. However, microdeletions in the male-specific portion of Yq are found in a significant proportion of men with low sperm count, ranging from cases of nonobstructive azoospermia (no sperm detectable in semen) to severe oligospermia (<5 million/mL; normal range, 20 to 40 million/mL). These findings suggest that one or more genes, termed *azoospermia factors (AZF),* are located on the Y chromosome, and three such regions on Yq (AZFa, AZFb, and AZFc) have been defined (see Fig. 6-9).

Genomic analysis of these microdeletions led to identification of a series of genes that appear to be important in spermatogenesis. For example, the 3.5-Mb-long AZFc deletion region contains seven different families of genes that are expressed only in the testis, including four copies of the *DAZ* genes (*deleted in azoospermia*) that encode nearly identical RNA-binding proteins expressed only in the premeiotic germ cells of the testis. De novo deletions of AZFc arise in approximately 1 in 4000 males and account for approximately 12% of azoospermic males and approximately 6% of males with severe oligospermia. Deletion of only two of the four *DAZ* genes has been associated with milder oligospermia. Similar to the other genomic disorders described earlier in this chapter, they are mediated by recombination between segmentally duplicated sequences (see Table 6-3). AZFa and AZFb deletions, although less common, also involve recombination. The Yq microdeletions are *not* syndromic, however; they are responsible only for a defect in spermatogenesis in otherwise normal males. The explanation is that all of the genes involved in the *AZF* deletions are expressed only in the testis and have no functions in other tissues or cell types.

Overall, approximately 2% of otherwise healthy males are infertile because of severe defects in sperm production, and it appears likely that de novo deletions or mutations of genes on Yq account for a significant proportion of these. Thus men with idiopathic infertility should be karyotyped, and Y chromosome molecular testing and genetic counseling may be appropriate before the initiation of assisted reproduction by intracytoplasmic sperm injection for such couples, mostly because of the risk for passing a Yq microdeletion responsible for infertility to the infertile couple's sons.

The X Chromosome

Aneuploidy for the X chromosome is among the most common of cytogenetic abnormalities. The relative tolerance of human development for X chromosome abnormalities can be explained in terms of **X chromosome inactivation,** the process by which most genes on one of the two X chromosomes in females are silenced epigenetically, introduced in Chapter 3. X inactivation and its consequences in relation to the inheritance of X-linked disorders are discussed in Chapter 7. Here we discuss the chromosomal and genomic mechanisms of X inactivation and their implications for human and medical genetics (see Box at the end of this section).

X Chromosome Inactivation

The principle of X inactivation is that in somatic cells in normal females (but not in normal males), one X chromosome is inactivated early in development, thus equalizing the expression of X-linked genes in the two sexes. In normal female development, because the choice of which X chromosome is to be inactivated is a random one that is then maintained clonally, females are mosaic with respect to X-linked gene expression (see Fig. 3-13).

There are many epigenetic features that distinguish the active and inactive X chromosomes in somatic cells (Table 6-5). These features can be useful diagnostically for identifying the inactive X chromosome(s) in clinical material. In patients with extra X chromosomes (whether male or female), any X chromosome in excess of one is inactivated (Fig. 6-12). Thus all diploid somatic cells in both males and females have a single active X chromosome, regardless of the total number of X or Y chromosomes present.

TABLE 6-5 Epigenetic and Chromosomal Features of X Chromosome Inactivation in Somatic Cells

Feature	Active X	Inactive X
Gene expression	Yes; similar to male X	Most genes silenced; ~15% expressed to some degree
Chromatin state	Euchromatin	Facultative heterochromatin; Barr body
Noncoding RNA	*XIST* gene silenced	*XIST* RNA expressed from Xi only; associates with Barr body
DNA replication timing	Synchronous with autosomes	Late-replicating in S phase
Histone variants	Similar to autosomes and male X	Enriched for macroH2A
Histone modifications	Similar to autosomes and male X	Enriched for heterochromatin marks; deficient in euchromatin marks

Xi, Inactive X.

Sexual phenotype	Karyotype	No. of active X's	No. of inactive X's
Male	46,XY; 47,XYY	1	0
	47,XXY; 48,XXYY	1	1
	48,XXXY; 49,XXXYY	1	2
	49,XXXXY	1	3
Female	45,X	1	0
	46,XX	1	1
	47,XXX	1	2
	48,XXXX	1	3
	49,XXXXX	1	4

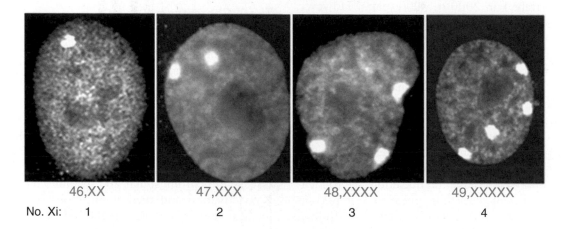

46,XX	47,XXX	48,XXXX	49,XXXXX
No. Xi: 1	2	3	4

Figure 6-12 Sex chromosome constitution and X chromosome inactivation. *Top,* In individuals with extra X chromosomes, any X in excess of one is inactivated, regardless of sex and regardless of the number of Y chromosomes present. Thus the number of inactive X chromosomes in diploid cells is always one less than the total number of X chromosomes. *Bottom,* Detection of inactive X chromosomes (Xi) in interphase nuclei from females with 46,XX, 47,XXX, 48,XXXX, and 49,XXXXX karyotypes. Regions of bright fluorescence indicate the presence of the histone variant macroH2A associated with inactive X chromosomes (see Table 6-5).

The X chromosome contains approximately 1000 genes, but not all of these are subject to inactivation. Notably, the genes that continue to be expressed, at least to some degree, from the inactive X are not distributed randomly along the X chromosome; many more genes "escape" inactivation on distal Xp (as many as 50%) than on Xq (just a few percent). This finding has important implications for genetic counseling in cases of partial X chromosome aneuploidy, because imbalance for genes on Xp may have greater clinical significance than imbalance for genes on Xq, where the effect is largely mitigated by X inactivation.

Patterns of X Inactivation. X inactivation is normally random in female somatic cells and leads to mosaicism for two cell populations expressing alleles from one or the other X (Fig. 6-13). Where examined, most females have approximately equal proportions of cells expressing alleles from the maternal or paternal X (i.e., approximately 50:50), and approximately 90% of phenotypically normal females fall within a distribution that extends from approximately 25:25 to approximately 75:25 (see Fig. 6-13). Such a distribution presumably reflects the expected range of outcomes for a random event (i.e., the choice of which X will be the inactive X) involving a relatively small number of cells during early embryogenesis. For individuals who are carriers for X-linked single-gene disorders (see Chapter 7), this X inactivation ratio can influence the clinical phenotype, depending on what proportion of cells in relevant tissues or cell types express the deleterious allele on the active X.

However, there are exceptions to the distribution expected for random X inactivation when the karyotype involves a **structurally abnormal X chromosome**. For example, in nearly all patients with unbalanced structural abnormalities of an X chromosome (including deletions, duplications, and isochromosomes), the structurally abnormal chromosome is always the inactive X. Because the initial inactivation event early in embryonic

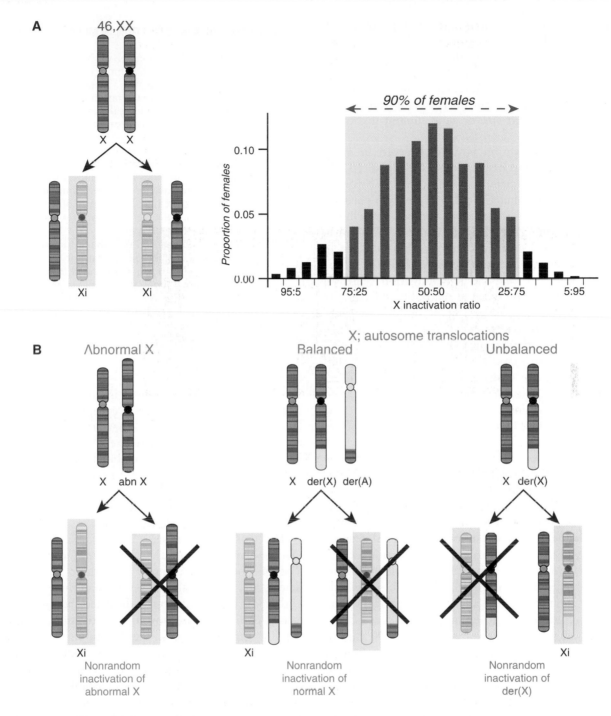

Figure 6-13 **X chromosome inactivation in karyotypes with normal or abnormal X chromosomes or X;autosome translocations. A,** Normal female cells (46,XX) undergo random X inactivation, resulting in a mosaic of two cell populations *(left)* in which either the paternal or maternal X is the inactive X (Xi, indicated by *shaded box*). In phenotypically normal females, the ratio of the two cell populations has a mode at 50:50, but with variation observed in the population *(right)*, some with an excess of cells expressing alleles from the paternal X and others with an excess of cells expressing alleles from the maternal X. **B,** Individuals carrying a structurally abnormal X (abn X) or X;autosome translocation in a balanced or unbalanced state show nonrandom X inactivation in which virtually all cells have the same X inactive. The other cell population is inviable or at a growth disadvantage because of genetic imbalance and is thus underrepresented or absent. der(X) and der(A) represent the two derivatives of the X;autosome translocation. *See Sources & Acknowledgments.*

development is likely random, the patterns observed after birth probably reflect secondary selection against genetically unbalanced cells that are inviable (see Fig. 6-13). Because of this preferential inactivation of the abnormal X, such X chromosome anomalies have less of an impact on phenotype than unbalanced abnormalities of similar size or gene content involving autosomes.

Nonrandom inactivation is also observed in most cases of **X;autosome translocations** (see Fig. 6-13). If such a translocation is balanced, the normal X chromosome is preferentially inactivated, and the two parts of the translocated chromosome remain active, again likely reflecting selection against cells in which critical autosomal genes have been inactivated. In the unbalanced offspring of a balanced carrier, however, only the translocation product carrying the **X inactivation center** is present, and this chromosome is invariably inactivated; the normal X is always active. These nonrandom patterns of inactivation have the general effect of minimizing, but not always eliminating, the clinical consequences of the particular chromosomal defect. Because patterns of X inactivation are strongly correlated with clinical outcome, determination of an individual's X inactivation pattern by cytological or molecular analysis (see Table 6-5) is indicated in all cases involving X;autosome translocations.

The X Inactivation Center. Inactivation of an X chromosome depends on the presence of the X inactivation center region (*XIC*) on that chromosome, whether it is a normal X chromosome or a structurally abnormal X. Detailed analysis of structurally abnormal, inactivated X chromosomes led to the identification of the *XIC* within an approximately 800-kb candidate region in proximal Xq, in band Xq13.2 (Fig. 6-14), which coordinates many, if not all, of the critical steps necessary to initiate and promulgate the silenced chromatin state along the near-entirety of the X chosen to become the inactive X. As introduced in Chapter 3, this complex series of events requires a noncoding RNA gene, *XIST,* that appears to be a key master regulatory locus for the onset of X inactivation. It is one of a suite of noncoding RNA genes in the interval, others of which may operate in the regulation of *XIST* expression and in other early events in the X inactivation process.

Cytogenetic Abnormalities of the Sex Chromosomes

Sex chromosome abnormalities are among the most common of all human genetic disorders, with an overall incidence of approximately 1 in 400 live births. Like abnormalities of the autosomes, they can be either numerical or structural and can be present in all cells or in mosaic form. As a group, disorders of the sex

---

SIGNIFICANCE OF X INACTIVATION IN MEDICAL GENETICS

Many of the underlying details of X inactivation are mechanistically similar to other, more localized epigenetic silencing systems (see Table 3-2). Nonetheless, there are a number of features of X inactivation that are of central importance to human and medical genetics:

- Its **chromosomal nature** reduces the impact of segmental- or whole-chromosome genetic imbalance, such that many numerical and structural abnormalities of the X chromosome are relatively less deleterious than comparable abnormalities of the autosomes.
- Its **random nature** and the resulting **clonal mosaicism** greatly influence the clinical phenotype of females who carry X-linked single-gene mutations on one of their X chromosomes (see Chapter 7).
- Its **dependence on the *XIC*** is required for normal XX female development, because even very small fragments of the X chromosome separated from the *XIC* can lead to severe phenotypic anomalies as a result of their expression from both copies of genes contained on the X fragment (see Fig. 6-14).

---

chromosomes tend to occur as isolated events without apparent predisposing factors, except for an effect of late maternal age in the cases that originate from errors of maternal meiosis I. There are a number of clinical indications that raise the possibility of a sex chromosome abnormality and thus the need for chromosomal or genomic studies. These indications include delay in onset of puberty, primary or secondary amenorrhea, infertility, and ambiguous genitalia.

The most common sex chromosome abnormalities involve aneuploidy for the X and/or Y chromosomes. The phenotypes associated with these chromosomal defects are, in general, less severe than those associated with comparable autosomal disorders because, as discussed earlier, X chromosome inactivation, as well as the low gene content of the Y, minimize the clinical consequences of sex chromosome imbalance. By far the most common sex chromosome defects in liveborn infants and in fetuses are the trisomic types (XXY, XXX, and XYY), but all three are rare in spontaneous abortions. In contrast, monosomy for the X (Turner syndrome) is less frequent in liveborn infants but is the most common chromosome anomaly reported in spontaneous abortions (see Table 5-2).

Sex Chromosome Aneuploidy

The incidence and major features of the four conditions associated with sex chromosome aneuploidy are compared in Tables 6-6 and 6-7. These well-defined syndromes are important causes of infertility, abnormal development, or both, and thus warrant a more detailed description. The effects of these chromosome abnormalities on development have been studied in long-term multicenter studies of hundreds of affected individuals,

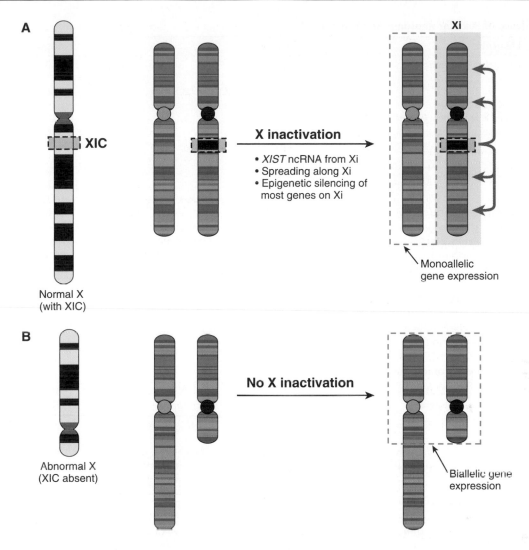

Figure 6-14 X chromosome inactivation and dependence on X inactivation center (XIC). **A,** On normal X chromosomes, XIC lies within an approximately 800-kb candidate region in Xq13.2 that contains a number of noncoding RNA (ncRNA) genes, including *XIST*, the master X inactivation control gene. In early development in XX embryos, the *XIST* RNA spreads along the length of one X, which will become the inactive X (Xi), with epigenetic silencing of most genes on that X chromosome, resulting in monoallelic expression of most, but not all X-linked genes. **B,** On structurally abnormal X chromosomes that lack the XIC, X inactivation cannot occur and genes present on the abnormal X are expressed biallelically. Although a fairly large abnormal X is shown here for illustrative purposes, in fact only very small such fragments are observed in female patients, who invariably display significant congenital anomalies, suggesting that biallelic expression of larger numbers of X-linked genes is inconsistent with normal development and is likely inviable.

some of whom have been monitored for more than 40 years. As a group, those with sex chromosome aneuploidy show reduced levels of psychosocial adaptation, educational achievement, occupational performance, and economic independence, and on average they score slightly lower on intelligence (IQ) tests than their peers. However, each group shows high variability, making it impossible to generalize to specific cases. In fact, the overall impression is a high degree of normalcy, particularly in adulthood, which is remarkable among those with major chromosomal anomalies. Because almost all patients with sex chromosome abnormalities have only mild developmental abnormalities, a parental

decision regarding potential termination of a pregnancy in which the fetus is found to have this type of defect can be a very difficult and even controversial one.

Here, we use **Klinefelter syndrome** to illustrate the major principles of sex chromosome aneuploidy. A more detailed presentation of **Turner syndrome** (45,X and its variants) can be found in the Cases (Case 47).

Klinefelter Syndrome (47,XXY). The phenotype of typical patients with Klinefelter syndrome is shown in Figure 6-15. Klinefelter patients are almost always infertile because of the failure of germ cell development, and patients are often identified clinically for the first time

TABLE 6-6 Incidence of Sex Chromosome Abnormalities

Sex	Disorder	Karyotype	Approximate Incidence
Male	Klinefelter syndrome	47,XXY	1/600 males
		48,XXXY	1/25,000 males
		Others (48,XXYY; 49,XXXYY; mosaics)	1/10,000 males
	47,XYY syndrome	47,XYY	1/1000 males
	Other X or Y chromosome abnormalities		1/1500 males
	XX testicular DSD	46,XX	1/20,000 males
		Overall incidence: 1/300 males	
Female	Turner syndrome	45,X	1/4000 females
		46,X,i(Xq)	1/50,000 females
		Others (deletions, mosaics)	1/15,000 females
	Trisomy X	47,XXX	1/1000 females
	Other X chromosome abnormalities		1/3000 females
	XY gonadal dysgenesis	46,XY	1/20,000 females
	Androgen insensitivity syndrome	46,XY	1/20,000 females
		Overall incidence: 1/650 females	

DSD, Disorder of sex development.
Data updated from Robinson A, Linden MG, Bender BG: Prenatal diagnosis of sex chromosome abnormalities. In Milunsky A, editor: *Genetic disorders of the fetus*, ed 4, Baltimore, 1998, Johns Hopkins University Press, pp 249-285.

TABLE 6-7 Features of Sex Chromosome Aneuploidy Conditions

Feature	47,XXY Klinefelter Syndrome	47,XYY	47,XXX Trisomy X	45,X Turner Syndrome
Prevalence	1 in 600 male births	1 in 1000 male births	1 in 1000 female births	1 in 2500 to 4000 female births
Clinical phenotype	Tall male; see Figure 6-15 and text	Tall, but otherwise typical male appearance	Hypotonia, delayed milestones; language and learning difficulties; tend to be taller than average	Short stature, webbed neck, lymphedema; risk for cardiac abnormalities
Cognition/intelligence	Verbal IQ reduced to low-normal range; educational difficulties	Verbal IQ reduced to low-normal range; language delay; reading difficulties	Normal to low-normal range (both verbal and performance IQ decreased)	Typically normal, but performance IQ lower than verbal IQ
Behavioral phenotype	No major disorders; tendency to poor social adjustments, but normal adult relationships	Subset with specific behavioral problems likely associated with lower IQ	Typically, no behavioral problems; some anxiety and low self-esteem; reduced social skills	Typically normal, but impaired social adjustment
Sex development/fertility	Hypogonadism, azoospermia, infertility	Normal	?Reduced fertility in some ?Premature ovarian failure	Gonadal dysgenesis, delayed maturation, infertility
Variant karyotypes	See Table 6-6		48,XXXX; 49,XXXXX Increased severity with additional X's	46,Xi(Xq); 45,X/46,XX mosaics; other mosaics

Summarized from Ross JL, Roeltgen DP, Kushner H, et al: Behavioral and social phenotypes in boys with 47,XYY syndrome or 47,XXY Klinefelter syndrome. *Pediatrics* 129:769-778, 2012; Pinsker JE: Turner syndrome: updating the paradigm of clinical care. *J Clin Endocrinol Metab* 97:E994-E1003, 2012; and AXYS, http://www.genetic.org.

because of infertility; as such, Klinefelter syndrome is classified among **disorders of sex development,** as we shall see in the next section. Klinefelter syndrome is relatively common among infertile males (approximately 4%) or males with oligospermia or azoospermia (approximately 10%). In adulthood, persistent androgen deficiency may result in decreased muscle tone, a loss of libido, and decreased bone mineral density.

The incidence of Klinefelter syndrome is estimated to be as high as 1 in 600 male births. Approximately half the cases result from nondisjunction in paternal meiosis I because of a failure of normal Xp/Yp recombination in the pseudoautosomal region. Among cases of maternal origin, most result from errors in maternal meiosis I; maternal age is increased in such cases. Approximately 15% of Klinefelter patients have mosaic

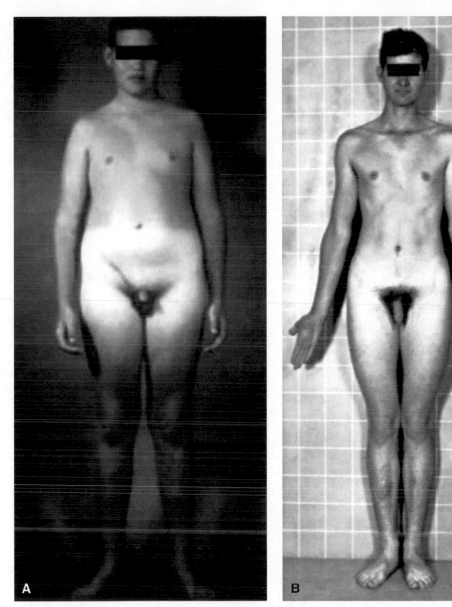

Figure 6-15 Phenotype of males with 47,XXY Klinefelter syndrome. The patients are tall and thin and have relatively long legs. They appear physically normal until puberty, when signs of hypogonadism become obvious. Puberty occurs at a normal age, but the testes remain small, and secondary sexual characteristics remain underdeveloped. Note narrow shoulders and chest. Gynecomastia is a feature of some Klinefelter males and is visible in the 16-year-old patient in **A**. *See Sources & Acknowledgments.*

karyotypes, most commonly 46,XY/47,XXY. As a group, such mosaic patients have variable phenotypes, and some may have normal testicular development.

Although there is wide phenotypic variation among patients with this and other sex chromosome aneuploidies, some consistent phenotypic differences have been identified between patients with Klinefelter syndrome and chromosomally normal males (see Table 6-7). Verbal comprehension and ability are below those of 46,XY males. Patients with Klinefelter syndrome have a several-fold increased risk for learning difficulties, especially in reading, that may require educational intervention. Language difficulties may lead to shyness,

unassertiveness, apparent immaturity, and an increased risk for depression. Although most Klinefelter males form normal adult relationships, many of the affected boys have relatively poor psychosocial adjustment. Because of the relatively mild yet variable phenotype, many cases are presumed to go undetected.

DISORDERS OF SEX DEVELOPMENT

Earlier in this chapter, we discussed the primary sex-determining role of the Y chromosome and the *SRY* gene. In this section, we examine the role of various genes in ovarian and testicular development and in the development of male and female external genitalia. Disorders of gonadal and sexual development can arise from errors at any of the major steps of normal sex determination outlined earlier (see Fig. 6-8). These conditions, ranging from gonadal abnormalities to complete incompatibility between chromosomal and phenotypic sex, are now collectively termed **disorders of sex development (DSD)**. They are among the most common birth defects; worldwide, 1 in 4500 babies are born with significant ambiguous genitalia, and DSDs are estimated to account for over 7% of all birth defects.

Although the chromosomal sex of an embryo is established at the time of fertilization, for some newborn infants, assignment of sex is difficult or impossible because the genitalia are ambiguous, with anomalies that tend to make them resemble in part those of the opposite chromosomal sex. Such anomalies may vary from mild hypospadias in males (a developmental anomaly in which the urethra opens on the underside of the penis or on the perineum) to an enlarged clitoris in females. In some patients, as we discuss later, *both* ovarian and testicular tissue is present. Abnormalities of either external or internal genitalia do not necessarily indicate a cytogenetic abnormality of the sex chromosomes but may reflect chromosomal changes elsewhere in the karyotype, single-gene defects, or nongenetic causes. Nonetheless, determination of the child's karyotype, frequently accompanied by chromosomal microarray, is an essential part of the investigation of such patients and can help guide both surgical and psychosocial management, as well as genetic counseling.

The detection of cytogenetic abnormalities, especially when seen in multiple patients, can also provide important clues about the location and nature of genes involved in sex determination and sex differentiation, some of which are listed in Table 6-8. DSDs can be classified into several major phenotypic and mechanistic groups, examples of which are discussed in the following sections. We focus on a few examples to illustrate the critical balance among various genes and their products that is necessary for normal gonadal and genital development in both males and females (see Box). These examples also reinforce the wide range of cytogenetic and genomic approaches—from standard karyotypes to FISH to microarrays to direct mutation analysis—needed for diagnosis, clinical and psychosocial management, and genetic counseling in these conditions.

GENE BALANCE AND DISORDERS OF SEX DEVELOPMENT

The discovery of different Y-linked, X-linked, and autosomal, chromosomal, genomic, and single-gene abnormalities in different patients underscores the finely tuned nature of the network of dosage-sensitive genes that control gonadal development. The right genes and their products have to be expressed in the right amounts at precisely the right time and in the right place in the developing embryo.

Imbalance in the expression of major genes in the sex development pathways can override the signals typical of the chromosomal sex, leading to testis formation, even in the absence of a Y chromosome, or to ovarian development, even in the presence of the Y. Mutations and/or dosage imbalance (duplications or deletions) of critical genes in these pathways can overcome chromosomal sex and lead to a mismatch between chromosomal and gonadal sex or between gonadal and phenotypic sex (Fig. 6-16).

Disorders of Gonadal Development

Gonadal dysgenesis refers to a progressive loss of germ cells, typically leading to underdeveloped and dysfunctional ("streak") gonads, with consequent failure to develop mature secondary sex characteristics. Gonadal dysgenesis is typically categorized according to the karyotype of a patient. **Complete gonadal dysgenesis (CGD)**—as in the case of XX males (now formally designated **46,XX testicular DSD**) or XY females (now

TABLE 6-8 Examples of Genes Involved in Disorders of Sex Development

Gene	Location	Genetic Abnormality	Phenotypic Sex, Disorder
46,XY Karyotype			
SRY	Yp11.3	*SRY* mutation	Female, XY gonadal dysgenesis
DAX1 (NR0B1)	Xp21.3	*DAX1* gene duplication	Female, XY gonadal dysgenesis
SOX9	17q24	*SOX9* mutation	Female, XY gonadal dysgenesis, with camptomelic dysplasia
NR5A1	9q33	*NRSA1* mutation	Ambiguous genitalia, XY partial gonadal dysgenesis
WNT4	1p35	*WNT4* gene duplication	Ambiguous genitalia, cryptorchidism
AR	Xq12	*AR* mutation	Female, complete or partial androgen insensitivity syndrome
46,XX Karyotype			
SRY	Yp11.3	*SRY* gene translocated to X	Male, XX (ovo)testicular DSD
SOX3	Xq27.1	*SOX3* gene duplication	Male, XX testicular DSD
SOX9	17q24	*SOX9* gene duplication	Male, XX testicular DSD
CYP21A2	6p21.3	*CYP21A2* mutation	Ambiguous genitalia, virilization, micropenis

DSD, Disorder of sex development.
Updated from Achermann JC, Hughes IA: Disorders of sex development. In Melmed S, Polonsky KS, Larsen PR, Kronenberg HM, editors: *Williams textbook of endocrinology*, ed 12, Philadelphia, 2011, WB Saunders, pp 886-934.

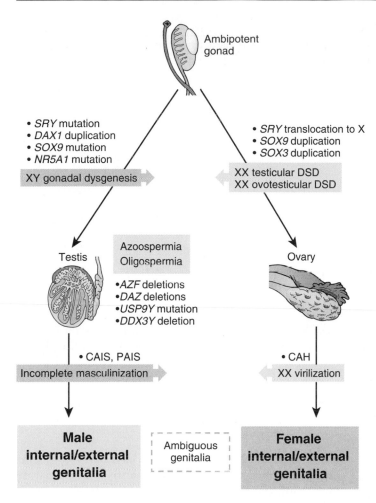

Figure 6-16 Disorders of sex development (DSDs), across the spectrum of developmental events in sex determination and gonadal differentiation (see Fig. 6-10). Selected DSDs are shown, along with particular gene mutations and genomic alterations that interfere with the primary effect of chromosomal sex (XX or XY) in sex development and shift—all or in part—sex development toward the opposite sex. These mutations, duplications, and deletions illustrate the role of gene balance and imbalance on development of gonadal sex, sex-specific differentiation, and phenotypic sex. See text and Tables 6-8 and 6-9. CAH, Congenital adrenal hyperplasia; CAIS, complete androgen insensitivity syndrome; PAIS, partial androgen insensitivity syndrome.

formally designated **46,XY CGD**)—is characterized by normal-appearing external genitalia of the opposite chromosomal sex. Cases with ambiguous external genitalia are said to have partial gonadal dysgenesis. Gonadal dysgenesis can also be associated with **sex chromosome DSDs**; it is a consistent feature of Turner syndrome (see Table 6-7), and patients with a 45,X/46,XY karyotype have **mixed gonadal dysgenesis**.

Various types of gonadal dysgenesis, their clinical phenotypes, and genetic causes are summarized in Table 6-9 and illustrated schematically in Figure 6-16.

Disorders Associated with a 46,XY Karyotype

We begin with DSDs associated with a 46,XY karyotype. The overall incidence of these conditions is approximately 1 in 20,000 live births. Although a number of cytogenetic or single-gene defects have been demonstrated, many such cases remain unexplained. Approximately 15% of patients with 46,XY CGD have deletions or mutations in the *SRY* gene that interfere with the normal male pathway. However, most females with a 46,XY karyotype have an apparently normal *SRY* gene.

The *DAX1* gene in Xp21.3 encodes a transcription factor that plays a dosage-sensitive role in determination of gonadal sex, implying a tightly regulated interaction between *DAX1* and *SRY*. Although production of *SRY* at a critical point in early development normally leads to testis formation, an excess of *DAX1* resulting from duplication of the gene can apparently suppress the normal male-determining function of *SRY*, leading to ovarian development (see Fig. 6-16).

A key master gene in gonadal development and the target of *SRY* signaling is the *SOX9* gene on chromosome 17. *SOX9* is normally expressed early in development in the genital ridge and is required for normal testis formation. Mutations in one copy of the *SOX9* gene, typically associated with a skeletal malformation disorder called **camptomelic dysplasia**, lead to complete gonadal dysgenesis in approximately 75% of 46,XY cases (see Table 6-8). In the absence of one copy of the *SOX9* gene, testes fail to form, and the ovarian pathway is followed instead. The phenotype of these patients suggests that the critical step for the male pathway is sufficient *SOX9* expression to drive the formation of testes, normally after up-regulation by the *SRY* gene. In 46,XY CGD, with either a mutation in *SRY* or a mutation in *SOX9*, the levels of *SOX9* expression remain too low for testis differentiation, allowing ovarian differentiation to ensue.

TABLE 6-9 Disorders of Sex Development and their Characteristics

Disorder	Gonadal Sex	Phenotypic Sex	Characteristics
Sex chromosome DSDs			
Klinefelter syndrome	Testes (dysgenetic)	Male	Gonadal dysgenesis; hypogonadism; azoospermia
Turner syndrome	Ovary (streak gonads)	Female	Gonadal dysgenesis; amenorrhea
46,XX testicular DSD	Testes (bilateral)	Normal male (≈80%) or ambiguous (≈20%)	Most present clinically after puberty with small testes, gynecomastia, azoospermia
46,XX ovotesticular DSD	Testicular and ovarian tissue (ovotestis or one of each)	Ambiguous	Uterus may be present; surgery often required to repair external genitalia; raised as male or female
46,XY DSD	Testes (dysgenetic)	Ambiguous	Variable müllerian structures; penoscrotal hypospadias; risk for gonadoblastoma; raised as male or female
46,XY complete gonadal dysgenesis	Undeveloped streak gonads; no sperm production	Female	Normal müllerian structures; risk for gonadoblastoma
46,XY partial gonadal dysgenesis	Regressed testes	Variable (male, female, or ambiguous)	Ambiguous external genitalia with or without müllerian structures; raised as male or female
45,X/46,XY mixed gonadal dysgenesis	Asymmetric (dysgenetic testis and streak gonad)	Variable (male, female, or ambiguous)	Variable phenotype, ranging from a typical (short) male to Turner syndrome female; risk for gonadoblastoma

DSD, Disorder of sex development.

Summarized from Achermann JC, Hughes IA: Disorders of sex development. In Melmed S, Polonsky KS, Larsen PR, Kronenberg HM, editors: *Williams textbook of endocrinology*, ed 12, Philadelphia, 2011, WB Saunders, pp 886-934; and Pagon RA, Adam MP, Bird TD, et al, editors: GeneReviews [Internet]. Seattle, 1993-2013, University of Washington, Seattle, http://www.ncbi.nlm.nih.gov/books/NBK1116/.

As many as 10% of patients with a range of 46,XY DSD phenotypes carry mutations in the *NR5A1* gene, which encodes a transcriptional regulator of a number of genes, including *SOX9* and *DAX1*. These mutations are associated with inadequate androgenization of external genitalia, leading to ambiguous genitalia, partial gonadal dysgenesis, and absent or rudimentary müllerian structures.

Disorders Associated with a 46,XX Karyotype

A series of phenotypes known as the **46,XX testicular DSDs** (previously termed *XX sex reversal*) are characterized by the presence of male external genitalia in individuals with an apparently normal 46,XX karyotype. The overall incidence is approximately 1 in 20,000.

Most patients have a normal male appearance at birth and are not diagnosed until puberty because of small testes, gynecomastia, and infertility, despite otherwise normal-appearing male genitalia and pubic hair (see Table 6-9). As described previously in the section on the Y chromosome, most of these patients are found to have a copy of a normal *SRY* gene translocated to an X chromosome as a result of aberrant recombination (see Fig. 6-11) (Case 41).

Those 46,XX males who lack an *SRY* gene, however, are a clinically more heterogeneous group. Approximately 15% to 20% of such patients are identifiable at birth because of ambiguous genitalia, including penoscrotal hypospadias and cryptorchidism (undescended testes); there are no identifiable müllerian structures, and their gender identity is male. A somewhat smaller percentage of patients, however, have *both* testicular and ovarian tissue, either as an ovotestis or as a separate ovary and testis, a condition known as **46,XX ovotesticular DSD** (formerly called true hermaphroditism).

Patients with either testicular DSD or ovotesticular DSD who lack a translocated *SRY* gene have been the subject of intense investigation to identify the responsible genetic defect(s). Duplications of at least two genes have been described, suggesting that increased levels of transcriptional regulators can overcome the absence of *SRY* and initiate the testis-specific pathway (see Table 6-8 and Fig. 6-16). Both gene duplications and regulatory mutations can increase the level of *SOX9* expression to bypass the requirement for *SRY*. Similarly, duplications of the X-linked *SOX3* gene, which is very closely related in sequence to the *SRY* gene, can stimulate increased *SOX9* expression, replacing the usual need for *SRY*.

Ovarian Development and Maintenance

A number of genes have been implicated in normal ovarian development through the study of DSDs (see Table 6-8). Thus ovarian development may not be the "default" pathway, as it is frequently described, but rather the result of balanced interactions among various genes, some of which normally stimulate the ovarian pathway and others of which normally inhibit factors involved in the opposing male pathway.

Ovarian maintenance typically lasts for up to five decades in normal females. Loss of normal ovarian function before the age of 40, as seen in approximately 1% of women, is considered **premature ovarian failure** (or **premature ovarian insufficiency**). It has long been thought that two X chromosomes are necessary for ovarian maintenance, because 45,X females, despite normal initiation of ovarian development in utero, are characterized by germ cell loss, oocyte degeneration, and ovarian dysgenesis. Further, patients with 47,XXX or with cytogenetic abnormalities involving Xq, as well as carriers of **fragile X syndrome** (Case 17), frequently show premature ovarian failure. Because many non-overlapping deletions on Xq show the same effect, this finding may reflect a need for two structurally normal X chromosomes in oogenesis or simply a requirement for multiple X-linked genes.

Nearly a dozen specific genes have been implicated in familial cases of premature ovarian failure and in various forms of 46,XX gonadal dysgenesis.

Disorders of Sex Development Involving Phenotypic Sex

Patients described earlier illustrate a mismatch between their chromosomal sex and their gonadal sex, frequently leading to gonadal dysgenesis (see Fig. 6-16). In contrast, individuals with 46,XX or 46,XY DSD have gonadal tissue that matches their chromosomal sex. However, their mismatch lies in the establishment of phenotypic sex: here, their internal and/or external genitalia show features that are contrary to those expected normally for those of the given chromosomal and gonadal sex (see Fig. 6-16). Thus patients with 46,XX DSD have a 46,XX karyotype with normal ovarian tissue but with ambiguous or male genitalia. And those with 46,XY DSD have a 46,XY karyotype and testicular tissue but with incompletely masculinized or female external genitalia. On this basis, patients of both types were thus previously described as having "pseudohermaphroditism," a term no longer in use.

Virilization of 46,XX Infants: Congenital Adrenal Hyperplasia

These patients include those who have 46,XX karyotypes with a normal uterus and ovaries but with ambiguous or male external genitalia due to excessive virilization. The majority of such patients have **congenital adrenal hyperplasia (CAH)**, an inherited disorder arising from specific defects in enzymes of the adrenal cortex required for cortisol biosynthesis and resulting in virilization of 46,XX infants. In addition to being a frequent cause of female virilization, CAH accounts for approximately half of all cases presenting with ambiguous external genitalia. Ovarian development is normal, but excessive production of androgens causes masculinization of the external genitalia, with clitoral enlargement and labial fusion to form a scrotum-like structure (Fig. 6-17).

Although any one of several enzymatic steps may be defective in CAH, by far the most common defect is

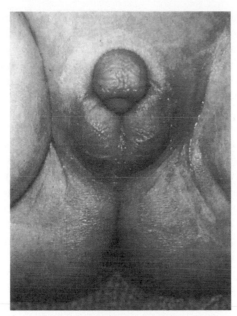

Figure 6-17 Masculinized external genitalia of a 46,XX infant caused by congenital adrenal hyperplasia (virilizing form). See text for discussion. *See Sources & Acknowledgments.*

deficiency of 21-hydroxylase, which has an incidence of approximately 1 in 12,500 births. Deficiency of 21-hydroxylase blocks the normal biosynthetic pathway of glucocorticoids and mineralocorticoids. This leads to overproduction of the precursors, which are then shunted into the pathway of androgen biosynthesis, causing abnormally high androgen levels in both XX and XY embryos. Whereas 46,XX infants with 21-hydroxylase deficiency are born with ambiguous genitalia, affected 46,XY infants have normal external genitalia and may go unrecognized in early infancy. Of patients with classic 21-hydroxylase deficiency, 25% have the simple virilizing type, and 75% have a salt-losing type due to a mineralocorticoid deficiency that is clinically more severe and may lead to neonatal death. A screening test developed to identify the condition in newborns is now in use in many countries (see Chapter 16). Prompt medical, surgical, and psychosocial management of 46,XX CAH patients is associated with improved fertility rates and normal female gender identity.

Incomplete Masculinization of 46,XY Infants: Androgen Insensitivity Syndrome

In addition to disorders of testis formation during embryological development, causes of DSD in 46,XY individuals include abnormalities of gonadotropins, inherited disorders of testosterone biosynthesis and metabolism, and abnormalities of androgen target cells. These disorders are heterogeneous both genetically and clinically, and in some cases they may correspond to milder manifestations of the same cause underlying ovotesticular DSD. Whereas the gonads are exclusively testes in 46,XY DSD, the genital ducts or external genitalia are incompletely masculinized (see Fig. 6-16).

There are several forms of androgen insensitivity that result in incomplete masculinization of 46,XY individuals. Here we illustrate the essential principles by considering the X-linked syndrome known as **androgen insensitivity syndrome** (once known as testicular feminization). As the original name indicates, testes are present either within the abdomen or in the inguinal canal, where they are sometimes mistaken for hernias in infants who otherwise appear to be normal females. Although the testes in these patients secrete androgen normally, end-organ unresponsiveness to androgens results from absence of androgen receptors in the appropriate target cells. The receptor protein, specified by the normal allele at the X-linked androgen receptor (AR) locus, has the role of forming a complex with testosterone and dihydrotestosterone. If the complex fails to form, the hormone fails to stimulate the transcription of target genes required for differentiation in the male direction. The molecular defect has been determined in many hundreds of cases and ranges from a complete deletion of the *AR* gene to point mutations in the androgen-binding or DNA-binding domains of the androgen receptor protein.

Affected individuals are chromosomal males (karyotype 46,XY) who have apparently normal female external genitalia but have a blind vagina and no uterus or uterine tubes. The incidence of androgen insensitivity is approximately 1 in 10,000 to 20,000 live births, and both complete and partial forms are known, depending on the severity of the genetic defect. In the complete form (Fig. 6-18), axillary and pubic hair are sparse or absent, and breast development occurs at the appropriate age, but without menses; primary amenorrhoea is frequently the presenting clinical finding that leads to a diagnosis. Gender assignment is typically not an issue, and psychosexual development and sexual function (except for fertility) are that of a typical 46,XX female.

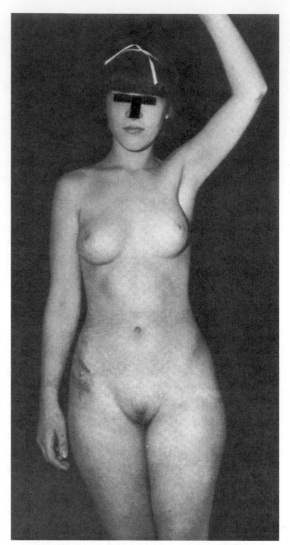

Figure 6-18 **Phenotype of a 46,XY individual with complete androgen insensitivity syndrome.** Note female body contours, breast development, absence of axillary hair, and sparse pubic hair. *See Sources & Acknowledgments.*

NEURODEVELOPMENTAL DISORDERS AND INTELLECTUAL DISABILITY

Lastly, we consider another class of disorders that like the disorders of sex development just discussed, frequently require a wide range of chromosomal and genomic approaches for diagnosis, management, and genetic counseling. Neurodevelopmental disorders are highly heterogeneous, encompassing impairments in cognition, communication, behavior, and motor functioning. Broadly considered, the category of neurodevelopmental disorders includes overlapping diagnoses such as **intellectual disability** (defined as impairment of cognitive and adaptive functions in childhood), **autism spectrum disorder (ASD)** (Case 5), and **attention deficit hyperactivity disorder (ADHD)**. This category can also include various neuropsychiatric conditions such as schizophrenia and bipolar disorder, complex traits of the type that are considered later in Chapter 8.

The overall incidence of intellectual disability and developmental delay is estimated to be at least 2% to 3%, whereas ASD affects as many as 1%. Determining the genetic cause of intellectual disability in most patients is a particular challenge, especially in the absence of other clinical clues or information about the specific gene or region of the genome responsible. Especially in sporadic cases without an obvious family history, a precise diagnosis can be helpful for clinical management and genetic counseling. Thus the full range of screening methods must be considered, including karyotyping and chromosomal microarrays, as well as whole-exome and whole-genome sequencing.

Genomic Imbalance in Neurodevelopmental Disorders

In large studies comparing diagnostic yield in this patient population, chromosomal microarray analysis detects

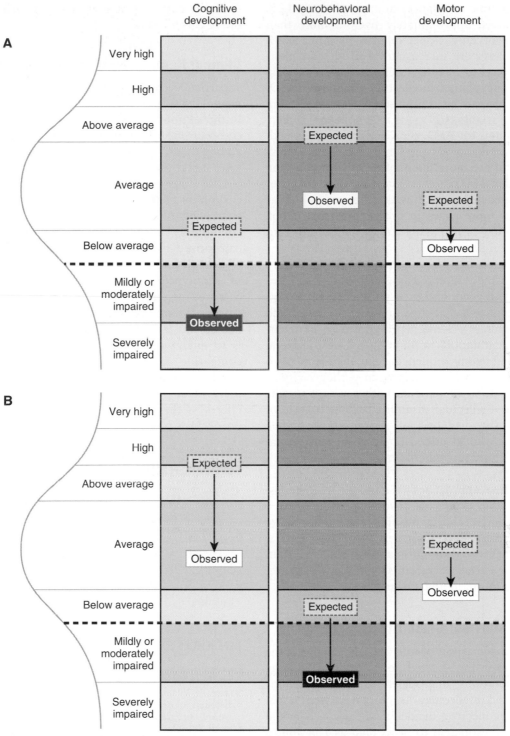

Figure 6-19 **Model for impact of genetic or genomic changes on an individual's cognitive, neurobehavioral, and motor development.** Here, the observed profile of abilities in probands (*solid boxes*) shows the deleterious effect of a copy number variant (CNV) on the predicted profile expected from familial background (*gray boxes*). The phenotypic effect of a particular CNV varies among the three elements of neurodevelopment. The *purple dotted line* represents the diagnostic threshold (2 SD below the mean). **A,** In this family, the deleterious effect of the CNV on quantitative cognitive traits (e.g., IQ) results in a diagnosis of intellectual disability, whereas neurobehavioral and motor features do not fall within the clinically impaired range. **B,** In contrast, in a different family, because of different familial norms, the deleterious effect of the same CNV leads to a diagnosis of a neurobehavioral disorder (e.g., schizophrenia), but without intellectual disability or motor impairment. *See Sources & Acknowledgments.*

pathogenic genomic imbalances in approximately 12% to 16% of cases, approximately fivefold more than G-banded karyotyping alone; on this basis, chromosomal microarrays are increasingly considered a first-tier clinical test to identify genomic imbalance in patients with unexplained intellectual disability or ASD. Although an increase in the presence of multiple rare copy number variants (CNVs) is true both for intellectual disability and for ASD, the CNVs in patients with intellectual disability tend to be larger and to encompass more genes and are more likely of de novo origin than those detected in ASD patients. Many hundreds of genes have been implicated to date, with estimates as high as a thousand or more genes in the genome that, when present in too few or too many copies, can lead to neurodevelopmental disorders.

Although screening for genomic imbalance due to CNVs is increasingly accepted as a diagnostic tool, identifying individual genes and their pathogenic mutations remains a significant challenge because of clinical and genetic heterogeneity. Some genes appear to be recurrent targets of mutation, accounting for up to several percent of cases; exome sequencing can identify de novo coding variants with likely or proven pathogenicity in approximately 15% of patients with severe, sporadic nonsyndromic intellectual disability and in cohorts of patients with the diagnosis of ASD. Whole-genome sequencing has also identified likely pathogenic mutations, either de novo or inherited, in ASD and in intellectual disability.

Although these approaches are valuable for gene discovery, large-scale exome or genome sequencing as a strategy for routine clinical testing will likely require substantial further reductions in cost, as well as improvements in the ability to distinguish a pathogenic mutation from the great excess of variants of unknown significance that are found in any single genome.

X-Linked Intellectual Disability

A long-appreciated aspect of intellectual disability is the excess of males in the affected population, and a large number of mutations, microdeletions, or duplications causing X-linked intellectual disability have been documented. The collective incidence of such X-linked defects has been estimated to be as high as 1 in 500 to 1000 live births.

The most common cause of X-linked intellectual disability is mutation in the *FMR1* gene in males with the **fragile X syndrome** (Case 17). However, nearly 100 other X-linked genes have been implicated in X-linked intellectual disability, mostly on the basis of large family studies. Chromosomal microarray analysis has identified presumptive causal CNVs and insertion-deletions in a further 10% of such families. In addition, exome-sequencing efforts summarized in the preceding section to identify de novo changes in patients with intellectual

disability have revealed an excess of such mutations on the X chromosome.

Clinical Heterogeneity and Diagnostic Overlap

A particular challenge for understanding neurodevelopmental disorders, their etiology, and their clinical course is the extraordinary degree of clinical heterogeneity, co-occurrence of symptoms, and diagnostic overlap among them. For cases due either to CNVs or to single-gene mutations, the same defect can lead to different clinical diagnoses in different cases and even in different family members—some with intellectual disability, some with ASD, and some with diagnosed psychiatric conditions. This heterogeneity and overlap, even when categorized by genetic/genomic diagnosis rather than clinical diagnosis, suggests the need for further study of genotype/phenotype correlations to meaningfully capture the broad range of phenotypes that might emerge among individuals with the same genetic disorder. As illustrated in Figure 6-19, one important factor is to analyze the effect of the CNV or mutation by comparing affected individuals to their unaffected family members (rather than to unrelated individuals in the general population), thus minimizing confounding effects of the wide range of cognitive and behavioral phenotypes observed even in the general population.

GENERAL REFERENCES

Achermann JC, Hughes IA: Disorders of sex development. In Melmed S, Polonsky KS, Larsen PR, Kronenberg HM, editors: *Williams textbook of endocrinology*, ed 12, Philadelphia, 2011, WB Saunders, pp 886–934.

Gardner RJM, Sutherland GR, Shaffer LG: *Chromosome abnormalities and genetic counseling*, ed 4, Oxford, England, 2012, Oxford University Press.

Moore KL, Persaud TVN, Torchia MG: *The developing human: clinically oriented embryology*, ed 9, Philadelphia, 2013, W.B. Saunders.

REFERENCES FOR SPECIFIC TOPICS

Bartolomei MS, Ferguson-Smith AC: Mammalian genomic imprinting, *Cold Spring Harb Perspect Biol* 3:a002592, 2011.

Baxter R, Vilain R: Translational genetics for diagnosis of human disorders of sex development, *Annu Rev Genomics Hum Genet* 14:371–392, 2013.

Cassidy SB, Schwartz S, Miller JL, et al: Prader-Willi syndrome, *Genet Med* 14:10–26, 2012.

Cooper GM, Coe BP, Girirajan S, et al: A copy number variation morbidity map of developmental delay, *Nat Genet* 43:838–846, 2011.

de Ligt J, Willemsen H, van Bon BWM, et al: Diagnostic exome sequencing in persons with severe intellectual disability, *N Engl J Med* 367:1921–1929, 2012.

Ellison JW, Rosenfeld JA, Shaffer LG: Genetic basis of intellectual disability, *Ann Rev Med* 64:441–450, 2013.

Gajecka M, MacKay KL, Shaffer LG: Monosomy 1p36 deletion syndrome, *Am J Med Genet Part C Semin Med Genet* 145C:346–356, 2007.

Higgins AW, Alkuraya FS, Bosco AF, et al: Characterization of apparently balanced chromosomal rearrangements from the Developmental Genome Anatomy Project, *Am J Hum Genet* 82:712–722, 2008.

Hughes IA, Davies JD, Bunch TI, et al: Androgen insensitivity syndrome, *Lancet* 380:1419–1428, 2012.

Hughes IA, Houk C, Ahmed SF, et al: Consensus statement on management of intersex disorders, *Arch Dis Child* 91:554–563, 2006.

Huguet G, Ey E, Bourgeron T: The genetic landscapes of autism spectrum disorders, *Ann Rev Genomics Hum Genet* 14:191–213, 2013.

Jiang Y, Yuen RKC, Jin X, et al: Detection of clinically relevant genetic variants in autism spectrum disorder by whole-genome sequencing, *Am J Hum Genet* 93:1–15, 2013.

Kaminsky EB, Kaul V, Paschall J, et al: An evidence-based approach to establish the functional and clinical significance of copy number variants in intellectual and developmental disabilities, *Genet Med* 13:777–784, 2011.

Korbel JO, Tirosh-Wagner T, Urban AE, et al: The genetic architecture of Down syndrome phenotypes revealed by high-resolution analysis of human segmental trisomies, *Proc Natl Acad Sci USA* 106:12031–12036, 2009.

Leggett V, Jacobs P, Nation K, et al: Neurocognitive outcomes of individuals with a sex chromosome trisomy: XXX, XYY, or XXY: a systematic review, *Devel Med Child Neurol* 52:119–129, 2010.

Mabb AM, Judson MC, Zylka MJ, et al: Angelman syndrome: insights into genomic imprinting and neurodevelopmental phenotypes, *Trends Neurosci* 34:293–303, 2011.

Moreno-De-Luca A, Myers SM, Challman TD, et al: Developmental brain dysfunction: revival and expansion of old concepts based on new genetic evidence, *Lancet Neurol* 12:406–414, 2013.

Morris JK, Alberman E, Mutton D, et al: Cytogenetic and epidemiological findings in Down syndrome: England and Wales 1989-2009, *Am J Med Genet A* 158A:1151–1157, 2012.

Najmabadi H, Hu H, Garshasbi M, et al: Deep sequencing reveals 50 novel genes for recessive cognitive disorders, *Nature* 478:57–63, 2011.

Silber SJ: The Y chromosome in the era of intracytoplasmic sperm injection, *Fertil Steril* 95:2439–2448, 2011.

Talkowski ME, Maussion G, Crapper L, et al: Disruption of a large intergenic noncoding RNA in subjects with neurodevelopmental disabilities, *Am J Hum Genet* 91:1128–1134, 2012.

Talkowski ME, Rosenfeld JA, Blumenthal I, et al: Sequencing chromosomal abnormalities reveals neurodevelopmental loci that confer risk across diagnostic boundaries, *Cell* 149:525–537, 2012.

Weischenfeldt J, Symmns O, Spitz F, et al: Phenotypic impact of genomic structural variation: insights from and for human disease, *Nat Rev Genet* 14:125–138, 2013.

Zufferey F, Sherr EH, Beckmann ND, et al: A 600 kb deletion syndrome at 16p11.2 leads to energy imbalance and neuropsychiatric disorders, *J Med Genet* 49:660–668, 2013.

PROBLEMS

1. In a woman with a 47,XXX karyotype, what types of gametes would theoretically be formed and in what proportions? What are the *theoretical* karyotypes and phenotypes of her progeny? What are the *actual* karyotypes and phenotypes of her progeny?

2. Individuals carrying a copy of the inv(9) described in the text are clinically normal. Provide two possible explanations.

3. The birth incidence rates of 47,XXY and 47,XYY males are approximately equal. Is this what you would expect on the basis of the possible origins of the two abnormal karyotypes? Explain.

4. How can a person with an XX karyotype differentiate as a phenotypically normal male?

5. A small centric ring X chromosome that lacks the X inactivation center is observed in a patient with short stature, gonadal dysgenesis, and intellectual disability. Because intellectual disability is not a typical feature of Turner syndrome, explain the presence of mental retardation with or without other associated physical anomalies in individuals with a 46,X,r(X) karyotype. In a prenatal diagnosis involving a different family, a somewhat larger ring that contains the X inactivation center is detected. What phenotype would you predict for the fetus in this pregnancy?

6. A baby girl with ambiguous genitalia is found to have 21-hydroxylase deficiency of the salt-wasting type. What karyotype would you expect to find? What is the disorder? What genetic counseling would you offer to the parents?

7. What are the expected clinical consequences of the following deletions? If the same amount of DNA is deleted in each case, why might the severity of each be different?
   a. 46,XX,del(13)(pter→p11.1:)
   b. 46,XY,del(Y)(pter→q12:)
   c. 46,XX,del(5)(p15)
   d. 46,XX,del(X)(q23q26)

8. Provide possible explanations for the fact that persons with X chromosome aneuploidy are clinically not completely normal.

9. In genetics clinic, you are counseling five pregnant women who inquire about their risk for having a Down syndrome fetus. What are their risks and why?
   a. a 23-year-old mother of a previous trisomy 21 child
   b. a 41-year-old mother of a previous trisomy 21 child
   c. a 27-year-old woman whose niece has Down syndrome
   d. a carrier of a 14;21 Robertsonian translocation
   e. a woman whose husband is a carrier of a 14;21 Robertsonian translocation

10. A young girl with Down syndrome is karyotyped and found to carry a 21q21q translocation. With use of standard cytogenetic nomenclature, what is her karyotype?

11. Paracentric inversions generally do not raise the problem of imbalance in offspring. Why not?

Patterns of Single-Gene Inheritance

In Chapter 1, we introduced and briefly characterized the three main categories of genetic disorders—single-gene, chromosomal, and complex. In this chapter, the typical patterns of transmission of single-gene disorders are discussed in detail, building on the mechanisms of gene and genome transmission presented generally in Chapters 2 and 3; the emphasis here is on the various inheritance patterns of genetic disease in families. Later, in Chapter 8, we will examine more complex patterns of inheritance, including multifactorial disorders that result from the interaction between variants at one or more genes, as well as environmental factors.

OVERVIEW AND CONCEPTS
Genotype and Phenotype

For autosomal loci (and X-linked loci in females), the **genotype** of a person at a locus consists of both of the alleles occupying that locus on the two homologous chromosomes (Fig. 7-1). Genotype should not be confused with **haplotype**, which refers to the set of alleles at two or more neighboring loci on *one* of the two homologous chromosomes. More broadly, the term *genotype* can refer to all of the allele pairs that collectively make up an individual's genetic constitution across the entire genome. **Phenotype**, as described initially in Chapter 3, is the expression of genotype as a morphological, clinical, cellular, or biochemical trait, which may be clinically observable or may only be detected by blood or tissue testing. The phenotype can be discrete—such as the presence or absence of a disease—or can be a measured quantity, such as body mass index or blood glucose levels. A phenotype may, of course, be either normal or abnormal in a given individual, but in this book, which emphasizes disorders of medical significance, the focus is on disease phenotypes—that is, genetic disorders.

When a person has a pair of identical alleles at a locus encoded in nuclear DNA, he or she is said to be **homozygous**, or a **homozygote**; when the alleles are different and one of the alleles is the wild-type allele, he or she is **heterozygous**, or a **heterozygote**. The term **compound heterozygote** is used to describe a genotype in which two different mutant alleles of a gene are present, rather than one wild-type and one mutant allele. These terms (*homozygous, heterozygous,* and *compound heterozygous*) can be applied either to a person or to a genotype.

In the special case in which a male has an abnormal allele for a gene located on the X chromosome and there is no other copy of the gene, he is neither homozygous nor heterozygous but is referred to as **hemizygous**. Mitochondrial DNA is still another special case. In contrast to the two copies of each gene per diploid cell, mitochondrial DNA molecules, and the genes encoded by the mitochondrial genome, are present in tens to thousands of copies per cell (see Chapter 2). For this reason, the terms *homozygous, heterozygous,* and *hemizygous* are not used to describe genotypes at mitochondrial loci.

A **single-gene disorder** is one that is determined primarily by the alleles at a single locus. The known single-gene diseases are listed in the late Victor A. McKusick's classic reference, *Mendelian Inheritance in Man*, which has been indispensable to medical geneticists for decades. These diseases follow one of the classic inheritance patterns in families (autosomal recessive, autosomal dominant, X-linked) and are therefore referred to as **mendelian** because, like the characteristics of the garden peas Gregor Mendel studied, they occur on average in fixed and predictable proportions among the offspring of specific types of matings.

A single abnormal gene or gene pair often produces multiple diverse phenotypic effects in multiple organ systems, with a variety of signs and symptoms occurring at different points during the life span. To cite just one example, individuals with a mutation in the *VHL* gene can have hemangioblastomas of the brain, spinal cord, and retina; renal cysts; pancreatic cysts; renal cell carcinoma; pheochromocytoma; and endolymphatic tumors of the inner ear; as well as tumors of the epididymis in males or of the broad ligament of the uterus in females—even though *all* of these disease manifestations stem from the same single mutation. Under these circumstances, the disorder is said to exhibit **pleiotropy** (from Greek *pleion* and *tropos*, more turns), and the expression of the gene defect is said to be **pleiotropic**. At present, for many pleiotropic disorders, the connection between the gene defect and the various manifestations is neither obvious nor well understood.

Single-gene disorders affect children disproportionately, but not exclusively. Serious single-gene disorders affect 1 in 300 neonates and are responsible for an estimated 7% of pediatric hospitalizations. Although less than 10% of single-gene disorders manifest after

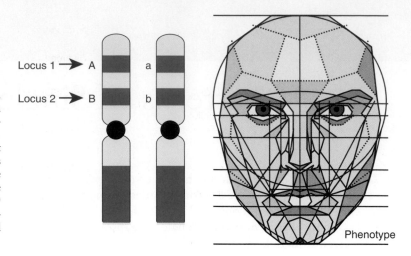

Locus 1 → A   a
Locus 2 → B   b

Phenotype

Figure 7-1 The concepts of genotype and phenotype. *(Left)* The genotype refers to information encoded in the genome. Diagram of one pair of homologous chromosomes and two loci on that chromosome, Locus 1 and Locus 2, in an individual who is heterozygous at both loci. He has alleles *A* and *a* at locus 1 and alleles *B* and *b* at locus 2. The locus 1 genotype is *Aa*, while the locus 2 genotype is *Bb*. The two haplotypes on these homologous chromosomes are *A-B* and *a-b*. *(Right)* The phenotype is the physical, clinical, cellular, or biochemical manifestation of the genotype, as illustrated here by morphometric aspects of an individual's face.

puberty, and only 1% occur after the end of the reproductive period, mendelian disorders are nonetheless important to consider in adult medicine. There are nearly 200 mendelian disorders whose phenotypes include common adult illnesses such as heart disease, stroke, cancer, and diabetes. Although mendelian disorders are by no means the major contributory factor in causing these common diseases in the population at large, they are important in individual patients because of their significance for the health of other family members and because of the availability of genetic testing and detailed management options for many of them.

Penetrance and Expressivity

For some genetic conditions, a disease-causing genotype is always fully expressed at birth as an abnormal phenotype. Clinical experience, however, teaches that other disorders are not expressed at all or may vary substantially in their signs and symptoms, clinical severity, or age of onset, even among members of a family who all share the same disease-causing genotype. Geneticists use distinct terms to describe such differences in clinical expression.

Penetrance is the probability that a mutant allele or alleles will have any phenotypic expression at all. When the frequency of expression of a phenotype is less than 100%—that is, when some of those who have the relevant genotype *completely* fail to express it—the disorder is said to show **reduced** or **incomplete penetrance**. Penetrance is all or nothing. It is the percentage of people at any given age with a predisposing genotype who are affected, regardless of the severity.

Penetrance of some disorders is age dependent; that is, it may occur any time, from early in intrauterine development all the way to the postreproductive years.

Some disorders are lethal prenatally, whereas others can be recognized prenatally (e.g., by ultrasonography; see Chapter 17) but are consistent with a liveborn infant; still others may be recognized only at birth (**congenital**).* Other disorders have their onset typically or exclusively in childhood or in adulthood. Even in these, however, and sometimes even in the same family, two individuals carrying the same disease-causing genotype may develop the disease at very different ages.

In contrast to penetrance, **expressivity** refers not to the presence or absence of a phenotype, but to the *severity of expression* of that phenotype among individuals with the same disease-causing genotype. When the severity of disease differs in people who have the same genotype, the phenotype is said to show **variable expressivity**. Even in the same family, two individuals carrying the same mutant genes may have some signs and symptoms in common, whereas their other disease manifestations may be quite different, depending on which tissues or organs happen to be affected. The challenge to the clinician caring for these families is to not miss very subtle signs of a disorder in a family member and, as a result, either mistake mild expressivity for lack of penetrance or infer that the individual does not have the disease-causing genotype.

PEDIGREES

Single-gene disorders are characterized by their patterns of transmission in families. To establish the pattern of transmission, a usual first step is to obtain information

---

*The terms *genetic* and *congenital* are frequently confused. Keep in mind that a genetic disorder is one that is determined by variation in genes, whereas a congenital disorder is simply one that is present at birth and may or may not have a genetic basis.

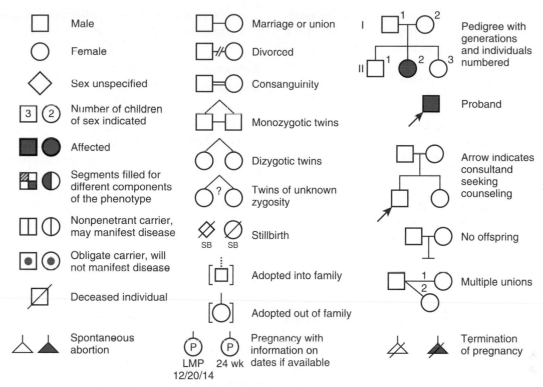

Figure 7-2 Symbols commonly used in pedigree charts. Although there is no uniform system of pedigree notation, the symbols used here are according to recent recommendations made by professionals in the field of genetic counseling.

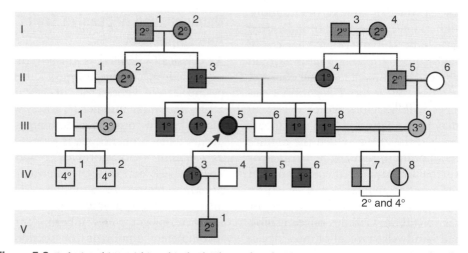

Figure 7-3 Relationships within a kindred. The proband, III-5 (*arrow*), represents an isolated case of a genetic disorder. She has four siblings, III-3, III-4, III-7, and III-8. Her partner/spouse is III-6, and they have three children (their F1 progeny). The proband has nine first-degree relatives (her parents, siblings, and offspring), nine second-degree relatives (grandparents, uncles and aunts, nieces and nephews, and grandchildren), two third-degree relatives (first cousins), and four fourth-degree relatives (first cousins once removed). IV-3, IV-5, and IV-6 are second cousins of IV-1 and IV-2. IV-7 and IV-8, whose parents are consanguineous, are doubly related to the proband: second-degree relatives through their father and fourth-degree relatives through their mother.

about the family history of the patient and to summarize the details in the form of a **pedigree**, a graphical representation of the family tree, with use of standard symbols (Fig. 7-2). The extended family depicted in such pedigrees is a **kindred** (Fig. 7-3). An affected individual

through whom a family with a genetic disorder is first brought to the attention of the geneticist (i.e., is ascertained) is the **proband, propositus,** or **index case.** The person who brings the family to attention by consulting a geneticist is referred to as the **consultand;** the consultand may be an affected individual or an unaffected relative of a proband. A family may have more than one proband, if they are ascertained through more than one source. Brothers and sisters are called **sibs** or **siblings**, and a family of sibs forms a **sibship**. Relatives are classified as **first degree** (parents, sibs, and offspring of the proband), **second degree** (grandparents and grandchildren, uncles and aunts, nephews and nieces, and half-sibs), or **third degree** (e.g., first cousins), and so forth, depending on the number of steps in the pedigree between the two relatives. The offspring of first cousins are second cousins, and a child of a first cousin is a "first cousin once removed" of his or her parents' first cousins. Couples who have one or more ancestors in common are **consanguineous**. If the proband is the only affected member in a family, he or she is an **isolated** case (see Fig. 7-3). If an isolated case is proven to be due to new mutation in the proband, it is referred to as a **sporadic** case. When there is a definitive diagnosis based on comparisons to other patients, well-established patterns of inheritance in other families with the same disorder can often be used as a basis for counseling, even if the patient is an isolated case in the family. Thus, even when a patient has no similarly affected relatives, it may still be possible to recognize that the disorder is genetic and determine the risk to other family members.

Examining a pedigree is an essential first step in determining the inheritance pattern of a genetic disorder in a family. There are, however, a number of situations that may make the inheritance pattern of an individual pedigree difficult to discern. The inheritance pattern in a family with a lethal disorder affecting a fetus early in pregnancy may be obscure because all that one observes are multiple miscarriages or reduced fertility. Conversely, for phenotypes with variable age of onset, an affected individual may have unaffected family members who have not yet reached the age at which the mutant gene reveals itself. In addition to reduced penetrance or variable expressivity that may mask the existence of relatives carrying the mutant genotype, the geneticist may lack accurate information about the presence of the disorder in relatives or about family relationships. Finally, with the small family size typical of most developed countries today, the patient may by chance alone be the only affected family member, making determination of any inheritance pattern very difficult.

MENDELIAN INHERITANCE

The patterns of inheritance shown by single-gene disorders in families depend chiefly on two factors:

- Whether the chromosomal location of the gene locus is on an autosome (chromosomes 1 to 22), on a sex chromosome (X and Y chromosomes), or in the mitochondrial genome
- Whether the phenotype is **dominant** (expressed when only one chromosome carries the mutant allele) or

recessive (expressed only when both chromosomes of a pair carry mutant alleles at a locus)

Autosomal, X-Linked, and Mitochondrial Inheritance

The different patterns of transmission of the autosomes, sex chromosomes, and mitochondria during meiosis result in distinctive inheritance patterns of mutant alleles on these different types of chromosome (see Chapter 2). Because only one of the two copies of each autosome passes into a single gamete during meiosis, males and females heterozygous for a mutant allele on an autosome have a 50% chance of passing that allele on to any offspring, regardless of the child's sex. Mutant alleles on an X chromosome, however, are not distributed equally to sons and daughters. Males pass their Y chromosome to their sons and their X to their daughters; they therefore *cannot* pass an allele on the X chromosome to their sons and *always* pass the allele to their daughters (unless it is at one of the pseudoautosomal loci; see Chapter 6). Because mitochondria are inherited from the mother only, regardless of the sex of the offspring, mutations in the mitochondrial genome are not inherited according to a mendelian pattern. Autosomal, X-linked, and mitochondrial inheritance will be discussed in the rest of the chapter that follows.

Dominant and Recessive Traits
Autosomal Loci

As classically defined, a phenotype is **recessive** if it is expressed *only* in homozygotes, hemizygotes, or compound heterozygotes, all of whom lack a wild-type allele, and *never* in heterozygotes, who do have a wild-type allele. In contrast, a **dominant** inheritance pattern occurs when a phenotype is expressed in heterozygotes *as well as* in homozygotes (or compound heterozygotes). For the vast majority of inherited dominant diseases, homozygotes or compound heterozygotes for mutant alleles at autosomal loci are more severely affected than are heterozygotes, an inheritance pattern known as **incompletely dominant** (or **semidominant**). Very few diseases are known in which homozygotes (or compound heterozygotes) show the same phenotype as heterozygotes; in such cases, the disorder is referred to as a **pure dominant** disease. Finally, if phenotypic expression of both alleles at a locus occurs in a compound heterozygote, inheritance is termed **codominant**.

ABO Blood Group. One medically important trait that demonstrates codominant expression is the ABO blood group system important in blood transfusion and tissue transplantation. The *A, B,* and *O* alleles at the *ABO* locus form a three-allele system in which two alleles (*A* and *B*) govern expression of either the A or B carbohydrate antigen on the surface of red cells as a codominant

TABLE 7-1 ABO Genotypes and Serum Reactivity

Genotype	Phenotype in RBCs	Reaction with Anti-A	Reaction with Anti-B	Antibodies in Serum
OO	O	–	–	Anti-A, anti-B
AA or AO	A	+	–	Anti-B
BB or BO	B	–	+	Anti-A
AB	AB	+	+	Neither

– Represents no reaction; + represents reaction. RBC, red blood cell.

trait; a third allele (O) results in expression of neither the A nor the B antigen and is recessive. The difference between the A and B antigen is which of two different sugar molecules makes up the terminal sugar on a cell surface glycoprotein called H. Whether the A or B form of the glycoprotein is made is specified by an enzyme encoded by the *ABO* gene that adds one or the other sugar molecule to the H antigen depending on which version of the enzyme is encoded by alleles at the *ABO* locus. There are therefore four phenotypes possible: O, A, B, and AB (Table 7-1). Type A individuals have antigen A on their red blood cells, type B individuals have antigen B, type AB individuals have both antigens, and type O individuals have neither.

A feature of the ABO groups not shared by other blood group systems is the reciprocal relationship, in an individual, between the antigens present on the red blood cells and the antibodies in the serum (see Table 7-1). When the red blood cells lack antigen A, the serum contains anti-A antibodies; when the cells lack antigen B, the serum contains anti-B. Formation of anti-A and anti-B antibodies in the absence of prior blood transfusion is believed to be a response to the natural occurrence of A-like and B-like antigens in the environment (e.g., in bacteria).

X-Linked Loci

For X-linked disorders, a condition expressed only in hemizygotes and *never* in heterozygotes has traditionally been referred to as an X-linked recessive, whereas a phenotype that is *always* expressed in heterozygotes as well as in hemizygotes has been called X-linked dominant. Because of epigenetic regulation of X-linked gene expression in carrier females due to X chromosome inactivation (introduced in Chapters 3 and 6), it can be difficult to determine phenotypically if a disease with an X-linked inheritance pattern is dominant or recessive, and some geneticists have therefore chosen not to use these terms when describing the inheritance of X-linked disease.

Strictly speaking, the terms *dominant* and *recessive* refer to the inheritance pattern of a phenotype rather than to the alleles responsible for that phenotype. Similarly, a gene is not dominant or recessive; it is the phenotype produced by a particular mutant allele in that gene that shows dominant or recessive inheritance.

AUTOSOMAL PATTERNS OF MENDELIAN INHERITANCE
Autosomal Recessive Inheritance

Autosomal recessive disease occurs only in individuals with two mutant alleles and no wild-type allele. Such homozygotes must have inherited a mutant allele from each parent, each of whom is (barring rare exceptions that we will consider later) a heterozygote for that allele.

When a disorder shows recessive inheritance, the mutant allele responsible generally reduces or eliminates the function of the gene product, a so-called **loss-of-function** mutation. For example, many recessive diseases are caused by mutations that impair or eliminate the function of an enzyme. The remaining normal gene copy in a heterozygote is able to compensate for the mutant allele and prevent the disease from occurring. However, when no normal allele is present, as in homozygotes or compound heterozygotes, disease occurs. Disease mechanisms and examples of recessive conditions are discussed in detail in Chapters 11 and 12.

Three types of matings can lead to homozygous offspring affected with an autosomal recessive disease. The most common mating by far is between two unaffected heterozygotes, who are often referred to as **carriers**. However, any mating in which each parent has at least one recessive allele can produce homozygous affected offspring. The transmission of a recessive condition can be followed if we symbolize the mutant recessive allele as *r* and its normal dominant allele as *R*.

As seen in the table, when both parents of an affected person are carriers, their children's risk for receiving a recessive allele is 50% from each parent. The chance of inheriting two recessive alleles and therefore being affected is thus $\frac{1}{2} \times \frac{1}{2}$ or 1 in 4 with each pregnancy. The 25% chance for two heterozygotes to have a child with an autosomal recessive disorder is independent of how many previous children there are who are either affected or unaffected. The proband may be the only affected family member, but if any others are affected, they are usually in the same sibship and not elsewhere in the kindred (Fig. 7-4).

Sex-Influenced Autosomal Recessive Disorders

Because males and females both have the same complement of autosomes, autosomal recessive disorders generally show the same frequency and severity in males and females. There are, however, exceptions. Some autosomal recessive diseases demonstrate a **sex-influenced phenotype**, that is, the disorder is expressed in both sexes but with different frequencies or severity. For example, **hereditary hemochromatosis** is an autosomal recessive phenotype that is 5 to 10 times more common in males than in females (Case 20). Affected individuals have enhanced absorption of dietary iron that can lead to iron overload and serious damage to the heart, liver, and pancreas. The lower incidence of the clinical disorder in homozygous females is believed to be due to their lower dietary iron intake, lower alcohol usage, and increased iron loss through menstruation.

Autosomal Recessive Inheritance

Carrier by Carrier		Parent 2 Genotype R/r Gametes		Risk for Disease
		R	r	
Parent 1 Genotype R/r Gametes	R	R/R	R/r	¼ Unaffected (R/R) ½ Unaffected carriers (R/r) ¼ Affected (r/r)
	r	R/r	r/r	

Carrier by Affected		Parent 2 Genotype r/r Gametes		Risk for Disease
		r	r	
Parent 1 Genotype R/r Gametes	R	R/r	R/r	½ Unaffected carriers (R/r) ½ Affected (r/r)
	r	r/r	r/r	

Affected by Affected		Parent 2 Genotype r/r Gametes		Risk for Disease
		r	r	
Parent 1 Genotype r/r Gametes	r	r/r	r/r	All affected (r/r)
	r	r/r	r/r	

The wild-type allele is denoted by uppercase R, a mutant allele by lowercase r.

Gene Frequency and Carrier Frequency

Mutant alleles responsible for a recessive disorder are generally rare, and so most people will not have even one copy of the mutant allele. Because an autosomal recessive disorder must be inherited from *both* parents, the risk that any carrier will have an affected child depends partly on the chance that his or her mate is also a carrier of a mutant allele for the condition. Thus knowledge of the carrier frequency of a disease is clinically important for genetic counseling.

The most common autosomal recessive disorder in white children is **cystic fibrosis (CF)** (Case 12), caused by mutations in the *CFTR* gene (see Chapter 12). Among white populations, approximately 1 child in 2000 has two mutant *CFTR* alleles and has the disease, from which we can infer that 1 in 23 individuals is a silent carrier who has no disease. (How one calculates heterozygote frequencies in autosomal recessive conditions will be addressed in Chapter 9.) Mutant alleles may be handed down from carrier to carrier for numerous generations without ever appearing in the homozygous state and causing overt disease. The presence of such hidden recessive genes is not revealed unless the carrier happens to mate with someone who also carries a mutant allele at the same locus and the two deleterious alleles are both inherited by a child.

Estimates of the number of deleterious alleles in each of our genomes range from 50 to 200 based on examining an individual's complete exome or genome sequence for clearly deleterious mutations in the coding regions of the genome (see Chapter 4). This estimate is imprecise, however. It may be an underestimate, because it does not include mutant alleles whose deleterious effect is not obvious from a simple examination of the DNA sequence. It may also, however, be an overestimate, because it includes mutations in many genes that are not known to cause disease.

Consanguinity

Because most mutant alleles are generally uncommon in the population, people with rare autosomal recessive disorders are typically **compound heterozygotes** rather than true homozygotes. One well-recognized exception

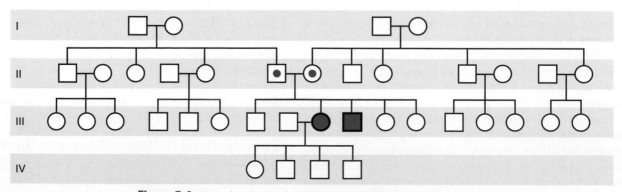

Figure 7-4 Typical pedigree showing autosomal recessive inheritance.

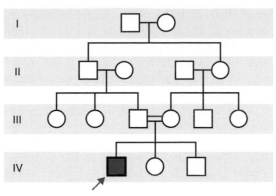

Figure 7-5 Pedigree in which parental consanguinity suggests autosomal recessive inheritance. *Arrow* indicates the proband.

to this rule occurs when an affected individual inherits the exact same mutant allele from both parents because the parents are consanguineous (i.e., they are related and carry the identical mutant allele inherited from a common ancestor). Finding consanguinity in the parents of a patient with a genetic disorder is strong evidence (although not proof) for the autosomal recessive inheritance of that condition. For example, the disorder in the pedigree in Figure 7-5 is likely to be an autosomal recessive trait, even though other information in the pedigree may seem insufficient to establish this inheritance pattern.

Consanguinity is more frequently found in the background of patients with very rare conditions than in those with more common recessive conditions. This is because it is less likely that two individuals mating at random in the population will both be carriers of a very rare disorder by chance alone than it is that they would both be carriers because they inherited the same mutant allele from a single common ancestor. For example, in **xeroderma pigmentosum** (Case 48), a very rare autosomal recessive condition of DNA repair (see Chapter 15), more than 20% of cases occur among the offspring of marriages between first cousins. In contrast, in more common recessive conditions, most cases of the disorder result from matings between *unrelated* persons, each of whom happens by chance to be a carrier. Thus most affected persons with a relatively common disorder, such as CF, are *not* the result of consanguinity, because the mutant allele is so common in the general population. How consanguinity is measured for different matings is described in Chapter 9.

The genetic risk to the offspring of marriages between related people is not as great as is sometimes imagined. For marriages between first cousins, the absolute risks of abnormal offspring, including not only known autosomal recessive diseases but also stillbirth, neonatal death, and congenital malformation, is 3% to 5%, approximately double the overall background risk of 2% to 3% for offspring born to any unrelated couple (see Chapter 16). Consanguinity at the level of third cousins or more remote relationships is not considered

to be genetically significant, and the increased risk for abnormal offspring is negligible in such cases.

The incidence of first-cousin marriage is low (≈1 to 10 per 1000 marriages) in many populations in Western societies today. However, it remains relatively common in some ethnic groups, for example, in families from rural areas of the Indian subcontinent, in other parts of Asia, and in the Middle East, where between 20% and 60% of all marriages are between cousins.

CHARACTERISTICS OF AUTOSOMAL RECESSIVE INHERITANCE

- An autosomal recessive phenotype, if not isolated, is typically seen only in the sibship of the proband, and not in parents, offspring, or other relatives.
- For most autosomal recessive diseases, males and females are equally likely to be affected.
- Parents of an affected child are asymptomatic carriers of mutant alleles.
- The parents of the affected person may in some cases be consanguineous. This is especially likely if the gene responsible for the condition is rare in the population.
- The recurrence risk for each sib of the proband is 1 in 4 (25%).

Autosomal Dominant Inheritance

More than half of all known mendelian disorders are inherited as autosomal dominant traits. The incidence of some autosomal dominant disorders can be high. For example, adult **polycystic kidney disease** (Case 37) occurs in 1 in 1000 individuals in the United States. Other autosomal dominant disorders show a high frequency only in certain populations from specific geographical areas: for example, the frequency of **familial hypercholesterolemia** (Case 16) is 1 in 100 for Afrikaner populations in South Africa and of **myotonic dystrophy** is 1 in 550 in the Charlevoix and Saguenay–Lac Saint Jean regions of northeastern Quebec. The burden of autosomal dominant disorders is further increased because of their hereditary nature; when they are transmitted through families, they raise medical and even social problems not only for individuals but also for whole kindreds, often through many generations.

The risk and severity of dominantly inherited disease in the offspring depend on whether one or both parents are affected and whether the trait is a pure dominant or is incompletely dominant. There are a number of different ways that one mutant allele can cause a dominantly inherited trait to occur in a heterozygote despite the presence of a normal allele. Disease mechanisms in various dominant conditions are discussed in Chapter 12.

Denoting *D* as the mutant allele and *d* as the wild-type allele, matings that produce children with an autosomal dominant disease can be between two heterozygotes (*D/d*) for the mutation or, more frequently, between a heterozygote for the mutation (*D/d*) and a homozygote for a normal allele (*d/d*).

As seen in the table, each child of a *D/d* by *d/d* mating has a 50% chance of receiving the affected parent's

Autosomal Dominant Inheritance

Affected by Unaffected		Parent 2 Genotype d/d Gametes		Risk for Disease
		d	d	
Parent 1 Genotype D/d Gametes	D	D/d	D/d	½ Affected (D/d) ½ Unaffected (d/d)
	d	d/d	d/d	

Affected by Affected		Parent 2 Genotype D/d Gametes		Risk for Disease
		D	d	
Parent 1 Genotype D/d Gametes	D	D/D	D/d	Strictly dominant ¾ Affected (D/D and D/d) ¼ Unaffected (d/d)
	d	D/d	d/d	Incompletely dominant ¼ Severely affected (D/D) ½ Affected (D/d) ¼ Unaffected (d/d)

The mutant allele causing dominantly inherited disease is denoted by uppercase D; the normal or wild-type allele is denoted by lowercase d.

abnormal allele *D* and a 50% chance of receiving the normal allele *d*. In the population as a whole, then, the offspring of *D/d* by *d/d* parents are approximately 50% *D/d* and 50% *d/d*. Of course, each pregnancy is an independent event, not governed by the outcome of previous pregnancies. Thus, within a family, the distribution of affected and unaffected children may be quite different from the theoretical expected ratio of 1:1, especially if the sibship is small. Typical autosomal dominant inheritance can be seen in the pedigree of a family with a dominantly inherited form of hereditary deafness (Fig. 7-6A).

In medical practice, homozygotes for dominant phenotypes are not often seen because matings that could produce homozygous offspring are rare. Again denoting the abnormal allele as *D* and the wild-type allele as *d*, the matings that can produce a *D/D* homozygote might theoretically be *D/d* by *D/d*, *D/D* by *D/d*, or *D/D* by *D/D*. In the case of two heterozygotes mating, three fourths of the offspring of a *D/d* by *D/d* mating will be affected to some extent and one fourth unaffected.

Pure Dominant Inheritance

As mentioned earlier, very few human disorders demonstrate a purely dominant pattern of inheritance. Even **Huntington disease** (Case 24), which is frequently considered to be a pure dominant because the disease is generally similar in the nature and severity of symptoms in heterozygotes and homozygotes, appears to have a somewhat accelerated time course from the onset of

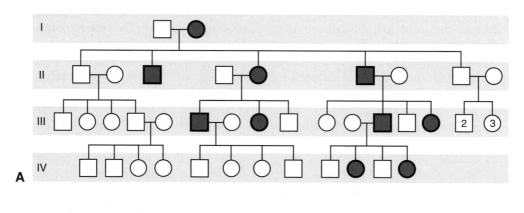

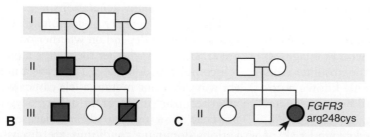

Figure 7-6 **A,** Pedigree showing typical inheritance of a form of adult-onset progressive sensorineural deafness (DFNA1) inherited as an autosomal dominant trait. **B,** Pedigree showing inheritance of achondroplasia, an incompletely dominant (or semidominant) trait. **C,** Pedigree showing a sporadic case of thanatophoric dwarfism, a genetic lethal, in the proband *(arrow)*.

disease to death in homozygous individuals compared with heterozygotes.

Incompletely Dominant Inheritance

As introduced in Chapter 4, **achondroplasia** (Case 2) is an incompletely dominant skeletal disorder of short-limbed dwarfism and large head caused by certain mutations in the fibroblast growth factor receptor 3 gene (*FGFR3*). Most achondroplasia patients have normal intelligence and lead normal lives within their physical capabilities. Marriages between two patients with achondroplasia are not uncommon. A pedigree of a mating between two individuals heterozygous for the most common mutation that causes achondroplasia is shown in Figure 7-6B. The deceased child, individual III-3, was a homozygote for the condition and had a disorder far more severe than in either parent, resulting in death soon after birth.

Sex-Limited Phenotype in Autosomal Dominant Disease

As discussed earlier for the autosomal recessive condition hemochromatosis, autosomal dominant phenotypes may also demonstrate a sex ratio that differs significantly from 1:1. Extreme divergence of the sex ratio is seen in sex-limited phenotypes, in which the defect is autosomally transmitted but expressed in only one sex. An example is **male-limited precocious puberty**, an autosomal dominant disorder in which affected boys develop secondary sexual characteristics and undergo an adolescent growth spurt at approximately 4 years of age (Fig. 7-7). In some families, the defect has been traced to mutations in the *LCGR* gene, which encodes the receptor for luteinizing hormone; these mutations constitutively activate the receptor's signaling action, even in the absence of its hormone. The defect shows no effect in heterozygous females. The pedigree in Figure 7-8 shows that, although the disease can be transmitted by unaffected (nonpenetrant carrier) females, it can also

be transmitted directly from father to son, showing that it is autosomal, not X-linked.

For disorders in which affected males do not reproduce, however, it is not always easy to distinguish sex-limited autosomal inheritance from X-linked inheritance because the critical evidence, absence of male-to-male transmission, cannot be provided. In that case, other lines of evidence, particularly gene mapping to learn

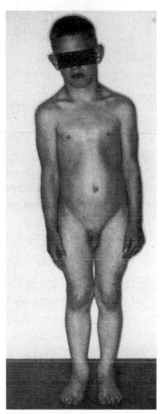

Figure 7-7 Male-limited precocious puberty, a sex-limited autosomal dominant disorder expressed exclusively in males. This child, at 4.75 years, is 120 cm in height (above the 97th percentile for his age). Note the muscle bulk and precocious development of the external genitalia. Epiphyseal fusion occurs at an early age, and affected persons are relatively short as adults.

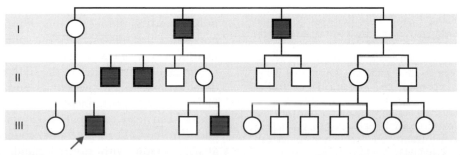

Figure 7-8 Part of a large pedigree of male-limited precocious puberty in the family of the child shown in Figure 7-7. This autosomal dominant disorder can be transmitted by affected males or by unaffected carrier females. Male-to-male transmission shows that inheritance is autosomal, not X-linked. Transmission of the trait through carrier females shows that inheritance cannot be Y-linked. *Arrow* indicates proband.

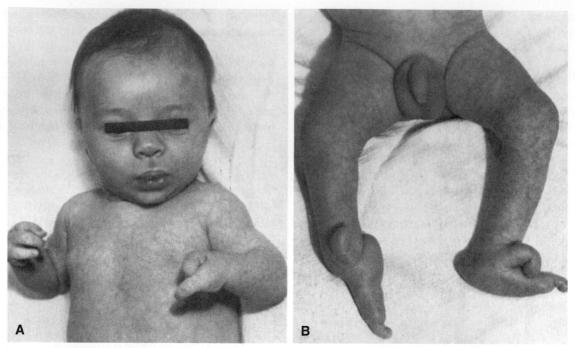

Figure 7-9 Split-hand deformity, an autosomal dominant trait involving the hands and feet, in a 3-month-old boy. A, Upper part of body. B, Lower part of body. *See Sources & Acknowledgments.*

whether the responsible gene maps to the X chromosome or to an autosome (see Chapter 10), can determine the pattern of inheritance and the consequent recurrence risk (see Box).

CHARACTERISTICS OF AUTOSOMAL DOMINANT INHERITANCE

- The phenotype usually appears in every generation, each affected person having an affected parent. Exceptions or apparent exceptions to this rule in clinical genetics are (1) cases originating from fresh mutations in a gamete of a phenotypically normal parent and (2) cases in which the disorder is not expressed (nonpenetrant) or is expressed only subtly in a person who has inherited the responsible mutant allele.
- Any child of an affected parent has a 50% risk for inheriting the trait. This is true for most families, in which the other parent is phenotypically normal. Because statistically each family member is the result of an "independent event," wide deviation from the expected 1:1 ratio may occur by chance in a single family.
- Phenotypically normal family members do not transmit the phenotype to their children. Failure of penetrance or subtle expression of a condition may lead to apparent exceptions to this rule.
- Males and females are equally likely to transmit the phenotype, to children of either sex. In particular, male-to-male transmission can occur, and males can have unaffected daughters.
- A significant proportion of isolated cases are sporadic cases due to new mutation. The less the fitness, the greater is the proportion of cases due to new mutation.

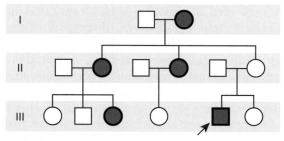

Figure 7-10 Pedigree of split-hand deformity demonstrating failure of penetrance in the mother of the proband *(arrow)* and his sister, the consultand. Reduced penetrance must be taken into account in genetic counseling.

Effect of Incomplete Penetrance, Variable Expressivity, and New Mutations on Autosomal Dominant Inheritance Patterns

Some of the difficulties raised by incomplete penetrance in fully understanding the inheritance of a disease phenotype are demonstrated by the **split-hand/foot malformation**, a type of ectrodactyly (Fig. 7-9). The split-hand malformation originates in the sixth or seventh week of development, when the hands and feet are forming. Failure of penetrance in pedigrees of split-hand malformation can lead to apparent skipping of generations, and this complicates genetic counseling because an at-risk person with normal hands may nevertheless carry the mutation for the condition and thus be capable of having children who are affected.

Figure 7-10 is a pedigree of split-hand deformity in which the unaffected sister of an affected man sought genetic counseling. Her mother is a nonpenetrant carrier of the split-hand mutation. The literature on split-hand deformity suggests that there is reduced penetrance of approximately 70% (i.e., only 70% of the people who have the mutation exhibit the clinical defect). Using this pedigree information to calculate conditional probabilities (as discussed further in Chapter 16), one can calculate that the risk that the consultand might herself be a nonpenetrant carrier is 23% and her chance of having a child with the abnormality is therefore approximately 8% (carrier risk × the risk for transmission × penetrance, or 23% × 50% × 70%).

An autosomal dominant inheritance pattern may also be obscured by variable expressivity. **Neurofibromatosis 1 (NF1)**, a common disorder of the nervous system, demonstrates both age-dependent penetrance and variable expressivity in a single family. Some adults may have only multiple flat, irregular pigmented skin lesions, known as café au lait spots, and small benign tumors (hamartomas) called Lisch nodules on the iris of the eye. Other family members can have these signs as well as multiple benign fleshy tumors (neurofibromas) in the skin. And, still others may have a much more severe phenotype, with intellectual disability, diffuse plexiform neurofibromas, or malignant tumors of nervous system or muscle in addition to the café au lait spots, Lisch nodules, and neurofibromas. Unless one looks specifically for mild manifestations of the disease in the relatives of the proband, heterozygous carriers may be incorrectly classified as unaffected, noncarriers.

Furthermore, the signs of NF1 may require many years to develop. For example, in the newborn period, less than half of all affected newborns show even the most subtle sign of the disease, an increased incidence of café au lait spots. Eventually, however, multiple café au lait spots and Lisch nodules do appear so that, by adulthood, heterozygotes always demonstrate some sign of the disease. The challenges for diagnosis and genetic counseling in NF1 are presented in (Case 34).

Finally, in classic autosomal dominant inheritance, every affected person in a pedigree has an affected parent, who also has an affected parent, and so on, as far back as the disorder can be traced (see Fig. 7-6A). In fact, however, many dominant conditions of medical importance occur because of a spontaneous, de novo mutation in a gamete inherited from a noncarrier parent (see Fig. 7-6C). An individual with an autosomal dominant disorder caused by new mutation will look like an isolated case, and his or her parents, aunts and uncles, and cousins will all be unaffected noncarriers. He or she will still be at risk for passing the mutation down to his or her own children, however. Once a new mutation has arisen, it will be transmitted to future generations following standard principles of inheritance, and, as we discuss in the next section, its survival in the population depends on the fitness of persons carrying it.

Relationship between New Mutation and Fitness in Autosomal Dominant Disorders

In many disorders, whether or not a condition demonstrates an obvious pattern of transmission in families depends on whether individuals affected by the disorder can reproduce. Geneticists coined the term **fitness** as a measure of the impact of a condition on reproduction. Fitness is defined as the ratio of the number of offspring of individuals affected with the condition who survive to reproductive age, compared to the number of offspring of individuals who do not carry the mutant allele. Fitness ranges from 0 (affected individuals never have children who survive to reproductive age) to 1 (affected individuals have the same number of offspring as unaffected controls). Although we will explore the impact of mutation, selection, and fitness on allele frequencies in greater detail in Chapter 9, here we discuss examples that illustrate the major concepts and range of impact of fitness on autosomal dominant conditions.

At one extreme are disorders that have a fitness of 0; patients with such disorders never reproduce, and the disorder is referred to as a **genetic lethal**. One example is the severe short-limb dwarfism syndrome known as **thanatophoric dwarfism** that occurs in heterozygotes for mutations in the *FGFR3* gene (see Fig. 7-6C). Thanatophoric dwarfism is lethal in the neonatal period, and therefore all probands with the disorder *must* be due to new mutations because these mutations cannot be transmitted to the next generation.

At the other extreme are disorders that have virtually normal reproductive fitness because of a late age of onset or a mild phenotype that does not interfere with reproduction. If the fitness is normal, the disorder will only rarely be the result of fresh mutation; a patient is much more likely to have inherited the disorder than to have a new mutant gene, and the pedigree is likely to show multiple affected individuals with clear-cut autosomal dominant inheritance. Late-onset **progressive hearing loss** is a good example of such an autosomal dominant condition, with a fitness of approximately 1 (see Fig. 7-6A). Thus there is an inverse relation between the fitness of a given autosomal dominant disorder and the proportion of all patients with the disorder who inherited the defective gene versus those who received it as a new mutation. The measurement of mutation frequency and the relation of mutation frequency to fitness will be discussed further in Chapter 9.

It is important to note that fitness is *not* simply a measure of physical or intellectual disability. Some individuals with an autosomal dominant disorder may appear phenotypically normal but have a fitness of 0; and at the other extreme, individuals may have normal or near-normal fitness, despite being affected by an autosomal dominant condition with an obvious and severe phenotype such as familial Alzheimer disease (Case 4).

X-LINKED INHERITANCE

In contrast to genes on the autosomes, genes on the X and Y chromosomes are distributed unequally to males and females in families. The patrilineal inheritance of the Y chromosome is straightforward. However, there are very few strictly Y-linked genes, almost all of which are involved in primary sex determination or the development of secondary male characteristics, as discussed in Chapter 6, and they will not be considered here. Approximately 800 protein-coding and 300 noncoding RNA genes have been identified on the X chromosome to date, of which over 300 genes are presently known to be associated with X-linked disease phenotypes. Phenotypes determined by genes on the X have a characteristic sex distribution and a pattern of inheritance that is usually easy to identify and easy to distinguish from the patterns of autosomal inheritance we just explored.

Because males have one X chromosome but females have two, there are only two possible genotypes in males and four in females with respect to mutant alleles at an X-linked locus. A male with a mutant allele at an X-linked locus is **hemizygous** for that allele, whereas females may be a homozygote for the wild-type allele, a homozygote for a mutant allele, a compound heterozygote for two different mutant alleles, or a heterozygous carrier of a mutant allele. For example, if X_H is the wild-type allele for an X-linked disease gene and a mutant allele, X_h, is the disease allele, the genotypes expected in males and females are as follows:

Genotypes and Phenotypes in X-linked Disease

	Genotypes	Phenotypes
Males	Hemizygous X_H	Unaffected
	Hemizygous X_h	Affected
Females	Homozygous X_H/X_H	Unaffected
	Heterozygous X_H/X_h	Carrier (may or may not be affected)
	Homozygous (or compound heterozygous) X_h/X_h	Affected

X Inactivation, Dosage Compensation, and the Expression of X-Linked Genes

As introduced in Chapters 3 and 6, X inactivation is a normal physiological process in which most of the genes on one of the two X chromosomes in normal females, but not the genes on the single X chromosome in males, are inactivated in somatic cells, thus equalizing the expression of most X-linked genes in the two sexes. The clinical relevance of X inactivation in X-linked diseases is profound. It leads to females having two cell populations, which express alleles of X-linked genes on one or the other of the two X chromosomes (see Fig. 3-13 and further discussion in Chapter 6). These two cell

populations are thus genetically identical but functionally distinct, and both cell populations in human females can be readily detected for some disorders. For example, in **Duchenne muscular dystrophy** (Case 14), female carriers exhibit typical mosaic expression of their dystrophin immunostaining (Fig. 7-11). Depending on the

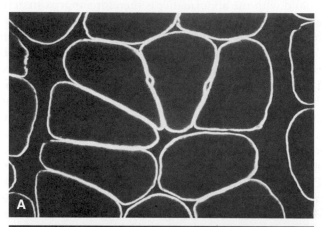
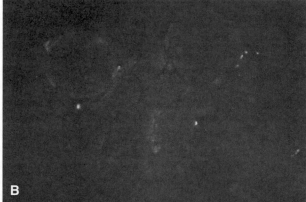
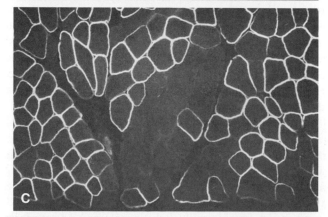

Figure 7-11 Immunostaining for dystrophin in muscle specimens. **A,** A normal female (×480). **B,** A male with Duchenne muscular dystrophy (DMD) (×480). **C,** A carrier female (×240). Staining creates the bright signals seen here encircling individual muscle fibers. Muscle from DMD patients lacks dystrophin staining. Muscle from DMD carriers exhibits both positive and negative patches of dystrophin immunostaining, representing fibers with either the normal or mutant allele on the active X. *See Sources & Acknowledgments.*

pattern of random X inactivation of the two X chromosomes, two female heterozygotes for an X-linked disease may have very different clinical presentations because they differ in the proportion of cells that have the mutant allele on the active X in a relevant tissue (as seen in **manifesting heterozygotes**, as described later).

Recessive and Dominant Inheritance of X-Linked Disorders

As mentioned earlier in this chapter, the use of the terms *dominant* and *recessive* is somewhat different in X-linked conditions than we just saw for autosomal disorders. So-called X-linked dominant and recessive patterns of inheritance are typically distinguished on the basis of the phenotype in heterozygous females. Some X-linked phenotypes are consistently apparent clinically, at least to some degree, in carriers and are thus referred to as dominant, whereas others typically are not and are considered to be recessive. The difficulty in classifying an X-linked disorder as dominant or recessive arises because females who are heterozygous for the same mutant allele in a family may or may not demonstrate the disease, depending on the pattern of random X inactivation and the proportion of the cells in pertinent tissues that have the mutant allele on the active or inactive X.

Nearly a third of X-linked disorders are penetrant in some but not all female heterozygotes and cannot be classified as either dominant or recessive. Even for disorders that can be so classified, they show incomplete penetrance that varies as a function of X inactivation patterns, not inheritance patterns. Because clinical expression of an X-linked condition does not depend strictly on the particular gene involved or even the particular mutation in the same family, some geneticists have recommended dispensing altogether with the terms *recessive* and *dominant* for X-linked disorders. Be that as it may, the terms are widely applied to X-linked disorders, and we will continue to use them, recognizing that they describe extremes of a continuum of penetrance and expressivity in female carriers of X-linked diseases.

X-Linked Recessive Inheritance

The inheritance of X-linked recessive phenotypes follows a well-defined and easily recognized pattern (Fig. 7-12 and Box). An X-linked recessive mutation is expressed phenotypically in all males who receive it, and, consequently, X-linked recessive disorders are generally restricted to males.

Hemophilia A is a classic X-linked recessive disorder in which the blood fails to clot normally because of a deficiency of factor VIII, a protein in the clotting cascade (**Case 21**). The hereditary nature of hemophilia and even its pattern of transmission have been recognized since ancient times, and the condition became known as

the "royal hemophilia" because of its occurrence among descendants of Britain's Queen Victoria, who was a carrier.

As in the earlier discussion, suppose X_h represents the mutant factor VIII allele causing hemophilia A, and X_H represents the normal allele. If a male with hemophilia mates with a normal female, all the sons receive their father's Y chromosome and a maternal X and are unaffected, but all the daughters receive the paternal X chromosome with its hemophilia allele and are obligate carriers. If a daughter of the affected male mates with an unaffected male, four genotypes are possible in the progeny, with equal probabilities:

The hemophilia of an affected grandfather, which did not appear in any of his own children, has a 50% chance of appearing in each son of his daughters. It will

X-Linked Recessive Inheritance

Affected Male by Noncarrier Female		Female Genotype X_H/X_H Gametes		Risk for Disease
		X_H	X_H	
Male Genotype X_h/Y Gametes	X_h	X_H/X_h	X_H/X_h	All female carriers (X_H/X_h)
	Y	X_H/Y	X_H/Y	All males unaffected (X_H/Y)

Unaffected Male by Carrier Female		Female Genotype X_H/X_h Gametes		Risk for Disease
		X_H	X_h	
Male Genotype X_H/Y Gametes	X_H	X_H/X_H	X_H/X_h	¼ Noncarrier female (X_H/X_H) ¼ Carrier female (X_H/X_h)
	Y	X_H/Y	X_h/Y	¼ Normal male (X_H/Y) ¼ Affected male (X_h/Y)

The wild-type allele at the X-linked hemophilia locus is denoted as X_H with an uppercase H, and the mutant allele is denoted as X_h with a lowercase h.

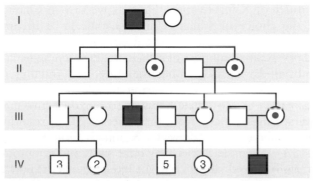

Figure 7-12 Pedigree pattern demonstrating an X-linked recessive disorder such as hemophilia A, transmitted from an affected male through females to an affected grandson and great-grandson.

not reappear among the descendants of his sons, however. A daughter of a carrier has a 50% chance of being a carrier herself (see Fig. 7-12). By chance, an X-linked recessive allele may be transmitted undetected through a series of female carriers before it is expressed in a male descendant.

CHARACTERISTICS OF X-LINKED RECESSIVE INHERITANCE

- The incidence of the trait is much higher in males than in females.
- Heterozygous females are usually unaffected, but some may express the condition with variable severity as determined by the pattern of X inactivation.
- The gene responsible for the condition is transmitted from an affected man through all his daughters. Any of his daughters' sons has a 50% chance of inheriting it.
- The mutant allele is never transmitted directly from father to son, but it is transmitted by an affected male to all his daughters.
- The mutant allele may be transmitted through a series of carrier females; if so, the affected males in a kindred are related through females.
- A significant proportion of isolated cases are due to new mutation.

Affected Females in X-linked Recessive Disease

Although X-linked conditions are classically seen only in males, they can be observed in females under two circumstances. In one, such a female can be homozygous for the relevant disease allele, although most X-linked diseases are so rare that this scenario is highly unlikely unless her parents are consanguineous. However, a few X-linked conditions, such as X-linked color blindness, are sufficiently common that such homozygotes are seen in female offspring of an affected father and a carrier mother.

More commonly, an affected female represents a carrier of a recessive X-linked allele who shows phenotypic expression of the disease and is referred to as a **manifesting heterozygote**. Whether a female carrier will be a manifesting heterozygote depends on a number of features of X inactivation. First, as we saw in Chapter 3, the choice of which X chromosome is to become inactive is a random one, but it occurs when there is a relatively small number of cells in the developing female embryo. By chance alone, therefore, the fraction of cells in various tissues of carrier females in which the normal or mutant allele happens to remain active may deviate substantially from the expected 50%, resulting in **unbalanced** or "skewed" X inactivation (see Fig. 6-13A). A female carrier may have signs and symptoms of an X-linked disorder if the skewed inactivation is unfavorable (i.e., a large majority of the active X chromosomes in pertinent tissues happen to contain the deleterious allele).

Favorably unbalanced or skewed inactivation, in which the mutant allele is found preferentially on the inactive X in some or all tissues of an unaffected heterozygous female, also occurs. Such skewed inactivation may simply be due to chance alone, as we just saw (albeit in the opposite direction). However, there are certain X-linked conditions in which there is reduced cell survival or a proliferative disadvantage for those cells that originally had the mutant allele on the active X early in development, resulting in a pattern of highly skewed inactivation that favors cells with the normal allele on the active X in relevant tissues. For example, highly skewed X inactivation is the rule in female carriers of certain **X-linked immunodeficiencies**, in whom only those early progenitor cells that happen to carry the normal allele on their active X chromosome can populate certain lineages in the immune system.

X-Linked Dominant Inheritance

As discussed earlier, an X-linked phenotype can be described as dominant if it is regularly expressed in heterozygotes. X-linked dominant inheritance can readily be distinguished from autosomal dominant inheritance by the lack of **male-to-male transmission**, which is impossible for X-linked inheritance because males transmit the Y chromosome, not the X, to their sons.

Thus the distinguishing feature of a fully penetrant X-linked dominant pedigree (Fig. 7-13) is that *all* the daughters and *none* of the sons of affected males are affected; if any daughter is unaffected or any son is affected, the inheritance must be autosomal, not X-linked. The pattern of inheritance through females is

X-Linked Dominant Inheritance

Unaffected Male by Affected Female		Female Genotype X_D/X_d Gametes		
		X_D	X_d	Risk for Disease
Male Genotype X_d/Y Gametes	X_d	X_D/X_d	X_d/X_d	¼ Affected females (X_D/X_d) ¼ Unaffected females (X_d/X_d)
	Y	X_D/Y	X_d/Y	¼ Affected males (X_D/Y) ¼ Unaffected males (X_d/Y)

Affected Male by Noncarrier Female		Female Genotype X_d/X_d Gametes		
		X_d	X_d	Risk for Disease
Male Genotype X_D/Y Gametes	X_D	X_D/X_d	X_D/X_d	All females affected (X_D/X_d)
	Y	X_d/Y	X_d/Y	All males unaffected (X_d/Y)

The wild-type allele at the hypophosphatemic rickets locus is denoted as X_d, and the mutant allele is denoted as X_D.

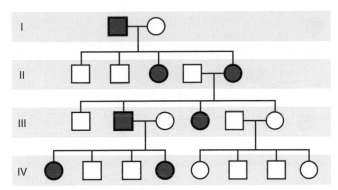

Figure 7-13 Pedigree pattern demonstrating X-linked dominant inheritance.

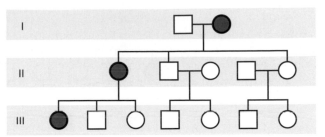

Figure 7-14 Pedigree pattern demonstrating X-linked dominant inheritance of a disorder that is lethal in males during the prenatal period.

no different from the autosomal dominant pattern; because females have a pair of X chromosomes just as they have pairs of autosomes, each child of an affected female has a 50% chance of inheriting the trait, regardless of sex. Across multiple families with an X-linked dominant disease, the expression is usually milder in heterozygous females, because the mutant allele is located on the inactive X chromosome in a proportion of their cells. Thus most X-linked dominant disorders are incompletely dominant, as is the case with most autosomal dominant disorders (see Box).

CHARACTERISTICS OF X-LINKED DOMINANT INHERITANCE

- Affected males with normal mates have no affected sons and no normal daughters.
- Both male and female offspring of female carriers have a 50% risk for inheriting the phenotype. The pedigree pattern is similar to that seen with autosomal dominant inheritance.
- Affected females are approximately twice as common as affected males, but affected females typically have milder (although variable) expression of the phenotype.
- One example of an X-linked dominant disorder is X-linked **hypophosphatemic rickets** (also known as vitamin D–resistant rickets), in which the ability of the kidney tubules to reabsorb filtered phosphate is impaired. This disorder fits the criterion of an X-linked dominant disorder in that both sexes are affected, although the serum phosphate level is less depressed and the rickets less severe in heterozygous females than in affected males.

X-Linked Dominant Disorders with Male Lethality

Although most X-linked conditions are typically apparent only in males, a few rare X-linked defects are expressed exclusively or almost exclusively in females. These X-linked dominant conditions are lethal in males before birth (Fig. 7-14). Typical pedigrees of these conditions show transmission by affected females, who

produce affected daughters, normal daughters, and normal sons in equal proportions (1:1:1); affected males are not seen.

Rett syndrome (Case 40) is a striking disorder that occurs nearly exclusively in females and meets all criteria for being an X-linked dominant disorder that is usually lethal in hemizygous males. The syndrome is characterized by normal prenatal and neonatal growth and development, followed by the rapid onset of neurological symptoms in affected girls. The disease mechanism is thought to reflect abnormalities in the regulation of a set of genes in the developing brain; the cause of male lethality is unknown but presumably reflects a requirement during early development for at least one functional copy of the *MECP2* gene on the X chromosome that is mutated in this syndrome.

X-Linked Dominant Disorders with Male Sparing

Other disorders are manifest only in carrier females because hemizygous males are largely spared the consequences of the mutation they carry. One such disorder is female-limited, X-linked **epilepsy and cognitive impairment**. Affected females are asymptomatic at birth and appear to be developing normally but then develop seizures, generally in the second year of life, after which development begins to regress. Most affected females go on to be developmentally delayed, which can vary from mild to severe. In contrast, male hemizygotes in the same families are completely unaffected (Fig. 7-15). The disorder is due to loss-of-function mutations in the protocadherin gene 19, an X-linked gene that encodes a cell surface molecule expressed on neurons in the central nervous system.

The explanation for this unusual pattern of inheritance is not clear. It is hypothesized that the epilepsy occurs in females because mosaicism for expression of protocadherin 19, resulting from random X inactivation in the brain, disrupts communication between groups of neurons with and without the cell surface protein. Neurons in males uniformly lack the cell surface molecule, but their brains are apparently spared cell-cell miscommunication by a different, compensating protocadherin.

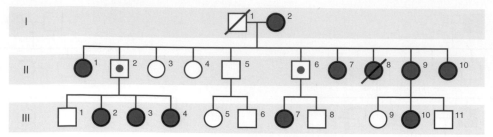

Figure 7-15 Pedigree pattern of familial female epilepsy and cognitive impairment, demonstrating its X-linked dominant inheritance with sparing of males hemizygous for a premature termination mutation in the protocadherin 19 gene.

Relationship between New Mutation and Fitness in X-Linked Disorders

Just as with autosomal dominant disorders, new mutations constitute a significant fraction of isolated cases of many X-linked diseases. Males carrying mutations causing X-linked disorders are exposed to selection that is complete for some disorders, partial for others, and absent for still others, depending on the fitness of the genotype. Males carrying mutant alleles for X-linked disorders such as **Duchenne muscular dystrophy** (Case 14), a disease of muscle that affects young boys, do not reproduce. Fitness of affected males is currently 0, although the situation may change as a result of advances in research aimed at therapy for affected boys (see Chapter 12). In contrast, patients with **hemophilia** (Case 21) also have reduced fitness, but the condition is not a genetic lethal: affected males have on average approximately 70% as many offspring as unaffected males do, and fitness of affected males is therefore approximately 0.70. This fitness may also increase with improvements in the treatment of this disease.

When fitness is reduced, the mutant alleles that these males carry are lost from the population. In contrast to autosomal dominant conditions, however, mutant alleles for X-linked diseases with reduced fitness may be partially or completely protected from selection when present in females. Thus, even in X-linked disorders with a fitness of 0, less than half of new cases will be due to new mutations. The overall incidence of the disease, then, will be determined both by the transmittal of a mutant allele from a carrier mother and by the rate of de novo mutations at the responsible locus. The balance between new mutation and selection will be discussed more fully from the population genetics perspective in Chapter 9.

PSEUDOAUTOSOMAL INHERITANCE

As we first saw in Chapter 2, meiotic recombination between X-linked loci only occurs between the two homologous X chromosomes and is therefore restricted to females. X-linked loci do not participate in meiotic recombination in males, who have a Y chromosome and

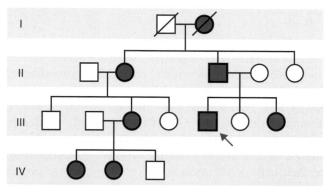

Figure 7-16 Pedigree showing inheritance of dyschondrosteosis due to mutations in *SHOX*, a pseudoautosomal gene on the X and Y chromosomes. The *arrow* shows a male who inherited the trait on his Y chromosome from his father. His father, however, inherited the trait on his X chromosome from his mother. *See Sources & Acknowledgments.*

only one X chromosome. There are, however, a small number of contiguous loci located at the tips of the p and q arms of the sex chromosomes that are homologous between the X and Y and do recombine between them in male meiosis. As a consequence, during spermatogenesis, a mutant allele at one of these loci on the X chromosome can be transferred onto the Y chromosome and passed on to male offspring, thereby demonstrating the male-to-male transmission characteristic of autosomal inheritance. Because these unusual loci on the X and Y demonstrate autosomal inheritance but are not located on an autosome, they are referred to as **pseudoautosomal** loci, and the segments of the X and Y chromosomes where they are located are referred to as the **pseudoautosomal regions**.

One example of a disease caused by a mutation at a pseudoautosomal locus is **dyschondrosteosis**, a dominantly inherited skeletal dysplasia with disproportionate short stature and deformity of the forearms. Although a greater prevalence of the disease in females as compared with males initially suggested an X-linked dominant disorder, the presence of male-to-male transmission clearly ruled out strict X-linked inheritance (Fig. 7-16). Mutations in the *SHOX* gene, located in the pseudoautosomal region on Xp and Yp, have been found responsible for this condition.

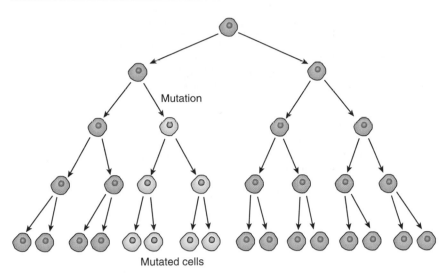

Figure 7-17 Schematic representation of a mutation occurring after conception, during mitotic cell divisions. Such a mutation can lead to a proportion of cells carrying the mutation—that is, to either somatic or germline mosaicism, depending on at what stage of embryonic or postnatal development the mutation occurred.

MOSAICISM

Although we are used to thinking of ourselves as being composed of cells that all carry exactly the same complement of genes and chromosomes, this is in reality an oversimplified view. Mosaicism is the presence in an individual or a tissue of at least two cell lineages that differ genetically but are derived from a single zygote. Mutations that occur after conception in a single cell in either prenatal or postnatal life can give rise to clones of cells genetically different from the original zygote because, given the nature of DNA replication, the mutation will persist in all the clonal descendants of that cell (Fig. 7-17). Mosaicism for numerical or structural abnormalities of chromosomes is a clinically important phenomenon (see Chapters 5 and 17), and somatic mutation is recognized as the major contributor to most types of cancer (see Chapter 15).

Mosaicism can affect any cells or tissue within a developing embryo or at any point after conception to adulthood, and it can be a diagnostic dilemma to determine just how widespread the mosaic pattern is. For example, the population of cells that carry a mutation in a mosaic pregnancy might be found only in extraembryonic tissue and not in the embryo proper (**confined placental mosaicism**; see Chapter 17), might be present in some tissues of the embryo but not in the gametes (pure **somatic mosaicism**), might be restricted to the gamete lineage only and nowhere else (pure **germline mosaicism**), or might be present in both somatic lineages and the germline—all depending on whether the mutation occurred before or after the separation of the inner cell mass, the germline cells, and the somatic cells during embryogenesis (see Chapter 17). Because there are approximately 30 mitotic divisions in the cells of the germline before meiosis in the female and several hundred in the male (see Chapter 2), there is ample opportunity for mutations to occur in germline cells

after the separation from somatic cells, resulting in pure gonadal mosaicism.

Determining whether mosaicism for a mutation is present only in the germline or only in somatic tissues may be difficult because failure to find a mutation in a subset of cells from a readily accessible somatic tissue (e.g., peripheral white blood cells, skin, or buccal cells) does not ensure that the mutation is not present elsewhere in the body, including the germline.

Segmental Mosaicism

A mutation affecting morphogenesis and occurring during embryonic development might be manifested as a segmental or patchy abnormality, depending on the stage at which the mutation occurred and the lineage of the somatic cell in which it originated. For example, **neurofibromatosis 1 (NF1)** (Case 34) is sometimes segmental, affecting only one part of the body. Segmental NF1 is caused by somatic mosaicism for a mutation that occurred after conception. Although the parents of such a patient would be unaffected and considered not at risk for transmitting the mutant gene, a patient with segmental NF1 could be at risk for having an affected child, whose phenotype would be typical for NF1, that is, *not* segmental. Whether the patient is at risk for transmitting the defect will depend on whether the mutation occurred before separation of germline cells from the somatic cell line that carries the mutation.

Germline Mosaicism

In pedigrees with germline mosaicism, unaffected individuals with no evidence of a disease-causing mutation in their genome (as evidenced by the failure to find the mutation in DNA extracted from their peripheral white blood cells) may still be at risk for having more than one child who inherited the mutation from them

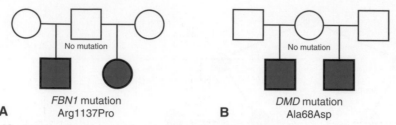

Figure 7-18 Pedigrees demonstrating two affected siblings with the autosomal dominant disorder Marfan syndrome (Family A) and the X-linked condition Becker muscular dystrophy (Family B). In Family A, the affected children have the same point mutation inherited from their father, who is unaffected and does not carry the mutation in DNA from examined somatic tissues. He must have been a mosaic for the *FBN1* mutation in his germline. In Family B, the affected children have the same point mutation inherited from their mother who is unaffected and does not carry the mutation in DNA from examined somatic tissues. She must have been a mosaic for the *DMD* mutation in her germline.

(Fig. 7-18). The existence of germline mosaicism means that geneticists and genetic counselors must be aware of the potential inaccuracy of assuming that normal examination results and normal gene test results of the parents of a child with an autosomal dominant or X-linked phenotype means the child must be a new mutation. The impact of this possibility on risk assessment will be discussed further in Chapter 16.

PARENT-OF-ORIGIN EFFECTS ON INHERITANCE PATTERNS
Unusual Inheritance Patterns due to Genomic Imprinting

According to Mendel's laws of heredity, a mutant allele of an autosomal gene is equally likely to be transmitted from a parent of either sex to an offspring of either sex; similarly, a female is equally likely to transmit a mutated X-linked gene to a child of either sex. Originally, little attention was paid to whether the sex of the parent had any effect on the *expression* of the genes each parent transmits. As discussed in Chapter 6, we now know, however, that in some genetic disorders, such as **Prader-Willi syndrome** (Case 38) and **Angelman syndrome**, the expression of the disease phenotype depends on whether the mutant allele has been inherited from the father or from the mother, a phenomenon known as **genomic imprinting**. The hallmark of genomic imprinting is that the sex of the parent who *transmits* the abnormality determines whether there is expression of the disorder in a child. This is very different from sex-limited inheritance (described earlier in this chapter), in which expression of the disease depends on the sex of the child who *inherits* the abnormality.

Imprinting can cause unusual inheritance patterns in pedigrees, in that a disorder can appear to be inherited in a dominant manner when transmitted from one parent, but not the other. For example, the **hereditary paragangliomas** (PGLs) are a group of autosomal dominant disorders in which multiple tumors develop in sympathetic and parasympathetic ganglia of the autonomic nervous system. Patients with paraganglioma can also develop a catecholamine-producing tumor known as a pheochromocytoma, either in the adrenal medulla or in sympathetic ganglia along the vertebral column. A pedigree of one type of PGL family is shown in Figure 7-19. The striking observation is that, although both males and females can be affected, this is only if they inherited the mutation from their father and not from their mother. A male heterozygote who has inherited his mutation from his mother will remain unaffected throughout life but is still at a 50% risk for transmitting the mutation to each of his children, who are then at high risk for developing the disease.

DYNAMIC MUTATIONS: UNSTABLE REPEAT EXPANSIONS

In all of the types of inheritance presented thus far in this chapter, the responsible mutation, once it occurs, is stable when it is transmitted from one generation to the next; that is, all affected members of a family share the identical inherited mutation. In contrast, an entirely different class of genetic disease has been recognized, diseases due to **dynamic mutations** that change from generation to generation (see Chapter 4). These conditions are characterized by an unstable expansion within the affected gene of a segment of DNA consisting of repeating units of three or more nucleotides that occur in tandem. Many such repeat units consist of three nucleotides, such as CAG or CCG, and the repeat will therefore be CAGCAGCAGCAG or CCGCCGCCG CCG. In general, genes associated with these diseases all have wild-type alleles that are polymorphic; that is, there is a variable number of repeat units in the normal population, as we saw in Chapter 4. As the gene is passed from generation to generation, however, the number of repeats can increase and undergo **expansion**, far beyond the normal polymorphic range, leading to abnormalities in gene expression and function. The discovery of this unusual group of conditions has dispelled the orthodox notions of germline stability and provided

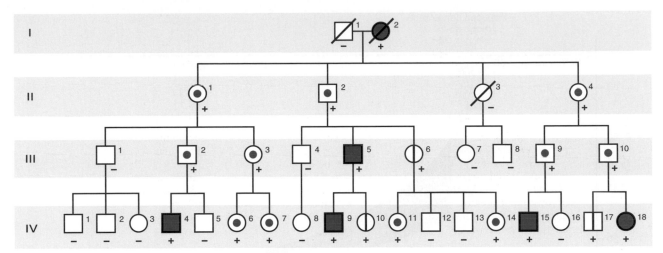

Figure 7-19 Pedigree of a family with paraganglioma syndrome 1 caused by a mutation in the *SDHD* gene. Individuals II-1, II-2, II-4, III-2, III-3, III-9, III-10, IV-6, IV-7, IV-11, and IV-14 each inherited the mutation from their mothers but are unaffected. However, when the males in this group pass on the mutation, those children can be affected. In addition to the imprinting, the family also demonstrates the effect of reduced and age-dependent penetrance in the children (III-6, IV-10, IV-17) of heterozygous fathers. The + and – symbols refer to the presence or absence of the *SDHD* mutation in this family.

a biological basis for peculiarities of familial transmission, discussed in the next section, that previously had no known mechanistic explanation.

More than a dozen diseases are known to result from unstable repeat expansions of this type. All of these conditions are primarily neurological. Here, we will review the inheritance patterns of two different unstable expansion diseases that illustrate the effects that different dynamic mutations can have on patterns of inheritance. A more complete description of the pathogenetic mechanisms of unstable repeat disorders is given in Chapter 12.

Polyglutamine Disorders

Several different neurological diseases share the property that the protein encoded by the gene mutated in each condition is characterized by a variable string of consecutive glutamine residues, the codon for which is the trinucleotide CAG. These so-called **polyglutamine disorders** result when an expansion of the CAG repeat leads to a protein with more glutamines than is compatible with normal function. **Huntington disease (HD)** is a well-known disorder that illustrates many of the common genetic features of the polyglutamine disorders caused by expansion of an unstable repeat (Case 24). The neuropathology is dominated by degeneration of the striatum and the cortex. Patients first present clinically in midlife and manifest a characteristic phenotype of motor abnormalities (chorea, dystonia), personality changes, a gradual loss of cognition, and ultimately death.

For a long time, HD was thought to be a typical autosomal dominant condition with age-dependent penetrance. The disease is transmitted from generation to

generation with a 50% risk to each offspring, and heterozygous and homozygous patients carrying the mutation have very similar phenotypes, although homozygotes may have a more rapid course of their disease. There are, however, obvious peculiarities in its inheritance that cannot be explained by simple autosomal dominant inheritance. First, the disease appears to develop at an earlier and earlier age as it is transmitted through the pedigree, a phenomenon referred to as **anticipation**. Second, anticipation seems to occur only when the mutant allele is transmitted by an affected father and not by an affected mother, a situation known as **parental transmission bias**.

The peculiarities of inheritance of HD are now readily explained by the discovery that the mutation is composed of an abnormally long CAG expansion in the coding region of the *HD* gene. Normal individuals carry alleles with between 9 and 35 CAG repeats in their *HD* gene, with the average being 18 or 19. Individuals affected with HD, however, have 40 or more repeats, with the average being around 46. Repeat numbers in the range of 40 to 50 usually result in disease later in life, which explains the age-dependent penetrance that is a hallmark of this condition. A borderline repeat number of 36 to 39, although usually associated with HD, can be found in a few individuals who show no signs of the disease even at a fairly advanced age. The age of onset varies with how many CAG repeats are present (Fig. 7-20).

How, then, does an individual come to have an expanded CAG repeat in his or her *HD* gene? First, he or she may inherit it from a parent who already has an expanded repeat beyond the normal range but has not yet developed the disease. Second, he or she may inherit

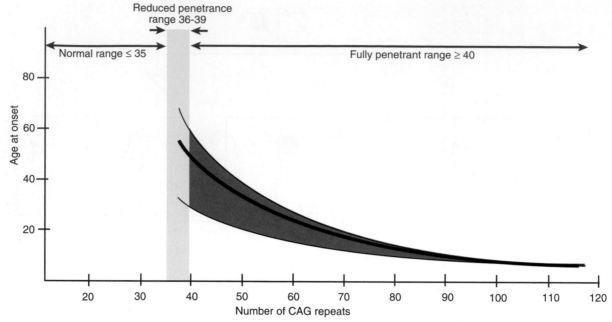

Figure 7-20 Graph correlating approximate age of onset of Huntington disease with the number of CAG repeats found in the *HD* gene. The *solid line* is the average age of onset, and the *shaded area* shows the range of age of onset for any given number of repeats. *See Sources & Acknowledgments.*

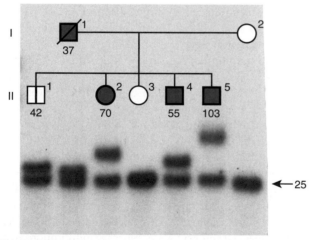

Figure 7-21 Pedigree of family with Huntington disease. Shown beneath the pedigree is a Southern blot analysis for CAG repeat expansions in the *HD* gene. In addition to a normal allele containing 25 CAG repeats, individual I-1 and his children, II-1, II-2, II-4, and II-5, are all heterozygous for expanded alleles, each containing a different number of CAG repeats. The repeat number is indicated below each individual. II-2, II-4, and II-5 are all affected; individual II-1 is unaffected at the age of 50 years but will develop the disease later in life. *See Sources & Acknowledgments.*

an expanded repeat from a parent with repeat length of 35 to 40, which may or may not cause disease in the parent's lifetime but may expand on transmission, resulting in earlier-onset disease in later generations (and thus explaining anticipation). For example, in the pedigree shown in Figure 7-21, individual I-1, now deceased, was diagnosed with HD at the age of 64 years and was heterozygous for an expanded allele with 37

CAG repeats and a normal, stable allele with 25 repeats. Four of his children inherited the unstable allele, with CAG repeat lengths ranging from 42 to more than 100 repeats. Finally, unaffected individuals may carry alleles with repeat lengths at the upper limit of the normal range (29 to 35 CAG repeats) that can expand during meiosis to 40 or more repeats. CAG repeat alleles at the upper limits of normal that do not cause disease but are capable of expanding into the disease-causing range are known as **premutations**.

Expansion in HD shows a paternal transmission bias and occurs most frequently during male gametogenesis, which is why the severe early-onset juvenile form of the disease, seen with the largest expansions (70 to 121 repeats), is always paternally inherited.

Fragile X Syndrome

The **fragile X syndrome** (Case 17) is the most common heritable form of moderate intellectual disability, one of many conditions now considered to be among the autism spectrum disorders. The name fragile X refers to a cytogenetic marker on the X chromosome at Xq27.3, a so-called **fragile site** induced in cultured cells in which the chromatin fails to condense properly during mitosis. The syndrome is inherited as an X-linked disorder with penetrance in females in the 50% to 60% range. The fragile X syndrome has a frequency of 1 in 3000 to 4000 male births and is so common that it requires consideration in the differential diagnosis of intellectual disability or autism in both males and females. Testing for the fragile X syndrome is among the most frequent

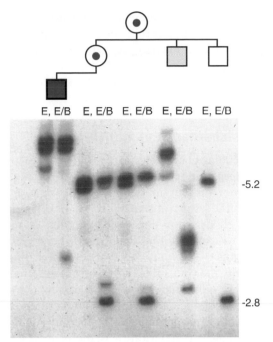

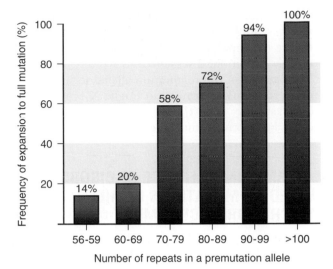

Figure 7-23 Frequency of expansion of a premutation triplet repeat in *FMR1* to a full mutation in oogenesis as a function of the length of the premutation allele carried by a heterozygous female. The risk for fragile X syndrome to her sons is approximately half this frequency, because there is a 50% chance a son will inherit the expanded allele. The risk for fragile X syndrome to her daughters is approximately one-fourth this frequency, because there is a 50% chance a daughter would inherit the full mutation, and penetrance of the full mutation in a female is approximately 50%. *See Sources & Acknowledgments.*

Figure 7-22 Southern blot DNA from the members of a family in which fragile X syndrome is segregating. In the family shown at the top, DNA samples were digested either with the endonuclease *EcoRI* alone (E) or with the combination of *EcoRI* and *BssH2* (B), an endonuclease that will not cut when the cytosines in its recognition sequence are methylated. *EcoRI* digestion normally yields a 5.2-kb fragment containing the region of the repeat, but the size of the fragment increases proportionately to the expansion of the triplet repeat. Digestion with *BssH2* along with *EcoRI* (E/B) will reduce the 5.2-kb fragment generated by *EcoRI* to a 2.8-kb fragment containing the repeats if the CGG repeats are unmethylated, as is the case on the active X chromosome in a female, or if the repeats are not expanded into the full mutation range (>200 repeats). *BssH2* cannot cut the 5.2-kb fragment coming from an inactive X or a fully expanded *FMR1* allele. The affected individual has a large *EcoRI* fragment, much greater than 5.2 kb, that contains the expanded CGG repeat and is resistant to *BssH2* digestion because it is mostly methylated. His mother has two fragments after *EcoRI* digestion, one normal in size and the other a few hundred base pairs larger, indicating she is a premutation carrier, as is her mother, the proband's grandmother. Upon double digestion, two fragments are seen, the normal at 2.8 kb and a premutation allele that is a few hundred base pairs larger. The proband has two uncles, one (shown in light blue) who appears mildly affected and has an expanded allele (based on *EcoRI* digestion) that is only partially methylated (based on *BssH2* digestion). The other uncle is a normal male with a normal sized, unmethylated allele. *See Sources & Acknowledgments.*

indications for genome analysis, genetic counseling, and prenatal diagnosis.

Like HD, fragile X syndrome is caused by an unstable repeat expansion. However, in this case, a massive expansion of a different triplet repeat, CGG, occurs in the 5′ untranslated region of a gene called *FMR1* (Fig. 7-22). The normal number of repeats is up to 55, whereas more than 200 (and even up to several thousand) repeats are found in patients with the "full" fragile X syndrome mutation. The syndrome is due to a lack of expression of the *FMR1* gene and failure to

produce the encoded protein. The expanded repeat leads to excessive methylation of cytosines in the promoter of *FMR1*; as discussed in Chapter 3, DNA methylation at CpG islands prevents normal promoter function and leads to gene silencing.

Triplet repeat numbers between 56 and 200 constitute an intermediate premutation stage of the fragile X syndrome. Expansions in this range are unstable when they are transmitted from mother to child and have an increasing tendency to undergo full expansion to more than 200 copies of the repeat during gametogenesis in the female (but almost never in the male), with the risk for expansion increasing dramatically with increasing premutation size (Fig. 7-23). The overall premutation frequency in females in the population is estimated to be greater than 1 in 200.

Similarities and Differences in Huntington Disease and Fragile X Pedigrees

A comparison of HD with the fragile X syndrome reveals some similarities but also many differences that illustrate many of the features of disorders due to dynamic mutations:

- Premutation expansions causing an increased risk for passing on full mutations are the rule in both of these disorders, and anticipation is commonly seen in both.
- However, the number of repeats in premutation alleles in HD is 29 to 35, far less than the 55 to 200 repeats in fragile X syndrome premutations.

- Premutation carriers for fragile X syndrome are at risk for adult-onset ataxia (in males) and ovarian failure (in females). But premutation carriers in HD are, by definition, disease-free.
- The expansion of premutation alleles occurs primarily in the female germline in fragile X syndrome; in contrast, the largest expansions causing juvenile-onset HD occur in the male germline.

MATERNAL INHERITANCE OF DISORDERS CAUSED BY MUTATIONS IN THE MITOCHONDRIAL GENOME

All of the patterns of inheritance described thus far are explained by mutations in the nuclear genome, in either autosomal or X-linked genes. However, some pedigrees of inherited diseases that do not show patterns of typical mendelian inheritance are caused by mutations in the mitochondrial genome and manifest strictly maternal inheritance. Disorders caused by mutations in mitochondrial DNA (mtDNA) demonstrate a number of unusual features that result from the unique characteristics of mitochondrial biology and function.

As introduced in Chapter 2, not all the RNA and protein synthesized in a cell are encoded in the DNA of the nucleus; a small but important fraction is encoded by genes in mtDNA. The mitochondrial genome consists of 37 genes that encode 13 subunits of enzymes involved in oxidative phosphorylation, as well as ribosomal RNAs and transfer RNAs required for translating the transcripts of the mitochondria-encoded polypeptides. Because mitochondria are essential to the normal functioning of nearly all cells, disruption of energy production of mutations in mtDNA often results in severe disease, affecting many different tissues. Thus pleiotropy is the rule, not the exception, in mitochondrial disorders.

More than 100 different rearrangements and 100 different point mutations have been identified in mtDNA that can cause a range of human diseases, often involving the central nervous and musculoskeletal systems, such as **myoclonic epilepsy with ragged-red fibers** (Case 33).

In this section, we will focus on the distinctive pattern of inheritance because of three unusual features of mitochondria: **maternal inheritance, replicative segregation,** and **homoplasmy** and **heteroplasmy.** The underlying mechanisms of mitochondrial disorders are discussed in more detail in Chapter 12.

Maternal Inheritance of mtDNA

The first defining characteristic of the genetics of mtDNA is its **maternal inheritance.** Sperm mitochondria are generally not present in the zygote, so that only the maternal mtDNA is transmitted to the next generation. Thus the children of a *female* who has a mtDNA mutation will inherit the mutation, whereas none of the offspring of a *male* carrying the same mutation will inherit the defective DNA. Pedigrees of such disorders are quite distinctive, as shown by the strictly maternal inheritance of a mtDNA mutation causing **Leber hereditary optic neuropathy** seen in Figure 7-24. Although maternal inheritance is the general expectation, at least one instance of paternal inheritance of mtDNA has occurred in a patient with a mitochondrial myopathy. Consequently, in patients with apparently sporadic mtDNA mutations, the rare occurrence of paternal mtDNA inheritance must be considered (see Box).

Replicative Segregation

A second feature of the mitochondrial genome is the stochastic nature of segregation during mitosis and meiosis. At cell division, the multiple copies of mtDNA in each of the mitochondria in each cell replicate and sort randomly among newly synthesized mitochondria, in stark contrast to the highly predictable and programmed segregation of the 46 nuclear chromosomes. The mitochondria themselves, in turn, are then distributed randomly between the two daughter cells. This process is known as **replicative segregation** and can result in significant variability in manifestations of mitochondrial disorders among different tissues and/or patients.

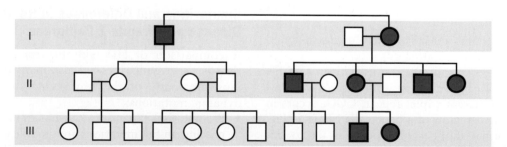

Figure 7-24 Pedigree of Leber hereditary optic neuropathy, a form of adult-onset blindness caused by a defect in mitochondrial DNA. Inheritance is only through the maternal lineage, in agreement with the known maternal inheritance of mitochondrial DNA. Note that no affected male transmits the disease.

Homoplasmy and Heteroplasmy

Finally, a distinctive feature of the genetics of mtDNA is seen when replicative segregation occurs in mitochondria containing both mutant and wild-type mitochondrial genomes. When a mutation first occurs in the mtDNA, it is present in only one of the mtDNA molecules in a mitochondrion. With cell division, all the mtDNAs replicate, the mitochondria undergo fission, and the mutant and wild-type DNA are distributed randomly into daughter organelles, which—simply by chance—may contain different proportions of wild-type and mutant mitochondrial genomes. The cell, which now contains mitochondria containing different mixtures of normal and mutant mtDNAs, in turn distributes those mitochondria randomly to its daughter cells. Daughter cells may thus receive a mixture of mitochondria, some with and some without the mutation (a situation known as **heteroplasmy**; Fig. 7-25). Occasionally, a daughter cell may receive, again by chance, mitochondria that contain a pure population of normal mtDNA or a pure population of mutant mtDNA (a situation known as **homoplasmy**). Because the phenotypic expression of a mutation in mtDNA depends on the relative proportions of normal and mutant mtDNA in the cells making up different tissues, reduced penetrance and variable expression are typical features of mitochondrial disorders (Case 33).

Maternal inheritance in the presence of heteroplasmy in the mother is associated with additional features of mtDNA genetics that are of medical significance. First, the number of mtDNA molecules within developing oocytes is reduced before being subsequently amplified to the huge total seen in mature oocytes. This restriction and subsequent amplification of mtDNA during oogenesis is termed the **mitochondrial genetic bottleneck**.

CHARACTERISTICS OF MITOCHONDRIAL INHERITANCE

- All children of females *homoplasmic* for a mutation will inherit the mutation; the children of males carrying a similar mutation almost always will not.
- Females *heteroplasmic* for *point mutations* and *duplications* will pass them on to all of their children. However, the fraction of mutant mitochondria in the offspring, and therefore the risk and severity of disease, can vary considerably, depending on the fraction of mutant mitochondria in their mother as well as on random chance operating on small numbers of mitochondria per cell at the oocyte bottleneck. Heteroplasmic *deletions* are generally not heritable.
- The fraction of mutant mitochondria in different tissues of an individual heteroplasmic for a mutation can vary tremendously, thereby causing a spectrum of disease among the members of a family in which there is heteroplasmy for a mitochondrial mutation. Pleiotropy and variable expressivity in different affected family members are also frequent.

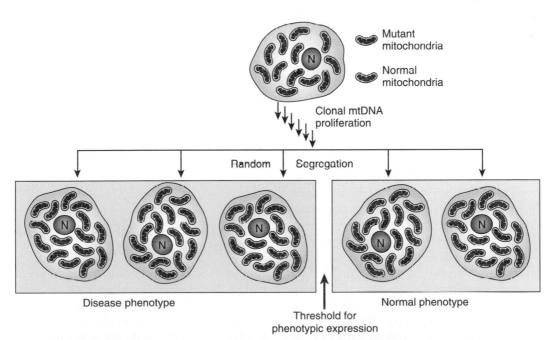

Figure 7-25 Replicative segregation of a heteroplasmic mitochondrial mutation. Random partitioning of mutant and wild-type mitochondria through multiple rounds of mitosis produces a collection of daughter cells with wide variation in the proportion of mutant and wild-type mitochondria carried by each cell. Cell and tissue dysfunction results when the fraction of mitochondria that are carrying a mutation exceeds a threshold level. mtDNA, Mitochondrial DNA; N, nucleus.

Consequently, the variability in the proportion of mutant mtDNA molecules seen in the offspring of a mother with heteroplasmy for a mtDNA mutation arises, at least in part, from the sampling of a reduced subset of the mtDNAs after the mitochondrial bottleneck that occurs in oogenesis. As might be expected, mothers with a high proportion of mutant mtDNA molecules are more likely to produce eggs with a higher proportion of mutant mtDNA and therefore are more likely to have clinically affected offspring than are mothers with a lower proportion.

CORRELATING GENOTYPE AND PHENOTYPE

An important component of medical genetics is identifying and characterizing the genotypes responsible for particular disease phenotypes. In doing so, it is important not to adhere to an overly simplistic view that each disease phenotype is caused uniquely by one particular mutation in a specific gene or that mutations in a particular gene always cause the same phenotype. In fact, there is often substantial heterogeneity in the complex relationship(s) among disease phenotypes, the genes that are mutated in those diseases, and the nature of the mutations found in those genes. Three main types of heterogeneity are distinguished, as will be illustrated in detail in Chapters 11 and 12. Here, we introduce them and outline their distinguishing features.

- **Allelic heterogeneity**, in which *different* mutations in a gene may produce the *same* phenotype
- **Locus heterogeneity**, in which mutations in *different* genes may cause the *same* phenotype
- **Clinical** or **phenotypic heterogeneity**, in which *different* mutations in a gene may result in *different* phenotypes.

Allelic Heterogeneity

Many loci possess more than one mutant allele; in fact, at a given locus, there may be several or many mutations in the population. Allelic heterogeneity may be responsible for differences in the severity or degree of pleiotropy demonstrated for a particular condition. As one example, more than 1000 different mutations have been found worldwide in the cystic fibrosis transmembrane conductance regulator gene (*CFTR*) among patients with CF (Case 12). Sometimes these different mutations result in clinically indistinguishable disorders. In other cases, different mutant alleles at the same locus produce a similar phenotype but along a continuum of severity. In autosomal recessive disorders, in particular, the fact that many patients are compound heterozygotes for two different alleles further adds to phenotypic variability of a disorder. For example, homozygotes or compound heterozygotes for many *CFTR* mutations have classic CF with pancreatic insufficiency, severe progressive lung disease, and congenital absence of the vas deferens in males, whereas other patients with combinations of other mutant alleles may have lung disease but normal pancreatic function, and still others will have only the abnormality of the male reproductive tract.

Allelic heterogeneity may also be manifest in the pattern of inheritance demonstrated for a particular condition. For example, in **retinitis pigmentosa**, a common cause of hereditary visual impairment due to photoreceptor degeneration, some mutations in the *ORP1* gene, encoding an oxygen-regulated photoreceptor protein, cause an autosomal recessive form of the disease, whereas others in the same gene result in an autosomal dominant form.

Locus Heterogeneity

Locus heterogeneity describes the situation in which clinically similar and even indistinguishable disorders may arise from mutations in different loci in different patients. For some phenotypes, pedigree analysis alone has been sufficient to demonstrate locus heterogeneity. Taking retinitis pigmentosa again as an example, it was recognized many years ago that the disease occurs in both autosomal and X-linked forms. Now, pedigree analysis combined with gene mapping has demonstrated that this single clinical entity can be caused by mutations in at least 56 different genes, 54 of which are autosomal and 2 of which are X-linked!

Clinical Heterogeneity

Different mutations in the same gene may produce very dissimilar phenotypes in different families, a phenomenon known as **clinical** or **phenotypic heterogeneity**. This situation occurs with mutations in the *LMNA* gene, which encodes a nuclear membrane protein. Different *LMNA* mutations have been associated with at least a half dozen phenotypically distinct disorders, including a form of muscular dystrophy, one form of hereditary dilated cardiomyopathy, one form of the Charcot-Marie-Tooth peripheral neuropathy, a disorder of adipose tissue called lipodystrophy, and the premature aging syndrome known as Hutchinson-Gilford progeria.

IMPORTANCE OF THE FAMILY HISTORY IN MEDICAL PRACTICE

Among medical specialties, medical genetics is distinctive in that it focuses not only on the patient but also on the entire family. A comprehensive family history is an important first step in the analysis of any disorder, whether or not the disorder is known to be genetic. As the late Barton Childs stated succinctly: "to fail to take a good family history is bad medicine." Despite the sophisticated cytogenetic, molecular, and genome testing now available to geneticists, an accurate family history (including the family pedigree) still remains a fundamental tool for all physicians and genetic counselors to use for determining the pattern of inheritance of a disorder in the family, forming a differential diagnosis, determining what genetic testing might be needed, and designing an individualized management and treatment plan for their patients. Furthermore, recognizing a familial component to a medical disorder allows the risk in other family members to be estimated so that proper management, prevention, and counseling can be offered to the patient *and* the family, as we will discuss in many of the chapters to follow.

GENERAL REFERENCES

Bennett RL, French KS, Resta RG, Doyle DL: Standardized human pedigree nomenclature: update and assessment of the recommendations of the National Society of Genetic Counselors, *J Genet Counsel* 17:424–433, 2008.

Online Mendelian Inheritance in Man, OMIM, Baltimore, Johns Hopkins University. Updated online at: http://omim.org/.

Rimoin DL, Pyeritz RE, Korf BR, editors: *Emery and Rimoin's essential medical genetics*, Oxford, 2013, Academic Press.

Scriver CR, Beaudet AL, Sly WS, et al, editors: *The metabolic and molecular bases of inherited disease*, ed 8, New York, 2000, McGraw-Hill. Updated online version available at: http://genetics.accessmedicine.com/.

PROBLEMS

1. Cathy is pregnant for the second time. Her first child, Donald, has cystic fibrosis (CF). Cathy has two brothers, Charles and Colin, and a sister, Cindy. Colin and Cindy are unmarried. Charles is married to an unrelated woman, Carolyn, and has a 2-year-old daughter, Debbie. Cathy's parents are Bob and Betty. Betty's sister Barbara is the mother of Cathy's husband, Calvin, who is 25. There is no previous family history of CF.
   a. Sketch the pedigree, using standard symbols.
   b. What is the pattern of transmission of CF, and what is the risk for CF for Cathy's next child?
   c. Which people in this pedigree are obligate heterozygotes?

2. George and Grace, who have normal hearing, have eight children; two of their five daughters and two of their three sons are congenitally deaf. Another couple, Harry and Helen, both with normal hearing, also have eight children; two of their six daughters and one of their two sons are deaf. A third couple, Gilbert and Gisele, who are congenitally deaf, have four children, also deaf. Their daughter Hedy marries Horace, a deaf son of George and Grace, and Hedy and Horace in turn have four deaf children. Their eldest son Isaac marries Ingrid, a daughter of Harry and Helen; although both Isaac and Ingrid are deaf, their six sons all have *normal* hearing. Sketch the pedigree and answer the following questions. (Hint: How many different types of congenital deafness are segregating in this pedigree?)
   a. State the probable genotypes of the children in the last generation.
   b. Why are all the children of Gilbert and Gisele and of Hedy and Horace deaf?

3. Consider the following situations:
   a. Retinitis pigmentosa occurs in X-linked and autosomal forms.
   b. Two parents each have a typical case of familial hypercholesterolemia diagnosed on the basis of hypercholesterolemia, arcus corneae, tendinous xanthomas, and demonstrated deficiency of low-density lipoprotein (LDL) receptors, together with a family history of the disorder; they have a child who has a very high plasma cholesterol level at birth and within a few years develops xanthomas and generalized atherosclerosis.

   c. A couple with normal vision, from an isolated community, have a child with autosomal recessive gyrate atrophy of the retina. The child grows up, marries another member (with normal vision) of the same community, and has a child with the same eye disorder.
   d. A child has severe neurofibromatosis 1 (NF1). Her father is phenotypically normal; her mother seems clinically normal but has several large café au lait spots and areas of hypopigmentation, and slit-lamp examination shows that she has a few Lisch nodules (hamartomatous growths on the iris).
   e. Parents of normal stature have a child with achondroplasia.
   f. An adult male with myotonic dystrophy has cataracts, frontal balding, and hypogonadism, in addition to myotonia.
   g. A man with vitamin D–resistant rickets transmits the condition to all his daughters, who have a milder form of the disease than their father has; none of his sons is affected. The daughters have approximately equal numbers of unaffected sons, affected sons, unaffected daughters, and affected daughters, the affected sons being more severely affected than their affected sisters.
   h. A boy has progressive muscular dystrophy with onset in early childhood and is wheelchair-bound by the age of 12 years. An unrelated man also has progressive muscular dystrophy but is still ambulant at the age of 30 years. Molecular analysis shows that both patients have large deletions in the dystrophin gene, which encodes the protein that is deficient or defective in the Duchenne and Becker types of muscular dystrophy.
   i. A patient with a recessive disorder is found to have inherited both copies of one chromosome from the same parent and no representative of that chromosome from the other parent.
   j. A child with maple syrup urine disease is born to parents who are first cousins.

   Which of the concepts listed here are illustrated by situations a to j?
   • Variable expressivity
   • Uniparental disomy
   • Consanguinity
   • Inbreeding

- X-linked dominant inheritance
- New mutation
- Allelic heterogeneity
- Locus heterogeneity
- Autosomal incompletely dominant trait
- Pleiotropy

4. Don and his maternal grandfather Barry both have hemophilia A. Don's partner Diane is his maternal aunt's daughter. Don and Diane have one son, Edward, and two daughters, Elise and Emily, all of whom have hemophilia A. They also have an unaffected daughter, Enid.
   a. Draw the pedigree.
   b. Why are Elise and Emily affected?
   c. What is the probability that a son of Elise would be hemophilic? What is the probability that her daughter would be hemophilic?
   d. What is the probability that a son of Enid would be hemophilic? A daughter?

5. A boy is born with a number of malformations but does not have a recognized syndrome. The parents are unrelated, and there is no family history of a similar condition. Which of the following conditions could explain this situation? Which are unlikely? Why?
   a. Autosomal dominant inheritance with new mutation
   b. Autosomal dominant inheritance with reduced penetrance
   c. Autosomal dominant inheritance with variable expressivity
   d. Autosomal recessive inheritance
   e. X-linked recessive inheritance
   f. Autosomal dominant inheritance, misattributed paternity
   g. Maternal ingestion of a teratogenic drug at a sensitive stage of embryonic development

6. A couple has a child with NF1. Both parents are clinically normal, and neither of their families shows a positive family history.
   a. What is the probable explanation for NF1 in their child?
   b. What is the risk for recurrence in other children of this couple?
   c. If the husband has another child by a different mother, what would the risk for NF1 be?
   d. What is the risk that any offspring of the affected child will also have NF1?

7. The consultand (*arrow*) wants to know her risk for having a child with a birth defect before starting her family because she and her husband are related (see pedigree). The family history reveals no known recessive disease. What is the chance that her child could be homozygous for a mutation for a recessive disorder carried by

one of the two common ancestors in generation I (the coefficient of inbreeding)? (Hint: The mutation could be on either of the two chromosomes in either of the two common ancestors.)

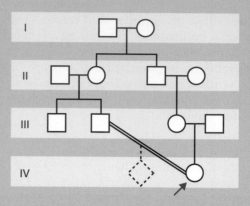

8. Given the pedigree below, what is/are the *most likely* inheritance pattern(s); *possible* but less likely inheritance pattern(s); *incompatible* inheritance pattern(s)? Patterns are autosomal recessive, autosomal dominant, X-linked recessive, X-linked dominant, mitochondrial. Justify your choices.

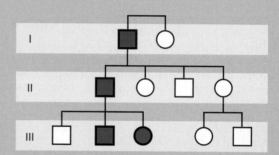

9. When a child is affected with an autosomal recessive condition, the assumption is that both parents are heterozygous carriers for the condition. Yet, new mutations occur all the time during the generation of gametes (see Chapter 4). Might not an individual have two mutant alleles for an autosomal recessive condition by virtue of inheriting one mutant allele from a carrier parent, whereas the other mutant allele arose de novo in a gamete that came from a parent who was not a carrier? Consider a child with cystic fibrosis. Calculate the odds (ratio of the probabilities) that both parents are carriers versus the probability that only the mother is a carrier and the sperm brought in a de novo mutation. Assume an average mutation rate of approximately 1×10^{-6} per male gamete per generation.

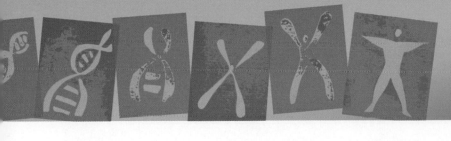
Complex Inheritance of Common Multifactorial Disorders

Common diseases such as congenital birth defects, myocardial infarction, cancer, neuropsychiatric disorders, diabetes, and Alzheimer disease cause morbidity and premature mortality in nearly two of every three individuals during their lifetimes (Table 8-1). Many of these diseases "run in families"—cases seem to cluster among the relatives of affected individuals more frequently than in the general population. However, their inheritance generally does not follow one of the mendelian patterns seen in the single-gene disorders described in Chapter 7. This is because such diseases rarely result simply from inheriting one or two alleles of major effect at a single locus, as is the case for dominant and recessive mendelian disorders. Instead, they are thought to result from *complex interactions* among a number of genetic variants that alter susceptibility to disease, combined with certain environmental exposures and perhaps chance events as well, all of which acting together may trigger, accelerate, or protect against the disease process. For this reason, these disorders are considered to be **multifactorial** in origin, and the familial clustering generates a pattern of inheritance that is referred to as **complex**.

The familial clustering and complex inheritance seen with multifactorial disorders can be explained by recognizing that family members share a greater proportion of their genetic information and environmental exposures than individuals chosen at random in the population. Thus the relatives of an affected individual are more likely to experience the same **gene-gene** and **gene environment interactions** that led to disease in the proband than are individuals who are unrelated to the proband.

In this chapter, we first address the question of how we infer that gene variants in the population predispose to such common diseases. We then describe how studies of familial aggregation and twin studies are used by geneticists to quantify the relative contributions of genetic variation and environment and show how these tools have been applied to multifactorial diseases. Finally, we devote the remainder of the chapter to describing a few examples of complex disorders where information is beginning to emerge about the specific nature of the genetic and environmental contributions to disease.

As we shall see in this chapter, the individual genes, the particular variants in those genes, and the environmental factors that interact with these variants have not yet been fully identified for the vast majority of common multifactorial diseases. A more detailed understanding of the approaches that geneticists use to identify the genetic factors underlying complex disease first requires a full appreciation of the distribution of genetic variation in different populations. This topic is presented in Chapter 9, after which we will turn, in Chapter 10, to discussion of the specific population-based epidemiological approaches that geneticists are using to identify the particular genes and the variants in those genes responsible for an increasing number of conditions with complex inheritance.

Ultimately, finding the genes and their variants that interact with the environment to contribute to susceptibility will give us a better understanding of the underlying processes leading to common multifactorial diseases and, perhaps, better tools for prevention or treatment.

QUALITATIVE AND QUANTITATIVE TRAITS

Multifactorial diseases with complex inheritance can be classified either as discrete **qualitative** traits or as continuous **quantitative** traits. A qualitative trait is the simpler of the two; a disease, such as lung cancer or rheumatoid arthritis, is either present or absent in an individual. Distinguishing between individuals who either have a disease or not is usually straightforward, but it may sometimes require detailed examination or specialized testing if the manifestations are subtle.

In contrast, a quantitative trait is a measurable physiological or biochemical quantity, such as height, blood pressure, serum cholesterol concentration, or body mass index, that varies among different individuals within a population. Although a quantitative trait varies continuously across a range of values, there are certain disease diagnoses, such as short stature, hypertension, hypercholesterolemia, or obesity, that are defined based on whether the value of the trait falls outside the so-called **normal range**, defined as an arbitrary interval around the population average. Frequently the normal range is derived by using the **normal distribution**, which

TABLE 8-1 Frequency of Different Types of Genetic Disease

Type	Incidence at Birth (per 1000)	Prevalence at Age 25 (per 1000)	Population Prevalence (per 1000)
Disorders due to genome and chromosome mutations	6	1.8	3.8
Disorders due to single-gene mutations	10	3.6	20
Disorders with multifactorial inheritance	≈50	≈50	≈600

Data from Rimoin DL, Connor JM, Pyeritz RE: *Emery and Rimoin's principles and practice of medical genetics*, ed 3, Edinburgh, 1997, Churchill Livingstone.

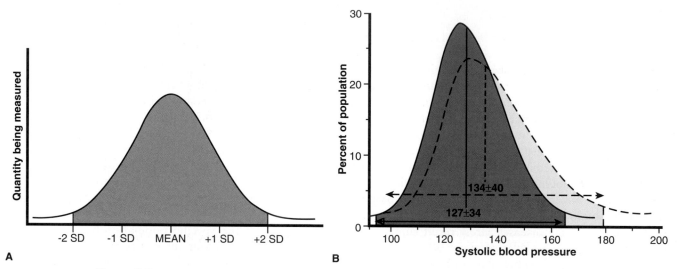

Figure 8-1 A, The normal gaussian distribution, with mean (average) and standard deviations (SDs) indicated. For many traits, the "normal" range is considered the mean ± 2 SD, as indicated by the *shaded* region. **B,** Distribution of systolic blood pressure in approximately 3300 men aged 40 to 45 (*solid line*) and approximately 2200 men aged 50 to 55 (*dotted line*). The mean and ± 2 SD are shown above *double-headed arrows. See Sources & Acknowledgments.*

is described in the next section, as an approximation for the distribution of the values of a quantitative trait in the population. Note that the term *normal* is used here in two different ways. Asserting that a physiological quantity has a "normal" distribution in the population and stating that an individual's value is in the "normal" range are different uses of the same word, one statistical and the other a measure of conformity to what is typically observed.

The Normal Distribution

As is often the case with physiological quantities, such as systolic blood pressure, a graph of the number (or the fraction) of individuals in the population (y-axis) having a particular quantitative value (x-axis) approximates the familiar, bell-shaped curve known as the **normal** (or **gaussian**) **distribution** (Fig. 8-1A). The position of the peak and the width of the curve of the normal distribution are governed by two quantities, the **mean** (μ) and the **variance** (σ^2), respectively. The mean is the arithmetic average of the values, and because more people have values for the trait near the average, the curve ordinarily has its peak at the mean value. The

variance (or its square root, σ, the **standard deviation,** abbreviated SD) is a measure of how much spread there is in the values to either side of the mean and therefore determines the breadth of the curve.

Any physiological quantity that can be measured across a sample of a population is a **quantitative phenotype,** and the mean and variance for that sample can be calculated and used to approximate the underlying mean and variance of the population from which the sample was drawn. For example, the systolic blood pressure of thousands of men in two different age-groups is shown in Figure 8-1B. The systolic blood pressure of the younger cohort is nearly symmetrical; in the older age-group, however, the curve becomes more "skewed" (asymmetrical), with more individuals with systolic blood pressures above the mean than below, indicating a tendency toward hypertension in that age-group.

The normal distribution provides guidelines for setting the limits of the normal range. A normal range is often defined as the values of a quantitative trait that are seen in approximately 95% of the population. Basic statistical theory states that when the values of a quantitative trait in a population follow the bell-shaped curve (i.e., are normally distributed), approximately 5% of the population will have measurements more than 2 SD above or below the population mean. For a given individual, however, it may still be perfectly "normal" (i.e., the individual is in good health), despite being a value outside the "normal" range.

FAMILIAL AGGREGATION AND CORRELATION

Allele Sharing among Relatives

The more closely related two individuals are in a family, the more alleles they have in common, inherited from their common ancestors (see Chapter 7). The most extreme examples of two individuals having alleles in common are identical (**monozygotic [MZ]**) twins (see later in this chapter), who have the same alleles at every locus. The next most closely related individuals in a family are **first-degree relatives**, such as a parent and child or a pair of sibs, including fraternal (**dizygotic [DZ]**) twins. In a parent-child pair, the child has exactly one allele out of two (50% of alleles) in common with each parent at every locus, that is, the allele the child inherited from that parent. Siblings (including DZ twins) also have 50% of their alleles in common with their other siblings, but this is only on average. This is because a pair of sibs inherits the same two alleles at a locus one fourth of the time, no alleles in common one fourth of the time, and one allele in common one half of the time (Fig. 8-2). At any one locus therefore, the average number of alleles an individual is expected to share with a sibling is given by:

$$\frac{1}{4} \text{ (2 alleles)} + \frac{1}{4} \text{ (0 allele)} + \frac{1}{2} \text{ (1 allele)} = 0.5 + 0 + 0.5 = 1 \text{ allele}$$

The more distantly related two members of a family are, the fewer alleles they will have in common, inherited from a common ancestor.

Familial Aggregation in Qualitative Traits

If certain alleles increase the chance of developing a disease, one would expect an affected individual to have a greater-than-expected number of affected relatives compared to what would be predicted from the frequency of the disease in the general population (**familial aggregation of disease**). This is because the more closely related the family members are to the affected relative, the more they will share the relevant alleles and the greater their chance of also being affected. Here, we will present two approaches to measuring familial aggregation: relative risk ratios and family history case-control studies.

Relative Risk Ratio

One way to measure familial aggregation of a disease is by comparing the frequency of the disease in the relatives of an affected proband with its frequency

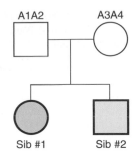

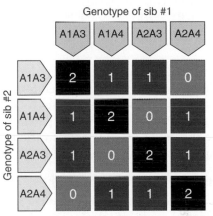

Figure 8-2 Allele sharing at an arbitrary locus between sibs concordant for a disease. The parents' genotypes are shown as *A1A2* for the father and *A3A4* for the mother. All four possible genotypes for sib #1 are given across the top of the table, and all four possible genotypes for sib #2 are given along the left side of the table. The numbers inside the boxes represent the number of alleles both sibs have in common for all 16 different combinations of genotypes for both sibs. For example, the upper left-hand corner has the number 2 because sib #1 and sib #2 both have the genotype *A1A3* and so have both A1 and A3 alleles in common. The bottom left-hand corner contains the number 0 because sib #1 has genotype *A1A3*, whereas sib #2 has genotype *A2A4*, so there are no alleles in common.

(prevalence) in the general population. The **relative risk ratio** λ_r (where the subscript "r" refers to relatives) is defined as:

$$\lambda_r = \frac{\text{Prevalence of the disease in the relatives of an affected person}}{\text{Prevalence of the disease in the general population}}$$

The value of λ_r as a measure of familial aggregation depends both on how frequently a disease is found to have recurred in a relative of an affected individual (the numerator) and on the population prevalence (the denominator); the larger λ_r is, the greater is the familial aggregation. The population prevalence enters into the calculation because the more common a disease is, the greater is the likelihood that aggregation may be just a coincidence based on drawing alleles from the overall gene pool rather than a result of sharing the alleles that predispose to disease because of familial inheritance. A value of $\lambda_r = 1$ indicates that a relative is no more likely to develop the disease than is any individual in the

TABLE 8-2 Risk Ratios λ_s for Siblings of Probands with Diseases with Familial Aggregation and Complex Inheritance

Disease	Relationship	λ_s
Schizophrenia	Siblings	12
Autism	Siblings	150
Manic-depressive (bipolar) disorder	Siblings	7
Type 1 diabetes mellitus	Siblings	35
Crohn disease	Siblings	25
Multiple sclerosis	Siblings	24

Data from Rimoin DL, Connor JM, Pyeritz RE: *Emery and Rimoin's principles and practice of medical genetics*, ed 3, Edinburgh, 1997, Churchill Livingstone; and King RA, Rotter JI, Motulsky AG: *The genetic basis of common diseases*, ed 2, Oxford, England, 2002, Oxford University Press.

population, whereas a value greater than 1 indicates that a relative is more likely to develop the disease. In practice, one measures λ for a particular class of relatives (e.g., r = s for sibs or r = p for parents). Examples of relative risk ratios determined for various diseases in samples of siblings (thus, λ_s) are shown in Table 8-2.

Family History Case-Control Studies

Another approach to assessing familial aggregation is the **case-control study**, in which patients with a disease (the cases) are compared with suitably chosen individuals without the disease (the controls), with respect to family history of disease (as well as other factors, such as environmental exposures, occupation, geographical location, parity, and previous illnesses). To assess a possible genetic contribution to familial aggregation of a disease, the frequency with which the disease is found in the extended families of the cases (**positive family history**) is compared with the frequency of positive family history among suitable controls, matched for age and ethnicity, but who do not have the disease. Spouses are often used as controls in this situation because they usually match the cases in age and ethnicity and share the same household environment. Other frequently used controls are patients with unrelated diseases matched for age, occupation, and ethnicity. Thus, for example, in a study of **multiple sclerosis (MS)**, approximately 3.5% of first-degree relatives of patients with MS also had MS, a prevalence that was much higher than among first-degree relatives of matched controls without MS (0.2%). Thus the odds of having a first-degree relative with MS were 18 times higher among MS patients than among controls. (In Chapter 10, we will discuss how one calculates odds ratios in case-control studies.) One can conclude therefore that substantial familial aggregation is occurring in MS, thereby providing evidence of a genetic predisposition to this disease.

Measuring the Genetic Contribution to Quantitative Traits

Just as a hereditary contribution to a disease increases familial aggregation for that disease, sharing of alleles that govern a particular quantitative trait affects the distribution of values of that trait in family members. The more sharing of alleles that govern a quantitative trait there is among relatives, the more similar the value of the trait will be among family members compared to what would be expected from the variance of the trait measured in the general population. The effect of genetic variation on quantitative traits is often measured and reported in two related ways: **correlation** between relatives and **heritability**.

Familial Correlation

The tendency for the values of a physiological measurement to be more similar among relatives than it is in the general population is measured by determining the degree of **correlation** of particular physiological quantities among relatives. The **coefficient of correlation** (symbolized by the letter r) is a statistical measure of correlation applied to a pair of measurements, such as, for example, a child's serum cholesterol level and that of a parent. Accordingly, a **positive correlation** would exist between the cholesterol measurements in a group of patients and the cholesterol levels in their relatives if it is found that the higher a patient's level, the proportionately higher is the level in the patient's relatives. When a correlation exists, a graph of values in the proband and his or her relatives, in which each point represents a proband-relative pair of values, will tend to cluster around a straight line. In such examples, the value of r can range from 0 when there is no correlation to +1 for perfect positive correlation. In the example of serum cholesterol, Figure 8-3 shows a modest positive correlation ($r = 0.294$) between serum cholesterol level of mothers aged 30 to 39 and those of their male children aged 4 to 9. In contrast, a **negative correlation** exists when the greater the increase in the patient's measurement, the lower the measurement is in the patient's relatives. The measurements are still correlated, but in the opposite direction. In such a case, the value of r can range from 0 to −1 for a perfect negative correlation.

Heritability

The concept of **heritability** of a quantitative trait (symbolized as H^2) was developed in an attempt to determine how much the genetic differences between individuals in a population contribute to variability of that trait in the population. H^2 is defined as *the fraction of the total phenotypic variance of a quantitative trait that is due to allelic variation in the broadest sense*, regardless of the mechanism by which the various alleles affect the phenotype. The higher the heritability, the greater is the contribution of genetic differences among people to the variability of the trait in the population. The value of H^2 varies from 0, if genotype contributes nothing to the total phenotypic variance in a population, to 1, if genotype is totally responsible for the phenotypic variance in that population.

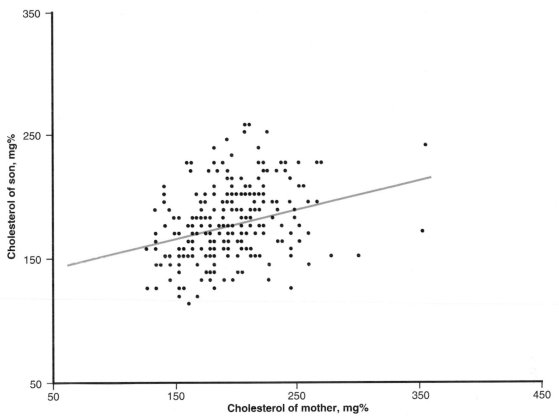

Figure 8-3 Plot of serum cholesterol levels in a group of mothers aged 30 to 39 and in their male children aged 4 to 9. Each dot represents a mother-son pair of measurements. The straight line is a "best fit" through the data points. *See Sources & Acknowledgments.*

Heritability of a human trait is a theoretical quantity that is usually estimated from the correlation between measurements of that trait among relatives of known degrees of relatedness, such as parents and children, siblings, or, as we shall see later in this chapter, twins.

DETERMINING THE RELATIVE CONTRIBUTIONS OF GENES AND ENVIRONMENT TO COMPLEX DISEASE

Distinguishing between Genetic and Environmental Influences Using Family Studies

For both qualitative and quantitative traits, similarities among family members are most likely the result of overlapping genotype and common exposure to nongenetic (i.e., environmental) factors such as socioeconomic status, local environment, dietary habits, or cultural behaviors, all of which are frequently shared among family members but are generally considered to be of nongenetic origin. Given evidence of familial aggregation of a disease or correlation of a quantitative trait, geneticists attempt to separate the relative contributions of genotype and environment to the phenotype using a variety of approaches. One approach is to

compare λ_r measurements or quantitative trait correlations between relatives who are of varying degrees of relatedness to the proband. For example, if genes predispose to a disease, one would expect λ_r to be greatest for MZ twins, to be somewhat smaller for first-degree relatives such as sibs or parent-child pairs, and to continue to decrease as allele-sharing decreases among the more distant relatives in a family (see Figure 7-3).

To illustrate this approach, consider **cleft lip with or without cleft palate**, or CL(P), one of the most common congenital malformations, affecting 1.4 per 1000 newborns worldwide. CL(P) originates as a failure of fusion of embryonic tissues that will go to make up the upper lip and the hard palate at approximately the 35th day of gestation. It is a multifactorial disorder with complex inheritance; for reasons that are not well understood, approximately 60% to 80% of those affected with CL(P) are males. Despite the similarity in names, CL(P) is usually etiologically distinct from isolated cleft palate (i.e., *without* cleft lip).

CL(P) is heterogeneous and includes forms in which the clefting is only one feature of a syndrome that includes other anomalies, known as **syndromic CL(P)**, as well as forms that are not associated with other birth defects, which are known as **nonsyndromic CL(P)**.

Syndromic CL(P) can be inherited as a mendelian single-gene disorder or can be caused by chromosome disorders (especially trisomy 13 and 4p⁻ deletion syndrome) (see Chapter 6) or teratogenic exposure (rubella embryopathy, thalidomide, or anticonvulsants) (see Chapter 14). Nonsyndromic CL(P) can also be inherited as a single-gene disorder but more commonly is a sporadic occurrence and demonstrates some degree of familial aggregation without an obvious mendelian inheritance pattern.

The risk for CL(P) in a child increases as a function of the number of relatives the child has who are affected with CL(P) and the more closely related they are to the child (Table 8-3). The simplest explanation for this is that the more closely related one is to the proband and, the more probands there are in the family, the more likely one is to share disease-susceptibility alleles with the probands; therefore one's risk for the disorder increases.

Another approach is to compare the disease relative risk ratio in *biological* relatives of the proband with that in *biologically unrelated* family members (e.g., adoptees or spouses), all living in the same household environment. Returning to MS, for example, λ_r is 190 for MZ twins and 20 to 40 for first-degree biological relatives (parents, children, and sibs). In contrast, λ_r is 1 for the *adopted* siblings of an affected individual, suggesting that most of the familial aggregation in MS is genetic rather than the result of a shared environment. A similar analysis can be carried out for quantitative traits such as blood pressure: *no* correlation exists between a child's blood pressure and that of his adopted siblings, in contrast to the positive correlation with blood pressure of biological siblings, all living in the same household.

Distinguishing between Genetic and Environmental Influences Using Twin Studies

Of all methods used to separate genetic and environmental influences, geneticists have relied most heavily on twin studies.

Twinning

MZ and DZ twins are "experiments of nature" that provide an excellent opportunity to separate environmental and genetic influences on phenotypes in humans. MZ twins arise from the cleavage of a single fertilized zygote into two separate zygotes early in embryogenesis (see Chapter 14). They occur in approximately 0.3% of all births, without significant differences among different ethnic groups. At the time the zygote cleaves in two, MZ twins start out with identical genotypes at every locus and are therefore often thought of as having identical genotypes and gene expression patterns.

In contrast, DZ twins arise from the simultaneous fertilization of two eggs by two sperm; genetically, DZ twins are siblings who share a womb and, like all

TABLE 8-3 Risk for Cleft Lip with or without Cleft Palate in a Child Depending on the Number of Affected Parents and Other Relatives

	Risk for CL(P) (%)		
	No. of Affected Parents		
Affected Relatives	0	1	2
None	0.1	3	34
One sibling	3	11	40
Two siblings	8	19	45
One sibling and one second-degree relative	6	16	43
One sibling and one third-degree relative	4	14	44

CL(P), Cleft lip with or without cleft palate.

siblings, share, on average, 50% of the alleles at all loci. DZ twins are of the same sex half the time and of opposite sex the other half. In contrast to MZ twins, DZ twins occur with a frequency that varies as much as fivefold in different populations, from a low of 0.2% among Asians to more than 1% of births in parts of Africa and among African Americans.

The striking difference between MZ and DZ twins in their genetic makeup is most easily seen by comparing the pattern for a type of so-called DNA fingerprint in twins (Fig. 8-4). This method of **DNA fingerprinting** is generated by simultaneously examining many DNA fragments of varying lengths that share a particular DNA sequence (minisatellite) and are located throughout the genome. MZ twins show an indistinguishable pattern, whereas many differences are seen between DZ twins, whether of same sex or not.

Disease Concordance in Monozygotic and Dizygotic Twins

When twins have the same disease, they are said to be **concordant** for that disorder. Conversely, when only one member of the pair of twins is affected and the other is not, the relatives are **discordant** for the disease. An examination of how frequently MZ twins are concordant for a disease is a powerful method for determining whether genotype alone is sufficient to produce a particular disease. The differences between a disease that is mendelian from one that shows complex inheritance are immediately evident. Using **sickle cell disease** (Case 42) as an example of a mendelian disorder, if one MZ twin has sickle cell disease, the other twin will always have the disease as well. In contrast, as an example of a multifactorial disorder, when one MZ twin has **type 1 diabetes mellitus** (previously known as insulin-dependent or juvenile diabetes) (Case 26), the other twin will also have type 1 diabetes in only approximately 40% of such twin pairs. *Disease concordance less than 100% in MZ twins is strong evidence that nongenetic factors play a role in the disease.* Such factors could include environmental influences, such as

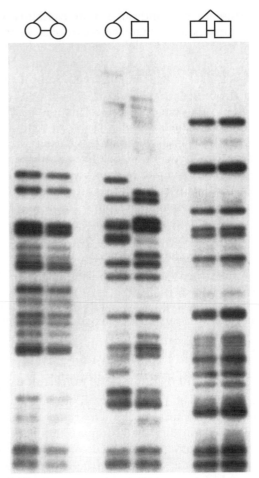

Figure 8-4 DNA fingerprinting of twins by detecting a variable number tandem repeat polymorphism, a class of polymorphism that has many alleles in loci around the genome due to variation in the number of copies repeated in tandem (see Chapter 4). Each pair of lanes contains DNA from a set of twins. The twins of the first and third sets have identical DNA fingerprints, indicating that they are identical (MZ) twins. The twins of the set in the middle have clearly distinguishable DNA fingerprints, confirming that they are fraternal (DZ) twins. *See Sources & Acknowledgments.*

TABLE 8-4 Concordance Rates in MZ and DZ Twins for Various Multifactorial Disorders

Disorder	Concordance (%)*	
	MZ	DZ
Nontraumatic epilepsy	70	6
Multiple sclerosis	18	2
Type 1 diabetes	40	5
Schizophrenia	46	15
Bipolar disease	62	8
Osteoarthritis	32	16
Rheumatoid arthritis	12	3
Psoriasis	72	15
Cleft lip with or without cleft palate	30	2
Systemic lupus erythematosus	22	0

*Rounded to the nearest percent.
DZ, Dizygotic; MZ, monozygotic.
Data from Rimoin DL, Connor JM, Pyeritz RE: *Emery and Rimoin's principles and practice of medical genetics*, ed 3, Edinburgh, 1997, Churchill Livingstone; King RA, Rotter JI, Motulsky AG: *The genetic basis of common diseases*, Oxford, England, 1992, Oxford University Press; and Tsuang MT: Recent advances in genetic research on schizophrenia. *J Biomed Sci* 5:28-30.

exposure to infection or diet, as well as other effects, such as somatic mutation, effects of aging, or epigenetic changes in gene expression in one twin compared with the other.

MZ and same-sex DZ twins share a common intra-uterine environment and sex and are usually reared together in the same household by the same parents. Thus a comparison of concordance for a disease between MZ and same-sex DZ twins shows how frequently disease occurs when relatives who experience the same prenatal and often the same postnatal environment have the same alleles at every locus (MZ twins), compared with only 50% of their alleles in common (DZ twins). *Greater concordance in MZ versus DZ twins is strong evidence of a genetic component to the disease,* as shown in Table 8-4 for a number of disorders.

Estimating Heritability from Twin Studies

Just as twin data may be used to assess the separate roles of genes and environment in qualitative disease traits, twins are also used to estimate the heritability of a quantitative trait using the correlation in the values of a physiological measurement in MZ and DZ twins. If one assumes that the alleles affecting the trait exert their effect additively (which is certainly overly simplistic and probably incorrect in many, if not all cases), MZ twins, who share 100% of their alleles, have twice the amount of allele sharing compared to that of DZ twins, who share 50% of their alleles on average. H^2, introduced earlier in this chapter, can therefore be approximated by taking twice the difference in the correlation coefficient r for a quantitative trait between MZ twins (r_{MZ}) and r between same-sex DZ twins (r_{DZ}) (as given by Falconer's formula):

$$H^2 = 2 \times (r_{MZ} - r_{DZ})$$

If the variability of the trait is determined chiefly by environment, the correlation within pairs of DZ twins will be similar to that seen between pairs of MZ twins; there will be little difference in the value of r for MZ and DZ twins. Thus, $r_{MZ} - r_{DZ} = \approx 0$, and H^2 will approach 0. At the other extreme, however, if the variability is determined exclusively by genetic makeup, the correlation coefficient r between MZ pairs will approach 1, whereas r between DZ twins will be half of that. Now, $r_{MZ} - r_{DZ} = \approx \frac{1}{2}$, and therefore H^2 will be approximately $2 \times (\frac{1}{2}) = 1$.

Twins Reared Apart

Although a rare occurrence, twins are sometimes separated at birth for social reasons and placed in different homes, thus providing an opportunity to observe individuals of identical or half-identical genotypes reared in different environments. Such studies have been used primarily in research in psychiatric disorders, substance abuse, and eating disorders, in which strong environmental influences within the family are believed to play a role in the development of disease. For example, in one study of obesity, the **body mass index** (BMI; weight/height^2, expressed in kg/m^2) was measured in MZ and DZ twins reared in the same household versus those reared apart (Table 8-5). Although the average BMI among MZ or DZ twins was similar, regardless of whether they were reared together or apart, the pairwise correlation for BMI between a pair of twins was much higher for the MZ than the DZ twins. Also interesting is that the higher correlation between MZ versus DZ twins was independent of whether the twins were reared together or apart, which suggests that genotype has a highly significant impact on adult weight and consequently on the risk for obesity and its complications.

Limitations of Familial Aggregation and Heritability Estimates from Family and Twin Studies
Potential Sources of Bias

There are a number of difficulties in measuring and interpreting λ_s. One is that studies of familial aggregation of disease are subject to various forms of **bias**. There is **ascertainment bias**, which arises when families with more than one affected sibling are more likely to come to a researcher's attention, thereby inflating the sibling recurrence risk λ_s. Ascertainment bias is also a problem for twin studies. Many studies rely on asking one twin with a particular disease to recruit the other twin to participate in a study (**volunteer-based ascertainment**), rather than ascertaining them first as twins through a twin registry and only then examining their health status (**population-based ascertainment**). Volunteer-based ascertainment can give biased results because twins, particularly MZ twins who may be emotionally close, are more likely to volunteer if they are concordant than if they are not, which inflates the concordance rate.

Similarly, because case-control studies of family history often rely, for practical reasons, on taking a history from the proband rather than examining all the relatives directly, there may be **recall bias**, in which a proband may be more likely to know of family members with the same or similar disease, than would the controls. Such biases will inflate the level of familial aggregation.

Other difficulties arise in measuring and interpreting heritability. The same trait may yield different measurements of heritability in different populations because of different allele frequencies or diverse environmental conditions. For example, heritability measurements of height would be lower when measured in a population with widespread famine that stunts growth in childhood as compared to the same population after food becomes plentiful. Heritability of a trait should therefore not be thought of as an intrinsic, universally applicable measure of "how genetic" the trait is, because it depends on the population and environment in which the estimate is being made. Although heritability estimates are still made in genetic research, most geneticists consider them to be only crude estimates of the role of genetic variation in causing phenotypic variation.

Potential Genetic or Epigenetic Differences

Despite the evident power of twin studies, one must caution against thinking of such studies as perfectly controlled experiments that compare individuals who share either half or all of their genetic variation and are exposed either to the same or to different environments. Studies of MZ twins assume the twins are genetically identical. Although this is mostly true, genotype and gene expression patterns may come to differ between MZ twins because of genetic or epigenetic changes that occur after the cleavage event that produced the MZ twin embryos. There are a number of ways that MZ twins may differ in their genotypes or patterns of gene expression. Genotype may differ due to somatic rearrangements and/or rare somatic mutations that occur after the cleavage event (see Chapter 3). Epigenetic changes may occur in response to environmental or stochastic factors, thus leading to differences in gene expression between MZ twins. (Female MZ twins have an additional source of variability, because of the stochastic nature of X inactivation patterns in various tissues, as presented in Chapter 6.)

TABLE 8-5 Pairwise Correlation of BMI between MZ and DZ Twins Reared Together and Apart

Twin Type	Rearing	Men			Women		
		No. of Pairs	BMI*	Pairwise Correlation	No. of Pairs	BMI*	Pairwise Correlation
Monozygotic	Apart	49	24.8 ± 2.4	0.70	44	24.2 ± 3.4	0.66
	Together	66	24.2 ± 2.9	0.74	88	23.7 ± 3.5	0.66
Dizygotic	Apart	75	25.1 ± 3.0	0.15	143	24.9 ± 4.1	0.25
	Together	89	24.6 ± 2.7	0.33	119	23.9 ± 3.5	0.27

*Mean ± 1 SD.
BMI, Body mass index; DZ, dizygotic; MZ, monozygotic.
Data from Stunkard A J, Harris JR, Pedersen NL, McClearn GE: The body-mass index of twins who have been reared apart. *N Engl J Med* 322:1483-1487, 1990.

Other Limitations

Another problem may arise when assuming that the environmental exposure of MZ and DZ twins has been held constant when they are reared together but not when twins are reared apart. Environmental exposures, including even intrauterine environment, may vary for twins reared in the same family. For example, MZ twins frequently share a placenta, and there may be a disparity between the twins in blood supply, intrauterine development, and birth weight. For late-onset diseases, such as neurodegenerative disease of late adulthood, the assumption that MZ and DZ twins are exposed to similar environments throughout their adult lives becomes less and less valid, and thus a difference in concordance provides less strong evidence for genetic factors in disease causation. Conversely, one assumes that by determining disease concordance in MZ twins reared apart, one is measuring the effect of different environments on the same genotype. However, the environment of twins reared apart may actually not be as different as one might suppose. *Thus no twin study is a perfectly controlled assessment of genetic versus environmental influence.*

Finally, caution is necessary when generalizing from twin studies. The most extreme situation would be when the phenotype being studied is only sometimes genetic in origin; that is, nongenetic phenocopies may exist. If genotype alone causes the disease in half the pairs of twins (MZ twin concordance of 100%) in your sample and a nongenetic phenocopy affects only one twin of the other half of twin pairs in your sample (MZ twin concordance of 0%), twin studies will show an intermediate level of 50% concordance that really applies to neither form of the disease.

EXAMPLES OF COMMON MULTIFACTORIAL DISEASES WITH A GENETIC CONTRIBUTION

In this section and the next, we turn to considering examples of several common conditions that illustrate general concepts of multifactorial disorders and their complex inheritance, as summarized here (see Box).

Multifactorial Congenital Malformations

Many common congenital malformations, occurring as isolated defects and not as part of a syndrome, are multifactorial and demonstrate complex inheritance (Table 8-6). Among these, **congenital heart malformations** are some of the most common and serve to illustrate the current state of understanding of other categories of congenital malformation.

Congenital heart defects (CHDs) occur at a frequency of approximately 4 to 8 per 1000 births. They are a heterogeneous group, caused in some cases by single-gene or chromosomal mechanisms and in others by exposure to teratogens, such as rubella infection or

CHARACTERISTICS OF INHERITANCE OF COMPLEX DISEASES

- Genetic variation contributes to diseases with complex inheritance, but these diseases are not single-gene disorders and do not demonstrate a simple mendelian pattern of inheritance.
- Diseases with complex inheritance often demonstrate familial aggregation because relatives of an affected individual are more likely to have disease-predisposing alleles in common with the affected person than with unrelated individuals.
- Diseases with complex inheritance are more common among the close relatives of a proband and become less common in relatives who are less closely related and therefore share fewer predisposing alleles. Greater concordance for disease is expected among monozygotic versus dizygotic twins.
- However, pairs of relatives who share disease-predisposing genotypes at relevant loci may still be discordant for phenotype (show lack of penetrance) because of the crucial role of nongenetic factors in disease causation. The most extreme examples of lack of penetrance despite identical genotypes are discordant monozygotic twins.

TABLE 8-6 Some Common Congenital Malformations with Multifactorial Inheritance

Malformation	Approximate Population Incidence (per 1000)
Cleft lip with or without cleft palate	0.4-1.7
Cleft palate	0.4
Congenital dislocation of hip	2*
Congenital heart defects	4-8
Ventricular septal defect	1.7
Patent ductus arteriosus	0.5
Atrial septal defect	1.0
Aortic stenosis	0.5
Neural tube defects	2-10
Spina bifida and anencephaly	Variable
Pyloric stenosis	1,[†] 5*

*Per 1000 males.
[†]Per 1000 females.
Data from Carter CO: Genetics of common single malformations. *Br Med Bull* 32:21-26, 1976; Nora JJ: Multifactorial inheritance hypothesis for the etiology of congenital heart diseases: the genetic environmental interaction. *Circulation* 38:604-617, 1968; and Lin AE, Garver KL: Genetic counseling for congenital heart defects. *J Pediatr* 113:1105-1109, 1988.

maternal diabetes. The cause is usually unknown, however, and the majority of cases are believed to be multifactorial in origin.

There are many types of CHDs, with different population incidences and empirical risks. It is known that when heart defects recur in a family, however, the affected children do not necessarily have exactly the same anatomical defect but instead show recurrence of lesions that are similar with regard to developmental mechanisms (see Chapter 14). By using developmental mechanisms as a classification scheme, five main groups of CHDs can be distinguished:

- Flow lesions
- Defects in cell migration
- Defects in cell death
- Abnormalities in extracellular matrix
- Defects in targeted growth

The subtype of congenital heart malformations known as flow lesions illustrates the familial aggregation and elevated risk for recurrence in relatives of an affected individual, all characteristic of a complex trait (Table 8-7). Flow lesions, which constitute

TABLE 8-7 Population Incidence and Recurrence Risks for Various Flow Lesions

Defect	Population Incidence (%)	Frequency in Sibs (%)	λ_s
Ventricular septal defect	0.17	4.3	25
Patent ductus arteriosus	0.083	3.2	38
Atrial septal defect	0.066	3.2	48
Aortic stenosis	0.044	2.6	59

approximately 50% of all CHDs, include hypoplastic left heart syndrome, coarctation of the aorta, atrial septal defect of the secundum type, pulmonary valve stenosis, a common type of ventricular septal defect, and other forms (Fig. 8-5). Up to 25% of patients with flow lesions, particularly tetralogy of Fallot, may have the deletion of chromosome region 22q11 seen in the **velo-cardiofacial syndrome** (see Chapter 6).

Certain isolated CHDs are inherited as multifactorial traits. Until more is known, the figures shown in Table 8-7 can be used as estimates of the recurrence risk for flow lesions in first-degree relatives. There is, however, a rapid falloff in risk (to levels not much higher than the population risk) in second- and third-degree relatives of index patients with flow lesions. Similarly, relatives of index patients with types of CHDs other than flow lesions can be offered reassurance that their risk is no greater than that of the general population. For further reassurance, many CHDs can now be assessed prenatally by ultrasonography (see Chapter 17).

Neuropsychiatric Disorders

Mental illnesses are some of the most common and perplexing of human diseases, affecting 4% of the

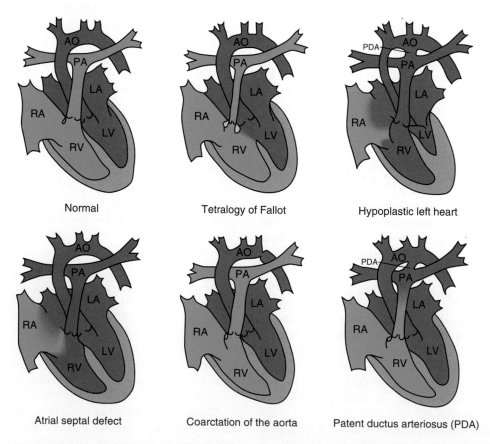

Figure 8-5 Diagram of various flow lesions seen in congenital heart disease. Blood on the left side of the circulation is shown in *red*, on the right side in *blue*. Abnormal admixture of oxygenated and deoxygenated blood is *purple*. AO, Aorta; LA, left atrium; LV, left ventricle; PA, pulmonary artery; RA, right atrium; RV, right ventricle.

TABLE 8-8 Recurrence Risks and Relative Risk Ratios in Schizophrenia Families

Relation to Individual Affected by Schizophrenia	Recurrence Risk (%)	λ_r
Child of two schizophrenic parents	46	23
Child	9-16	11.5
Sibling	8-14	11
Nephew or niece	1-4	2.5
Uncle or aunt	2	2
First cousin	2-6	4
Grandchild	2-8	5

TABLE 8-9 Recurrence Risks and Relative Risk Ratios in Bipolar Disorder Families

Relation to Individual Affected with Bipolar Disease	Recurrence Risk (%)*	λ_r
Child of two parents with bipolar disease	50-70	75
Child	27	34
Sibling	20-30	31
Second-degree relative	5	6

*Recurrence of bipolar, unipolar, or schizoaffective disorder.

human population worldwide. The annual cost in medical care and social services exceeds $150 billion in the United States alone. Among the most severe of the mental illnesses are schizophrenia and bipolar disease (manic-depressive illness).

Schizophrenia affects 1% of the world's population. It is a devastating psychiatric illness, with onset commonly in late adolescence or young adulthood, and is characterized by abnormalities in thought, emotion, and social relationships, often associated with delusional thinking and disordered mood. A genetic contribution to schizophrenia is supported by both twin and family aggregation studies. MZ concordance in schizophrenia is estimated to be 40% to 60%; DZ concordance is 10% to 16%. The recurrence risk ratio is elevated in first- and second-degree relatives of schizophrenic patients (Table 8-8).

Although there is considerable evidence of a genetic contribution to schizophrenia, only a subset of the genes and alleles that predispose to the disease has been identified to date. A major exception is the small percentage (<2%) of all schizophrenia that is found in individuals with interstitial deletions of particular chromosomes, such as the 22q11 deletion responsible for the velocardiofacial syndrome. It is estimated that 25% of patients with 22q11 deletions develop schizophrenia, even in the absence of many or most of the other physical signs of the syndrome. The mechanism by which a deletion of 3 Mb of DNA on 22q11 (see Fig. 6-5) causes mental illness in patients with this syndrome is unknown. Chromosomal microarrays have been used to scan the entire genome for other deletions and duplications, many too small to be detectable by standard cytogenetic approaches, as introduced in Chapter 5. These studies have revealed numerous deletions and duplications (copy number variants [CNVs]) throughout the genome in both normal individuals and individuals with a variety of psychiatric and neurodevelopmental disorders (see Chapter 6). In particular, small (1- to 1.5-Mb) interstitial deletions at 1q21.1, 15q11.2, and 15q13.3 have been implicated repeatedly in a small fraction of patients with schizophrenia. For the vast majority of patients with schizophrenia, however, genetic lesions are not known, and counseling therefore relies on empirical risk figures (see Table 8-8).

Bipolar disease is predominantly a mood disorder in which episodes of mood elevation, grandiosity, high-risk dangerous behavior, and inflated self-esteem (mania) alternate with periods of depression, decreased interest in what are normally pleasurable activities, feelings of worthlessness, and suicidal thinking. The prevalence of bipolar disease is 0.8%, approximately equal to that of schizophrenia, with a similar age at onset. The seriousness of this condition is underscored by the high (10% to 15%) rate of suicide in affected patients.

A genetic contribution to bipolar disease is strongly supported by twin and family aggregation studies. MZ twin concordance is 40% to 60%; DZ twin concordance is 4% to 8%. Disease risk is also elevated in relatives of affected individuals (Table 8-9). One striking aspect of bipolar disease in families is that the condition has variable expressivity; some members of the same family demonstrate classic bipolar illness, others have depression alone (unipolar disorder), and others carry a diagnosis of a psychiatric syndrome that involves both thought and mood (**schizoaffective disorder**). Even less is known about genes and alleles that predispose to bipolar disease than is known for schizophrenia; in particular, although an increase in de novo deletions or duplications has been identified in bipolar psychosis, *recurrent* CNVs involving particular regions of the genome have not been identified. Counseling therefore typically relies on empirical risk figures (see Table 8-9).

Coronary Artery Disease

Coronary artery disease (CAD) kills approximately 500,000 individuals in the United States yearly and is one of the most frequent causes of morbidity and mortality in the developed world. CAD due to atherosclerosis is the major cause of the nearly 1,500,000 cases of myocardial infarction (MI) and the more than 200,000 deaths from acute MI occurring annually. In the aggregate, CAD costs more than $143 billion in health care expenses alone each year in the United States, not including lost productivity. For unknown reasons, males are at higher risk for CAD both in the general population and within affected families.

Family studies have repeatedly supported a role for heredity in CAD, particularly when it occurs in relatively young individuals. The pattern of increased risk suggests that when the proband is female or young, there is likely to be a greater genetic contribution to MI in the family, thereby increasing the risk for disease in the proband's relatives. For example, the recurrence risk (Table 8-10) in male first-degree relatives of a female proband is sevenfold greater than that in the general population, compared with the 2.5-fold increased risk in female relatives of a male proband. When the proband is young (<55 years) and female, the risk for CAD is more than 11 times greater than that of the general population. Having multiple relatives affected at a young age increases risk substantially as well. Twin studies also support a role for genetic variants in CAD (Table 8-11).

A few mendelian disorders leading to CAD are known. **Familial hypercholesterolemia** (Case 16), an autosomal dominant defect of the low-density lipoprotein (LDL) receptor discussed in Chapter 12, is one of the most common of these but accounts for only approximately 5% of survivors of MI. Most cases of CAD show multifactorial inheritance, with both nongenetic and genetic predisposing factors. There are many stages in the evolution of atherosclerotic lesions in the coronary artery. What begins as a fatty streak in the intima of the artery evolves into a fibrous plaque containing smooth muscle, lipid, and fibrous tissue. These intimal plaques become vascular and may bleed, ulcerate, and calcify, thereby causing severe vessel narrowing as well as providing fertile ground for thrombosis, resulting in sudden, complete occlusion and MI. Given the many stages in the evolution of atherosclerotic lesions in the coronary artery, it is not surprising that many genetic differences affecting the various pathological processes involved could predispose to or protect from CAD (Fig. 8-6; also see Box). Additional risk factors for CAD include other disorders that are themselves multifactorial with genetic components, such as hypertension, obesity, and diabetes mellitus. The metabolic and physiological derangements represented by these disorders also contribute to enhancing the risk for CAD. Finally, diet, physical activity, systemic inflammation, and smoking are environmental factors that also play a major role in influencing the risk for CAD. Given all the different processes, metabolic derangements, and environmental factors that contribute to the development of CAD, it is easy to imagine that genetic susceptibility to CAD could be a complex multifactorial condition (see Box).

GENES AND GENE PRODUCTS INVOLVED IN THE STEPWISE PROCESS OF CORONARY ARTERY DISEASE

A large number of genes and gene products have been suggested and, in some cases, implicated in promoting one or more of the developmental stages of coronary artery disease. These include genes involved in the following:

- Serum lipid transport and metabolism—cholesterol, apolipoprotein E, apolipoprotein C-III, the low-density lipoprotein (LDL) receptor, and lipoprotein(a)—as well as total cholesterol level. Elevated LDL cholesterol level and decreased high-density lipoprotein cholesterol level, both of which elevate the risk for coronary artery disease, are themselves quantitative traits with significant heritabilities of 40% to 60% and 45% to 75%, respectively.
- Vasoactivity, such as angiotensin-converting enzyme
- Blood coagulation, platelet adhesion, and fibrinolysis, such as plasminogen activator inhibitor 1, and the platelet surface glycoproteins Ib and IIIa
- Inflammatory and immune pathways
- Arterial wall components

TABLE 8-10 Risk for Coronary Artery Disease in Relatives of a Proband

Proband	Increased Risk for CAD in a Family Member*
Male	3-fold in male first-degree relatives
	2.5-fold in female first-degree relatives
Female	7-fold in male first-degree relatives
Female <55 years of age	11.4-fold in male first-degree relatives
Two male relatives <55 years of age	13-fold in first-degree relatives

*Relative to the risk in the general population.
CAD, Coronary artery disease.
Data from Silberberg JS: Risk associated with various definitions of family history of coronary heart disease. *Am J Epidemiol* 147:1133-1139, 1998.

TABLE 8-11 Twin Concordance Rates and Relative Risks for Fatal Myocardial Infarction When Proband Had Early Fatal Myocardial Infarction*

Sex of the Twins	Concordance MZ Twins	Increased Risk[†] in a MZ Twin	Concordance DZ Twins	Increased Risk[†] in a DZ Twin
Male	0.39	6- to 8-fold	0.26	3-fold
Female	0.44	15-fold	0.14	2.6-fold

*Early myocardial infarction defined as age <55 years in males, age <65 years in females.
[†]Relative to the risk in the general population.
DZ, Dizygotic; MZ, monozygotic.
Data from Marenberg ME: Genetic susceptibility to death from coronary heart disease in a study of twins. *N Engl J Med* 330:1041-1046, 1994.

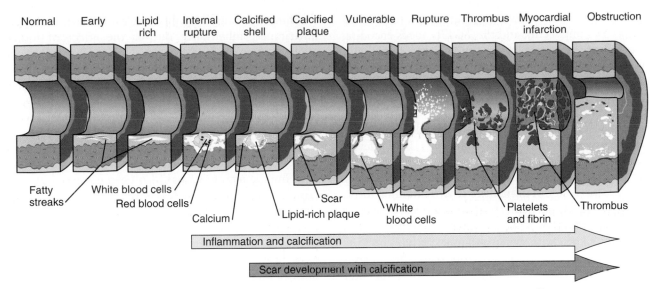

| Normal | Early | Lipid rich | Internal rupture | Calcified shell | Calcified plaque | Vulnerable | Rupture | Thrombus | Myocardial infarction | Obstruction |

Inflammation and calcification

Scar development with calcification

Figure 8-6 Sections of coronary artery demonstrating the steps leading to coronary artery disease. Genetic and environmental factors operating at any or all of the steps in this pathway can contribute to the development of this complex, common disease. *See Sources & Acknowledgments.*

CAD is often an incidental finding in family histories of patients with other genetic diseases. In view of the high recurrence risk, physicians and genetic counselors may need to consider whether first-degree relatives of patients with CAD should be evaluated further and offered counseling and therapy, even when CAD is not the primary genetic problem for which the patient or relative has been referred. Such an evaluation is clearly indicated when the proband is young, particularly if the proband is female.

EXAMPLES OF MULTIFACTORIAL TRAITS FOR WHICH SPECIFIC GENETIC AND ENVIRONMENTAL FACTORS ARE KNOWN

Up to this point, we have described some of the epidemiological approaches involving family and twin studies that are used to assess the extent to which there may be a genetic contribution to a complex trait. It is important to realize, however, that studies of familial aggregation, disease concordance, or heritability do *not* specify how many loci there are, which loci and alleles are involved, or how a particular genotype and set of environmental influences interact to cause a disease or to determine the value of a particular physiological measurement. In most cases, all we can show is that there is some genetic contribution and estimate its magnitude. There are, however, a few multifactorial diseases with complex inheritance for which we have begun to identify the genetic and, in some cases, environmental factors responsible for increasing disease susceptibility. We give a few examples in the next part of this chapter, illustrating increasing levels of complexity.

Modifier Genes in Mendelian Disorders

As discussed in Chapter 7, allelic variation at a single locus can explain variation in the phenotype in many single-gene disorders. However, even for well-characterized mendelian disorders known to be due to defects in a single gene, variation at other gene loci may impact some aspect of the phenotype, illustrating features therefore of complex inheritance.

In **cystic fibrosis (CF)** (Case 12), for example, whether or not a patient has pancreatic insufficiency requiring enzyme replacement can be explained largely by which mutant alleles are present in the *CFTR* gene (see Chapter 12). The correlation is imperfect, however, for other phenotypes. For example, the variation in the degree of pulmonary disease seen in CF patients remains unexplained by allelic heterogeneity. It has been proposed that the genotype at other genetic loci could act as **genetic modifiers**, that is, genes whose alleles have an effect on the severity of pulmonary disease seen in CF patients. For example, reduction in forced expiratory volume after 1 second (FEV$_1$), calculated as a percentage of the value expected for CF patients (a CF-specific FEV$_1$ percent), is a quantitative trait commonly used to measure deterioration in pulmonary function in CF patients. A comparison of CF-specific FEV$_1$ percent in affected MZ versus affected DZ twins provides an estimate of the heritability of the severity of lung disease in CF patients of approximately 50%. This value is independent of the specific *CFTR* allele(s) (because both kinds of twins will have the same CF mutations).

Two loci harboring alleles responsible for modifying the severity of pulmonary disease in CF are known:

MBL2, a gene that encodes a serum protein called mannose-binding lectin; and the *TGFB1* locus encoding the cytokine transforming growth factor β (TGFβ). Mannose-binding lectin is a plasma protein in the innate immune system that binds to many pathogenic organisms and aids in their destruction by phagocytosis and complement activation. A number of common alleles that result in reduced blood levels of the lectin exist at the *MBL2* locus in European populations. Lower levels of mannose-binding lectin appear associated with worse outcomes for CF lung disease, perhaps because low levels of lectin result in difficulties with containing respiratory pathogens, particularly *Pseudomonas.* Alleles at the *TGFB1* locus that result in higher TGFβ production are also associated with worse outcome, perhaps because TGFβ promotes lung scarring and fibrosis after inflammation. Thus both *MBL2* and *TGFB1* are **modifier genes,** variants at which—while they do not cause CF—can modify the clinical phenotype associated with disease-causing alleles at the *CFTR* locus.

Digenic Inheritance

The next level of complexity is a disorder determined by the additive effect of the genotypes at two or more loci. One clear example of such a disease phenotype has been found in a few families of patients with a form of retinal degeneration called **retinitis pigmentosa (RP)** (Fig. 8-7). Affected individuals in these families are heterozygous for mutant alleles at two *different* loci (**double heterozygotes**). One locus encodes the photoreceptor membrane protein peripherin and the other encodes a related photoreceptor membrane protein called Rom1. Heterozygotes for only one or the other of these mutations in these families are unaffected. Thus the RP in

this family is caused by the simplest form of multigenic inheritance, inheritance due to the effect of mutant alleles at two loci, without any known environmental factors that influence disease occurrence or severity. The proteins encoded by these two genes are likely to have overlapping physiological function because they are both located in the stacks of membranous disks found in retinal photoreceptors. It is the additive effect of having an abnormality in two proteins with overlapping function that produces disease.

A multigenic model has also been observed in a few families with **Bardet-Biedl syndrome,** a rare birth defect characterized by obesity, variable degrees of intellectual disability, retinal degeneration, polydactyly, and genitourinary malformations. Fourteen different genes have been found in which mutations cause the syndrome. Although inheritance is clearly autosomal recessive in most families, a few families appear to demonstrate digenic inheritance, in which the disease occurs only when an individual is homozygous for mutations at one of these 14 loci and is heterozygous for a mutation at another of the loci.

Gene-Environment Interactions in Venous Thrombosis

Another example of gene-gene interaction predisposing to disease is found in the group of conditions referred to as hypercoagulability states, in which venous or arterial clots form inappropriately and cause life-threatening complications of **thrombophilia** (Case 46). With hypercoagulability, however, there is a third factor, an environmental influence that in the presence of the predisposing genetic factors, increases the risk for disease even more.

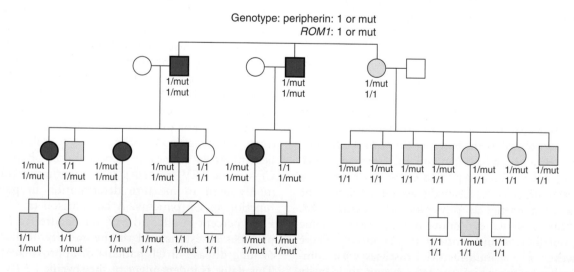

Figure 8-7 Pedigree of a family with retinitis pigmentosa due to digenic inheritance. *Dark blue* symbols are affected individuals. Each individual's genotypes at the peripherin locus (*first line*) and *ROM1* locus (*second line*) are written below each symbol. The normal allele is 1; the mutant allele is mut. *Light blue* symbols are unaffected, despite carrying a mutation in one or the other gene. *See Sources & Acknowledgments.*

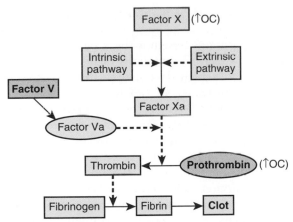

Figure 8-8 The clotting cascade relevant to factor V Leiden and prothrombin variants. Once factor X is activated, through either the intrinsic or extrinsic pathway, activated factor V promotes the production of the coagulant protein thrombin from prothrombin, which in turn cleaves fibrinogen to generate fibrin required for clot formation. Oral contraceptives (OC) increase blood levels of prothrombin and factor X as well as a number of other coagulation factors. The hypercoagulable state can be explained as a synergistic interaction of genetic and environmental factors that increase the levels of factor V, prothrombin, factor X and others to promote clotting. Activated forms of coagulation proteins are indicated by the letter *a*. *Solid arrows* are pathways; *dashed arrows* are stimulators.

One such disorder is **idiopathic cerebral vein thrombosis,** a disease in which clots form in the venous system of the brain, causing catastrophic occlusion of cerebral veins in the absence of an inciting event such as infection or tumor. It affects young adults, and although quite rare (<1 per 100,000 in the population), it carries a high mortality rate (5% to 30%). Three relatively common factors—two genetic and one environmental—that lead to abnormal coagulability of the clotting system are each known to individually increase the risk for cerebral vein thrombosis (Fig. 8-8):

- A missense variant in the gene for the clotting factor, factor V
- A variant in the 3′ untranslated region (UTR) of the gene for the clotting factor prothrombin
- The use of oral contraceptives

A polymorphic allele of factor V, **factor V Leiden** (FVL), in which arginine is replaced by glutamine at position 506 (Arg506Gln), has a frequency of approximately 2.5% in white populations but is rarer in other population groups. This alteration affects a cleavage site used to degrade factor V, thereby making the protein more stable and able to exert its procoagulant effect for a longer duration. Heterozygous carriers of FVL, approximately 5% of the white population, have a risk for cerebral vein thrombosis that, although still quite low, is sevenfold higher than that in the general population; homozygotes have a risk that is eightyfold higher.

The second genetic risk factor, a mutation in the **prothrombin** gene, changes a G to an A at position 20210 in the 3′ UTR of the gene (prothrombin

g.20210G>A). Approximately 2.4% of white individuals are heterozygotes, but it is rare in other ethnic groups. This change appears to increase the level of prothrombin mRNA, resulting in increased translation and elevated levels of the protein. Being heterozygous for the prothrombin 20210G>A allele raises the risk for cerebral vein thrombosis three to sixfold.

Finally, the use of **oral contraceptives** containing synthetic estrogen increases the risk for thrombosis fourteen- to twentytwofold, independent of genotype at the factor V and prothrombin loci, probably by increasing the levels of many clotting factors in the blood. Although using oral contraceptives and being heterozygous for FVL cause only a modest increase in risk compared with either factor alone, oral contraceptive use in a heterozygote for prothrombin 20210G>A raises the relative risk for cerebral vein thrombosis 30- to 150-fold!

There is also interest in the role of FVL and prothrombin 20210G>A alleles in **deep venous thrombosis (DVT)** of the lower extremities, a condition that occurs in approximately 1 in 1000 individuals per year, far more common than idiopathic cerebral venous thrombosis. Mortality due to DVT (primarily due to pulmonary embolus) can be up to 10%, depending on age and the presence of other medical conditions. Many environmental factors are known to increase the risk for DVT and include trauma, surgery (particularly orthopedic surgery), malignant disease, prolonged periods of immobility, oral contraceptive use, and advanced age.

The FVL allele increases the relative risk for a first episode of DVT sevenfold in heterozygotes; heterozygotes who use oral contraceptives see their risk increased thirtyfold compared with controls. Heterozygotes for prothrombin 20210G>A also have an increase in their relative risk for DVT of twofold to threefold. Notably, double heterozygotes for FVL and prothrombin 20210G>A have a relative increased risk of twentyfold— a risk approaching a few percent of the population.

Thus each of these three factors, two genetic and one environmental, on its own increases the risk for an abnormal hypercoagulable state; having two or all three of these factors at the same time raises the risk even more, to the point that thrombophilia screening programs for selected populations of patients may be indicated in the future.

Multiple Coding and Noncoding Elements in Hirschsprung Disease

A more complicated set of interacting genetic factors has been described in the pathogenesis of a developmental abnormality of the enteric nervous system in the gut known as **Hirschsprung disease (HSCR)** (Case 22). In HSCR, there is complete absence of some or all of the intrinsic ganglion cells in the myenteric and submucosal plexuses of the colon. An aganglionic colon is incapable of peristalsis, resulting in severe constipation, symptoms of intestinal obstruction, and massive dilatation of the colon (megacolon) proximal to the aganglionic segment. The disorder affects approximately 1 in 5000 newborns of European ancestry but is twice as common among Asian infants. HSCR occurs as an isolated birth defect 70% of the time, as part of a chromosomal syndrome 12% of the time, and as one element of a broad constellation of congenital abnormalities in the remainder of cases. Among patients with HSCR as an isolated birth defect, 80% have only a single, *short* aganglionic segment of colon at the level of the rectum (hence, HSCR-S), whereas 20% have aganglionosis of a *long* segment of colon, the entire colon or, occasionally, the entire colon plus the ileum as well (hence, HSCR-L).

Familial HSCR-L is often characterized by patterns of inheritance that suggest dominant or recessive inheritance, but consistently with reduced penetrance. HSCR-L is most commonly caused by loss-of-function missense or nonsense mutations in the *RET* gene, which encodes RET, a receptor tyrosine kinase. A small minority of families have mutations in genes encoding ligands that bind to RET, but with even lower penetrance than those families with *RET* mutations.

HSCR-S is the more common type of HSCR and has many of the characteristics of a disorder with complex genetics. The relative risk ratio for sibs, λ_s, is very high (approximately 200), but MZ twins do not show perfect concordance and families do not show any obvious mendelian inheritance pattern for the disorder. When pairs of siblings concordant for HSCR-S were analyzed genome-wide to see which loci and which sets of alleles at these loci each sib had in common with an affected brother or sister, alleles at three loci (including *RET*) were found to be significantly shared, suggesting gene-gene interactions and/or multigenic inheritance; indeed, most of the concordant sibpairs were found to share alleles at all three loci. Although the non-*RET* loci have yet to be identified, Figure 8-9 illustrates the range of interactions necessary to account for much of the penetrance of HSCR in even this small cohort of patients.

HSCR mutations have now been described at over a dozen loci, with *RET* mutations being by far the most common. The current data suggest that the *RET* gene is implicated in nearly all HSCR patients and, in particular, have pointed to two interacting **noncoding regulatory variants** near the *RET* gene, one in a potent gut enhancer with a binding site for the relevant transcription factor SOX10 and the other at an even more distant noncoding site some 125 kb upstream of the *RET* transcription start site. Thus HSCR-S is a multifactorial disease that results from mutations in or near the *RET* locus, perturbing the normally tightly controlled process of enteric nervous system development, combined with mutations at a number of other loci, both known and still unknown. Current genomic approaches of the type discussed in Chapter 10 suggest the possibility that many dozens of additional genes could be involved.

The identification of common, low-penetrant variants in noncoding elements serves to illustrate that the gene variants responsible for modifying expression of a multifactorial trait may be subtle in how they exert their effects on gene expression and, as a consequence, on disease penetrance and expressivity. It is also sobering to realize that the underlying genetic mechanisms for this relatively well defined congenital malformation have turned out to be so surprisingly complex; still, they are likely to be far simpler than are the mechanisms involved in the more common complex diseases, such as diabetes.

Type 1 Diabetes Mellitus

A common complex disease for which some of the underlying genetic architecture is being delineated is diabetes mellitus. Diabetes occurs in two major forms: **type 1 (T1D)** (sometimes referred to as insulin-dependent; IDDM) **(Case 26)** and **type 2 (T2D)** (sometimes referred

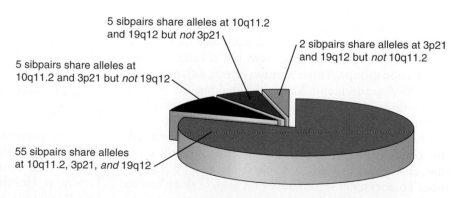

Figure 8-9 Patterns of allele sharing among sibpairs concordant for Hirschsprung disease, divided according to the number of loci for which the sibs show allele sharing. The three loci are located at 10q11.2 (the *RET* locus), 3p21, and 19q12. *See Sources & Acknowledgments.*

5 sibpairs share alleles at 10q11.2 and 19q12 but *not* 3p21

2 sibpairs share alleles at 3p21 and 19q12 but *not* 10q11.2

5 sibpairs share alleles at 10q11.2 and 3p21 but *not* 19q12

55 sibpairs share alleles at 10q11.2, 3p21, *and* 19q12

Loci showing allele sharing in 67 sibpairs concordant for Hirschsprung disease

to as non–insulin-dependent; NIDDM) (Case 35), representing approximately 10% and 88% of all cases, respectively. Familial aggregation is seen in both types of diabetes, but in any given family, usually only T1D or T2D is present. They differ in typical onset age, MZ twin concordance, and association with particular genetic variants at particular loci. Here, we focus on T1D to illustrate the major features of complex inheritance in diabetes.

T1D has an incidence in the white population of approximately 2 per 1000 (0.2%), but this is lower in African and Asian populations. It usually manifests in childhood or adolescence. It results from autoimmune destruction of the β cells of the pancreas, which normally produce insulin. A large majority of children who will go on to have T1D develop multiple autoantibodies early in childhood against a variety of endogenous proteins, including insulin, well before they develop overt disease.

There is strong evidence for genetic factors in T1D: concordance among MZ twins is approximately 40%, which far exceeds the 5% concordance in DZ twins. The lifetime risk for T1D in siblings of an affected proband is approximately 7%, resulting in an estimated λ_s of ≈ 35. However, the earlier the age of onset of the T1D in the proband, the greater is λ_s.

The Major Histocompatibility Complex

The major genetic factor in T1D is the major histocompatibility complex (MHC) locus, which spans some 3 Mb on chromosome 6 and is the most highly polymorphic locus in the human genome, with over 200 known genes (many involved in immune functions) and well over 2000 alleles known in populations around the globe (Fig. 8-10). On the basis of structural and functional differences, two major subclasses, the class I and class II genes, correspond to the **human leukocyte antigen (HLA)** genes, originally discovered by virtue of their importance in tissue transplantation between unrelated individuals. The HLA class I (HLA-A, HLA-B, HLA-C) and class II (HLA-DR, HLA-DQ, HLA-DP) genes encode cell surface proteins that play a critical role in the presentation of antigen to lymphocytes, which cannot recognize and respond to an antigen unless it is complexed with an HLA molecule on the surface of an antigen-presenting cell. Within the MHC, the HLA class I and class II genes are by far the most highly polymorphic loci (see Fig. 8-10).

The original studies showing an association between T1D and alleles designated as *HLA-DR3* and *HLA-DR4* relied on a serological method in use at that time for distinguishing between different HLA alleles, one that was based on immunological reactions in a test tube. This method has long been superseded by direct determination of the DNA sequence of different alleles, and sequencing of the MHC in a large number of individuals has revealed that the serologically determined

"alleles" associated with T1D are not single alleles at all (see Box). Both *DR3* and *DR4* can be subdivided into a dozen or more alleles located at a locus now termed *HLA-DRB1*.

HUMAN LEUKOCYTE ANTIGEN ALLELES AND HAPLOTYPES

The human leukocyte antigen (HLA) system can be confusing at first because the nomenclature used to define and describe different HLA alleles has undergone a fundamental change with the advent of widespread DNA sequencing of the major histocompatibility complex (MHC). According to the older system of HLA nomenclature, the different alleles were distinguished from one another serologically. However, as the genes responsible for encoding the class I and class II MHC chains were identified and sequenced (see Fig. 8-10), single HLA alleles initially defined serologically were shown to consist of multiple alleles defined by different DNA sequence variants even within the same serological allele. The 100 serological specificities at *HLA-A, B, C, DR, DQ,* and *DP* loci now comprise more than 1300 alleles defined at the DNA sequence level! For example, what used to be a single *B27* allele defined serologically, is now referred to as *HLA-B*2701, HLA-B*2702,* and so on, based on DNA-based genotyping.

The set of HLA alleles at the different class I and class II loci on a given chromosome together form a **haplotype**. Within any one ethnic group, some HLA alleles and haplotypes are found commonly; others are rare or never seen. The differences in the distribution and frequency of the alleles and haplotypes within the MHC are the result of complex genetic, environmental, and historical factors at play in each of the different populations. The extreme levels of polymorphism at HLA loci and their resulting haplotypes have been extraordinarily useful for identifying associations of particular variants with specific diseases (see Chapter 10), many of which (as one might predict) are **autoimmune disorders**, associated with an abnormal immune response apparently directed against one or more self-antigens resulting from polymorphism in immune response genes.

Furthermore, it is now clear that the association between certain *DRB1* alleles and T1D is due, in part, to alleles at two other class II loci, *DQA1* and *DQB1*, located approximately 80 kb away from *DRB1*, that form a particular combination of alleles with each other—that is, a **haplotype**—that is typically inherited as a unit (due to linkage disequilibrium; see Chapter 10). *DQA1* and *DQB1* encode the α and β chains of the class II DQ protein. Certain combinations of alleles at these three loci form a haplotype that *increases* the risk for T1D more than elevenfold over that for the general population, whereas other combinations of alleles *reduce* the risk fiftyfold. The DQB1*0303 allele contained in this protective haplotype results in the amino acid

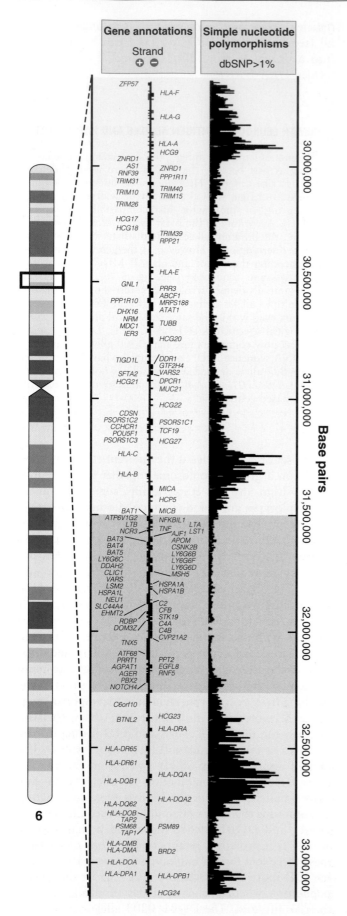

Figure 8-10 Genomic landscape of the major histocompatibility complex (MHC). The classic MHC is shown on the short arm of chromosome 6, comprising the class I region (*yellow*) and class II region (*blue*), both enriched in human leukocyte antigen (HLA) genes. Sequence-level variation is shown for single nucleotide polymorphisms (SNPs) found with at least 1% frequency. Remarkably high levels of polymorphism are seen in regions containing the classic HLA genes where variation is enriched in coding exons involved in defining the antigen-binding cleft. Other genes (*pink*) in the MHC region show lower levels of polymorphism. dbSNP, minor allele frequency in the Single Nucleotide Polymorphism database. *See Sources & Acknowledgments.*

aspartic acid at position 57 of the DQB1 product, whereas other amino acids at this position (alanine, valine, or serine) confer susceptibility. In fact, approximately 90% of patients with T1D are homozygous for *DQB1* alleles that do *not* encode aspartic acid at position 57. It is likely that differences in antigen binding, determined by which amino acid is at position 57, contribute directly to the autoimmune response that destroys the insulin-producing cells of the pancreas. Other loci and alleles in the MHC, however, are also important, as can be seen from the fact that some patients with T1D do have an aspartic acid at this position.

Genes Other Than Class II Major Histocompatibility Complex Loci in Type 1 Diabetes

The MHC haplotype alone accounts for only a portion of the genetic contribution to the risk for T1D in siblings of a proband. Family studies in T1D (Table 8-12) suggest that even when siblings share the same MHC class II haplotypes, the risk for disease is only approximately 17%, still well below the MZ twin concordance rate of approximately 40%. Thus there must be other genes, elsewhere in the genome, that contribute to the development of T1D, assuming that MZ twins and sibs have similar environmental exposures. Indeed, genetic association studies (to be described in Chapter 10) indicate that variation at nearly 50 different loci around the genome can increase susceptibility to T1D, although most have very small effects on increasing disease susceptibility.

It is important to stress, however, that genetic factors alone do not cause T1D because the MZ twin concordance rate is only approximately 40%, not 100%. Until a more complete picture develops of the genetic and nongenetic factors that cause T1D, risk counseling using HLA haplotyping must remain empirical (see Table 8-12).

Alzheimer Disease

Alzheimer disease (AD) (Case 4) is a fatal neurodegenerative disease that affects 1% to 2% of the United

TABLE 8-12 Empirical Risks for Counseling in Type 1 Diabetes

Relationship to Affected Individual	Risk for Development of Type 1 Diabetes (%)
None	0.2
MZ twin	40
Sibling	7
Sibling with no DR haplotypes in common	1
Sibling with 1 DR haplotype in common	5
Sibling with 2 DR haplotypes in common	17*
Child	4
Child of affected mother	3
Child of affected father	5

*20%-25% for particular shared haplotypes.
MZ, Monozygotic.

TABLE 8-13 Cumulative Age- and Sex-Specific Risks for Alzheimer Disease and Dementia

Time Interval Past 65 Years of Age	Risk for Development of AD (%)	Risk for Development of Any Dementia (%)
65 to 80 years		
Male	6.3	10.9
Female	12	19
65 to 100 years		
Male	25	32.8
Female	28.1	45

AD, Alzheimer disease.
Data from Seshadri S, Wolf PA, Beiser A, et al: Lifetime risk of dementia and Alzheimer's disease. The impact of mortality on risk estimates in the Framingham Study. *Neurology* 49:1498-1504, 1997.

States population. It is the most common cause of dementia in older adults and is responsible for more than half of all cases of dementia. As with other dementias, patients experience a chronic, progressive loss of memory and other cognitive functions, associated with loss of certain types of cortical neurons. Age, sex, and family history are the most significant risk factors for AD. Once a person reaches 65 years of age, the risk for any dementia, and AD in particular, increases substantially with age and female sex (Table 8-13).

AD can be diagnosed definitively only postmortem, on the basis of neuropathological findings of characteristic protein aggregates (β-amyloid plaques and neurofibrillary tangles; see Chapter 12). The most important constituent of the plaques is a small (39– to 42–amino acid) peptide, Aβ, derived from cleavage of a normal neuronal protein, the amyloid protein precursor. The secondary structure of Aβ gives the plaques the staining characteristics of amyloid proteins.

In addition to three rare autosomal dominant forms of the disease (see Chapter 12), in which disease onset is in the third to fifth decade, there is a common form of AD with onset after the age of 60 years (late onset).

TABLE 8-14 Association of Apolipoprotein E ε4 Allele with Alzheimer Disease*

Genotype	Frequency			
	United States		Japan	
	AD	Control	AD	Control
ε4/ε4; ε4/ε3; or ε4/ε2	0.64	0.31	0.47	0.17
ε3/ε3; ε2/ε3; or ε2/ε2	0.36	0.69	0.53	0.83

*Frequency of genotypes with and without the ε4 allele among Alzheimer disease (AD) patients and controls from the United States and Japan.

This form has no obvious mendelian inheritance pattern but does show familial aggregation and an elevated relative risk ratio ($\lambda_s = \approx 4$) typical of disorders with complex inheritance. Twin studies have been inconsistent but suggest MZ concordance of approximately 50% and DZ concordance of approximately 18%.

The ε4 Allele of Apolipoprotein E

The major locus with alleles found to be significantly associated with common late-onset AD is *APOE*, which encodes apolipoprotein E. Apolipoprotein E is a protein component of the LDL particle and is involved in clearing LDL through an interaction with high-affinity receptors in the liver. Apolipoprotein E is also a constituent of amyloid plaques in AD and is known to bind the Aβ peptide. The *APOE* gene has three alleles, ε2, ε3, and ε4, due to substitutions of arginine for two different cysteine residues in the protein (see Chapter 12).

When the genotypes at the *APOE* locus were analyzed in AD patients and controls, a genotype with at least one ε4 allele was found two to three times more frequently among patients compared with controls in both the general U.S. and Japanese populations (Table 8-14), with much less of an association in Hispanic and African American populations. Even more striking is that the risk for AD appears to increase further if both *APOE* alleles are ε4, through an effect on the age at onset of AD; patients with two ε4 alleles have an earlier onset of disease than those with only one. In a study of patients with AD and unaffected controls, the age at which AD developed in the affected patients was earliest for ε4/ε4 homozygotes, next for ε4/ε3 heterozygotes, and significantly less for the other genotypes (Fig. 8-11).

In the population in general, the risk for developing AD by age 80 is approaching 10%. The ε4 allele is clearly a predisposing factor that increases the risk for development of AD by shifting the age at onset to an earlier age, such that ε3/ε4 heterozygotes have a 40% risk for developing the disease, and ε4/ε4 have a 60% risk by age 85. Despite this increased risk, other genetic and environmental factors must be important because a significant proportion of ε3/ε4 and ε4/ε4 individuals live to extreme old age with no evidence of AD. There are also reports of association between the presence of the ε4 allele and neurodegenerative disease after traumatic head injury (as

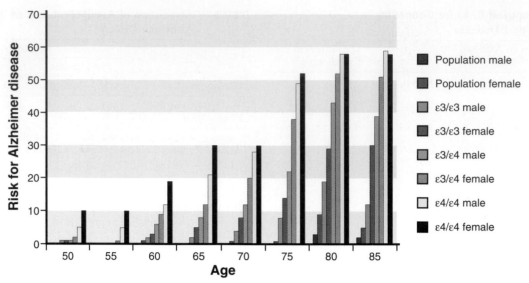

Figure 8-11 Chance of developing Alzheimer disease as a function of age for different *APOE* genotypes for each sex. At one extreme is the ε4/ε4 homozygote, who has ≈40% chance of remaining free of the disease by the age of 85 years, whereas an ε3/ε3 homozygote has ≈70% to ≈90% chance of remaining disease free at the age of 85 years, depending on the sex. General population risk is also shown for comparison. *See Sources & Acknowledgments.*

seen in professional boxers, football players, and soldiers who have suffered blast injuries), indicating that at least one environmental factor, brain trauma, can interact with the ε4 allele in the pathogenesis of AD.

The ε4 variant of *APOE* represents a prime example of a predisposing allele: *it predisposes to a complex trait in a powerful way but does not predestine any individual carrying the allele to the disease.* Additional genes as well as environmental effects are also clearly involved; although several of these appear to have a significant effect, most remain to be identified. In general, testing of asymptomatic people for the *APOE* ε4 allele remains inadvisable because knowing that one is a heterozygote or homozygote for the ε4 allele does not mean one will develop AD, nor are there any interventions currently known that can affect the chance one will or will not develop AD (see Chapter 18).

THE CHALLENGE OF MULTIFACTORIAL DISEASE WITH COMPLEX INHERITANCE

The greatest challenge facing medical genetics and genomic medicine going forward is unraveling the complex interactions between the variants at multiple loci and the relevant environmental factors that underlie the susceptibility to common multifactorial disease. This area of research is the central focus of the field of population-based **genetic epidemiology** (to be discussed more fully in Chapter 10). The field is developing rapidly, and it is clear that the genetic contribution to many more complex diseases in humans will be

elucidated in the coming years. Such understanding will, in time, allow the development of novel preventive and therapeutic measures for the common disorders that cause such significant morbidity and mortality in the population.

GENERAL REFERENCES

Chakravarti A, Clark AG, Mootha VK: Distilling pathophysiology from complex disease genetics, *Cell* 155:21–26, 2013.
Rimoin DL, Pyeritz RE, Korf BR: *Emery and Rimoin's essential medical genetics*, Waltham, MA, 2013, Academic Press (Elsevier).
Scott W, Ritchie M: *Genetic analysis of complex disease*, ed 3, Hoboken, NJ, 2014, John Wiley and Sons.

REFERENCES FOR SPECIFIC TOPICS

Amiel J, Sproat-Emison E, Garcia-Barcelo M, et al: Hirschsprung disease, associated syndromes, and genetics: a review, *J Med Genet* 45:1–14, 2008.
Bertram L, Lill CM, Tanzi RE: The genetics of Alzheimer disease: back to the future, *Neuron* 68:270–281, 2010.
Concannon P, Rich SS, Nepom GT: Genetics of type 1A diabetes, *N Engl J Med* 360:1646–1664, 2009.
Emison ES, Garcia-Barcelo M, Grice EA, et al: Differential contributions of rare and common, coding and noncoding Ret mutations to multifactorial Hirschsprung Disease liability, *Am J Hum Genet* 87:60–74, 2010.
Malhotra D, McCarthy S, Michaelson JJ, et al: High frequencies of de novo CNVs in bipolar disorder and schizophrenia, *Neuron* 72:951–963, 2011.
Segal JB, Brotman DJ, Necochea AJ, et al: Predictive value of Factor V Leiden and prothrombin G20210A in adults with venous thromboembolism and in family members of those with a mutation, *JAMA* 301:2472–2485, 2009.
Trowsdale J, Knight JC: Major histocompatibility complex genomics and human disease, *Annu Rev Genomics Hum Genet* 14:301–323, 2013.

PROBLEMS

1. For a certain malformation, the recurrence risk in sibs and offspring of affected persons is 10%, the risk in nieces and nephews is 5%, and the risk in first cousins is 2.5%.
   a. Is this more likely to be an autosomal dominant trait with reduced penetrance or a multifactorial trait? Explain.
   b. What other information might support your conclusion?

2. A large sex difference in affected persons is often a clue to X-linked inheritance. How would you establish that pyloric stenosis is multifactorial rather than X-linked?

3. A series of children with a particular congenital malformation includes both boys and girls. In all cases, the parents are normal. How would you determine whether the malformation is more likely to be multifactorial than autosomal recessive?

CHAPTER 9

Genetic Variation in Populations

We have explored in previous chapters the nature of genetic and genomic variation and mutation and the inheritance of different alleles in families. Throughout, we have alluded to differences in the frequencies of different alleles in different populations, whether assessed by examining different single nucleotide polymorphisms (SNPs), indels, or copy number variants (CNVs) in the genomes of many thousands of individuals studied worldwide (see Chapter 4) or inferred by ascertaining individuals with specific phenotypes and genetic disorders among populations around the globe (see Chapters 7 and 8). Here we consider in greater detail the genetics of populations and the principles that influence the frequency of genotypes and phenotypes in those populations.

Population genetics is the quantitative study of the distribution of genetic variation in populations and of how the frequencies of genes and genotypes are maintained or change over time both within and between populations. Population genetics is concerned both with genetic factors, such as mutation and reproduction, and with environmental and societal factors, such as selection and migration, which together determine the frequency and distribution of alleles and genotypes in families and communities. A mathematical description of the behavior of alleles in populations is an important element of many disciplines, including anthropology, evolutionary biology, and human genetics. At present, human geneticists use the principles and methods of population genetics to address many unanswered questions concerning the history and genetic structure of human populations, the flow of alleles between populations and between generations, and, importantly, the optimal methods for identifying genetic susceptibilities to common diseases, as we introduced in Chapter 8. In the practice of medical genetics, population genetics provides knowledge about various disease genes that are common in different populations. Such information is needed for clinical diagnosis and genetic counseling, including determining the allele frequencies required for accurate risk calculations.

In this chapter, we describe the central, organizing concept of population genetics, **Hardy-Weinberg equilibrium**; we consider its assumptions and the factors that may cause true or apparent deviation from equilibrium in real as opposed to idealized populations. Finally, we provide some insight into how differences in allelic variant or disease gene frequencies arise among members of different, more or less genetically isolated groups.

GENOTYPES AND PHENOTYPES IN POPULATIONS

Allele and Genotype Frequencies in Populations

To illustrate the relationship between allele and genotype frequencies in populations, we begin with an important example of a common autosomal trait governed by a single pair of alleles. Consider the gene *CCR5*, which encodes a cell surface cytokine receptor that serves as an entry point for certain strains of the human immunodeficiency virus (HIV), which causes the **acquired immunodeficiency syndrome** (**AIDS**). A 32-bp deletion in this gene results in an allele (ΔCCR5) that encodes a nonfunctional protein due to a frameshift and premature termination. Individuals homozygous for the ΔCCR5 allele do not express the receptor on the surface of their immune cells and, as a consequence, are resistant to HIV infection. Loss of function of *CCR5* appears to be a benign trait, and its only known phenotypic consequence is resistance to HIV infection. A sampling of 788 individuals from Europe illustrates the distribution of individuals who were homozygous for the wild-type *CCR5* allele, homozygous for the ΔCCR5 allele, or heterozygous (Table 9-1).

On the basis of the observed genotype frequencies, we can directly determine the allele frequencies by simply counting the alleles. In this context, when we refer to the population frequency of an allele, we are considering a hypothetical **gene pool** as a collection of all the alleles at a particular locus for the entire population. For autosomal loci, the size of the gene pool at one locus is twice the number of individuals in the population because each autosomal genotype consists of two alleles; that is, a ΔCCR5/ΔCCR5 individual has two ΔCCR5 alleles, and a *CCR5*/ΔCCR5 individual has one of each. In this example, then, the observed frequency of the *CCR5* allele is:

$$\frac{(2 \times 647) + (1 \times 134)}{788 \times 2} = 0.906$$

Similarly, one can calculate the frequency of the ΔCCR5 allele as 0.094, by adding up how many ΔCCR5 alleles

TABLE 9-1 Genotype Frequencies for the Wild-Type *CCR5* Allele and the Δ*CCR5* Deletion Allele

Genotype	Number of Individuals	Observed Genotype Frequency	Allele	Derived Allele Frequencies
CCR5/CCR5	647	0.821		
CCR5/Δ*CCR5*	134	0.168	*CCR5*	0.906
Δ*CCR5*/Δ*CCR5*	7	0.011	Δ*CCR5*	0.094
Total	788	1.000		

Data from Martinson JJ, Chapman NH, Rees DC, et al: Global distribution of the *CCR5* gene 32-basepair deletion. *Nat Genet* 16:100-103, 1997.

are present $[(2 \times 7) + (1 \times 134)] = 148$ out of a total of 1576 alleles in this sample], resulting in a Δ*CCR5* allele frequency of $148/1576 = 0.094$. Alternatively (and more simply), one can subtract the frequency of the normal *CCR5* allele, 0.906, from 1, because the frequencies of the two alleles must add up to 1, resulting in a Δ*CCR5* allele frequency of 0.094.

The Hardy-Weinberg Law

As we have just shown with the *CCR5* example, we can use a sample of individuals with known genotypes in a population to derive estimates of the allele frequencies by simply counting the alleles in individuals with each genotype. How about the converse? Can we calculate the proportion of the population with various genotypes once we know the allele frequencies? Deriving genotype frequencies from allele frequencies is not as straightforward as counting because we actually do not know in advance how the alleles are distributed among homozygotes and heterozygotes. If a population meets certain assumptions (see later), however, there is a simple mathematical equation for calculating genotype frequencies from allele frequencies. This equation is known as the **Hardy-Weinberg law**. This law, the cornerstone of population genetics, was named for Godfrey Hardy, an English mathematician, and Wilhelm Weinberg, a German physician, who independently formulated it in 1908.

The Hardy-Weinberg law has two critical components. The first is that under certain ideal conditions (see Box), a simple relationship exists between allele frequencies and genotype frequencies in a population. Suppose p is the frequency of allele A, and q is the frequency of allele a in the gene pool. Assume alleles combine into genotypes randomly; that is, mating in the population is completely at random with respect to the genotypes at this locus. The chance that two A alleles will pair up to give the AA genotype is then p^2; the chance that two a alleles will come together to give the aa genotype is q^2; and the chance of having one A and one a pair, resulting in the Aa genotype, is $2pq$ (the factor 2 comes from the fact that the A allele could be inherited from the mother and the a allele from the father, or vice versa). *The Hardy-Weinberg law states that the frequency of the three genotypes AA, Aa, and aa is given by the terms of the binomial expansion of* $(p + q)^2 = p^2 + 2pq + q^2$. This law applies to all autosomal loci and to the X chromosome in females, but not to X-linked loci in males who have only a single X chromosome.

THE HARDY-WEINBERG LAW

The Hardy-Weinberg law rests on these assumptions:
- The population under study is large, and matings are random with respect to the locus in question.
- Allele frequencies remain constant over time because of the following:
  - There is no appreciable rate of new mutation.
  - Individuals with all genotypes are equally capable of mating and passing on their genes; that is, there is no selection against any particular genotype.
  - There has been no significant immigration of individuals from a population with allele frequencies very different from the endogenous population.

A population that reasonably appears to meet these assumptions is considered to be in **Hardy-Weinberg equilibrium**.

The law can be adapted for genes with more than two alleles. For example, if a locus has three alleles, with frequencies p, q, and r, the genotypic distribution can be determined from $(p + q + r)^2$. In general terms, the genotype frequencies for any known number of alleles a_n with allele frequencies p_1, p_2, ... p_n can be derived from the terms of the expansion of $(p_1 + p_2 + ... p_n)^2$.

A second component of the Hardy-Weinberg law is that if allele frequencies do not change from generation to generation, the proportion of the genotypes will not change either; that is, the *population genotype frequencies from generation to generation will remain constant, at equilibrium, if the allele frequencies p and q remain constant*. More specifically, when there is random mating in a population that is at equilibrium and genotypes AA, Aa, and aa are present in the proportions $p^2:2pq:q^2$, then genotype frequencies in the next generation will remain in the same relative proportions, $p^2:2pq:q^2$. Proof of this equilibrium is shown in Table 9-2. It is important to note that Hardy-Weinberg equilibrium does not specify any particular values for p and q; whatever allele frequencies happen to be present in the population will result in genotype frequencies of $p^2:2pq:q^2$, and these relative genotype frequencies will

TABLE 9-2 Frequencies of Mating Types and Offspring for a Population in Hardy-Weinberg Equilibrium with Parental Genotypes in the Proportion $p^2 : 2pq : q^2$

Types of Matings			Offspring		
Mother	Father	Frequency	*AA*	*Aa*	*aa*
AA	*AA*	$p^2 \times p^2 = p^4$	p^4		
AA	*Aa*	$p^2 \times 2pq = 2p^3q$	½ $(2p^3q)$	½ $(2p^3q)$	
Aa	*AA*	$2pq \times p^2 = 2p^3q$	½ $(2p^3q)$	½ $(2p^3q)$	
AA	*aa*	$p^2 \times q^2 = p^2q^2$		p^2q^2	
aa	*AA*	$q^2 \times p^2 = p^2q^2$		p^2q^2	
Aa	*Aa*	$2pq \times 2pq = 4p^2q^2$	¼ $(4p^2q^2)$	½ $(4p^2q^2)$	¼ $(4p^2q^2)$
Aa	*aa*	$2pq \times q^2 = 2pq^3$		½ $(2pq^3)$	½ $(2pq^3)$
aa	*Aa*	$q^2 \times 2pq = 2pq^3$		½ $(2pq^3)$	½ $(2pq^3)$
aa	*aa*	$q^2 \times q^2 = q^4$			q^4

Sum of *AA* offspring $= p^4 + p^3q + p^3q + p^2q^2 = p^2(p^2 + 2pq + q^2) = p^2(p + q)^2 = p^2$. (Remember that $p + q = 1$.)

Sum of *Aa* offspring $= p^3q + p^3q + p^2q^2 + p^2q^2 + 2p^2q^2 + pq^3 + pq^3 = 2pq(p^2 + 2pq + q^2) = 2pq(p + q)^2 = 2pq$.

Sum of *aa* offspring $= p^2q^2 + pq^3 + pq^3 + q^4 = q^2(p^2 + 2pq + q^2) = q^2(p + q)^2 = q^2$.

remain constant from generation to generation as long as the allele frequencies remain constant and the other conditions introduced in the Box are met.

Applying the Hardy-Weinberg formula to the *CCR5* example given earlier, with relative frequencies of the two alleles in the population of 0.906 (for the wild-type allele *CCR5*) and 0.094 (for Δ*CCR5*), then the Hardy-Weinberg law states that the relative proportions of the three combinations of alleles (genotypes) are $p^2 = 0.906 \times 0.906 = 0.821$ (for an individual having two wild-type *CCR5* alleles), $q^2 = 0.094 \times 0.094 = 0.009$ (for two Δ*CCR5* alleles), and $2pq = (0.906 \times 0.094) + (0.094 \times 0.906) = 0.170$ (for one *CCR5* and one Δ*CCR5* allele). When these genotype frequencies, which were *calculated* by the Hardy-Weinberg law, are applied to a population of 788 individuals, the derived numbers of people with the three different genotypes (647 : 134 : 7) are, in fact, identical to the actual *observed* numbers in Table 9-1. As long as the assumptions of the Hardy-Weinberg law are met in a population, we would expect these genotype frequencies (0.821 : 0.170 : 0.009) to remain constant generation after generation in that population.

The Hardy-Weinberg Law in Autosomal Recessive Disease

The major practical application of the Hardy-Weinberg law in medical genetics is in genetic counseling for autosomal recessive disorders. For a disease such as **phenylketonuria (PKU)**, there are hundreds of different mutant alleles with frequencies that vary among different population groups defined by geography and/or ethnicity (see Chapter 12). Affected individuals can be homozygotes for the same mutant allele but, more often than not, are compound heterozygotes for different mutant alleles (see Chapter 7). For many disorders, however, it is convenient to consider all disease-causing alleles together and treat them as a single mutant allele, with frequency q, even when there is significant allelic heterogeneity in disease-causing alleles. Similarly, the

combined frequency of all wild-type or normal alleles, p, is given by $1 - q$.

Suppose we would like to know the frequency of all disease-causing PKU alleles in a population for use in genetic counseling, for example, to inform couples of their risk for having a child with PKU. If we were to attempt to determine the frequency of disease-causing PKU alleles directly from genotype frequencies, as we did in the earlier example of the Δ*CCR5* allele, we would need to know the frequency of heterozygotes in the population, a frequency that cannot be measured directly because of the recessive nature of PKU; heterozygotes are asymptomatic silent carriers (see Chapter 7), and their frequency in the population (i.e., $2pq$) cannot be reliably determined directly from phenotype.

However, the frequency of affected homozygotes/compound heterozygotes for disease-causing alleles in the population (i.e., q^2) *can* be determined directly, by counting the number of babies with PKU born over a given period of time and identified through newborn screening programs (see Chapter 18), divided by the total number of babies screened during that same period of time. Now, using the Hardy-Weinberg law, we can calculate the mutant allele frequency (q) from the observed frequency of homozygotes/compound heterozygotes alone (q^2), thereby providing an estimate ($2pq$) of the frequency of heterozygotes for use in genetic counseling.

To illustrate this example further, consider a population in Ireland, where the frequency of PKU is approximately 1 per 4500. If we group all disease-causing alleles together and treat them as a single allele with frequency q, then the frequency of affected individuals $q^2 = 1/4500$. From this, we calculate $q = 0.015$, and thus $2pq = 0.029$. The carrier frequency for all disease-causing alleles lumped together in the Irish population is therefore approximately 3%. For an individual known to be a carrier of PKU through the birth of an affected child in the family, there would then be an approximately 3% chance that he or she would find a new mate of Irish ethnicity who would also be a carrier, and this estimate could be used to provide genetic counseling. Note, however, that this estimate applies only to the population in question; if the new mate was not from Ireland, but from Finland, where the frequency of PKU is much lower ($\approx$1 per 200,000), his or her chance of being a carrier would be only 0.6%.

In this example, we lumped all PKU-causing alleles together for the purpose of estimating q. For other disorders, however, such as hemoglobin disorders that we will consider in Chapter 11, different mutant alleles can lead to very different diseases, and therefore it would make no sense to group all mutant alleles together, even when the same locus is involved. Instead, the frequency of alleles leading to different phenotypes (such as sickle cell anemia and β-thalassemia in the case of different mutant alleles at the β-globin locus) is calculated separately.

The Hardy-Weinberg Law in X-Linked Disease

Recall from Chapter 7 that, for X-linked genes, there are three female genotypes but only two possible male genotypes. To illustrate gene frequencies and genotype frequencies when the gene of interest is X-linked, we use the trait known as X-linked **red-green color blindness,** which is caused by mutations in the series of visual pigment genes on the X chromosome. We use color blindness as an example because, as far as we know, it is not a deleterious trait (except for possible difficulties with traffic lights), and color blind persons are not subject to selection. As discussed later, allowing for the effect of selection complicates estimates of gene frequencies.

In this example, we use the symbol cb for all the mutant color blindness alleles and the symbol + for the wild-type allele, with frequencies q and p, respectively (Table 9-3). The frequencies of the two alleles can be determined directly from the incidence of the corresponding phenotypes in *males* by simply counting the alleles. Because females have two X chromosomes, their genotypes are distributed like autosomal genotypes, but because color blindness alleles are recessive, the normal homozygotes and heterozygotes are typically not distinguishable. As shown in Table 9-3, the frequency of color blindness in females is much lower than that in males.

Less than 1% of females are color blind, but nearly 15% are carriers of a mutant color blindness allele and have a 50% chance of having a color blind son with each male pregnancy.

FACTORS THAT DISTURB HARDY-WEINBERG EQUILIBRIUM

Underlying the Hardy-Weinberg law and its use are a number of assumptions (see Box, earlier), not all of which can be met (or reasonably inferred to be met) by all populations. The first is that the population under study is large and that mating is random. However, a very small population in which random events can radically alter an allele frequency may not meet this first assumption. This first assumption is also breached when the population contains subgroups whose members choose to marry within their own subgroup rather than the population at large. The second assumption is that allele frequencies do not change significantly over time. This requires that there is no migration in or out of the population by groups whose allele frequencies at a locus of interest are radically different from the allele frequencies in the population as a whole. Similarly, selection for or against particular alleles, or the addition of new alleles to the gene pool due to mutations, will break the assumptions of the Hardy-Weinberg law.

In practice, some of these violations are more damaging than others to the application of the law to human populations. As shown in the sections that follow, violating the assumption of random mating can cause large deviations from the frequency of individuals homozygous for an autosomal recessive condition that we might expect from population allele frequencies. On the other hand, changes in allele frequency due to mutation, selection, or migration usually cause more minor and subtle deviations from Hardy-Weinberg equilibrium. Finally, when Hardy-Weinberg equilibrium does not hold for a particular disease allele at a particular locus, it may be instructive to investigate *why* the allele and its associated genotypes are not in equilibrium because this may provide clues about the pathogenesis of the condition or point to historical events that have affected the frequency of alleles in different population groups over time.

TABLE 9-3 X-Linked Genes and Genotype Frequencies (Color Blindness)

Sex	Genotype	Phenotype	Incidence (Approximate)
Male	X^+	Normal color vision	$p = 0.92$
	X^{cb}	Color blind	$q = 0.08$
Female	X^+/X^+	Normal (homozygote)	$p^2 = (0.92)^2 = 0.8464$
	X^+/X^{cb}	Normal (heterozygote)	$2pq = 2(0.92)(0.08) = 0.1472$
		Normal (combined)	$p^2 + 2pq = 0.9936$
	X^{cb}/X^{cb}	Color blind	$q^2 = (0.08)^2 = 0.0064$

Exceptions to Large Populations with Random Mating

As introduced earlier, the principle of random mating is that for any locus, an individual of a given genotype has a purely random probability of mating with an individual of any other genotype, the proportions being determined only by the relative frequencies of the different genotypes in the population. One's choice of mate, however, may not be at random. In human populations, nonrandom mating may occur because of three distinct but related phenomena: **stratification**, **assortative mating**, and **consanguinity**.

Stratification

Stratification describes a population in which there are a number of subgroups that have—for a variety of historical, cultural, or religious reasons—remained relatively genetically separate during modern times. Worldwide, there are numerous stratified populations; for example, the United States population is stratified into many subgroups, including whites of northern or southern European ancestry, African Americans, and numerous Native American, Asian, and Hispanic groups. Similarly stratified populations exist in other parts of the world as well, either currently or in the recent past, such as Sunni and Shia Muslims, Orthodox Jews, French-speaking Canadians, or different castes in India. When mate selection in a population is restricted for any reason to members of one particular subgroup, and that subgroup happens to have a variant allele with a higher frequency than in the population as a whole, the result will be an apparent excess of homozygotes in the overall population beyond what one would predict from allele frequencies in the population as a whole if there were truly random mating.

To illustrate this point, suppose a population contains a minority group, constituting 10% of the population, in which a mutant allele for an autosomal recessive disease has a frequency $q_{min} = 0.05$ and the wild-type allele has frequency $p_{min} = 0.95$. In the remaining majority 90% of the population, the mutant allele is nearly absent (i.e., q_{maj} is ≈ 0 and $p_{maj} = 1$). An example of just such a situation is the African American population of the United States and the mutant allele at the β-globin locus responsible for **sickle cell disease** (Case 42). The overall frequency of the disease allele in the total population, q_{pop}, is therefore equal to $0.1 \times 0.05 = 0.005$, and, simply applying the Hardy-Weinberg law, the frequency of the disease in the population as a whole would be predicted to be $q^2_{pop} = (0.005)^2 = 2.5 \times 10^{-5}$ if mating were perfectly random throughout the entire population. If, however, individuals belonging to the minority group were to mate exclusively with other members of that same minority group (an extreme situation that does not apply in reality), then the frequency of affected individuals in the minority group would be

$(q^2_{min}) = (0.05)^2 = 0.0025$. Because the minority group is one tenth of the entire population, the frequency of disease in the total population is $0.0025/10 = 2.5 \times 10^{-4}$, or 10-fold higher than the calculated $q^2_{pop} = 2.5 \times 10^{-5}$ obtained by naively applying the Hardy-Weinberg law to the population as a whole without consideration of stratification.

By way of comparison, stratification has no effect on the frequency of autosomal dominant disease and would have only a minor effect on the frequency of X-linked disease by increasing the small number of females homozygous for the mutant allele.

Assortative Mating

Assortative mating is the choice of a mate because the mate possesses some particular trait. Assortative mating is usually positive; that is, people tend to choose mates who resemble themselves (e.g., in native language, intelligence, stature, skin color, musical talent, or athletic ability). To the extent that the characteristic shared by the partners is genetically determined, the overall genetic effect of positive assortative mating is an increase in the proportion of the homozygous genotypes at the expense of the heterozygous genotype.

A clinically important aspect of assortative mating is the tendency to choose partners with similar medical problems, such as congenital deafness or blindness or exceptionally short stature. In such a case, the expectations of Hardy-Weinberg equilibrium do not apply because the genotype of the mate at the disease locus is not determined by the allele frequencies found in the general population. For example, consider **achondroplasia** (Case 2), an autosomal dominant form of skeletal dysplasia with a population incidence of 1 per 15,000 to 1 per 40,000 live births. Offspring homozygous for the achondroplasia mutation have a severe, lethal form of skeletal dysplasia that is almost never seen unless both parents have achondroplasia and are thus heterozygous for the mutation. This would be highly unlikely to occur by chance, except for assortative mating among those with achondroplasia.

When mates have autosomal recessive disorders caused by the same mutation or by allelic mutations in the same gene, all of their offspring will also have the disease. Importantly, however, not all cases of blindness, deafness, or short stature have the same genetic basis; many families have been described, for example, in which two parents with albinism have had children with normal pigmentation, or two deaf parents have had hearing children, because of locus heterogeneity (discussed in Chapter 7). Even if there is locus heterogeneity with assortative mating, however, the chance that two individuals are carrying mutations in the same disease locus is increased over what it would be under true random mating, and therefore the risk for the disorder in their offspring is also increased. Although the long-term population effect of this kind of positive assortative mating on disease gene frequencies is insignificant, a specific family may find itself at very high genetic risk that would not be predicted from strict application of the Hardy-Weinberg law.

Consanguinity and Inbreeding

Consanguinity, like stratification and positive assortative mating, brings about an increase in the frequency of autosomal recessive disease by increasing the frequency with which carriers of an autosomal recessive disorder mate. Unlike the disorders in stratified populations, in which each subgroup is likely to have a high frequency of a few alleles, the kinds of recessive disorders seen in the offspring of related parents may be very rare and unusual in the population as a whole because consanguineous mating allows an uncommon allele inherited from a heterozygous common ancestor to become homozygous. A similar phenomenon is seen in **genetic isolates,** small populations derived from a limited number of common ancestors who tended to mate only among themselves. Mating between two apparently "unrelated" individuals in a genetic isolate may have the same risk for certain recessive conditions as that observed in consanguineous marriages because the individuals are both carriers by inheritance from common ancestors of the isolate, a phenomenon known as **inbreeding**.

For example, among Ashkenazi Jews in North America, mutant alleles for **Tay-Sachs disease** (GM$_2$ gangliosidosis) (Case 43), discussed in detail in Chapter 12, are relatively more common than in other ethnic groups. The frequency of Tay-Sachs disease is 100 times higher in Ashkenazi Jews (1 per 3600) than in most other populations (1 per 360,000). Thus the Tay-Sachs carrier frequency among Ashkenazi Jews is approximately 1 in 30 ($q^2 = 1/3600$, $q = 1/60$, $2pq = \approx 1/30$) as compared to a carrier frequency of approximately 1 in 300 in non-Ashkenazi individuals.

Exceptions to Constant Allele Frequencies
Effect of Mutation

We have shown that nonrandom mating can substantially upset the relative frequency of various genotypes predicted by the Hardy-Weinberg law, even within the time of a single generation. In contrast, changes in allele frequency due to selection or mutation usually occur slowly, in small increments, and cause much less deviation from Hardy-Weinberg equilibrium, at least for recessive diseases.

The rates of new mutations (see Chapter 4) are generally well below the frequency of heterozygotes for autosomal recessive diseases; the addition of new mutant alleles to the gene pool thus has little effect in the short term on allele frequencies for such diseases. In addition, most deleterious recessive alleles are hidden in asymptomatic heterozygotes and thus are not subject to

selection. As a consequence, selection is not likely to have major short-term effects on the allele frequency of these recessive alleles. Therefore, to a first approximation, Hardy-Weinberg equilibrium may apply even for alleles that cause severe autosomal recessive disease.

Importantly, however, for dominant or X-linked disease, mutation and selection *do* perturb allele frequencies from what would be expected under Hardy-Weinberg equilibrium, by substantially reducing or increasing certain genotypes.

Selection and Fitness

The molecular and genomic basis for mutation was considered earlier in Chapter 4. Here we examine the concept of **fitness**, the chief factor that determines whether a mutation is eliminated immediately, becomes stable in the population, or even becomes, over time, the predominant allele at the locus concerned. The frequency of an allele in a population at any given time represents a balance between the rate at which mutant alleles appear through mutation and the effects of selection. If either the mutation rate or the effectiveness of selection is altered, the allele frequency is expected to change.

Whether an allele is transmitted to the succeeding generation depends on its fitness, *f,* which is a measure of the number of offspring of affected persons who survive to reproductive age, compared with an appropriate control group. If a mutant allele is just as likely as the normal allele to be represented in the next generation, *f* equals 1. If an allele causes death or sterility, selection acts against it completely, and *f* equals 0. Values between 0 and 1 indicate transmission of the mutation, but at a rate that is less than that of individuals who do not carry the mutant allele.

A related parameter is the **coefficient of selection, *s*,** which is a measure of the *loss* of fitness and is defined as 1 − *f,* that is, the proportion of mutant alleles that are *not* passed on and are therefore lost as a result of selection. In the genetic sense, a mutation that prevents reproduction by an adult is just as "lethal" as one that causes a very early miscarriage of an embryo, because in neither case is the mutation transmitted to the next generation. Fitness is thus the outcome of the joint effects of survival and fertility. When a genetic disorder limits reproduction so severely that the fitness is zero (i.e., *s* = 1), it is thus referred to as a **genetic lethal**. In the biological sense, fitness has no connotation of superior endowment except in a single respect: comparative ability to contribute alleles to the next generation.

Selection in Recessive Disease. Selection against harmful recessive mutations has far less effect on the population frequency of the mutant allele than does selection against dominant mutations because only a small proportion of the genes are present in homozygotes and are therefore exposed to selective forces. Even if there were complete selection against homozygotes ($f = 0$), as in many lethal autosomal recessive conditions, it would take many generations to reduce the gene frequency appreciably because most of the mutant alleles are carried by heterozygotes with normal fitness. For example, as we saw previously in this chapter, the frequency of mutant alleles causing Tay-Sachs disease, q, can be as high as 1.5% in Ashkenazi Jewish populations. Given this value of q, we can estimate that approximately 3% of such populations ($2 \times p \times q$) are heterozygous and carry one mutant allele, whereas only 1 individual per 3600 (q^2) is a homozygote with two mutant alleles. The proportion of all mutant alleles found in homozygotes in such a population is thus given by:

$$\frac{2 \times 0.00028}{(2 \times 0.00028) + (1 \times 0.03)} = \approx 0.0183$$

Thus, less than 2% of all the mutant alleles in the population are in affected homozygotes and would therefore be exposed to selection in the absence of effective treatment.

Reduction or removal of selection against an autosomal recessive disorder by successful medical treatment (e.g., as in the case of PKU [see Chapter 12]) would have just as slow an effect on increasing the gene frequency over many generations. Thus *as long as mating is random, genotypes in autosomal recessive diseases can be considered to be in Hardy-Weinberg equilibrium, despite selection against homozygotes for the recessive allele.* Thus the mathematical relationship between genotype and allele frequencies described in the Hardy-Weinberg law holds for most practical purposes in recessive disease.

Selection in Dominant Disorders.

In contrast to recessive mutant alleles, dominant mutant alleles are exposed *directly* to selection. Consequently, the effects of selection and mutation are more obvious and can be more readily measured for dominant traits. A genetic lethal dominant allele, if fully penetrant, will be exposed to selection in heterozygotes, thus removing all alleles responsible for the disorder in a single generation. Several human diseases are thought or known to be autosomal dominant traits with zero or near-zero fitness and thus always result from new rather than inherited autosomal dominant mutations (Table 9-4), a point of great significance for genetic counseling. In some, the genes and specific mutant alleles are known, and family studies show new mutations in the affected individuals that were not inherited from the parents. In other conditions, the genes are not known, but a paternal age effect (see Chapter 4) has been seen, suggesting (but not proving) that a de novo mutation in the paternal germline is a possible cause of the disorder. The implication

TABLE 9-4 Examples of Disorders Occurring as Sporadic Conditions due to New Mutations with Zero Fitness

Disorder	Description
Atelosteogenesis	Early lethal form of short-limbed skeletal dysplasia
Cornelia de Lange syndrome	Intellectual disability, micromelia, synophrys, and other abnormalities; can be caused by mutation in the *NIPBL* gene
Osteogenesis imperfecta, type II	Perinatal lethal type, with a defect in type I collagen (*COL1A1*, *COL1A2*) (see Chapter 12)
Thanatophoric dysplasia	Early lethal form of skeletal dysplasia due to de novo mutations in the *FGFR3* gene (see Fig. 7-6C)

for genetic counseling is that the parents of a child with an autosomal dominant but genetically lethal condition will typically have a very low risk for recurrence in subsequent pregnancies because the condition would generally require another independent mutation to recur. A caveat to keep in mind, however, is the possibility of germline mosaicism, as we saw in Chapter 7 (see Fig. 7-18).

Mutation and Selection Balance in Dominant Disease.

If a dominant disease is deleterious but not lethal, affected persons may reproduce but will nevertheless contribute fewer than the average number of offspring to the next generation; that is, their fitness, f, will be reduced. Such a mutation will be lost through selection at a rate proportional to the reduced fitness of heterozygotes. The frequency of the mutant alleles responsible for the disease in the population therefore represents a balance between loss of mutant alleles through the effects of selection and gain of mutant alleles through recurrent mutation. A stable allele frequency will be reached at whatever level balances the two opposing forces: one (selection) that removes mutant alleles from the gene pool and one (de novo mutation) that adds new ones back. The mutation rate per generation, μ, at a disease locus must be sufficient to account for that fraction of all the mutant alleles (allele frequency q) that are lost by selection from each generation. Thus,

$$\mu = sq$$

As an illustration of this relationship, in achondroplasia, the fitness of affected patients is not zero, but they have only approximately one fifth as many children as people of normal stature in the population. Thus their average fitness, f, is 0.20, and the coefficient of selection, s, is $1 - f$, or 0.80. In the subsequent generation, then, only 20% of current achondroplasia alleles are passed on from the current generation to the next. Because the frequency of achondroplasia appears stable from generation to generation, new mutations must be responsible for replacing the 80% of mutant genes in the population lost through selection.

If the fitness of affected persons suddenly improved (e.g., because of medical advances), the observed incidence of the disease in the population would be predicted to increase and reach a new equilibrium. **Retinoblastoma** (Case 39) and other dominant embryonic tumors with childhood onset are examples of conditions that now have a greatly improved prognosis, with a predicted consequence of increased disease frequency in the population. Allele frequency, mutation rate, and fitness are related; thus, if any two of these three characteristics are known, the third can be estimated.

Mutation and Selection Balance in X-Linked Recessive Mutations.

For those X-linked phenotypes of medical interest that are recessive, or nearly so, selection occurs in hemizygous males and not in heterozygous females, except for the small proportion of females who are manifesting heterozygotes with reduced fitness (see Chapter 7). In this brief discussion, however, we assume that heterozygous females have normal fitness.

Because males have one X chromosome and females two, the pool of X-linked alleles in the entire population's gene pool is partitioned at any given time, with one third of mutant alleles present in males and two thirds in females. As we saw in the case of autosomal dominant mutations, mutant alleles lost through selection must be replaced by recurrent new mutations to maintain the observed disease incidence. If the incidence of a serious X-linked disease is not changing and selection is operating against (and only against) hemizygous males, the mutation rate, μ, must equal the coefficient of selection, s (i.e., the proportion of mutant alleles that are *not* passed on), times q, the allele frequency, adjusted by a factor of 3 because *selection is operating only on the third of the mutant alleles in the population that are present in males at any time*. Thus,

$$\mu = sq/3$$

For an X-linked genetic lethal disease, $s = 1$, and one third of all copies of the mutant gene responsible are lost from each generation and must, in a stable equilibrium, be replaced by de novo mutations. Therefore, in such disorders, one third of all persons who have X-linked lethal disorders are predicted to carry a new mutation, and their genetically normal mothers have a low risk for having subsequent children with the same disorder (again, assuming the absence of germline mosaicism). The remaining two thirds of the mothers of individuals with an X-linked lethal disorder would be carriers, with a 50% risk for having another affected son. However, the prediction that two thirds of the mothers of individuals with an X-linked lethal disorder are carriers of a disease-causing mutation is based on the assumption that mutation rates in males and in females are equal. It can be shown that if the mutation rate in males is much greater than in females, then the chance of a new mutation in the egg is very low, and most of the mothers of affected children will be carriers, having inherited the mutation as a new mutation from their unaffected fathers and then passing it on to their affected children. The effect on genetic counseling of differences in the rate of disease-causing mutations in male and female gametes will be discussed in Chapter 16.

In less severe disorders, such as **hemophilia A** (Case 21), the proportion of affected individuals representing new mutations is less than one third (currently approximately 15%). Because the treatment of hemophilia is improving rapidly, the total frequency of mutant alleles can be expected to rise relatively rapidly and to reach a new equilibrium. Assuming (as seems reasonable) that the mutation rate at this locus stays the same over time, the *proportion* of hemophiliacs who result from a new mutation will decrease, but the overall *incidence* of the disease will increase. Such a change would have significant implications for genetic counseling for this disorder (see Chapter 16).

Genetic Drift

Chance events can have a much greater effect on allele frequencies in a small population than in a large one. For example, when a new mutation occurs in a small population, its frequency is represented by only one copy among all the copies of that gene in the population. Random effects of environment or other chance occurrences that are *independent* of the genotype (i.e., events that *occur for reasons unrelated to whether an individual is carrying the mutant allele*) can produce significant changes in the frequency of the disease allele when the population is small. Such chance occurrences disrupt Hardy-Weinberg equilibrium and cause the allele frequency to change from one generation to the next. This phenomenon, known as **genetic drift**, can explain how allele frequencies can change as a result of chance. During the next few generations, although the population size of the new group remains small, there may be considerable fluctuation in gene frequency until allele frequencies come to a new equilibrium as the population increases in size. In contrast to **gene flow** (see next section), in which allele frequencies change because of the mixing of previously distinct populations, the mechanism of genetic drift is simply chance operating on a small population.

Founder Effect.

One special form of genetic drift is referred to as **founder effect**. When a small subpopulation breaks off from a larger population, the gene frequencies in the small population may be different from those of the population from which it originated because the new group contains a small, random sample of the parent group and, by chance, may not have the same gene frequencies as the parent group. If one of the original founders of a new group just happens to carry a relatively rare allele, that allele will have a far higher frequency than it had in the larger group from which the new group was derived.

Migration and Gene Flow

Migration can change allele frequency by the process of **gene flow**, defined as the slow diffusion of genes across a barrier. Gene flow usually involves a large population and a gradual change in gene frequencies. The genes of migrant populations with their own characteristic allele frequencies are gradually merged into the gene pool of the population into which they have migrated, a process referred to as **genetic admixture**. The term *migration* is used here in the broad sense of crossing a reproductive barrier, which may be racial, ethnic, or cultural and not necessarily geographical and requiring physical movement from one region to another. Some examples of admixture reflect well-known and well-documented events in human history (e.g., the African diaspora from the 15th to the 19th century), whereas others can only be inferred from the genomic study of variation in ancient DNA samples (see Box).

Returning to the example of the 32-bp deletion allele of the *CCR5* cytokine receptor gene, ΔCCR5, the frequency of this allele has been studied in many populations all over the world. The frequency of the ΔCCR5 allele is highest, up to 18%, in parts of northwestern Europe and then declines along a gradient into eastern and southern Europe, falling to a few percent in the Middle East and the Indian subcontinent. The ΔCCR5 allele is virtually absent from Africa and the Far East. The best interpretation of the current geographical distribution of the ΔCCR5 allele is that the mutation originated in northern Europe and then underwent both positive selection and gene flow over long distances (Fig. 9-1).

ETHNIC DIFFERENCES IN THE FREQUENCY OF VARIOUS GENETIC DISEASES

The previous discussion of the Hardy-Weinberg law explained how, at equilibrium, genotype frequencies are determined by allele frequencies and remain stable from generation to generation, assuming the allele frequencies in a large, isolated, randomly mating population remain constant. However, there is a problem of interest to human geneticists that the Hardy-Weinberg law does

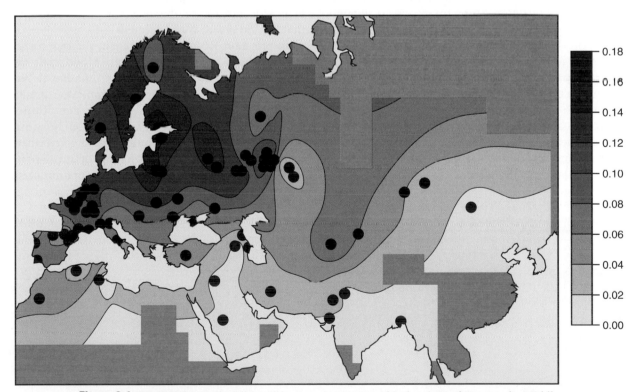

Figure 9-1 The frequency of ΔCCR5 alleles in various geographical regions of Europe, the Middle East, and the Indian subcontinent. The various allele frequencies are shown with color coding provided on the right. *Black dots* indicate the locations where allele frequencies were sampled; the rest of the frequencies were then interpolated in the regions between where direct sampling was done. *Gray areas* are regions where there were insufficient data to estimate allele frequencies. *See Sources & Acknowledgments.*

ANCIENT MIGRATIONS AND GENE FLOW

A fascinating example of gene flow during human prehistory comes from the sequencing of DNA samples obtained from the bones of three Neanderthals who died approximately 38,000 years ago in Europe. The most recent common ancestors of Neanderthals and *Homo sapiens* lived in Africa over 200,000 years ago, well before the migration of Neanderthals out of Africa to settle in Europe and the Middle East. An analysis of the sequence of Neanderthal DNA revealed that approximately 1% to 4% of the DNA of modern Europeans and Asians, but not of Africans, matches Neanderthal DNA. A variety of statistical techniques indicate that the introduction of Neanderthal DNA likely occurred approximately 50,000 years ago, well after the migration of modern humans out of Africa into Europe and beyond, which explains why traces of the Neanderthal genome are not present in modern Africans.

The analysis of individual Neanderthal genomes and their comparison to genomes of modern human populations promises to provide clues about characteristic differences between these groups, as well as about the frequency of possible disease genes or alleles that were more or less common in these ancient populations compared to different modern human populations.

not address: Why are allele frequencies different in different populations in the first place? In particular, for the medical geneticist, why are some mutant alleles that are clearly deleterious more common in certain population groups than in others? We address these issues in the rest of this chapter.

Differences in frequencies of alleles that cause genetic disease are of particular interest to the medical geneticist and genetic counselor because they cause different disease risks in specific population groups. Well-known examples include Tay-Sachs disease in people of Ashkenazi Jewish ancestry, sickle cell disease in African Americans, and hemolytic disease of the newborn and PKU in white populations (Table 9-5).

The Rh System

One clinically important example of marked differences in allele frequencies is seen with the Rh blood group. The Rh blood group is very important clinically because of its role in hemolytic disease of the newborn and in transfusion incompatibilities. In simplest terms, the population is separated into Rh-positive individuals, who express, on their red blood cells, the antigen Rh D, a polypeptide encoded by the *RHD* gene, and Rh-negative individuals, who do not express this antigen. Being Rh-negative is therefore inherited as an autosomal recessive trait in which the Rh-negative phenotype occurs in individuals homozygous or compound heterozygous for nonfunctional alleles of the *RHD* gene. The frequency of Rh-negative individuals varies enormously in different ethnic groups (see Table 9-5).

Hemolytic Disease of the Newborn Caused by Rh Incompatibility

The chief significance of the Rh system is that Rh-negative persons can readily form anti-Rh antibodies after exposure to Rh-positive red blood cells. Normally, during pregnancy, small amounts of fetal blood cross the placental barrier and reach the maternal bloodstream. If the mother is Rh-negative and the fetus Rh-positive, the mother will form antibodies that return to the fetal circulation and damage the fetal red blood cells, causing hemolytic disease of the newborn with consequences that can be severe if not treated.

In pregnant Rh-negative women, the risk for immunization by Rh-positive fetal red blood cells can be minimized with an injection of Rh immune globulin at 28 to 32 weeks of gestation and again after pregnancy. Rh immune globulin serves to clear any Rh-positive fetal cells from the mother's circulation before she is sensitized. Rh immune globulin is also given after miscarriage, termination of pregnancy, or invasive procedures such as chorionic villus sampling or amniocentesis,

TABLE 9-5 Incidence, Gene Frequency, and Heterozygote Frequency for Selected Autosomal Disorders in Different Populations

Disorder	Population	Incidence	Allele Frequency	Heterozygote Frequency
Recessive		q^2	q	$2pq$
Sickle cell anemia (*S/S* genotype)	U.S. African American	1 in 400	0.05	1 in 11
	Hispanic American	1 in 40,000	0.005	1 in 101
Rh (all Rh-negative alleles)	U.S. white	1 in 6	0.41	≈1 in 2
	U.S. African Americans	1 in 14	0.26	≈2 in 5
	Japanese	1 in 200	0.071	≈1 in 8
Phenylketonuria (all mutant alleles)	Scotland	1 in 5300	0.014	1 in 37
	Finland	1 in 200,000	0.002	1 in 250
	Japan	1 in 109,000	0.003	1 in 166
Dominant		$2pq + q^2$	q	
Familial hypercholesterolemia	Isolate in Quebec, Canada	1 in 122	0.004	—
	Afrikaner, South Africa	1 in 70	0.007	—
	U.S. population	1 in 500	0.001	—
Myotonic dystrophy	Isolate in Quebec, Canada	1 in 475	0.0011	—
	Europe	1 in 25,000	0.00002	—

in case any Rh-positive cells gained access to the mother's circulation. The discovery of the Rh system and its role in hemolytic disease of the newborn has been a major contribution of genetics to medicine. At one time ranking as the most common human genetic disease among individuals of European ancestry, hemolytic disease of the newborn caused by Rh incompatibility is now relatively rare, but only because obstetricians remain vigilant, identify at-risk patients, and routinely give them Rh immune globulin to prevent sensitization.

Ethnic Differences in Disease Frequencies

A number of factors discussed earlier in this chapter are thought to explain how differences in alleles and allele frequencies among ethnic groups develop. One is the lack of gene flow due to genetic isolation, so that a mutation in one group would not have an opportunity to be spread through matings to other groups. Other factors are **genetic drift**, including nonrandom distribution of alleles among the individuals who founded particular subpopulations (**founder effect**), and **heterozygote advantage** under environmental conditions that favor the reproductive fitness of carriers of deleterious mutations. Specific examples of these are illustrated in the next section. However, in many cases, we do not have a clear explanation for how these differences developed.

Founder Effect

One extreme example of a difference in the incidence of genetic disease among different ethnic groups is the high incidence of **Huntington disease** (Case 24) among the indigenous inhabitants around Lake Maracaibo, Venezuela, that resulted from the introduction of a Huntington disease mutation into this genetic isolate. There are numerous other examples of founder effect involving other disease alleles in genetic isolates throughout the world, such as the French-Canadian population of Canada, which has high frequencies of certain disorders that are rare elsewhere. For example, hereditary **type I tyrosinemia** is an autosomal recessive condition that causes hepatic failure and renal tubular dysfunction due to deficiency of fumarylacetoacetase, an enzyme in the degradative pathway of tyrosine. The disease frequency is 1 in 685 in the Saguenay–Lac-Saint-Jean region of Quebec, but only 1 in 100,000 in other populations. As predicted for a founder effect, 100% of the mutant alleles in the Saguenay–Lac-Saint-Jean patients are due to the same mutation.

Thus one of the outcomes of the founder effect and genetic drift is that each population may be characterized by its own particular mutant alleles, as well as by an increase or decrease in specific diseases. The relative mobility of most present-day populations, in comparison with their ancestors of only a few generations ago, may reduce the effect of genetic drift in the future while increasing the effect of gene flow.

Positive Selection for Heterozygotes (Heterozygote Advantage)

Although certain mutant alleles may be deleterious in homozygotes, there may be environmental conditions in which heterozygotes for some diseases have *increased* fitness not only over homozygotes for the mutant allele but even over homozygotes for the normal allele. This situation is termed **heterozygote advantage**. Even a slight heterozygote advantage can lead to an increase in frequency of an allele that is severely detrimental in homozygotes, because heterozygotes greatly outnumber homozygotes in the population. A situation in which selective forces operate both to maintain a deleterious allele and to remove it from the gene pool is described as a **balanced polymorphism**.

Malaria and Hemoglobinopathies. A well-known example of heterozygote advantage is resistance to malaria in heterozygotes for the mutation in **sickle cell disease** (Case 42). The sickle cell allele in the β-globin gene has reached its highest frequency in certain regions of West Africa, where heterozygotes are more fit than either type of homozygote because heterozygotes are relatively more resistant to the malarial organism. In regions where malaria is endemic, normal homozygotes are susceptible to malaria; many become infected and are severely, even fatally, affected, leading to reduced fitness. Sickle cell homozygotes are even more seriously disadvantaged, with a low relative fitness that approaches zero because of their severe hematological disease, discussed more fully in Chapter 11. Heterozygotes for sickle cell disease have red cells that are inhospitable to the malaria parasite but do not undergo sickling under normal environmental conditions; the heterozygotes are thus relatively more fit than homozygotes for the normal β-globin allele and reproduce at a higher rate. Thus, over time, the sickle cell mutant allele has reached a frequency as high as 0.15 in some areas of West Africa that are endemic for malaria, far higher than could be accounted for by recurrent mutation alone.

The heterozygote advantage in sickle cell disease demonstrates how violating one of the fundamental assumptions of Hardy-Weinberg equilibrium—that allele frequencies are not significantly altered by selection—causes the mathematical relationship between allele and genotype frequencies to diverge from what is expected under the Hardy-Weinberg law (see Box).

Change in selective pressures would be expected to lead to a rapid change in the relative frequency of the sickle cell allele. Today, in fact, major efforts are being made to eradicate the mosquito responsible for transmitting the disease in malarial areas; in addition, many sickle cell heterozygotes live in nonmalarial regions. There is evidence that in the African American population in the United States, the frequency of the sickle cell gene may already be falling from its high level in the

BALANCED SELECTION AND THE HARDY-WEINBERG LAW

Consider two alleles at the β-globin gene, the normal *A* allele and the mutant *S* allele, which give rise to three genotypes: *A/A* (normal), *A/S* (heterozygous carriers), and *S/S* (sickle cell disease). In a study of 12,387 individuals from an adult West African population, the three genotypes were detected in the proportions 9365 *A/A*:2993 *A/S*:29 *S/S*.

By counting the *A* and *S* alleles in these three genotypes, one can determine the allele frequencies in this population to be $p = 0.877$ for the *A* allele and $q = 0.123$ for the *S* allele. Under Hardy-Weinberg equilibrium, the ratio of genotypes is determined by $p^2:2pq:q^2$ and should therefore be 9527 *A/A*:2672 *A/S*:188 *S/S*. In this West African population, the observed number of *A/S* individuals exceeds what was predicted assuming Hardy-Weinberg equilibrium, whereas the observed number of *S/S* homozygotes is far below what was predicted, reflecting balanced selection at this locus. This example of the sickle cell allele illustrates how the forces of selection, operating both negatively on the relatively rare *S/S* genotype but also positively on the more common *A/S* genotype, cause a deviation from Hardy-Weinberg equilibrium in a population.

original African population of several generations ago, although other factors, such as the admixture of alleles from non-African populations into the African American gene pool, may also be playing a role. Some other deleterious alleles, including those responsible for **thalassemia** (Case 44), and **glucose-6-phosphate dehydrogenase deficiency** (Case 19), are also thought to be maintained at their present high frequencies in certain populations because of the protection that they provide against malaria.

Balanced Selection in Other Infectious Diseases

The effects of balanced selection in malaria are also apparent in other infectious diseases. For example, many Africans and African Americans with the severe renal disease known as **focal segmental glomerulosclerosis** are homozygotes for certain variant alleles in the coding region of the *APOL1* gene that encodes the apolipoprotein L1. Apolipoprotein L1 is a serum factor that kills the trypanosome parasite *Trypanosoma brucei* that causes trypanosomiasis (sleeping sickness). The same variants that increase the risk for severe kidney disease in homozygotes tenfold over the rest of the population protect heterozygotes carrying these variants against strains of trypanosomes (*T. brucei rhodesiense*) that have developed resistance to wild-type apolipoprotein L1. As a result, the frequency of heterozygous carriers for these variant alleles can be as high as approximately 45% in parts of Africa in which the rhodesiense trypanosomiasis is endemic.

Drift Versus Heterozygote Advantage

Determining whether drift or heterozygote advantage is responsible for the increased frequency of some deleterious alleles in certain populations is not always straightforward, because it involves the integration of modern genetic data and public health with the historical record of population movement and ancient diseases. The environmental selective pressure responsible for heterozygote advantage may have been operating in the past and not be identifiable in modern times. As seen in Figure 9-1, the northwest to southeast gradient in the frequency of the Δ*CCR5* allele reflects major differences in the frequency of this allele in different ethnic groups. For example, the highest frequency of the Δ*CCR5* allele, seen among Ashkenazi Jews, is 0.21, and it is nearly that high in Iceland and the British Isles. The moderate variation in allele frequencies across Europe is most consistent with genetic drift acting on a neutral polymorphism.

However, the overall elevation of allele frequencies in Europe (relative to non-European populations) is more suggestive of positive selection in response to some infectious agent. Although the current AIDS pandemic is too recent to have affected gene frequencies through selection, it is possible that a different selective factor (perhaps another infectious disease such as smallpox or bubonic plague) may have elevated the frequency of the Δ*CCR5* allele in northern European populations during a period of intense selection many generations ago. Thus geneticists continue to debate whether genetic drift or heterozygote advantage (or both) adequately accounts for the unusually high frequencies that some deleterious alleles achieve in some populations.

GENETICS AND ANCESTRY

Ancestry Informative Markers

Although the approximately 20,000 coding genes and their location and order on the chromosomes are nearly identical in all humans, we saw in Chapter 4 that humans as a whole have tens of millions of different alleles, ranging from changes in single base pairs (SNPs) to large genomic variants (CNVs or indels) hundreds of kilobases in size, that underlie extensive polymorphism among individuals. Many of the alleles found in one population are found in all human populations, at similar frequencies around the globe.

However, following a period of explosive population growth, today's human species of more than 7 billion members are derived from much smaller subpopulations, which, until quite recently, existed as separate subpopulations or ethnic groups, with different geographical origins and population histories that resulted in restricted mating between the subgroups. Different alleles arose from random mutation in humans who lived in these small isolated settlements; most of these would be expected to confer no selective advantage or disadvantage and are therefore selectively neutral. For the population geneticist and anthropologist, selectively neutral genetic markers provide a means of tracing human history. The interactions of genetic drift, selection due to environmental factors, and gene flow brought about by migration and intermarriage have different effects at loci around the genome: they may equalize allele frequencies throughout many subpopulations, may cause major differences in frequency between populations, or may cause certain alleles to be restricted to just one population.

Alleles that show large differences in allele frequency among populations originating in different parts of the world are referred to as **ancestry informative markers (AIMs).** Sets of AIMs have been identified whose frequencies differ among populations derived from widely separated geographical origins (e.g., European, African, Far East Asian, Middle Eastern, Native American, and Pacific Islanders). They are therefore useful as markers for charting human migration patterns, for documenting historical admixture between or among populations, and for determining the degree of genetic diversity among identifiable population subgroups. Studies of hundreds of thousands of AIMs from across the genome have been used to distinguish and determine the genome-wide relationships among many different population groups, including communities of Jews in Europe, Africa, Asia, and the Americas; dozens of distinct Native American populations from South America, North America, and Siberia; and many castes and tribal groups in India. Figure 9-2 illustrates this type of analysis to establish that Hispanics as a group are genetically very heterogeneous, with ancestors from many parts of the world.

Although there are millions of variants with different allele frequencies that can distinguish different population groups, genotyping as few as just a few hundred or a thousand SNPs in an individual is sufficient to identify the likely proportion of his or her genome contributed by ancestors from these different continental populations and to infer, therefore, the likely geographical origin(s) of that individual's ancestors. For example, Figure 9-3 shows the results from several hundred individuals from Puerto Rico, whose individual genomes can be shown to consist of various proportions of

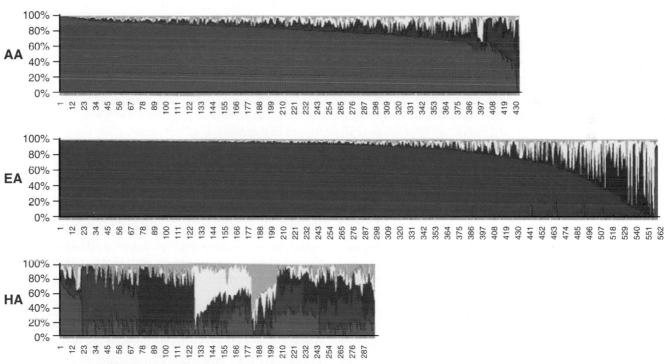

Figure 9-2 Mixed ancestry of a group of Americans who self-identify as African American (AA), European American (EA), and Hispanic American (HA) using ancestry informative markers. Each vertical line represents one individual ((totaling hundreds, as shown by the numbers), and subjects are displayed according to the predominant ancestry contribution to their genomes. Different colors indicate origin from a different geographical origin, as inferred from AIMs, as follows: Africa (*blue*), Europe (*red*), Middle East (*purple*), Central Asia (*yellow*), Far East Asia (*cyan*), Oceania (*amber*), and America (*green*). Most African Americans have genomes of predominantly African origin (*blue*), and most European Americans have genomes of predominantly European origin (*red*), although there is a range of ancestry contribution among different subjects. In contrast, Hispanic Americans are a more heterogeneous group, and most individuals have genomes with significant contributions from four or five different origins. *See Sources & Acknowledgments.*

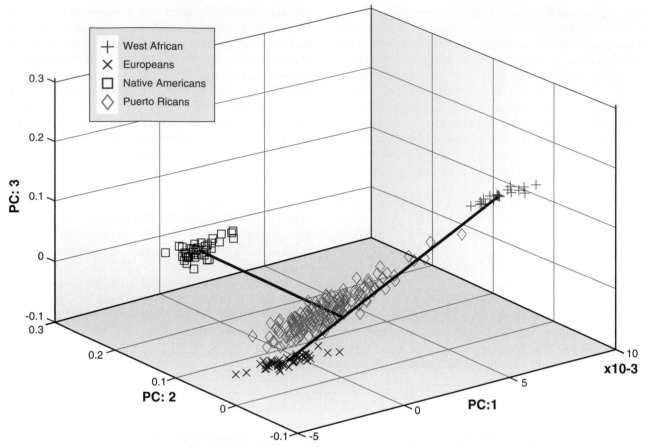

Figure 9-3 Ancestry contributions in an admixed Puerto Rican population. Three-dimensional display of the similarity of genomes from 192 Puerto Ricans to West African, European, and Native American control genomes, using a statistical measure known as principal components (PC) analysis. The PCs shown on the three axes correspond to groups of ancestry informative markers that distinguish the populations in question. The West African, European, and Native American genomes cluster in three distinct locations by PC analysis. The analysis demonstrates that the Puerto Rican genomes are heterogeneous; some individuals have genomes of predominantly European origin, and others have a much greater contribution from West Africa, whereas there is much less contribution from the Americas. *See Sources & Acknowledgments.*

African and European heritage, with much less Native American genetic heritage. Dozens of companies now offer ancestry testing services to consumers. Although there is disagreement about the scientific, medical, or anthropological value of the information for most individuals, the availability of ancestry testing has attracted widespread attention from those with interests in their family histories or diasporic heritage.

Population Genetics and Race

Population genetics uses quantitative methods to explain why and how differences in the frequency of genetic disease and the alleles responsible for them arose among different individuals and ethnic groups. What population genetics does *not* do, however, is provide a biological foundation for the concept of "race."

In one sense, racial distinctions are both "real" and widely used (and misused); they are social constructs that can have a profound impact on the health of

individuals experiencing racial categorization in their day-to-day lives. Physicians must pay attention to the social milieu their patients navigate, and the impact of racial categorization on the health and well-being of their patients must be taken into account if physicians are to understand and respond to patient needs (see Box).

From the scientific point of view, however, race is a fiction. Racial categories are constructed using poorly defined criteria that subdivide humankind using physical appearance (i.e., skin color, hair texture, and facial structures) combined with social characteristics that have their origins in the geographical, historical, cultural, religious, and linguistic backgrounds of the community in which an individual was born and raised. Although some of these distinguishing characteristics have a basis in the differences in the alleles carried by individuals of different ancestry, others likely have little or no basis in genetics. Racial categorization has been widely used in the past in medicine as the basis for making a number

ANCESTRY AND HEALTH

The significance of genetic ancestry for the practice of medicine reflects the role that allelic variants with different frequencies in different populations have on various clinically relevant functions. Although this area of study is still in its early stages, it is already clear that including assessment of genetic ancestry can provide useful information to improve prognostic predictions compared to those that depend only on self-declared racial or ethnic identity.

For example, when examined with panels of AIMs, the genomes of individuals who self-identify as African American contain DNA that ranges from less than 10% to more than 95% of African origin. For genetically determined traits influenced by ancestry, the effect of a particular single nucleotide polymorphism(s) on gene function will depend then on whether the responsible allele(s) is of African or European origin, a determination of genetic origin that is distinct from one's self-identification as an African American.

As an illustration of this point, a study of lung function to classify disease severity in a group of African American asthmatics showed that predictions of the overall degree of lung impairment in these patients were more accurate when genetic ancestry was considered, rather than relying solely on self-reported race. Disease classification (i.e., lung function in the "normal" range or not) could be misclassified in up to 5% of patients when ancestry information was omitted.

In addition to the potential for clinical management, AIMs testing can also be useful in research for identifying the particular genes and variants that are responsible for genetic diseases and other complex traits that differ markedly in incidence between different ethnic or geographical groups. Successful examples of this powerful approach are described in Chapter 10.

of assumptions concerning an individual's genetic makeup. Knowing the frequencies of alleles of relevance to health and disease in different populations around the globe is valuable for alerting a physician to an increased likelihood for disease based on an individual's genetic ancestry. However, with the expansion of individualized genetic medicine, it is hoped that more and more of the variants that contribute to disease will be assessed directly rather than having ethnicity or "race" used as a surrogate for an accurate genotype.

GENERAL REFERENCES

Li CC: *First course in population genetics*, Pacific Grove, CA, 1975, Boxwood Press.
Nielsen R, Slatkin M: *An introduction to population genetics*, Sunderland, MA, 2013, Sinauer Associates, Inc.

REFERENCES FOR SPECIFIC TOPICS

Behar DM, Yunusbayev B, Metspalu M, et al: The genome-wide structure of the Jewish people, *Nature* 466:238–242, 2010.
Corona E, Chen R, Sikora M, et al: Analysis of the genetic basis of disease in the context of worldwide human relationships and migration, *PLoS Genet* 9:e1003447, 2013.
Kumar R, Seibold MA, Aldrich MC, et al: Genetic ancestry in lung-function predictions, *N Engl J Med* 363:321–330, 2010.
Reich D, Patterson N, Campbell D, et al: Reconstructing Native American population history, *Nature* 488:370–374, 2012.
Reich D, Thangaraj K, Patterson N, et al: Reconstructing Indian population history, *Nature* 461:489–494, 2009.
Royal CD, Novembre J, Fullerton SM, et al: Inferring genetic ancestry: opportunities, challenges and implications, *Am J Hum Genet* 86:661–673, 2010.
Sankararaman S, Mallick S, Dannemann M, et al: The genomic landscape of Neanderthal ancestry in present-day humans, *Nature* 507:354–357, 2014.

PROBLEMS

1. A short tandem repeat DNA polymorphism consists of five different alleles, each with a frequency of 0.20. What proportion of individuals would be expected to be heterozygous at this locus? What if the five alleles have frequency of 0.40, 0.30, 0.15, 0.10, and 0.05?

2. If the allele frequency for Rh-negative is 0.26 in a population, what fraction of first pregnancies would sensitize the mother (assume Hardy-Weinberg equilibrium)? If no prophylaxis were given, what fraction of second pregnancies would be at risk for hemolytic disease of the newborn due to Rh incompatibility?

3. In a population at equilibrium, three genotypes are present in the following proportions: A/A, 0.81; A/a, 0.18; a/a, 0.01.
   a. What are the frequencies of A and a?
   b. What will their frequencies be in the next generation?
   c. What proportion of all matings in this population are $A/a \times A/a$?

4. In a screening program to detect carriers of β-thalassemia in an Italian population, the carrier frequency was found to be approximately 4%. Calculate:
   a. The frequency of the β-thalassemia allele (assuming that there is only one common β-thalassemia mutation in this population)
   b. The proportion of matings in this population that could produce an affected child
   c. The incidence of affected fetuses or newborns in this population
   d. The incidence of β-thalassemia among the offspring of couples both found to be heterozygous

5. Which of the following populations is in Hardy-Weinberg equilibrium?
   a. A/A, 0.70; A/a, 0.21; a/a, 0.09.
   b. For the MN blood group polymorphism, with two codominant alleles, M and N: (i) M, 0.33; MN, 0.34; N, 0.33. (ii) 100% MN.
   c. A/A, 0.32; A/a, 0.64; a/a, 0.04.
   d. A/A, 0.64; A/a, 0.32; a/a, 0.04.

What explanations could you offer to explain the frequencies in those populations that are not in equilibrium?

6. You are consulted by a couple, Abby and Andrew, who tell you that Abby's sister Anna has Hurler syndrome (a mucopolysaccharidosis) and that they are concerned that they themselves might have a child with the same disorder. Hurler syndrome is an autosomal recessive condition with a population incidence of approximately 1 in 90,000 in your community.

   a. If Abby and Andrew are not consanguineous, what is the risk that Abby and Andrew's first child will have Hurler syndrome?

   b. If they are first cousins, what is the risk?

   c. How would your answers to these questions differ if the disease in question were cystic fibrosis instead of Hurler syndrome?

7. In a certain population, each of three serious neuromuscular disorders—autosomal dominant facioscapulohumeral muscular dystrophy, autosomal recessive Friedreich ataxia, and X-linked Duchenne muscular dystrophy—has a population frequency of approximately 1 in 25,000.

   a. What are the gene frequency and the heterozygote frequency for each of these?

   b. Suppose that each one could be treated, so that selection against it is substantially reduced and affected individuals can have children. What would be the effect on the gene frequencies in each case? Why?

8. As discussed in this chapter, the autosomal recessive condition tyrosinemia type I has an observed incidence of 1 in 685 individuals in one population in the province of Quebec, but an incidence of approximately 1 in 100,000 elsewhere. What is the frequency of the mutant tyrosinemia allele in these two groups? Suggest two possible explanations for the difference in allele frequencies between the population in Quebec and populations elsewhere.

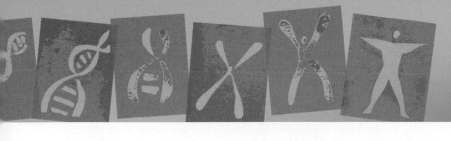

Identifying the Genetic Basis for Human Disease

This chapter provides an overview of how geneticists study families and populations to identify genetic contributions to disease. Whether a disease is inherited in a recognizable mendelian pattern, as illustrated in Chapter 7, or just occurs at a higher frequency in relatives of affected individuals, as explored in Chapter 8, it is the different genetic and genomic variants carried by affected family members or affected individuals in the population that either cause disease directly or influence their susceptibility to disease. Genome research has provided geneticists with a catalogue of all known human genes, knowledge of their location and structure, and an ever-growing list of tens of millions of variants in DNA sequence found among individuals in different populations. As we saw in previous chapters, some of these variants are common, others are rare, and still others differ in frequency among different ethnic groups. Whereas some variants clearly have functional consequences, others are certainly neutral. For most, their significance for human health and disease is unknown.

In Chapter 4, we dealt with the effect of mutation, which alters one or more genes or loci to generate variant alleles and polymorphisms. And in Chapters 7 and 8, we examined the role of genetic factors in the pathogenesis of various mendelian or complex disorders. In this chapter, we discuss how geneticists go about discovering the particular genes implicated in disease and the variants they contain that underlie or contribute to human diseases, focusing on three approaches.

- The first approach, **linkage analysis**, is *family-based*. Linkage analysis takes explicit advantage of family pedigrees to follow the inheritance of a disease among family members and to test for consistent, repeated coinheritance of the disease with a *particular genomic region* or even with a *specific variant or variants*, whenever the disease is passed on in a family.
- The second approach, **association analysis**, is *population-based*. Association analysis does not depend explicitly on pedigrees but instead takes advantage of the entire history of a population to look for increased or decreased frequency of a *particular allele* or *set of alleles* in a sample of affected individuals taken from the population, compared with a control set of unaffected people from that same population. It is

particularly useful for complex diseases that do not show a mendelian inheritance pattern.
- The third approach involves direct **genome sequencing** of affected individuals and their parents and/or other individuals in the family or population. This approach is particularly useful for rare mendelian disorders in which linkage analysis is not possible because there are simply not enough such families to do linkage analysis or because the disorder is a genetic lethal that always results from new mutations and is never inherited. In these situations, sequencing the genome (or just the coding exons of every gene, the **exome**) of an affected individual and sifting through the resulting billions (or in the case of the exome, tens of millions) of bases of DNA has been successfully used to find the gene responsible for the disorder. This new approach takes advantage of recently developed technology that has reduced the cost of DNA sequencing a millionfold from what it was when the original reference genome was being prepared during the Human Genome Project.

Use of linkage, association, and sequencing to map and identify disease genes has had an enormous impact on our understanding of the pathogenesis and pathophysiology of many diseases. In time, knowledge of the genetic contributions to disease will also suggest new methods of prevention, management, and treatment.

GENETIC BASIS FOR LINKAGE ANALYSIS AND ASSOCIATION

A fundamental feature of human biology is that each generation reproduces by combining haploid gametes containing 23 chromosomes, resulting from independent assortment and recombination of homologous chromosomes (see Chapter 2). To understand fully the concepts underlying genetic linkage analysis and tests for association, it is necessary to review briefly the behavior of chromosomes and genes during meiosis as they are passed from one generation to the next. Some of this information repeats the classic material on gametogenesis presented in Chapter 2, illustrating it with new information that has become available as a result of the

Human Genome Project and its applications to the study of human variation.

Independent Assortment and Homologous Recombination in Meiosis

During meiosis I, homologous chromosomes line up in pairs along the meiotic spindle. The paternal and maternal homologues exchange homologous segments by crossing over and creating new chromosomes that are a "patchwork" consisting of alternating portions of the grandmother's chromosomes and the grandfather's chromosomes (see Fig. 2-15). In the family illustrated in Figure 10-1, examples of recombined chromosomes are shown in the offspring (generation II) of the couple in generation I. Also shown is that the individual in generation III inherits a maternal chromosome that contains segments derived from all four of his maternal grandparents' chromosomes. The creation of such patchwork chromosomes emphasizes the notion of human genetic individuality: each chromosome inherited by a child from a parent is never exactly the same as either of the two copies of that chromosome in the parent.

Although any two homologous chromosomes generally look identical under the microscope, they differ substantially at the DNA sequence level. As discussed in Chapter 4, these differences at the same position (locus) on a pair of homologous chromosomes are **alleles.** Alleles that are common (generally considered to

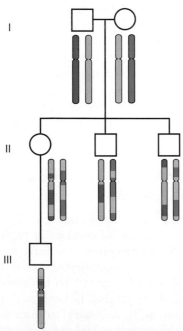

Figure 10-1 The effect of recombination on the origin of various portions of a chromosome. Because of crossing over in meiosis, the copy of the chromosome the boy (generation III) inherited from his mother is a mosaic of segments of all four of his grandparents' copies of that chromosome.

be those carried by approximately 2% or more of the population) constitute a **polymorphism,** and linkage analysis in families (as we will explore later in the chapter) requires following the inheritance of specific alleles as they are passed down in a family. Allelic variants on homologous chromosomes allow geneticists to trace each segment of a chromosome inherited by a particular child to determine if and where recombination events have occurred along the homologous chromosomes. Several tens of millions of genetic markers are available to serve as genetic markers for this purpose. It is a truism now in human genetics to say that it is essentially always possible to determine with confidence, through a series of analyses outlined in this chapter, whether a given allele or segment of the genome in a patient has been inherited from his or her father or mother. This advance—a singular product of the Human Genome Project—is an essential feature of genetic analysis to determine the precise genetic basis of disease.

Alleles at Loci on Different Chromosomes Assort Independently

Assume there are two polymorphic loci, 1 and 2, on different chromosomes, with alleles A and a at locus 1 and alleles B and b at locus 2 (Fig. 10-2). Suppose an individual's genotype at these loci is Aa and Bb; that is, she is heterozygous at both loci, with alleles A and B inherited from her father and alleles a and b inherited from her mother. The two different chromosomes will line up on the metaphase plate at meiosis I in one of two combinations with equal likelihood. After recombination and chromosomal segregation are complete, there will be four possible combinations of alleles, AB, ab, Ab, and aB, in a gamete; each combination is as likely to occur as any other, a phenomenon known as **independent assortment.** Because AB gametes contain only her paternally derived alleles, and ab gametes only her maternally derived alleles, these gametes are designated **parental.** In contrast, Ab or aB gametes, each containing one paternally derived allele and one maternally derived allele, are termed **nonparental** gametes. On average, half (50%) of gametes will be parental (AB or ab) and 50% nonparental (Ab or aB).

Alleles at Loci on the Same Chromosome Assort Independently If at Least One Crossover between Them Always Occurs

Now suppose that an individual is heterozygous at two loci 1 and 2, with alleles A and B paternally derived and a and b maternally derived, but the loci are on the *same* chromosome (Fig. 10-3). Genes that reside on the same chromosome are said to be **syntenic** (literally, "on the same thread"), regardless of how close together or how far apart they lie on that chromosome.

How will these alleles behave during meiosis? We know that between one and four crossovers occur between homologous chromosomes during meiosis I

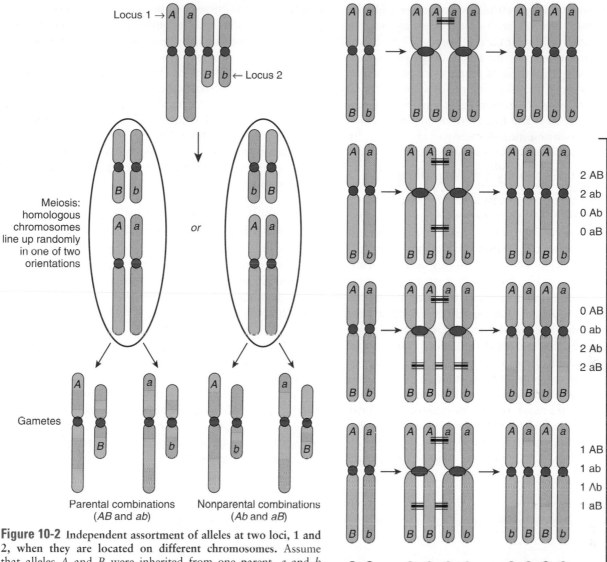

Figure 10-2 Independent assortment of alleles at two loci, 1 and 2, when they are located on **different chromosomes.** Assume that alleles *A* and *B* were inherited from one parent, *a* and *b* from the other. The two chromosomes can line up on the metaphase plate in meiosis I in one of two equally likely combinations, resulting in independent assortment of the alleles on these two chromosomes.

when there are two chromatids per homologous chromosome. If no crossing over occurs within the segment of the chromatids between the loci 1 and 2 (and ignoring whatever happens in segments outside the interval between these loci), then the chromosomes we see in the gametes will be *AB* and *ab*, which are the same as the original parental chromosomes; a parental chromosome is therefore a **nonrecombinant** chromosome. If crossing over occurs at least once in the segment between the loci, the resulting chromatids may be either nonrecombinant or *Ab* and *aB*, which are not the same as the parental chromosomes; such a nonparental chromosome is therefore a **recombinant** chromosome (shown in Fig. 10-3). One, two, or more recombinations occurring between two loci at the four-chromatid stage result in gametes that are 50% nonrecombinant (parental) and

Figure 10-3 Crossing over between homologous chromosomes (*black horizontal lines*) in meiosis is shown between chromatids of two homologous chromosomes on the *left*. Crossovers result in new combinations of maternally and paternally derived alleles on the recombinant chromosomes present in gametes, shown on the *right*. If no crossing over occurs in the interval between loci 1 and 2, only parental (nonrecombinant) allele combinations, *AB* and *ab*, occur in the offspring. If one or two crossovers occur in the interval between the loci, half the gametes will contain a nonrecombinant combination of alleles and half the recombinant combination. The same is true if more than two crossovers occur between the loci (not illustrated here). NR, Nonrecombinant; R, recombinant.

50% recombinant (nonparental), which is precisely the same proportions one sees with independent assortment of alleles at loci on different chromosomes. Thus, if two syntenic loci are sufficiently far apart on the same chromosome to ensure that there is going to be at least one crossover between them in every meiosis, the ratio of recombinant to nonrecombinant genotypes will be, on average, 1:1, just as if the loci were on separate chromosomes and assorting independently.

Recombination Frequency and Map Distance
Frequency of Recombination as a Measure of Distance between Loci

Suppose now that two loci are on the same chromosome but are either far apart, very close together, or somewhere in between (Fig. 10-4). As we just saw, when the loci are far apart (see Fig. 10-4A), at least one crossover will occur in the segment of the chromosome between loci 1 and 2, and there will be gametes of both the nonrecombinant genotypes *AB* and *ab* and recombinant genotypes *Ab* and *aB*, in equal proportions (on average) in the offspring. On the other hand, if two loci are so close together on the same chromosome that crossovers *never* occur between them, there will be no recombination; the nonrecombinant genotypes (parental chromosomes *AB* and *ab* in Fig. 10-4B) are transmitted together *all* of the time, and the frequency of the recombinant genotypes *Ab* and *aB* will be 0. In between these two

extremes is the situation in which two loci are far enough apart that one recombination between the loci occurs in some meioses but not in others (see Fig. 10-4C). In this situation, we observe nonrecombinant combinations of alleles in the offspring when no crossover occurred and recombinant combinations when a recombination has occurred, but the frequency of recombinant chromosomes at the two loci will fall between 0% and 50%. The crucial point is that *the closer together two loci are, the smaller the recombination frequency, and the fewer recombinant genotypes are seen in the offspring.*

Detecting Recombination Events Requires Heterozygosity and Knowledge of Phase

Detecting the recombination events between loci requires that (1) a parent be heterozygous (**informative**) at both loci and (2) we know which allele at locus 1 is on the same chromosome as which allele at locus 2. In an individual who is heterozygous at two syntenic loci, one with alleles *A* and *a*, the other *B* and *b*, which allele at the first locus is on the same chromosome with which allele at the second locus defines what is referred to as the **phase** (Fig. 10-5). The set of alleles on the same homologue (*A* and *B*, or *a* and *b*) are said to be in **coupling** (or *cis*) and form what is referred to as a **haplotype** (see Chapters 7 and 8). In contrast, alleles on the different homologues (*A* and *b*, or *a* and *B*) are in **repulsion** (or *trans*) (see Fig. 10-5).

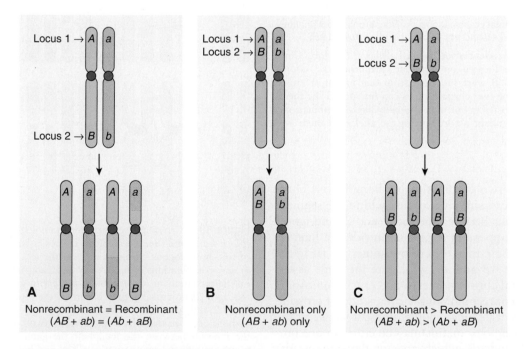

Figure 10-4 Assortment of alleles at two loci, 1 and 2, when they are located on the same chromosome. **A,** The loci are far apart and at least one crossover between them is likely to occur in every meiosis. **B,** The loci are so close together that crossing over between them is not observed, regardless of the presence of crossovers elsewhere on the chromosome. **C,** The loci are close together on the same chromosome but far enough apart that crossing over occurs in the interval between the two loci only in some meioses but not in most others.

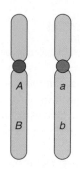

In coupling: *A* and *B*   *a* and *b*
In repulsion: *a* and *B*   *A* and *b*

Figure 10-5 Possible phases of alleles *A* and *a* and alleles *B* and *b*.

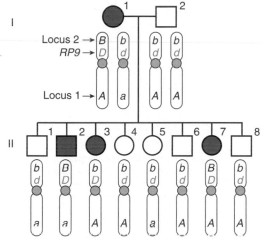

Figure 10-6 Coinheritance of the gene for an autosomal dominant form of retinitis pigmentosa (RP), with marker locus 2 and not with marker locus 1. Only the mother's contribution to the children's genotypes is shown. The mother (I-1) is affected with this dominant disease and is heterozygous at the *RP9* locus (*Dd*) as well as at loci 1 and 2. She carries the *A* and *B* alleles on the same chromosome as the mutant *RP9* allele (*D*). The unaffected father is homozygous normal (*dd*) at the *RP9* locus as well as at the two marker loci (*AA* and *bb*); his contributions to his offspring are not considered further. Two of the three affected offspring have inherited the *B* allele at locus 2 from their mother, whereas individual II-3 inherited the *b* allele. The five unaffected offspring have also inherited the *b* allele. Thus seven of eight offspring are nonrecombinant between the *RP9* locus and locus 2. However, individuals II-2, II-4, II-6, and II-8 are recombinant for *RP9* and locus 1, indicating that meiotic crossover has occurred between these two loci.

Figure 10-6 shows a pedigree of a family with multiple individuals affected by autosomal dominant **retinitis pigmentosa (RP)**, a degenerative disease of the retina that causes progressive blindness in association with abnormal retinal pigmentation. As shown, individual I-1 is heterozygous at both marker locus 1 (with alleles *A* and *a*) and marker locus 2 (with alleles *B* and *b*), as well as heterozygous for the disorder (*D* is the dominant disease allele, *d* is the recessive normal allele). The alleles

A-D-B form one haplotype, and *a-d-b* the other. Because we know her spouse is homozygous at all three loci and can only pass on the *a*, *b*, and *d* alleles, we can easily determine which alleles the children received from their mother and thus trace the inheritance of her RP-causing allele or her normal allele at that locus, as well as the alleles at both marker loci in her children. Close inspection of Figure 10-6 allows one to determine whether each child has inherited a recombinant or a nonrecombinant haplotype from the mother.

However, if the mother (I-1) had been homozygous *bb* at locus 2, then all children would inherit a maternal *b* allele, regardless of whether they received a mutant *D* or normal *d* allele at the *RP9* locus. Because she is not informative at locus 2 in this scenario, it would be impossible to determine whether recombination had occurred. Similarly, if the information provided for the family in Figure 10-6 was simply that individual I-1 was heterozygous, *Bb,* at locus 2 and heterozygous for an autosomal dominant form of RP, but the phase was not known, one could not determine which of her children were nonrecombinant between the *RP9* locus and locus 2 and which of her children were recombinant. Thus determination of who is or is not a recombinant requires that we know whether the *B* or *b* allele at locus 2 was on the same chromosome as the mutant *D* allele for RP in individual I-1 (see Fig. 10-6).

Linkage and Recombination Frequency

Linkage is the term used to describe a departure from the independent assortment of two loci, or, in other words, the tendency for alleles at loci that are close together on the same chromosome to be transmitted together, as an intact unit, through meiosis. Analysis of linkage depends on determining the frequency of recombination as a measure of how close two loci are to each other on a chromosome. A common notation for recombination frequency (as a proportion, not a percentage) is the Greek letter theta, θ, where θ varies from 0 (no recombination at all) to 0.5 (independent assortment). If two loci are so close together that $\theta = 0$ between them (as in Fig. 10-4B), they are said to be **completely linked**; if they are so far apart that $\theta = 0.5$ (as in Fig. 10-4A), they are assorting independently and are **unlinked**. In between these two extremes are various degrees of linkage.

Genetic Maps and Physical Maps

The **map distance** between two loci is a *theoretical* concept that is based on *actual* data—the extent of observed recombination, θ, between the loci. Map distance is measured in units called **centimorgans** (cM), defined as the genetic length over which, on average, one crossover occurs in 1% of meioses. (The centimorgan is $\frac{1}{100}$ of a "morgan," named after Thomas Hunt Morgan, who first observed genetic recombination in the fruit fly *Drosophila*.) Therefore a recombination fraction of 1% (i.e., θ = 0.01) translates approximately into a map distance of 1 cM. As we discussed before in this chapter, the recombination frequency between two loci increases proportionately with the distance between two loci only up to a point because, once markers are far enough apart that at least one recombination will always occur, the observed recombination frequency will equal 50% (θ = 0.5), no matter how far apart physically the two loci are.

To accurately measure true genetic map distance between two widely spaced loci, therefore, one has to use markers spaced at short genetic distances (1 cM or less) in the interval between these two loci, and then add up the values of θ between the intervening markers, because the values of θ between pairs of closely neighboring markers will be good approximations of the genetic distances between them. Using this approach, the genetic length of an entire human genome has been measured and, interestingly, found to differ between the sexes. When measured in female meiosis, genetic length of the human genome is approximately 60% greater (≈4596 cM) than when it is measured in male meiosis (2868 cM), and this sex difference is consistent and uniform across each autosome. The sex-averaged genetic length of the entire haploid human genome, which is estimated to contain approximately 3.3 billion base pairs of DNA, or ≈3300 Mb (see Chapter 2), is 3790 cM, for an average of approximately 1.15 cM/Mb. The reason for the observed increased recombination per unit length of DNA in females compared with males is unknown, although one might speculate that it has to do with the increased opportunity for crossing over afforded by the many years that female gamete precursors remain in meiosis I before ovulation (see Chapter 2).

Pairwise measurements of recombination between genetic markers separated by 1 Mb or more gives a fairly constant ratio of genetic distance to physical distance of approximately 1 cM/Mb. However, when recombination is measured at much higher resolution, such as between markers spaced less than 100 kb apart, recombination per unit length becomes nonuniform and can range over four orders of magnitude (0.01 to 100 cM/Mb). When viewed on the scale of a few tens of kilobase pairs of DNA, the apparent linear relationship between physical distance in base pairs and recombination between polymorphic markers located millions of base pairs of DNA apart is, in fact, the result of an averaging of so-called **hot spots of recombination** interspersed among regions of little or no recombination. Hot spots occupy only approximately 6% of sequence in the genome and yet account for approximately 60% of all the meiotic recombination in the human genome. The biological basis for these recombination hot spots is unknown. The impact of this nonuniformity of recombination at high resolution is discussed next, as we address the phenomenon of linkage disequilibrium.

Linkage Disequilibrium

It is generally the case that the alleles at two loci will not show any preferred phase in the population if the loci are linked but at a distance of 0.1 to 1 cM or more. For example, suppose loci 1 and 2 are 1 cM apart. Suppose further that allele A is present on 50% of the chromosomes in a population and allele a on the other 50% of chromosomes, whereas at locus 2, a disease susceptibility allele S is present on 10% of chromosomes and the protective allele s is on 90% (Fig. 10-7). Because the frequency of the A-S haplotype, freq(A-S), is simply

Linkage equilibrium: Haplotype frequencies are as expected from allele frequencies

		Allele frequencies at locus 2	
		freq(S) = 0.1	freq(s) = 0.9
Allele frequencies at locus 1	freq(A) = 0.5	Haplotype A-S freq(A-S) = 0.05	Haplotype A-s freq(A-s) = 0.45
	freq(a) = 0.5	Haplotype a-S freq(a-S) = 0.05	Haplotype a-s freq(a-s) = 0.45

A

Linkage disequilibrium: Haplotype frequencies diverge from what is expected from allele frequencies

		Allele frequencies at locus 2	
		freq(S) = 0.1	freq(s) = 0.9
Allele frequencies at locus 1	freq(A) = 0.5	Haplotype A-S freq(A-S) = 0	Haplotype A-s freq(A-s) = 0.5
	freq(a) = 0.5	Haplotype a-S freq(a-S) = 0.1	Haplotype a-s freq(a-s) = 0.4

B

Partial linkage disequilibrium: Haplotype frequencies are rarer than expected from allele frequencies

		Allele frequencies at locus 2	
		freq(S) = 0.1	freq(s) = 0.9
Allele frequencies at locus 1	freq(A) = 0.5	Haplotype A-S freq(A-S) = 0.01	Haplotype A-s freq(A-s) = 0.49
	freq(a) = 0.5	Haplotype a-S freq(a-S) = 0.09	Haplotype a-s freq(a-s) = 0.41

C

Figure 10-7 Tables demonstrating how the same allele frequencies can result in different haplotype frequencies indicative of linkage equilibrium, strong linkage disequilibrium, or partial linkage disequilibrium. **A,** Under **linkage equilibrium,** haplotype frequencies are as expected from the product of the relevant allele frequencies. **B,** Loci 1 and 2 are located very close to one another, and alleles at these loci show strong **linkage disequilibrium.** Haplotype A-S is absent and a-s is less frequent (0.4 instead of 0.45) compared to what is expected from allele frequencies. **C,** Alleles at loci 1 and 2 show **partial linkage disequilibrium.** Haplotypes, A-S and a-s are underrepresented compared to what is expected from allele frequencies. Note that the allele frequencies for A and a at locus 1 and for S and s at locus 2 are the same in all three tables; it is the way the alleles are distributed in haplotypes, shown in the central four cells of the table, that differ.

the product of the frequencies of the two alleles—freq(A) × freq(S) = 0.5 × 0.1 = 0.05, the alleles are said to be in **linkage equilibrium** (see Fig. 10-7A). That is, the frequencies of the four possible haplotypes, A-S, A-s, a-S, and a-s follow directly from the allele frequencies of A, a, S, and s.

However, as we examine haplotypes involving loci that are very close together, we find that knowing the allele frequencies for these loci individually does *not* allow us to predict the four haplotype frequencies. The frequency of any one of the haplotypes, freq(A-S) for example, may *not* be equal to the product of the frequencies of the individual alleles that make up that haplotype; in this situation, freq(A-S) ≠ freq(A) × freq(S), and the alleles are thus said to be in **linkage disequilibrium (LD)**. The deviation ("delta") between the expected and actual haplotype frequencies is called D and is given by:

$$D = freq(A\text{-}S) \times freq(a\text{-}s) - freq(A\text{-}s) \times freq(a\text{-}S)$$

D ≠ 0 is equivalent to saying the alleles are in LD, whereas D = 0 means the alleles are in linkage equilibrium.

Examples of LD are illustrated in Figures 10-7B and 10-7C. Suppose one discovers that *all* chromosomes carrying allele S also have allele a, whereas none has allele A (see Fig. 10-7B). Then allele S and allele a are said to be in complete LD. As a second example, suppose the A-S haplotype is present on only 1% of chromosomes in the population (see Fig. 10-7C). The A-S haplotype has a frequency much below what one would expect on the basis of the frequencies of alleles A and S in the population as a whole, and D < 0, whereas the haplotype a-S has a frequency much greater than expected and D > 0. In other words, chromosomes carrying the susceptibility allele S are enriched for allele a at the expense of allele A, compared with chromosomes that carry the protective allele s. Note, however, that the individual allele frequencies are unchanged; it is only how they are distributed into haplotypes that differs, and this is what determines if there is LD.

Linkage Disequilibrium Has Both Biological and Historical Causes

What causes LD? When a disease allele first enters the population (by mutation or by immigration of a founder who carries the disease allele), the particular set of alleles at polymorphic loci linked to (i.e., syntenic with) the disease locus constitutes a **disease-containing haplotype** in which the disease allele is located (Fig. 10-8). The degree to which this original disease-containing haplotype will persist over time depends in part on the

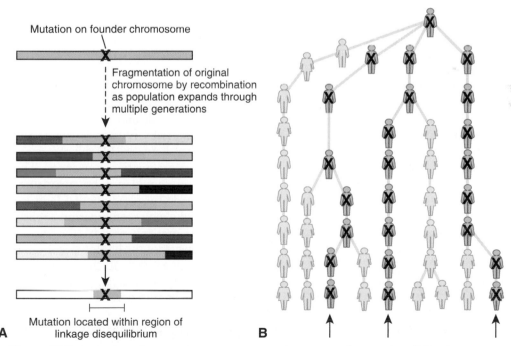

Figure 10-8 A, With each generation, meiotic recombination exchanges the alleles that were initially present at polymorphic loci on a chromosome on which a disease-associated mutation arose (▬) for other alleles present on the homologous chromosome. Over many generations, the only alleles that remain in coupling phase with the mutation are those at loci so close to the mutant locus that recombination between the loci is very rare. These alleles are in linkage disequilibrium with the mutation and constitute a disease-associated haplotype. **B,** Affected individuals in the current generation (*arrows*) carry the mutation (**X**) in linkage disequilibrium with the disease-associated haplotype (*individuals in blue*). Depending on the age of the mutation and other population genetic factors, a disease-associated haplotype ordinarily spans a region of DNA of a few kb to a few hundred kb. *See Sources & Acknowledgments.*

probability that recombination moves the disease allele *off* of the original haplotype and *onto* chromosomes with different sets of alleles at these linked loci. The speed with which recombination will move the disease allele onto a new haplotype depends on a number of factors:

- The number of generations (and therefore the number of opportunities for recombination) since the mutation first appeared.
- The frequency of recombination per generation between the loci. The smaller the value of θ, the greater is the chance that the disease-containing haplotype will persist intact.
- Processes of natural selection for or against particular haplotypes. If a haplotype combination undergoes either positive selection (and therefore is preferentially passed on) or experiences negative selection (and therefore is less readily passed on), it will be either overrepresented or underrepresented in that population.

Measuring Linkage Disequilibrium

Although conceptually valuable, the discrepancy, D, between the expected and observed frequencies of haplotypes is not a good way to quantify LD because it varies not only with degree of LD but also with the allele frequencies themselves. To quantify varying degrees of LD, therefore, geneticists often use a measure derived from D, referred to as D′ (see Box). D′ is designed to vary from 0, indicating linkage equilibrium, to a maximum of ±1, indicating very strong LD. Because LD is a result not only of genetic distance but also of the amount of time during which recombination had a chance to occur and the possible effects of selection for or against particular haplotypes, different populations living in different environments and with different histories can have different values of D′ between the same two alleles at the same loci in the genome.

$$D' = D/F$$

where $D = \text{freq}(A\text{-}S) \times \text{freq}(a\text{-}s) - \text{freq}(A\text{-}s) \times \text{freq}(a\text{-}S)$

and F is a correction factor that helps account for the allele frequencies.

The value of F depends on whether D itself is a positive or negative number.

F = the smaller of freq(*A*) × freq(*s*) or freq(*a*) × freq(*S*) if D > 0

F = the smaller of freq(*A*) × freq(*S*) or freq(*a*) × freq(*s*) if D < 0

Clusters of Alleles Form Blocks Defined by Linkage Disequilibrium

Analysis of pairwise measurements of D′ for neighboring variants, particularly single nucleotide polymorphism

(SNPs), across the genome reveals a complex genetic architecture for LD. Contiguous SNPs can be grouped into clusters of varying size in which the SNPs in any one cluster show high levels of LD with each other but not with SNPs outside that cluster (Fig. 10-9). For example, the nine polymorphic loci in cluster 1 (see Fig. 10-9A), each consisting of two alleles, have the potential to generate $2^9 = 512$ different haplotypes; yet, *only five haplotypes constitute 98% of all haplotypes seen.* The absolute values of |D′| between SNPs within the cluster are well above 0.8. Clusters of loci with alleles in high LD across segments of only a few kilobase pairs to a few dozen kilobase pairs are termed **LD blocks.**

The size of an LD block encompassing alleles at a particular set of polymorphic loci is not identical in all populations. African populations have smaller blocks, averaging 7.3 kb per block across the genome, compared with 16.3 kb in Europeans; Chinese and Japanese block sizes are comparable to each other and are intermediate, averaging 13.2 kb. This difference in block size is almost certainly the result of the smaller number of generations since the founding of the non-African populations compared with populations in Africa, thereby limiting the time in which there has been opportunity for recombination to break up regions of LD.

Is there a biological basis for LD blocks, or are they simply genetic phenomena reflecting human (and genome) history? It appears that biology does contribute to LD block structure in that the boundaries between LD blocks often coincide with meiotic recombination hot spots, discussed earlier (see Fig. 10-9C). Such recombination hot spots would break up any haplotypes spanning them into two shorter haplotypes more rapidly than average, resulting in linkage *equilibrium* between SNPs on one side and the other side of the hot spot. The correlation is by no means exact, and many apparent boundaries between LD blocks are not located over evident recombination hot spots. This lack of perfect correlation should not be surprising, given what we have already surmised about LD: it is affected not only by how likely a recombination event is (i.e., where the hot spots are) but also by the age of the population, the frequency of the haplotypes originally present in the founding members of that population, and whether there has been either positive or negative selection for particular haplotypes.

MAPPING HUMAN DISEASE GENES
Why Map Disease Genes?

In clinical medicine, a disease state is defined by a collection of phenotypic findings seen in a patient or group of patients. Designating such a disease as "genetic"—and thus inferring the existence of a gene responsible for or contributing to the disease—comes from detailed genetic analysis, applying the principles outlined in Chapters 7 and 8. However, surmising the existence of

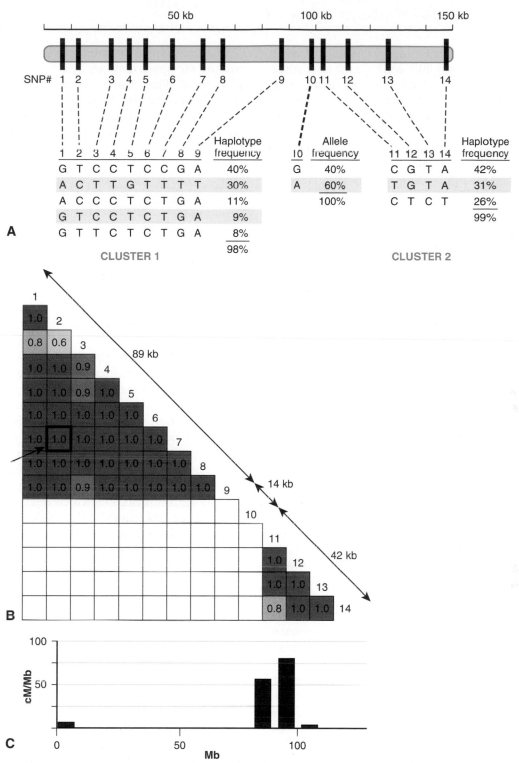

Figure 10-9 **A,** A 145-kb region of chromosome 4 containing 14 single nucleotide polymorphism (SNPs). In cluster 1, containing SNPs 1 through 9, five of the $2^9 = 512$ theoretically possible haplotypes are responsible for 98% of all the haplotypes in the population, reflecting substantial linkage disequilibrium (LD) among these SNP loci. Similarly, in cluster 2, only three of the $2^4 = 16$ theoretically possible haplotypes involving SNPs 11 to 14 represent 99% of all the haplotypes found. In contrast, alleles at SNP 10 are found in linkage equilibrium with the SNPs in cluster 1 and cluster 2. **B,** A schematic diagram in which each *red box* contains the pairwise measurement of the degree of LD between two SNPs (e.g., the *arrow* points to the box, *outlined in black*, containing the value of D' for SNPs 2 and 7). The higher the degree of LD, the darker the color in the box, with maximum D' values of 1.0 occurring when there is complete LD. Two LD blocks are detectable, the first containing SNPs 1 through 9, and the second SNPs 11 through 14. Between blocks, the 14-kb region containing SNP 10 shows no LD with neighboring SNPs 9 or 11 or with any of the other SNP loci. **C,** A graph of the ratio of map distance to physical distance (cM/Mb), showing that a recombination hot spot is present in the region between SNP 10 and cluster 2, with values of recombination that are fifty- to sixtyfold above the average of approximately 1.15 cM/Mb for the genome. *See Sources & Acknowledgments.*

a gene or genes in such a way does *not* tell us which of the perhaps 40,000 to 50,000 coding and noncoding genes in the genome is involved, what the function of that gene or genes might be, or how that gene or genes cause or contribute to the disease.

Disease gene mapping is often a critical first step in identifying the gene or genes in which variants are responsible for causing or increasing susceptibility to disease. Mapping the gene focuses attention on a region of the genome in which to carry out a systematic analysis of all the genes in that region to find the mutations or variants that contribute to the disease. Once the gene is identified that harbors the DNA variants responsible for either causing a mendelian disorder or increasing susceptibility to a genetically complex disease, the full spectrum of variation in that gene can be studied. In this way, we can determine the degree of allelic heterogeneity, the penetrance of different alleles, whether there is a correlation between certain alleles and various aspects of the phenotype (genotype-phenotype correlation), and the frequency of disease-causing or predisposing variants in various populations.

Other patients with the same or similar disorders can be examined to see whether or not they also harbor mutations in the same gene, which would indicate there is locus heterogeneity for a particular disorder. Once the gene and variants in that gene are identified in affected individuals, highly specific methods of diagnosis, including prenatal diagnosis, and carrier screening can be offered to patients and their families.

The variants associated with disease can then be modeled in other organisms, which allows us to use powerful genetic, biochemical, and physiological tools to better understand how the disease comes about. Finally, armed with an understanding of gene function and how the alleles associated with disease affect that function, we can begin to develop specific therapies, including gene replacement therapy, to prevent or ameliorate the disorder. Indeed, *much of the material in the next few chapters about the etiology, pathogenesis, mechanism, and treatment of various diseases begins with gene mapping.* Here, we examine the major approaches used to discover genes involved in genetic disease, as outlined at the beginning of this chapter.

Mapping Human Disease Genes by Linkage Analysis

Determining Whether Two Loci Are Linked

Linkage analysis is a method of mapping genes that uses studies of recombination in families to determine whether two genes show linkage when passed on from one generation to the next. We use information from the known or suspected mendelian inheritance pattern (dominant, recessive, X-linked) to determine which of the individuals in a family have inherited a recombinant or a nonrecombinant chromosome.

To decide whether two loci are linked and, if so, how close or far apart they are, we rely on two pieces of information. First, using the family data in hand, we need to estimate θ, the recombination frequency between the two loci, because that will tell us how close or far apart they are. Next, we need to ascertain whether θ is statistically significantly different from 0.5, because determining whether two loci are linked is equivalent to asking whether the recombination fraction between them differs significantly from the 0.5 fraction expected for unlinked loci. Estimating θ and, at the same time, determining the statistical significance of any deviation of θ from 0.5, relies on a statistical tool called the **likelihood ratio** (as discussed later in the Chapter).

Linkage analysis begins with a set of actual family data with N individuals. Based on a mendelian inheritance model, count the number of chromosomes, r, that show recombination between the allele causing the disease and alleles at various polymorphic loci around the genome (so-called "markers"). The number of chromosomes that do not show a recombination is therefore N − r. The recombination fraction θ can be considered to be the unknown probability, with each meiosis, that a recombination will occur between the two loci; the probability that no recombination occurs is therefore 1 − θ. Because each meiosis is an independent event, one multiplies the probability of a recombination, θ, or of no recombination, (1 − θ), for each chromosome. The formula for the likelihood (which is just the probability) of observing this number of recombinant and nonrecombinant chromosomes when θ is unknown is therefore given by $\{N!/r!(N − r)!\}\theta^r (1 − \theta)^{(N−r)}$. (The factorial term, $N!/r!(N − r)!$, is necessary to account for all the possible birth orders in which the recombinant and nonrecombinant children can appear in the pedigree). Calculate a second likelihood based on the null hypothesis that the two loci are unlinked, that is, make θ = 0.50. The ratio of the likelihood of the family data supporting linkage with unknown θ to the likelihood that the loci are unlinked is the odds in favor of linkage and is given by:

$$\frac{\text{Likelihood of the data if loci were linked at distance } \theta}{\text{Likelihood of the data if loci were unlinked } (\theta = 0.5)} =$$

$$\frac{\{N!/r!(N−r)!\}\theta^r(1−\theta)^{(N−r)}}{\{N!/r!(N−r)!\}(\frac{1}{2})^r(\frac{1}{2})^{(N−r)}}$$

Fortunately, the factorial terms are always the same in the numerator and denominator of the likelihood ratio, and therefore they cancel each other out and can be ignored. If θ = 0.5, the numerator and denominator are the same and the odds equal 1.

Statistical theory tells us that when the value of the likelihood ratio for all values of θ between 0 and 0.5 are calculated, the value of θ that gives the greatest value of this likelihood ratio is, in fact, the best estimate of the recombination fraction you can make given the data and is referred to as θ_{max}. By convention, the computed likelihood ratio for different values of θ is usually expressed as the $\log_{10}$ and is called the **LOD score** (Z) where LOD stands for "*L*ogarithm of the *OD*ds." The use of logarithms allows likelihood ratios calculated from different families to be combined by simple addition instead of having to multiply them together.

How is LOD score analysis actually carried out in families with mendelian disorders? (See Box this page) Return to the family shown in Figure 10-6, in which the mother has an autosomal dominant form of **retinitis pigmentosa**. There are dozens of different forms of this disease, many of which have been mapped to specific sites within the genome and the genes for which have now been identified. Typically, when a new family comes to clinical attention, one does not know which form of RP a patient has. In this family, the mother is also heterozygous for two marker loci on chromosome 7, locus 1 in distal 7q and locus 2 in 7p14. Suppose we know (from other family data) that the disease allele D is in coupling with allele A at locus 1 and allele B at locus 2. Given this phase, one can see that there has been recombination between RP and locus 2 in only one of her eight children, her daughter II-3. The alleles at the disease locus, however, show no tendency to follow the alleles at locus 1 or alleles at any of the other hundreds of marker loci tested on the other autosomes. Thus, although the RP locus involved in this family could in principle have mapped anywhere in the human genome, one now begins to suspect on the basis of the linkage data that the responsible RP locus lies in the region of chromosome 7 near marker locus 2.

To provide a quantitative assessment of this suspicion, suppose we let θ be the "true" recombination fraction between RP and locus 2, the fraction we would see if we had unlimited numbers of offspring to test. The likelihood ratio for this family is therefore

$$\frac{(\theta)^1(1-\theta)^7}{(\frac{1}{2})^1(\frac{1}{2})^7}$$

and reaches a maximum LOD score of $Z_{max} = 1.1$ at $\theta_{max} = 0.125$.

The value of θ that maximizes the likelihood ratio, θ_{max}, may be the best estimate one can make for θ given the data, but how good an estimate is it? The magnitude of the LOD score provides an assessment of how good an estimate of θ_{max} you have made. *By convention, a LOD score of +3 or greater (equivalent to greater than 1000:1 odds in favor of linkage) is considered firm evidence that two loci are linked—that is, that θ_{max} is statistically significantly different from 0.5.* In our RP example, $\frac{7}{8}$ of the offspring are nonrecombinant and $\frac{1}{8}$ are recombinant. The $\theta_{max} = 0.125$, but the LOD score is only 1.1, enough to raise a suspicion of linkage but insufficient to prove linkage because Z_{max} falls far short of 3.

LINKAGE ANALYSIS OF MENDELIAN DISEASES

Linkage analysis is used when there is a particular mode of inheritance (autosomal dominant, autosomal recessive, or X-linked) that explains the inheritance pattern.

LOD score analysis allows mapping of genes in which mutations cause diseases that follow mendelian inheritance.

The LOD score gives both:
- A best estimate of the recombination frequency, θ_{max}, between a marker locus and the disease locus; and
- An assessment of how strong the evidence is for linkage at that value of θ_{max}. Values of the LOD score Z above 3 are considered strong evidence.

Linkage at a particular θ_{max} of a disease gene locus to a marker with known physical location implies that the disease gene locus must be near the marker. The smaller the θ_{max} is, the closer the disease locus is to the linked marker locus.

Combining LOD Score Information across Families

In the same way that each meiosis in a family that produces a nonrecombinant or recombinant offspring is an independent event, so too are the meioses that occur in different families. We can therefore multiply the likelihoods in the numerators and denominators of each family's likelihood odds ratio together. Suppose two additional families with RP were studied and one showed no recombination between locus 2 and RP in four children and the other showed no recombination in five children. The individual LOD scores can be generated for each family and added together (Table 10-1). Because the maximum LOD score Z_{max} exceeds 3 at $\theta_{max} = \approx 0.06$, the RP gene in this group of families is linked to locus 2 at a recombination distance of ≈ 0.06. Because the genomic location of marker locus 2 is known to be at 7p14, the RP in this family can be mapped to the 7p14 region and likely involves the *RP9* gene, one of the already identified loci for a form of autosomal dominant RP.

If, however, some of the families being used for the study were to have RP due to mutations at a different locus, the LOD scores between families would diverge, with some showing a trend to being positive at small values of θ and others showing strongly negative LOD scores at these values. Thus, in linkage analysis involving more than one family, unsuspected locus heterogeneity can obscure what may be real evidence for linkage in a subset of families.

Phase-Known and Phase-Unknown Pedigrees

In the RP example just discussed, we assumed that we knew the phase of marker alleles on chromosome 7 in the affected mother in that family. Let us now look at the implications of knowing phase in more detail.

TABLE 10-1 LOD Score for Three Families with Retinitis Pigmentosa

	0.00	0.01	0.05	0.06	0.07	0.10	0.125	0.20	0.30	0.40
Family 1	—	0.38	0.95	1.00	1.03	1.09	**1.1**	1.03	0.80	0.46
Family 2	**1.2**	1.19	1.11	1.10	1.08	1.02	0.97	0.82	0.58	0.32
Family 3	**1.5**	1.48	1.39	1.37	1.35	1.28	1.22	1.02	0.73	0.39
Total	—	3.05	3.45	**3.47**	3.46	3.39	3.29	2.87	2.11	1.17

Individual Z_{max} for each family is shown in **bold**. The overall $Z_{max} = 3.47$ at $\theta_{max} = 0.06$.

Consider the three-generation family with autosomal dominant **neurofibromatosis, type 1 (NF1)** (Case 34) in Figure 10-10. The affected mother, II-2, is heterozygous at both the NF1 locus *(D/d)* and a marker locus *(A/a)*, but (as shown in Fig. 10-10A) we have no genotype information on her parents. The two affected children received the *A* alleles along with the *D* disease allele, and the one unaffected child received the *a* allele along with the normal *d* allele. Without knowing the phase of these alleles in the mother, either all three offspring are recombinants or all three are nonrecombinants. Because both possibilities are equally likely in the absence of any other information, we consider the phase on her two chromosomes to be *D-a* and *d-A* half of the time and *D-A* and *d-a* the other half (which assumes the alleles in these haplotypes are in linkage equilibrium). To calculate the overall likelihood of this pedigree, we then add the likelihood calculated assuming one phase in the mother to the likelihood calculated assuming the other phase. Therefore, the overall likelihood $= \frac{1}{2}\theta^0(1-\theta)^3 + \frac{1}{2}(\theta^3)(1-\theta)^0$ and the likelihood ratio for this pedigree, then, is:

$$\frac{\frac{1}{2}(1-\theta)^3(\theta^0) + \frac{1}{2}(\theta^3)(1-\theta)^0}{\frac{1}{8}}$$

giving a maximum LOD score of $Z_{max}= 0.602$ at $\theta_{max} = 0$.

If, however, additional genotype information in the maternal grandfather I-1 becomes available (as in Fig. 10-10B), the phase can now be determined to be *D-A* (i.e., the NF1 allele *D* was in coupling with the *A* in individual II-2). In light of this new information, the three children can now be scored definitively as nonrecombinants, and we no longer have to consider the possibility of the opposite phase. The numerator of the likelihood ratio now becomes $(1 - \theta)^3(\theta^0)$ and the maximum LOD score $Z_{max} = 0.903$ at $\theta_{max} = 0$. Thus knowing the phase increases the power of the data available to test for linkage.

Mapping Human Disease Genes by Association
Designing an Association Study

An entirely different approach to identification of the genetic contribution to disease relies on finding *particular alleles* that are associated with the disease in a sample from the population. In contrast to linkage analysis, this approach does not depend upon there being a mendelian inheritance pattern and is therefore better suited for discovering the genetic contributions to disorders with complex inheritance (see Chapter 8). The presence of a particular allele at a locus at increased or decreased frequency in affected individuals compared with controls is known as a **disease association**. There are two commonly used study designs for association studies:

- **Case-control studies.** Individuals *with* the disease are selected in a population, a matching group of controls *without* disease are then selected, and the genotypes of individuals in the two groups are determined and used to populate a two-by-two table (see below).
- **Cross-sectional or cohort studies.** A random sample of the entire population is chosen and then analyzed for whether they have (cross-sectional) or, after being followed over time, develop (cohort) a particular disease; the genotypes of everyone in the study

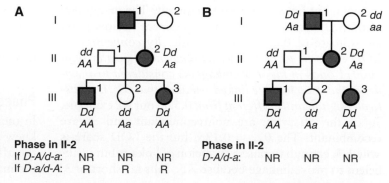

Figure 10-10 Two pedigrees of autosomal dominant neurofibromatosis, type 1 (NF1). **A,** Phase of the disease allele *D* and marker alleles *A* and *a* in individual II-2 is unknown. **B,** Availability of genotype information for generation I allows a determination that the disease allele *D* and marker allele *A* are in coupling in individual II-2. NR, Non-recombinant; R, recombinant.

population are determined. The numbers of individuals with and without disease and with and without an allele (or genotype or haplotype) of interest are used to fill out the cells of a two-by-two table.

Odds Ratios and Relative Risks

The two different types of association studies report the strength of the association, using either the odds ratio or relative risk.

In a **case-control study**, the frequency of a *particular allele or haplotype* (e.g., for a human leukocyte antigen [HLA] haplotype or a particular SNP allele or SNP haplotype) is compared between the selected affected and unaffected individuals, and an association between disease and genotype is then calculated by an **odds ratio (OR)**.

	Patients	Controls	Totals
With genetic marker*	a	b	a + b
Without genetic marker	c	d	c + d
Totals	a + c	b + d	

*A genetic marker can be an allele, a genotype, or a haplotype.

Using the two-by-two table, the odds of an allele carrier developing the disease is the ratio (a/b) of the number of allele carriers who develop the disease (a) to the number of allele carriers who do not develop the disease (b). Similarly, the odds of a noncarrier developing the disease is the ratio (c/d) of noncarriers who develop the disease (c) divided by the number of noncarriers who do not develop the disease (d). The disease odds ratio is then the ratio of these odds.

$$OR = \frac{\frac{a}{b}}{\frac{c}{d}} = \frac{ad}{bc}$$

An OR that differs from 1 means there is an association of disease risk with the genetic marker, whereas OR = 1 means there is no association.

Alternatively, if the association study was designed as a **cross-sectional or cohort study**, the strength of an association can be measured by the **relative risk (RR)**. The RR is the ratio of the proportion of those with the disease who carry a particular allele ([a/(a + b)]) to the proportion of those without the disease who carry that allele ([c/(c + d)]).

$$RR = \frac{\frac{a}{a+b}}{\frac{c}{c+d}}$$

Again, an RR that differs from 1 means there is an association of disease risk with the genetic marker, whereas RR = 1 means there is no association. (The relative risk RR introduced here should not be confused

with λ_r, the **risk ratio in relatives**, which was discussed in Chapter 8. λ_r is the prevalence of a particular disease phenotype in an affected individual's relatives versus that in the general population.)

For diseases that are rare (i.e., a < b and c < d), a case-control design with calculation of the OR is best, because any random sample of a population is unlikely to contain sufficient numbers of affected individuals to be suitable for a cross-sectional or cohort study design. Note, however, that when a disease is rare and calculating an OR in a case-control study is the only practical approach, OR is a good approximation for an RR. (Examine the formula for RR and convince yourself that, when a < b and c < d, (a + b) ≈ b and (c + d) ≈ d, and thus RR ≈ OR.)

The information obtained in an association study comes in two parts. The first is **the magnitude of the association** itself: the further the RR or OR diverges from 1, the greater is the effect of the genetic variant on the association. However, an OR or RR for an association is a statistical measure and requires a test of statistical significance. The **significance** of any association can be assessed by simply asking with a chi-square test if the frequencies of the allele (a, b, c, and d in the two-by-two table) differ significantly from what would be expected if there were no association (i.e., if the OR or RR were equal to 1.0). A common way of expressing whether there is statistical significance to an estimate of OR or RR is to provide a 95% (or 99%) **confidence interval**. The confidence interval is the range within which one would expect the OR or RR to fall 95% (or 99%) of the time by chance alone in a sample taken from the population. If a confidence interval excludes the value 1.0, then the OR or RR deviates significantly from what would be expected if there were no association with the marker locus being tested, and the null hypothesis of no association can be rejected at the corresponding significance level. (Later in this chapter we will explain why a level of 0.05 or 0.01 is inadequate for assessing statistical significance when *multiple* marker loci in the genome are tested simultaneously for association.)

To illustrate these approaches, we first consider a case-control study of **cerebral vein thrombosis (CVT)**, which we introduced in Chapter 8. In this study, suppose a group of 120 patients with CVT and 120 matched controls were genotyped for the 20210G>A allele in the prothrombin gene (see Chapter 8).

	Patients with CVT	Controls without CVT	Totals
20210G>A allele present	23	4	27
20210G>A allele absent	97	116	213
Total	120	120	240

CVT, Cerebral vein thrombosis.

Because this is a case-control study, we will calculate an odds ratio: OR = (23/4)/(97/116) = ≈6.9 with 95% confidence limits of 2.3 to 20.6. There is clearly a substantial effect size of 6.9 and 95% confidence limits that exclude 1.0, thereby demonstrating that *there is a strong and statistically significant association between the 20210G>A allele and CVT*. Stated simply, individuals carrying the prothrombin 20210G>A allele have nearly seven times greater odds of having the disease than those who do not carry this allele.

To illustrate a longitudinal cohort study in which RR, instead of an OR, can be calculated, consider **statin-induced myopathy**, a rare but well-recognized adverse drug reaction that can develop in some individuals during statin therapy to lower cholesterol. In one study, subjects enrolled in a cardiac protection study were randomized to receive 40 mg of the statin drug simvastatin or placebo. Over 16,600 participants exposed to the statin were genotyped for a variant (Val174Ala) in the *SLCO1B1* gene, which encodes a hepatic drug transporter, and were watched for development of the adverse drug response. Out of the entire genotyped group exposed to the statin, 21 developed myopathy. Examination of their genotypes showed that the RR for developing myopathy associated with the presence of the Val174Ala allele is approximately 2.6, with 95% confidence limits of 1.3 to 5.1. Thus here *there is a statistically significant association between the Val174Ala allele and statin-induced myopathy*; those carrying this allele are at moderately increased risk for developing this adverse drug reaction relative to those who do not carry this allele.

One common misconception concerning an association study is that the more significant the *P* value, the stronger is the association. In fact, a significant *P* value for an association does *not* provide information concerning the magnitude of the effect of an associated allele on disease susceptibility. Significance is a statistical measure that describes how likely it is that the population sample used for the association study could have yielded an observed OR or RR that differs from 1.0 simply by chance alone. In contrast, the actual magnitude of the OR or RR—how *far* it diverges from 1.0—is a measure of the impact a particular variant (or genotype or haplotype) has on increasing or decreasing disease susceptibility.

Genome-Wide Association Studies
The Haplotype Map (HapMap)

For many years, association studies for human disease genes were limited to particular sets of variants in restricted sets of genes chosen either for convenience or because they were thought to be involved in a pathophysiological pathway relevant to a disease and thus appeared to be logical **candidate genes** for the disease under investigation. Thus many such association studies were undertaken before the Human Genome Project era with use of the HLA or blood group loci, for example, because these loci were highly polymorphic and easily genotyped in case-control studies. Ideally, however, one would like to be able to test systematically for an association between any disease of interest and *every* one of the tens of millions of rare and common alleles in the genome in an unbiased fashion without any preconception of what genes and genetic variants might be contributing to the disease.

Association analyses on a genome scale are referred to as **genome-*w*ide *a*ssociation *s*tudies**, known by their acronym **GWAS**. Such an undertaking for *all* known variants is impractical for many reasons but can be approximated by genotyping cases and controls for a mere 300,000 to 1 million individual variants located throughout the genome to search for association with the disease or trait in question. The success of this approach depends on exploiting LD because, as long as a variant responsible for altering disease susceptibility is in LD with one or more of the genotyped variants within an LD block, a positive association should be detectable between that disease and the alleles in the LD block.

Developing such a set of markers led to the launch of the **Haplotype Mapping (HapMap) Project**, one of the biggest human genomics efforts to follow completion of the Human Genome Project. The HapMap Project began in four geographically distinct groups—a primarily European population, a West African population, a Han Chinese population, and a population from Japan—and included collecting and characterizing millions of SNP loci and developing methods to genotype them rapidly and inexpensively. Since that time, whole-genome sequencing has been applied to many populations in what is referred to as the **1000 Genomes Project**, resulting in a massive expansion in the database of DNA variants available for GWAS with different populations around the globe.

Gene Mapping by Genome-Wide Association Studies in Complex Traits

The purpose of the HapMap was not just to gather basic information about the distribution of LD across the human genome. Its primary purpose was to provide a powerful new tool for finding the genetic variants that contribute to human disease and other traits by making possible an approximation to an idealized, full-scale, genome-wide association. The driving principle behind this approach is a straightforward one: *detecting an association with alleles within an LD block pinpoints the genomic region within the block as being likely to contain the disease-associated allele*. Consequently, although the approach does not typically pinpoint the *actual* variant that is responsible functionally for the association with disease, this region will be the place to focus additional studies to find the allelic variant that *is* functionally involved in the disease process itself.

Historically, detailed analysis of conditions associated with high-density variants in the class I and class II HLA regions (see Fig. 8-10) have exemplified this approach (see Box). However, with the tens of millions of variants now available in different populations, this approach can be broadened to examine the genetic basis of virtually *any* complex disease or trait. Indeed, to date, thousands of GWAS have uncovered an enormous number of naturally occurring variants associated with a variety of genetically complex multifactorial diseases, ranging from diabetes and inflammatory bowel disease to rheumatoid arthritis and cancer, as well as for traits such as stature and pigmentation. Research to uncover the underlying biological basis for these associations will be ongoing for years to come.

HUMAN LEUKOCYTE ANTIGEN AND DISEASE ASSOCIATION

Among more than a thousand genome-trait or genome-disease associations from around the genome, the region with the highest concentration of associations to different phenotypes is the human leukocyte antigen (HLA) region. In addition to the association of specific alleles and haplotypes to **type 1 diabetes** discussed in Chapter 8, association of various HLA polymorphisms has been demonstrated for a wide range of conditions, most but not all of which are **autoimmune**, that is, associated with an abnormal immune response apparently directed against one or more self-antigens. These associations are thought to be related to variation in the immune response resulting from polymorphism in immune response genes.

The functional basis of most HLA-disease associations is unknown. HLA molecules are integral to T-cell recognition of antigens. Different polymorphic HLA alleles are thought to result in structural variation in these cell surface molecules, leading to differences in the capacity of the proteins to interact with antigen and the T-cell receptor in the initiation of an immune response, thereby affecting such critical processes as immunity against infections and self-tolerance to prevent autoimmunity.

Ankylosing spondylitis, a chronic inflammatory disease of the spine and sacroiliac joints, is one example. More than 95% of those with ankylosing spondylitis are *HLA-B27*-positive; the risk for developing ankylosing spondylitis is at least 150 times higher for people who have certain *HLA-B27* alleles than for those who do not. These alleles lead to HLA-B27 heavy chain misfolding and inefficient antigen presentation.

In other disorders, the association between a particular HLA allele or haplotype and a disease is not due to functional differences in immune response genes themselves. Instead, the association is due to a particular allele being present at a very high frequency on chromosomes that also happen to contain disease-causing mutations in another gene within the major histocompatibility complex region. One example is **hemochromatosis,** a common disorder of iron overload. More than 80% of patients with hemochromatosis are homozygous for a common mutation, Cys282Tyr, in the hemochromatosis gene (*HFE*) and have *HLA-A*0301* alleles at their *HLA-A* locus. The association is not the result of *HLA-A*0301*, however. *HFE* is involved with iron transport or metabolism in the intestine; *HLA-A*, as a class I immune response gene, has no effect on iron transport. The association is due to proximity of the two loci and LD between the Cys282Tyr *HFE* mutation and the *A*0301* allele at *HLA-A*.

Pitfalls in Design and Analysis of GWAS

Association methods are powerful tools for pinpointing precisely the genes that contribute to genetic disease by demonstrating not only the genes but also the particular alleles responsible. They are also relatively easy to perform because one needs samples only from a set of unrelated affected individuals and controls and does not have to carry out laborious family studies and collection of samples from many members of a pedigree.

Association studies must be interpreted with caution, however. One serious limitation of association studies is the problem of totally artifactual association caused by **population stratification** (see Chapter 9). If a population is stratified into separate subpopulations (e.g., by ethnicity or religion) and members of one subpopulation rarely mate with members of other subpopulations, then a disease that happens to be more common in one subpopulation for whatever reason can appear (incorrectly) to be associated with any alleles that also happen to be more common in that subpopulation than in the population as a whole. Factitious association due to population stratification can be minimized, however, by careful selection of matched controls. In particular, one form of quality control is to make sure the cases and controls have similar frequencies of alleles whose frequencies are known to differ markedly between populations (**ancestry informative markers,** as we discussed in Chapter 9). If the frequencies seen in cases and controls are similar, then unsuspected or cryptic stratification is unlikely.

In addition to the problem of stratification producing false-positive associations, false-positive results in GWAS can also arise if an inappropriately lax test for statistical significance is applied. This is because as the number of alleles being tested for a disease association increases, the chance of finding associations *by chance alone* also increases, a concept in statistics known as the problem of **multiple hypothesis testing**. To understand why the cut-off for statistical significance must be much more stringent when multiple hypotheses are being tested, imagine flipping a coin 50 times and having it come up heads 40 times. Such a highly unusual result has a probability of occurring of only once in approximately 100,000 times. However, if the same experiment were repeated a million times, chances are greater than 99.999% that *at least* one coin flip experiment out of the million performed will result in 40 or more heads!

Thus even rare events that occur by chance alone in an experiment become frequent when the experiment is repeated over and over again. This is why when testing for an association with hundreds of thousands to millions of variants across the genome, tens of thousands of variants could appear associated with $P < 0.05$ *by chance alone*, making a typical cutoff for statistical significance of $P < 0.05$ far too low to point to a true association. Instead, a significance level of $P < 5 \times 10^{-8}$ is considered to be more appropriate for GWAS that tests hundreds of thousands to millions of variants. Even with appropriately stringent cutoffs for genome-wide significance, however, false-positive results due to chance alone will still occur. To take this into account, a properly performed GWAS usually include a **replication study** in a different, completely independent group of individuals to show that alleles near the same locus are associated. A caveat, however, is that alleles that show association may be different in different ethnic groups.

Finally, it is important to emphasize that if an association is found between a disease and a polymorphic marker allele that is part of a dense haplotype map, one *cannot* infer there is a functional role for that marker allele in increasing disease susceptibility. Because of the nature of LD, *all* alleles in LD with an allele at a locus involved in the disease will show an apparently positive association, whether or not they have any functional relevance in disease predisposition. An association based on LD is still quite useful, however, because in order for the polymorphic marker alleles to appear associated, the associated polymorphic marker alleles would likely sit within an LD block that also harbors the actual disease locus.

A comparison of the characteristics, strengths, and weaknesses of linkage and association methods for disease gene mapping are summarized in the Box.

FROM GENE MAPPING TO GENE IDENTIFICATION

The application of gene mapping to medical genetics using the approaches outlined in the previous section has met with many spectacular successes. This strategy has led to the identification of the genes associated with thousands of mendelian disorders and a growing number of genes and alleles associated with genetically complex disorders. The power of these approaches has increased enormously with the introduction of highly efficient and less expensive technologies for genome analysis.

In this section, we describe how genetic and genomic methods led to the identification of the genes involved in two disorders, one first using linkage analysis and LD to narrow down the location of the gene responsible for

COMPARISON OF LINKAGE AND ASSOCIATION METHODS

Linkage	Association
• Follows inheritance of a disease trait and regions of the genome from individual to individual in family pedigrees	• Tests for altered frequency of particular alleles or haplotypes in affected individuals compared with controls in a population
• Looks for regions of the genome harboring disease alleles; uses polymorphic variants only as a way of marking which region an individual has inherited from which parent	• Examines particular alleles or haplotypes for their contribution to the disease
• Uses hundreds to thousands of polymorphic markers across the genome	• Uses anywhere from a few markers in targeted genes to hundreds of thousands of markers for genome-wide analyses
• Not designed to find the specific variant responsible for or predisposing to the disease; can only demarcate where the variant can be found within (usually) one or a few megabases	• Can occasionally pinpoint the variant that is actually functionally responsible for the disease; more frequently, defines a disease-containing haplotype over a 1- to 10-kb interval (usually)
• Relies on recombination events occurring in families during only a few generations to allow measurement of the genetic distance between a disease gene and polymorphic markers on chromosomes	• Relies on finding a set of alleles, including the disease gene, that remained together for many generations because of a *lack* of recombination events among the markers
• Requires sampling of families, not just people affected by the disease	• Can be carried out on case-control or cohort samples from populations
• Loses power when disease has complex inheritance with substantial lack of penetrance	• Is sensitive to population stratification artifact, although this can be controlled by proper case-control designs or the use of family-based approaches
• Most often used to map disease-causing mutations with strong enough effects to cause a mendelian inheritance pattern	• Is the best approach for finding variants with small effect that contribute to complex traits

the common autosomal recessive disease **cystic fibrosis (CF)** (Case 12) and one using GWAS to find multiple allelic variants in genes that increase susceptibility to **age-related macular degeneration (AMD)** (Case 3), a devastating disorder that robs older adults of their vision.

Gene Finding in a Common Mendelian Disorder by Linkage Mapping

Example: Cystic Fibrosis

Because of its relatively high frequency, particularly in white populations, and the nearly total lack of understanding of the abnormalities underlying its pathogenesis, CF represented a prime candidate for identifying the gene responsible by using linkage to find the gene's location, rather than using any information on the disease process itself. DNA samples from nearly 50 CF families were analyzed for linkage between CF and hundreds of DNA markers throughout the genome until linkage of CF to markers on the long arm of chromosome 7 was finally identified. Linkage to additional DNA markers in 7q31-q32 narrowed the localization of the CF gene to an approximately 500-kb region of chromosome 7.

Linkage Disequilibrium in Cystic Fibrosis. At this point, however, an important feature of CF genetics emerged: even though the closest linked markers were still some distance from the CF gene, it became clear that there was significant LD between the disease locus and a particular haplotype at loci tightly linked to the disease. Regions with the greatest degree of LD were analyzed for gene sequences, leading to the isolation of the CF gene in 1989. As described in detail in Chapter 12, the gene responsible, which was named the cystic fibrosis transmembrane conductance regulator (*CFTR*), showed an interesting spectrum of mutations. A 3-bp deletion (ΔF508) that removed a phenylalanine at position 508 in the protein was found in approximately 70% of all mutant CF alleles in northern European populations but never among normal alleles at this locus. Although subsequent studies have demonstrated many hundreds of mutant *CFTR* alleles worldwide, it was the high frequency of the ΔF508 mutation in the families used to map the CF gene and the LD between it and alleles at polymorphic marker loci nearby that proved so helpful in the ultimate identification of the *CFTR* gene.

Mapping of the CF locus and cloning of the *CFTR* gene made possible a wide range of research advances and clinical applications, from basic pathophysiology to molecular diagnosis for genetic counseling, prenatal diagnosis, animal models, and finally current ongoing attempts to treat the disorder (see Chapter 12).

Finding the Genes Contributing to a Complex Disease by Genome-Wide Association

Example: Age-Related Macular Degeneration

AMD is a progressive degenerative disease of the portion of the retina responsible for central vision. It causes blindness in 1.75 million Americans older than 50 years. The disease is characterized by the presence of drusen, which are clinically visible, discrete extracellular deposits of protein and lipids behind the retina in the region of the macula (Case 3). Although there is ample evidence for a genetic contribution to the disease, most individuals with AMD are not in families in which there is a likely mendelian pattern of inheritance. Environmental contributions are also important, as shown by the increased risk for AMD in cigarette smokers compared with nonsmokers.

Initial case-control GWAS of AMD revealed association of two common SNPs near the complement factor H (*CFH*) gene. The most frequent at-risk haplotype containing these alleles was seen in 50% of cases versus only 29% of controls (OR = 2.46; 95% confidence interval [CI], 1.95 to 3.11). Homozygosity for this haplotype was found in 24.2% of cases, compared to only 8.3% of the controls (OR = 3.51; 95% CI, 2.13-5.78). A search through the SNPs within the LD block containing the AMD-associated haplotype revealed a nonsynonymous SNP in the *CFH* gene that substituted a histidine for tyrosine at position 402 of the CFH protein (Tyr402His). The Tyr402His alteration, which has an allele frequency of 26% to 29% in white and African populations, showed an even stronger association with AMD than did the two SNPs that showed an association in the original GWAS.

Given that drusen contain complement factors and that CFH is found in retinal tissues around drusen, it is believed that the Tyr402His variant is less protective against the inflammation that is thought to be responsible for drusen formation and retinal damage. Thus Tyr402His is likely to be the variant at the *CFH* locus responsible for increasing the risk for AMD.

More recent GWAS of AMD using more than 7600 cases and more than 50,000 controls and millions of variants genome-wide have revealed that alleles at a minimum of 19 loci are associated with AMD, with genome-wide significance of $P < 5 \times 10^{-8}$. A popular way to summarize GWAS in graphic form is to plot the $-\log_{10}$ significance levels for each associated variant in what is referred to as a "Manhattan plot," because it is thought to bear a somewhat fanciful similarity to the skyline of New York City (Fig. 10-11). The ORs for AMD of these variants range from a high of 2.76 for a gene of unknown function, *ARMS2,* and 2.48 for *CFH* to 1.1 for many other genes involved in multiple pathways, including the complement system, atherosclerosis, blood vessel formation, and others.

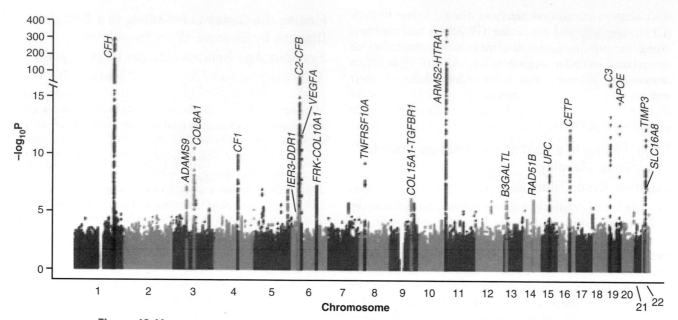

Figure 10-11 "Manhattan plot" of genome-wide association studies (GWAS) of age-related macular degeneration using approximately 1 million genome-wide single nucleotide polymorphism (SNP) alleles located along all 22 autosomes on the x-axis. Each *blue dot* represents the statistical significance (expressed as $-\log_{10}(P)$ plotted on the y-axis), confirming a previously known association; *green dots* are the statistical significance for novel associations. The discontinuity in the y-axis is needed because some of the associations have extremely small P values $< 1 \times 10^{-16}$. *See Sources & Acknowledgments.*

In this example of AMD, a complex disease, GWAS led to the identification of strongly associated, common SNPs that in turn were in LD with a common coding SNP in the gene that appears to be the functional variant involved in the disease. This discovery in turn led to the identification of other SNPs in the complement cascade and elsewhere that can also predispose to or protect against the disease. Taken together, these results give important clues to the pathogenesis of AMD and suggest that the complement pathway might be a fruitful target for novel therapies. Equally interesting is that GWAS revealed that a novel gene of unknown function, *ARMS2*, is also involved, thereby opening up an entirely new line of research into the pathogenesis of AMD.

Importance of Associations Discovered with GWAS

There is vigorous debate regarding the interpretation of GWAS results and their value as a tool for human genetic studies. The debate arises primarily from a misunderstanding of what an OR or RR means. It is true that many properly executed GWAS yield significant associations, but of very modest effect size (similar to the OR of 1.1 just mentioned for AMD). In fact, significant associations of smaller and smaller effect size have become more common as larger and larger sample sizes are used that allow detection of statistically significant genome-wide associations with smaller and smaller ORs or RRs. This has led to the suggestion that GWAS are of little value because the effect size of the association, as measured by

OR or RR, is too small for the gene and pathway implicated by that variant to be important in the pathogenesis of the disease. This is faulty reasoning on two accounts.

First, ORs are a measure of the impact of a specific allele (e.g., the *CFH* Tyr402His allele for AMD) on complex pathogenetic pathways, such as the alternative complement pathway of which CFH is a component. The subtlety of that impact is determined by how that allele perturbs the biological function of the gene in which it is located, and not by whether the gene harboring that allele might be important in disease pathogenesis. In autoimmune disorders, for example, studies of patients with a number of different autoimmune disorders, such as **rheumatoid arthritis, systemic lupus erythematosus,** and **Crohn disease,** reveal modest associations, but with some of the same variants, suggesting there are common pathways leading to these distinct but related diseases that will likely be quite illuminating in studies of their pathogenesis (see Box).

Second, even if the effect size of any one variant is small, GWAS demonstrate that many of these disorders are indeed extremely polygenic, even more polygenic than previously suspected, with thousands of variants, most of which contribute only a little (ORs between 1.01 and 1.1) to disease susceptibility by themselves but, in the aggregate, account for a substantial fraction of the observed clustering of these diseases within certain families (see Chapter 8).

Although the observation of modest effect size for most alleles found by GWAS is correct, it misses a critical and perhaps most fundamental finding of GWAS: *the genetic architecture of some of the most common complex diseases studied to date may involve hundreds to thousands of loci harboring variants of small effect in many genes and pathways.* These genes and pathways are important to our understanding of how complex diseases occur, even if each allele exerts only subtle effects on gene regulation or protein function and has only a modest effect on disease susceptibility on a per allele basis.

Thus GWAS remain an important human genetics research tool for dissecting the many contributions to complex disease, regardless of whether or not the individual variants found to be associated with the disease substantially raise the risk for the disease in individuals carrying those alleles (see Chapter 16). We expect that many more genetic variants responsible for complex diseases will be successfully identified by genome-wide association and that deep sequencing of the regions showing disease associations should uncover the variants or collections of variants functionally responsible for disease associations. Such findings should provide us with powerful insights and potential therapeutic targets for many of the common diseases that cause so much morbidity and mortality in the population.

FINDING GENES RESPONSIBLE FOR DISEASE BY GENOME SEQUENCING

Thus far in this chapter, we have focused on two approaches to map and then identify genes involved in disease, linkage analysis and GWAS. Now we turn to a third approach, involving direct genome sequencing of affected individuals and their parents and/or other individuals in the family or population.

The development of vastly improved methods of DNA sequencing, which has cut the cost of sequencing six orders of magnitude from what was spent generating the Human Genome Project's reference sequence, has opened up new possibilities for discovering the genes and mutations responsible for disease, particularly in the case of rare mendelian disorders. As introduced in Chapter 4, these new technologies make it possible to generate a **whole-genome sequence** (WGS) or, in what may be a cost-effective compromise, the sequence of only the approximately 2% of the genome containing the exons of genes, referred to as a **whole-exome sequence** (WES).

Filtering Whole-Genome Sequence or Whole-Exome Sequence Data to Find Potential Causative Variants

As an example of what is now possible, consider a family "trio" consisting of a child affected with a rare

FROM GWAS TO PHEWAS

In genome-wide association studies (GWAS), one explores the genetic basis for a given phenotype, disease, or trait by searching for associations with large, unbiased collections of DNA markers from the entire genome. But can one do the reverse? Can one uncover the potential *phenotypic links* associated with genome variants by searching for associations with large, unbiased collections of phenotypes from the entire "phenome?" Thus far, the results of this approach appear to be highly promising.

In an approach dubbed **phenome-wide association studies (PheWAS)**, genetic variants are tested for association, not just with a particular phenotype of interest (say, rheumatoid arthritis or systolic blood pressure above 160 mm Hg), but with *all* medically relevant phenotypes and laboratory values found in **electronic medical records (EMRs)**. In this way, one can seek novel and unanticipated associations in an unbiased manner, using search algorithms, billing codes, and open text mining to query all electronic entries, which are fast becoming available for health records in many countries.

As an illustration of this approach, SNPs for a major class II *HLA-DRB1* haplotype (as described in Chapter 8) were screened against over 4800 phenotypes in EMRs from over 4000 patients; this PheWAS detected association not only with multiple sclerosis (as expected from previous studies), but also with alcohol-induced cirrhosis of the liver, erythematous conditions such as rosacea, various benign neoplasms, and several dozen other phenotypes.

Although the potential of PheWAS is just being realized, such unbiased interrogation of vast clinical data sets may allow discovery of previously unappreciated comorbidities and/or less common side effects or drug-drug interactions in patients receiving prescribed drugs.

disorder and his parents. WGS is performed for all three, yielding typically over 4 million differences compared to the human genome reference sequence (see Chapter 4). Which of these variants is responsible for the disease? Extracting useful information from this massive amount of data relies on creating a variant filtering scheme based on a variety of reasonable assumptions about which variants are more likely to be responsible for the disease.

One example of a filtering scheme that can be used to sort through these variants is shown in Figure 10-12.

1. *Location with respect to protein-coding genes.* Keep variants that are within or near exons of protein-coding genes, and discard variants deep within introns or intergenic regions. It is possible, of course, that the responsible mutation might lie in a noncoding RNA gene or in regulatory sequences located some distance from a gene, as introduced in Chapter 3. However, these are currently more difficult to assess, and thus, as a simplifying assumption, it is reasonable to focus initially on protein-coding genes.

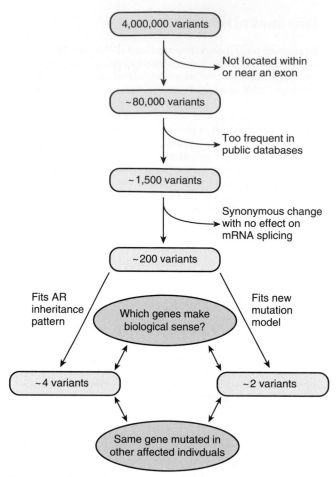

Figure 10-12 Representative filtering scheme for reducing the millions of variants detected in whole-genome sequencing of a family consisting of two unaffected parents and an affected child to a small number that can be assessed for biological and disease relevance. The initial enormous collection of variants is winnowed down into smaller and smaller bins by applying filters that remove variants that are unlikely to be causative based on assuming that variants of interest are likely to be located near a gene, will disrupt its function, and are rare. Each remaining candidate gene is then assessed for whether the variants in that gene are inherited in a manner that fits the most likely inheritance pattern of the disease, whether a variant occurs in a candidate gene that makes biological sense given the phenotype in the affected child, and whether other affected individuals also have mutations in that gene. AR, Autosomal recessive; mRNA, messenger RNA.

2. *Population frequency.* Keep rare variants from step 1, and discard common variants with allele frequencies greater than 0.05 (or some other arbitrary number between 0.01 and 0.1), because common variants are highly unlikely to be responsible for a disease whose population prevalence is much less than the q^2 predicted by Hardy-Weinberg equilibrium (see Chapter 9).

3. *Deleterious nature of the mutation.* Keep variants from step 2 that cause nonsense or nonsynonymous changes in codons within exons, cause frameshift mutations, or alter highly conserved splice sites, and discard synonymous changes that have no predicted effect on gene function.

4. *Consistency with likely inheritance pattern.* If the disorder is considered most likely to be autosomal recessive, keep any variants from step 3 that are found in both copies of a gene in an affected child. The child need not be homozygous for the same deleterious variant but could be a compound heterozygote for two different deleterious mutations in the same gene (see Chapter 7). If the hypothesized mode of inheritance is correct, then the parents should both be heterozygous for the variants. If there were consanguinity in the parents, the candidate genes and variants might be further filtered by requiring that the child be a true homozygote for the same mutation derived from a single common ancestor (see Chapter 9). If the disorder is severe and seems more likely to be a new dominant mutation, because unaffected parents rarely if ever have more than one affected child, keep variants from step 3 that are de novo changes in the child and are not present in either parent.

In the end, millions of variants can be filtered down to a handful occurring in a small number of genes. Once the filtering reduces the number of genes and alleles to a manageable number, they can be assessed for other characteristics. First, do any of the genes have a known function or tissue expression pattern that would be expected if it were the potential disease gene? Is the gene involved in other disease phenotypes, or does it have a role in pathways with other genes in which mutations can cause similar or different phenotypes? Finally, is this same gene mutated in other patients with the disease? Finding mutations in one of these genes in other patients would then confirm this was the responsible gene in the original trio.

In some cases, one gene from the list in step 4 may rise to the top as a candidate because its involvement makes biological or genetic sense or it is known to be mutated in other affected individuals. In other cases, however, the gene responsible may turn out to be entirely unanticipated on biological grounds or may not be mutated in other affected individuals because of locus heterogeneity (i.e., mutations in other as yet undiscovered genes can cause a similar disease).

Such variant assessments require extensive use of public genomic databases and software tools. These include the human genome reference sequence, databases of allele frequencies, software that assesses how deleterious an amino acid substitution might be to gene function, collections of known disease-causing mutations, and databases of functional networks and biological pathways. The enormous expansion of this information over the past few years has played a crucial role in facilitating gene discovery of rare mendelian disorders.

Example: Identification of the Gene Mutated in Postaxial Acrofacial Dysostosis

The WGS approach just outlined was used in the study of a family in which two siblings affected with a rare congenital malformation known as **postaxial acrofacial dysostosis (POAD)** were born to two unaffected, unrelated parents. Patients with this disorder have small jaws, missing or poorly developed digits on the ulnar sides of their hands, underdevelopment of the ulna, cleft lip, and clefts (colobomas) of the eyelids. The disorder was thought to be autosomal recessive because the parents of an affected child in some other families are consanguineous, and there are a few families, like the one here, with multiple affected siblings born to unaffected parents—both findings that are hallmarks of recessive inheritance (see Chapter 7). This small family alone was clearly inadequate for linkage analysis. Instead, all four members of the family had their entire genomes sequenced and analyzed.

From an initial list of more than 4 million variants and assuming autosomal recessive inheritance of the disorder in both affected children, a filtering scheme similar to that described earlier yielded only four possible genes. One of these, *DHODH*, was also shown to be mutated in two other unrelated patients with POAD, thereby confirming this gene was responsible for the disorder in these families. *DHODH* encodes dihydroorotate dehydrogenase, a mitochondrial enzyme involved in pyrimidine biosynthesis, and was not suspected on biological grounds to be the gene responsible for this malformation syndrome.

Applications of Whole-Genome Sequence or Whole-Exome Sequence in Clinical Settings

Since the application of WGS or WES to rare mendelian disorders was first described in 2009, many hundreds of such disorders have been studied and the causative mutations found in over 300 previously unrecognized disease genes. Although the genome sequencing approach may miss certain categories of mutation that are difficult to detect routinely by sequencing alone (e.g., deletions or copy number variants) or that are difficult or impossible to recognize with our current understanding (e.g., noncoding mutations or regulatory mutations in intergenic regions), many groups report up to 25% to 40% success rates in identifying a causative mutation. These discoveries not only provide information useful for genetic counseling in the families involved, but also may inform clinical management and the potential development of effective treatments.

It is anticipated that the success rate of this approach will only increase as the costs of sequencing continue to fall and as our ability to interpret the likely functional consequences of sequence changes in the genome improves.

GENERAL REFERENCES

Altshuler D, Daly MJ, Lander ES: Genetic mapping in human disease, *Science* 322:881–888, 2008.
Manolio TA: Genomewide association studies and assessment of the risk of disease, *N Engl J Med* 363:166–176, 2010.
Risch N, Merikangas K: The future of genetic studies of complex human diseases, *Science* 273:1516-1517, 1996.
Terwilliger JD, Ott J: *Handbook of human genetic linkage*, Baltimore, 1994, Johns Hopkins University Press.

REFERENCES FOR SPECIFIC TOPICS

Abecasis GR, Auton A, Brooks LD, et al: An integrated map of genetic variation from 1,092 human genomes, *Nature* 491:56–65, 2012.
Bainbridge MN, Wiszniewski W, Murdock DR, et al: Whole-genome sequencing for optimized patient management, *Science Transl Med* 3:87re3, 2011.
Bush WS, Moore JH: Genome-wide association studies, *PLoS Computational Biol* 8:e1002822, 2012.
Denny JC, Bastarache L, Ritchie MD, et al: Systematic comparison of phenome-wide association study of electronic medical record data and genome-wide association data, *Nat Biotechnol* 31:1102–1110, 2013.
Fritsche LG, Chen W, Schu M, et al: Seven new loci associated with age-related macular degeneration, *Nat Genet* 17:1783–1786, 2013.
Gonzaga-Jauregui C, Lupski JR, Gibbs RA: Human genome sequencing in health and disease, *Annu Rev Med* 63:35–61, 2012.
Hindorff LA, MacArthur J, Morales J, et al: A catalog of published genome-wide association studies. Available at: www.genome.gov/gwastudies. Accessed February 1, 2015.
International HapMap Consortium: A second generation human haplotype map of over 3.1 million SNPs, *Nature* 449:851–861, 2007.
Kircher M, Witten DM, Jain P, et al: A general framework for estimating the relative pathogenicity of human genetic variants, *Nat Genet* 46:310–315, 2014.
Koboldt DC, Steinberg KM, Larson DE, et al: The next-generation sequencing revolution and its impact on genomics, *Cell* 155:27–38, 2013.
Manolio TA: Bringing genome-wide association findings into clinical use, *Nat Rev Genet* 14:549–558, 2014.
Matise TC, Chen F, Chen W, et al: A second-generation combined linkage-physical map of the human genome, *Genome Res* 17:1783–1786, 2007.
Roach JC, Glusman G, Smit AF, et al: Analysis of genetic inheritance in a family quartet by whole-genome sequencing, *Science* 328:636–639, 2010.
Robinson PC, Brown MA: Genetics of ankylosing spondylitis, *Mol Immunol* 57:2–11, 2014.
SEARCH Collaborative Group: *SLCO1B1* variants and statin-induced myopathy—a genomewide study, *N Engl J Med* 359:789–799, 2008.
Stahl EA, Wegmann D, Trynka G, et al: Bayesian inference analyses of the polygenic architecture of rheumatoid arthritis, *Nature Genet* 44:4383–4391, 2012.
Yang Y, Muzny DM, Reid JG, et al: Clinical whole-exome sequencing for the diagnosis of mendelian disorders, *N Engl J Med* 369:1502–1511, 2013.

PROBLEMS

1. The Huntington disease (HD) locus was found to be tightly linked to a DNA polymorphism on chromosome 4. In the same study, however, linkage was ruled out between HD and the locus for the MNSs blood group polymorphism, which also maps to chromosome 4. What is the explanation?

2. LOD scores (Z) between a polymorphism in the α-globin locus on the short arm of chromosome 16 and an autosomal dominant disease was analyzed in a series of British and Dutch families, with the following data:

θ	0.00	0.01	0.10	0.20	0.30	0.40
Z	$-\infty$	23.4	24.6	19.5	12.85	5.5

$Z_{max} = 25.85$ at $\theta_{max} = 0.05$

How would you interpret these data? Why is the value of Z given as $-\infty$ at $\theta = 0$?

In a subsequent study, a large family from Sicily with what looks like the same disease was also investigated for linkage to α-globin, with the following results:

θ	0.00	0.10	0.20	0.30	0.40
LOD scores (Z)	$-\infty$	-8.34	-3.34	-1.05	-0.02

How would you interpret the data in this second study?

3. This pedigree was obtained in a study designed to determine whether a mutation in a gene for γ-crystallin, one of the major proteins of the eye lens, may be responsible for an autosomal dominant form of cataract. The filled-in symbols in the pedigree indicate family members with cataracts. The letters indicate three alleles at the polymorphic γ-crystallin locus on chromosome 2. If you examine each affected person who has passed on the cataract to his or her children, how many of these represent a meiosis that is informative for linkage between the cataract and γ-crystallin? In which individuals is the phase known between the cataract mutation and the γ-crystallin alleles? Are there any meioses in which a crossover must have occurred to explain the data? What would you conclude about linkage between the cataract and γ-crystallin from this study? What additional studies might be performed to confirm or reject the hypothesis?

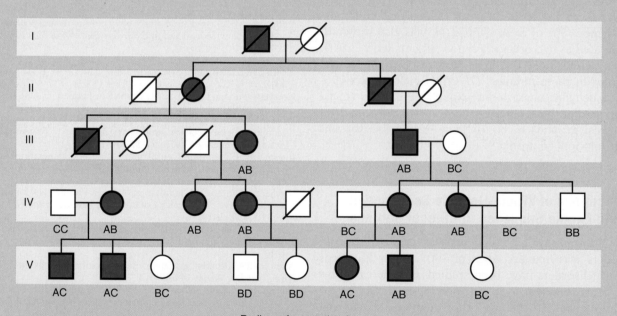

Pedigree for question 3

4. The following pedigree shows an example of molecular diagnosis in Wiskott-Aldrich syndrome, an X-linked immunodeficiency, by use of a linked DNA polymorphism with a map distance of approximately 5 cM between the polymorphic locus and the Wiskott-Aldrich syndrome gene.
   a. What is the likely phase in the carrier mother? How did you determine this? What diagnosis would you make regarding the current prenatal diagnosis if it were a male fetus?
   b. The maternal grandfather now becomes available for DNA testing and shows allele B at the linked locus. How does this finding affect your determination of phase in the mother? What diagnosis would you make now in regard to the current prenatal diagnosis?

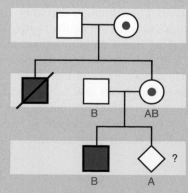

Pedigree for question 4

5. Review the pedigree in Figure 10-10B. If the unaffected grandmother, I-2, had been an *A/a* heterozygote, would it be possible to determine the phase in the affected parent, individual II-2?

6. In the pedigree below, showing a family with X-linked hemophilia A, can you determine the phase of the mutant factor VIII gene (*h*) and the normal allele (*H*) with respect to polymorphic alleles *M* and *m* in the mother of the two affected boys?

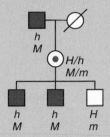

    Pedigree of X-linked hemophilia. The affected grandfather in the first generation has the disease (mutant allele *h*) and allele *M* at a polymorphic locus on the X chromosome.

7. Calculate D′ for the three scenarios listed in Figure 10-7.

8. Relative risk calculations are used for cohort studies and not case-control studies. To demonstrate why this is the case, imagine a case-control study for the effect of a genetic variant on disease susceptibility. The investigator has ascertained as many affected individuals (a + c) as possible and then arbitrarily chooses a set of (b + d) controls. They are genotyped as to whether a variant is present: a/(a + c) of the affected have the variant, whereas b/(b + d) of the controls have the variant.

	Disease Present	Disease Absent
Variant present	a	b
Variant absent	c	d
	a + c	b + d

Calculate the odds ratio and relative risk for the association between the variant being present and the disease being present.

Now, imagine the investigator arbitrarily decided to use three times as many unaffected individuals, 3 × (b + d), as controls. The investigator has every right to do so because it is a case-control study and the numbers of affected and unaffected are not determined by the prevalence of the disease in the population being studied, as they would be in a cohort study. Assume the distribution of the variant remains the same in this control group as with the smaller control group that is, 3b/[3 × (b + d)] = b/(b + d) carrying the allele.

	Disease Present	Disease Absent
Variant present	a	3b
Variant absent	c	3d
	a + c	3 × (b + d)

Recalculate the OR and RR with this new control group. Do the same when an arbitrary control group is an n-tuple of the original control group; that is, the size of the control group is n × (b + d).

Which of these measures, OR or RR, does not change when different, arbitrarily sized control groups are used?

CHAPTER 11

The Molecular Basis of Genetic Disease

General Principles and Lessons from the Hemoglobinopathies

The term *molecular disease*, introduced over six decades ago, refers to disorders in which the primary disease-causing event is an alteration, either inherited or acquired, affecting a gene(s), its structure, and/or its expression. In this chapter, we first outline the basic genetic and biochemical mechanisms underlying monogenic or single-gene disorders. We then illustrate them in the context of their molecular and clinical consequences using inherited diseases of hemoglobin—the hemoglobinopathies—as examples. This overview of mechanisms is expanded in Chapter 12 to include other genetic diseases that illustrate additional principles of genetics in medicine.

A genetic disease occurs when an alteration in the DNA of an essential gene changes the amount or function, or both, of the gene products—typically messenger RNA (mRNA) and protein but occasionally specific noncoding RNAs (ncRNAs) with structural or regulatory functions. Although almost all known single-gene disorders result from mutations that affect the function of a protein, a few exceptions to this generalization are now known. These exceptions are diseases due to mutations in ncRNA genes, including microRNA (miRNA) genes that regulate specific target genes, and mitochondrial genes that encode transfer RNAs (tRNAs; see Chapter 12). It is essential to understand genetic disease at the molecular and biochemical levels, because this knowledge is the foundation of rational therapy. In this chapter, we restrict our attention to diseases caused by defects in protein-coding genes; the study of phenotype at the level of proteins, biochemistry, and metabolism constitutes the discipline of **biochemical genetics.**

By 2014, the online version of *Mendelian Inheritance in Man* listed over 5500 phenotypes for which the molecular basis is known, largely phenotypes with autosomal and X-linked inheritance. Although it is impressive that the basic molecular defect has been found in so many disorders, it is sobering to realize that the pathophysiology is not entirely understood for *any* genetic disease. **Sickle cell disease** (Case 42), discussed later in this chapter, is among the best characterized of all inherited disorders, but even here, knowledge is incomplete—despite its being the first molecular disease to be recognized, more than 65 years ago.

THE EFFECT OF MUTATION ON PROTEIN FUNCTION

Mutations involving protein-coding genes have been found to cause disease through one of four different effects on protein function (Fig. 11-1). The most common effect by far is a **loss of function** of the mutant protein. Many important conditions arise, however, from other mechanisms: a **gain of function**, the acquisition of a **novel property** by the mutant protein, or the expression of a gene at the wrong time (**heterochronic expression**) and/or in the wrong place (**ectopic expression**).

Loss-of-Function Mutations

The loss of function of a gene may result from alteration of its coding, regulatory, or other critical sequences due to nucleotide substitutions, deletions, insertions, or rearrangements. A loss of function due to deletion, leading to a reduction in gene dosage, is exemplified by the α-thalassemias (Case 44), which are most commonly due to deletion of α-globin genes (see later discussion); by chromosome-loss diseases (Case 27), such as monosomies like **Turner syndrome** (see Chapter 6) (Case 47); and by acquired somatic mutations—often deletions—that occur in tumor-suppressor genes in many cancers, such as **retinoblastoma** (Case 39) (see Chapter 15). Many other types of mutations can also lead to a complete loss of function, and all are illustrated by the β-thalassemias (Case 44) (see later discussion), a group of hemoglobinopathies that result from a reduction in the abundance of β-globin, one of the major adult hemoglobin proteins in red blood cells.

The severity of a disease due to loss-of-function mutations generally correlates with the amount of function lost. In many instances, the retention of even a

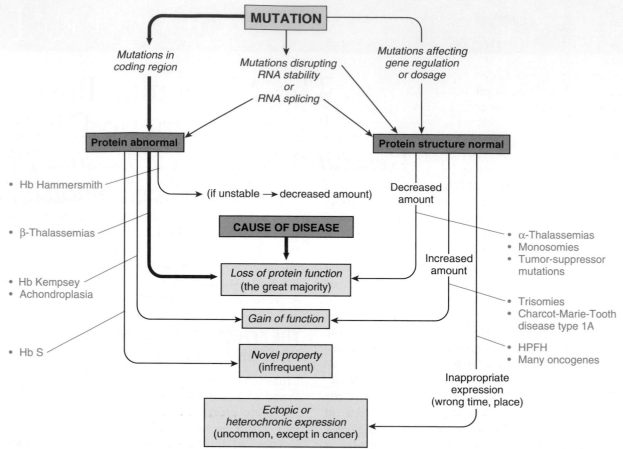

Figure 11-1 A general outline of the mechanisms by which disease-causing mutations produce disease. Mutations in the coding region result in structurally abnormal proteins that have a loss or gain of function or a novel property that causes disease. Mutations in noncoding sequences are of two general types: those that alter the stability or splicing of the messenger RNA (mRNA) and those that disrupt regulatory elements or change gene dosage. Mutations in regulatory elements alter the abundance of the mRNA or the time or cell type in which the gene is expressed. Mutations in either the coding region or regulatory domains can decrease the amount of the protein produced. HPFH, Hereditary persistence of fetal hemoglobin.

small percent of residual function by the mutant protein greatly reduces the severity of the disease.

Gain-of-Function Mutations

Mutations may also enhance one or more of the normal functions of a protein; in a biological system, however, more is not necessarily better, and disease may result. It is critical to recognize when a disease is due to a gain-of-function mutation because the treatment must necessarily differ from disorders due to other mechanisms, such as loss-of-function mutations. Gain-of-function mutations fall into two broad classes:

- **Mutations that increase the production of a normal protein.** Some mutations cause disease by increasing the synthesis of a normal protein in cells in which the protein is *normally present*. The most common mutations of this type are due to increased gene dosage, which generally results from duplication of part or all of a chromosome. As discussed in Chapter 6, the classic example is **trisomy 21 (Down syndrome)**,

which is due to the presence of three copies of chromosome 21. Other important diseases arise from the increased dosage of single genes, including one form of familial Alzheimer disease due to a duplication of the amyloid precursor protein (βAPP) gene (see Chapter 12), and the peripheral nerve degeneration **Charcot-Marie-Tooth disease type 1A (Case 8)**, which generally results from duplication of only one gene, the gene for peripheral myelin protein 22 (*PMP22*).

- **Mutations that enhance one normal function of a protein.** Rarely, a mutation in the coding region may increase the ability of each protein molecule to perform one or more of its normal functions, even though this increase is detrimental to the overall physiological role of the protein. For example, the missense mutation that creates hemoglobin Kempsey locks hemoglobin into its high oxygen affinity state, thereby reducing oxygen delivery to tissues. Another example of this mechanism occurs in the form of short stature called **achondroplasia (Case 2)**.

Novel Property Mutations

In a few diseases, a change in the amino acid sequence confers a novel property on the protein, without necessarily altering its normal functions. The classic example of this mechanism is **sickle cell disease** (Case 42), which, as we will see later in this chapter, is due to an amino acid substitution that has *no* effect on the ability of sickle hemoglobin to transport oxygen. Rather, unlike normal hemoglobin, sickle hemoglobin chains aggregate when they are deoxygenated and form abnormal polymeric fibers that deform red blood cells. That novel property mutations are infrequent is not surprising, because most amino acid substitutions are either neutral or detrimental to the function or stability of a protein that has been finely tuned by evolution.

Mutations Associated with Heterochronic or Ectopic Gene Expression

An important class of mutations includes those that lead to inappropriate expression of the gene at an abnormal time or place. These mutations occur in the regulatory regions of the gene. Thus cancer is frequently due to the abnormal expression of a gene that normally promotes cell proliferation—an **oncogene**—in cells in which the gene is not normally expressed (see Chapter 15). Some mutations in hemoglobin regulatory elements lead to the continued expression in adults of the γ-globin gene, which is normally expressed at high levels only in fetal life. Such γ-globin gene mutations cause a benign phenotype called the **hereditary**

persistence of fetal hemoglobin (Hb F), as we explore later in this chapter.

HOW MUTATIONS DISRUPT THE FORMATION OF BIOLOGICALLY NORMAL PROTEINS

Disruptions of the normal functions of a protein that result from the various types of mutations outlined earlier can be well exemplified by the broad range of diseases due to mutations in the globin genes, as we will explore in the second part of this chapter. To form a biologically active protein (such as the hemoglobin molecule), information must be transcribed from the nucleotide sequence of the gene to the mRNA and then translated into the polypeptide, which then undergoes progressive stages of maturation (see Chapter 3). Mutations can disrupt any of these steps (Table 11-1). As we shall see next, abnormalities in five of these stages are illustrated by various hemoglobinopathies; the others are exemplified by diseases to be presented in Chapter 12.

THE RELATIONSHIP BETWEEN GENOTYPE AND PHENOTYPE IN GENETIC DISEASE

Variation in the clinical phenotype observed in an inherited disease may have any of three genetic explanations, namely:

- allelic heterogeneity
- locus heterogeneity, or
- the effect of modifier genes

TABLE 11-1 The Eight Steps at Which Mutations Can Disrupt the Production of a Normal Protein

Step	Disease Example
Transcription	**Thalassemias** due to reduced or absent production of a globin mRNA because of deletions or mutations in regulatory or splice sites of a globin gene
	Hereditary persistence of fetal hemoglobin, which results from increased postnatal transcription of one or more γ-globin genes
Translation	**Thalassemias** due to nonfunctional or rapidly degraded mRNAs with nonsense or frameshift mutations
Polypeptide folding	More than 70 **hemoglobinopathies** are due to abnormal hemoglobins with amino acid substitutions or deletions that lead to unstable globins that are prematurely degraded (e.g., Hb Hammersmith)
Post-translational modification	**I-cell disease,** a lysosomal storage disease that is due to a failure to add a phosphate group to mannose residues of lysosomal enzymes. The mannose 6-phosphate residues are required to target the enzymes to lysosomes (see Chapter 12)
Assembly of monomers into a holomeric protein	Types of **osteogenesis imperfecta** in which an amino acid substitution in a procollagen chain impairs the assembly of a normal collagen triple helix (see Chapter 12)
Subcellular localization of the polypeptide or the holomer	**Familial hypercholesterolemia mutations** (class 4), in the carboxyl terminus of the LDL receptor, that impair the localization of the receptor to clathrin-coated pits, preventing the internalization of the receptor and its subsequent recycling to the cell surface (see Chapter 12)
Cofactor or prosthetic group binding to the polypeptide	Types of **homocystinuria** due to poor or absent binding of the cofactor (pyridoxal phosphate) to the cystathionine synthase apoenzyme (see Chapter 12)
Function of a correctly folded, assembled, and localized protein produced in normal amounts	Diseases in which the mutant protein is normal in nearly every way, except that one of its critical biological activities is altered by an amino acid substitution (e.g., in **Hb Kempsey,** impaired subunit interaction locks hemoglobin into its high oxygen affinity state)

LDL, Low-density lipoprotein; mRNA, messenger RNA.

TABLE 11-2 Types of Heterogeneity Associated with Genetic Disease

Type of Heterogeneity	Definition	Example
Genetic heterogeneity		
Allelic heterogeneity	The occurrence of more than one allele at a locus	α-Thalassemia
		β-Thalassemia
Locus heterogeneity	The association of more than one locus with a clinical phenotype	Thalassemia can result from mutations in either the α-globin or β-globin genes
Clinical or phenotypic heterogeneity	The association of more than one phenotype with mutations at a single locus	Sickle cell disease and β-thalassemia each result from distinct β-globin gene mutations

Each of these types can be illustrated by mutations in the α-globin or β-globin genes (Table 11-2).

Allelic Heterogeneity

Genetic heterogeneity is most commonly due to the presence of multiple alleles at a single locus, a situation referred to as allelic heterogeneity (see Chapter 7 and Table 11-1). In many instances, there is a clear **genotype-phenotype correlation** between a specific allele and a specific phenotype. The most common explanation for the effect of allelic heterogeneity on the clinical phenotype is that alleles that confer more **residual function** on the mutant protein are often associated with a milder form of the principal phenotype associated with the disease. In some instances, however, alleles that confer some residual protein function are associated with only one or a subset of the complete set of phenotypes seen with a missing or completely nonfunctional allele (frequently termed a *null allele*). As we will explore more fully in Chapter 12, this situation prevails with certain variants of the **cystic fibrosis** gene, *CFTR*; these variants lead to a phenotypically different condition, **congenital absence of the vas deferens**, but not to the other manifestations of cystic fibrosis.

A second explanation for allele-based variation in phenotype is that the variation may reflect the specific property of the protein that is most perturbed by the mutation. This situation is well illustrated by **Hb Kempsey**, a β-globin allele that maintains the hemoglobin in a high oxygen affinity structure, causing polycythemia because the reduced peripheral delivery of oxygen is misinterpreted by the hematopoietic system as being due to an inadequate production of red blood cells.

The biochemical and clinical consequences of a specific mutation in a protein are often unpredictable. Thus no one would have foreseen that the β-globin allele associated with sickle cell disease would lead to the formation of globin polymers that deform erythrocytes to a sickle cell shape (see later in this chapter). Sickle cell disease is highly unusual in that it results *only* from a single specific mutation—the Glu6Val substitution in the β-globin chain—whereas most disease phenotypes can arise from any of a number or many substitutions, usually loss-of-function mutations, in the affected protein.

Locus Heterogeneity

Genetic heterogeneity also arises when mutations in more than one locus can result in a specific clinical condition, a situation termed *locus heterogeneity* (see Chapter 7). This phenomenon is illustrated by the finding that thalassemia can result from mutations in either the β-globin or α-globin chain (see Table 11-2). Once locus heterogeneity has been documented, careful comparison of the phenotype associated with each gene sometimes reveals that the phenotype is not as homogeneous as initially believed.

Modifier Genes

Sometimes even the most robust genotype-phenotype relationships are found not to hold for a specific patient. Such phenotypic variation can, in principle, be ascribed to environmental factors or to the action of other genes, termed *modifier genes* (see Chapter 8). To date, only a few modifier genes for human monogenic disorders have been identified, although one anticipates that there will be numerous examples as our understanding of the basis for disease increases. One example described later in this chapter is seen in β-thalassemia homozygotes (carrying mutations at the β-globin locus) who also inherit an α-thalassemia variant at the α-globin locus.

THE HEMOGLOBINS

To illustrate in greater detail the concepts introduced in the first section of this chapter, we now turn to disorders of human hemoglobins—referred to as hemoglobinopathies—the most common single-gene diseases in humans. These disorders cause substantial morbidity, and the World Health Organization estimates that more than 5% of the world's population are carriers of genetic variants for clinically important disorders of hemoglobin. They are also important because their molecular and biochemical pathology is better understood than perhaps that of any other group of genetic diseases. Before the hemoglobinopathies are

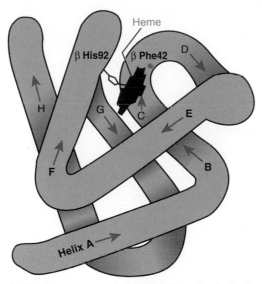

Figure 11-2 **The structure of a hemoglobin subunit.** Each subunit has eight helical regions, designated A to H. The two most conserved amino acids are shown: His92, the histidine to which the iron of heme is covalently linked; and Phe42, the phenylalanine that wedges the porphyrin ring of heme into the heme "pocket" of the folded protein. See discussion of Hb Hammersmith and Hb Hyde Park, which have substitutions for Phe42 and His92, respectively, in the β-globin molecule.

discussed in depth, it is important to briefly introduce the normal aspects of the globin genes and hemoglobin biology.

Structure and Function of Hemoglobin

Hemoglobin is the oxygen carrier in vertebrate red blood cells. Each hemoglobin molecule consists of four subunits: two α-globin chains and two β- (or β-like) globin chains. Each subunit is composed of a polypeptide chain, globin, and a prosthetic group, heme, which is an iron-containing pigment that combines with oxygen to give the molecule its oxygen-transporting ability (Fig. 11-2). The predominant adult human hemoglobin, Hb A, has an $\alpha_2\beta_2$ structure in which the four chains are folded and fitted together to form a globular tetramer.

As with all proteins that have been strongly conserved throughout evolution, the *tertiary structure* of globins is constant; virtually all globins have seven or eight helical regions (depending on the chain) (see Fig. 11-2). Mutations that disrupt this tertiary structure invariably have pathological consequences. In addition, mutations that substitute a highly conserved amino acid or that replace one of the nonpolar residues, which form the hydrophobic shell that excludes water from the interior of the molecule, are likely to cause a hemoglobinopathy (see Fig. 11-2). Like all proteins, globin has sensitive areas, in which mutations cannot occur without affecting function, and insensitive areas, in which variations are more freely tolerated.

The Globin Genes

In addition to Hb A, with its $\alpha_2\beta_2$ structure, there are five other normal human hemoglobins, each of which has a tetrameric structure like that of Hb A in consisting of two α or α-like chains and two non-α chains (Fig. 11-3A). The genes for the α and α-like chains are clustered in a tandem arrangement on chromosome 16. Note that there are two *identical* α-globin genes, designated α1 and α2, on each homologue. The β- and β-like globin genes, located on chromosome 11, are close family members that, as described in Chapter 3, undoubtedly arose from a common ancestral gene (see Fig. 11-3A). Illustrating this close evolutionary relationship, the β- and δ-globins differ in only 10 of their 146 amino acids.

Developmental Expression of Globin Genes and Globin Switching

The expression of the various globin genes changes during development, a process referred to as **globin switching** (see Fig. 11-3B). Note that the genes in the α- and β-globin clusters are arranged in the same transcriptional orientation and, remarkably, the genes in each cluster are situated in the same order in which they are expressed during development. The temporal switches of globin synthesis are accompanied by changes in the principal site of erythropoiesis (see Fig. 11-3B). Thus the three embryonic globins are made in the yolk sac from the third to eighth weeks of gestation, but at approximately the fifth week, hematopoiesis begins to move from the yolk sac to the fetal liver. Hb F ($\alpha_2\gamma_2$), the predominant hemoglobin throughout fetal life, constitutes approximately 70% of total hemoglobin at birth. In adults, however, Hb F represents less than a few percent of the total hemoglobin, although this can vary from less than 1% to approximately 5% in different individuals.

β-chain synthesis becomes significant near the time of birth, and by 3 months of age, almost all hemoglobin is of the adult type, Hb A ($\alpha_2\beta_2$) (see Fig. 11-3B). In diseases due to mutations that decrease the abundance of β-globin, such as β-thalassemia (see later section), strategies to increase the normally small amount of γ globin (and therefore of Hb F ($\alpha_2\gamma_2$)) produced in adults are proving to be successful in ameliorating the disorder (see Chapter 13).

The Developmental Regulation of β-Globin Gene Expression: The Locus Control Region

Elucidation of the mechanisms that control the expression of the globin genes has provided insight into both normal and pathological biological processes. The expression of the β-globin gene is only partly controlled by the promoter and two enhancers in the immediate

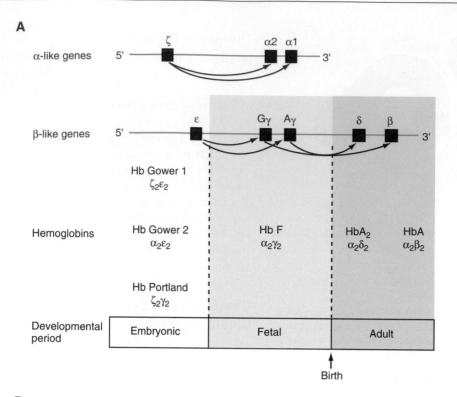

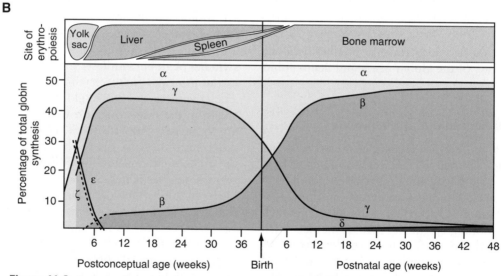

Figure 11-3 Organization of the human globin genes and hemoglobins produced in each stage of human development. **A,** The α-like genes are on chromosome 16, the β-like genes on chromosome 11. The *curved arrows* refer to the switches in gene expression during development. **B,** Development of erythropoiesis in the human fetus and infant. Types of cells responsible for hemoglobin synthesis, organs involved, and types of globin chain synthesized at successive stages are shown. *See Sources & Acknowledgments.*

flanking DNA (see Chapter 3). A critical requirement for additional regulatory elements was first suggested by the identification of a unique group of patients who had no gene expression from *any* of the genes in the β-globin cluster, even though the genes themselves (including their individual regulatory elements) were intact. These informative patients were found to have large deletions upstream of the β-globin complex, deletions that removed an approximately 20-kb domain called the **locus control region (LCR),** which begins approximately

6 kb upstream of the ε-globin gene (Fig. 11-4). Although the resulting disease, εγδβ-thalassemia, is described later in this chapter, these patients demonstrate that the LCR is required for the expression of all the genes in the β-globin cluster.

The LCR is defined by five so-called DNase I hypersensitive sites (see Fig. 11-4), genomic regions that are unusually open to certain proteins (such as the enzyme DNase I) that are used experimentally to reveal potential regulatory sites. Within the context of the epigenetic

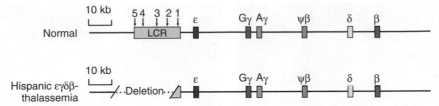

Figure 11-4 The β-globin locus control region (LCR). Each of the five regions of open chromatin (*arrows*) contains several consensus binding sites for both erythroid-specific and ubiquitous transcription factors. The precise mechanism by which the LCR regulates gene expression is unknown. Also shown is a deletion of the LCR that has led to εγδβ-thalassemia, which is discussed in the text. *See Sources & Acknowledgments.*

packaging of chromatin (see Chapter 3), these sites configure an open chromatin state at the locus in erythroid cells, the role of which is to maintain an open chromatin configuration at the locus, a configuration that gives transcription factors access to the regulatory elements that mediate the expression of each of the β-globin genes (see Chapter 3). The LCR, along with its associated DNA-binding proteins, interacts with the genes of the β-globin locus to form a nuclear domain called the **active chromatin hub**, where β-globin gene expression takes place. The sequential switching of gene expression that occurs among the five members of the β-globin gene complex during development results from the sequential association of the active chromatin hub with the different genes in the cluster as the hub moves from the most proximal gene in the complex (the ε-globin gene in embryos) to the most distal (the δ- and β-globin genes in adults).

The clinical significance of the LCR is threefold. First, as mentioned, patients with deletions of the LCR fail to express the genes of the β-globin cluster. Second, components of the LCR are likely to be essential in gene therapy (see Chapter 13) for disorders of the β-globin cluster, so that the therapeutic normal copy of the gene in question is expressed at the correct time in life and in the appropriate tissue. And third, knowledge of the molecular mechanisms that underlie globin switching may make it feasible to up-regulate the expression of the γ-globin gene in patients with β-thalassemia (who have mutations only in the β-globin gene), because Hb F ($\alpha_2\gamma_2$) is an effective oxygen carrier in adults who lack Hb A ($\alpha_2\beta_2$) (see Chapter 13).

Gene Dosage, Developmental Expression of the Globins, and Clinical Disease

The differences both in the gene dosage of the α- and β-globins (four α-globin and two β-globin genes per diploid genome), and in their patterns of expression during development, are important to an understanding of the pathogenesis of many hemoglobinopathies. Mutations in the β-globin gene are more likely to cause disease than are α-chain mutations because a single β-globin gene mutation affects 50% of the β chains, whereas a single α-chain mutation affects only 25% of

the α chains. On the other hand, β-globin mutations have no prenatal consequences because γ-globin is the major β-like globin before birth, with Hb F constituting 75% of the total hemoglobin at term (see Fig. 11-3B). In contrast, because α chains are the only α-like components of hemoglobins 6 weeks after conception, α-globin mutations cause severe disease in both fetal and postnatal life.

THE HEMOGLOBINOPATHIES

Hereditary disorders of hemoglobin can be divided into the following three broad groups, which in some instances overlap:

- **Structural variants**, which alter the amino acid sequence of the globin polypeptide, altering properties such as its ability to transport oxygen, or reducing its stability. *Example*: Sickle cell disease (Case 42), due to a mutation that makes deoxygenated β-globin relatively insoluble, changing the shape of the red cell (Fig. 11-5).
- **Thalassemias**, which are diseases that result from the decreased abundance of one or more of the globin chains (Case 44). The decrease can result from decreased production of a globin chain or, less commonly, from a structural variant that destabilizes the chain. The resulting imbalance in the ratio of the α:β chains underlies the pathophysiology of these conditions. *Example*: promoter mutations that decrease expression of the β-globin mRNA to cause β-thalassemia.
- **Hereditary persistence of fetal hemoglobin**, a group of *clinically benign* conditions that impair the perinatal switch from γ-globin to β-globin synthesis. *Example*: a deletion, found in African Americans, that removes both the δ- and β-globin genes but leads to continued postnatal expression of the γ-globin genes, to produce Hb F, which is an effective oxygen transporter (see Fig. 11-3).

Hemoglobin Structural Variants

Most variant hemoglobins result from point mutations in one of the globin structural genes. More than 400 abnormal hemoglobins have been described, and

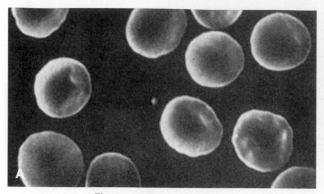

Figure 11-5 Scanning electron micrographs of red cells from a patient with sickle cell disease. A, Oxygenated cells are round and full. B, The classic sickle cell shape is produced only when the cells are in the deoxygenated state. *See Sources & Acknowledgments.*

TABLE 11-3 The Major Classes of Hemoglobin Structural Variants

Variant Class*	Amino Acid Substitution	Pathophysiological Effect of Mutation	Inheritance
Hb S	β chain: Glu6Val	Deoxygenated Hb S polymerizes → sickle cells → vascular occlusion and hemolysis	AR
Hb Hammersmith	β chain: Phe42Ser	An unstable Hb → Hb precipitation → hemolysis; also low oxygen affinity	AD
Hb Hyde Park (a Hb M)	β chain: His92Tyr	The substitution makes oxidized heme iron resistant to methemoglobin reductase → Hb M, which cannot carry oxygen → cyanosis (asymptomatic)	AD
Hb Kempsey	β chain: Asp99Asn	The substitution keeps the Hb in its high oxygen affinity structure → less oxygen to tissues → polycythemia	AD
Hb E	β chain: Glu26Lys	The mutation → an abnormal Hb *and* decreased synthesis (abnormal RNA splicing) → mild thalassemia[†] (see Fig. 11-11)	AR

*Hemoglobin variants are often named after the home town of the first patient described.
[†]Additional β-chain structural variants that cause β-thalassemia are depicted in Table 11-5.
AD, Autosomal dominant; AR, autosomal recessive; Hb M, methemoglobin; see text.

approximately half of these are clinically significant. The hemoglobin structural variants can be separated into the following three classes, depending on the clinical phenotype (Table 11-3):

- Variants that cause **hemolytic anemia**, most commonly because they make the hemoglobin tetramer unstable.
- Variants with **altered oxygen transport,** due to increased or decreased oxygen affinity or to the formation of methemoglobin, a form of globin incapable of reversible oxygenation.
- Variants due to mutations in the coding region that cause **thalassemia** because they reduce the abundance of a globin polypeptide. Most of these mutations impair the rate of synthesis of the mRNA or otherwise affect the level of the encoded protein.

Hemolytic Anemias

Hemoglobins with Novel Physical Properties: Sickle Cell Disease.
Sickle cell hemoglobin is of great clinical importance in many parts of the world. The disease results from a single nucleotide substitution that changes the codon of the sixth amino acid of β-globin from glutamic acid to valine (GAG → GTG: Glu6Val; see Table 11-3). Homozygosity for this mutation is the cause of **sickle cell disease** (Case 42). The disease has a characteristic geographical distribution, occurring most frequently in equatorial Africa and less commonly in the Mediterranean area and India and in countries to which people from these regions have migrated. Approximately 1 in 600 African Americans is born with this disease, which may be fatal in early childhood, although longer survival is becoming more common.

Clinical Features. Sickle cell disease is a severe autosomal recessive hemolytic condition characterized by a tendency of the red blood cells to become grossly abnormal in shape (i.e., take on a sickle shape) under conditions of low oxygen tension (see Fig. 11-5). Heterozygotes, who are said to have **sickle cell trait**, are generally clinically normal, but their red cells can sickle when they are subjected to very low oxygen pressure in vitro. Occasions when this occurs are uncommon, although heterozygotes appear to be at risk for splenic infarction, especially at high altitude (for example in airplanes with reduced cabin pressure) or when exerting themselves to extreme levels in athletic competition.

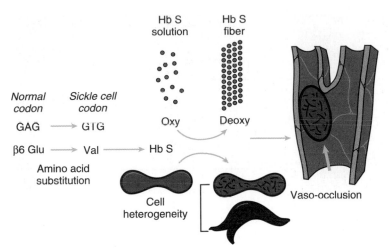

Figure 11-6 The pathogenesis of sickle cell disease. *See Sources & Acknowledgments.*

The heterozygous state is present in approximately 8% of African Americans, but in areas where the sickle cell allele (β^S) frequency is high (e.g., West Central Africa), up to 25% of the newborn population are heterozygotes.

The Molecular Pathology of Hb S. Nearly 60 years ago, Ingram discovered that the abnormality in sickle cell hemoglobin was a replacement of one of the 146 amino acids in the β chain of the hemoglobin molecule. All the clinical manifestations of sickle cell hemoglobin are consequences of this single change in the β-globin gene. Ingram's discovery was the first demonstration *in any organism* that a mutation in a structural gene could cause an amino acid substitution in the corresponding protein. Because the substitution is in the β-globin chain, the formula for sickle cell hemoglobin is written as $\alpha_2\beta_2^S$ or, more precisely, $\alpha_2^A\beta_2^S$. A heterozygote has a mixture of the two types of hemoglobin, A and S, summarized as $\alpha_2^A\beta_2^A/\alpha_2^A\beta_2^S$, as well as a hybrid hemoglobin tetramer, written as $\alpha_2^A\beta^A\beta^S$. Strong evidence indicates that the sickle mutation arose in West Africa but that it also occurred independently elsewhere. The β^S allele has attained high frequency in malarial areas of the world because it confers protection against malaria in heterozygotes (see Chapter 9).

Sickling and Its Consequences. The molecular and cellular pathology of sickle cell disease is summarized in Figure 11-6. Hemoglobin molecules containing the mutant β-globin subunits are normal in their ability to perform their principal function of binding oxygen (provided they have not polymerized, as described next), but in deoxygenated blood, they are only one fifth as soluble as normal hemoglobin. Under conditions of low oxygen tension, this relative insolubility of deoxyhemoglobin S causes the sickle hemoglobin molecules to aggregate in the form of rod-shaped polymers or fibers (see Fig. 11-5). These molecular rods distort the $\alpha_2\beta_2^S$ erythrocytes to a sickle shape that prevents them from squeezing single file through capillaries, as do normal red cells, thereby blocking blood flow and causing local ischemia.

They may also cause disruption of the red cell membrane (hemolysis) and release of free hemoglobin, which can have deleterious effects on the availability of vasodilators, such as nitric oxide, thereby exacerbating the ischemia.

Modifier Genes Determine the Clinical Severity of Sickle Cell Disease. It has long been known that a strong modifier of the clinical severity of sickle cell disease is the patient's level of Hb F ($\alpha_2\gamma_2$), higher levels being associated with less morbidity and lower mortality. The physiological basis of the ameliorating effect of Hb F is clear: Hb F is a perfectly adequate oxygen carrier in postnatal life and also inhibits the polymerization of deoxyhemoglobin S.

Until recently, however, it was not certain whether the variation in Hb F expression was heritable. Genome-wide association studies (GWAS) (see Chapter 10) have demonstrated that single nucleotide polymorphisms (SNPs) at three loci—the γ-globin gene and two genes that encode transcription factors, *BCL11A* and *MYB*—account for 40% to 50% of the variation in the levels of Hb F in patients with sickle cell disease. Moreover, the Hb F–associated SNPs are also associated with the painful clinical episodes thought to be due to capillary occlusion caused by sickled red cells (Fig. 11-6).

The genetically driven variations in the level of Hb F are also associated with variation in the clinical severity of β-thalassemia (discussed later) because the reduced abundance of β-globin (and thus of Hb A [$\alpha_2\beta_2$]) in that disease is partly alleviated by higher levels of γ-globin and thus of Hb F ($\alpha_2\gamma_2$). The discovery of these genetic modifiers of Hb F abundance not only explains much of the variation in the clinical severity of sickle cell disease and β-thalassemia, but it also highlights a general principle introduced in Chapter 8: *modifier genes can play a major role in determining the clinical and physiological severity of a single-gene disorder.*

BCL11A, a Silencer of γ-Globin Gene Expression in Adult Erythroid Cells. The identification of genetic modifiers of Hb F levels, particularly *BCL11A*, has

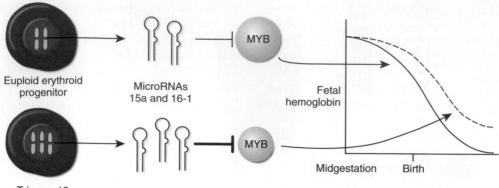

Figure 11-7 A model demonstrating how elevations of microRNAs 15a and 16-1 in trisomy 13 can result in elevated fetal hemoglobin expression. Normally, the basal level of these microRNAs can moderate expression of targets such as the *MYB* gene during erythropoiesis. In the case of trisomy 13, elevated levels of these microRNAs results in additional down-regulation of *MYB* expression, which in turn results in a delayed switch from fetal to adult hemoglobin and persistent expression of fetal hemoglobin. *See Sources & Acknowledgments.*

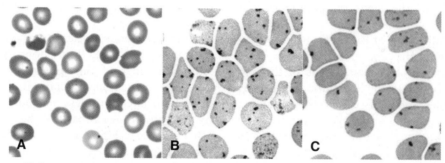

Figure 11-8 Visualization of one pathological effect of the deficiency of β chains in β-thalassemia: the precipitation of the excess normal α chains to form a Heinz body in the red blood cell. Peripheral blood smear and Heinz body preparation. A-C, The peripheral smear (**A**) shows "bite" cells with pitted-out semicircular areas of the red blood cell membrane as a result of removal of Heinz bodies by macrophages in the spleen, causing premature destruction of the red cell. The Heinz body preparation (**B**) shows increased Heinz bodies in the same specimen when compared to a control (**C**). *See Sources & Acknowledgments.*

great therapeutic potential. The product of the *BCL11A* gene is a transcription factor that normally silences γ-globin expression, thus shutting down Hb F production postnatally. Accordingly, drugs that suppress *BCL11A* activity postnatally, thereby *increasing* the expression of Hb F, might be of great benefit to patients with sickle cell disease and β-thalassemia (see Chapter 13), disorders that affect millions of individuals worldwide. Small molecule screening programs to identify potential drugs of this type are now underway in many laboratories.

Trisomy 13, MicroRNAs, and MYB, Another Silencer of γ-Globin Gene Expression. The indication from GWAS that MYB is an important regulator of γ-globin expression has received further support from an unexpected direction, studies investigating the basis for the persistent increased postnatal expression of Hb F that is observed in patients with trisomy 13 (see Chapter 6). Two miRNAs, miR-15a and miR-16-1, directly target the 3' untranslated region (UTR) of the *MYB* mRNA, thereby reducing *MYB* expression. The genes for these two miRNAs are located on chromosome 13; their extra dosage in trisomy 13 is predicted to reduce *MYB* expression below normal levels, thereby partly relaxing the postnatal suppression of γ-globin gene expression normally mediated by the MYB protein, and leading to increased expression of Hb F (Fig. 11-7).

Unstable Hemoglobins. The unstable hemoglobins are due largely to point mutations that cause denaturation of the hemoglobin tetramer in mature red blood cells. The denatured globin tetramers are insoluble and precipitate to form inclusions (Heinz bodies) that contribute to damage of the red cell membrane and cause the hemolysis of mature red blood cells in the vascular tree (Fig. 11-8, showing a Heinz body due to β-thalassemia).

The amino acid substitution in the unstable hemoglobin **Hb Hammersmith** (β-chain Phe42Ser; see Table

11-3) leads to denaturation of the tetramer and consequent hemolysis. This mutation is particularly notable because the substituted phenylalanine residue is one of the two amino acids that are conserved in all globins in nature (see Fig. 11-2). It is therefore not surprising that substitutions of this phenylalanine produce serious disease. In normal β-globin, the bulky phenylalanine wedges the heme into a "pocket" in the folded β-globin monomer. Its replacement by serine, a smaller residue, creates a gap that allows the heme to slip out of its pocket. In addition to its instability, Hb Hammersmith has a low oxygen affinity, which causes cyanosis in heterozygotes.

In contrast to mutations that destabilize the *tetramer*, other variants destabilize the globin *monomer* and never form the tetramer, causing chain imbalance and thalassemia (see following section).

Variants with Altered Oxygen Transport

Mutations that alter the ability of hemoglobin to transport oxygen, although rare, are of general interest because they illustrate how a mutation can impair one function of a protein (in this case, oxygen binding and release) and yet leave the other properties of the protein relatively intact. For example, the mutations that affect oxygen transport generally have little or no effect on hemoglobin stability.

Methemoglobins. Oxyhemoglobin is the form of hemoglobin that is capable of reversible oxygenation; its heme iron is in the reduced (or ferrous) state. The heme iron tends to oxidize spontaneously to the ferric form and the resulting molecule, referred to as methemoglobin, is incapable of reversible oxygenation. If significant amounts of methemoglobin accumulate in the blood, cyanosis results. Maintenance of the heme iron in the reduced state is the role of the enzyme methemoglobin reductase. In several mutant globins (either α or β), substitutions in the region of the heme pocket affect the heme-globin bond in a way that makes the iron resistant to the reductase. Although heterozygotes for these mutant hemoglobins are cyanotic, they are asymptomatic. The homozygous state is presumably lethal. One example of a β-chain methemoglobin is **Hb Hyde Park** (see Table 11-3), in which the conserved histidine (His92 in Fig. 11-2) to which heme is covalently bound has been replaced by tyrosine (His92Tyr).

Hemoglobins with Altered Oxygen Affinity. Mutations that alter oxygen affinity demonstrate the importance of subunit interaction for the normal function of a multimeric protein such as hemoglobin. In the Hb A tetramer, the α:β interface has been highly conserved throughout evolution because it is subject to significant movement between the chains when the hemoglobin shifts from the oxygenated (relaxed) to the deoxygenated (tense) form of the molecule. Substitutions in residues at this interface, exemplified by the β-globin mutant **Hb Kempsey** (see Table 11-3), prevent the normal oxygen-related movement between the chains; the mutation "locks" the hemoglobin into the high oxygen affinity state, thus reducing oxygen delivery to tissues and causing polycythemia.

Thalassemia: An Imbalance of Globin-Chain Synthesis

The thalassemias (from the Greek *thalassa*, sea, and *haema*, blood) are collectively the most common human single-gene disorders in the world (Case 44). They are a heterogeneous group of diseases of hemoglobin synthesis in which mutations reduce the synthesis or stability of either the α-globin or β-globin chain to cause **α-thalassemia** or **β-thalassemia**, respectively. The resulting imbalance in the ratio of the α:β chains underlies the pathophysiology. The chain that is produced at the normal rate is in relative excess; in the absence of a complementary chain with which to form a tetramer, the excess normal chains eventually precipitate in the cell, damaging the membrane and leading to premature red blood cell destruction. The excess β or β-like chains are insoluble and precipitate in both red cell precursors (causing ineffective erythropoiesis) and in mature red cells (causing hemolysis) because they damage the cell membrane. The result is a lack of red cells (anemia) in which the red blood cells are both hypochromic (i.e., pale red cells) and microcytic (i.e., small red cells).

The name *thalassemia* was first used to signify that the disease was discovered in persons of Mediterranean origin. Both α-thalassemia and β-thalassemia, however, have a high frequency in many populations, although α-thalassemia is more prevalent and more widely distributed. The high frequency of thalassemia is due to the protective advantage against malaria that it confers on carriers, analogous to the heterozygote advantage of sickle cell hemoglobin carriers (see Chapter 9). There is a characteristic distribution of the thalassemias in a band around the Old World—in the Mediterranean, the Middle East, and parts of Africa, India, and Asia.

An important clinical consideration is that alleles for both types of thalassemia, as well as for structural hemoglobin abnormalities, not uncommonly coexist in an individual. As a result, clinically important interactions may occur among different alleles of the same globin gene or among mutant alleles of different globin genes.

The α-Thalassemias

Genetic disorders of α-globin production disrupt the formation of *both* fetal and adult hemoglobins (see Fig. 11-3) and therefore cause intrauterine as well as postnatal disease. In the absence of α-globin chains with which to associate, the chains from the β-globin cluster are free to form a homotetrameric hemoglobin.

Hemoglobin with a γ_4 composition is known as **Hb Bart's,** and the β_4 tetramer is called Hb H. Because neither of these hemoglobins is capable of releasing oxygen to tissues under normal conditions, they are completely ineffective oxygen carriers. Consequently, infants with severe α-thalassemia and high levels of Hb Bart's (γ_4) suffer severe intrauterine hypoxia and are born with massive generalized fluid accumulation, a condition called **hydrops fetalis.** In milder α-thalassemias, an anemia develops because of the gradual precipitation of the Hb H (β_4) in the erythrocyte. The formation of Hb H inclusions in mature red cells and the removal of these inclusions by the spleen damages the cells, leading to their premature destruction.

Deletions of the α-Globin Genes. The most common forms of α-thalassemia are the result of gene deletions. The high frequency of deletions in mutants of the α chain and not the β chain is due to the presence of the two identical α-globin genes on each chromosome 16 (see Fig. 11-3A); the intron sequences within the two α-globin genes are also similar. This arrangement of tandem homologous α-globin genes facilitates misalignment due to homologous pairing and subsequent recombination between the $\alpha1$ gene domain on one chromosome and the corresponding $\alpha2$ gene region on the other (Fig. 11-9). Evidence supporting this pathogenic mechanism is provided by reports of rare normal individuals with a triplicated α-globin gene complex.

Deletions or other alterations of one, two, three, or all four copies of the α-globin genes cause a proportionately severe hematological abnormality (Table 11-4).

The α-thalassemia trait, caused by deletion of two of the four α-globin genes, is distributed throughout the world. However, the homozygous deletion type of α-thalassemia, involving all four copies of α-globin and leading to Hb Bart's (γ_4) and hydrops fetalis, is largely restricted to Southeast Asia. In this population, the high frequency of hydrops fetalis due to α-thalassemia can be explained by the nature of the deletion responsible. Individuals with two normal and two mutant α-globin genes are said to have α-thalassemia trait, which can result from either of two genotypes ($--/\alpha\alpha$ or $-\alpha/-\alpha$), differing in whether or not the deletions are in *cis* or in *trans*. Heterozygosity for deletion of both copies of the α-globin gene in *cis* ($--/\alpha\alpha$ genotype) is relatively common among Southeast Asians, and offspring of two carriers of this deletion allele may consequently receive two $--/--$ chromosomes. In other groups, however, α-thalassemia trait is usually the result of the *trans* $-\alpha/-\alpha$ genotype, which cannot give rise to $--/--$ offspring.

In addition to α-thalassemia mutations that result in deletion of the α-globin genes, mutations that delete only the LCR of the α-globin complex have also been found to cause α-thalassemia. In fact, similar to the observations discussed earlier with respect to the β-globin LCR, such deletions were critical for demonstrating the

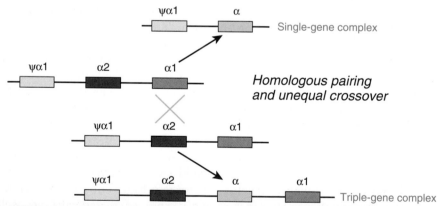

Figure 11-9 The probable mechanism underlying the most common form of α-thalassemia, which is due to deletions of one of the two α-globin genes on a chromosome 16. Misalignment, homologous pairing, and recombination between the $\alpha1$ gene on one chromosome and the $\alpha2$ gene on the homologous chromosome result in the deletion of one α-globin gene.

TABLE 11-4 Clinical States Associated with α-Thalassemia Genotypes

Clinical Condition	Number of Functional α Genes	α-Globin Gene Genotype	α-Chain Production
Normal	4	$\alpha\alpha/\alpha\alpha$	100%
Silent carrier	3	$\alpha\alpha/\alpha-$	75%
α-Thalassemia trait (mild anemia, microcytosis)	2	$\alpha-/\alpha-$ or $\alpha\alpha/--$	50%
Hb H (β_4) disease (moderately severe hemolytic anemia)	1	$\alpha-/--$	25%
Hydrops fetalis or homozygous α-thalassemia (Hb Bart's: γ_4)	0	$--/--$	0%

existence of this regulatory element at the α-globin locus.

Other Forms of α-Thalassemia. In all the classes of α-thalassemia described earlier, deletions in the α-globin genes or mutations in their *cis*-acting sequences account for the reduction of α-globin synthesis. Other types of α-thalassemia occur much less commonly. One important rare form of α-thalassemia is **ATR-X syndrome**, which is associated with both α-thalassemia and intellectual disability and illustrates the importance of epigenetic packaging of the genome in the regulation of gene expression (see Chapter 3). The X-linked *ATRX* gene encodes a chromatin remodeling protein that functions, in *trans*, to activate the expression of the α-globin genes. The ATRX protein belongs to a family of proteins that function within large multiprotein complexes to change DNA topology. ATR-X syndrome is one of a growing number of monogenic diseases that result from mutations in chromatin remodeling proteins.

ATR-X syndrome was initially recognized as unusual because the first families in which it was identified were northern Europeans, a population in which the deletion forms of α-thalassemia are uncommon. In addition, all affected individuals were males who also had severe X-linked intellectual disability together with a wide range of other abnormalities, including characteristic facial features, skeletal defects, and urogenital malformations. This diversity of phenotypes suggests that ATRX regulates the expression of numerous other genes besides the α-globins, although these other targets are presently unknown.

In patients with ATR-X syndrome, the reduction in α-globin synthesis is due to increased accumulation at the α-globin gene cluster of a histone variant (see Chapter 3) called macroH2A, an accumulation that reduces α-globin gene expression and causes α-thalassemia. All the mutations identified to date in the *ATRX* gene in ATR-X syndrome are partial loss-of-function mutations, leading to mild hematological defects compared with those seen in the classic forms of α-thalassemia.

In patients with ATR-X syndrome, abnormalities in DNA methylation patterns indicate that the ATRX protein is also required to establish or maintain the methylation pattern in certain domains of the genome, perhaps by modulating the access of the DNA methyltransferase enzyme to its binding sites. This finding is noteworthy because mutations in another gene, *MECP2*, which encodes a protein that binds to methylated DNA, cause **Rett syndrome** (Case 40) by disrupting the epigenetic regulation of genes in regions of methylated DNA, leading to neurodevelopmental regression. Normally, ATRX and the MeCP2 protein interact, and the impairment of this interaction due to *ATRX* mutations may contribute to the intellectual disability seen in ATR-X syndrome.

The β-Thalassemias

The β-thalassemias share many features with α-thalassemia. In β-thalassemia, the decrease in β-globin production causes a hypochromic, microcytic anemia and an imbalance in globin synthesis due to the excess of α chains. The excess α chains are insoluble and precipitate (see Fig. 11-8) in both red cell precursors (causing ineffective erythropoiesis) and mature red cells (causing hemolysis) because they damage the cell membrane. In contrast to α-globin, however, the β chain is important only in the postnatal period. Consequently, the onset of β-thalassemia is not apparent until a few months after birth, when β-globin normally replaces γ-globin as the major non-α chain (see Fig. 11-3B), and only the synthesis of the major adult hemoglobin, Hb A, is reduced. The level of Hb F is increased in β-thalassemia, not because of a reactivation of the γ-globin gene expression that was switched off at birth, but because of selective survival and perhaps also increased production of the minor population of adult red blood cells that contain Hb F.

In contrast to α-thalassemia, the β-thalassemias are usually due to single base pair substitutions rather than to deletions (Table 11-5). In many regions of the world where β-thalassemia is common, there are so many different β-thalassemia mutations that persons carrying two β-thalassemia alleles are more likely to be **genetic compounds** (i.e., carrying two different β-thalassemia alleles) than to be true homozygotes for one allele. Most individuals with two β-thalassemia alleles have **thalassemia major**, a condition characterized by severe anemia and the need for lifelong medical management. When the β-thalassemia alleles allow so little production of β-globin that no Hb A is present, the condition is designated $β^0$-thalassemia. If some Hb A is detectable, the patient is said to have $β^+$-thalassemia. Although the severity of the clinical disease depends on the combined effect of the two alleles present, survival into adult life was, until recently, unusual.

Infants with homozygous β-thalassemia present with anemia once the postnatal production of Hb F decreases, generally before 2 years of age. At present, treatment of the thalassemias is based on correction of the anemia and the increased marrow expansion by blood transfusion and on control of the consequent iron accumulation by the administration of chelating agents. Bone marrow transplantation is effective, but this is an option only if an HLA-matched family member can be found.

Carriers of one β-thalassemia allele are clinically well and are said to have **thalassemia minor**. Such individuals have hypochromic, microcytic red blood cells and may have a slight anemia that can be misdiagnosed initially as iron deficiency. The diagnosis of thalassemia minor can be supported by hemoglobin electrophoresis, which generally reveals an increase in the level of Hb A_2 ($α_2δ_2$) (see Fig. 11-3A). In many countries,

TABLE 11-5   The Molecular Basis of Some Causes of Simple β-Thalassemia

Type	Example						Phenotype	Affected Population
Defective mRNA Synthesis								
RNA splicing defects (see Fig. 11-11C)	Abnormal acceptor site of intron 1: AG → GG						β^0	Black
Promoter mutants	Mutation in the ATA box −31 −30 −29 −28 −31 −30 −29 −28 A T A A → G T A A						β^+	Japanese
Abnormal RNA cap site	A → C transversion at the mRNA cap site						β^+	Asian
Polyadenylation signal defects	AATAAA → AACAAA						β^+	Black
Nonfunctional mRNAs								
Nonsense mutations	Codon 39 gln → stop CAG → UAG						β^0	Mediterranean (especially Sardinia)
Frameshift mutations	Codon 16 (1-bp deletion) *Normal* trp gly lys val asn 15 16 17 18 19 UGG GGC AAG GUG AAC UGG GCA AGG UGA *Mutant* trp ala arg stop						β^0	Indian
Coding Region Mutations That Also Alter Splicing*								
Synonymous mutations	Codon 24 gly → gly GGU → GGA						β^+	Black

*One other hemoglobin structural variant that causes β-thalassemia is shown in Table 11-3.

mRNA, Messenger RNA.

Derived in part from Weatherall DJ, Clegg JB, Higgs DR, Wood WG: The hemoglobinopathies. In Scriver CR, Beaudet AL, Sly WS, Valle D, editors: *The metabolic and molecular bases of inherited disease*, ed 7, New York, 1995, McGraw-Hill, pp 3417-3484; and Orkin SH: Disorders of hemoglobin synthesis: the thalassemias. In Stamatoyannopoulos G, Nienhuis AW, Leder P, Majerus PW, editors: *The molecular basis of blood diseases*, Philadelphia, 1987, WB Saunders, pp 106-126.

thalassemia heterozygotes are sufficiently numerous to require diagnostic distinction from iron deficiency anemia and to be a relatively common source of referral for prenatal diagnosis of affected homozygous fetuses (see Chapter 17).

α-Thalassemia Alleles as Modifier Genes of β-Thalassemia.

One of the best examples in human genetics of a modifier gene comes from the fact that both β-thalassemia and α-thalassemia alleles may be present in a population. In such populations, β-thalassemia homozygotes may also inherit an α-thalassemia allele. The clinical severity of the β-thalassemia is sometimes ameliorated by the presence of the α-thalassemia allele, which acts as a modifier gene: the imbalance of globin chain synthesis that occurs in β-thalassemia, due to the relative excess of α chains, is reduced by the decrease in α-chain production that results from the α-thalassemia mutation.

β-Thalassemia, Complex Thalassemias, and Hereditary Persistence of Fetal Hemoglobin.

Almost every type of mutation known to reduce the synthesis of an mRNA or protein has been identified as a cause of β-thalassemia. The following overview of these genetic defects is therefore instructive about mutational mechanisms in general, describing in particular the molecular basis of one of the most common and severe genetic diseases in the world. Mutations of the β-globin gene complex are separated into two broad groups with different clinical phenotypes. One group of defects, which accounts for the great majority of patients, impairs the production of β-globin alone and causes **simple β-thalassemia**. The second group of mutations consists of large deletions that cause the **complex thalassemias**, in which the β-globin gene as well as one or more of the other genes—or the LCR—in the β-globin cluster is removed. Finally, some deletions within the β-globin cluster do not cause thalassemia but rather a benign phenotype termed the **hereditary persistence of fetal hemoglobin** (i.e., the persistence of γ-globin gene expression throughout adult life) that informs us about the regulation of globin gene expression.

Molecular Basis of Simple β-Thalassemia.

Simple β-thalassemia results from a remarkable diversity of molecular defects, predominantly point mutations, in the β-globin gene (Fig. 11-10; see Table 11-5). Most mutations causing simple β-thalassemia lead to a decrease in the abundance of the β-globin mRNA and

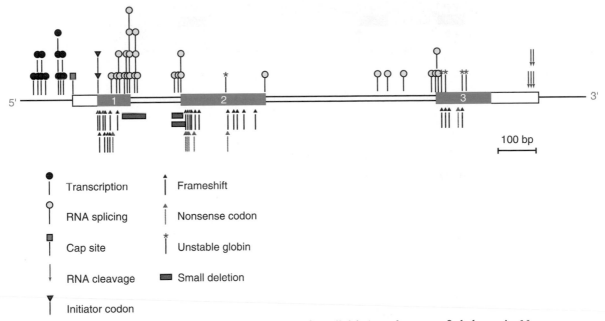

Figure 11-10 Representative point mutations and small deletions that cause β-thalassemia. Note the distribution of mutations throughout the gene and that the mutations affect virtually every process required for the production of normal β-globin. More than 100 different β-globin point mutations are associated with simple β-thalassemia. *See Sources & Acknowledgments.*

include promoter mutants, RNA splicing mutants (the most common), mRNA capping or tailing mutants, and frameshift or nonsense mutations that introduce premature termination codons within the coding region of the gene. A few hemoglobin structural variants also impair processing of the β-globin mRNA, as exemplified by Hb E (described later).

RNA Splicing Mutations. Most β-thalassemia patients with a decreased abundance of β-globin mRNA have abnormalities in RNA splicing. More than two dozen defects of this type have been described, and their combined clinical burden is substantial. These mutations have also acquired high visibility because their effects on splicing are often unexpectedly complex, and analysis of the mutant mRNAs has contributed extensively to knowledge of the sequences critical to normal RNA processing (introduced in Chapter 3). The splice defects are separated into three groups (Fig. 11-11), depending on the region of the unprocessed RNA in which the mutation is located.

- **Splice junction mutations** include mutations at the 5′ donor or 3′ acceptor splice junctions of the introns or in the consensus sequences surrounding the junctions. The critical nature of the conserved GT dinucleotide at the 5′ intron donor site and of the AG at the 3′ intron acceptor site (see Chapter 3) is demonstrated by the complete loss of normal splicing that results from mutations in these dinucleotides (see Fig. 11-11B). The inactivation of the normal acceptor site elicits the use of other acceptor-like sequences elsewhere in the RNA precursor molecule. These alternative sites are termed **cryptic splice sites** because they are normally not used by the splicing apparatus if the

correct site is available. Cryptic donor or acceptor splice sites can be found in either exons or introns.

- **Intron mutations** result from defects within an intron cryptic splice site that enhances the use of the cryptic site by making it more similar or identical to the normal splice site. The "activated" cryptic site then competes with the normal site, with variable effectiveness, thereby reducing the abundance of the normal mRNA by decreasing splicing from the correct site, which remains perfectly intact (see Fig. 11-11C). Cryptic splice site mutations are often "leaky," which means that some use of the normal site occurs, producing a β⁺-thalassemia phenotype.

- **Coding sequence changes that also affect splicing** result from mutations in the open reading frame that may or may not alter the amino acid sequence but that activate a cryptic splice site in an exon (see Fig. 11-11D). For example, a mild form of β⁺-thalassemia results from a mutation in codon 24 (see Table 11-5) that activates a cryptic splice site but does not change the encoded amino acid (both GGT and GGA code for glycine [see Table 3-1]); this is an example of a **synonymous mutation** that is *not* neutral in its effect.

Nonfunctional mRNAs. Some mRNAs are nonfunctional and cannot direct the synthesis of a complete polypeptide because the mutation generates a premature stop codon, which prematurely terminates translation. Two β-thalassemia mutations near the amino terminus exemplify this effect (see Table 11-5). In one (Gln-39Stop), the failure in translation is due to a single nucleotide substitution that creates a **nonsense mutation**. In the other, a **frameshift mutation** results from a

A

Normal splicing pattern

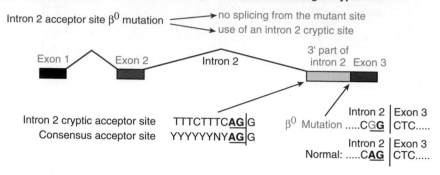

B

Mutation destroying a normal splice acceptor site and activating a cryptic site

Intron 2 acceptor site β⁰ mutation → no splicing from the mutant site
→ use of an intron 2 cryptic site

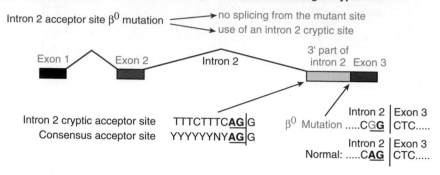

	Intron 2	Exon 3
Intron 2 cryptic acceptor site	TTTCTTTC**AG**	G
Consensus acceptor site	YYYYYYNY**AG**	G

β⁰ Mutation	Intron 2	Exon 3
	.....CG**G**	CTC.....
Normal:	Intron 2	Exon 3
	.....C**AG**	CTC.....

C

Mutation creating a new splice acceptor site in an intron

Intron 1 bp 110 β⁺ mutation → reduced use of unaffected normal site
in a cryptic acceptor site → preferred use of mutant site

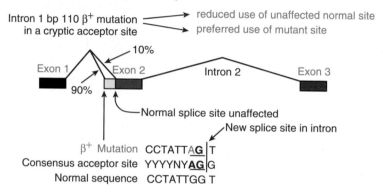

Normal splice site unaffected

New splice site in intron

β⁺ Mutation	CCTATT**AG**	T
Consensus acceptor site	YYYYNY**AG**	G
Normal sequence	CCTATTGG	T

D

Mutation enhancing a cryptic splice donor site in an exon

Hb E: Exon 1 mutation in a cryptic donor site → reduced use of normal site
→ moderate use of cryptic site

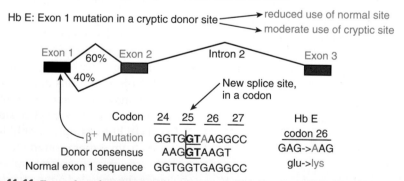

New splice site, in a codon

Codon	24	25	26	27

β⁺ Mutation	GGTG**GT**AAGGCC
Donor consensus	AAG**GT**AAGT
Normal exon 1 sequence	GGTGGTGAGGCC

Hb E
codon 26
GAG->AAG
glu->lys

Figure 11-11 Examples of mutations that disrupt normal splicing of the β-globin gene to cause β-thalassemia. A, Normal splicing pattern. **B,** An intron 2 mutation (IVS2-2A>G) in the normal splice acceptor site aborts normal splicing. This mutation results in the use of a cryptic acceptor site in intron 2. The cryptic site conforms perfectly to the consensus acceptor splice sequence (where Y is either pyrimidine, T or C). Because exon 3 has been enlarged at its 5′ end by inclusion of intron 2 sequences, the abnormal alternatively spliced messenger RNA (mRNA) made from this mutant gene has lost the correct open reading frame and cannot encode β-globin. **C,** An intron 1 mutation (G > A in base pair 110 of intron 1) activates a cryptic acceptor site by creating an AG dinucleotide and increasing the resemblance of the site to the consensus acceptor sequence. The globin mRNA thus formed is elongated (19 extra nucleotides) at the 5′ side of exon 2; a premature stop codon is introduced into the transcript. A β⁺ thalassemia phenotype results because the correct acceptor site is still used, although at only 10% of the wild-type level. **D,** In the Hb E defect, the missense mutation (Glu26Lys) in codon 26 in exon 1 activates a cryptic donor splice site in codon 25 that competes effectively with the normal donor site. Moderate use is made of this alternative splicing pathway, but the majority of RNA is still processed from the correct site, and mild β⁺ thalassemia results.

single base pair deletion early in the open reading frame that removes the first nucleotide from codon 16, which normally encodes glycine; in the mutant reading frame that results, a premature stop codon is quickly encountered downstream, well before the normal termination signal. Because no β-globin is made from these alleles, both of these types of nonfunctional mRNA mutations cause β⁰-thalassemia in the homozygous state. In some instances, frameshifts near the carboxyl terminus of the protein allow most of the mRNA to be translated normally or to produce elongated globin chains, resulting in a variant hemoglobin rather than β^0-thalassemia.

In addition to ablating the production of the β-globin polypeptide, nonsense codons, including the two described earlier, often lead to a reduction in the abundance of the mutant mRNA; indeed, the mRNA may be undetectable. The mechanisms underlying this phenomenon, called **nonsense-mediated mRNA decay**, appears to be restricted to nonsense codons located more than 50 bp upstream of the final exon-exon junction.

Defects in Capping and Tailing of β-Globin mRNA. Several β^+-thalassemia mutations highlight the critical nature of post-transcriptional modifications of mRNAs. For example, the 3′ UTR of almost all mRNAs ends with a polyA sequence, and if this sequence is not added, the mRNA is unstable. As introduced in Chapter 3, polyadenylation of mRNA first requires enzymatic cleavage of the mRNA, which occurs in response to a signal for the cleavage site, AAUAAA, that is found near the 3′ end of most eukaryotic mRNAs. Patients with a substitution that changes the signal sequence to AACAAA produced only a minor fraction of correctly polyadenylated β-globin mRNA.

Hemoglobin E: A Variant Hemoglobin with Thalassemia Phenotypes

Hb E is probably the most common structurally abnormal hemoglobin in the world, occurring at high frequency in Southeast Asia, where there are at least 1 million homozygotes and 30 million heterozygotes. Hb E is a β-globin variant (Glu26Lys) that reduces the rate of synthesis of the mutant β chain and is another example of a coding sequence mutation that also impairs normal splicing by activating a cryptic splice site (see Fig. 11-10D). Although Hb E homozygotes are asymptomatic and only mildly anemic, individuals who are genetic compounds of Hb E and another β-thalassemia allele have abnormal phenotypes that are largely determined by the severity of the other allele.

Complex Thalassemias and the Hereditary Persistence of Fetal Hemoglobin

As mentioned earlier, the large deletions that cause the **complex thalassemias** remove the β-globin gene plus one or more other genes—or the LCR—from the β-globin cluster. Thus, affected individuals have reduced expression of β-globin and one or more of the other β-like chains. These disorders are named according to the genes deleted, for example, $(\delta\beta)^0$-thalassemia or $({}^A\gamma\delta\beta)^0$-thalassemia, and so on (Fig. 11-12). Deletions that remove the β-globin LCR start approximately 50 to 100 kb upstream of the β-globin gene cluster and extend

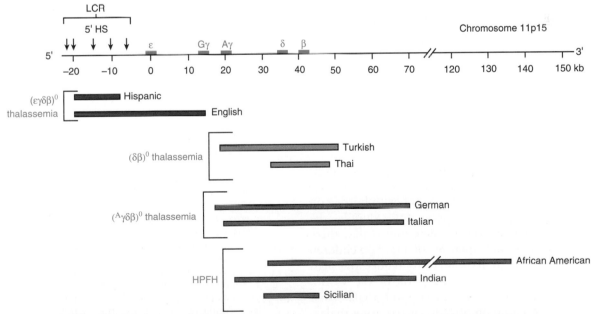

Figure 11-12 Location and size of deletions of various $(\varepsilon\gamma\delta\beta)^0$-thalassemia, $(\delta\beta)^0$-thalassemia, $({}^A\gamma\delta\beta)^0$-**thalassemia, and HPFH mutants.** Note that deletions of the locus control region (LCR) abrogate the expression of all genes in the β-globin cluster. The deletions responsible for δβ-thalassemia, $^A\gamma\delta\beta$-thalassemia, and HPFH overlap (see text). HPFH, Hereditary persistence of fetal hemoglobin; HS, hypersensitive sites. *See Sources & Acknowledgments.*

3′ to varying degrees. Although some of these deletions (such as the Hispanic deletion shown in Fig. 11-12) leave all or some of the genes at the β-globin locus completely intact, they ablate expression from the entire cluster to cause $(\epsilon\gamma\delta\beta)^0$-thalassemia. Such mutations demonstrate the total dependence of gene expression from the β-globin gene cluster on the integrity of the LCR (see Fig. 11-4).

A second group of large β-globin gene cluster deletions of medical significance are those that leave at least one of the γ genes intact (such as the English deletion in Fig. 11-12). Patients carrying such mutations have one of two clinical manifestations, depending on the deletion: either $\delta\beta^0$-thalassemia or a benign condition called hereditary persistence of fetal hemoglobin (HPFH) that is due to disruption of the perinatal switch from γ-globin to β-globin synthesis. Homozygotes with either of these conditions are viable because the remaining γ gene or genes are still active after birth, instead of switching off as would normally occur. As a result, Hb F $(\alpha_2\gamma_2)$ synthesis continues postnatally at a high level and compensates for the absence of Hb A.

The clinically innocuous nature of HPFH that results from the substantial production of γ chains is due to a higher level of Hb F in heterozygotes (17% to 35% Hb F) than is generally seen in $\delta\beta^0$-thalassemia heterozygotes (5% to 18% Hb F). Because the deletions that cause $\delta\beta^0$-thalassemia overlap with those that cause HPFH (see Fig. 11-12), it is not clear why patients with HPFH have higher levels of γ gene expression. One possibility is that some HPFH deletions bring enhancers closer to the γ-globin genes. Insight into the role of regulators of Hb F expression, such as BCL11A and MYB (see earlier discussion), has been partly derived from the study of patients with complex deletions of the β-globin gene cluster. For example, the study of several individuals with HPFH due to rare deletions of the β-globin gene cluster identified a 3.5-kb region, near the 5′ end of the δ-globin gene, that contains binding sites for BCL11A, the critical silencer of Hb F expression in the adult.

Public Health Approaches to Preventing Thalassemia

Large-Scale Population Screening. The clinical severity of many forms of thalassemia, combined with their high frequency, imposes a tremendous health burden on many societies. In Thailand alone, for example, the World Health Organization has determined that there are between half and three quarters of a million children with severe forms of thalassemia. To reduce the high incidence of the disease in some parts of the world, governments have introduced successful thalassemia control programs based on offering or requiring thalassemia carrier screening of individuals of childbearing age in the population (see Box). As a result of such programs, in many parts of the Mediterranean the birth rate of affected newborns has been reduced by as much

as 90% through programs of education directed both to the general population and to health care providers. In Sardinia, a program of voluntary screening, followed by testing of the extended family once a carrier is identified, was initiated in 1975.

ETHICAL AND SOCIAL ISSUES RELATED TO POPULATION SCREENING FOR β-THALASSEMIA*

Approximately 70,000 infants are born worldwide each year with β-thalassemia, at high economic cost to health care systems and at great emotional cost to affected families.

To identify individuals and families at increased risk for the disease, screening is done in many countries. National and international guidelines recommend that screening not be compulsory and that education and genetic counseling should inform decision making.

Widely differing cultural, religious, economic, and social factors significantly influence the adherence to guidelines. For example:

In Greece, screening is voluntary, available both premaritally and prenatally, requires informed consent, is widely advertised by the mass media and in military and school programs, and is accompanied by genetic counseling for carrier couples.

In Iran and Turkey, these practices differ only in that screening is mandatory premaritally (but in all countries with mandatory screening, carrier couples have the right to marry if they wish).

In Taiwan, antenatal screening is available and voluntary, but informed consent is not required and screening is currently not accompanied by educational programs or genetic counseling.

In the United Kingdom, screening is offered to all pregnant women, but public awareness is poor, and the screening is questionably voluntary because many if not most women tested are unaware they have been screened until they are found to be carriers. In some UK programs, women are not given the results of the test.

Major obstacles to more effective population screening for β-thalassemia

The principal obstacles include the facts that pregnant women feel overwhelmed by the array of tests offered to them, many health professionals have insufficient knowledge of genetic disorders, appropriate education and counseling are costly and time-consuming, it is commonly misunderstood that informing a women about a test is equivalent to giving consent, and the effectiveness of mass education varies greatly, depending on the community or country.

The effectiveness of well-executed β-thalassemia screening programs

In populations where β-thalassemia screening has been effectively implemented, the reduction in the incidence of the disease has been striking. For example, in Sardinia, screening between 1975 and 1995 reduced the incidence from 1 per 250 to 1 per 4000 individuals. Similarly, in Cyprus, the incidence of affected births fell from 51 in 1974 to none up to 2007.

*Based on Cousens NE, Gaff CL, Metcalfe SA, et al: Carrier screening for β-thalassaemia: a review of international practice, *Eur J Hum Genet* 18:1077-1083, 2010.

Screening Restricted to Extended Families. In developing countries, the initiation of screening programs for thalassemia is a major economic and logistical challenge. Recent work in Pakistan and Saudi Arabia, however, has demonstrated the effectiveness of a screening strategy that may be broadly applicable in countries where consanguineous marriages are common. In the Rawalpindi region of Pakistan, β-thalassemia was found to be largely restricted to a specific group of families that came to attention because there was an identifiable index case (see Chapter 7). In 10 extended families with such an index case, testing of almost 600 persons established that approximately 8% of the married couples examined consisted of two carriers, whereas no couple at risk was identified among 350 randomly selected pregnant women and their partners outside of these 10 families. All carriers reported that the information provided was used to avoid further pregnancy if they already had two or more healthy children or, in the case of couples with only one or no healthy children, for prenatal diagnosis. Although the long-term impact of this program must be established, extended family screening of this type may contribute importantly to the control of recessive diseases in parts of the world where a cultural preference for consanguineous marriage is present. In other words, because of consanguinity, disease gene variants are "trapped" within extended families, so that an affected child is an indicator of an extended family at high risk for the disease.

The initiation of carrier testing and prenatal diagnosis programs for thalassemia requires not only the education of the public and of physicians but also the establishment of skilled central laboratories and the consensus of the population to be screened (see Box). Whereas population-wide programs to control thalassemia are inarguably less expensive than the cost of caring for a large population of affected individuals over their lifetimes, the temptation for governments or physicians to pressure individuals into accepting such programs must be avoided. The autonomy of the individual in reproductive decision making, a bedrock of modern bioethics, and the cultural and religious views of their communities must both be respected.

GENERAL REFERENCES

Higgs DR, Engel JD, Stamatoyannopoulos G: Thalassaemia, *Lancet* 379:373–383, 2012.

Higgs DR, Gibbons RJ: The molecular basis of α-thalassemia: a model for understanding human molecular genetics, *Hematol Oncol Clin North Am* 24:1033–1054, 2010.

McCavit TL: Sickle cell disease, *Pediatr Rev* 33:195–204, 2012.

Roseff SD: Sickle cell disease: a review, *Immunohematology* 25:67–74, 2009.

Weatherall DJ: The role of the inherited disorders of hemoglobin, the first "molecular diseases," in the future of human genetics, *Annu Rev Genomics Hum Genet* 14:1–24, 2013.

REFERENCES FOR SPECIFIC TOPICS

Bauer DE, Orkin SH: Update on fetal hemoglobin gene regulation in hemoglobinopathies, *Curr Opin Pediatr* 23:1–8, 2011.

Ingram VM: Specific chemical difference between the globins of normal human and sickle-cell anaemia haemoglobin, *Nature* 178:792–794, 1956.

Ingram VM: Gene mutations in human haemoglobin: the chemical difference between normal and sickle cell haemoglobin, *Nature* 180:326–328, 1957.

Kervestin S, Jacobson A: NMD, a multifaceted response to premature translational termination, *Nat Rev Mol Cell Biol* 13:700–712, 2012.

Pauling L, Itano HA, Singer SJ, et al: Sickle cell anemia, a molecular disease, *Science* 110:543–548, 1949.

Sankaran VG, Lettre G, Orkin SH, et al: Modifier genes in Mendelian disorders: the example of hemoglobin disorders, *Ann N Y Acad Sci* 1214:47–56, 2010.

Steinberg MH, Sebastiani P: Genetic modifiers of sickle cell disease, *Am J Hematol* 87:795–803, 2012.

Weatherall DJ: The inherited diseases of hemoglobin are an emerging global health burden, *Blood* 115:4331–4336, 2010.

PROBLEMS

1. A child dies of hydrops fetalis. Draw a pedigree with genotypes that illustrates to the carrier parents the genetic basis of the infant's thalassemia. Explain why a Melanesian couple whom they met in the hematology clinic, who both also have the α-thalassemia trait, are unlikely to have a similarly affected infant.

2. Why are most β-thalassemia patients likely to be genetic compounds? In what situations might you anticipate that a patient with β-thalassemia would be likely to have two identical β-globin alleles?

3. Tony, a young Italian boy, is found to have moderate β-thalassemia, with a hemoglobin concentration of 7 g/dL (normal amounts are 10 to 13 g/dL). When you perform a Northern blot of his reticulocyte RNA, you unexpectedly find three β-globin mRNA bands, one of normal size, one larger than normal, and one smaller than normal.

   What mutational mechanisms could account for the presence of three bands like this in a patient with β-thalassemia? In *this* patient, the fact that the anemia is mild suggests that a significant fraction of normal β-globin mRNA is being made. What types of mutation would allow this to occur?

4. A man is heterozygous for Hb M Saskatoon, a hemoglobinopathy in which the normal amino acid His is replaced by Tyr at position 63 of the β chain. His mate is heterozygous for Hb M Boston, in which His is replaced by Tyr at position 58 of the α chain. Heterozygosity for either of these mutant alleles produces methemoglobinemia. Outline the possible genotypes and phenotypes of their offspring.

5. A child has a paternal uncle and a maternal aunt with sickle cell disease; both of her parents do not. What is the probability that the child has sickle cell disease?

6. A woman has sickle cell trait, and her mate is heterozygous for Hb C. What is the probability that their child has no abnormal hemoglobin?

7. Match the following:

_____ complex β-thalassemia	1. detectable Hb A
_____ β⁺-thalassemia	2. three
_____ number of α-globin genes missing in Hb H disease	3. β-thalassemia
	4. α-thalassemia
_____ two different mutant alleles at a locus	5. high-level β-chain expression
_____ ATR-X syndrome	6. α-thalassemia trait
_____ insoluble β chains	7. compound heterozygote
_____ number of α-globin genes missing in hydrops fetalis with Hb Bart's	8. δβ genes deleted
	9. four
_____ locus control region	10. mental retardation
_____ α–/α– genotype	
_____ increased Hb A₂	

8. Mutations in noncoding sequences may change the number of protein molecules produced, but each protein molecule made will generally have a normal amino acid sequence. Give examples of some exceptions to this rule, and describe how the alterations in the amino acid sequence are generated.

9. What are some possible explanations for the fact that thalassemia control programs, such as the successful one in Sardinia, have not reduced the birth rate of newborns with severe thalassemia to zero? For example, in Sardinia from 1999 to 2002, approximately two to five such infants were born each year.

CHAPTER 12

The Molecular, Biochemical, and Cellular Basis of Genetic Disease

In this chapter, we extend our examination of the molecular and biochemical basis of genetic disease beyond the hemoglobinopathies to include other diseases and the abnormalities in gene and protein function that cause them. In Chapter 11, we presented an outline of the general mechanisms by which mutations cause disease (see Fig. 11-1) and reviewed the steps at which mutations can disrupt the synthesis or function of a protein (see Table 11-2). Those outlines provide a framework for understanding the pathogenesis of all genetic disease. However, mutations in other classes of proteins often disrupt cell and organ function by processes that differ from those illustrated by the hemoglobinopathies, and we explore them in this chapter.

To illustrate these other types of disease mechanisms, we examine here well-known disorders such as **phenylketonuria, cystic fibrosis, familial hypercholesterolemia, Duchenne muscular dystrophy,** and **Alzheimer disease.** In some instances, less common disorders are included because they best demonstrate a specific principle. The importance of selecting representative disorders becomes apparent when one considers that to date, mutations in almost 3000 genes have been associated with a clinical phenotype. In the coming decade, one anticipates that many more of the approximately 20,000 to 25,000 coding genes in the human genome will be shown to be associated with both monogenic and genetically complex diseases.

DISEASES DUE TO MUTATIONS IN DIFFERENT CLASSES OF PROTEINS

Proteins carry out an astounding number of different functions, some of which are presented in Figure 12-1. Mutations in virtually every functional class of protein can lead to genetic disorders. In this chapter, we describe important genetic diseases that affect representative proteins selected from the groups shown in Figure 12-1; many other of the proteins listed, as well as the diseases associated with them, are described in the Cases section.

Housekeeping Proteins and Specialty Proteins in Genetic Disease

Proteins can be separated into two general classes on the basis of their pattern of expression: *housekeeping proteins,* which are present in virtually every cell and have fundamental roles in the maintenance of cell structure and function; and tissue-specific *specialty proteins,* which are produced in only one or a limited number of cell types and have unique functions that contribute to the individuality of the cells in which they are expressed. Most cell types in humans express 10,000 to 15,000 protein-coding genes. Knowledge of the tissues in which a protein is expressed, particularly at high levels, is often useful in understanding the pathogenesis of a disease.

Two broad generalizations can be made about the relationship between the site of a protein's expression and the site of disease.

- *First* (and somewhat intuitively), mutation in a tissue-specific protein most often produces a disease restricted to that tissue. However, there may be secondary effects on other tissues, and in some cases mutations in tissue-specific proteins may cause abnormalities primarily in organs that do not express the protein at all; ironically, the tissue expressing the mutant protein may be left entirely unaffected by the pathological process. This situation is exemplified by **phenylketonuria,** discussed in depth in the next section. Phenylketonuria is due to the absence of phenylalanine hydroxylase (PAH) activity in the liver, but it is the brain (which expresses very little of this enzyme), and not the liver, that is damaged by the high blood levels of phenylalanine resulting from the lack of hepatic PAH. Consequently, one cannot necessarily infer that disease in an organ results from mutation in a gene expressed principally or only in that organ, or in that organ at all.
- *Second,* although housekeeping proteins are expressed in most or all tissues, the clinical effects of mutations in housekeeping proteins are frequently limited to one or just a few tissues, for at least two reasons. In

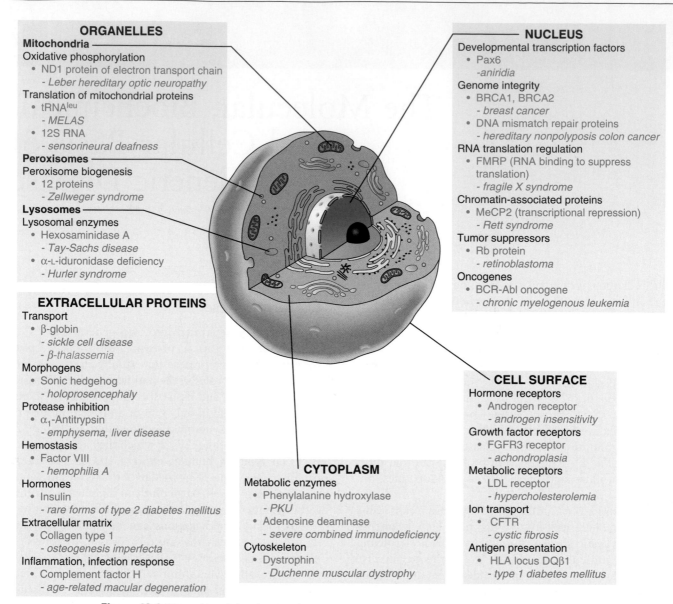

ORGANELLES

Mitochondria
Oxidative phosphorylation
- ND1 protein of electron transport chain
  - *Leber hereditary optic neuropathy*
Translation of mitochondrial proteins
- tRNA^{leu}
  - *MELAS*
- 12S RNA
  - *sensorineural deafness*

Peroxisomes
Peroxisome biogenesis
- 12 proteins
  - *Zellweger syndrome*

Lysosomes
Lysosomal enzymes
- Hexosaminidase A
  - *Tay-Sachs disease*
- α-L-iduronidase deficiency
  - *Hurler syndrome*

EXTRACELLULAR PROTEINS
Transport
- β-globin
  - *sickle cell disease*
  - *β-thalassemia*
Morphogens
- Sonic hedgehog
  - *holoprosencephaly*
Protease inhibition
- α$_1$-Antitrypsin
  - *emphysema, liver disease*
Hemostasis
- Factor VIII
  - *hemophilia A*
Hormones
- Insulin
  - *rare forms of type 2 diabetes mellitus*
Extracellular matrix
- Collagen type 1
  - *osteogenesis imperfecta*
Inflammation, infection response
- Complement factor H
  - *age-related macular degeneration*

NUCLEUS
Developmental transcription factors
- Pax6
  - *-aniridia*
Genome integrity
- BRCA1, BRCA2
  - *breast cancer*
- DNA mismatch repair proteins
  - *hereditary nonpolyposis colon cancer*
RNA translation regulation
- FMRP (RNA binding to suppress translation)
  - *fragile X syndrome*
Chromatin-associated proteins
- MeCP2 (transcriptional repression)
  - *Rett syndrome*
Tumor suppressors
- Rb protein
  - *retinoblastoma*
Oncogenes
- BCR-Abl oncogene
  - *chronic myelogenous leukemia*

CELL SURFACE
Hormone receptors
- Androgen receptor
  - *androgen insensitivity*
Growth factor receptors
- FGFR3 receptor
  - *achondroplasia*
Metabolic receptors
- LDL receptor
  - *hypercholesterolemia*
Ion transport
- CFTR
  - *cystic fibrosis*
Antigen presentation
- HLA locus DQβ1
  - *type 1 diabetes mellitus*

CYTOPLASM
Metabolic enzymes
- Phenylalanine hydroxylase
  - *PKU*
- Adenosine deaminase
  - *severe combined immunodeficiency*
Cytoskeleton
- Dystrophin
  - *Duchenne muscular dystrophy*

Figure 12-1 Examples of the classes of proteins associated with diseases with a strong genetic component (most are monogenic), and the part of the cell in which those proteins normally function. CFTR, Cystic fibrosis transmembrane regulator; FMRP, fragile X mental retardation protein; HLA, human leukocyte antigen; LDL, low-density lipoprotein; MELAS, mitochondrial encephalomyopathy with lactic acidosis and strokelike episodes; PKU, phenylketonuria.

most such instances, a single or a few tissue(s) may be affected because the housekeeping protein in question is normally expressed abundantly there and serves a specialty function in that tissue. This situation is illustrated by **Tay-Sachs disease,** as discussed later; the mutant enzyme in this disorder is hexosaminidase A, which is expressed in virtually all cells, but its absence leads to a fatal neurodegeneration, leaving non-neuronal cell types unscathed. In other instances, another protein with overlapping biological activity may also be expressed in the unaffected tissue, thereby lessening the impact of the loss of function of the mutant gene, a situation known as **genetic redundancy.** Unexpectedly, even mutations in

genes that one might consider as essential to every cell, such as actin, can result in viable offspring.

DISEASES INVOLVING ENZYMES

Enzymes are the catalysts that mediate the efficient conversion of a substrate to a product. The diversity of substrates on which enzymes act is huge. Accordingly, the human genome contains more than 5000 genes that encode enzymes, and there are hundreds of human diseases—the so-called **enzymopathies**—that involve enzyme defects. We first discuss one of the best-known groups of inborn errors of metabolism, the **hyperphenylalaninemias.**

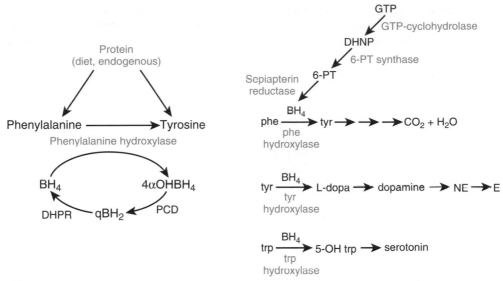

Figure 12-2 The biochemical pathways affected in the hyperphenylalaninemias. BH_4, tetrahydrobiopterin; $4\alpha OHBH_4$, 4α-hydroxytetrahydrobiopterin; qBH_2, quinonoid dihydrobiopterin, the oxidized product of the hydroxylation reactions, which is reduced to BH_4 by dihydropteridine reductase (DHPR); PCD, pterin 4α-carbinolamine dehydratase; phe, phenylalanine; tyr, tyrosine; trp, tryptophan; GTP, guanosine triphosphate; DHNP, dihydroneopterin triphosphate; 6-PT, 6-pyruvoyltetrahydropterin; L-dopa, L-dihydroxyphenylalanine; NE, norepinephrine; E, epinephrine; 5-OH trp, 5-hydroxytryptophan.

TABLE 12-1 Locus Heterogeneity in the Hyperphenylalaninemias

Biochemical Defect	Incidence/10^6 Births	Enzyme Affected	Treatment
Mutations in the Gene Encoding Phenylalanine Hydroxylase			
Classic PKU	5-350 (depending on the population)	PAH	Low-phenylalanine diet*
Variant PKU	Less than classic PKU	PAH	Low-phenylalanine diet (less restrictive than that required to treat PKU*
Non-PKU hyperphenylalaninemia	15-75	PAH	None, or a much less restrictive low-phenylalanine diet*
Mutations in Genes Encoding Enzymes of Tetrahydrobiopterin Metabolism			
Impaired BH_4 recycling	<1	PCD DHPR	Low-phenylalanine diet + L-dopa, 5-HT, carbidopa (+ folinic acid for DHPR patients)
Impaired BH_4 synthesis	<1	GTP-CH 6-PTS	Low-phenylalanine diet + L-dopa, 5-HT, carbidopa and pharmacological doses of BH_4

*BH_4 supplementation may increase the PAH activity of some patients in each of these three groups.

BH_4, Tetrahydrobiopterin; DHPR, dihydropteridine reductase; GTP-CH, guanosine triphosphate cyclohydrolase; 5-HT, 5-hydroxytryptophan; PAH, phenylalanine hydroxylase; PCD, pterin 4α-carbinolamine dehydratase; PKU, phenylketonuria; 6-PTS, 6-pyruvoyltetrahydropterin synthase.

Aminoacidopathies

The Hyperphenylalaninemias

The abnormalities that lead to an increase in the blood level of phenylalanine, most notably PAH deficiency or phenylketonuria (PKU), illustrate almost every principle of biochemical genetics related to enzyme defects. The biochemical causes of hyperphenylalaninemia are illustrated in Figure 12-2, and the principal features of the diseases associated with the biochemical defect at the five known hyperphenylalaninemia loci are presented in Table 12-1. All the genetic disorders of phenylalanine metabolism are inherited as autosomal recessive conditions and are due to loss-of-function mutations in the gene encoding PAH or in genes required for the synthesis or reutilization of its cofactor, tetrahydrobiopterin (BH_4).

Phenylketonuria. Classic PKU is the epitome of the enzymopathies. It results from mutations in the gene encoding PAH, which converts phenylalanine to tyrosine (see Fig. 12-2 and Table 12-1). The discovery of PKU in 1934 marked the first demonstration of a genetic defect as a cause of intellectual disability. Because patients with PKU cannot degrade phenylalanine, it accumulates in body fluids and damages the developing central nervous system in early childhood. A small fraction of phenylalanine is metabolized to produce increased amounts of phenylpyruvic acid, the keto acid responsible for the name of the disease. Ironically, although the enzymatic defect has been known for many decades, the precise pathogenetic mechanism(s) by which increased phenylalanine damages the brain is still uncertain. Importantly, the neurological damage is largely avoided by reducing the dietary intake of phenylalanine. The management of PKU is a paradigm of the treatment of the many metabolic diseases whose outcome can be improved by preventing accumulation of an enzyme substrate and its derivatives; this therapeutic principle is described further in Chapter 13.

MUTANT ENZYMES AND DISEASE: GENERAL CONCEPTS

The following concepts are fundamental to the understanding and treatment of enzymopathies.

* **Inheritance patterns**

    Enzymopathies are almost always recessive or X-linked (see Chapter 7). Most enzymes are produced in quantities significantly in excess of minimal biochemical requirements, so that heterozygotes (typically with approximately 50% of residual activity) are clinically normal. In fact, many enzymes may maintain normal substrate and product levels with activities of less than 10%, a point relevant to the design of therapeutic strategies (e.g., **homocystinuria** due to cystathionine synthase deficiency—see Chapter 13). The enzymes of porphyrin synthesis are exceptions (see discussion of acute intermittent porphyria in main text, later).

* **Substrate accumulation or product deficiency**

    Because the function of an enzyme is to convert a substrate to a product, all of the pathophysiological consequences of enzymopathies can be attributed to the accumulation of the substrate (as in PKU), to the deficiency of the product (as in *glucose-6-phosphate dehydrogenase deficiency* (Case 19), or to some combination of the two (Fig. 12-3).

* **Diffusible versus macromolecular substrates**

    An important distinction can be made between enzyme defects in which the substrate is a small molecule (such as phenylalanine, which can be readily distributed throughout body fluids by diffusion or transport) and defects in which the substrate is a macromolecule (such as a mucopolysaccharide, which remains trapped within its organelle or cell). The pathological change of the macromolecular diseases, such as *Tay-Sachs disease*, is confined to the tissues in which the substrate accumulates. In contrast, the site of the disease in the small molecule disorders is often unpredictable, because the unmetabolized substrate or its derivatives can move freely throughout the body, damaging cells that may normally have no relationship to the affected enzyme, as in PKU.

* **Loss of multiple enzyme activities**

    A patient with a single-gene defect may have a loss of function in more than one enzyme. There are several possible mechanisms: the enzymes may use the same cofactor (e.g., *BH₄ deficiency*); the enzymes may share a common subunit or an activating, processing, or

Variant Phenylketonuria and Nonphenylketonuria Hyperphenylalaninemia. Whereas PKU results from a virtual absence of PAH activity (less than 1% of that in controls), less severe phenotypes, designated non-PKU hyperphenylalaninemia and variant PKU (see Table 12-1), result when the mutant PAH enzyme has some residual activity. The fact that a very small amount of residual enzyme activity can have a large impact on phenotype is another general principle of the enzymopathies (see Box).

Variant PKU includes patients who require only some dietary phenylalanine restriction but to a lesser degree than that required in classic PKU, because their increases in blood phenylalanine levels are more moderate and less damaging to the brain. In contrast to classic PKU,

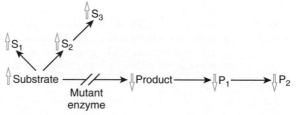

Figure 12-3 A model metabolic pathway showing that the potential effects of an enzyme deficiency include accumulation of the substrate (S) or derivatives of it (S_1, S_2, S_3) and deficiency of the product (P) or compounds made from it (P_1, P_2). In some cases, the substrate derivatives are normally only minor metabolites that may be formed at increased rates when the substrate accumulates (e.g., phenylpyruvate in phenylketonuria).

stabilizing protein (e.g., the *GM₂ gangliosidoses*); the enzymes may all be processed by a common modifying enzyme, and in its absence, they may be inactive, or their uptake into an organelle may be impaired (e.g., *I-cell disease,* in which failure to add mannose 6-phosphate to many lysosomal enzymes abrogates the ability of cells to recognize and import the enzymes); and a group of enzymes may be absent or ineffective if the organelle in which they are normally found is not formed or is abnormal (e.g., *Zellweger syndrome,* a disorder of peroxisome biogenesis).

* **Phenotypic homology**

    The pathological and clinical features resulting from an enzyme defect are often shared by diseases due to deficiencies of other enzymes that function in the same area of metabolism (e.g., the *mucopolysaccharidoses*) as well as by the different phenotypes that can result from partial versus complete defects of one enzyme. Partial defects often present with clinical abnormalities that are a subset of those found with the complete deficiency, although the etiological relationship between the two diseases may not be immediately obvious. For example, partial deficiency of the purine enzyme hypoxanthine-guanine phosphoribosyltransferase causes only hyperuricemia, whereas a complete deficiency causes hyperuricemia as well as a profound neurological disease, *Lesch-Nyhan syndrome,* which resembles cerebral palsy.

in which the plasma phenylalanine levels are greater than 1000 µmol/L when the patient is receiving a normal diet, **non-PKU hyperphenylalaninemia** is defined by plasma phenylalanine concentrations above the upper limit of normal (120 µmol/L), but less than the levels seen in classic PKU. If the increase in non-PKU hyperphenylalaninemia is small (<400 µmol/L), no treatment is required; these individuals come to clinical attention only because they are identified by newborn screening (see Chapter 17). Their normal phenotype has been the best indication of the "safe" level of plasma phenylalanine that must not be exceeded in treating classic PKU. The association of these three clinical phenotypes with mutations in the *PAH* gene is a clear example of allelic heterogeneity leading to clinical heterogeneity (see Table 12-1).

Allelic and Locus Heterogeneity in the Hyperphenylalaninemias

Allelic Heterogeneity in the* PAH *Gene. A striking degree of allelic heterogeneity at the *PAH* locus—more than 700 different mutations worldwide—has been identified in patients with hyperphenylalaninemia associated with classic PKU, variant PKU, and non-PKU hyperphenylalaninemia (see Table 12-1). Seven mutations account for a majority of known mutant alleles in populations of European descent, whereas six others represent the majority of *PAH* mutations in Asian populations (Fig. 12-4). The remaining disease-causing mutations are individually rare. To record and make this information publicly available, a *PAH* database has been developed by an international consortium.

The allelic heterogeneity at the *PAH* locus has major clinical consequences. Most important is the fact that most hyperphenylalaninemic subjects are **compound heterozygotes** (i.e., they have two different disease-causing alleles) (see Chapter 7). This allelic heterogeneity accounts for much of the enzymatic and phenotypic heterogeneity observed in this patient population. Thus, mutations that eliminate or dramatically reduce PAH activity generally cause classic PKU, whereas greater residual enzyme activity is associated with milder phenotypes. However, homozygous patients with certain PAH mutations have been found to have phenotypes ranging all the way from classic PKU to non-PKU hyperphenylalaninemia. Accordingly, it is now clear that other unidentified biological variables—undoubtedly including modifier genes—generate variation in the phenotype seen with any specific genotype. This lack of a strict genotype-phenotype correlation, initially somewhat surprising, is now recognized to be a common feature of many single-gene diseases and highlights the fact that even monogenic traits like PKU are not genetically "simple" disorders.

Defects in Tetrahydrobiopterin Metabolism. In 1% to 3% of hyperphenylalaninemic patients, the *PAH* gene is

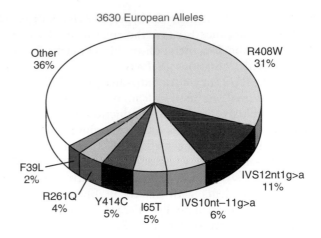

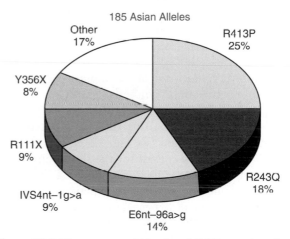

Figure 12-4 The nature and identity of *PAH* mutations in populations of European and Asian descent (the latter from China, Korea, and Japan). The one-letter amino acid code is used (see Table 3-1). *See Sources & Acknowledgments.*

normal, and the hyperphenylalaninemia results from a defect in one of the steps in the biosynthesis or regeneration of BH_4, the cofactor for PAH (see Table 12-1 and Fig. 12-2). The association of a single biochemical phenotype, such as hyperphenylalaninemia, with mutations in different genes, is an example of locus heterogeneity (see Table 11-1). The proteins encoded by genes that manifest locus heterogeneity generally act at different steps in a single biochemical pathway, another principle of genetic disease illustrated by the genes associated with hyperphenylalaninemia (see Fig. 12-2). BH_4-deficient patients were first recognized because they developed profound neurological problems in early life, despite the successful administration of a low-phenylalanine diet. This poor outcome is due in part to the requirement for the BH_4 cofactor of two other enzymes, tyrosine hydroxylase and tryptophan hydroxylase. These hydroxylases are critical for the synthesis of the monoamine neurotransmitters dopamine, norepinephrine, epinephrine, and serotonin (see Fig. 12-2).

The locus heterogeneity of hyperphenylalaninemia is of great significance because the treatment of patients with a defect in BH$_4$ metabolism differs markedly from subjects with mutations in *PAH,* in two ways. First, because the PAH enzyme of individuals with BH$_4$ defects is itself normal, its activity can be restored by large doses of oral BH$_4$, leading to a reduction in their plasma phenylalanine levels. This practice highlights the principle of product replacement in the treatment of some genetic disorders (see Chapter 13). Consequently, phenylalanine restriction can be significantly relaxed in the diet of patients with defects in BH$_4$ metabolism, and some patients actually tolerate a normal (i.e., a phenylalanine-unrestricted) diet. Second, one must also try to normalize the neurotransmitters in the brains of these patients by administering the products of tyrosine hydroxylase and tryptophan hydroxylase, L-dopa and 5-hydroxytryptophan, respectively (see Fig. 12-2 and Table 12-1).

Remarkably, mutations in sepiapterin reductase, an enzyme in the BH$_4$ synthesis pathway, do not cause hyperphenylalaninemia. In this case, only dopa-responsive dystonia is seen, due to impaired synthesis of dopamine and serotonin (see Fig. 12-2). It is thought that alternative pathways exist for the final step in BH$_4$ synthesis, bypassing the sepiapterin reductase deficiency in peripheral tissues, an example of genetic redundancy.

For these reasons, all hyperphenylalaninemic infants must be screened to determine whether their hyperphenylalaninemia is the result of an abnormality in PAH or in BH$_4$ metabolism. The hyperphenylalaninemias thus illustrate the critical importance of obtaining a specific molecular diagnosis in all patients with a genetic disease phenotype—the underlying genetic defect may not be what one first suspects, and the treatment can vary accordingly.

Tetrahydrobiopterin Responsiveness in PAH Mutations.

Many hyperphenylalaninemia patients with mutations in the *PAH* gene (rather than in BH$_4$ metabolism) will also respond to large oral doses of BH$_4$ cofactor, with a substantial decrease in plasma phenylalanine. BH$_4$ supplementation is therefore an important adjunct therapy for PKU patients of this type, allowing them a less restricted dietary intake of phenylalanine. The patients most likely to respond are those with significant residual PAH activity (i.e., patients with variant PKU and non-PKU hyperphenylalaninemia), but even a minority of patients with classic PKU are also responsive. The presence of residual PAH activity does not, however, necessarily guarantee an effect of BH$_4$ administration on plasma phenylalanine levels. Rather, the degree of BH$_4$ responsiveness will depend on the specific properties of each mutant PAH protein, reflecting the allelic heterogeneity underlying *PAH* mutations.

The provision of increased amounts of a cofactor is a general strategy that has been employed for the treatment of many inborn errors of enzyme metabolism, as discussed further in Chapter 13. In the general case, a cofactor comes into contact with the protein component of an enzyme (termed an *apoenzyme*) to form the active *holoenzyme,* which consists of both the cofactor and the otherwise inactive apoenzyme. Illustrating this strategy, BH$_4$ supplementation has been shown to exert its beneficial effect through one or more mechanisms, all of which result from the increased amount of the cofactor that is brought into contact with the mutant PAH apoenzyme. These mechanisms include stabilization of the mutant enzyme, protection of the enzyme from degradation by the cell, and increase in the cofactor supply for a mutant enzyme that has a low affinity for BH$_4$.

Newborn Screening.
PKU is the prototype of genetic diseases for which mass newborn screening is justified (see Chapter 18) because it is relatively common in some populations (up to approximately 1 in 2900 live births), mass screening is feasible, failure to treat has severe consequences (profound developmental delay), and treatment is effective if begun early in life. To allow time for the postnatal increase in blood phenylalanine levels to occur, the test is performed after 24 hours of age. Blood from a heel prick is assayed in a central laboratory for blood phenylalanine levels and measurement of the phenylalanine-to-tyrosine ratio. Positive test results must be confirmed quickly because delays in treatment beyond 4 weeks postnatally have profound effects on intellectual outcome.

Maternal Phenylketonuria.
Originally, the low-phenylalanine diet was discontinued in mid-childhood for most patients with PKU. Subsequently, however, it was discovered that almost all offspring of women with PKU not receiving treatment are clinically abnormal; most are severely delayed developmentally, and many have microcephaly, growth impairment, and malformations, particularly of the heart. As predicted by principles of mendelian inheritance, all of these children are heterozygotes. Thus their neurodevelopmental delay is not due to their own genetic constitution but to the highly teratogenic effect of elevated levels of phenylalanine in the maternal circulation. Accordingly, it is imperative that women with PKU who are planning pregnancies commence a low-phenylalanine diet before conceiving.

Lysosomal Storage Diseases: A Unique Class of Enzymopathies

Lysosomes are membrane-bound organelles containing an array of hydrolytic enzymes involved in the degradation of a variety of biological macromolecules. Mutations in these hydrolases are unique because they lead to the accumulation of their substrates inside the lysosome, where the substrates remain trapped because their large size prevents their egress from the organelle. Their accumulation and sometimes toxicity interferes with normal cell function, eventually causing cell death. Moreover, the substrate accumulation underlies one uniform clinical feature of these diseases—their unrelenting progression. In most of these conditions, substrate storage increases the mass of the affected tissues and organs. When the brain is affected, the picture is one of neurodegeneration. The clinical phenotypes are very distinct and often make the diagnosis of a storage disease straightforward. More than 50 lysosomal hydrolase or lysosomal membrane transport deficiencies, almost all inherited as autosomal recessive conditions, have been described. Historically, these diseases were untreatable. However, **bone marrow transplantation** and **enzyme replacement therapy** have dramatically improved the prognosis of these conditions (see Chapter 13).

Tay-Sachs Disease

Tay-Sachs disease (Case 43) is one of a group of heterogeneous lysosomal storage diseases, the GM_2 gangliosidoses, that result from the inability to degrade a sphingolipid, GM_2 ganglioside (Fig. 12-5). The biochemical lesion is a marked deficiency of hexosaminidase A (hex A). Although the enzyme is ubiquitous, the disease has its clinical impact almost solely on the brain, the predominant site of GM_2 ganglioside synthesis. Catalytically active hex A is the product of a three-gene system (see Fig. 12-5). These genes encode the α and β subunits of the enzyme (the *HEXA* and *HEXB* genes, respectively) and an activator protein that must associate with the substrate and the enzyme before the enzyme can cleave the terminal N-acetyl-β-galactosamine residue from the ganglioside.

The clinical manifestations of defects in the three genes are indistinguishable, but they can be differentiated by enzymatic analysis. Mutations in the *HEXA* gene affect the α subunit and disrupt hex A activity to cause Tay-Sachs disease (or less severe variants of hex A deficiency). Defects in the *HEXB* gene or in the gene encoding the activator protein impair the activity of both hex A and hex B (see Fig. 12-5) to produce Sandhoff disease or activator protein deficiency (which is very rare), respectively.

The clinical course of Tay-Sachs disease is tragic. Affected infants appear normal until approximately 3 to 6 months of age but then gradually undergo progressive neurological deterioration until death at 2 to 4 years. The effects of neuronal death can be seen directly in the form of the so-called cherry-red spot in the

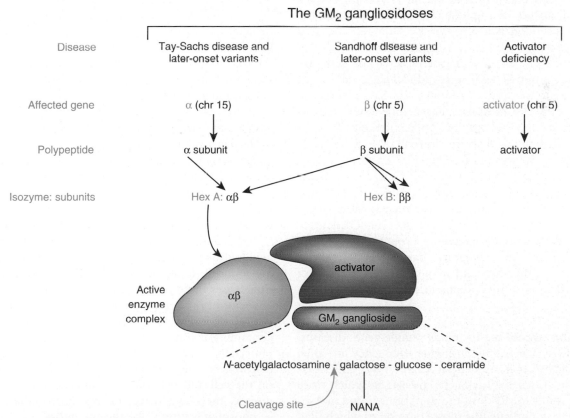

Figure 12-5 The three-gene system required for hexosaminidase A activity and the diseases that result from defects in each of the genes. The function of the activator protein is to bind the ganglioside substrate and present it to the enzyme. Hex A, Hexosaminidase A; hex B, hexosaminidase B; NANA, N-acetyl neuraminic acid. *See Sources & Acknowledgments.*

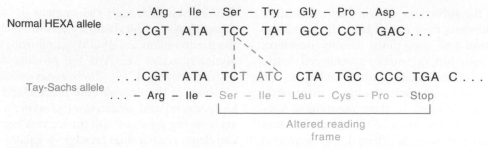

Figure 12-6 Four-base insertion (TATC) in the hexosaminidase A (hex A) gene in Tay-Sachs disease, leading to a frameshift mutation. This mutation is the major cause of Tay-Sachs disease in Ashkenazi Jews. No detectable hex A protein is made, accounting for the complete enzyme deficiency observed in these infantile-onset patients.

retina (Case 43). In contrast, *HEXA* alleles associated with some residual activity lead to later-onset forms of neurological disease, with manifestations including lower motor neuron dysfunction and ataxia due to spinocerebellar degeneration. In contrast to the infantile disease, vision and intelligence usually remain normal, although psychosis develops in one third of these patients. Finally, **pseudodeficiency alleles** (discussed next) do not cause disease at all.

Hex A Pseudodeficiency Alleles and Their Clinical Significance. An unexpected consequence of screening for Tay-Sachs carriers in the Ashkenazi Jewish population was the discovery of a unique class of hex A alleles, the so-called pseudodeficiency alleles. Although the two pseudodeficiency alleles are clinically benign, individuals identified as pseudodeficient in screening tests are genetic compounds with a pseudodeficiency allele on one chromosome and a common Tay-Sachs mutation on the other chromosome. These individuals have a low level of hex A activity (approximately 20% of controls) that is adequate to prevent GM2 ganglioside accumulation in the brain. The importance of hex A pseudodeficiency alleles is twofold. First, they complicate prenatal diagnosis because a pseudodeficient fetus could be incorrectly diagnosed as affected. More generally, the recognition of the hex A pseudodeficiency alleles indicates that screening programs for other genetic diseases must recognize that comparable alleles may exist at other loci and may confound the correct characterization of individuals in screening or diagnostic tests.

Population Genetics. In many single-gene diseases, some alleles are found at higher frequency in some populations than in others (see Chapter 9). This situation is illustrated by Tay-Sachs disease, in which three alleles account for 99% of the mutations found in Ashkenazi Jewish patients, the most common of which (Fig. 12-6) accounts for 80% of cases. Approximately 1 in 27 Ashkenazi Jews is a carrier of a Tay-Sachs allele, and the incidence of affected infants is 100 times higher than

in other populations. A founder effect or heterozygote advantage is the most likely explanation for this high frequency (see Chapter 9). Because most Ashkenazi Jewish carriers will have one of the three common alleles, a practical benefit of the molecular characterization of the disease in this population is the degree to which carrier screening has been simplified.

Altered Protein Function due to Abnormal Post-translational Modification
A Loss of Glycosylation: I-Cell Disease

Some proteins have information contained in their primary amino acid sequence that directs them to their subcellular residence, whereas others are localized on the basis of post-translational modifications. This latter mechanism is true of the acid hydrolases found in lysosomes, but this form of cellular trafficking was unrecognized until the discovery of **I-cell disease**, a severe autosomal recessive lysosomal storage disease. The disorder has a range of phenotypic effects involving facial features, skeletal changes, growth retardation, and intellectual disability and survival of less than 10 years (Fig. 12-7). The cytoplasm of cultured skin fibroblasts from I-cell patients contains numerous abnormal lysosomes, or inclusions, (hence the term *inclusion cells* or *I cells*).

In I-cell disease, the cellular levels of many lysosomal acid hydrolases are greatly diminished, and instead they are found in excess in body fluids. This unusual situation arises because the hydrolases in these patients have not been properly modified post-translationally. A typical hydrolase is a glycoprotein, the sugar moiety containing mannose residues, some of which are phosphorylated. The mannose-6-phosphate residues are essential for recognition of the hydrolases by receptors on the cell and lysosomal membrane surface. In I-cell disease, there is a defect in the enzyme that transfers a phosphate group to the mannose residues. The fact that many enzymes are affected is consistent with the diversity of clinical abnormalities seen in these patients.

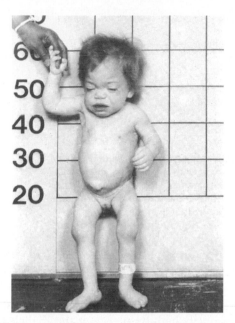

Figure 12-7 I-cell disease facies and habitus in an 18-month-old girl. *See Sources & Acknowledgments.*

Gains of Glycosylation: Mutations That Create New (Abnormal) Glycosylation Sites

In contrast to the failure of protein glycosylation exemplified by I-cell disease, it has been shown that an unexpectedly high proportion (approximately 1.5%) of the missense mutations that cause human disease may be associated with abnormal gains of *N*-glycosylation due to mutations creating new consensus *N*-glycosylation sites in the mutant proteins. That such novel sites can actually lead to inappropriate glycosylation of the mutant protein, with pathogenic consequences, is highlighted by the rare autosomal recessive disorder, **mendelian susceptibility to mycobacterial disease (MSMD)**.

MSMD patients have defects in any one of a number of genes that regulate the defense against some infections. Consequently, they are susceptible to disseminated infections upon exposure to moderately virulent mycobacterial species, such as the bacillus Calmette-Guérin (BCG) used throughout the world as a vaccine against tuberculosis, or to nontuberculous environmental bacteria that do not normally cause illness. Some MSMD patients carry missense mutations in the gene for interferon-γ receptor 2 (*IFNGR2*) that generate novel *N*-glycosylation sites in the mutant IFNGR2 protein. These novel sites lead to the synthesis of an abnormally large, overly glycosylated receptor. The mutant receptors reach the cell surface but fail to respond to interferon-γ. Mutations leading to gains of glycosylation have also been found to lead to a loss of protein function in several other monogenic disorders. The discovery that removal of the abnormal polysaccharides restores function to the mutant IFNGR2 proteins in MSMD offers hope that disorders of this type

may be amenable to chemical therapies that reduce the excessive glycosylation.

Loss of Protein Function due to Impaired Binding or Metabolism of Cofactors

Some proteins acquire biological activity only after they associate with cofactors, such as BH_4 in the case of PAH, as discussed earlier. Mutations that interfere with cofactor synthesis, binding, transport, or removal from a protein (when ligand binding is covalent) are also known. For many of these mutant proteins, an increase in the intracellular concentration of the cofactor is frequently capable of restoring some residual activity to the mutant enzyme, for example by increasing the stability of the mutant protein. Consequently, enzyme defects of this type are among the most responsive of genetic disorders to specific biochemical therapy because the cofactor or its precursor is often a water-soluble vitamin that can be administered safely in large amounts (see Chapter 13).

Impaired Cofactor Binding: Homocystinuria due to Cystathionine Synthase Deficiency

Homocystinuria due to cystathionine synthase deficiency (Fig. 12-8) was one of the first aminoacidopathies to be recognized. The clinical phenotype of this autosomal recessive condition is often dramatic. The most common features include dislocation of the lens, intellectual disability, osteoporosis, long bones, and thromboembolism of both veins and arteries, a phenotype that can be confused with **Marfan syndrome**, a disorder of connective tissue (Case 30). The accumulation of homocysteine is believed to be central to most, if not all, of the pathology.

Homocystinuria was one of the first genetic diseases shown to be vitamin responsive; pyridoxal phosphate is the cofactor of the enzyme, and the administration of large amounts of pyridoxine, the vitamin precursor of the cofactor, often ameliorates the biochemical abnormality and the clinical disease (see Chapter 13). In many patients, the affinity of the mutant enzyme for pyridoxal phosphate is reduced, indicating that altered conformation of the protein impairs cofactor binding.

Not all cases of homocystinuria result from mutations in cystathionine synthase. Mutations in five different enzymes of cobalamin (vitamin B_{12}) or folate metabolism can also lead to increased levels of homocysteine in body fluids. These mutations impair the provision of the vitamin B_{12} cofactor, methylcobalamin (methyl-B_{12}), or of methyl-H_4-folate (see Fig. 12-8) and thus represent another example (like the defects in BH_4 synthesis that lead to hyperphenylalaninemia) of genetic diseases due to defects in the biogenesis of enzyme cofactors. The clinical manifestation of these disorders is variable but includes megaloblastic anemia, developmental delay, and failure to thrive. These conditions, all

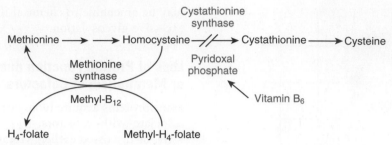

Figure 12-8 **Genetic defects in pathways that impinge on cystathionine synthase, or in that enzyme itself, and cause homocystinuria.** Classic homocystinuria is due to defective cystathionine synthase. Several different defects in the intracellular metabolism of cobalamins (not shown) lead to a decrease in the synthesis of methylcobalamin (methyl-B_{12}) and thus in the function of methionine synthase. Defects in methylene-H_4-folate reductase (not shown) decrease the abundance of methyl-H_4-folate, which also impairs the function of methionine synthase. Some patients with cystathionine synthase abnormalities respond to large doses of vitamin B_6, increasing the synthesis of pyridoxal phosphate, thereby increasing cystathionine synthase activity and treating the disease (see Chapter 13).

of which are autosomal recessive, are often partially or completely treatable with high doses of vitamin B_{12}.

Mutations of an Enzyme Inhibitor: α₁-Antitrypsin Deficiency

α_1-Antitrypsin (α1AT) deficiency is an important autosomal recessive condition associated with a substantial risk for chronic obstructive lung disease (emphysema) (Fig. 12-9) and cirrhosis of the liver. The α1AT protein belongs to a major family of protease inhibitors, the *ser*ine *pro*tease *in*hibitors or serpins; *SERPINA1* is the formal gene name. Notwithstanding the specificity suggested by its name, α1AT actually inhibits a wide spectrum of proteases, particularly elastase released from neutrophils in the lower respiratory tract.

In white populations, α1AT deficiency affects approximately 1 in 6700 persons, and approximately 4% are carriers. A dozen or so α1AT alleles are associated with an increased risk for lung or liver disease, but only the

Z allele (Glu342Lys) is relatively common. The reason for the relatively high frequency of the Z allele in white populations is unknown, but analysis of DNA haplotypes suggests a single origin with subsequent spread throughout northern Europe. Given the increased risk for emphysema, α1AT deficiency is an important public health problem, affecting an estimated 60,000 persons in the United States alone.

The α1AT gene is expressed principally in the liver, which normally secretes α1AT into plasma. Approximately 17% of Z/Z homozygotes present with neonatal jaundice, and approximately 20% of this group subsequently develop cirrhosis. The liver disease associated with the Z allele is thought to result from a novel property of the mutant protein—its tendency to aggregate, trapping it within the rough endoplasmic reticulum (ER) of hepatocytes. The molecular basis of the Z protein aggregation is a consequence of structural changes in the protein that predispose to the formation of long beadlike necklaces of mutant α1AT polymers.

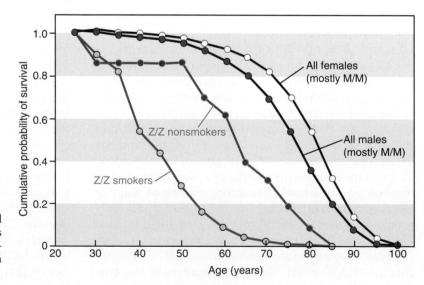

Figure 12-9 **The effect of smoking on the survival of patients with α₁-antitrypsin deficiency.** The curves show the cumulative probability of survival to specified ages of smokers, with or without α_1-antitrypsin deficiency. *See Sources & Acknowledgments.*

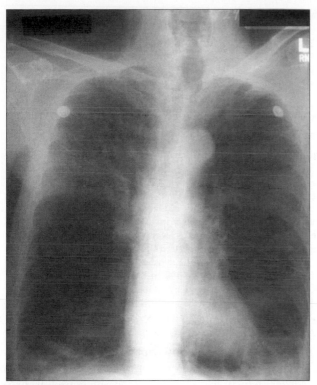

Figure 12-10 A posteroanterior chest radiograph of an individual carrying two Z alleles of the α1AT gene, showing the hyperinflation and basal hyperlucency characteristic of emphysema. *See Sources & Acknowledgments.*

Thus, like the sickle cell disease mutation in β-globin (see Chapter 11), the Z allele of α1AT is a clear example of a mutation that confers a novel property on the protein (in both of these examples, a tendency to aggregate) (see Fig. 11-1).

Both sickle cell disease and the α1AT deficiency associated with homozygosity for the Z allele are examples of inherited **conformational diseases.** These disorders occur when a mutation causes the shape or size of a protein to change in a way that predisposes it to self-association and tissue deposition. Notably, some fraction of the mutant protein is invariably correctly folded in these disorders, including α1AT deficiency. Note that not all conformational diseases are single-gene disorders, as illustrated, for example, by nonfamilial Alzheimer disease (discussed later) and prion diseases.

The lung disease associated with the Z allele of α1AT deficiency is due to the alteration of the normal balance between elastase and α1AT, which allows progressive degradation of the elastin of alveolar walls (Fig. 12-10). Two mechanisms contribute to the elastase:α1AT imbalance. First, the block in the hepatic secretion of the Z protein, although not complete, is severe, and Z/Z patients have only approximately 15% of the normal plasma concentration of α1AT. Second, the Z protein has only approximately 20% of the ability of the normal α1AT protein to inhibit neutrophil elastase. The infusion of normal α1AT is used in some patients to augment the level of α1AT in the plasma, to rectify the elastase:α1AT imbalance. At present, it is still uncertain whether progression of the lung disease is slowed by α1AT augmentation.

α₁-Antitrypsin Deficiency as an Ecogenetic Disease

The development of lung or liver disease in subjects with α1AT deficiency is highly variable, and although no modifier genes have yet been identified, a major environmental factor, cigarette smoke, dramatically influences the likelihood of emphysema. The impact of smoking on the progression of the emphysema is a powerful example of the effect that environmental factors may have on the phenotype of a monogenetic disease. Thus, for persons with the Z/Z genotype, survival after 60 years of age is approximately 60% in nonsmokers but only approximately 10% in smokers (see Fig. 12-9). One molecular explanation for the effect of smoking is that the active site of α1AT, at methionine 358, is oxidized by both cigarette smoke and inflammatory cells, thus reducing its affinity for elastase by 2000-fold.

The field of **ecogenetics,** illustrated by α1AT deficiency, is concerned with the interaction between environmental factors and different human genotypes. This area of medical genetics is likely to be one of increasing importance as genotypes are identified that entail an increased risk for disease on exposure to certain environmental agents (e.g., drugs, foods, industrial chemicals, and viruses). At present, the most highly developed area of ecogenetics is that of **pharmacogenetics,** presented in Chapter 16.

Dysregulation of a Biosynthetic Pathway: Acute Intermittent Porphyria

Acute intermittent porphyria (AIP) is an autosomal dominant disease associated with intermittent neurological dysfunction. The primary defect is a deficiency of porphobilinogen (PBG) deaminase, an enzyme in the biosynthetic pathway of heme, required for the synthesis of both hemoglobin and hepatic cytochrome p450 drug-metabolizing enzymes (Fig. 12-11). All individuals with AIP have an approximately 50% reduction in PBG deaminase enzymatic activity, whether their disease is clinically latent (90% of patients throughout their lifetime) or clinically expressed (approximately 10%). This reduction is consistent with the autosomal dominant inheritance pattern (see Chapter 7). Homozygous deficiency of PBG deaminase, a critical enzyme in heme biosynthesis, would presumably be incompatible with life. AIP illustrates one molecular mechanism by which an autosomal dominant disease may manifest only episodically.

The pathogenesis of the nervous system disease is uncertain but may be mediated directly by the increased

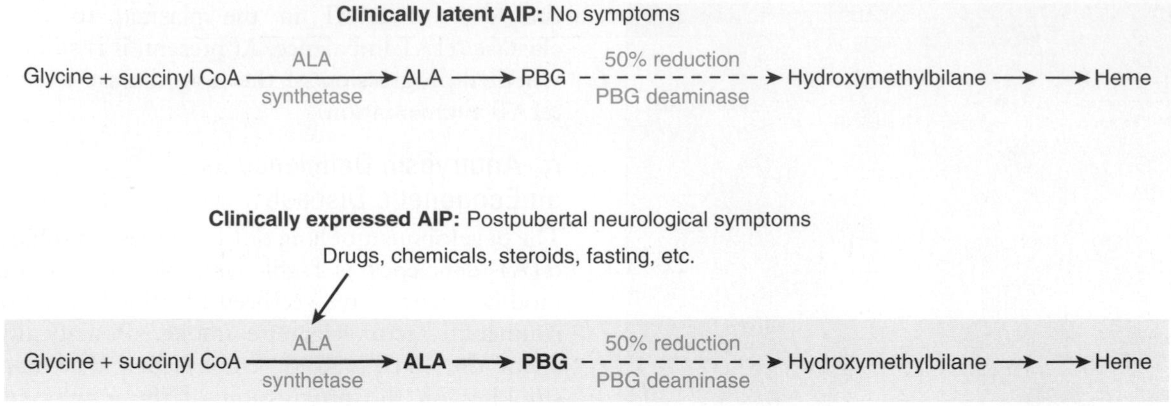

Clinically latent AIP: No symptoms

Glycine + succinyl CoA $\xrightarrow[\text{synthetase}]{\text{ALA}}$ ALA $\longrightarrow$ PBG $\dashrightarrow[\text{PBG deaminase}]{\text{50\% reduction}}$ Hydroxymethylbilane $\longrightarrow \longrightarrow$ Heme

Clinically expressed AIP: Postpubertal neurological symptoms

Drugs, chemicals, steroids, fasting, etc.

Glycine + succinyl CoA $\xrightarrow[\text{synthetase}]{\text{ALA}}$ **ALA** $\longrightarrow$ **PBG** $\dashrightarrow[\text{PBG deaminase}]{\text{50\% reduction}}$ Hydroxymethylbilane $\longrightarrow \longrightarrow$ Heme

Figure 12-11 The pathogenesis of acute intermittent porphyria (AIP). Patients with AIP who are either clinically latent or clinically affected have approximately half the control levels of porphobilinogen (PBG) deaminase. When the activity of hepatic δ-aminolevulinic acid (ALA) synthase is increased in carriers by exposure to inducing agents (e.g., drugs, chemicals), the synthesis of ALA and PBG is increased. The residual PBG deaminase activity (approximately 50% of controls) is overloaded, and the accumulation of ALA and PBG causes clinical disease. CoA, Coenzyme A. *See Sources & Acknowledgments.*

levels of δ-aminolevulinic acid (ALA) and PBG that accumulate due to the 50% reduction in PBG deaminase (see Fig. 12-11). The peripheral, autonomic, and central nervous systems are all affected, and the clinical manifestations are diverse. Indeed, this disorder is one of the great mimics in clinical medicine, with manifestations ranging from acute abdominal pain to psychosis.

Clinical crises in AIP are elicited by a variety of precipitating factors: drugs (most prominently the barbiturates, and to this extent, AIP is a **pharmacogenetic disease;** see Chapter 18); some steroid hormones (clinical disease is rare before puberty or after menopause); and catabolic states, including reducing diets, intercurrent illnesses, and surgery. The drugs provoke the clinical manifestations by interacting with drug-sensing nuclear receptors in hepatocytes, which then bind to transcriptional regulatory elements of the ALA synthetase gene, increasing the production of both ALA and PBG. In normal individuals the drug-related increase in ALA synthetase is beneficial because it increases heme synthesis, allowing greater formation of hepatic cytochrome P450 enzymes that metabolize many drugs. In patients with AIP, however, the increase in ALA synthetase causes the accumulation of ALA and PBG because of the 50% reduction in PBG deaminase activity (see Fig. 12-11). The fact that half of the normal activity of PBG deaminase is inadequate to cope with the increased requirement for heme synthesis in some situations accounts for both the dominant inheritance of the condition and the episodic nature of the clinical illness.

DEFECTS IN RECEPTOR PROTEINS

The recognition of a class of diseases due to defects in receptor molecules began with the identification by Goldstein and Brown of the low-density lipoprotein (LDL) receptor as the polypeptide affected in the most common form of familial hypercholesterolemia. This disorder, which leads to a greatly increased risk for myocardial infarction, is characterized by elevation of plasma cholesterol carried by LDL, the principal cholesterol transport protein in plasma. Goldstein and Brown's discovery has cast much light on normal cholesterol metabolism and on the biology of cell surface receptors in general. LDL receptor deficiency is representative of a number of disorders now recognized to result from receptor defects.

Familial Hypercholesterolemia: A Genetic Hyperlipidemia

Familial hypercholesterolemia is one of a group of metabolic disorders called the hyperlipoproteinemias. These diseases are characterized by elevated levels of plasma lipids (cholesterol, triglycerides, or both) carried by apolipoprotein B (apoB)-containing lipoproteins. Other monogenic hyperlipoproteinemias, each with distinct biochemical and clinical phenotypes, have also been recognized.

In addition to mutations in the LDL receptor gene (Table 12-2), abnormalities in three other genes can also lead to familial hypercholesterolemia (Fig. 12-12). Remarkably, all four of the genes associated with familial hypercholesterolemia disrupt the function or abundance either of the LDL receptor at the cell surface or of apoB, the major protein component of LDL and a ligand for the LDL receptor. Because of its importance, we first review familial hypercholesterolemia due to mutations in the LDL receptor. We also discuss mutations in the *PCSK9* protease gene; although gain-of-function mutations in this gene cause hypercholesterolemia, the greater importance of *PCSK9* lies in the fact

TABLE 12-2 Four Genes Associated with Familial Hypercholesterolemia

Mutant Gene Product	Pattern of Inheritance	Effect of Disease-Causing Mutations	Typical LDL Cholesterol Level (Normal Adults: ≈120 mg/dL)
LDL receptor	Autosomal dominant	Loss of function	Heterozygotes: 350 mg/dL Homozygotes: 700 mg/dL
Apoprotein B-100	Autosomal dominant*	Loss of function	Heterozygotes: 270 mg/dL Homozygotes: 320 mg/dL
ARH adaptor protein	Autosomal recessive†	Loss of function	Homozygotes: 470 mg/dL
PCSK9 protease	Autosomal dominant	Gain of function	Heterozygotes: 225 mg/dL

*Principally in individuals of European descent.
†Principally in individuals of Italian and Middle Eastern descent.
LDL, Low-density lipoprotein.
Partly modified from Goldstein JL, Brown MS: The cholesterol quartet. *Science* 292:1310–1312, 2001.

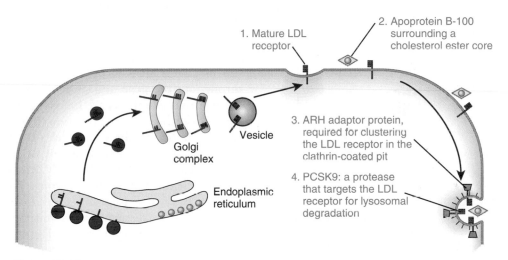

Figure 12-12 The four proteins associated with familial hypercholesterolemia. The low-density lipoprotein (LDL) receptor binds apoprotein B-100. Mutations in the LDL receptor-binding domain of apoprotein B-100 impair LDL binding to its receptor, reducing the removal of LDL cholesterol from the circulation. Clustering of the LDL receptor–apoprotein B-100 complex in clathrin-coated pits requires the ARH adaptor protein, which links the receptor to the endocytic machinery of the coated pit. Homozygous mutations in the ARH protein impair the internalization of the LDL:LDL receptor complex, thereby impairing LDL clearance. PCSK9 protease activity targets LDL receptors for lysosomal degradation, preventing them from recycling back to the plasma membrane (see text).

that several common loss-of-function sequence variants *lower* plasma LDL cholesterol levels, conferring substantial *protection* from coronary heart disease.

Familial Hypercholesterolemia due to Mutations in the LDL Receptor

Mutations in the LDL receptor gene (*LDLR*) are the most common cause of familial hypercholesterolemia **(Case 16)**. The receptor is a cell surface protein responsible for binding LDL and delivering it to the cell interior. Elevated plasma concentrations of LDL cholesterol lead to premature atherosclerosis (accumulation of cholesterol by macrophages in the subendothelial space of major arteries) and increased risk for heart attack and stroke in both untreated heterozygote and homozygote carriers of mutant alleles. Physical stigmata of familial hypercholesterolemia include xanthomas (cholesterol deposits in skin and tendons) **(Case 16)** and premature

arcus corneae (deposits of cholesterol around the periphery of the cornea). Few diseases have been as thoroughly characterized; the sequence of pathological events from the affected locus to its effect on individuals and populations has been meticulously documented.

Genetics. Familial hypercholesterolemia due to mutations in the *LDLR* gene is inherited as an autosomal semidominant trait. Both homozygous and heterozygous phenotypes are known, and a clear gene dosage effect is evident; the disease manifests earlier and much more severely in homozygotes than in heterozygotes, reflecting the greater reduction in the number of LDL receptors and the greater elevation in plasma LDL cholesterol (Fig. 12-13). Homozygotes may have clinically significant coronary heart disease in childhood and, if untreated, few live beyond the third decade. The heterozygous form of the disease, with a population frequency

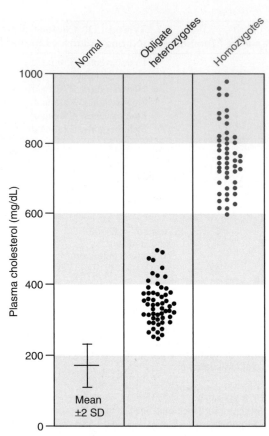

Figure 12-13 Gene dosage in low-density lipoprotein (LDL) deficiency. Shown is the distribution of total plasma cholesterol levels in 49 patients homozygous for deficiency of the LDL receptor, their parents (obligate heterozygotes), and normal controls. *See Sources & Acknowledgments.*

of approximately 2 per 1000, is one of the most common single-gene disorders. Heterozygotes have levels of plasma cholesterol that are approximately twice those of controls (see Fig. 12-13). Because of the inherited nature of familial hypercholesterolemia, it is important to make the diagnosis in the approximately 5% of survivors of premature (<50 years of age) myocardial infarction who are heterozygotes for an LDL receptor defect. It is important to stress, however, that, among those in the general population with plasma cholesterol concentrations above the 95th percentile for age and sex, only approximately 1 in 20 has familial hypercholesterolemia; most such individuals have an uncharacterized hypercholesterolemia due to multiple common genetic variants, as presented in Chapter 8.

Cholesterol Uptake by the LDL Receptor. Normal cells obtain cholesterol from either de novo synthesis or the uptake from plasma of exogenous cholesterol bound to lipoproteins, especially LDL. The majority of LDL uptake is mediated by the LDL receptor, which recognizes apoprotein B-100, the protein moiety of LDL. LDL receptors on the cell surface are localized to invaginations (coated pits) lined by the protein clathrin

(Fig. 12-14). Receptor-bound LDL is brought into the cell by endocytosis of the coated pits, which ultimately evolve into lysosomes in which LDL is hydrolyzed to release free cholesterol. The increase in free intracellular cholesterol reduces endogenous cholesterol formation by suppressing the rate-limiting enzyme of the synthetic pathway, 3-hydroxy-3-methylglutaryl coenzyme A (HMG CoA) reductase. Cholesterol not required for cellular metabolism or membrane synthesis may be re-esterified for storage as cholesteryl esters, a process stimulated by the activation of acyl coenzyme A : cholesterol acyltransferase (ACAT). The increase in intracellular cholesterol also reduces synthesis of the LDL receptor (see Fig. 12-14).

Classes of Mutations in the LDL Receptor

More than 1100 different mutations have been identified in the *LDLR* gene, and these are distributed throughout the gene and protein sequence. Not all of the reported mutations are functionally significant, and some disturb receptor function more severely than others. The great majority of alleles are single nucleotide substitutions, small insertions, or deletions; structural rearrangements account for only 2% to 10% of the *LDLR* alleles in most populations. The mature LDL receptor has five distinct structural domains that for the most part have distinguishable functions that mediate the steps in the life cycle of an LDL receptor, shown in Figure 12-14. Analysis of the effect on the receptor of mutations in each domain has played an important role in defining the function of each domain. These studies exemplify the important contribution that genetic analysis can make in determining the structure-function relationships of a protein.

Fibroblasts cultured from affected patients have been used to characterize the mutant receptors and the resulting disturbances in cellular cholesterol metabolism. *LDLR* mutations can be grouped into six classes, depending on which step of the normal cellular itinerary of the receptor is disrupted by the mutation (see Fig. 12-14).

- Class 1 mutations are *null alleles* that prevent the synthesis of any detectable receptor; they are the most common type of disease-causing mutations at this locus. In the remaining five classes, the receptor is synthesized normally, but its function is impaired.
- Mutations in class 2 (like those in classes 4 and 6) define features of the polypeptide critical to its subcellular localization. The relatively common class 2 mutations are designated *transport-deficient* because the LDL receptors accumulate at the site of their synthesis, the ER, instead of being transported to the Golgi complex. These alleles are predicted to prevent proper folding of the protein, an apparent requisite for exit from the ER.
- Class 3 mutant receptors reach the cell surface but are *incapable of binding LDL.*

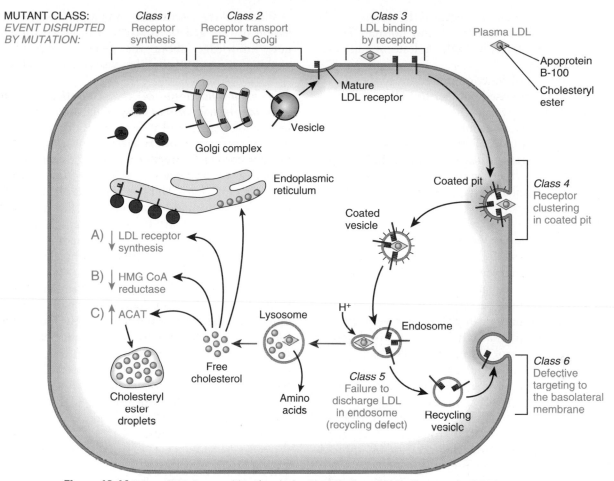

Figure 12-14 The cell biology and biochemical role of the low-density lipoprotein (LDL) receptor and the six classes of mutations that alter its function. After synthesis in the endoplasmic reticulum (ER), the receptor is transported to the Golgi apparatus and subsequently to the cell surface. Normal receptors are localized to clathrin-coated pits, which invaginate, creating coated vesicles and then endosomes, the precursors of lysosomes. Normally, intracellular accumulation of free cholesterol is prevented because the increase in free cholesterol (A) decreases the formation of LDL receptors, (B) reduces de novo cholesterol synthesis, and (C) increases the storage of cholesteryl esters. The biochemical phenotype of each class of mutant is discussed in the text. ACAT, Acyl coenzyme A:cholesterol acyltransferase; HMG CoA reductase, 3-hydroxy-3-methylglutaryl coenzyme A reductase. *See Sources & Acknowledgments.*

- Class 4 mutations *impair localization* of the receptor to the coated pit, and consequently the bound LDL is not internalized. These mutations alter or remove the cytoplasmic domain at the carboxyl terminus of the receptor, demonstrating that this region normally targets the receptor to the coated pit.
- Class 5 mutations are *recycling-defective alleles.* Receptor recycling requires the dissociation of the receptor and the bound LDL in the endosome. Mutations in the epidermal growth factor precursor homology domain prevent the release of the LDL ligand. This failure leads to degradation of the receptor, presumably because an occupied receptor cannot return to the cell surface.
- Class 6 mutations lead to *defective targeting* of the mutant receptor to the basolateral membrane, a process that depends on a sorting signal in the

cytoplasmic domain of the receptor. Mutations affecting the signal can mistarget the mutant receptor to the apical surface of hepatic cells, thereby impairing the recycling of the receptor to the basolateral membrane and leading to an overall reduction of endocytosis of the LDL receptor.

The PCSK9 Protease, a Potential Drug Target for Lowering LDL Cholesterol

Rare cases of autosomal dominant familial hypercholesterolemia have been found to result from gain-of-function missense mutations in the gene encoding PCSK9 protease (proprotein convertase subtilisin/kexin type 9). The role of PCSK9 is to target the LDL receptor for lysosomal degradation, thereby reducing receptor abundance at the cell surface (see Fig. 12-12). Consequently, the increase in PSCK9 activity associated with gain-of-function mutations reduces the levels of the LDL receptor at the cell surface below normal, leading to increased blood levels of LDL cholesterol and coronary heart disease.

Conversely, loss-of-function mutations in the *PCSK9* gene result in an increased number of LDL receptors at the cell surface by decreasing the activity of the protease. More receptors increase cellular uptake of LDL cholesterol, lowering cholesterol and providing protection against coronary artery disease. Notably, the complete absence of PCSK9 activity in the few known individuals with two *PCSK9* null alleles appears to have no adverse clinical consequences.

Some PCSK9 Sequence Variants Protect against Coronary Heart Disease.

The link between monogenic familial hypercholesterolemia and the *PCSK9* gene suggested that common sequence variants in *PCSK9* might be linked to very high or very low LDL cholesterol levels in the general population. Importantly, several *PCSK9* sequence variants are strongly linked to low levels of plasma LDL cholesterol (Table 12-3). For example, in the African American population one of two *PCSK9* nonsense variants is found in 2.6% of all subjects; each variant is associated with a mean reduction in LDL cholesterol of approximately 40%. This reduction in LDL cholesterol has a powerful protective effect against coronary artery disease, reducing the risk by approximately 90%; only approximately 1% of African American subjects carrying one of these two *PCSK9* nonsense variants developed coronary artery disease over a 15-year period, compared to almost 10% of individuals without either variant. A missense allele (Arg46Leu) is more common in white populations (3.2% of subjects) but appears to confer only approximately a 50% reduction in coronary heart disease. These findings have major public health implications because they suggest that modest but lifelong reductions in plasma LDL cholesterol levels of 20 to 40 mg/dL would significantly decrease the incidence of coronary heart disease in the population. The strong protective effect of *PCSK9* loss-of-function alleles, together with the apparent absence of any clinical sequelae in subjects with a total absence of PCSK9 activity, has made PCSK9 a strong candidate target for drugs that inactivate or diminish the activity of the enzyme.

Finally, these discoveries emphasize how the investigation of rare genetic disorders can lead to important new knowledge about the genetic contribution to common genetically complex diseases.

Clinical Implications of the Genetics of Familial Hypercholesterolemia.

Early diagnosis of the familial hypercholesterolemias is essential both to permit the prompt application of cholesterol-lowering therapies to prevent coronary artery disease and to initiate genetic screening of first-degree relatives. With appropriate drug therapy, familial hypercholesterolemia heterozygotes have a normal life expectancy. For homozygotes, onset of coronary artery disease can be remarkably delayed by plasma apheresis (which removes the hypercholesterolemic plasma), but will ultimately require liver transplantation.

Finally, the elucidation of the biochemical basis of familial hypercholesterolemia has had a profound impact on the treatment of the vastly more common forms of sporadic hypercholesterolemia by leading to the development of the statin class of drugs that inhibit de novo cholesterol biosynthesis (see Chapter 13). Newer therapies include monoclonal antibodies that directly target PCSK9, which lower LDL cholesterol by an additional 60% in clinical trials.

TRANSPORT DEFECTS

Cystic Fibrosis

Since the 1960s, cystic fibrosis (CF) has been one of the most publicly visible of all human monogenic diseases (Case 12). It is the most common fatal autosomal recessive genetic disorder of children in white populations, with an incidence of approximately 1 in 2500 white births (and thus a carrier frequency of approximately 1 in 25), whereas it is much less prevalent in other ethnic groups, such as African Americans (1 in 15,000 births) and Asian Americans (1 in 31,000 births). The isolation of the CF gene (called *CFTR*, for *CF transmembrane regulator*) (see Chapter 10) more than 25 years ago was one of the first illustrations of the power of molecular genetic and genomic approaches to identify disease genes. Physiological analyses have shown that the CFTR protein is a regulated chloride

TABLE 12-3 *PCSK9* Variants Associated with Low LDL Cholesterol Levels

Sequence Variant	Population Frequency	LDL Cholesterol Level (Normal ≤ ≈100 mg/dL)	Impact on Incidence of Coronary Heart Disease
Null or dominant negative alleles	Rare genetic compounds, one dominant negative heterozygote	7-16 mg/dL	Unknown, but likely to greatly reduce risk
Tyr142Stop or Cys679Stop	African American heterozygotes: 2.6%	Mean: 28% (38 mg/dL)	90% reduction
Arg46Leu	White heterozygotes: 3.2%	Mean: 15% (20 mg/dL)	50% reduction

LDL, Low-density lipoprotein.

Derived from Cohen JC, Boerwinkle E, Mosley TH, Hobbs H: Sequence variants in *PCSK9*, low LDL, and protection against coronary heart disease, *N Engl J Med* 354:1264–1272, 2006.

channel located in the apical membrane of the epithelial cells affected by the disease.

The Phenotypes of Cystic Fibrosis. The lungs and exocrine pancreas are the principal organs affected by CF **(Case 12)**, but a major diagnostic feature is increased sweat sodium and chloride concentrations (often first noted when parents kiss their infants). In most CF patients, the diagnosis is initially based on the clinical pulmonary or pancreatic findings and on an elevated level of sweat chloride. Less than 2% of patients have normal sweat chloride concentration despite an otherwise typical clinical picture; in these cases, molecular analysis can be used to ascertain whether they have mutations in the *CFTR* gene.

The pancreatic defect in CF is a maldigestion syndrome due to the deficient secretion of pancreatic enzymes (lipase, trypsin, chymotrypsin). Approximately 5% to 10% of patients with CF have enough residual pancreatic exocrine function for normal digestion and are designated *pancreatic sufficient.* Moreover, patients with CF who are pancreatic sufficient have better growth and overall prognosis than the majority, who are *pancreatic insufficient.* The clinical heterogeneity of the pancreatic disease is at least partly due to allelic heterogeneity, as discussed later.

Many other phenotypes are observed in CF patients. For example, neonatal lower intestinal tract obstruction (**meconium ileus**) occurs in 10% to 20% of CF newborns. The genital tract is also affected; females with CF have some reduction in fertility, but more than 95% of CF males are infertile because they lack the vas deferens, a phenotype known as **congenital bilateral absence of the vas deferens (CBAVD).** In a striking example of allelic heterogeneity giving rise to a partial phenotype, it has been found that some infertile males who are otherwise well (i.e., have no pulmonary or pancreatic disease) have CBAVD associated with specific mutant alleles in the *CFTR* gene. Similarly, some individuals with **idiopathic chronic pancreatitis** are carriers of mutations in *CFTR*, yet lack other clinical signs of CF.

The *CFTR* Gene and Protein. The *CFTR* gene has 27 exons and spans approximately 190 kb of DNA. The CFTR protein encodes a large integral membrane protein of approximately 170 kD (Fig. 12-15). The protein belongs to the so-called ABC (*A*TP [adenosine triphosphate]–*b*inding *c*assette) family of transport proteins. At least 22 ABC transporters have been implicated in mendelian disorders and complex trait phenotypes.

The CFTR chloride channel has five domains, shown in Figure 12-15: two membrane-spanning domains,

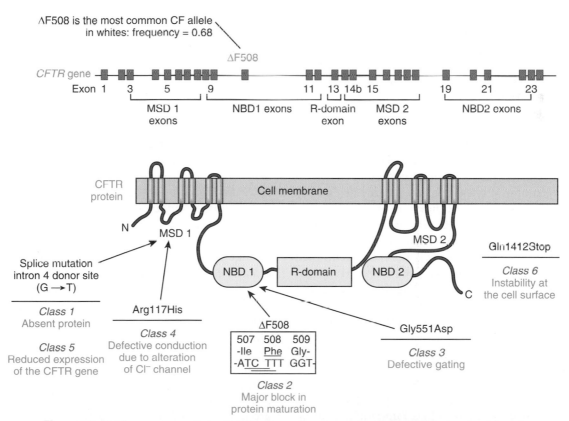

Figure 12-15 **The structure of the *CFTR* gene and a schematic of the CFTR protein.** Selected mutations are shown. The exons, introns, and domains of the protein are not drawn to scale. ΔF508 results from the deletion of TCT or CTT, replacing the Ile codon with ATT, and deleting the Phe codon. CF, Cystic fibrosis; MSD, membrane-spanning domain; NBD, nucleotide-binding domain; R-domain, regulatory domain. *See Sources & Acknowledgments.*

each with six transmembrane sequences; two nucleotide (ATP)-binding domains; and a regulatory domain with multiple phosphorylation sites. The importance of each domain is demonstrated by the identification of CF-causing missense mutations in each of them (see Fig. 12-15). The pore of the chloride channel is formed by the 12 transmembrane segments. ATP is bound and hydrolyzed by the nucleotide-binding domains, and the energy released is used to open and close the channel. Regulation of the channel is mediated, at least in part, by phosphorylation of the regulatory domain.

The Pathophysiology of Cystic Fibrosis. CF is due to abnormal fluid and electrolyte transport across epithelial apical membranes. This abnormality leads to disease in the lung, pancreas, intestine, hepatobiliary tree, and male genital tract. The physiological abnormalities have been most clearly elucidated for the sweat gland. The loss of CFTR function means that chloride in the duct of the sweat gland cannot be reabsorbed, leading to a reduction in the electrochemical gradient that normally drives sodium entry across the apical membrane. This defect leads, in turn, to the increased chloride and sodium concentrations in sweat. The effects on electrolyte transport due to the abnormalities in the CFTR protein have also been carefully studied in airway and pancreatic epithelia. In the lung, the hyperabsorption of sodium and reduced chloride secretion result in a depletion of airway surface liquid. Consequently, the mucous layer of the lung may become adherent to cell surfaces, disrupting the cough and cilia-dependent clearance of mucus and providing a niche favorable to *Pseudomonas aeruginosa*, the major cause of chronic pulmonary infection in CF.

The Genetics of Cystic Fibrosis

Mutations in the Cystic Fibrosis Transmembrane Regulator Polypeptide. The most common CF mutation is a deletion of a phenylalanine residue at position 508 (ΔF508) in the first ATP-binding fold (NBD1; see Fig. 12-15), accounting for approximately 70% of all CF alleles in white populations. In these populations, only seven other mutations are more frequent than 0.5%, and the remainder are each quite rare. Mutations of all types have been identified, but the largest single group (nearly half) are missense substitutions. The remainder are point mutations of other types, and less than 1% are genomic rearrangements. Although nearly 2000 *CFTR* gene sequence variants have been associated with disease, the actual number of missense mutations that are disease-causing is uncertain because few have been subjected to functional analysis. However, a new project called the Clinical and Functional Translation of CFTR (CFTR2 project; cftr2.org) has succeeded in assigning pathogenicity to more than 125 *CFTR* mutations, which together account for at least 96% of all *CFTR* alleles worldwide.

Although the specific biochemical abnormalities associated with most CF mutations are not known, six general classes of dysfunction of the CFTR protein have been identified to date. Alleles representative of each class are shown in Figure 12-15.

- Class 1 mutations are null alleles—no CFTR polypeptide is produced. This class includes alleles with premature stop codons or that generate highly unstable RNAs. Because CFTR is a glycosylated membrane-spanning protein, it must be processed in the endoplasmic reticulum and Golgi apparatus to be glycosylated and secreted.
- Class 2 mutations impair the folding of the CFTR protein, thereby arresting its maturation. The ΔF508 mutant typifies this class; this misfolded protein cannot exit from the endoplasmic reticulum. However, the biochemical phenotype of the ΔF508 protein is complex, because it also exhibits defects in stability and activation in addition to impaired folding.
- Class 3 mutations allow normal delivery of the CFTR protein to the cell surface, but disrupt its function (see Fig. 12-15). The prime example is the Gly551Asp mutation that impedes the opening and closing of the CFTR ion channel at the cell surface. This mutation is particularly notable because, although it constitutes only approximately 2% of *CFTR* alleles, the drug ivacaftor has been shown to be remarkably effective in correcting the function of the mutant Gly551Asp protein at the cell surface, resulting in both physiological and clinical improvements (see Chapter 13).
- Class 4 mutations are located in the membrane-spanning domains and, consistent with this localization, have defective chloride ion conduction.
- Class 5 mutations reduce the number of *CFTR* transcripts.
- Class 6 mutant proteins are synthesized normally but are unstable at the cell surface.

A Cystic Fibrosis Genocopy: Mutations in the Epithelial Sodium Channel Gene SCNN1. Although *CFTR* is the only gene that has been associated with classic CF, several families with nonclassic presentations (including CF-like pulmonary infections, less severe intestinal disease, elevated sweat chloride levels) have been found to carry mutations in the epithelial sodium channel gene *SCNN1*, a so-called **genocopy**, that is, a phenotype that, although genetically distinct, has a very closely related phenotype. This finding is consistent with the functional interaction between the CFTR protein and the epithelial sodium channel. Its main clinical significance, at present, is the demonstration that patients with nonclassic CF display locus heterogeneity and that if *CFTR* mutations are not identified in a particular case, abnormalities in *SCNNI* must be considered.

Genotype-Phenotype Correlations in Cystic Fibrosis. Because all patients with the classic form of CF appear to have mutations in the *CFTR* gene, clinical heterogeneity in CF must arise from allelic heterogeneity, from the effects of other modifying loci, or from nongenetic factors. Independent of the *CFTR* alleles that a particular patient may have, a significant genetic contribution from other (modifier) genes to several CF phenotypes has been recognized, with effects on lung function, neonatal intestinal obstruction, and diabetes.

Two generalizations have emerged from the genetic and clinical analysis of patients with CF. First, the specific *CFTR* genotype is a *good* predictor of exocrine pancreatic function. For example, patients homozygous for the common ΔF508 mutation or for predicted null alleles generally have pancreatic insufficiency. On the other hand, alleles that allow the synthesis of a partially functional CFTR protein, such as Arg117His (see Fig. 12-15), tend to be associated with pancreatic sufficiency.

Second, however, the specific *CFTR* genotype is a *poor* predictor of the severity of pulmonary disease. For example, among patients homozygous for the ΔF508 mutation, the severity of lung disease is variable. One reason for this poor phenotype-genotype correlation is inherited variation in the gene encoding transforming growth factor β1 (TGFβ1), as also discussed in Chapter 8. Overall, the evidence indicates that *TGFB1* alleles that increase TGFβ1 expression lead to more severe CF lung disease, perhaps by modulating tissue remodeling and inflammatory responses. Other genetic modifiers of CF lung disease, including alleles of the interferon-related developmental regulator 1 gene (*IFRD1*) and the interleukin-8 gene (*IL8*), may act by influencing the ability of the CF lung to tolerate infection. Similarly, a few modifier genes have been identified for other CF-related phenotypes, including diabetes, liver disease, and meconium ileus.

The Cystic Fibrosis Gene in Populations. At present, it is not possible to account for the high *CFTR* mutant allele frequency of 1 in 50 that is observed in white populations (see Chapter 9). The disease is much less frequent in nonwhites, although it has been reported in Native Americans, African Americans, and Asians (e.g., approximately 1 in 90,000 Hawaiians of Asian descent). The ΔF508 allele is the only one found to date that is common in virtually all white populations, but its frequency among all mutant alleles varies significantly in different European populations, from 88% in Denmark to 45% in southern Italy.

In populations in which the ΔF508 allele frequency is approximately 70% of all mutant alleles, approximately 50% of patients are homozygous for the ΔF508 allele; an additional 40% are genetic compounds for ΔF508 and another mutant allele. In addition, approximately 70% of CF carriers have the ΔF508 mutation. As noted earlier, except for ΔF508, other mutations at the *CFTR* locus are rare, although

in specific populations, some alleles are relatively common.

Population Screening. The complex issues raised by considering population screening for diseases such as CF are discussed in Chapter 18. At present, CF meets most of the criteria for a newborn screening program, except it is not yet clear that early identification of affected infants significantly improves long-term prognosis. Nevertheless, the advantages of early diagnosis (such as improved nutrition from the provision of pancreatic enzymes) have led some jurisdictions to implement newborn screening programs. It is generally agreed that universal screening for carriers should not be considered until at least 90% of the mutations in a population can be detected. Although population screening for couples has been underway in the United States for several years, the sensitivity of carrier screening for CF has only recently surpassed 90%.

Genetic Analysis of Families of Patients and Prenatal Diagnosis. The high frequency of the ΔF508 allele is useful when CF patients without a family history present for DNA diagnosis. The identification of the ΔF508 allele, in combination with a panel of 127 common mutations suggested by the American College of Medical Genetics, can be used to predict the status of family members for confirmation of disease (e.g., in a newborn or a sibling with an ambiguous presentation), carrier detection, and prenatal diagnosis. Given the vast knowledge of CF mutations in many populations, direct mutation detection is the method of choice for genetic analysis. Nevertheless, if linkage is used in the absence of knowing the specific mutation, accurate diagnosis is possible in virtually all families. For fetuses with a 1-in-4 risk, prenatal diagnosis by DNA analysis at 10 to 12 weeks, with tissue obtained by chorionic villus biopsy, is the method of choice (see Chapter 17).

Molecular Genetics and the Treatment of Cystic Fibrosis. Historically, the treatment of CF has been directed toward controlling pulmonary infection and improving nutrition. Increasing knowledge of the molecular pathogenesis has made it possible to design pharmacological interventions, including the drug ivacaftor, that modulate CFTR function in some patients (see Chapter 13). Alternatively, gene transfer therapy may be possible in CF, but there are many difficulties.

DISORDERS OF STRUCTURAL PROTEINS
The Dystrophin Glycoprotein Complex: Duchenne, Becker, and Other Muscular Dystrophies

Like CF, **Duchenne muscular dystrophy (DMD)** has long received attention from the general and medical

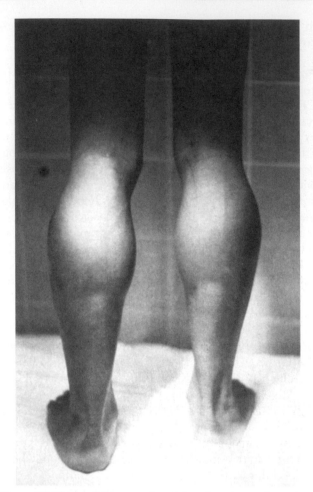

Figure 12-16 Pseudohypertrophy of the calves due to the replacement of normal muscle tissue with connective tissue and fat in an 8-year-old boy with Duchenne muscular dystrophy. *See Sources & Acknowledgments.*

communities as a relatively common, severe, and progressive muscle-wasting disease with relentless clinical deterioration (Case 14). The isolation of the gene affected in this X-linked disorder and the characterization of its protein (named dystrophin because of its association with DMD) have given insight into every aspect of the disease, greatly improved the genetic counseling of affected families, and suggested strategies for treatment. The study of dystrophin led to the identification of a major complex of other muscular dystrophy–associated muscle membrane proteins, the dystrophin glycoprotein complex (DGC), described later in this section.

The Clinical Phenotype of Duchenne Muscular Dystrophy. Affected boys are normal for the first year or two of life but develop muscle weakness by 3 to 5 years of age (Fig. 12-16), when they begin to have difficulty climbing stairs and rising from a sitting position. The child is typically confined to a wheelchair by the age of 12 years. Although DMD is currently incurable, recent advances in the management of pulmonary and cardiac complications (which were leading causes of death in

DMD boys) have changed the disease from a life-limiting to a life-threatening disorder. In the preclinical and early stages of the disease, the serum level of creatine kinase is grossly elevated (50 to 100 times the upper limit of normal) because of its release from diseased muscle. The brain is also affected; on average, there is a moderate decrease in IQ of approximately 20 points.

The Clinical Phenotype of Becker Muscular Dystrophy. Becker muscular dystrophy (BMD) is also due to mutations in the dystrophin gene, but the BMD alleles produce a much milder phenotype. Patients are said to have BMD if they are still walking at the age of 16 years. There is significant variability in the progression of the disease, and some patients remain ambulatory for many years. In general, patients with BMD carry mutated alleles that maintain the reading frame of the protein and thus express some dystrophin, albeit often an altered product at reduced levels. Dystrophin is generally demonstrable in the muscle of patients with BMD (Fig. 12-17). In contrast, patients with DMD have little or no detectable dystrophin.

The Genetics of Duchenne Muscular Dystrophy and Becker Muscular Dystrophy

Inheritance. DMD has an incidence of approximately 1 in 3300 live male births, with a calculated mutation rate of 10^{-4}, an order of magnitude higher than the rate observed in genes involved in most other genetic diseases (see Chapter 4). In fact, given a production of approximately 8×10^7 sperm per day, a normal male produces a sperm with a new mutation in the *DMD* gene every 10 to 11 seconds! In Chapter 7, DMD was presented as a typical X-linked recessive disorder that is lethal in males, so that one third of cases are predicted to be due to new mutations and two thirds of patients have carrier mothers (see also Chapter 16). The great majority of carrier females have no clinical manifestations, although approximately 70% have slightly elevated levels of serum creatine kinase. In accordance with random inactivation of the X chromosome (see Chapter 6), however, the X chromosome carrying the normal *DMD* allele appears to be inactivated above a critical threshold of cells in some female heterozygotes. Nearly 20% of adult female carriers have some muscle weakness, whereas in 8%, life-threatening cardiomyopathy and serious proximal muscle disability occur. In rare instances, females have been described with DMD. Some have X;autosome translocations (see Chapter 6), whereas others have only one X chromosome (Turner syndrome) with a *DMD* mutation on that chromosome.

BMD accounts for approximately 15% of the mutations at the locus. An important genetic distinction between these allelic phenotypes is that whereas DMD is a genetic lethal, the reproductive fitness of males with BMD is high (up to approximately 70% of normal), so that they can transmit the mutant gene to their

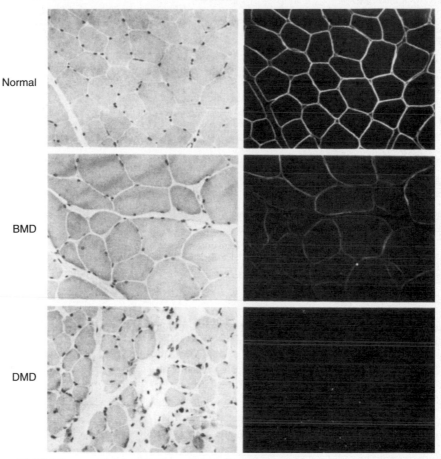

Figure 12-17 Microscopic visualization of the effect of mutations in the dystrophin gene in a patient with Becker muscular dystrophy (BMD) and a patient with Duchenne muscular dystrophy (DMD). *Left column,* Hematoxylin and eosin staining of muscle. *Right column,* Immunofluorescence microscopy staining with an antibody specific to dystrophin. Note the localization of dystrophin to the myocyte membrane in normal muscle, the reduced quantity of dystrophin in BMD muscle, and the complete absence of dystrophin from the myocytes of the DMD muscle. The amount of connective tissue between the myocytes in the DMD muscle is increased. *See Sources & Acknowledgments.*

daughters. Consequently, and in contrast to DMD, a high proportion of BMD cases are inherited, and relatively few (only approximately 10%) represent new mutations.

The DMD *Gene and Its Product.* The most remarkable feature of the *DMD* gene is its size, estimated to be 2300 kb, or 1.5% of the entire X chromosome. This huge gene is among the largest known in any species, by an order of magnitude. The high mutation rate can be at least partly explained by the fact that the locus is a large target for mutation but, as described later, it is also structurally prone to deletion and duplication. The *DMD* gene is complex, with 79 exons and seven tissue-specific promoters. In muscle, the large (14-kb) dystrophin transcript encodes a huge 427-kD protein (Fig. 12-18). In accordance with the clinical phenotype, the protein is most abundant in skeletal and cardiac muscle, although many tissues express at least one dystrophin isoform.

The Molecular and Physiological Defects in Becker Muscular Dystrophy and Duchenne Muscular Dystrophy. The most common molecular defects in patients with DMD are deletions (60% of alleles) (see Figs. 12-18 and 12-19), which are not randomly distributed. Rather, they are clustered in either the 5′ half of the gene or in a central region that encompasses an apparent deletion hot spot (see Fig. 12-18). The mechanism of deletion in the central region is unknown, but it appears to involve the tertiary structure of the genome and, in some cases, recombination between *Alu* repeat sequences (see Chapter 2) in large central introns. Point mutations account for approximately one third of the alleles and are randomly distributed throughout the gene.

The absence of dystrophin in DMD destabilizes the myofiber membrane, increasing its fragility and allowing increased Ca^{++} entry into the cell, with subsequent activation of inflammatory and degenerative pathways. In addition, the chronic degeneration of myofibers even-

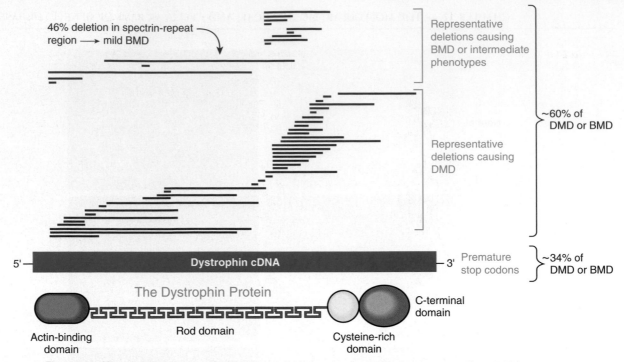

46% deletion in spectrin-repeat region → mild BMD

Representative deletions causing BMD or intermediate phenotypes

~60% of DMD or BMD

Representative deletions causing DMD

5' **Dystrophin cDNA** 3'

Premature stop codons

~34% of DMD or BMD

The Dystrophin Protein

C-terminal domain

Actin-binding domain

Rod domain

Cysteine-rich domain

Figure 12-18 A representation of the full-length dystrophin protein, the corresponding cDNA, and the distribution of representative deletions in patients with Becker muscular dystrophy (BMD) and Duchenne muscular dystrophy (DMD). Partial duplications of the gene (not shown) account for approximately 6% of DMD or BMD alleles. The actin-binding domain links the protein to the filamentous actin cytoskeleton. The rod domain presumably acts as a spacer between the N-terminal and C-terminal domains. The cysteine-rich domain mediates protein-protein interactions. The C-terminal domain, which associates with a large transmembrane glycoprotein complex (see Fig. 12-19), is also found in three dystrophin-related proteins (DRPs): utrophin (DRP-1), DRP-2, and dystrobrevin. The protein domains are not drawn to scale.

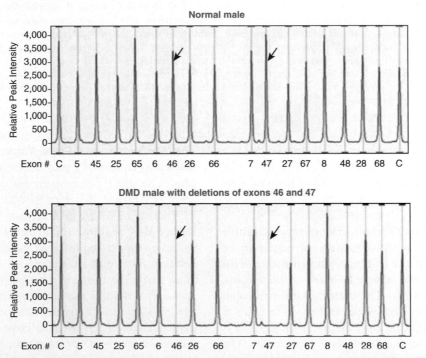

Normal male

Relative Peak Intensity

Exon # C 5 45 25 65 6 46 26 66   7 47 27 67 8 48 28 68 C

DMD male with deletions of exons 46 and 47

Relative Peak Intensity

Exon # C 5 45 25 65 6 46 26 66   7 47 27 67 8 48 28 68 C

Figure 12-19 Diagnosis of Duchenne muscular dystrophy (DMD) involves screening for deletions and duplications by a procedure called multiplex ligation-dependent probe amplification (MLPA). MLPA allows the simultaneous analysis of all 79 exons of the *DMD* gene in a single DNA sample and can detect exon deletions and duplications in males or females. Each amplification peak represents a single *DMD* gene exon, after separation of the amplification products by capillary electrophoresis. *Top panel,* The amplification profiles of 16 exons of a normal male sample. Control (C) DNAs are included at each end of the scan. The MLPA DNA fragments elute according to size, which is why the exons are not numbered sequentially. *Bottom panel,* The corresponding amplification profile from a DMD patient with a deletion of exons 46 and 47. *See Sources & Acknowledgments.*

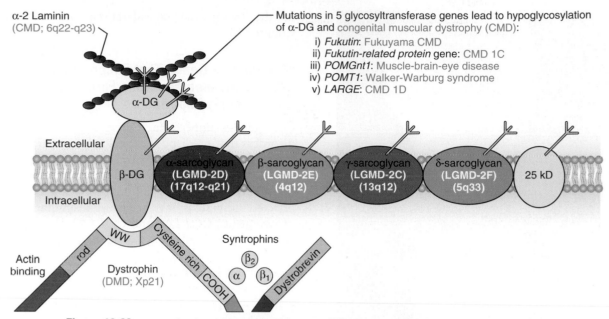

Figure 12-20 In muscle, dystrophin links the extracellular matrix (laminin) to the actin cytoskeleton. Dystrophin interacts with a multimeric complex composed of the dystroglycans (DG), the sarcoglycans, the syntrophins, and dystrobrevin. The α,β-dystroglycan complex is a receptor for laminin and agrin in the extracellular matrix. The function of the sarcoglycan complex is uncertain, but it is integral to muscle function; mutations in the sarcoglycans have been identified in limb girdle muscular dystrophies (LGMDs) types 2C, 2D, 2E, and 2F. Mutations in laminin type 2 (merosin) cause a congenital muscular dystrophy (CMD). The branched structures represent glycans. The WW domain of dystrophin is a tryptophan-rich, protein-binding motif.

tually exhausts the pool of myogenic stem cells that are normally activated to regenerate muscle. This reduced regenerative capacity eventually leads to the replacement of muscle with fat and fibrotic tissue.

The Dystrophin Glycoprotein Complex (DGC). Dystrophin is a structural protein that anchors the DGC at the cell membrane. The DGC is a veritable constellation of polypeptides associated with a dozen genetically distinct muscular dystrophies (Fig. 12-20). This complex serves several major functions. First, it is thought to be essential for the maintenance of muscle membrane integrity, by linking the actin cytoskeleton to the extracellular matrix. Second, it is required to position the proteins in the complex at the sarcolemma. Although the function of many of the proteins in the complex is unknown, their association with diseases of muscle indicates that they are essential components of the complex. Mutations in several of these proteins cause autosomal recessive limb girdle muscular dystrophies and other congenital muscular dystrophies (Fig. 12-20).

That each component of the DGC is affected by mutations that cause other types of muscular dystrophies highlights the principle that no protein functions in isolation but rather is a component of a biological pathway or a multiprotein complex. Mutations in the genes encoding other components of a pathway or a complex often lead to genocopies, much as we saw previously in the case of CF.

Post-translational Modification of the Dystrophin Glycoprotein Complex. Five of the muscular dystrophies associated with the DGC result from mutations in glycosyltransferases, leading to hypoglycosylation of α-dystroglycan (see Fig. 12-20). That five proteins are required for the post-translational modification of one other polypeptide testifies to the critical nature of glycosylation to the function of α-dystroglycan in particular but, more generally, to the importance of post-translational modifications for the normal function of most proteins.

Clinical Applications of Gene Testing in Muscular Dystrophy

Prenatal Diagnosis and Carrier Detection. With gene-based technologies, accurate carrier detection and prenatal diagnosis are available for most families with a history of DMD. In the 60% to 70% of families in whom the mutation results from a deletion or duplication, the presence or absence of the defect can be assessed by examination of fetal DNA using methods that assess the gene's genomic continuity and size (see Fig. 12-19). In most other families, point mutations can be identified by sequencing of the coding region and intron-exon boundaries. Because the disease has a very high frequency of new mutations and is not manifested in carrier females, approximately 80% of Duchenne boys are born into families with no previous history of the disease (see Chapter 7). Thus the incidence of DMD will

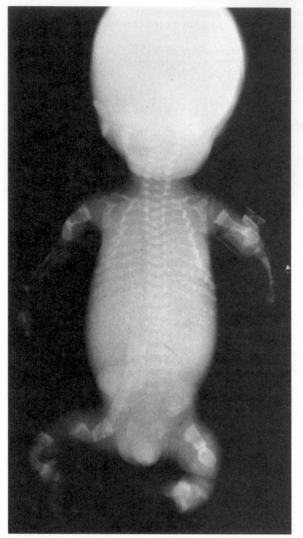

Figure 12-21 Radiograph of a premature (26 weeks' gestation) infant with the perinatal lethal form (type II) of osteogenesis imperfecta. The skull is relatively large and unmineralized and was soft to palpation. The thoracic cavity is small, the long bones of the arms and legs are short and deformed, and the vertebral bodies are flattened. All the bones are undermineralized. *See Sources & Acknowledgments.*

not decrease substantially until universal prenatal or preconception screening for the disease is possible.

Maternal Mosaicism. If a boy with DMD is the first affected member of his family, and if his mother is not found to carry the mutation in her lymphocytes, the usual explanation is that he has a new mutation at the *DMD* locus. However, approximately 5% to 15% of such cases appear to be due to maternal germline mosaicism, in which case the recurrence risk is significant (see Chapter 7).

Therapy. At present, only symptomatic treatment is available for DMD. The possibilities for rational therapy for DMD have greatly increased with the understanding of the normal role of dystrophin in the myocyte. Some

of the therapeutic considerations are discussed in Chapter 13.

Mutations in Genes That Encode Collagen or Other Components of Bone Formation: Osteogenesis Imperfecta

Osteogenesis imperfecta (OI) is a group of inherited disorders that predispose to skeletal deformity and easy fracturing of bones, even with little trauma (Fig. 12-21). The combined incidence of all forms of the disease is approximately 1 per 10,000. Approximately 95% of affected individuals have heterozygous mutations in one of two genes, *COL1A1* and *COL1A2*, that encode the chains of type I collagen, the major protein in bone. A remarkable degree of clinical variation has been recognized, from lethality in the perinatal period to only a mild increase in fracture frequency. The clinical heterogeneity is explained by both locus and allelic heterogeneity; the phenotypes are influenced by which chain of type I procollagen is affected and according to the type and location of the mutation at the locus. The major phenotypes and genotypes associated with mutations in the type I collagen genes are outlined in Table 12-4.

Normal Collagen Structure and Its Relationship to Osteogenesis Imperfecta

It is important to appreciate the major features of normal type I collagen to understand the pathogenesis of OI. The type I procollagen molecule is formed from two proα1(I) chains (encoded by *COL1A1*) and one similar but distinct proα2(I) chain (encoded by *COL1A2*) (Fig. 12-22).

Proteins composed of subunits, like collagen, are often subject to mutations that prevent subunit association by altering the subunit interfaces. The triple helical (collagen) section is composed of 338 tandemly arranged Gly-X-Y repeats; proline is often in the X position, and hydroxyproline or hydroxylysine is often in the Y position. Glycine, the smallest amino acid, is the only residue compact enough to occupy the axial position of the helix, and consequently, mutations that substitute other residues for those glycines are highly disruptive to the helical structure.

Several features of procollagen maturation are of special significance to the pathophysiology of OI. First, the assembly of the individual proα chains into the trimer begins at the carboxyl terminus, and triple helix formation progresses toward the amino terminus. Consequently, mutations that alter residues in the carboxyl-terminal part of the triple helical domain are more disruptive because they interfere earlier with the propagation of the triple helix (Fig. 12-23). Second, the post-translational modification (e.g., proline or lysine hydroxylation; hydroxylysyl glycosylation) of procollagen continues on any part of a chain not assembled into the triple helix. Thus, when triple helix assembly is

TABLE 12-4 Summary of the Genetic, Biochemical, and Molecular Features of the Types of Osteogenesis Imperfecta due to Mutations in Type 1 Collagen Genes

Type	Phenotype	Inheritance	Biochemical Defect	Gene Defect
Defective Production of Type I Collagen*				
I	**Mild:** blue sclerae, brittle bones but no bone deformity	Autosomal dominant	All the collagen made is normal (i.e., solely from the normal allele), but the quantity is *reduced* by half	Largely null alleles that impair the production of proα1(I) chains, such as defects that interfere with mRNA synthesis
Structural Defects in Type I Collagen				
II	**Perinatal lethal:** severe skeletal abnormalities, dark sclerae, death within 1 month (see Fig. 12-21)	Autosomal dominant (new mutation)	Production of *abnormal* collagen molecules due to substitution of the glycine in Gly-X-Y of the triple helical domain located, in general, throughout the protein	Missense mutations in the glycine codons of the genes for the α1 and α2 chains
III	**Progressive deforming:** with blue sclerae; fractures, often at birth; progressive bone deformity, limited growth	Autosomal dominant[†]		
IV	**Normal sclerae, deforming:** mild-moderate bone deformity, short stature fractures	Autosomal dominant		

*A few patients with type I disease have substitutions of glycine in one of the type I collagen chains.
[†]Rare cases are autosomal recessive.
mRNA, Messenger RNA.
Modified from Byers PH: Disorders of collagen biosynthesis and structure. In Scriver CR, Beaudet AL, Sly WS, Valle D, editors: *The metabolic basis of inherited disease,* ed 6, New York, 1989, McGraw-Hill, pp 2805–2842; and Byers PH: Brittle bones—fragile molecules: disorders of collagen structure and expression. *Trends Genet* 6:293–300, 1990.

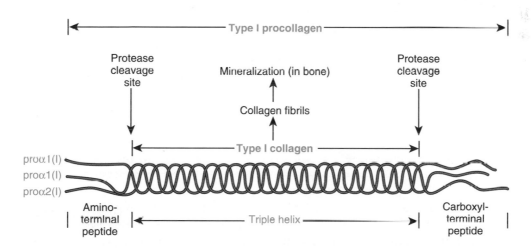

Figure 12-22 The structure of type I procollagen. Each collagen chain is made as a procollagen triple helix that is secreted into the extracellular space. The amino- and carboxyl-terminal domains are cleaved extracellularly to form collagen; mature collagen fibrils are then assembled and, in bone, mineralized. Note that type I procollagen is composed of two proα1(I) chains and one proα2(I) chain. *See Sources & Acknowledgments.*

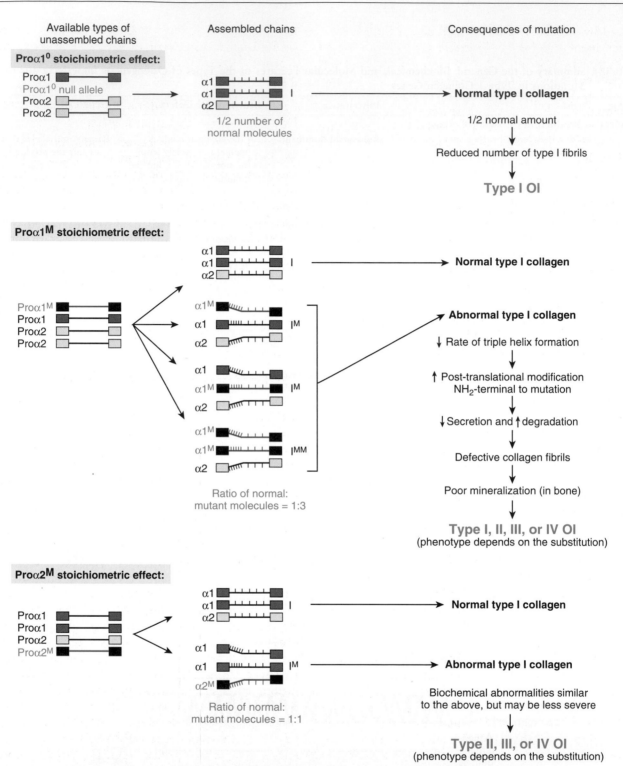

Figure 12-23 The pathogenesis of the major classes of type I procollagen mutants. *Column 1,* The types of procollagen chains available for assembly into a triple helix. Although there are two α1 and two α2 collagen genes/genome, as implied in the left column, twice as many α1 collagen molecules are produced, compared to α2 collagen molecules, as shown in the central column. *Column 2,* The effect of type I procollagen stoichiometry on the ratio of normal to defective collagen molecules formed in mutants with proα1 chain versus proα2 chain mutations. The small vertical bars on each procollagen chain indicate post-translational modifications (see text). *Column 3,* The effect of mutations on the biochemical processing of collagen. OI, Osteogenesis imperfecta; Proα1^M, a proα1 chain with a missense mutation; Proα2^M, a proα2 chain with a missense mutation; Proα1^0, a proα1 chain null allele. OI, Osteogenesis imperfecta.

slowed by a mutation, the unassembled sections of the chains amino-terminal to the defect are modified excessively, which slows their secretion into the extracellular space. Overmodification may also interfere with the formation of collagen fibrils. As a result of all of these abnormalities, the number of secreted collagen molecules is reduced, and many of them are abnormal. In bone, the abnormal chains and their reduced number lead to defective mineralization of collagen fibrils (see Fig. 12-21).

Molecular Abnormalities of Collagen in Osteogenesis Imperfecta

More than 2000 different mutations affecting the synthesis or structure of type I collagen have been found in individuals with OI. The clinical heterogeneity of this disease reflects even greater heterogeneity at the molecular level (see Table 12-4). For the type I collagen genes, the mutations fall into two general classes, those that reduce the *amount* of type I procollagen made and those that alter the *structure* of the molecules assembled.

Type I: Diminished Collagen Production. Most individuals with OI type I have mutations that result in production by cells of approximately half the normal amount of type I procollagen. Most of these mutations result in premature termination codons in one *COL1A1* allele that render the mRNA from that allele untranslatable. Because type I procollagen molecules must have two proα1(I) chains to assemble into a triple helix, loss of half the mRNA leads to production of half the normal quantity of type I procollagen molecules, although these molecules are normal (see Fig. 12-23). Missense mutations can also give rise to this milder form of OI when the amino acid change is located in the amino terminus. This is because amino terminal substitutions tend to be less disruptive of collagen chain assembly, which can still initiate as usual at the carboxy terminus.

Types II, III, and IV: Structurally Defective Collagens. The type II, III, and IV phenotypes of OI usually result from mutations that produce structurally abnormal proα1(I) or proα2(I) chains (see Fig. 12-23 and Table 12-4). Most of these patients have substitutions in the triple helix that replace a glycine with a more bulky residue that disrupts formation of the triple helix. The specific collagen affected, the location of the substitution, and the nature of the substituting residue are all important phenotypic determinants, but some generalizations about the phenotype likely to result from a specific substitution are nevertheless possible. Thus substitutions in the proα1(I) chain are more prevalent in patients with OI types III and IV and are more often lethal. In either chain, replacement of glycine (a neutral residue) with a charged residue (aspartic acid, glutamic acid, arginine) or large residue (tryptophan) is usually very disruptive and often associated with a severe (type II) phenotype (see Fig. 12-23). Sometimes, a specific substitution is associated with more than one phenotype, an outcome that is likely to reflect the influence of powerful modifier genes.

Novel Forms of Osteogenesis Imperfecta That Do Not Result from Collagen Mutations

Three additional forms of clinically defined OI (types V, VI, and VII) do not result from mutations in type I collagen genes but involve defects in other genes. These 5% of OI subjects with normal collagen genes have either dominant mutations in the *IFITM5* gene (encoding interferon-induced transmembrane protein 5) or biallelic mutations in any of almost a dozen other genes that encode proteins that regulate osteoblast development and facilitate bone formation or that mediate collagen assembly by interacting with collagens during synthesis and secretion. These genes include, for example, *WNT1*, which encodes a secreted signaling protein, and *BMP1*, which encodes bone morphogenetic protein 1, an inducer of cartilage formation.

The Genetics of Osteogenesis Imperfecta

As just discussed, most of the mutations in type I collagen genes that cause OI act in a dominant manner. This group of disorders illustrates the genetic complexities that result when mutations alter structural proteins, particularly those composed of multiple different subunits, or alter proteins that are involved in the folding and transport of collagens to their place of action.

The relatively mild phenotype and dominant inheritance of OI type I are consistent with the fact that although only half the normal number of molecules is made, they are of normal quality (see Fig. 12-23). The more severe consequences of producing structurally defective proα1(I) chains from one allele (compared with producing no chains) partly reflect the stoichiometry of type I collagen, which contains two proα1(I) chains and one proα2(I) chain (see Fig. 12-23). Accordingly, if half the proα1(I) chains are abnormal, three of four type I molecules have at least one abnormal chain; in contrast, if half the proα2(I) chains are defective, only one in two molecules is affected. Mutations such as the proα1(I) missense allele (proα1^M) shown in Figure 12-23 are thus **dominant negative alleles** because they impair the contribution of both the normal proα1(I) chains and the normal proα2(I) chains. In other words, the effect of the mutant allele is amplified because of the trimeric nature of the collagen molecule. Consequently, in dominantly inherited diseases such as OI, it is actually better to have a mutation that generates no gene product than one that produces an abnormal gene product. The biochemical mechanism in OI by which the dominant negative effect of dominant negative alleles of the *COL1A1* genes is exerted is one of the best understood in all of human genetics (see Case 8 and Case 30 for other examples of dominant negative alleles).

Although mutations that produce structurally abnormal proα2(I) chains reduce the number of normal type I collagen molecules by half, this reduction is nevertheless sufficient, in the case of some mutations, to cause the severe perinatal lethal phenotype (see Table 12-4). Most infants with OI type II, the perinatal lethal form, have a de novo dominant mutation, and consequently there is a very low likelihood of recurrence in the family. In occasional families, however, more than one sibling is affected with OI type II. Such recurrences are usually due to parental germline mosaicism, as described in Chapter 7.

Clinical Management. If a patient's molecular defect can be determined, increasing knowledge of the correlation between OI genotypes and phenotypes has made it possible to predict, at least to some extent, the natural history of the disease. The treatment of children with the more clinically significant forms of OI is based on physical medicine approaches to increase ambulation and mobility, often in the context of treatment with parenteral bisphosphonates, a class of drugs that act by decreasing bone resorption, to increase bone density and reduce fracture rate. These drugs appear to be less effective in individuals with the recessive forms of OI. The development of better and targeted drugs is a critical issue to improve care.

NEURODEGENERATIVE DISORDERS

Until recently, the biochemical and molecular mechanisms underlying almost all neurodegenerative diseases were completely obscure. In this section, we discuss three different conditions, each with a different genetic and genomic basis and illustrating different mechanisms of pathogenesis:
- Alzheimer disease
- Disorders of mitochondrial DNA
- Diseases due to the expansion of unstable repeat sequences

Alzheimer Disease

One of the most common adult-onset neurodegenerative conditions is **Alzheimer disease (AD)** (Case 4), introduced in Chapter 8 in the context of complex genetic disorders. AD generally manifests in the sixth to ninth decades, but there are monogenic forms that often present earlier, sometimes as soon as the third decade. The clinical picture of AD is characterized by a progressive deterioration of memory and of higher cognitive functions, such as reasoning, in addition to behavioral changes. These abnormalities reflect degeneration of neurons in specific regions of the cerebral cortex and hippocampus. AD affects approximately 1.4% of persons in developed countries and is responsible for at least 100,000 deaths per year in the United States alone.

The Genetics of Alzheimer Disease

The lifetime risk for AD in the general population is 12.1% in men and 20.3% in women by age 85. Most of the increased risk in relatives of affected individuals is not due to mendelian inheritance; rather, as described in Chapter 8, this familial aggregation results from a complex genetic contribution involving one or more incompletely penetrant genes that act independently, from multiple interacting genes, or from some combination of genetic and environmental factors.

Approximately 7% to 10% of patients, however, do have a monogenic highly penetrant form of AD that is inherited in an autosomal dominant manner. In the 1990s, four genes associated with AD were identified (Table 12-5). Mutations in three of these genes—encoding the β-amyloid precursor protein (βAPP), presenilin 1, and presenilin 2—lead to autosomal dominant AD. The fourth gene, *APOE*, encodes apolipoprotein E (apoE), the protein component of several plasma lipoproteins. Mutations in *APOE* are not associated with monogenic AD. Rather, as we saw in Chapter 8, the ε4 allele of *APOE* modestly increases susceptibility to nonfamilial AD and influences the age at onset of at least some of the monogenic forms (see later).

The identification of the four genes associated with AD has provided great insight not only into the pathogenesis of monogenic AD but also, as is commonly the case in medical genetics, into the mechanisms that underlie the more common form, nonfamilial or sporadic AD. Indeed, overproduction of one proteolytic product of βAPP, called the Aβ peptide, appears to be at the center of AD pathogenesis, and the currently available experimental evidence suggests that the βAPP, presenilin 1, and presenilin 2 proteins all play a direct role in the pathogenesis of AD.

The Pathogenesis of Alzheimer Disease: β-Amyloid Peptide and Tau Protein Deposits

The most important pathological abnormalities of AD are the deposition in the brain of two fibrillary proteins, β-amyloid peptide (Aβ) and tau protein. The Aβ peptide is generated from the larger βAPP protein (see Table 12-5), as discussed in the next section, and is found in extracellular amyloid or senile plaques in the extracellular space of AD brains. Amyloid plaques contain other proteins besides the Aβ peptide, notably apoE (see Table 12-5). Tau is a microtubule-associated protein expressed abundantly in neurons of the brain. Hyperphosphorylated forms of tau compose the neurofibrillary tangles that, in contrast to the extracellular amyloid plaques, are found *within* AD neurons. The tau protein normally promotes the assembly and stability of microtubules, functions that are diminished by phosphorylation. Although the formation of tau neurofibrillary tangles appears to be one of the causes of the neuronal degeneration in AD, mutations in the tau gene are associated

TABLE 12-5 Genes and Proteins Associated with Inherited Susceptibility to Alzheimer Disease

Gene	Inheritance	% of FAD	Protein	Normal Function	Role in FAD
PSEN1	AD	50%	Presenilin 1 (PS1): A 5 to 10 membrane-spanning domain protein found in cell types both inside and outside the brain	Unknown, but required for γ-secretase cleavage of βAPP.	May participate in the abnormal cleavage at position 42 of βAPP and its derivative proteins. More than 100 mutations identified in Alzheimer disease.
PSEN2	AD	1%-2%	Presenilin 2 (PS2): Structure similar to PS1, maximal expression outside the brain.	Unknown, likely to be similar to PS1.	At least 5 missense mutations identified.
APP	AD	1%-2%	Amyloid precursor protein (βAPP): An intracellular transmembrane protein. Normally, βAPP is cleaved endoproteolytically within the transmembrane domain (see Fig. 12-24), so that little of the β-amyloid peptide (Aβ) is formed.	Unknown.	β-Amyloid peptide (Aβ) is the principal component of senile plaques. Increased Aβ production, especially of the $A\beta_{42}$ form, is a key pathogenic event. Approximately 10 mutations have been identified in FAD.
APOE	See Table 12-6	NA	Apolipoprotein E (apoE): A protein component of several plasma lipoproteins. The apoE protein is imported into the cytoplasm of neurons from the extracellular space.	Normal function in neurons is unknown. Outside the brain, apoE participates in lipid transport between tissues and cells. Loss of function causes one form (type III) of hyperlipoproteinemia.	An Alzheimer disease susceptibility gene (see Table 12-6). ApoE is a component of senile plaques.

AD, Autosomal dominant; FAD, familial Alzheimer disease; NA, not applicable.

Data derived from St. George-Hyslop PH, Farrer LA: Alzheimer's disease and the fronto-temporal dementias: diseases with cerebral deposition of fibrillar proteins. In Scriver CR, Beaudet AL, Sly WS, Valle D, editors: *The molecular and metabolic bases of inherited disease*, ed 8, New York, 2000, McGraw-Hill; and Martin JB. Molecular basis of the neurodegenerative disorders. *N Engl J Med* 340:1970–1980, 1999.

not with AD but with another autosomal dominant dementia, **frontotemporal dementia.**

The Amyloid Precursor Protein Gives Rise to the β-Amyloid Peptide

The major features of the βAPP and its corresponding gene are summarized in Table 12-5. βAPP is a single-pass intracellular transmembrane protein found in endosomes, lysosomes, the ER and the Golgi apparatus. It is subject to three distinct proteolytic fates, depending on the relative activity of three different proteases: α-secretase and β-secretase, which are cell surface proteases; and γ-secretase, which is an atypical protease that cleaves membrane proteins within their transmembrane domains. The predominant fate of approximately 90% of βAPP is cleavage by the α-secretase (Fig. 12-24), an event that precludes the formation of the Aβ peptide, because α-secretase cleaves within the Aβ peptide domain. The other approximately 10% of βAPP is cleaved by the β- and γ-secretases to form either the nontoxic $A\beta_{40}$ peptide or the $A\beta_{42}$ peptide. The $A\beta_{42}$ peptide is thought to be neurotoxic because it is more prone to aggregation than its $A\beta_{40}$ counterpart, a feature that makes AD a conformational disease like α1AT deficiency (described previously in this chapter). Normally, little $A\beta_{42}$ peptide is produced, and the factors

that determine whether γ-secretase cleavage will produce the $A\beta_{40}$ or $A\beta_{42}$ peptide are not well defined.

In monogenic AD due to missense substitutions in the gene encoding βAPP (*APP*), however, several mutations lead to the relative overproduction of the $A\beta_{42}$ peptide. This increase leads to accumulation of the neurotoxic $A\beta_{42}$, an occurrence that appears to be the central pathogenic event of all forms of AD, monogenic or sporadic. Consistent with this model is the fact that patients with Down syndrome, who possess three copies of the *APP* gene (which is on chromosome 21), typically develop the neuropathological changes of AD by 40 years of age. Moreover, mutations in the AD genes presenilin 1 and presenilin 2 (see Table 12-5) also lead to increased production of $A\beta_{42}$. Notably, the amount of the neurotoxic $A\beta_{42}$ peptide is increased in the serum of individuals with mutations in the βAPP, presenilin 1, and presenilin 2 genes; furthermore, in cultured cell systems, the expression of mutant βAPP, presenilin 1, and presenilin 2 increases the relative production of $A\beta_{42}$ peptide by twofold to tenfold.

The central role of the $A\beta_{42}$ peptide in AD is highlighted by the discovery of a coding mutation (Ala673Thr) in the *APP* gene (Fig. 12-25) that protects against both AD and cognitive decline in older adults. The protective effect is likely due to reduced formation

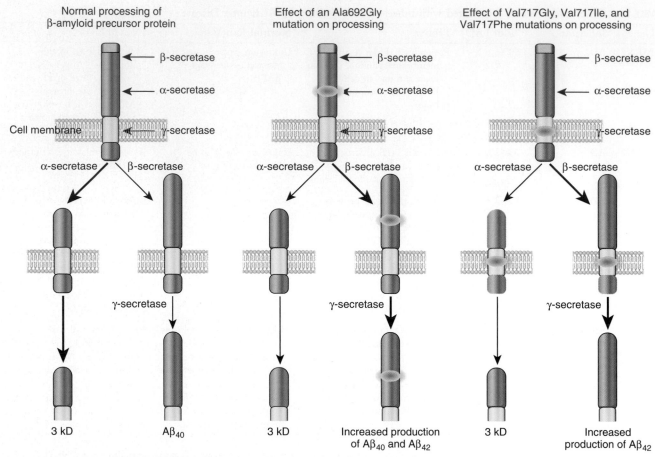

Normal processing of
β-amyloid precursor protein

Effect of an Ala692Gly
mutation on processing

Effect of Val717Gly, Val717Ile, and
Val717Phe mutations on processing

Figure 12-24 The normal processing of β-amyloid precursor protein (βAPP)and the effect on processing of missense mutations in the βAPP gene associated with familial Alzheimer disease. The *ovals* show the locations of the missense mutations. *See Sources & Acknowledgments.*

of the $A\beta_{42}$ peptide, reflecting the proximity of Thr673 to the β-secretase cleavage site (see Fig. 12-25).

The Presenilin 1 and 2 Genes

The genes encoding presenilin 1 and presenilin 2 (see Table 12-5) were identified in families with autosomal dominant AD. Presenilin 1 is required for γ-secretase cleavage of βAPP derivatives. Indeed, some evidence suggests that presenilin 1 is a critical cofactor protein of γ-secretase. The mutations in presenilin 1 associated with AD, through an unclear mechanism, increase production of the $A\beta_{42}$ peptide. A major difference between presenilin 1 and presenilin 2 mutations is that the age at onset with the latter is much more variable (presenilin 1, 35 to 60 years; presenilin 2, 40 to 85 years); indeed, in one family, an asymptomatic octogenarian carrying a presenilin 2 mutation transmitted the disease to his offspring. The basis of this variation is partly dependent on the number of *APOE* ε4 alleles (see Table 12-5 and later discussion) carried by individuals with a presenilin 2 mutation; two ε4 alleles are associated with an earlier age at onset than one allele, and one confers an earlier onset than other *APOE* alleles.

The APOE Gene is an Alzheimer Disease Susceptibility Locus

As presented in Chapter 8, the ε4 allele of the *APOE* gene is a major risk factor for the development of AD. The role for *APOE* as a major AD susceptibility locus was suggested by multiple lines of evidence, including linkage to AD in late-onset families, increased association of the ε4 allele with AD patients compared with controls, and the finding that apoE binds to the Aβ peptide. The *APOE* protein has three common forms encoded by corresponding *APOE* alleles (Table 12-6). The ε4 allele is significantly overrepresented in patients with AD (≈40% vs. ≈15% in the general population) and is associated with an early onset of AD (for ε4/ε4 homozygotes, the age at onset of AD is approximately 10 to 15 years earlier than in the general population; see Chapter 8). Moreover, the relationship between the ε4 allele and the disease is dose-dependent; two copies of ε4 are associated with an earlier age at onset (mean onset before 70 years) than with one copy (mean onset after 70 years) (see Fig. 8-11 and Table 8-14). In contrast, the ε2 allele has a protective effect and

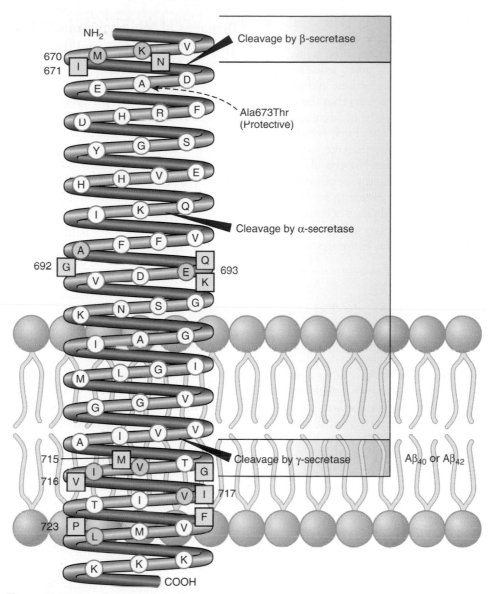

Figure 12-25 The topology of the amyloid precursor protein (βAPP), its nonamyloidogenic cleavage by α-secretase, and its alternative cleavage by putative β-secretase and γ-secretase to generate the amyloidogenic β amyloid peptide (Aβ). Letters are the single-letter code for amino acids in β-amyloid precursor protein, and numbers show the position of the affected amino acid. Normal residues involved in missense mutations are shown in *highlighted circles*, whereas the amino acid residues representing various missense mutations are shown in *boxes*. The mutated amino acid residues are near the sites of β-, α-, and γ-secretase cleavage (*black arrowheads*). The mutations lead to the accumulation of toxic peptide $A\beta_{42}$ rather than the wild-type $A\beta_{40}$ peptide. The location of the protective allele Ala673Thr is indicated by the *dashed arrow*. *See Sources & Acknowledgments.*

TABLE 12-6 Amino Acid Substitutions Underlying the Three Common Apolipoprotein E Polymorphisms

Allele	ε2	ε3	ε4
Residue 112	Cys	Cys	Arg
Residue 158	Cys	Arg	Arg
Frequency in white populations	10%	65%	25%
Frequency in patients with Alzheimer disease	2%	58%	40%
Effect on Alzheimer disease	Protective	None known	30%-50% of the genetic risk for Alzheimer disease

These figures are estimates, with differences in allele frequencies that vary with ethnicity in control populations, and with age, gender, and ethnicity in Alzheimer disease subjects.

Data derived from St. George Hyslop PH, Farrer LA, Goedert M: Alzheimer disease and the frontotemporal dementias: diseases with cerebral deposition of fibrillar proteins. In Valle D, Beaudet AL, Vogelstein B, et al, editors: The online metabolic & molecular bases of inherited disease (OMMBID). Available at: http://www.ommbid.com/.

correspondingly is more common in elderly subjects who are unaffected by AD (see Table 12-6).

The mechanisms underlying these effects are not known, but apoE polymorphisms may influence the processing of βAPP and the density of amyloid plaques in AD brains. It is also important to note that the *APOE* ε4 allele is not only associated with an increased risk for AD; carriers of ε4 alleles can also have poorer neurological outcomes after head injury, stroke, and other neuronal insults. Although carriers of the *APOE* ε4 allele have a clearly increased risk for development of AD, there is currently no role for screening for the presence of this allele in healthy individuals; such testing has poor positive and negative predictive values and would therefore generate highly uncertain estimates of future risk for AD (see Chapter 18).

Other Genes Associated with AD

One significant modifier of AD risk, the *TREM2* gene (which encodes the so-called *t*riggering *r*eceptor *e*xpressed on *m*yeloid cells 2), was identified by whole-exome and whole-genome sequencing in families with multiple individuals affected with AD. Several moderately rare missense coding variants in this gene are associated with a fivefold increase in risk for late-onset AD, making *TREM2* mutations the second most common contributor to classic late-onset AD after *APOE ε4*. Statistical analyses suggest that an additional four to eight genes may significantly modify the risk for AD, but their identity remains obscure.

Although case-control association studies (see Chapter 10) of candidate genes with hypothetical functional links to the known biology of AD have suggested more than 100 genes in AD, only one such candidate gene, *SORL1* (sortilin-related receptor 1), has been robustly implicated. Single nucleotide polymorphisms (SNPs) in the *SORL1* gene confer a moderately increased relative risk for AD of less than 1.5. The *SORL1*-encoded protein affects the processing of APP and favors the production of the neurotoxic $A\beta_{42}$ peptide from βAPP.

Genome-wide association studies analyses (see Chapter 10), on the other hand, have greatly expanded the number of genes believed to be associated with AD, identifying at least nine novel SNPs associated with a predisposition to nonfamilial late-onset forms of AD. The genes implicated by these SNPs and their causal role(s) in AD are presently uncertain.

Overall, it is becoming clear that genetic variants alter the risk for AD in at least two general ways: first, by modulating the production of Aβ, and second, through their impact on other processes, including the regulation of innate immunity, inflammation, and the resecretion of protein aggregates. These latter variants likely modulate AD risk by altering the flux through downstream pathways in response to a given load of Aβ.

Diseases of Mitochondrial DNA (mtDNA)
The mtDNA Genome and the Genetics of mtDNA Diseases

The general characteristics of the mtDNA genome and the features of the inheritance of disorders caused by mutations in this genome were first described in Chapters 2 and 7 but are reviewed briefly here. The small circular mtDNA chromosome is located inside mitochondria and contains only 37 genes (Fig. 12-26). Most cells have at least 1000 mtDNA molecules, distributed among hundreds of individual mitochondria, with multiple copies of mtDNA per mitochondrion. In addition to encoding two types of ribosomal RNA (rRNA) and 22 transfer RNAs (tRNAs), mtDNA encodes 13 proteins that are subunits of oxidative phosphorylation.

Mutations in mtDNA can be inherited maternally (see Chapter 7) or acquired as somatic mutations. The diseases that result from mutations in mtDNA show distinctive patterns of inheritance due to three features of mitochondrial chromosomes:

- Replicative segregation
- Homoplasmy and heteroplasmy
- Maternal inheritance

Replicative segregation refers to the fact that the multiple copies of mtDNA in each mitochondrion replicate and sort randomly among newly synthesized mitochondria, which in turn are distributed randomly between the daughter cells (see Fig. 7-25). **Homoplasmy** is the situation in which a cell contains a pure population of normal mtDNA or of mutant mtDNA, whereas **heteroplasmy** describes the presence of a mixture of mutant and normal mtDNA molecules within a cell. Thus the phenotype associated with a mtDNA mutation will depend on the relative proportion of normal and mutant mtDNA in the cells of a particular tissue (see Fig. 7-25). As a result, mitochondrial disorders are generally characterized by reduced penetrance, variable expression, and pleiotropy. The **maternal inheritance** of mtDNA (discussed in greater detail in Chapter 7; see Fig. 7-24) reflects the fact that sperm mitochondria are generally eliminated from the embryo, so that mtDNA is almost always inherited entirely from the mother; paternal inheritance of mtDNA disease is highly unusual and has been well documented in only one instance.

The 74 polypeptides of the oxidative phosphorylation complex not encoded in the mtDNA are encoded by the nuclear genome, which contains the genes for most of the estimated 1500 mitochondrial proteins. To date, more than 100 nuclear genes are associated with disorders of the respiratory chain. Thus diseases of oxidative phosphorylation arise not only from mutations in the mitochondrial genome but also from mutations in nuclear genes that encode oxidative phosphorylation components. Furthermore, the nuclear genome encodes up to 200 proteins required for the maintenance and expression of mtDNA genes or for the assembly of

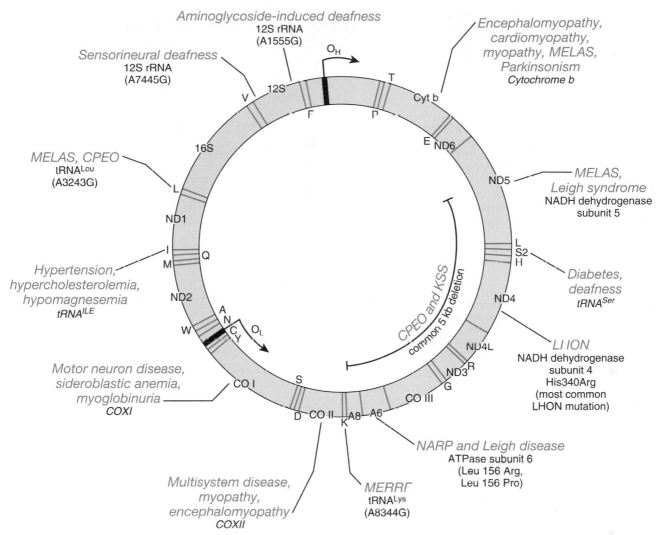

Figure 12-26 Representative disease-causing mutations and deletions in the human mtDNA genome, shown in relation to the location of the genes encoding the 22 transfer RNAs (tRNAs), 2 ribosomal RNAs (rRNAs), and 13 proteins of the oxidative phosphorylation complex. Specific alleles are indicated when they are the predominant or only alleles associated with the phenotype or particular features of it. O_H and O_L are the origins of replication of the two DNA strands, respectively; 12S, 12S ribosomal RNA; 16S, 16S ribosomal RNA. The locations of each of the tRNAs are indicated by the single-letter code for their corresponding amino acids. The 13 oxidative phosphorylation polypeptides encoded by mitochondrial DNA (mtDNA) include components of complex I: NADH dehydrogenase (ND1, ND2, ND3, ND4, ND4L, ND5, and ND6); complex III: cytochrome b (cyt b); complex IV: cytochrome c oxidase I or cytochrome c (COI, COII, COIII); and complex V: ATPase 6 and 8 (A6, A8). The disease abbreviations used in this figure (e.g., MELAS, MERRF, LHON) are explained in Table 12-7. CPEO, Chronic progressive external ophthalmoplegia; NARP, neuropathy, ataxia, and retinitis pigmentosa. *See Sources & Acknowledgments.*

oxidative phosphorylation protein complexes. Mutations in many of these nuclear genes can also lead to disorders with the phenotypic characteristics of mtDNA diseases, but of course the patterns of inheritance in these cases are those typically seen with nuclear genome mutations (see Chapter 7).

Mutations in mtDNA and Disease

The sequence of the mtDNA genome and the presence of pathogenic mutations in mtDNA have been known for over three decades. Unexpected and still unexplained, however, is the fact that the mtDNA genome mutates at a rate approximately tenfold greater than does nuclear DNA. The range of clinical disease resulting from mtDNA mutations is diverse (Fig. 12-27), although neuromuscular disease predominates. More than 100 different rearrangements and approximately 100 different point mutations that are disease-related have been identified in mtDNA. The prevalence of mtDNA mutations has been shown, in at least one

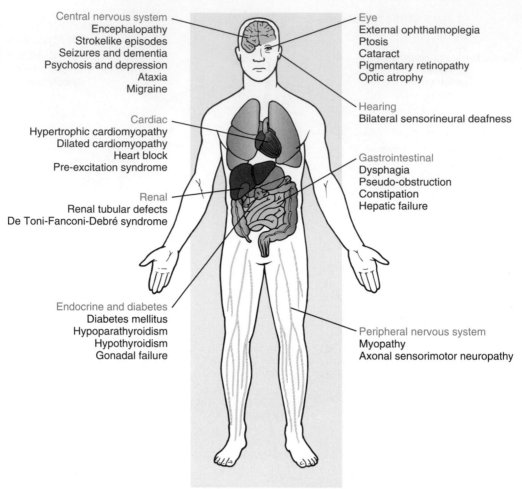

Central nervous system
Encephalopathy
Strokelike episodes
Seizures and dementia
Psychosis and depression
Ataxia
Migraine

Eye
External ophthalmoplegia
Ptosis
Cataract
Pigmentary retinopathy
Optic atrophy

Hearing
Bilateral sensorineural deafness

Cardiac
Hypertrophic cardiomyopathy
Dilated cardiomyopathy
Heart block
Pre-excitation syndrome

Gastrointestinal
Dysphagia
Pseudo-obstruction
Constipation
Hepatic failure

Renal
Renal tubular defects
De Toni-Fanconi-Debré syndrome

Endocrine and diabetes
Diabetes mellitus
Hypoparathyroidism
Hypothyroidism
Gonadal failure

Peripheral nervous system
Myopathy
Axonal sensorimotor neuropathy

Figure 12-27 The range of affected tissues and clinical phenotypes associated with mutations in mitochondrial DNA (mtDNA). *See Sources & Acknowledgments.*

population, to be approximately 1 per 8000. Representative mutations and the diseases associated with them are presented in Figure 12-26 and Table 12-7. In general, as illustrated in the sections to follow, three types of mutations have been identified in mtDNA: rearrangements that generate deletions or duplications of the mtDNA molecule; point mutations in tRNA or rRNA genes that impair mitochondrial protein synthesis; and missense mutations in the coding regions of genes that alter the activity of an oxidative phosphorylation protein.

Deletions of mtDNA and Disease. In most cases, mtDNA deletions that cause disease, such as **Kearns-Sayre syndrome** (see Table 12-7), are inherited from an unaffected mother, who carries the deletion in her oocytes but generally not elsewhere, an example of gonadal mosaicism. Under these circumstances, disorders caused by mtDNA deletions appear to be sporadic, because oocytes carrying the deletion are relatively rare. In approximately 5% of cases, the mother may be affected and transmit the deletion. The reason for the low frequency of transmission is uncertain, but it may simply reflect the fact that women with a high proportion of the deleted mtDNAs in their germ cells have such a severe phenotype that they rarely reproduce.

The importance of deletions in mtDNA as a cause of disease has recently been highlighted by the discovery that *somatic* mtDNA deletions are common in dopaminergic neurons of the substantia nigra, both in normal aging individuals and perhaps to a greater extent in individuals with **Parkinson disease**. The deletions that have occurred in individual neurons from Parkinson disease patients have been shown to be unique, indicating that clonal expansion of the different mtDNA deletions occurred in each cell. These findings indicate that somatic deletions of the mtDNA may contribute to the loss of dopaminergic neurons in the aging substantia nigra and raise the possibility that the common sporadic form of Parkinson disease results from a greater than normal accumulation of deleted mtDNA molecules in the substantia nigra, with a consequently more severe impairment of oxidative phosphorylation. At present, the mechanisms leading to the deletions and their clonal expansions are entirely unclear.

TABLE 12-7 Representative Examples of Disorders due to Mutations in Mitochondrial DNA and Their Inheritance

Disease	Phenotypes—Largely Neurological	Most Frequent Mutation in mtDNA Molecule	Homoplasmy vs. Heteroplasmy	Inheritance
Leber hereditary optic neuropathy (LHON)	Rapid onset of blindness in young adult life due to optic nerve atrophy; some recovery of vision, depending on the mutation. Strong sex bias: ≈50% of male carriers have visual loss vs. ≈10% of females.	Substitution 1178A>G in the ND4 subunit of complex I of the electron transport chain; this mutation, with two others, accounts for more than 90% of cases.	Largely homoplasmic	Maternal
Leigh syndrome	Early-onset progressive neurodegeneration with hypotonia, developmental delay, optic atrophy, and respiratory abnormalities	Point mutations in the ATPase subunit 6 gene	Heteroplasmic	Maternal
MELAS	Myopathy, mitochondrial encephalomyopathy, lactic acidosis, and stroke like episodes; may present only as diabetes mellitus and deafness	Point mutations in tRNA$^{leu(UUR)}$, a mutation hot spot, most commonly 3243A>G	Heteroplasmic	Maternal
MERRF **(Case 33)**	Myoclonic epilepsy with ragged-red muscle fibers, myopathy, ataxia, sensorineural deafness, dementia	Point mutations in tRNA^{lys}, most commonly 8344A>G	Heteroplasmic	Maternal
Deafness	Progressive sensorineural deafness, often induced by aminoglycoside antibiotics; nonsyndromic sensorineural deafness	1555A>G mutation in the 12S rRNA gene	Homoplasmic	Maternal
		7445A>G mutation in the 12S rRNA gene	Homoplasmic	Maternal
Kearns-Sayre syndrome (KSS)	Progressive myopathy, progressive external ophthalmoplegia of early onset, cardiomyopathy, heart block, ptosis, retinal pigmentation, ataxia, diabetes	The ≈5-kb large deletion (see Fig. 12-26)	Heteroplasmic	Generally sporadic, likely due to maternal gonadal mosaicism

mtDNA, Mitochondrial DNA; rRNA, ribosomal RNA; tRNA, transfer RNA.

Mutations in tRNA and rRNA Genes of the Mitochondrial Genome. Mutations in the noncoding tRNA and rRNA genes of mtDNA are of general significance because they illustrate that not all disease-causing mutations in humans occur in genes that encode proteins **(Case 33)**. More than 90 pathogenic mutations have been identified in 20 of the 22 tRNA genes of the mtDNA, and they are the most common cause of oxidative phosphorylation abnormalities in humans (see Fig. 12-26 and Table 12-7). The resulting phenotypes are those generally associated with mtDNA defects. The tRNA mutations include 18 substitutions in the tRNA$^{leu(UUR)}$ gene, some of which, like the common 3243A>G mutation, cause a phenotype referred to as **MELAS**, an acronym for *m*itochondrial *e*ncephalomyopathy with *l*actic *a*cidosis and *s*trokelike episodes (see Fig. 12-26 and Table 12-7); others are associated predominantly with myopathy. An example of a 12S rRNA mutation is a homoplasmic substitution (see Table 12-7) that causes **sensorineural prelingual deafness** after exposure to aminoglycoside antibiotics (see Fig. 12-26).

The Phenotypes of Mitochondrial Disorders
Oxidative Phosphorylation and mtDNA Diseases. Mitochondrial mutations generally affect those tissues that depend on intact oxidative phosphorylation to satisfy high demands for metabolic energy. This phenotypic focus reflects the central role of the oxidative phosphorylation complex in the production of cellular energy. Consequently, decreased production of ATP characterizes many diseases of mtDNA and is likely to underlie the cell dysfunction and cell death that occur in mtDNA diseases. The evidence that mechanisms other than decreased energy production contribute to the pathogenesis of mtDNA diseases is either indirect or weak, but the generation of reactive oxygen species as a byproduct of faulty oxidative phosphorylation may also contribute to the pathology of mtDNA disorders. A substantial body of evidence indicates that there is a **phenotypic threshold effect** associated with mtDNA heteroplasmy (see Fig. 7-25); a critical threshold in the proportion of mtDNA molecules carrying the detrimental mutation must be exceeded in cells from the affected tissue before clinical disease becomes apparent. The threshold appears to be approximately 60% for disorders due to deletions in mtDNA and approximately 90% for diseases due to other types of mutations.

The neuromuscular system is the one most commonly affected by mutations in mtDNA; the consequences can include encephalopathy, myopathy, ataxia, retinal degeneration, and loss of function of the external ocular muscles. Mitochondrial myopathy is characterized by so-called ragged-red (muscle) fibers, a histological phenotype due to the proliferation of structurally and biochemically abnormal mitochondria in muscle fibers. The spectrum of mitochondrial disease is broad and, as illustrated in Figure 12-27, may include liver dysfunction, bone marrow failure, pancreatic islet cell deficiency and diabetes, deafness, and other disorders.

HETEROPLASMY AND MITOCHONDRIAL DISEASE

Heteroplasmy accounts for three general characteristics of genetic disorders of mtDNA that are of importance to their pathogenesis.
- *First,* female carriers of heteroplasmic mtDNA point mutations or of mtDNA duplications usually transmit some mutant mtDNAs to their offspring.
- *Second,* the fraction of mutant mtDNA molecules inherited by each child of a carrier mother is very variable. This is because the number of mtDNA molecules within each oocyte is reduced before being subsequently amplified to the huge total seen in mature oocytes. This restriction and subsequent amplification of mtDNA during oogenesis is termed the **mitochondrial genetic bottleneck.** Consequently, the variability in the percentage of mutant mtDNA molecules seen in the offspring of a mother carrying a mtDNA mutation arises, at least in part, from the sampling of only a subset of the mtDNAs during oogenesis.
- *Third,* despite the variability in the degree of heteroplasmy arising from the bottleneck, mothers with a high proportion of mutant mtDNA molecules are more likely to have clinically affected offspring than are mothers with a lower proportion, as one would predict from the random sampling of mtDNA molecules through the bottleneck. Nevertheless, even women carrying low proportions of pathogenic mtDNA molecules have some risk for having an affected child because the bottleneck can lead to the sampling and subsequent expansion, by chance, of even a rare mutant mtDNA species.

Unexplained and Unexpected Phenotypic Variation in mtDNA Diseases. As seen in Table 12-7, heteroplasmy is the rule for many mtDNA diseases. Heteroplasmy leads to an unpredictable and variable fraction of mutant mtDNA being present in any particular tissue, undoubtedly accounting for much of the pleiotropy and variable expressivity of mtDNA mutations (see Box). An example is provided by what appears to be the most common mtDNA mutation, the 3243A>G substitution in the tRNA$^{leu(UUR)}$ gene just mentioned in the context of the MELAS phenotype. This mutation leads predominantly to diabetes and deafness in some families, whereas in others it causes a disease called **chronic progressive external ophthalmoplegia.** Moreover, a very small fraction (<1%) of diabetes mellitus in the general population, particularly in Japanese, has been attributed to the 3243A>G substitution.

It is likely that much of the phenotypic variation observed among patients with mutations in mitochondrial genes will be explained by the fact that the proteins within mitochondria are remarkably heterogeneous between tissues, differing on average by approximately 25% between any two organs. This molecular heterogeneity is reflected in biochemical heterogeneity. For example, whereas much of the energy generated by brain mitochondria derives from the oxidation of ketones, skeletal muscle mitochondria preferentially use fatty acids as their fuel.

Interactions between the Mitochondrial and Nuclear Genomes

Because both the nuclear and mitochondrial genomes contribute polypeptides to oxidative phosphorylation, it is not surprising that the phenotypes associated with mutations in the nuclear genes are often indistinguishable from those due to mtDNA mutations. Moreover, mtDNA depends on many nuclear genome–encoded proteins for its replication and the maintenance of its integrity. Genetic evidence has highlighted the direct nature of the relationship between the nuclear and mtDNA genomes. The first indication of this interaction was provided by the identification of the syndrome of **autosomally transmitted deletions in mtDNA.** Mutations in at least two genes have been associated with this phenotype. The protein encoded by one of these genes, amusingly called Twinkle, appears to be a DNA primase or helicase. The product of the second gene is a mitochondrial-specific DNA polymerase γ, whose loss of function is associated with both dominant and recessive multiple deletion syndromes.

A second autosomal disorder, the **mtDNA depletion syndrome,** is the result of mutations in any of six nuclear genes that lead to a reduction in the number of copies of mtDNA (both per mitochondrion and per cell) in various tissues. Several of the affected genes encode proteins required to maintain nucleotide pools or to metabolize nucleotides appropriately in the mitochondrion. For example, both myopathic and hepatocerebral phenotypes result from mutations in the nuclear genes for mitochondrial thymidine kinase and deoxyguanosine kinase. Because mutations in the six genes identified to date account for only a minority of affected individuals, additional genes must also be involved in this disorder.

Apart from the insights that these rare disorders provide into the biology of the mitochondrion, the identification of the affected genes facilitates genetic counseling and prenatal diagnosis in some families and suggests, in some instances, potential treatments. For example, the blood thymidine level is markedly increased in thymidine phosphorylase deficiency, suggesting that lowering thymidine levels might have therapeutic benefits if an excess of substrate rather than a deficiency of the product plays a major role in the pathogenesis of the disease.

Nuclear Genes Can Modify the Phenotype of mtDNA Diseases.

Although heteroplasmy is a major source of phenotypic variability in mtDNA diseases (see Box), additional factors, including alleles at nuclear loci, must also play a role. Strong evidence for the existence of such factors is provided by families carrying mutations associated with **Leber hereditary optic neuropathy** (LHON; see Table 12-7), which is generally homoplasmic (thus ruling out heteroplasmy as the explanation for the observed phenotypic variation). LHON is expressed phenotypically as rapid, painless bilateral loss of central vision due to optic nerve atrophy in young adults (see Table 12-7 and Fig. 12-26). Depending on the mutation, there is often some recovery of vision, but the pathogenic mechanisms of the optic nerve damage are unclear.

There is a striking and unexplained increase in the penetrance of the disease in males; approximately 50% of male carriers but only approximately 10% of female carriers of a LHON mutation develop symptoms. The variation in penetrance and the male bias of the LHON phenotype are determined by a haplotype on the short arm of the X chromosome. The gene at this nuclear-encoded modifier locus has not yet been identified, but it is contained, notably, in a haplotype that is common in the general population. When the protective haplotype is transmitted from a typically unaffected mother to individuals who have inherited the LHON mtDNA mutation from that mother, the phenotype is substantially ameliorated. Thus males who carry the high-risk X-linked haplotype as well as a LHON mtDNA mutation (other than the one associated with the most severe LHON phenotype [see Table 12-7]) are thirty-fivefold more likely to develop visual failure than those who carry the low-risk X-linked haplotype. These observations are of general significance because they demonstrate the powerful effect that modifier loci can have on the phenotype of a monogenic disease.

Diseases due to the Expansion of Unstable Repeat Sequences

The inheritance pattern of diseases due to unstable repeat expansions was presented in Chapter 7, with emphasis on the unusual genetics of this unique group of almost 20 disorders. These features include the unstable and dynamic nature of the mutations, which are due to the expansion, within the *transcribed* region of the affected gene, of repeated sequences such as the codon for glutamine (CAG) in **Huntington disease** (Case 24) and most of a group of neurodegenerative disorders called the **spinocerebellar ataxias,** or due to the expansion of trinucleotides in *noncoding* regions of RNAs, including CGG in **fragile X**

syndrome (Case 17), GAA in **Friedreich ataxia,** and CUG in **myotonic dystrophy 1** (Fig. 12-28).

Although the initial nucleotide repeat diseases to be described are all due to the expansion of three nucleotide repeats, other disorders have now been found to result from the expansion of longer repeats; these include a tetranucleotide (CCTG) in **myotonic dystrophy 2** (a close genocopy of myotonic dystrophy 1) and a pentanucleotide (ATTCT) in **spinocerebellar atrophy 10**. Because the affected gene is passed from generation to generation, the number of repeats may expand to a degree that is pathogenic, ultimately interfering with normal gene expression and function. The intergenerational expansion of the repeats accounts for the phenomenon of **anticipation**, the appearance of the disease at an earlier age as it is transmitted through a family. The biochemical mechanism most commonly proposed to underlie the expansion of unstable repeat sequences is slipped mispairing (Fig. 12-29). Remarkably, the repeat expansions appear to occur both in proliferating cells such as spermatogonia (during meiosis) and in nonproliferating somatic cells such as neurons. Consequently, expansion can occur, depending on the disease, during both DNA replication (as shown in Fig. 12-29) and genome maintenance (i.e., DNA repair).

The clinical phenotypes of Huntington disease and fragile X syndrome are presented in Chapter 7 and in their respective Cases. For reasons that are gradually becoming apparent, particularly in the case of fragile X syndrome, diseases due to the expansion of unstable repeats are primarily neurological; the clinical presentations include ataxia, cognitive defects, dementia, nystagmus, parkinsonism, and spasticity. Nevertheless, other systems are sometimes involved, as illustrated by some of the diseases discussed here.

The Pathogenesis of Diseases due to Unstable Repeat Expansions

Diseases of unstable repeat expansion are diverse in their pathogenic mechanisms and can be divided into three classes, considered in turn in the sections to follow.

- *Class 1*: diseases due to the expansion of noncoding repeats that cause a loss of protein expression
- *Class 2*: disorders resulting from expansions of noncoding repeats that confer novel properties on the RNA
- *Class 3*: diseases due to repeat expansion of a codon such as CAG (for glutamine) that confers novel properties on the affected protein

Class 1: Diseases due to the Expansion of Noncoding Repeats That Cause a Loss of Protein Expression

Fragile X Syndrome.

In the X-linked fragile X syndrome, the expansion of the CGG repeat in the 5′ untranslated region (UTR) of the *FMR1* gene to more than 200 copies leads to excessive methylation of

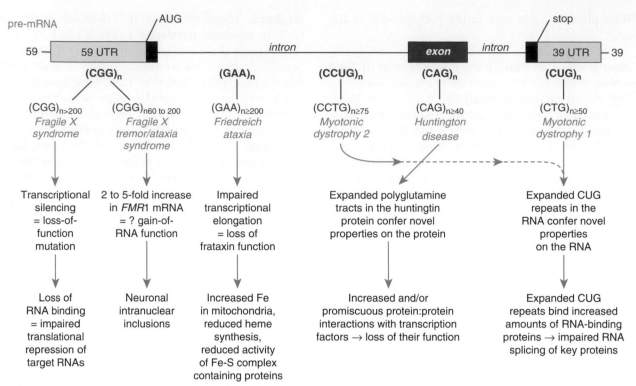

Figure 12-28 The locations of the trinucleotide repeat expansions and the sequence of each trinucleotide in five representative trinucleotide repeat diseases, shown on a schematic of a generic pre–messenger RNA (mRNA). The minimal number of repeats in the DNA sequence of the affected gene associated with the disease is also indicated. The effect of the expansion on the mutant RNA or protein is also indicated. *See Sources & Acknowledgments.*

cytosines in the promoter, an epigenetic modification of the DNA that silences transcription of the gene (see Figs. 7-22 and 12-28). Remarkably, the epigenetic silencing appears to be mediated by the mutant *FMR1* mRNA itself. The initial step in the silencing of *FMR1* results from the *FMR1* mRNA, containing the transcribed CGG repeat, hybridizing with the complementary CGG-repeat sequence of the *FMR1* gene, to form an RNA:DNA duplex. The mechanisms that subsequently maintain the silencing of the *FMR1* gene are unknown. The loss of the fragile X mental retardation protein (FMRP) is the cause of the intellectual disability and learning deficits and the non-neurological features of the clinical phenotype, including macroorchidism and connective tissue dysplasia **(Case 17)**. FMRP is an RNA-binding protein that associates with polyribosomes to suppress the translation of proteins from its RNA targets. These targets appear to be involved in cytoskeletal structure, synaptic transmission, and neuronal maturation, and the disruption of these processes is likely to underlie the intellectual disability and learning abnormalities seen in fragile X patients. For example, FMRP appears to regulate the translation of proteins required for the formation of synapses because the brains of individuals with the fragile X syndrome have increased density of abnormally long, immature dendritic spines. Moreover, FMRP localizes to dendritic spines, where at least one of its roles is to regulate

synaptic plasticity, the capacity to alter the strength of a synaptic connection, a process critical to learning and memory.

Fragile X Tremor/Ataxia Syndrome. Remarkably, the pathogenesis of disease in individuals with less pronounced CGG repeat expansion (60 to 200 repeats) in the *FMR1* gene, causing the clinically distinct **fragile X tremor/ataxia syndrome (FXTAS)**, is entirely different from that of the fragile X syndrome itself. Although decreased translational efficiency impairs the expression of the FMRP protein in FXTAS, this reduction cannot be responsible for the disease because males with full mutations and virtually complete loss of function of the *FMR1* gene never develop FXTAS. Rather, the evidence suggests that FXTAS results from the twofold to fivefold increased levels of the *FMR1* mRNA present in these patients, representing a gain-of-function mutation. This pathogenic RNA leads to the formation of intranuclear neuronal inclusions, the cellular signature of the disease.

Class 2: Disorders Resulting from Expansions of Noncoding Repeats That Confer Novel Properties on the RNA

Myotonic Dystrophy. Myotonic dystrophy 1 (DM1) is an autosomal dominant condition with the most pleiotropic phenotype of all the unstable repeat expansion

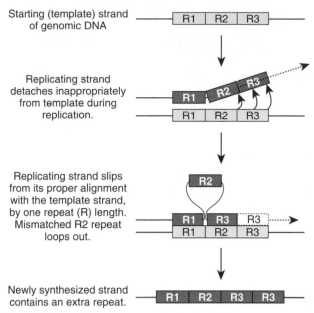

Starting (template) strand of genomic DNA

Replicating strand detaches inappropriately from template during replication.

Replicating strand slips from its proper alignment with the template strand, by one repeat (R) length. Mismatched R2 repeat loops out.

Newly synthesized strand contains an extra repeat.

Figure 12-29 The slipped mispairing mechanism thought to underlie the expansion of unstable repeats, such as the (CAG)$_n$ repeat found in Huntington disease and the spinocerebellar ataxias. An insertion occurs when the newly synthesized strand aberrantly dissociates from the template strand during replication synthesis. When the new strand reassociates with the template strand, the new strand may slip back to align out of register with an incorrect repeat copy. Once DNA synthesis is resumed, the misaligned molecule will contain one or more extra copies of the repeat (depending on the number of repeat copies that slipped out in the misalignment event).

disorders. In addition to myotonia, it is characterized by muscle weakness and wasting, cardiac conduction defects, testicular atrophy, insulin resistance, and cataracts; there is also a congenital form with intellectual disability. The disease results from a CTG expansion in the 3′ UTR of the *DMPK* gene, which encodes a protein kinase (see Fig. 12-28). Myotonic dystrophy 2 (DM2) is also an autosomal dominant trait and shares most of the clinical features of DM1, except that there is no associated congenital presentation. DM2 is due to the expansion of a CCTG tetranucleotide in the first intron of the gene encoding zinc finger protein 9 (see Fig. 12-28). The strikingly similar phenotypes of DM1 and DM2 suggest that they have a common pathogenesis. Because the unstable expansions occur within the non coding regions of two different genes that encode unrelated proteins, the CTG trinucleotide expansion itself (and the resulting expansion of CUG in the mRNA) is thought to underlie an RNA-mediated pathogenesis.

What is the mechanism by which large tracts of the CUG trinucleotide, in the noncoding region of genes, lead to the DM1 and DM2 phenotypes? The pathogenesis appears to result from the binding of the CUG repeats to RNA-binding proteins. Consequently, the pleiotropy that typifies the disease may reflect the broad

array of RNA-binding proteins to which the CUG repeats bind. Many of the RNA-binding proteins sequestered by the excessive number of CUG repeats are regulators of splicing, and indeed more than a dozen distinct pre-mRNAs have been shown to have splicing alterations in patients with DM1, including cardiac troponin T (which might account for the cardiac abnormalities) and the insulin receptor (which may explain the insulin resistance). Thus the myotonic dystrophies are referred to as **spliceopathies**. Even though our knowledge of the abnormal processes underlying DM1 and DM2 is still incomplete, these molecular insights offer the hope that a rational small molecule therapy might be developed.

Class 3: Diseases due to Repeat Expansion of a Codon That Confers Novel Properties on the Affected Protein

Huntington Disease. Huntington disease is an autosomal dominant neurodegenerative disorder associated with chorea, athetosis (uncontrolled writhing movements of the extremities), loss of cognition, and psychiatric abnormalities (Case 24). The pathological process is caused by the expansion—to more than 40 repeats—of the codon CAG in the *HD* gene, resulting in long polyglutamine tracts in the mutant protein, huntingtin (see Figs. 7-20 and 7-21). The bulk of evidence suggests that the mutant proteins with expanded polyglutamine sequences are novel property mutants (see Chapter 11), the expanded tract conferring novel features on the protein that damage specific populations of neurons and produce neurodegeneration by unique toxic mechanisms. The most striking cellular hallmark of the disease is the presence of insoluble aggregates of the mutant protein (as well as other polypeptides) clustered in nuclear inclusions in neurons. The aggregates are thought to result from normal cellular responses to the misfolding of huntingtin that results from the polyglutamine expansion. Dramatic as these inclusions are, however, their formation may actually be protective rather than pathogenic.

A unifying model of the neuronal death mediated by polyglutamine expansion in huntingtin is not at hand. Many cellular processes have been shown to be disrupted by mutant huntingtin in its soluble or its aggregated form, including transcription, vesicular transport, mitochondrial fission, and synaptic transmission and plasticity. Ultimately, the most critical and primary events in the pathogenesis will be identified, perhaps guided by genetic analyses that lead to correction of the phenotype. For example, it has been found that mutant huntingtin abnormally associates with a mitochondrial fission protein, GTPase dynamin-related protein 1 (DRP1) in Huntington disease patients, leading to multiple mitochondrial abnormalities. Remarkably, in mice, these defects are rescued by reducing DRP1 GTPase activity, suggesting both that DRP1 as a therapeutic target for the disorder and that mitochondrial abnormalities play important roles in Huntington disease.

CONCLUDING COMMENTS

Despite the substantial progress in our understanding of the molecular events that underlie the pathology of the unstable repeat expansion diseases, we are only beginning to dissect the pathogenic complexity of these important conditions. It is clear that the study of animal models of these disorders is providing critical insights into these disorders, insights that will undoubtedly lead to therapies to prevent or to reverse the pathogenesis of these slowly developing disorders in the near future. We begin to explore the concepts relevant to the treatment of disease in the next chapter.

GENERAL REFERENCES

Hamosh A: Online mendelian inheritance in man, OMIM. McKusick-Nathans Institute of Genetic Medicine, Baltimore, MD, Johns Hopkins University. Available at http://omim.org/.

Lupski JR, Stankiewicz P, editors: *Genomic disorders: the genomic basis of disease*, Totowa, NJ, 2006, Humana Press.

Pagon RA, Adam MP, Bird TD, et al: GeneReviews. Expert-authored summaries about diagnosis, management and genetic counseling for specific inherited conditions, University of Washington. Available at http://www.ncbi.nlm.nih.gov/books/NBK1116/.

Rimoin DL, Connor JM, Pyeritz RE, et al: *Emery and Rimoin's essential medical genetics*, Waltham, MA, 2013, Academic Press (Elsevier).

Strachan T, Read AP: *Human molecular genetics*, ed 4, New York, 2010, Garland Science.

Valle D, Beaudet AL, Vogelstein B, et al, editors: The online metabolic & molecular bases of inherited disease (OMMBID), http://www.ommbid.com.

REFERENCES TO SPECIFIC TOPICS

Bettens K, Sleegers K, Van Broeckhoven C: Genetic insights in Alzheimer's disease, *Lancet Neurol* 12:92–104, 2013.

Blau N, Hennermann JB, Langenbeck U, et al: Diagnosis, classification, and genetics of phenylketonuria and tetrahydrobiopterin (BH$_4$) deficiencies, *Mol Genet Metab* 104:S2–S9, 2011.

Byers PH, Pyott SM: Recessively inherited forms of osteogenesis imperfecta, *Ann Rev Genet* 46:475–497, 2012.

Chamberlin JS: Duchenne muscular dystrophy models show their age, *Cell* 143:1040–1042, 2010.

Chillon M, Casals T, Mercier B, et al: Mutations in the cystic fibrosis gene in patients with congenital absence of the vas deferens, *N Engl J Med* 332:1475–1480, 1995.

Colak D, Zaninovic N, Cohen MS, et al: Promoter-bound trinucleotide repeat mRNA drives epigenetic silencing in fragile X syndrome, *Science* 343:1002–1005, 2014.

Cutting GR: Modifier genes in Mendelian disorders: the example of cystic fibrosis, *Ann N Y Acad Sci* 1214:57–69, 2010.

Flanigan KM: The muscular dystrophies, *Semin Neurol* 32:255–263, 2012.

Fong LG, Young SG: PCSK9 function and physiology, *J Lipid Res* 49:1152–1156, 2008.

Goldstein JL, Brown MS: Molecular medicine: the cholesterol quartet, *Science* 292:1310–1312, 2001.

Gu YY, Harley ITW, Henderson LB, et al: IFRD1 polymorphisms in cystic fibrosis with potential link to altered neutrophil function, *Nature* 458:1039–1042, 2009.

Janciauskiene SM, Bals R, Koczulla R, et al: The discovery of alpha1-antitrypsin and its role in health and disease, *Respir Med* 105:1129–1139, 2011.

Jonsson T, Atwal JK, Steinberg S, et al: A mutation in APP protects against Alzheimer's disease and age-related cognitive decline, *Nature* 488:96–99, 2012.

Kathiresan S, Melander O, Guiducci C, et al: Six new loci associated with blood low-density lipoprotein cholesterol, high-density lipoprotein cholesterol or triglycerides in humans, *Nat Genet* 40:189–197, 2008.

Koopman WJ, Willems PH, Smeitink JA: Monogenic mitochondrial disorders, *N Engl J Med* 366:1132–1141, 2012.

Laine CM, Joeng KS, Campeau PM, et al: WNT1 mutations in early-onset osteoporosis and osteogenesis imperfecta, *N Engl J Med* 368:1809–1816, 2013.

Lopez CA, Cleary JD, Pearson CE: Repeat instability as the basis for human diseases and as a potential target for therapy, *Nat Rev Mol Cell Biol* 11:165–170, 2010.

Moskowitz SM, James F, Chmiel JF, et al: Clinical practice and genetic counseling for cystic fibrosis and CFTR-related disorders, *GeneTests* 10:851–868, 2008.

Raal FJ, Santos ED: Homozygous familial hypercholesterolemia: current perspectives on diagnosis and treatment, *Atherosclerosis* 223:262–268, 2012.

Ramsey BW, Banks-Schlegel S, Accurso FJ, et al: Future directions in early cystic fibrosis lung disease research: an NHLBI workshop report, *Am J Respir Crit Care Med* 185:887–892, 2012.

Schon EA, DiMauro S, Hirano M: Human mitochondrial DNA: roles of inherited and somatic mutations, *Nat Rev Genet* 13:878–890, 2012.

Selkoe DJ: Alzheimer's disease, *Cold Spring Harb Perspect Biol* 3:a004457, 2011.

Sosnay PR, Siklosi KR, Van Goor F, et al: Defining the disease liability of mutations in the cystic fibrosis transmembrane conductance regulator gene, *Nature Genet* 45:1160–1167, 2013.

Vafai SB, Mootha VK: Mitochondrial disorders as windows into an ancient organelle, *Nature* 491:374–383, 2012.

Zoghbi HY, Orr HT: Pathogenic mechanisms of a polyglutamine-mediated neurodegenerative disease, spinocerebellar ataxia type 1, *J Biol Chem* 284:7425–7429, 2009.

USEFUL WEBSITES

Mutation Databases

Clinical and functional translation of CFTR (CFTR2 project).
  http://www.cftr2.org/
Collagen mutation database.
  http://www.le.ac.uk/genetics/collagen/
Cystic fibrosis and *CFTR* gene mutation database.
  http://www.gene.sickkids.on.ca/cftr/
Human mitochondrial genome database.
  http://www.gen.emory.edu/mitomap.html
Phenylalanine hydroxylase mutation database.
  http://www.pahdb.mcgill.ca
The Human Gene Mutation Database.
  http://www.hgmd.cf.ac.uk/ac/index.php

PROBLEMS

1. One mutant allele at the LDL receptor locus (leading to familial hypercholesterolemia) encodes an elongated protein that is approximately 50,000 Da larger than the normal 120,000-Da receptor. Indicate at least three mechanisms that could account for this abnormality. Approximately how many extra nucleotides would need to be translated to add 50,000 Da to the protein?

2. Are autosomal dominant *PSCK9* gain-of-function mutations that cause familial hypercholesterolemia deficiency phenocopies, or genocopies, of familial hypercholesterolemia due to autosomal dominant mutations in the LDL receptor gene? Explain your answer.

3. In discussing the nucleotide changes found to date in the coding region of the CF gene, we stated that some of the changes (the missense changes) found so far are only "putative" disease-causing mutations. What criteria would one need to fulfill before knowing that a nucleotide change is pathogenic and not a benign polymorphism?

4. Johnny, 2 years of age, is failing to thrive. Investigations show that although he has clinical findings of CF, his sweat chloride concentration is normal. The sweat chloride concentration is normal in less than 2% of patients with CF. His pediatrician and parents want to know if DNA analysis can determine whether he indeed has CF.
   a. Would DNA analysis be useful in this case? Briefly outline the steps involved in obtaining a DNA diagnosis for CF.
   b. If he has CF, what is the probability that he is homozygous for the ΔF508 mutation? (Assume that 85% of CF mutations could be detected at the time you are consulted and that his parents are from northern Europe, where the ΔF508 allele has a frequency of 0.70.)
   c. If he does not have the ΔF508 mutation, does this disprove the diagnosis? Explain.

5. James is the only person in his kindred affected by DMD. He has one unaffected brother, Joe. DNA analysis shows that James has a deletion in the *DMD* gene and that Joe has received the same maternal X chromosome, but one without a deletion. What genetic counseling would you give the parents regarding the recurrence risk for DMD in a future pregnancy?

6. *DMD* has a high mutation rate but shows no ethnic variation in frequency. Use your knowledge of the gene and the genetics of DMD to suggest why this disorder is equally common in all populations.

7. A 3½-year-old girl, T.N., has been noted to have increasing difficulty standing up after sitting on the floor. Her serum level of creatine kinase is grossly elevated. Although a female, the presumptive clinical diagnosis is Duchenne muscular dystrophy. Females with DMD are rare. Identify three mechanisms of mutation that could account for the occurrence of DMD in a female.

8. In patients with osteogenesis imperfecta, explain why the missense mutations at glycine positions in the triple helix of type I collagen are confined to a limited number of other amino acid residues (Ala, Ser, Cys, Arg, Val, Asp).

9. Glucose-6-phosphate dehydrogenase (G6PD) is encoded by an X-linked gene. G6PD loss-of-function mutations can lead to hemolysis on exposure to some drugs, fava beans, and other compounds (see Chapter 18). Electrophoresis of red blood cell hemolysates shows that some females have two G6PD bands, but males have a single band. Explain this observation and the possible pathological and genetic significance of the finding of two bands in an African American female.

10. A 2 year-old infant, the child of first-cousin parents, has unexplained developmental delay. A survey of various biochemical parameters indicates that he has a deficiency of four lysosomal enzymes. Explain how a single autosomal recessive mutation might cause the loss of function of four enzyme activities. Why is it most likely that the child has an autosomal recessive condition, if he has a genetic condition at all?

11. The effect of a dominant negative allele illustrates one general mechanism by which mutations in a protein cause dominantly inherited disease. What other mechanism is commonly associated with dominance in genes encoding the subunits of multimeric proteins?

12. The clinical effects of mutations in a housekeeping protein are frequently limited to one or a few tissues, often tissues in which the protein is abundant and serves a specialty function. Identify and discuss examples that illustrate this generalization, and explain why they fit it.

13. The relationship between the site at which a protein is expressed and the site of pathological change in a genetic disease may be unpredictable. In addition, the tissue that lacks the mutant protein may even be left unaffected by disease. Give examples of this latter phenomenon and discuss them.

14. The two pseudodeficiency alleles of hex A are Arg-247Trp and Arg249Trp. What is the probable reason that the missense substitutions of these alleles are so close together in the protein?

15. Why are gain-of-function mutations in proteins, as seen with the autosomal dominant *PCSK9* mutations that cause hypercholesterolemia, almost always missense mutations?

16. What are the possible explanations for the presence of three predominant alleles for Tay-Sachs disease in Ashkenazi Jews? Does the presence of three alleles, and the relatively high frequency of Tay-Sachs disease in this population, necessarily accord with a heterozygote advantage hypothesis or a founder effect hypothesis?

17. All of the known loci associated with Alzheimer disease do not account for the implied genetic risk. Identify at least three other sources of genetic variation that may account for the genetic contribution to AD.

18. Propose a molecular therapy that might counteract the effect of the CUG expansions in the RNAs of myotonic dystrophy 1 and 2 and that would reduce the binding of RNA-binding proteins to the CUG repeats. Anticipate some possible undesirable effects of your proposed therapy.

CHAPTER 13

The Treatment of Genetic Disease

The understanding of genetic disease at a molecular level, as presented in Chapters 11 and 12, is the foundation of rational therapy. In the coming decades, increasing annotation of the human genome sequence and the catalogue of human genes, as well as gene, RNA, and protein therapy, will have an enormous impact on the treatment of genetic conditions and other disorders. In this chapter, we review established therapies as well as new strategies for treating genetic disease. Our emphasis will be on therapies that reflect the genetic approach to medicine, and our focus is on single-gene diseases, rather than genetically complex disorders.

The objective of treating genetic disease is to eliminate or ameliorate the effects of the disorder, not only on the patient but also on his or her family. The importance of educating the patient is paramount—not only to achieve understanding of the disease and its treatment, but also to ensure compliance with therapy that may be inconvenient and lifelong. The family must be informed about the risk that the disease may occur in other members. Thus genetic counseling is a major component of the management of hereditary disorders and will be dealt with separately, in Chapter 16.

For single-gene disorders due to loss-of-function mutations, treatment is directed to replacing the defective protein, improving its function, or minimizing the consequences of its deficiency. Replacement of the defective gene product (RNA or protein) may be achieved by direct administration, cell or organ transplantation, or gene therapy. In principle, gene therapy or gene editing will be the preferred mode of treatment of some and perhaps many single-gene diseases, once these approaches become routinely safe and effective. However, even when copies of a normal gene can be transferred into the patient to effect permanent cure, the family will need ongoing genetic counseling, carrier testing, and prenatal diagnosis, in many cases for several generations.

Recent discoveries promise many more exciting and dramatic therapies for genetic disease. These achievements include the first cures of inherited disorders using gene therapy, the development of novel small molecule therapies that can restore activity to mutant proteins, and the ability to prevent the clinical manifestations of previously lethal disorders, including lysosomal storage diseases, by protein replacement therapy.

THE CURRENT STATE OF TREATMENT OF GENETIC DISEASE

Genetic disease can be treated at any level from the mutant gene to the clinical phenotype (Fig. 13-1). Treatment at the level of the clinical phenotype includes all the medical or surgical interventions that are not unique to the management of genetic disease. Throughout this chapter, we describe the rationale for treatment at each of these levels. For diseases in which the biochemical or genetic defect is known, the approximate frequency with which the most common strategies are employed is shown in Figure 13-2. The current treatments are not necessarily mutually exclusive, although only gene therapy, gene editing, or cell transplantation can potentially provide cures.

Although powerful advances are being made, the overall treatment of single-gene diseases is presently deficient. A 25-year longitudinal survey of the effectiveness of treatment of 57 inborn errors of metabolism, reflecting the state of the field up to 2008, is shown in Figure 13-3. Note, however, that inborn errors are a group of diseases for which treatment is advanced, in general, compared to most other types of genetic disorders such as those due, for example, to chromosomal abnormalities, imprinting defects, or copy number variation. An encouraging trend over past decades is that treatment is more likely to be successful if the basic biochemical defect is known. In one study, for example, although treatment increased life span in only 15% of all single-gene diseases studied, life span was improved by approximately 50% in the subset of 57 inborn errors in which the cause was known; significant improvements were also observed for other phenotypes, including growth, intelligence, and social adaptation. Thus research to elucidate the genetic and biochemical bases of hereditary disease has a major impact on the clinical outcome.

The improving but still unsatisfactory state of treatment of monogenic diseases is due to numerous factors, including the following:

- **Gene not identified or pathogenesis not understood.** Although more than 3000 genes have been associated with monogenic diseases, the affected gene is still unknown in more than half of these disorders. This fraction will decrease dramatically over the next

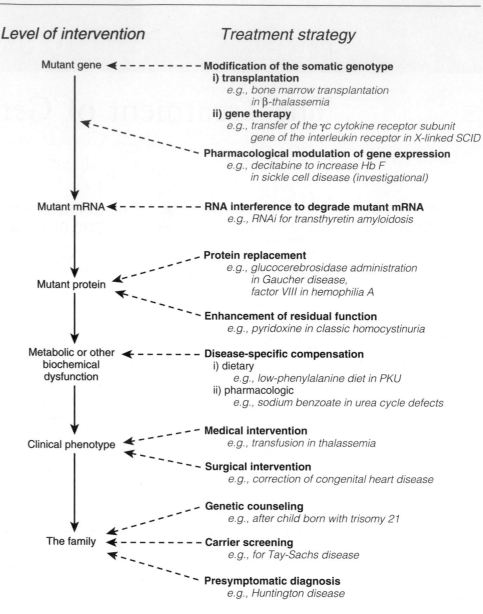

Figure 13-1 The various levels of treatment that are relevant to genetic disease, with the corresponding strategies used at each level. For each level, a disease discussed in the book is given as an example. All the therapies listed are used clinically in many centers, unless indicated otherwise. Hb F, Fetal hemoglobin; mRNA, messenger RNA; PKU, phenylketonuria; RNAi, RNA interference; SCID, severe combined immunodeficiency. *See Sources & Acknowledgments.*

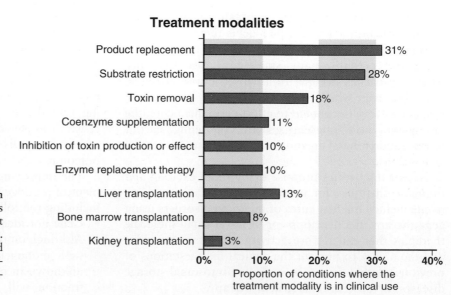

Figure 13-2 Treatment modalities for inborn errors of metabolism. This figure represents the findings of an analysis of the treatment efficacy of 57 inborn errors of metabolism. The total of the nine different approaches used exceeds 100% because more than one treatment can sometimes be used for a given condition. *See Sources & Acknowledgments.*

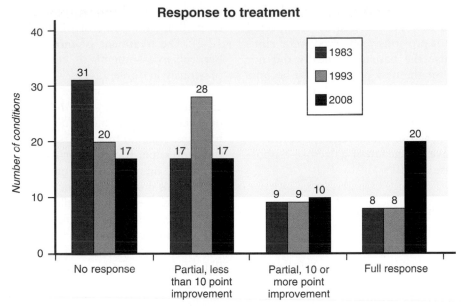

Figure 13-3 The effect of treatment of 57 genetic diseases in which the affected gene or biochemical function is known and for which sufficient information was available for analysis in 2008. A quantitative phenotype scoring system was used to evaluate the efficacy of the therapies. The fraction of treatable diseases will have increased to a small extent since this 2008 survey because of the increasing success of enzyme replacement and a few other treatments, including gene therapy. *See Sources & Acknowledgments.*

TABLE 13-1 Examples of Prenatal Medical Treatment of Monogenic Disorders

Disease	Treatment
Biotinidase deficiency	Prenatal biotin administration
Cobalamin-responsive methylmalonic aciduria	Prenatal maternal cobalamin administration
Congenital adrenal hyperplasia	Dexamethasone, a cortisol analogue
Phosphoglycerate dehydrogenase (PGDH) deficiency, a disorder of L-serine synthesis	Prenatal L-serine administration

decade because of the impact of whole-genome and whole-exome sequencing. However, even when the mutant gene in known, knowledge of the pathophysiological mechanism is often inadequate and can lag well behind gene discovery. In phenylketonuria (PKU), for example, despite decades of study, the mechanisms by which the elevation in phenylalanine impairs brain development and function are still poorly understood (see Chapter 12).

- **Prediagnostic fetal damage.** Some mutations act early in development or cause irreversible pathological changes before they are diagnosed. These problems can sometimes be anticipated if there is a family history of the genetic disease or if carrier screening identifies couples at risk. In some cases, prenatal treatment is possible (Table 13-1).
- **Severe phenotypes are less amenable to intervention.** The initial cases of a disease to be recognized are

usually the most severely affected, but they are often less amenable to treatment. In such individuals, the mutation frequently leads to the absence of the encoded protein or to a severely compromised mutant protein with no residual activity. In contrast, when the mutation is less disruptive, the mutant protein may retain some residual function and it may be possible to increase the small amount of function sufficiently to have a therapeutic effect, as described later.

- **The challenge of dominant negative alleles.** For some dominant disorders, the mutant protein interferes with the function of the normal allele. The challenge is to decrease the expression or impact of the mutant allele or its encoded mutant protein specifically, without disrupting expression or function of the normal allele or its normal protein.

SPECIAL CONSIDERATIONS IN TREATING GENETIC DISEASE

Long-Term Assessment of Treatment Is Critical

For treating monogenetic diseases, long-term evaluation of cohorts of treated individuals, often over decades, is critical for several reasons. First, treatment initially judged as successful may eventually be revealed to be imperfect; for example, although well-managed children with PKU have escaped severe retardation and have normal or nearly normal IQs (see later), they often manifest subtle learning disorders and behavioral disturbances that impair their academic performance in later years.

Second, successful treatment of the pathological changes in one organ may be followed by unexpected problems in tissues not previously observed to be clinically involved, because the patients typically did not survive long enough for the new phenotype to become evident. **Galactosemia,** a well-known inborn error of carbohydrate metabolism, illustrates this point. This disorder results from an inability to metabolize galactose, a component of lactose (milk sugar), because of the autosomal recessive deficiency of galactose-1-phosphate uridyltransferase (GALT)

$$\text{Galactose-1-phosphate} \xrightarrow{\text{GALT}} \text{UDP galactose}$$

Affected infants are usually normal at birth but develop gastrointestinal problems, cirrhosis of the liver, and cataracts in the weeks after they are given milk. The pathogenesis is thought to be due to the negative impact of galactose-1-phosphate accumulation on other critical enzymes. If not recognized, galactosemia causes severe intellectual disability and is often fatal. Complete removal of milk from the diet, however, can protect against most of the harmful consequences, although, as with PKU, learning disabilities are now recognized to be common, even in well-treated patients. Moreover, despite conscientious treatment, most females with galactosemia have ovarian failure that appears to result from continued galactose toxicity.

Another example is provided by hereditary retinoblastoma (Case 39) due to germline mutations in the **retinoblastoma** *(RB1)* gene (see Chapter 15). Patients successfully treated for the eye tumor in the first years of life are unfortunately at increased risk for development of other independent malignant neoplasms, particularly osteosarcoma, after the first decade of life. Ironically, therefore, treatment that successfully prolongs life provides an opportunity for the manifestation of a previously unrecognized phenotype.

In addition, therapy that is free of side effects in the short term may be associated with serious problems in the long term. For example, clotting factor infusion in hemophilia (Case 21) sometimes results in the formation of antibodies to the infused protein, and blood transfusion in thalassemia (Case 44) invariably produces iron overload, which must then be managed by the administration of iron-chelating agents, such as deferoxamine.

Genetic Heterogeneity and Treatment

The optimal treatment of single-gene defects requires an unusual degree of diagnostic precision; one must often define not only the biochemical abnormality, but also the specific gene that is affected. For example, as we saw in Chapter 12, hyperphenylalaninemia can result from mutations in either the phenylalanine hydroxylase *(PAH)* gene or in one of the genes that encodes the

enzymes required for the synthesis of tetrahydrobiopterin (BH$_4$), the cofactor of the PAH enzyme (see Fig. 12-2). The treatment of these two different causes of hyperphenylalaninemia is entirely different, as shown previously in Table 12-1.

Allelic heterogeneity (see Chapter 7) may also have critical implications for therapy. Some alleles may produce a protein that is decreased in abundance but has some residual function, so that strategies to increase the expression, function, or stability of such a partially functional mutant protein may correct the biochemical defect. This situation is again illustrated by some patients with hyperphenylalaninemia due to mutations in the *PAH* gene; the mutations in some patients lead to the formation of a mutant PAH enzyme whose activity can be increased by the administration of high doses of the BH$_4$ cofactor (see Chapter 12). Of course, if a patient carries two alleles with no residual function, nothing will be gained by increasing the abundance of the mutant protein. One of the most striking examples of the importance of knowing the specific mutant allele in a patient with a genetic disease is exemplified by cystic fibrosis (CF); the drug ivacaftor (Kalydeco) is presently approved for treating CF patients carrying any one of only nine of the many hundreds of *CFTR* missense alleles.

TREATMENT BY THE MANIPULATION OF METABOLISM

Presently, the most successful disease-specific approach to the treatment of genetic disease is directed at the metabolic abnormality in inborn errors of metabolism. The principal strategies used to manipulate metabolism in the treatment of this group of diseases are listed in Table 13-2. The necessity for patients with pharmacogenetic diseases, such as glucose-6-phosphate dehydrogenase deficiency, to avoid certain drugs and chemicals is described in Chapter 18.

Substrate Reduction

As illustrated by the damaging effects of hyperphenylalaninemia in PKU, enzyme deficiencies may lead to substrate accumulation, with pathophysiological consequences (see Chapter 12). Strategies to prevent the accumulation of the offending substrate have been one of the most effective methods of treating genetic disease. The most common approach is to reduce the dietary intake of the substrate or of a precursor of it, and presently several dozen disorders—most involving amino acid catabolic pathways—are managed in this way. The drawback is that severe lifelong restriction of dietary protein intake is often necessary, requiring strict adherence to an artificial diet that is onerous for the family as well as for the patient. Nutrients such as 20 essential amino acids cannot be withheld entirely, however; their intake must be sufficient for anabolic needs such as protein synthesis.

TABLE 13-2 **Treatment of Genetic Disease by Metabolic Manipulation**

Type of Metabolic Intervention	Substance or Technique	Disease
Avoidance	Antimalarial drugs	G6PD deficiency
	Isoniazid	Slow acetylators
Dietary restriction	Phenylalanine	PKU
	Galactose	Galactosemia
Replacement	Thyroxine	Monogenic forms of congenital hypothyroidism
	Biotin	Biotinidase deficiency
Diversion	Sodium benzoate	Urea cycle disorders
	Drugs that sequester bile acids in the intestine (e.g., colesevelam)	Familial hypercholesterolemia heterozygotes
Enzyme inhibition	Statins	Familial hypercholesterolemia heterozygotes
Receptor antagonism	Losartan (investigational)	Marfan syndrome
Depletion	LDL apheresis (direct removal of LDL from plasma)	Familial hypercholesterolemia homozygotes

G6PD, Glucose-6-phosphate dehydrogenase; LDL, low-density lipoprotein; PKU, phenylketonuria.
Updated from Rosenberg LE: Treating genetic diseases: lessons from three children. *Pediatr Res* 27:S10–S16, 1990.

A diet restricted in phenylalanine largely circumvents the neurological damage in classic PKU (see Chapter 12). Phenylketonuric children are normal at birth because the maternal enzyme protects them during prenatal life. Treatment is most effective if begun promptly after diagnosis by newborn screening. Without treatment, irreversible developmental delay occurs, the degree of intellectual deficit being directly related to the delay in commencing the low-phenylalanine diet. It is now recommended that patients with PKU remain on a low-phenylalanine diet for life because neurological and behavioral abnormalities develop in many (although perhaps not all) patients if the diet is stopped. However, even PKU patients who have been effectively treated throughout life may have neuropsychological deficits (e.g., impaired conceptual, visual-spatial, and language skills), despite their having normal intelligence as measured by IQ tests. Nonetheless, treatment produces results vastly superior to the severe developmental delay that occurs without treatment. As discussed in Chapter 12, continued phenylalanine restriction is particularly important in women with PKU during pregnancy to prevent prenatal damage to the fetus, even though the fetus is highly unlikely to be affected by PKU.

Replacement

The provision of essential metabolites, cofactors, or hormones whose deficiency is due to a genetic disease is simple in concept and often simple in application. Some of the most successfully treated single-gene defects belong to this category. A prime example is provided by **congenital hypothyroidism**, of which 10% to 15% of cases are monogenic in origin. Monogenic congenital hypothyroidism can result from mutations in any one of numerous genes encoding proteins required for the development of the thyroid gland or the biosynthesis or metabolism of thyroxine. Because congenital hypothyroidism from

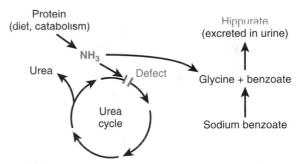

Figure 13-4 The strategy of metabolite diversion. In this example, ammonia cannot be removed by the urea cycle because of a genetic defect of a urea cycle enzyme. The administration of sodium benzoate diverts ammonia to glycine synthesis, and the nitrogen moiety is subsequently excreted as hippurate.

all causes is common (approximately 1 in 4000 neonates), neonatal screening is conducted in many countries so that thyroxine administration may be initiated soon after birth to prevent the severe intellectual defects that are otherwise inevitable (see Chapter 18).

Diversion

Diversion therapy is the enhanced use of alternative metabolic pathways to reduce the concentration of a harmful metabolite. A major use of this strategy is in the treatment of the **urea cycle disorders** (Fig. 13-4). The function of the urea cycle is to convert ammonia, which is neurotoxic, to urea, a benign end product of protein catabolism excreted in urine. If the cycle is disrupted by an enzyme defect such as **ornithine transcarbamylase deficiency** (Case 36), the consequent hyperammonemia can be only partially controlled by dietary protein restriction. Blood ammonia levels can be reduced to normal, however, by the diversion of ammonia to metabolic pathways that are normally of minor significance, leading to the synthesis of harmless compounds. Thus,

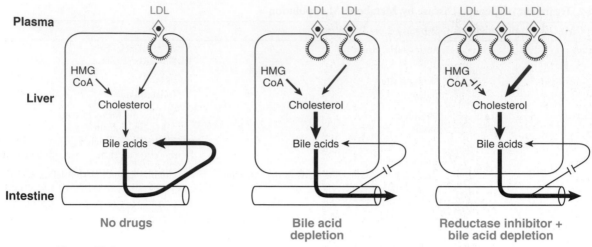

Figure 13-5 Rationale for the combined use of a reagent that sequesters bile acids, such as colesevelam, together with an inhibitor of 3-hydroxy-3-methylglutaryl coenzyme A reductase (HMG CoA reductase) in the treatment of familial hypercholesterolemia heterozygotes. LDL, Low-density lipoprotein. *See Sources & Acknowledgments.*

the administration to hyperammonemic patients of large quantities of sodium benzoate forces the ligation of ammonia with glycine to form hippurate, which is excreted in urine (see Fig. 13-4). Glycine synthesis is thereby increased, and for each mole of glycine formed, one mole of ammonia is consumed.

A comparable approach is used to reduce cholesterol levels in *heterozygotes* for **familial hypercholesterolemia** (Case 16) (see Chapter 12). If bile acids are sequestered in the intestine by the oral administration of a compound such as colesevelam and then excreted in feces rather than being reabsorbed, bile acid synthesis from cholesterol increases (Fig. 13-5). The reduction in hepatic cholesterol levels leads to increased production of low-density lipoprotein (LDL) receptors from their single normal LDL receptor gene, increased hepatic uptake of LDL-bound cholesterol, and lower levels of plasma LDL cholesterol. This treatment significantly reduces plasma cholesterol levels because 70% of all LDL receptor uptake of cholesterol occurs in the liver. An important general principle is illustrated by this example: autosomal dominant diseases may sometimes be treated by increasing the expression of the normal allele.

Enzyme Inhibition

The pharmacological inhibition of enzymes is sometimes used to reduce the impact of metabolic abnormalities in treating inborn errors. This principle is also illustrated by the treatment of heterozygotes of familial hypercholesterolemia. If a statin, a class of drugs that are powerful inhibitors of 3-hydroxy-3-methylglutaryl coenzyme A reductase, or HMG CoA reductase (the rate-limiting enzyme of cholesterol synthesis), is used to decrease hepatic de novo cholesterol synthesis in these

patients, the liver compensates by increasing the synthesis of LDL receptors from the remaining intact LDL receptor allele. The increase in LDL receptors typically lowers plasma LDL cholesterol levels by 40% to 60% in familial hypercholesterolemia heterozygotes; used together with colesevelam, the effect is synergistic, and even greater decreases can be achieved (see Fig. 13-5).

Receptor Antagonism

In some instances, the pathophysiology of an inherited disease results from the increased and inappropriate activation of a biochemical or signaling pathway. In such cases, one therapeutic approach is to antagonize critical steps in the pathway. A powerful example is provided by an investigational treatment of an autosomal dominant connective tissue disorder, **Marfan syndrome** (Case 30). The disease results from mutations in the gene that encodes fibrillin 1, an important structural component of the extracellular matrix. The syndrome is characterized by many connective tissue abnormalities, such as aortic aneurysm, pulmonary emphysema, and eye-lens dislocation (Fig. 13-6).

Unexpectedly, the pathophysiology of Marfan syndrome is only partially explained by the impact of the reduction in fibrillin-1 microfibrils on the structure of the extracellular matrix. Rather, it has been found that a major function of microfibrils is to regulate signaling by the transforming growth factor β (TGF-β), by binding TGF-β to the large latent protein complex of TGF-β. The decreased abundance of microfibrils in Marfan syndrome leads to an increase in the local abundance of unbound TGF-β and in local activation of TGF-β signaling. This increased TGF-β signaling has been suggested to underlie the pathogenesis of many of the phenotypes of Marfan syndrome, particularly the progressive

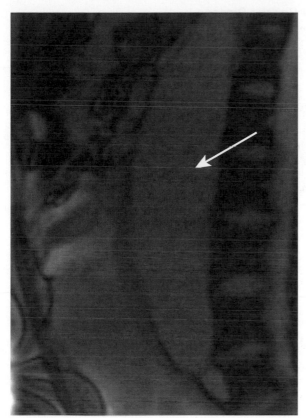

Figure 13-6 Magnetic resonance image (MRI) of the abdominal aorta of a 29-year-old pregnant woman with Marfan syndrome. The massive dilatation of the abdominal aorta is indicated by the *arrow. See Sources & Acknowledgments.*

dilation of the aortic root, and aortic aneurysm and dissection, the major cause of death in this disorder. Moreover, a recently recognized group of other vasculopathies, such as nonsyndromic forms of thoracic aortic aneurysm, has also proved to be driven by altered TGF-β signaling.

Angiotensin II signaling is known to increase TGF-β activity and the angiotensin II type 1 receptor antagonist, losartan, a widely used antihypertensive agent, has been shown to attenuate TGF-β signaling by decreasing the transcription of genes encoding TGF-β ligands, receptor subunits, and activators. Treatment with losartan has been found to decrease substantially the rate of aortic root dilation in initial clinical trials of Marfan syndrome patients, an effect that appears to be largely due to decreased TGF-β signaling.

The novel use of a U.S. Food and Drug Administration (FDA) approved drug, losartan, to treat a rare inherited disease, Marfan syndrome, is likely to represent a paradigm that will be repeated regularly in the future, as small molecule chemical screens to identify compounds with therapeutic potential—often including the thousands of FDA approved drugs—are undertaken to identify safe, effective treatments for other uncommon genetic disorders.

Depletion

Genetic diseases characterized by the accumulation of a harmful compound are sometimes treated by direct removal of the compound from the body. This principle is illustrated by the treatment of *homozygous familial hypercholesterolemia*. In this instance, for patients whose LDL levels cannot be lowered by other approaches, a procedure called apheresis is used to remove LDL from the circulation. Whole blood is removed from the patient, LDL is removed from plasma by any one of several methods, and the plasma and blood cells are returned to the patient. The use of phlebotomy to alleviate the iron accumulation of **hereditary hemochromatosis** (Case 20) provides another example of depletion therapy.

TREATMENT TO INCREASE THE FUNCTION OF THE AFFECTED GENE OR PROTEIN

The growth in knowledge of the molecular pathophysiology of monogenic diseases has been accompanied by a small but promising increase in therapies that—at the level of DNA, RNA, or protein—increase the function of the gene affected by the mutation. Some of the novel treatments have led to striking improvement in the lives of affected individuals, an outcome that, until recently, would have seemed fanciful. An overview of the molecular treatment of single-gene diseases is presented in Figure 13-7. These molecular therapies represent one facet of the important paradigm embraced by the concept of **personalized** or **precision medicine**. The term *precision medicine* is a general one used to describe the diagnosis, prevention, and treatment of a disease—tailored to individual patients—based on a profound understanding of the mechanisms that underlie its etiology and pathogenesis.

Treatment at the Level of the Protein

In many situations, if a mutant protein product is made, it may be possible to increase its function. For example, the stability or function of a mutant protein with *some* residual function may be further increased. With enzymopathies, the improvement in function obtained by this approach is usually very small, on the order of a few percent, but this increment is often all that is required to restore biochemical homeostasis.

Enhancement of Mutant Protein Function with Small Molecule Therapy

Small molecules are compounds with molecular weights in the few hundreds to thousands. They include vitamins, nonpeptide hormones, and indeed most drugs, whether synthesized by organic chemists or isolated from nature. A new strategy for identifying potential drugs is to use high-throughput screening of chemical

The Molecular Treatment of Genetic Disease

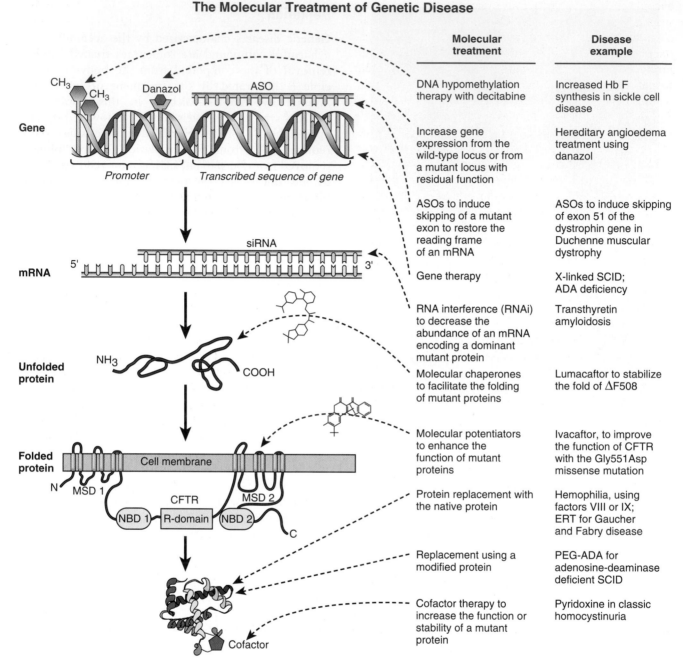

	Molecular treatment	Disease example
	DNA hypomethylation therapy with decitabine	Increased Hb F synthesis in sickle cell disease
	Increase gene expression from the wild-type locus or from a mutant locus with residual function	Hereditary angioedema treatment using danazol
	ASOs to induce skipping of a mutant exon to restore the reading frame of an mRNA	ASOs to induce skipping of exon 51 of the dystrophin gene in Duchenne muscular dystrophy
	Gene therapy	X-linked SCID; ADA deficiency
	RNA interference (RNAi) to decrease the abundance of an mRNA encoding a dominant mutant protein	Transthyretin amyloidosis
	Molecular chaperones to facilitate the folding of mutant proteins	Lumacaftor to stabilize the fold of ΔF508
	Molecular potentiators to enhance the function of mutant proteins	Ivacaftor, to improve the function of CFTR with the Gly551Asp missense mutation
	Protein replacement with the native protein	Hemophilia, using factors VIII or IX; ERT for Gaucher and Fabry disease
	Replacement using a modified protein	PEG-ADA for adenosine-deaminase deficient SCID
	Cofactor therapy to increase the function or stability of a mutant protein	Pyridoxine in classic homocystinuria

Figure 13-7 **The molecular treatment of inherited disease.** Each molecular therapy is discussed in the text. ADA, Adenosine deaminase; ASO, antisense oligonucleotide; ERT, enzyme replacement therapy; Hb F, fetal hemoglobin; mRNA, messenger RNA; MSD, membrane-spanning domain; NBD, nucleotide-binding domain; PEG, polyethylene glycol; SCID, severe combined immunodeficiency; siRNA, small interfering RNA.

compound libraries, often containing tens of thousands of known chemicals, against a drug target, such as the protein whose function is disrupted by a mutation. As we will discuss, two drugs that are now FDA approved for the treatment of some patients with CF, and another that is investigational, were discovered using such high-throughput screens. Progress in the development of these drugs represents a new frontier with great potential for the treatment of genetic disease.

Small Molecule Therapy to Allow Skipping over Nonsense Codons. Nonsense mutations account for 11% of defects in the human genome. Approximately 9% of all *CFTR* alleles are nonsense mutations, and approximately 50% of Ashkenazi Jewish patients with CF carry at least one *CFTR* allele with a premature stop codon (e.g., Arg553Stop). A potentially ideal therapeutic approach (other than gene therapy) for patients with a nonsense mutation would be a safe drug that encourages the translational apparatus to misread the stop codon by a transfer RNA (tRNA) that is near-cognate to the stop codon tRNA. If the amino acid thereby inserted into the polypeptide by that tRNA still produces a functional protein, the activity of the protein would be restored. An event of this type, for example, would convert the *CFTR* Arg553Stop mutation to 553Tyr, a substitution that generates a CFTR peptide with nearly normal properties. High-throughput chemical screens for a drug of this type identified ataluren (PTC124), and evidence suggests that it is most effective in allowing read-through of TGA nonsense codons. Moreover, studies in model organisms have firmly demonstrated that it can correct the mutant phenotype of some nonsense mutations. Ataluren has not been established to be clinically effective, but a Phase III clinical trial in CF patients carrying at least one nonsense mutation showed a promising trend toward statistically significant improvement in lung function, and a follow-up trial is underway. Even if ataluren proves ineffective in humans, thousands of other small molecules are being examined in laboratories around the world to identify novel nontoxic compounds that facilitate the skipping of nonsense codons, not only for the treatment of CF but also for Duchenne muscular dystrophy patients carrying nonsense codons, as well as other diseases. Safe, effective drugs of this type will have a major impact on the treatment of inherited disease.

Small Molecules to Correct the Folding of Mutant Membrane Proteins: Pharmacological Chaperones

Some mutations in membrane proteins may disrupt their ability to fold, pass through the endoplasmic reticulum, and be trafficked to the plasma membrane. These mutant proteins are recognized by the cellular protein quality control machinery, trapped in the endoplasmic reticulum, and prematurely degraded by the proteosome. The ΔF508 deletion of the CFTR protein—which constitutes 65% of all CF mutations worldwide—is perhaps the best-known example (see Fig. 12-15) of a mutation that impairs trafficking of a membrane protein. If the folding/trafficking defect could be overcome to increase the abundance of CFTR channels at the apical surface of the cell by 20% to 25%, it is thought that a clinical benefit would be obtained, because once the ΔF508 CFTR protein reaches the cell surface, it is an effective Cl⁻ channel.

Small molecule screens to identify compounds that can serve as a chaperone to prevent misfolding and correct the ΔF508 CFTR trafficking defect in in vitro assay systems have identified lumacaftor (VX-809) as an effective, although incomplete, corrector of this specific CFTR mutant polypeptide (see Fig. 13-7). Lumacaftor interacts directly with the mutant CFTR to stabilize its three-dimensional structure, specifically correcting the underlying trafficking defect and enhancing Cl⁻ transport. Although monotherapy with lumacaftor

had no clinical benefits, a recently completed Phase III clinical trial using lumacaftor together with another small molecule, ivacaftor (VX-770), discussed later, showed significant improvements in lung function in homozygous ΔF508 *CFTR* patients. This finding is notable because it is the first treatment shown to have a favorable impact on the primary biochemical defect in patients carrying the most common *CFTR* allele, ΔF508. Ongoing studies of the long-term effectiveness and safety of the lumacaftor-ivacaftor combination therapy are in progress. Irrespective of their success, this example is a milestone in medical genetics, because it establishes the principle that molecular chaperones can have clinical benefits in the treatment of monogenic disease.

Small Molecules to Increase the Function of Correctly Trafficked Mutant Membrane Proteins.

Amino acid substitutions in membrane proteins may not disrupt the trafficking of the mutant polypeptide to the plasma membrane, but rather interfere with its function at the cell surface. Small molecule screens for new treatments for CF have also led this area of drug discovery. Screens for so-called potentiators—molecules that could enhance the function of mutant CFTR proteins that are correctly positioned at the cell surface—identified ivacaftor (VX-770), which improves the Cl⁻ transport of some mutant CFTR proteins, such as the Gly551Asp *CFTR* missense mutation (see Fig. 12-15) that inactivates anion transport; this allele is carried by 4% to 5% of all CF patients. In one clinical trial, patients carrying at least one Gly551Asp allele experienced a significant improvement in lung function (Fig. 13-8), weight gain, respiratory symptoms, and a decline in sweat Cl⁻. Ivacaftor is presently FDA approved for the treatment of eight other CFTR missense mutations, and more alleles will certainly be added to this group. Although fewer than 200 CF patients in the United States have one of these eight alleles, the allele-specific indications for ivacaftor treatment highlight both the benefits and dilemmas of personalized medicine for genetic disease: effective drugs can be discovered, but they may be effective only in a relatively small numbers of individuals. Moreover, at present ivacaftor is extremely expensive, costing approximately $300,000 per year.

Small Molecules to Enhance the Function of Mutant Enzymes: Vitamin-Responsive Inborn Errors of Metabolism.

The biochemical abnormalities of a number of inherited metabolic diseases may respond, sometimes dramatically, to the administration of large amounts of the vitamin cofactor of the enzyme impaired by the mutation (Table 13-3). In fact, the vitamin-responsive inborn errors are among the most successfully treated of all genetic diseases. The vitamins used are remarkably nontoxic, generally allowing the safe administration of amounts 100 to 500 times greater

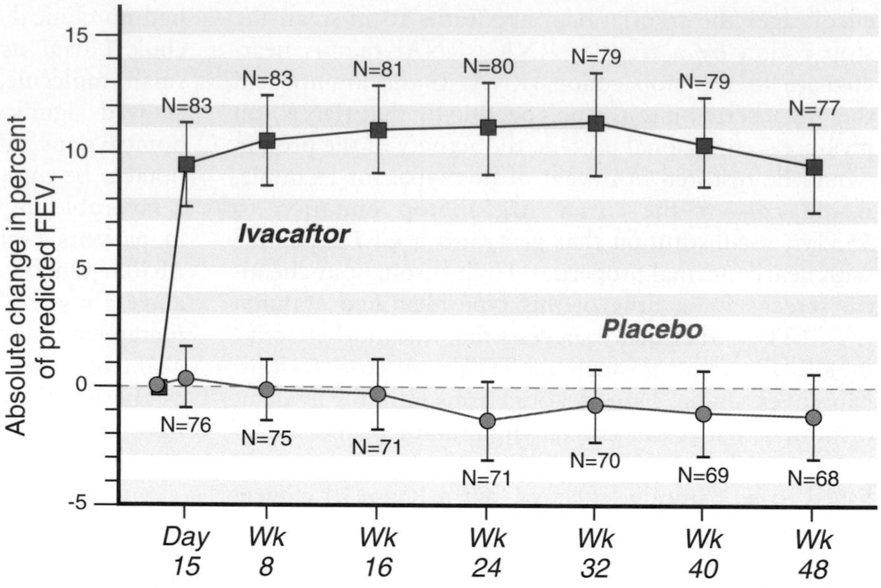

Figure 13-8 The effect of ivacaftor (Kalydeco) on lung function of cystic fibrosis patients carrying at least one Gly551Asp CFTR allele. The figure shows the absolute mean change from baseline in the percent of predicted forced expiratory volume in 1 second (FEV$_1$) through week 48 of a clinical trial. N refers to the number of subjects studied at each time point during the trial. *See Sources & Acknowledgments.*

TABLE 13-3 Treatment of Genetic Disease at the Level of the Mutant Protein

Strategy	Example	Status
Enhancement of Mutant Protein Function		
Small molecules that facilitate translational "skipping" over mutant stop codons	Ataluren in the 10% of cystic fibrosis patients with nonsense mutations in the *CFTR* gene	Investigational in CF: confirmatory Phase III clinical trial was begun in 2014
Small molecule "correctors" that increase the trafficking of the mutant protein through the ER to the plasma membrane	Lumacaftor (VX-809) to increase the abundance of the ΔF508 mutant CFTR protein at the apical membrane of epithelial cells in CF patients	Investigational: very promising improvements in lung function in ΔF508 homozygotes, when used in combination with ivacaftor; expensive
Small molecule "potentiators" that increase the function at the cell membrane of correctly trafficked membrane proteins	Ivacaftor (VX-770) used alone to enhance the function of specific mutant CFTR proteins at the epithelial apical membrane	FDA approved for the treatment of CF patients carrying specific alleles; expensive
Vitamin cofactor administration to increase the residual activity of the mutant enzyme	Vitamin B$_6$ for pyridoxine-responsive homocystinuria	Treatment of choice in the 50% of cystathionine synthase patients who are responsive
Protein Augmentation		
Replacement of an extracellular protein	Factor VIII in hemophilia A	Well-established, effective, safe
Extracellular replacement of an intracellular protein	Polyethylene glycol–modified adenosine deaminase (PEG-ADA) in ADA deficiency	Well-established, safe, and effective, but costly; now used principally to stabilize patients before gene therapy or HLA-matched bone marrow transplantation
Replacement of an intracellular protein—cell targeting	β-glucocerebrosidase in non-neuronal Gaucher disease	Established; biochemically and clinically effective; expensive

ADA, Adenosine deaminase; CF, cystic fibrosis; ER, endoplasmic reticulum; FDA, U.S. Food and Drug Administration; HLA, human leukocyte antigen; PEG, polyethylene glycol.

than those required for normal nutrition. In **homocystinuria** due to cystathionine synthase deficiency (see Fig. 12-8), for example, approximately 50% of patients respond to the administration of high doses of pyridoxine (vitamin B$_6$, the precursor of pyridoxal phosphate, the cofactor for the enzyme), an example—as we saw earlier in the case of BH$_4$ administration in PKU—of cofactor responsiveness in a metabolic disease. In most of these responsive patients, homocystine completely

disappears from the plasma, even though the increase in hepatic cystathionine synthase activity is usually only a fewfold, from 1.5% to 4.5% of control activity. The increased pyridoxal phosphate concentrations may stabilize the mutant enzyme or overcome reduced affinity of the mutant enzyme for the cofactor (Fig. 13-9). In any case, vitamin B$_6$ treatment substantially improves the clinical course of the disease in responsive patients. Nonresponsive patients generally carry null alleles and

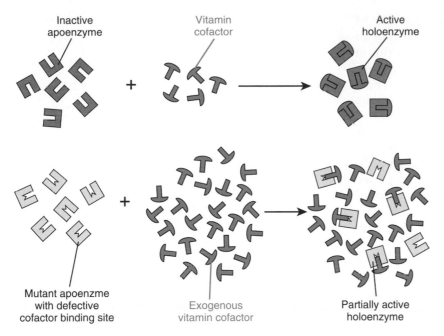

Figure 13-9 The mechanism of response of a mutant apoenzyme to the administration of its cofactor at high doses. Vitamin-responsive enzyme defects are often due to mutations that reduce the normal affinity (*top*) of the enzyme protein (apoenzyme) for the cofactor needed to activate it. In the presence of the high concentrations of the cofactor that result from the administration of up to 500 times the normal daily requirement, the mutant enzyme acquires a small amount of activity sufficient to restore biochemical normalcy. *See Sources & Acknowledgments.*

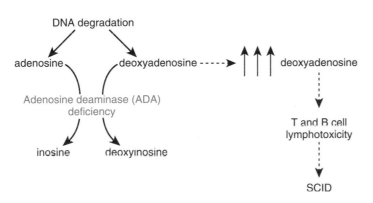

Figure 13-10 Adenosine deaminase (ADA) converts adenosine to inosine and deoxyadenosine to deoxyinosine. In ADA deficiency, deoxyadenosine accumulation in lymphocytes is lymphotoxic, killing the cells by impairing DNA replication and cell division to cause severe combined immunodeficiency (SCID).

therefore have no residual cystathionine synthase activity to augment.

Protein Augmentation

The principal types of protein augmentation are summarized in Table 13-3. Protein augmentation is a routine therapeutic approach in only a few diseases, all involving proteins whose principal site of action is in the plasma or extracellular fluid. The prime example is the prevention or arrest of bleeding episodes in patients with **hemophilia (Case 21)** by the infusion of plasma fractions enriched for the appropriate factor. The decades of experience with this disease illustrate the problems that can be anticipated as new strategies for replacing other, particularly intracellular, polypeptides are attempted. These problems include the difficulty and cost of procuring sufficient amounts of the protein to treat all patients at the optimal frequency, the need to administer the protein at a frequency consistent with its half-life (only 8 to 10 hours for factor VIII), and the formation of neutralizing antibodies in some patients (5% of classic hemophiliacs).

Enzyme Replacement Therapy: Extracellular Administration of an Intracellular Enzyme

Adenosine Deaminase Deficiency. Adenosine deaminase (ADA) is a critical enzyme of purine metabolism that catalyzes the deamination of adenosine to inosine and of deoxyadenosine to deoxyinosine (Fig. 13-10). The pathology of ADA deficiency, an autosomal recessive disease, results entirely from the accumulation of toxic purines, particularly deoxyadenosine, in lymphocytes. A profound failure of both cell-mediated (T-cell) and humoral (B-cell) immunity results, making ADA deficiency one cause of **severe combined immunodeficiency (SCID)**. Untreated patients die of infection within the first 2 years of life. The long-term treatment of ADA deficiency is rapidly evolving, with gene therapy (see later section) now a strong alternative to bone marrow transplantation from a fully human leukocyte antigen (HLA) compatible donor. The administration of a modified form of the bovine ADA enzyme, described in the next section, is no longer a first choice for long-term management, but it is an effective stabilizing measure in the short term until these other treatments can be used.

Modified Adenosine Deaminase. The infusion of bovine ADA modified by the covalent attachment of an inert polymer, polyethylene glycol (PEG), is superior in several ways to the use of the unmodified ADA enzyme. First, PEG-ADA largely protects the patient from a neutralizing antibody response (which would remove the ADA from plasma). Second, the modified enzyme remains in the extracellular fluid where it can degrade toxic purines. Third, the plasma half-life of PEG-ADA is 3 to 6 days, much longer than the half-life of unmodified ADA. Although the near-normalization of purine metabolism obtained with PEG-ADA does not completely correct immune function (most patients remain T lymphopenic), immunoprotection is restored, with dramatic clinical improvement.

The general principles exemplified by the use of PEG-ADA are that (1) proteins can be chemically modified to improve their effectiveness as pharmacological reagents, and (2) an enzyme that is normally located inside the cell can be effective extracellularly if its substrate is in equilibrium with the extracellular fluid and if its product can be taken up by the cells that require it.

Enzyme Replacement Therapy: Targeted Augmentation of an Intracellular Enzyme. Enzyme replacement therapy (ERT) is now established therapy for six lysosomal storage diseases, with clinical trials being conducted for several others. Non-neuronal (type 1) **Gaucher disease** was the first lysosomal storage disease for which ERT was shown to be effective. It is the most prevalent lysosomal storage disorder, affecting up to 1 in 450 Ashkenazi Jews and 1 in 40,000 to 100,000 individuals in other populations (Case 18). This autosomal recessive condition results from deficiency of β-glucocerebrosidase. Loss of this enzyme activity leads to the accumulation of its substrate, the complex lipid glucocerebroside, in the lysosome, where it is normally degraded. The lysosomal accumulation of glucocerebroside, particularly in the macrophages and monocytes of the reticuloendothelial system, leads to gross enlargement of the liver and spleen. Bone marrow is slowly replaced by lipid-laden macrophages (Gaucher cells), leading to anemia and thrombocytopenia. The bone lesions cause episodic pain, osteonecrosis, and substantial morbidity.

More than 5000 patients with non-neuronal Gaucher disease have been treated worldwide with β-glucocerebrosidase ERT, with dramatic clinical benefits. The increase in the hemoglobin level of one patient, a response that is representative of the effectiveness of this treatment, is shown in Figure 13-11. Overall, this therapy also reduces the enlargement of liver and spleen, increases the platelet count, accelerates growth, and improves the characteristic skeletal abnormalities and bone density. Early treatment is most effective in preventing irreversible damage to bones and liver.

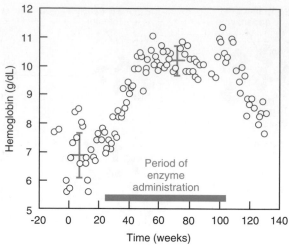

Figure 13-11 The effect of weekly intravenous infusions of modified glucocerebrosidase on the hemoglobin concentration of a child with non-neuronal (type 1) Gaucher disease. A review of the response of more than 1000 patients indicates that this response is representative. Treatment was begun at 4 years of age and continued for 18 months. The therapy was accompanied by an increased platelet count and radiological improvement in the bone abnormalities. The hematological parameters returned to pretreatment levels when the infusions were stopped. *See Sources & Acknowledgments.*

The success of ERT for non-neuronopathic Gaucher disease provides guidance in the development of enzyme and protein replacement therapy for other lysosomal storage disorders, and perhaps other classes of diseases as well, for several reasons. First, this use of ERT highlights the importance of understanding the biology of the relevant cell types. As demonstrated by I-cell disease (see Chapter 12), lysosomal hydrolases such as β-glucocerebrosidase contain post-translationally added mannose sugars that target the enzyme to the macrophage through a mannose receptor on the plasma membrane. Once bound, the enzyme is internalized and delivered to the lysosome. Thus, β-glucocerebrosidase ERT in Gaucher disease targets the protein both to a particular relevant cell and to a specific intracellular address, in this case the macrophage and the lysosome, respectively.

Second, the human enzyme can be produced in abundance from cultured cells expressing the glucocerebrosidase gene, a key factor because this treatment, given as twice-monthly infusions, must be continuous. Only approximately 1% to 5% of the normal intracellular enzyme activity is required to correct the biochemical abnormalities in this and other lysosomal storage disorders. Third, the administered β-glucocerebrosidase is not recognized as a foreign antigen because patients with non-neuronal Gaucher disease have small amounts of residual enzyme activity. Unfortunately, however, because β-glucocerebrosidase does not cross the blood-brain barrier, ERT cannot treat the neuronopathic forms of Gaucher disease. Although ERT for any lysosomal disease is very expensive, its success has been a

TABLE 13-4 Treatment by Modification of the Genome or its Expression

Type of Modification	Example	Status
Pharmacological modulation of gene expression	Decitabine therapy to stimulate γ-globin (and thus Hb F) synthesis in sickle cell disease	Effective in increasing Hb F levels; concerns about cytotoxicity drive the search for safer but effective cytidine analogues.
RNA interference (RNAi) to reduce the abundance of a toxic or dominant negative protein	RNAi for transthyretin amyloidosis	Successful Phase I clinical trial completed
Induction of exon skipping	Use of antisense oligonucleotides to induce skipping of exon 51 in Duchenne muscular dystrophy	Investigational; clinical trials offer cautious optimism.
Gene editing	CRISPR/Cas9 inactivation of the *CCR5* gene in CD4 T cells of HIV-infected individuals	Investigational; Phase I trial successful
Partial modification of the somatic genotype	Bone marrow transplantation in β-thalassemia	Curative with HLA-matched donor; good results overall
By transplantation	Bone marrow transplantation in storage diseases (e.g., Hurler syndrome)	Excellent results in some diseases, even if the brain is affected, such as Hurler syndrome
	Cord blood stem cell transplantation for presymptomatic Krabbe disease; Hurler syndrome	Excellent results for these two disorders.
	Liver transplantation in α₁-antitrypsin deficiency	Up to 80% survival over 5 yr for genetic liver disease
By gene transfer into somatic tissues (see Table 13-5)	See Table 13-5.	See Table 13-5.

cas, CRISPR-associated; CRISPR, clustered regularly interspaced short palindromic repeats; Hb F, fetal hemoglobin; HLA, human leukocyte antigen.

tremendous advance in the treatment of monogenic disorders. It has established the feasibility of directing an intracellular enzyme to its physiologically relevant location to produce clinically significant effects.

Modulation of Gene Expression

Decades ago, the idea that one might treat a genetic disease through the use of drugs that modulate gene expression would have seemed fanciful. Increasing knowledge of the normal and pathological bases of gene expression, however, has made this approach feasible. Indeed, it seems likely that this strategy will become only more widely used as our understanding of gene expression, and how it might be manipulated, increases.

Increasing Gene Expression from the Wild-Type or Mutant Locus

Therapeutic effects can be obtained by increasing the amount of messenger RNA (mRNA) transcribed from the wild-type locus associated with a dominant disease or from the mutant locus, if the mutant protein retains some function (Table 13-4; see Fig. 13-7). An effective therapy of this type is used to manage **hereditary angioedema,** a rare but potentially fatal autosomal dominant condition due to mutations in the gene encoding the complement 1 (C1) esterase inhibitor. Affected individuals are subject to unpredictable episodes, of widely varying severity, of submucosal and subcutaneous edema. Attacks that involve the upper respiratory tract can be fatal. Because of the rapid and unpredictable

nature of the attacks, long-term prophylaxis with attenuated androgens, particularly danazol, is often employed. Danazol significantly increases the abundance of the C1 esterase inhibitor mRNA by modulating transcription of the gene, presumably from both the normal and mutant loci. In the great majority of patients, the frequency of serious attacks is dramatically reduced, although long-term androgen administration is not free of side effects.

Increasing Gene Expression from a Locus Not Affected by the Disease

A related therapeutic strategy is to increase the expression of a normal gene that compensates for the effect of mutation at another locus. This approach is extremely promising in the management of sickle cell disease (Case 42) and β-thalassemia (Case 44), for which drugs that induce **DNA hypomethylation** are being used to increase the abundance of fetal hemoglobin (Hb F) (see Chapter 11), which normally constitutes less than 1% of total hemoglobin in adults. Sickle cell disease causes illness because of both the anemia and the sickling of red blood cells (see Chapter 11); the increase in the level of Hb F ($\alpha_2\gamma_2$) benefits these patients because Hb F is a perfectly adequate oxygen carrier in postnatal life and because the polymerization of deoxyhemoglobin S is inhibited by Hb F. In β-thalassemia, Hb F restores the imbalance between α and non–α-globin chains (see Chapter 11), substituting Hb F ($\alpha_2\gamma_2$) for Hb A ($\alpha_2\beta_2$).

The normal postnatal decrease in the expression of the γ-globin gene is at least partly due to methylation of

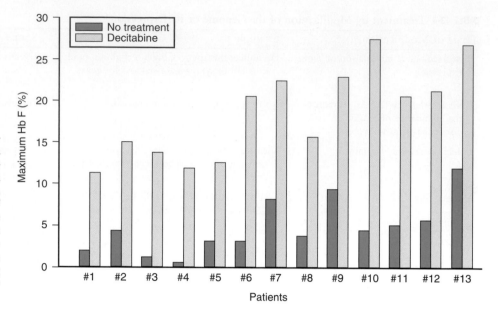

Figure 13-12 The effect of the cytosine analogue decitabine, a DNA hypomethylating agent, on the percentage of fetal hemoglobin (Hb F) in 13 patients with sickle cell disease, compared with their level of Hb F without any treatment. Note the wide variation between patients in the levels of Hb F without treatment. Every patient shown had a significant increase in Hb F during decitabine therapy. *See Sources & Acknowledgments.*

CpG residues (see Chapter 3) in the promoter region of the gene. Methylation of the promoter is inhibited if a cytidine analogue such as decitabine (5-aza-2′-deoxycytidine) is incorporated into DNA instead of cytidine. The inhibition of methylation is associated with substantial increases in γ-globin gene expression and, accordingly, in the proportion of Hb F in blood. Both patients with sickle cell anemia and patients with some forms of β-thalassemia treated with decitabine uniformly display increases in Hb F to levels that are likely to have a significant positive impact on morbidity and mortality (Fig. 13-12). The use of inhibitors of γ-globin gene methylation is evolving rapidly, and more effective inhibitors of methylation, with fewer side effects, are likely to be developed.

As described earlier, any approach that allows a patient with β-thalassemia or sickle cell anemia to retain Hb F expression is likely to be very beneficial to the patient. The BCL11A protein, described in Chapter 11, is a *trans*-acting effector of hemoglobin switching that turns off γ-globin production postnatally but nevertheless allows β-globin gene expression. Genome editing (see later) in hematopoietic stem cells (HSCs) is currently being explored as a method to delete an erythroid enhancer of the *BCL11A* gene, thereby blocking its expression in the erythroid cell lineage. As a result, hemoglobin switching from Hb F to Hb A would not occur, and patients would retain Hb F instead of a hemoglobin containing a mutant β-thalassemia or sickle cell allele.

Reducing the Expression of a Dominant Mutant Gene Product: Small Interfering RNAs

The pathology of some inherited diseases results from the presence of a mutant protein that is toxic to the cell, as seen with proteins with expanded polyglutamine tracts

(see Chapter 12), as in **Huntington disease (Case 24)**, or with disorders such as the inherited amyloidoses. The autosomal dominant disorder **transthyretin amyloidosis** is the result of any of more than 100 missense mutations in transthyretin, a protein produced mainly in liver, that transports retinol (one form of vitamin A) and thyroxine in body fluids. The major phenotypes are amyloidotic polyneuropathy, due to deposition of the amyloid in peripheral nerves (causing intractable peripheral sensory neuropathy and autonomic neuropathy), and amyloidotic cardiomyopathy, due to its deposition in the heart. Both disorders greatly shorten the life span, and the only current treatment is hepatic transplantation.

A promising therapy, however, is provided by a technology called **RNA interference (RNAi)**, which can mediate the degradation of a specific target RNA, such as that encoding transthyretin. Briefly, short RNAs that correspond to specific sequences of the targeted RNA (see Fig. 13-7)—termed **small interfering RNAs (siRNAs)**—are introduced into cells by, for example, lipid nanoparticles or viral vectors. Strands of the interfering RNA, approximately 21 nucleotides long, bind to the target RNA and initiate its cleavage. A Phase I clinical trial using an siRNA (encapsulated in injected lipid nanoparticles) directed against transthyretin, led to a 56% to 67% reduction in transthyretin levels by the 28th day of study, with no significant toxicity. This trial established proof of concept for RNAi treatment of an inherited disease, an approach that will undoubtedly be applied to other diseases where elimination of the mutant gene product is the goal.

Induction of Exon Skipping

Exon skipping refers to the use of molecular interventions to exclude an exon from a pre-mRNA that encodes

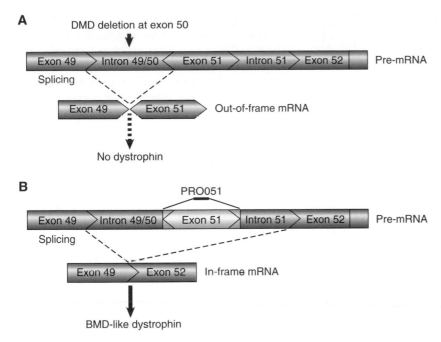

A

DMD deletion at exon 50

Exon 49 > Intron 49/50 < Exon 51 > Intron 51 > Exon 52 | Pre-mRNA

Splicing

Exon 49 > Exon 51 | Out-of-frame mRNA

No dystrophin

B

PRO051

Exon 49 > Intron 49/50 < Exon 51 > Intron 51 > Exon 52 | Pre-mRNA

Splicing

Exon 49 > Exon 52 | In-frame mRNA

BMD-like dystrophin

Figure 13-13 Schematic representation of exon skipping. In a patient with Duchenne muscular dystrophy (DMD) who has a deletion of exon 50, an out-of-frame transcript is generated in which exon 49 is spliced to exon 51 (**A**). As a result, a stop codon is generated in exon 51, which prematurely aborts dystrophin synthesis. The sequence-specific binding of the exon-internal antisense oligonucleotide PRO051 interferes with the correct inclusion of exon 51 during splicing, so that the exon is actually skipped (**B**). This restores the open reading frame of the transcript and allows the synthesis of a dystrophin similar to that in patients with Becker muscular dystrophy (BMD). mRNA, Messenger RNA. *See Sources & Acknowledgments.*

a reading frame–disrupting mutation, thereby rescuing expression of the mutant gene. If the number of nucleotides in the excluded exon is a multiple of three, no frame shift will occur and, if the resulting polypeptide with the deleted amino acids retains sufficient function, a therapeutic benefit will result. The most widely studied method of inducing exon skipping is through the use of **antisense oligonucleotides (ASOs)**, which are synthetic 15- to 35-nucleotide single-stranded molecules that can hybridize to specific corresponding sequences in a pre-mRNA (see Fig. 13-7). The clearest example of the potential of this strategy is provided by Duchenne muscular dystrophy (DMD) (see Chapter 12) (Case 14).

The goal of exon skipping in DMD is to convert a DMD mutation into an in-frame counterpart that generates a functional dystrophin, just as the deletions that allow the production of a partially functioning dystrophin are associated with the milder phenotype of Becker muscular dystrophy (see Fig. 12-18). The distribution of DMD mutations is nonrandomly distributed in the gene (see Chapter 12), and thus, remarkably, the skipping of just exon 51 alone would restore the dystrophin reading frame of an estimated 13% of all DMD patients (Fig. 13-13). This exon has therefore been the major focus of exon-skipping drug development. Several clinical trials have established that ASOs that cause skipping of exon 51 can produce significant increases in the number of dystrophin-positive muscle fibers of DMD patients. Moreover, one trial demonstrated stabilization of patient walking ability, but the treatment group was small and must be studied in a larger number of subjects. Irrespective of the specific challenges posed by DMD, it will be surprising if exon-skipping strategies

do not ultimately play a significant role in the therapy of some inherited disorders.

Gene Editing

Over the last decade, molecular biologists have developed methods to introduce site-specific genomic sequence changes into the DNA of intact organisms, including primates. The correction of a mutant gene sequence in its natural DNA context, in a sufficient number of target cells, would be an ideal treatment. This new technology, termed **genome editing,** uses engineered endonucleases containing a DNA-binding domain that will recognize a specific sequence in the genome, such as the sequence in which a missense mutation is embedded. Subsequently, a nuclease domain creates a double-stranded break, and cellular mechanisms for homology-directed repair (HDR) then repair the break (see Chapter 4), introducing the wild-type nucleotide to replace the mutant one. The template for the HDR must be based on a matching homologous wild-type DNA template that is introduced into the target cells before editing. The most widely used editing approach at present is the *c*lustered *r*egularly *i*nterspaced *s*hort *p*alindromic *r*epeats (CRISPR)/CRISPR-associated (Cas) 9 system, commonly referred to as CRISPR/Cas9.

In humans, genome editing offers possibilities for the correction of genetic defects in their natural genomic landscape, without the risks associated with the semi-random vector integration of some viral vectors used in gene therapy (see later section). The first clinical use of this technology was a Phase I (safety) clinical trial reported in 2014. This study took advantage of the knowledge that a naturally occurring deletion in *CCR5*, the gene that encodes the cell membrane coreceptor for human immunodeficiency virus (HIV), renders homozygous carriers resistant to HIV infection but does not impair CD4 T-cell function (see Chapter 9). When CD4 T cells taken from HIV-infected patients were treated with an adenoviral vector expressing a nuclease designed to generate a null allele of the *CCR5* gene, and then reinfused into the patient, the CCR5 gene was "knocked out" in 11% to 28% of the CD4 T cells in these patients; the modified cells had a half-life of almost 1 year, and HIV RNA became undetectable in one of four patients who could be evaluated. This study demonstrates the great clinical potential of gene editing.

A major concern whose real dimensions are presently unknown is that the endonucleases can have off-target effects, which could cause mutations elsewhere in the genome. Nevertheless, considerable optimism is justified in thinking that this technology can be extended to the correction of mutations in the cells of individuals with genetic diseases in the future, including, for example, bone marrow stem cells for the treatment of inherited blood and immune system disorders (see later discussion).

Modification of the Somatic Genome by Transplantation

Transplanted cells retain the genotype of the donor, and consequently transplantation can be regarded as a form of gene transfer therapy because it leads to a modification of the somatic genome. There are two general indications for the use of transplantation in the treatment of genetic disease. First, cells or organs may be transplanted to introduce wild-type copies of a gene into a patient with mutations in that gene. This is the case, for example, in homozygous familial hypercholesterolemia (see Chapter 12), for which liver transplantation is an effective but high-risk procedure. The second and more common indication is for cell replacement, to compensate for an organ damaged by genetic disease (for example, a liver that has become cirrhotic in α_1-antitrypsin (deficiency). Some examples of the uses of transplantation in genetic disease are provided in Table 13-4.

Stem Cell Transplantation

Stem cells are defined by two properties: (1) their ability to proliferate to form the differentiated cell types of a tissue in vivo; and (2) their ability to self-renew—that is, to form another stem cell. Embryonic stem cells, which can give rise to the whole organism, are discussed in Chapter 14.

Only three types of stem cells are in clinical use at present: **hematopoietic stem cells (HSCs)**, which can reconstitute the blood system after bone marrow transplantation; **corneal stem cells**, which are used to regenerate the corneal epithelium, and **skin stem cells**. These cells are derived from immunologically compatible donors. The possibility that other types of stem cells will be used clinically in the future is enormous because stem cell research is one of the most active and promising areas of biomedical investigation. Although it is easy to overstate the potential of such treatment, optimism about the long-term future of stem cell therapy is justified.

Hematopoietic Stem Cell Transplantation in Nonstorage Diseases. In addition to its extensive application in the management of cancer, HSC transplantation using bone marrow stem cells is the treatment of choice for a selected group of monogenic immune deficiency disorders, including SCID of any type. Its role in the management of genetic disease in general, however, is less certain and under careful evaluation. For example, excellent outcomes have been obtained with allogenic HSC transplantation in the treatment of children with β-thalassemia and sickle cell disease. Nevertheless, for each disease that bone marrow transplantation might benefit, its outcomes must be evaluated for many years and weighed against the results obtained with other therapies.

Hematopoietic Stem Cell Transplantation for Lysosomal Storage Diseases

Transplantation of Hematopoietic Stem Cells from Bone Marrow. Bone marrow stem cell transplants are effective in correcting lysosomal storage in many tissues including, in some diseases, the brain, through the two mechanisms depicted in Figure 13-14. First, the transplanted cells are a source of lysosomal enzymes that can be transferred to other cells through the extracellular fluid, as discussed in Chapter 12 for I-cell disease. Because bone marrow–derived cells constitute approximately 10% of the total cell mass of the body, the quantitative impact of enzymes transferred from them may be significant. Second, the mononuclear phagocyte system in tissues is derived from bone marrow stem cells so that, after bone marrow transplantation, this system is of donor origin throughout the body. Of special note are the brain perivascular microglial cells, whose bone marrow origin may partially account for the correction of nervous system abnormalities by bone marrow transplantation in some storage disorders, as we will see next in the case of **Hurler syndrome**, a lysosomal storage disease due to α-l-Iduronidase **deficiency**.

Bone marrow transplantation corrects or reduces the visceral abnormalities of many storage diseases. For example, a normalization or reduction in the size of the enlarged liver, spleen, and heart seen in Hurler syndrome can be achieved, and improvements in upper airway obstruction, joint mobility, and corneal clouding are also obtained. Most rewarding, however, has been the impact of transplantation on the neurological component of this disease. Patients who have good

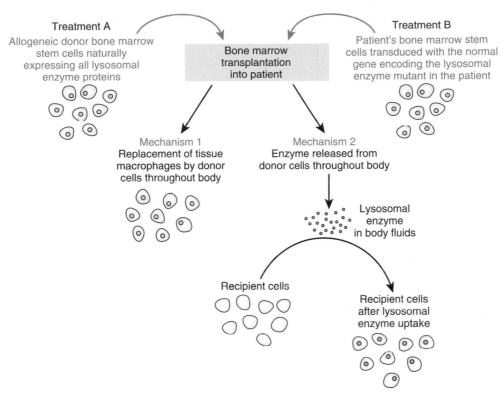

Figure 13-14 The two major mechanisms by which bone marrow transplantation or gene transfer into bone marrow may reduce the substrate accumulation in lysosomal storage diseases. In the case of either treatment, bone marrow transplantation from an allogeneic donor (A) or genetic correction of the patient's own bone marrow stem cells by gene transfer (B), the bone marrow stem cell progeny, now expressing the relevant lysosomal enzyme, expand to repopulate the monocyte-macrophage system of the patient (mechanism 1). In addition, lysosomal enzymes are released from the bone marrow cells derived from the donor or from the genetically modified marrow cells of the patient and taken up by enzyme-deficient cells from the extracellular fluid (mechanism 2).

developmental indices before transplantation, and who receive transplants before 24 months of age, continue to develop cognitively after transplantation, in contrast to the inexorable loss of intellectual function that otherwise occurs. Interestingly, a gene dosage effect is manifested in the donor marrow; children who receive cells from *homozygous* normal donors appear to be more likely to retain fully normal intelligence than do the recipients of *heterozygous* donor cells.

Transplantation of Hematopoietic Stem Cells from Placental Cord Blood. The discovery that placental cord blood is a rich source of HSCs is beginning to make a substantial impact on the treatment of genetic disease. The use of placental cord blood has three great advantages over bone marrow as a source of transplantable HSCs. First, recipients are more tolerant of histoincompatible placental blood than of other allogeneic donor cells. Thus engraftment occurs even if as many as three HLA antigens, cell surface markers encoded by the major histocompatibility complex (see Chapter 8), are mismatched between the donor and the recipient. Second, the wide availability of placental cord blood, together with the increased tolerance of

histoincompatible donor cells, greatly expands the number of potential donors for any recipient. This feature is of particular significance to patients from minority ethnic groups, for whom the pool of potential donors is relatively small. Third, the risk for graft-versus-host disease is substantially reduced with use of placental cord blood cells. Cord blood transplantation from unrelated donors appears to be as effective as bone marrow transplantation from a matched donor for the treatment of Hurler syndrome (Fig. 13-15).

Liver Transplantation

For some metabolic liver diseases, liver transplantation is the only treatment of known benefit. For example, the chronic liver disease associated with CF or α1AT deficiency can be treated only by liver transplantation, and together these two disorders account for a large fraction of all the liver transplants performed in the pediatric population. Liver transplantation has now been undertaken for more than two dozen genetic diseases. At present, the 5-year survival rate of all children who receive liver transplants is in the range of 70% to 85%. For almost all of these patients, the quality of life

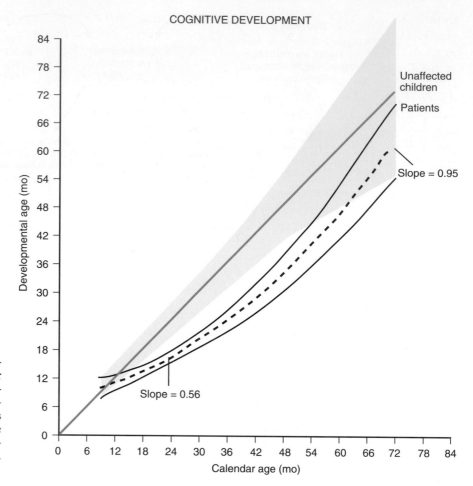

COGNITIVE DEVELOPMENT

Figure 13-15 Preservation of neurocognitive development in children with Hurler syndrome treated by cord blood transplantation. The figure displays the mean cognitive growth curve for transplanted patients compared with unaffected children. The *thin black lines* represent the 95% confidence interval for transplanted patients. *See Sources & Acknowledgments.*

is generally much improved, the specific metabolic abnormality necessitating the transplant is corrected, and in those conditions in which hepatic damage has occurred (such as α1AT deficiency), the provision of healthy hepatic tissue restores growth and normal pubertal development.

The Problems and the Future of Transplantation

Two major problems limit the wider use of transplantation for the treatment of genetic disease. First, the mortality after transplantation is still significant, and the morbidity from superimposed infection due to the requirement for immunosuppression and graft-versus-host disease is substantial. Nevertheless, the ultimate goal of transplantation research—transplantation without immunosuppression—comes incrementally closer. The increased tolerance of the recipient to cord blood transplants, compared with bone marrow–derived donor cells, exemplifies the advances in this area.

The second problem with transplantation is the finite supply of organs, cord blood being a singular exception. For example, for all indications, including genetic disease, more than 6000 liver transplants are performed annually in the United States alone, but more than double that number are added to the waiting list each year. In addition, it remains to be demonstrated that transplanted organs are generally capable of functioning normally for a lifetime.

One solution to these difficulties involves the combination of stem cell *and* either genome editing or gene therapy. Here, a patient's own stem cells would be cultured in vitro and either transfected by gene therapy with the gene of interest or corrected by CRISPR/Cas9 editing and returned to the patient to repopulate the affected tissue with genetically restored cells. The identification of stem cells in a variety of adult human tissues and recent advances in gene transfer therapy offer great hope for this strategy.

Induced Pluripotent Stem Cells. The recently developed ability to induce the formation of pluripotent stem cells (**iPSCs**) from somatic cells has the potential to provide the optimal solution to both of the challenges of transplantation posed earlier. In this approach somatic cells, such as skin fibroblasts, would be taken from a patient in need of a transplant, and induced to form differentiated cells of the organ of interest. For example, the loss-of-function mutation in the α1-antitrypsin gene in the fibroblasts cultured from a patient with α1AT deficiency (see Chapter 12) could be corrected, either by gene editing (see earlier section) or gene therapy (see later section); the corrected cells could then be induced to form liver-specific iPSCs, which could then be transplanted into the liver of the patient to differentiate into hepatocytes. Alternatively, mature hepatocytes derived in vitro from the genetically corrected iPSCs could be transplanted. The great merit of this approach is that the genetically corrected liver cells are derived from the patient's own genome, thus evading immunological rejection of the transplanted cells as well as graft-versus-host disease. Experimental work in animal models has established that this strategy is capable of correcting inherited disorders. Substantial hurdles with iPSCs must first be overcome, however, including establishing the safety of transplanting cells derived by iPSC methodology and preventing epigenetic modifications in the derived cell type that are not characteristic of wild-type cells of the tissue of interest.

GENE THERAPY

Gene therapy is the introduction of a biologically active gene into a cell to achieve a therapeutic benefit. In 2012, the first gene therapy product was licensed in the United States and Europe for the treatment of lipoprotein lipase deficiency, and gene therapy has now been shown to be effective or extremely promising in clinical trials for almost a dozen inherited diseases, some of which are outlined in Table 13-5. These recent successes firmly establish that the treatment of genetic disease at its most

TABLE 13-5 Examples of Inherited Diseases Treated by Gene Therapy of Somatic Tissues

Disease	Affected Protein (Gene)	Vector, Cell Transduced	Outcome
X-linked SCID	γc-cytokine receptor subunit of several interleukin receptors (*IL2RG*)	Retroviral vector Allogenic hematopoietic stem cells	Significant clinical improvement in 27 of 32 patients, 5 of whom developed a leukemia-like disorder that was treatable in 4
SCID due to ADA deficiency	Adenosine deaminase (*ADA*)	Retroviral vector Allogenic hematopoietic stem cells	29 of 40 treated patients are off PEG-ADA enzyme replacement therapy
X-linked adrenoleukodystrophy	A peroxisomal adenosine triphosphate–binding cassette transporter (*ABCD1*)	Lentiviral vector Autologous hematopoietic stem cells	Apparent arrest of cerebral demyelination in the two boys studied
Lipoprotein lipase deficiency	Lipoprotein lipase (*LPL*)	Adeno-associated virus vector injected intramuscularly	Decreased frequency of pancreatitis in affected individuals
Metachromatic leukodystrophy	Arylsulfatase A (*ARSA*)	Lentiviral vector expressing supraphysiological levels of ARSA Autologous hematopoietic stem cells	Apparent arrest of neurodegeneration in three patients, with no genotoxic effects. Long-term follow-up is required to know the true safety and efficacy of the treatment.
Wiskott-Aldrich syndrome	WAS protein, a regulator of actin polymerization in hematopoietic cells (*WAS*)	Lentiviral vector Autologous hematopoietic stem cells	Marked immunological, hematological, and clinical improvement in the first three patients treated.
Hemophilia B	Factor IX (*F9*)	Adeno-associated virus vector Patients received a single IV injection	Stable expression of factor IX at 1%-7% of normal levels up to 3 years post-treatment; 4 of 6 patients able to stop prophylactic factor IX treatment.
β-Thalassemia	β-Globin (*HBA1*)	Lentiviral vector Autologous hematopoietic stem cells	A single patient, with compound β^E/β^0-thalassemia. Stable Hb levels of 9-10 g/dL, but only a third of the total Hb originated from the vector (see text).
Leber congenital amaurosis (one form)	RPE65, a protein required for the cycling of retinoids (vitamin A metabolites) to photoreceptors (*RPE65*)	Adeno-associated virus vector Retinal pigment epithelial cells	Initially improved vision in many patients in the first trials, but the evidence now suggests, unexpectedly, that the photoreceptor (PR) degeneration continues nevertheless. The cause of this PR death is unknown.

ADA, Adenosine deaminase; Hb, hemoglobin; IV, intravenous; PEG, polyethylene glycol; SCID, severe combined immunodeficiency; WAS, Wiskott-Aldrich syndrome.

fundamental level—the gene—will be increasingly feasible. The goal of gene therapy is to transfer the therapeutic gene early enough in the life of the patient to prevent the pathogenetic events that damage cells. Moreover, correction of the *reversible* features of genetic diseases should also be possible for many conditions.

In this section, we outline the potential, methods, and probable limitations of gene transfer for the treatment of human genetic disease. The minimal requirements that must be met before the use of gene transfer can be considered for the treatment of a genetic disorder are presented in the Box.

General Considerations for Gene Therapy

In the treatment of inherited disease, the most common use of gene therapy will be the introduction of functional copies of the relevant gene into the appropriate target cells of a patient with a loss-of-function mutation (because most genetic diseases result from such mutations).

In these instances, precisely *where* the transferred gene inserts into the genome of a cell would, in principle, generally not be important (see later discussion). If gene editing (see earlier discussion and Table 13-4) to treat inherited disease becomes possible, then correction of the defect in the mutant gene in its normal genomic context would be ideal and would alleviate concerns such as the activation of a nearby oncogene by the regulatory activity of a viral vector, or the inactivation of a tumor suppressor due to insertional mutagenesis by the vector. In some long-lived types of cells, stable, long-term expression may not require integration of the introduced gene into the host genome. For example, if the transferred gene is stabilized in the form of an **episome** (a stable nuclear but nonchromosomal DNA molecule, such as that formed by an adeno-associated viral vector, discussed later), and if the target cell is long-lived (e.g., T cells, neurons, myocytes, hepatocytes), then long-term expression can occur without integration.

Gene therapy may also be undertaken to inactivate the product of a dominant mutant allele whose abnormal product causes the disease. For example, vectors carrying siRNAs (see earlier section) could, in principle, be used to mediate the selective degradation of a mutant mRNA encoding a dominant negative proα1(I) collagen that causes osteogenesis imperfecta (see Chapter 12).

Gene Transfer Strategies

An appropriately engineered gene may be transferred into target cells by one of two general strategies (Fig. 13-16). The first involves introduction of the gene into cells that have been cultured from the patient ex vivo (that is, outside the body) and then reintroduction of the cells to the patient after the gene transfer. In the second approach, the gene is injected directly in vivo

ESSENTIAL REQUIREMENTS OF GENE THERAPY FOR AN INHERITED DISORDER

- **Identity of the molecular defect**
  The identity of the affected gene must be known.
- **A functional copy of the gene**
  A complementary DNA (cDNA) clone of the gene or the gene itself must be available. If the gene or cDNA is too large for the current generation of vectors, a functional version of the gene from which nonessential components have been removed to reduce its size may suffice.
- **An appropriate vector**
  The most commonly used vectors at present are derived from the adeno-associated viruses (AAVs) or retroviruses, including lentivirus.
- **Knowledge of the pathophysiological mechanism**
  Knowledge of the pathophysiological mechanism of the disease must be sufficient to suggest that the gene transfer will ameliorate or correct the pathological process and prevent, slow, or reverse critical phenotypic abnormalities. Loss-of-function mutations require replacement with a functional gene; for diseases due to dominant negative alleles, inactivation of the mutant gene or its products will be necessary.
- **Favorable risk-to-benefit ratio**
  A substantial disease burden and a favorable risk-to-benefit ratio, in comparison with alternative therapies, must be present.
- **Appropriate regulatory components for the transferred gene**
  Tight regulation of the level of gene expression is relatively unimportant in some diseases and critical in others. In thalassemia, for example, overexpression of the transferred gene would cause a new imbalance of globin chains in red blood cells, whereas low levels of expression would be ineffective. In some enzymopathies, a few percent of normal expression may be therapeutic, and abnormally high levels of expression may have no adverse effect.
- **An appropriate target cell**
  Ideally, the target cell must have a long half-life or good replicative potential in vivo. It must also be accessible for direct introduction of the gene or, alternatively, it must be possible to deliver sufficient copies of the gene to it (e.g., through the bloodstream) to attain a therapeutic benefit. The feasibility of gene therapy is often enhanced if the target cell can be cultured in vitro to facilitate gene transfer into it; in this case, it must be possible to introduce a sufficient number of the recipient cells into the patient and have them functionally integrate into the relevant organ.
- **Strong evidence of efficacy and safety**
  Cultured cell and animal studies must indicate that the vector and gene construct are both effective and safe. The ideal precedent is to show that the gene therapy is effective, benign, and enduring in a large animal genetic model of the disease in question. At present, however, large animal models exist for only a few monogenic diseases. Genetically engineered or spontaneous mutant mouse models are much more widely available.
- **Regulatory approval**
  Protocol review and approval by an institutional review board are essential. In most countries, human gene therapy trials are also subject to oversight by a governmental agency.

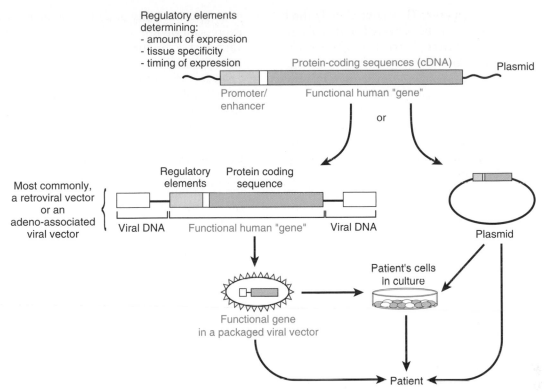

Figure 13-16 **The two major strategies used to transfer a gene to a patient.** For patients with a genetic disease, the most common approach is to construct a viral vector containing the human complementary DNA (cDNA) of interest and to introduce it directly into the patient or into cells cultured from the patient that are then returned to the patient. The viral components at the ends of the molecule are required for the integration of the vector into the host genome. In some instances, the gene of interest is placed in a plasmid, which is then used for the gene transfer.

into the tissue or extracellular fluid of interest (from which it is taken up by the target cells). In some cases, it may be desirable to target the vector to a specific cell type; this is usually achieved by modifying the coat of a viral vector so that only the designated cells bind the viral particles.

The Target Cell

The ideal target cells are stem cells (which are self-replicating) or progenitor cells taken from the patient (thereby eliminating the risk for graft-versus-host disease); both cell types have substantial replication potential. Introduction of the gene into stem cells can result in the expression of the transferred gene in a large population of daughter cells. At present, bone marrow is the only tissue whose stem cells have been successfully targeted as recipients of transferred genes. Genetically modified bone marrow stem cells have been used to cure two forms of SCID, as discussed later. Gene transfer therapy into blood stem cells is also likely to be effective for the treatment of hemoglobinopathies and storage diseases for which bone marrow transplantation has been effective, as discussed earlier.

An important logistical consideration is the number of cells into which the gene must be introduced in

order to have a significant therapeutic effect. To treat PKU, for example, the approximate number of liver cells into which the phenylalanine hydroxylase gene would have to be transferred is approximately 5% of the hepatocyte mass, or approximately 10^{10} cells, although this number could be much less if the level of expression of the transferred gene is higher than wild type. A much greater challenge is gene therapy for muscular dystrophies, for which the gene must be inserted into a significant fraction of the huge number of myocytes in the body in order to have therapeutic efficacy.

DNA Transfer into Cells: Viral Vectors

The ideal vector for gene therapy would be safe, readily made, and easily introduced into the appropriate target tissue, and it would express the gene of interest for life. Indeed, no single vector is likely to be satisfactory in all respects for all types of gene therapy, and a repertoire of vectors will probably be required. Here, we briefly review three of the most widely used classes of viral vectors, those derived from **retroviruses, adeno-associated viruses (AAVs),** and **adenoviruses.**

One of the most widely used classes of vectors is derived from retroviruses, simple RNA viruses that can integrate into the host genome. They contain only three structural genes, which can be removed and replaced with the gene to be transferred (see Fig. 13-16). The current generation of retroviral vectors has been engineered to render them incapable of replication. In addition, they are nontoxic to the cell, and only a low number of copies of the viral DNA (with the transferred gene) integrate into the host genome. Moreover, the integrated DNA is stable and can accommodate up to 8 kb of added DNA, commodious enough for many genes that might be transferred. A major limitation of many retroviral vectors, however, is that the target cell must undergo division for integration of the virus into the host DNA, limiting the use of such vectors in nondividing cells such as neurons. In contrast, **lentiviruses,** the class of retroviruses that includes HIV, are capable of DNA integration in nondividing cells, including neurons. Lentiviruses have the additional advantage of not showing preferential integration into any specific gene locus, thus reducing the chances of activating an oncogene in a large number of cells.

AAVs do not elicit strong immunological responses, a great advantage that enhances the longevity of their expression. Moreover, they infect dividing or nondividing cells to remain in a predominantly episomal form that is stable and confers long-term expression of the transduced gene. A disadvantage is that the current AAV vectors can accommodate inserts of up to only 5 kb, which is smaller than many genes in their natural context.

The third group of viral vectors, adenovirus-derived vectors, can be obtained at high titer, will infect a wide variety of dividing or nondividing cell types, and can accommodate inserts of 30 to 35 kb. However, in addition to other limitations, they have been associated with at least one death in a gene therapy trial through the elicitation of a strong immune response. At present their use is restricted to gene therapy for cancer.

Risks of Gene Therapy

Gene therapy for the treatment of human disease has risks of three general types:

- **Adverse response to the vector or vector-disease combination.** Principal among the concerns is that the patient will have an adverse reaction to the vector or the transferred gene. Such problems should be largely anticipated with appropriate animal and preliminary human studies.
- **Insertional mutagenesis causing malignancy.** The second concern is insertional mutagenesis, that is, that the transferred gene will integrate into the patient's DNA and activate a proto-oncogene or disrupt a tumor suppressor gene, leading possibly to cancer (see Chapter 15). The illicit expression of an oncogene is less likely to occur with the current generation of viral vectors, which have been altered to minimize the ability of their promoters to activate the expression of adjacent host genes. Insertional inactivation of a tumor suppressor gene is likely to be infrequent and, as such, is an acceptable risk in diseases for which there is no therapeutic alternative.
- **Insertional inactivation of an essential gene.** A third risk—that insertional inactivation could disrupt a gene essential for viability—will, in general, be without significant effect because such lethal mutations are expected to be rare and will kill only single cells. Although vectors appear to somewhat favor insertion into transcribed genes, the chance that the same gene will be disrupted in more than a few cells is extremely low. The one exception to this statement applies to the germline; an insertion into a gene in the germline could create a dominant disease-causing mutation that might manifest in the treated patient's offspring. Such events, however, are likely to be rare and the risk acceptable because it would be difficult to justify withholding, on this basis, carefully planned and reviewed trials of gene therapy from patients who have no other recourse. Moreover, the problem of germline modification by disease treatment is not confined to gene therapy. For example, most chemotherapy used in the treatment of malignant disease is mutagenic, but this risk is accepted because of the therapeutic benefits.

Diseases That Have Been Amenable to Gene Therapy

Although nearly a dozen single-gene diseases have been shown to improve with gene therapy, a large number of other monogenic disorders are potential candidates for this strategy, including retinal degenerations; hematopoietic conditions, such as sickle cell anemia and thalassemia; and disorders affecting liver proteins, such as PKU, urea cycle disorders, familial hypercholesterolemia, and α1AT deficiency. Here we discuss several disorders in which gene therapy has been clearly effective, but which also highlight some of the challenges associated with this therapeutic approach.

Severe X-Linked Combined Immunodeficiency

The SCIDs are due to mutations in genes required for lymphocyte maturation. Affected individuals fail to thrive and die early in life of infection because they lack functional B and T lymphocytes. The most common form of the disease, X-linked SCID, results from mutations in the X-linked gene (*IL2RG*) encoding the γc-cytokine receptor subunit of several interleukin receptors. The receptor deficiency causes an early block in T- and natural killer–lymphocyte growth, survival, and differentiation and is associated with severe infections, failure to thrive, and death in infancy or early childhood if left untreated. This condition was chosen for a gene therapy trial for two principal reasons. First, bone marrow transplantation cures the disease, indicating that the restoration of lymphocyte expression of *IL2RG* can reverse the pathophysiological changes. Second, it was believed that so-called transduced cells carrying the transferred gene would have a selective survival advantage over untransduced cells.

The outcome of trials of X-linked SCID has been dramatic and resulted, in 2000, in the first gene therapy cure of a patient with a genetic disease. Subsequent confirmation has been obtained in most patients in subsequent clinical trials (see Table 13-5). Bone marrow stem cells from the patients were infected in culture (ex vivo) with a retroviral vector that expressed the γc cytokine subunit cDNA. A selective advantage was conferred on the transduced cells by the gene transfer. Transduced T cells and natural killer cells populated the blood of treated patients, and the T cells appeared to behave normally. Although the frequency of transduced B cells was low, adequate levels of serum immunoglobulin and antibody levels were obtained. Dramatic clinical improvement occurred, with resolution of protracted diarrhea and skin lesions and restoration of normal growth and development. These initial trials demonstrated the great potential of gene therapy for the correction of inherited disease.

This highly promising outcome, however, came at the cost of induction of a leukemia-like disorder in 5 of the 20 treated patients, who developed an extreme lymphocytosis resembling T-cell acute lymphocytic leukemia; 4 of them are now well after treatment of the leukemia. The malignancy was due to insertional mutagenesis: the retroviral vector inserted into the *LMO2* locus, causing aberrant expression of the *LMO2* mRNA, which encodes a component of a transcription factor complex that mediates hematopoietic development. Consequently, trials using integrating vectors in hematopoietic cells must now monitor insertion sites and survey for clonal proliferation. Current-generation vectors are designed to avoid this mutagenic effect by using strategies such as including a self-inactivating or "suicide" gene cassette in the vector to eliminate clones of malignant cells. At this point, bone marrow stem cell transplantation remains the treatment of choice for those children with SCID fortunate enough to have a donor with an HLA-identical match. For patients without such a match, autologous transplantation of hematopoietic stem and progenitor cells, in which the genetic defect has been corrected by gene therapy, offers a lifesaving alternative, but one that may not be without risk.

Metachromatic Leukodystrophy

Metachromatic leukodystrophy (MLD) is an autosomal recessive neurodegenerative disorder that, in the late infantile form, is generally fatal by 5 years of age. It results from mutations in the gene, *ARSA*, that encodes arylsulfatase A, a lysosomal enzyme that degrades sulfatides that are neurotoxic, leading to demyelination in the central and peripheral nervous system. As described earlier, HSC transplantation is an effective treatment of some lysosomal storage diseases because some of the donor-derived macrophages and microglia can enter the central nervous system, scavenge the stored material (such as sulfatide in MLD), and release lysosomal enzymes that are taken up by the mutant cells of the patient. HSC transplants have not been successful for MLD, however, a failure thought to be due to a level of *ARSA* expression from the transplanted cells that is too low to have a therapeutic effect.

In an apparently successful treatment, the autologous HSCs of three patients with MLD were transduced with a lentiviral vector that was engineered to produce above-normal levels of arylsulfatase A from a functional *ARSA* gene, and the genetically corrected HSCs were then engrafted (Fig. 13-17). Although more than 36,000 different lentiviral integration sites were examined, no evidence of genotoxicity was observed, suggesting that lentiviral vectors can be effective in the gene therapy of HSCs. Dramatically, disease progression was arrested, at least up to 24 months after treatment, but long-term follow-up will be required to establish that the effect of the gene therapy is benign and enduring.

Hemophilia B

Hemophilia B is an X-linked disorder of coagulation caused by mutations in the *F9* gene, leading to a deficiency or dysfunction of clotting factor IX (Case 21). The disease is characterized by bleeding into soft tissues, muscles, and weight-bearing joints, and occurs within hours to days after trauma. Severely affected subjects, with less than 1% of normal levels of factor IX, have frequent bleeding that causes crippling joint disease and early death. Prophylactic—but not curative—treatment with intravenous factor IX concentrate several times a week is expensive and leads to the generation of inhibitory antibodies.

In 2011, the first successful gene therapy treatment of hemophilia B was reported in six patients using an AAV8 vector that is tropic for hepatocytes, where factor IX is normally produced. After a single infusion of the AAV8-*F9* vector, four patients were able to discontinue prophylactic factor IX infusions, whereas the other two tolerated longer intervals between infusions. The two patients who received the highest dose of the vector had transient asymptomatic increases in liver enzyme levels—which resolved with steroid treatment—indicating that immune-related side effects must remain a concern in future studies. Unfortunately, the AAV vectors cannot accommodate the gene for factor VIII, so that other vectors will have to be developed for hemophilia A patients. Apart from this limitation of cargo size, however, AAV-mediated gene therapy targeted to hepatocytes may be applicable to any genetic disease in which production of the protein in the liver is the desired goal.

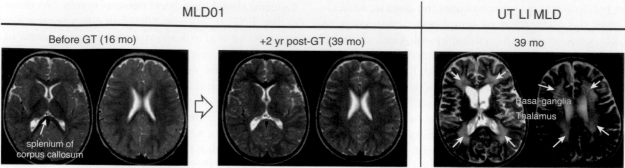

Figure 13-17 Clinical follow-up of a metachromatic leukodystrophy (MLD) patient after hematopoietic stem cell gene therapy (GT) with the arylsulfatase A gene. Magnetic resonance images from patient MLD01 before gene therapy and 2 years after treatment. The brain of this patient appeared largely normal 2 years after treatment. In contrast, the brain of an untreated, age matched late infantile MLD patient (UT LI MLD) showed severe demyelination associated with diffuse atrophy. In MLD01 images, a small area of hyperintensity is present within the splenium of the corpus callosum (*white arrow*). This area appeared at the 12-month follow-up and remained stable thereafter. In UT LI MLD images, extensive, diffuse symmetrical hyperintensities with typical striped "tigroid pattern" (*white arrows*) are seen within periventricular white matter, corpus callosum, external and internal capsules, and cerebellar deep white matter. Severe diffuse brain atrophy involving basal ganglia and thalamus, which show a T2 hypointense signal, is also present. *See Sources & Acknowledgments.*

β-Thalassemia

The hemoglobinopathies are the most common genetic defects in the world (see Chapter 11), but at present they are incurable except by HSC transplantation from a matched donor. Consequently, the development of effective, safe, and affordable gene therapy for these disorders, the most common being sickle cell disease and the α- and β-thalassemias, would be a medical triumph.

In 2010, the first successful gene therapy trial for a hemoglobinopathy was reported, in a single patient with β-thalassemia who was transfusion-dependent, with hemoglobin levels of only 4 to 6 g/dL. This individual was a genetic compound of β^E and β^0 alleles, the β^E allele generating a mutant β-globin of decreased abundance, with the β^0 allele being a null. The patient's HSCs were transduced with a lentiviral vector containing a β-globin gene. The patient became transfusion-independent, with hemoglobin levels ranging from 9 to 10 g/dL, although the vector-encoded hemoglobin accounted for only approximately one third of the total, the remainder being the mutant Hb E and Hb F. Unexpectedly, the increase in normal β-globin expression was largely attributable to one bone marrow cell clone, in which the lentiviral vector integrated into a gene encoding a transcriptional regulator called HMGA2. This integration activated expression in erythroid cells of a truncated form of HMGA2, an event that confounded the interpretation of the result, because the extent to which the clonal dominance of cells expressing the truncated HMGA2 accounted for the therapeutic benefits of the gene therapy is unclear.

This study offers great promise but highlights the potential risks associated with the random insertion of viral vectors in the genome. Much current research is therefore devoted to the development of safer gene delivery vectors, including modified lentiviral vectors.

The Prospects for Gene Therapy

To date, almost 2000 clinical gene therapy trials (approximately two thirds of which are for cancer) have been undertaken worldwide to evaluate both the safety and efficacy of this long-promised and conceptually promising technology. Approximately 180 of these trials were for the treatment of monogenic diseases. The exciting results obtained with gene therapy to date, albeit with small numbers of patients and only a few diseases, validates the optimism behind this immense effort. Although the breadth of applications remains uncertain, it is to be hoped that over the next few decades, gene therapy for both monogenic and genetically complex diseases will contribute to the management of many disorders, both common and rare.

PRECISION MEDICINE: THE PRESENT AND FUTURE OF THE TREATMENT OF MENDELIAN DISEASE

The treatment of single-gene diseases embodies the concept of precision medicine tailored to the individual patient as deeply as any other area of medical treatment. Knowledge of the specific mutant sequence in an individual is central to many of the targeted therapies described in this chapter. The promise of gene therapy for an individual with a mendelian disorder must be based on the identification of the mutant gene in each affected individual and on the design of a vector that will deliver the therapeutic gene to the targeted tissue. Similarly, approaches based on gene editing require knowledge of the specific mutation to be corrected.

Beyond this, however, precision medicine will frequently require knowledge of the precise mutant allele and of its specific effect on the mRNA and protein. In many cases, the exact nature of the mutation will define the drug that will bind to a specific regulatory sequence to enhance or reduce the expression of a gene. In other cases, the mutation will dictate the sequence of an allele-specific oligonucleotide to mediate the skipping of an exon with a premature termination codon, or of an siRNA to suppress a dominant negative allele. A compendium of small molecules will gradually become available to suppress particular stop codons, to act as chaperones that will rescue mutant proteins from misfolding and proteosomal degradation, or to potentiate the activity of mutant proteins.

Genetic treatment is not only becoming more and more creative, it is becoming more and more precise. The future promises not only a longer life for many patients, but a life of vastly better quality.

GENERAL REFERENCES

Campeau PM, Scriver CR, Mitchell JJ: A 25-year longitudinal analysis of treatment efficacy in inborn errors of metabolism, *Mol Genet Metab* 95:11–16, 2008.

Dietz HC: New therapeutic approaches to mendelian disorders, *N Engl J Med* 363:852–863, 2010.

Valle D, Beaudet AL, Vogelstein B, et al, editors: The online metabolic and molecular bases of inherited disease, 2014. Available at http://ommbid.mhmedical.com/book.aspx?bookID=474.

REFERENCES FOR SPECIFIC TOPICS

Arora N, Daley GQ: Pluripotent stem cells in research and treatment of hemoglobinopathies, *Cold Spring Harb Perspect Med* 2:a011841, 2012.

Bélanger-Quintana A, Burlina A, Harding CO, et al: Up to date knowledge on different treatment strategies for phenylketonuria, *Mol Genet Metabolism* 104:S19–S25, 2011.

Biffi A, Montini E, Lorioli L, et al: Lentiviral hematopoietic stem cell gene therapy benefits metachromatic leukodystrophy, *Science* 341:1233158, 2013. doi:10.1126/science.1233158.

Cathomen T, Ehl S: Translating the genomic revolution—targeted genome editing in primates, *N Engl J Med* 370:2342–2345, 2014.

Coelho T, Adams D, Silva A, et al: Safety and efficacy of RNAi therapy for transthyretin amyloidosis, *N Engl J Med* 369(9):818–829, 2013.

Daley GQ: The promise and perils of stem cell therapeutics, *Cell Stem Cell* 10:740–749, 2012.

Desnick RJ, Schuchman EH: Enzyme replacement therapy for lysosomal diseases: lessons from 20 years of experience and remaining challenges, *Annu Rev Genomics Hum Genet* 13:307–335, 2012.

de Souza N: Primer: genome editing with engineered nucleases, *Nat Methods* 9:27, 2012.

Dong A, Rivella S, Breda L: Gene therapy for hemoglobinopathies: progress and challenges, *Trans Res* 161:293–306, 2013.

Gaspar HB, Qasim W, Davies EG, et al: How I treat severe combined immunodeficiency, *Blood* 122:3749–3758, 2013.

Gaziev J, Lucarelli G: Hematopoietic stem cell transplantation for thalassemia, *Curr Stem Cell Res Ther* 6:162–169, 2011.

Goemans NM, Tulinius M, van den Akker JT: Systemic administration of PRO051 in Duchenne's muscular dystrophy, *N Engl J Med* 364:1513–1522, 2011.

Groenink M, den Hartog AW, Franken R, et al: Losartan reduces aortic dilatation rate in adults with Marfan syndrome: a randomized controlled trial, *Eur Heart J* 34:3491–3500, 2013.

Hanna JH, Saha K, Jaenisch R: Pluripotency and cellular reprogramming: facts, hypotheses, unresolved issues, *Cell* 143:508–525, 2010.

Hanrahan JW, Sampson HM, Thomas DY: Novel pharmacological strategies to treat cystic fibrosis, *Trends Pharmacol Sci* 34:119–125, 2013.

High KA: Gene therapy in clinical medicine. In Longo D, Fauci A, Kasper D, et al, editors: *Harrison's principles of internal medicine*, ed 19, New York, 2015, McGraw-Hill, in press.

Huang R, Southall N, Wang Y, et al: The NCGC Pharmaceutical Collection: A comprehensive resource of clinically approved drugs enabling repurposing and chemical genomics, *Sci Transl Med* 3:80ps16, 2011.

Jarmin S, Kymalainen H, Popplewell L, et al: New developments in the use of gene therapy to treat Duchenne muscular dystrophy, *Expert Opin Biol Ther* 14:209–230, 2014.

Johnson SM, Connelly S, Fearns C, et al: The transthyretin amyloidoses: from delineating the molecular mechanism of aggregation linked to pathology to a regulatory agency approved drug, *J Mol Biol* 421:185–203, 2012.

Li M, Suzuki K, Kim NY, et al: A cut above the rest: targeted genome editing technologies in human pluripotent stem cells, *J Biol Chem* 289:4594–4599, 2014.

Mukherjee S, Thrasher AJ: Gene therapy for primary immunodeficiency disorders: progress, pitfalls and prospects, *Gene* 525:174–181, 2013.

Nathwani AC, Tuddenham EGD, Rangarajan S: Adenovirus-associated virus vector–mediated gene transfer in hemophilia B, *N Engl J Med* 365:2357–2365, 2011.

Okam MM, Ebert BL: Novel approaches to the treatment of sickle cell disease: the potential of histone deacetylase inhibitors, *Expert Rev Hematol* 5:303–311, 2012.

Otsuru S, Gordon PL, Shimono K, et al: Transplanted bone marrow mononuclear cells and MSCs impart clinical benefit to children with osteogenesis imperfecta through different mechanisms, *Blood* 120:1933–1941, 2012.

Peltz SW, Morsy M, Welch EW, et al: Ataluren as an agent for therapeutic nonsense suppression, *Annu Rev Med* 64:407–425, 2013.

Perrine SP, Pace BS, Faller DV: Targeted fetal hemoglobin induction for treatment of beta hemoglobinopathies, *Hematol Oncol Clin North Am* 28:233–248, 2014.

Prasad VK, Kurtzberg J: Cord blood and bone marrow transplantation in inherited metabolic diseases: scientific basis, current status and future directions, *Br J Haematol* 148:356–372, 2009.

Ramsey BW, Davies J, McElvaney NG, et al: A CFTR potentiator in patients with cystic fibrosis and the G551D mutation, *N Engl J Med* 365:1663–1672, 2011.

Robinton DA, Daley GQ: The promise of induced pluripotent stem cells in research and therapy, *Nature* 481:295–305, 2012.

Sander JD, Joung JK: CRISPR-Cas systems for editing, regulating and targeting genomes, *Nat Biotechnol* 32:347–355, 2014.

Southwell AL, Skotte NH, Bennett CF, et al: Antisense oligonucleotide therapeutics for inherited neurodegenerative diseases, *Trends Mol Med* 18:634–643, 2012.

Tebas P, Stein D, Tang WW, et al: Gene editing of CCR5 in autologous CD4 T cells of persons infected with HIV, *N Engl J Med* 370:901–910, 2014.

van Ommen G-JB, Aartsma-Rus A: Advances in therapeutic RNA-targeting, *Trends Mol Med* 18:634–643, 2012.

Verma IM: Gene therapy that works, *Science* 341:853–855, 2013.

Xu J, Peng C, Sankaran VG, et al: Correction of sickle cell disease in adult mice by interference with fetal hemoglobin silencing, *Science* 334:993–996, 2011.

USEFUL WEBSITES

Registry and results database of publicly and privately supported clinical studies of human participants conducted around the world: https://clinicaltrials.gov/

Gene Therapy Clinical Trials Worldwide: http://www.wiley.com/legacy/wileychi/genmed/clinical/

PROBLEMS

1. X-linked chronic granulomatous disease (CGD) is an uncommon disorder characterized by a defect in host defense that leads to severe, recurrent, and often fatal pyogenic infections beginning in early childhood. The X-linked *CGD* locus encodes the heavy chain of cytochrome b, a component of the oxidase that generates superoxide in phagocytes. Because interferon-γ (IFN-γ) is known to enhance the oxidase activity of normal phagocytes, IFN-γ was administered to boys with X-linked CGD to see whether their oxidase activity increased. Before treatment, the phagocytes of *some* less severely affected patients had small but detectable bursts of oxidase activity (unlike those of severely affected patients), suggesting that increased activity in these less severely affected subjects is the result of greater production of cytochrome b from the affected locus. In these less severe cases, IFN-γ increased the cytochrome b content, superoxide production, and killing of *Staphylococcus aureus* in the granulocytes. The IFN-γ effect was associated with a definite increase in the abundance of the cytochrome b chain. Presumably, the cytochrome b polypeptide of these patients is partially functional, and increased expression of the residual function improved the physiological defect. Describe the genetic differences that might account for the fact that the phagocytes of some patients with X-linked CGD respond to IFN-γ in vitro and others do not.

2. Identify some of the limitations on the types of proteins that can be considered for extracellular replacement therapy, as exemplified by polyethylene glycol–adenosine deaminase (PEG-ADA). What makes this approach inappropriate for phenylalanine hydroxylase deficiency? If Tay-Sachs disease caused only liver disease, would this strategy succeed? If not, why?

3. A 3-year-old girl, Rhonda, has familial hypercholesterolemia due to a deletion of the 5′ end of each of her low-density lipoprotein (LDL) receptor genes that removed the promoter and the first two exons. (Rhonda's parents are second cousins.) You explain to the parents that she will require plasmapheresis every 1 to 2 weeks for years. At the clinic, however, they meet another family with a 5-year-old boy with the same disease. The boy has been treated with drugs with some success. Rhonda's parents want to know why she has not been offered similar pharmacological therapy. Explain.

4. What classes of mutations are likely to be found in homocystinuric patients who are not responsive to the administration of large doses of pyridoxine (vitamin B_6)? How might you explain the fact that Tom is completely responsive, whereas his first cousin Allan has only a partial reduction in plasma homocystine levels when he is given the same amount of vitamin B_6?

5. You have isolated the gene for phenylalanine hydroxylase (PAH) and wish ultimately to introduce it into patients with PKU. Your approach will be to culture cells from the patient, introduce a functional version of the gene into the cells, and reintroduce the cells into the patient.
   a. What DNA components do you need to make a functional PAH protein in a gene transfer experiment?
   b. Which tissues would you choose in which to express the enzyme, and why? How does this choice affect your gene construct in (a)?
   c. You introduce your version of the gene into fibroblasts cultured from a skin biopsy specimen from the patient. Northern (RNA) blot analysis shows that the messenger RNA (mRNA) is present in normal amounts and is the correct size. However, no PAH protein can be detected in the cells. What kinds of abnormalities in the transferred gene would explain this finding?
   d. You have corrected all the problems identified in (c). On introducing the new version of the gene into the cultured cells, you now find that the PAH protein is present in great abundance, and when you harvest the cells and assay the enzyme (in the presence of all the required components), normal activity is obtained. However, when you add ^{3}H-labeled phenylalanine to the cells in culture, no ^{3}H-labeled tyrosine is formed (in contrast, some cultured liver cells produce a large quantity of ^{3}H-labeled tyrosine in this situation). What are the most likely explanations for the failure to form ^{3}H-tyrosine? How does this result affect your gene therapy approach to patients?
   e. You have developed a method to introduce your functional version of the gene directly into a large proportion of the hepatocytes of patients with PAH deficiency. Unexpectedly, you find that much lower levels of PAH enzymatic activity are obtained in patients in whom significant amounts of the inactive PAH homodimer were detectable in hepatocytes before treatment than in patients who had no detectable PAH polypeptide before treatment. How can you explain this result? How might you overcome the problem?

6. Both alleles of an autosomal gene that is mutant in your patient produce a protein that is decreased in abundance but has residual function. What therapeutic strategies might you consider in such a situation?

7. A Phase III clinical trial is undertaken to evaluate the effectiveness of a small molecule drug that facilitates skipping over nonsense mutation codons. The drug had been shown in earlier trials to have a modest but significant clinical effect in patients with cystic fibrosis with at least one *CFTR* nonsense mutation. Two cystic fibrosis (CF) patients each have a nonsense mutation in one *CFTR* allele, but at different locations in the reading frame. One patient responds to the drug, whereas the other does not. Discuss how the location of the nonsense mutation in the predicted reading frame of the protein could account for this differential response.

CHAPTER 14

Developmental Genetics and Birth Defects

Knowledge of the principles and concepts of developmental genetics, including the mechanisms and pathways responsible for normal human development in utero, is essential for the practitioner who seeks to develop a rational approach to the diagnostic evaluation of a patient with a birth defect. With an accurate diagnostic assessment in hand, the practitioner can make predictions about prognosis, recommend management options, and provide an accurate recurrence risk for the parents and other relatives of the affected child. In this chapter, we provide an overview of the branch of medicine concerned with birth defects and review basic mechanisms of embryological development, with examples of some of these mechanisms and pathways in detail. We present examples of birth defects that result from abnormalities in these processes. And finally, we show how an appreciation of developmental biology is essential for understanding prenatal diagnosis (see Chapter 17) and stem cell therapy as applied to regenerative medicine (see Chapter 13).

DEVELOPMENTAL BIOLOGY IN MEDICINE
The Public Health Impact of Birth Defects

The medical impact of birth defects is considerable. In 2013, the most recent year for which final statistics are available, the infant mortality rate in the United States was 5.96 infant deaths per 1000 live births; more than 20% of infant deaths were attributed to birth defects, that is, abnormalities (often referred to as **anomalies**) that are present at birth in the development of organs or other structures. Another 20% of infant deaths may be attributed to complications of prematurity, which can be considered a failure of maintenance of the maternal-fetal developmental environment. Therefore *nearly half of the deaths of infants are caused by derangements of normal development.* In addition to mortality, congenital anomalies are a major cause of long-term morbidity, intellectual disability, and other dysfunctions that limit the productivity of affected individuals.

Developmental anomalies certainly have a major impact on public health. Genetic counseling and prenatal diagnosis, with the option to continue or to terminate a pregnancy, are important for helping individuals faced with a risk for serious birth defects in their offspring improve their chances of having healthy children (see Chapter 17). Physicians and other health care professionals must be careful, however, not to limit the public health goal of reducing disease solely to preventing the birth of children with anomalies through voluntary pregnancy termination. Primary prevention of birth defects can be accomplished. For example, recommendations to supplement prenatal folic acid intake, which markedly reduces the incidence of neural tube defects, and public health campaigns that focus on preventing teratogenic effects of alcohol during pregnancy, are successful public health approaches to the prevention of birth defects that do not depend on prenatal diagnosis and elective abortion. In the future, it is hoped that our continued understanding of the developmental processes and pathways that regulate them will lead to therapies that may improve the morbidity and mortality associated with birth defects.

Dysmorphology and Mechanisms That Cause Birth Defects

Dysmorphology is the study of congenital birth defects that alter the shape or form of one or more parts of the body of a newborn child. Researchers attempt to understand the contribution of both abnormal genes and nongenetic, environmental influences to birth defects, as well as how those genes participate in conserved developmental pathways. The objectives of the medical geneticist who sees a child with birth defects are:
- to diagnose a child with a birth defect,
- to suggest further diagnostic evaluations,
- to give prognostic information about the range of outcomes that could be expected,
- to develop a plan to manage the expected complications,
- to provide the family with an understanding of the causation of the malformation, and
- to give recurrence risks to the parents and other relatives.

To accomplish these diverse and demanding objectives, the clinician must acquire and organize data from

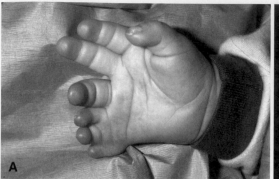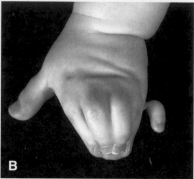

Figure 14-1 Polydactyly and syndactyly malformations. **A,** Insertional polydactyly. This patient has heptadactyly with insertion of a digit in the central ray of the hand and a supernumerary postaxial digit. This malformation is typically associated with metacarpal fusion of the third and fourth digits. Insertional polydactyly is common in patients with Pallister-Hall syndrome. **B,** Postaxial polydactyly with severe cutaneous syndactyly of digits two through five. This type of malformation is seen in patients with Greig cephalopolysyndactyly syndrome. *See Sources & Acknowledgments.*

the patient, the family history, and published clinical and basic science literature. Medical geneticists work closely with specialists in pediatric surgery, neurology, rehabilitation medicine, and the allied health professions to provide ongoing care for children with serious birth defects.

Malformations, Deformations, and Disruptions

Medical geneticists divide birth defects into three major categories: **malformations, deformations,** and **disruptions.** We will illustrate the difference between these three categories with examples of three distinct birth defects, all involving the limbs.

Malformations result from *intrinsic* abnormalities in one or more genetic programs operating in development. An example of a malformation is the extra fingers in the disorder known as **Greig cephalopolysyndactyly** (Fig. 14-1). This syndrome, discussed later in the chapter, results from loss-of-function mutations in a gene for a transcription factor, GLI3, which is one component of a complex network of transcription factors and signaling molecules that interact to cause the distal end of the human upper limb bud to develop into a hand with five digits. Because malformations arise from intrinsic defects in genes that specify a series of developmental steps or programs, and because such programs are often used more than once in different parts of the embryo or fetus at different stages of development, a malformation in one part of the body is often but not always associated with malformations elsewhere as well.

In contrast to malformations, **deformations** are caused by *extrinsic* factors impinging physically on the fetus during development. They are especially common during the second trimester of development when the fetus is constrained within the amniotic sac and uterus. For example, contractions of the joints of the extremities, known as **arthrogryposes,** in combination with deformation of the developing skull, occasionally

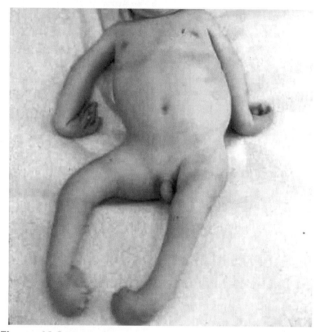

Figure 14-2 Deformation known as congenital arthrogryposis seen with a condition referred to as amyoplasia. There are multiple, symmetrical joint contractures due to abnormal muscle development caused by severe fetal constraint in a pregnancy complicated by oligohydramnios. Intelligence is generally normal, and orthopedic rehabilitation is often successful. *See Sources & Acknowledgments.*

accompany constraint of the fetus due to twin or triplet gestations or prolonged leakage of amniotic fluid (Fig. 14-2). Most deformations apparent at birth either resolve spontaneously or can be treated by external fixation devices to reverse the effects of the instigating cause.

Disruptions, the third category of birth defect, result from destruction of irreplaceable normal fetal tissue. Disruptions are more difficult to treat than deformations because they involve actual loss of normal tissue. Disruptions may be the result of vascular insufficiency, trauma, or teratogens. One example is **amnion disruption,** the partial amputation of a fetal limb associated with strands of amniotic tissue. Amnion disruption is often recognized clinically by the presence of partial and irregular digit amputations in conjunction with constriction rings (Fig. 14-3).

The pathophysiological concepts of malformations, deformations, and disruptions are useful clinical guides to the recognition, diagnosis, and treatment of birth defects, but they sometimes overlap. For example,

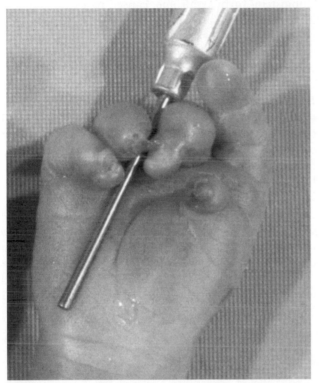

Figure 14-3 Disruption of limb development associated with amniotic bands. This 26-week fetus shows nearly complete disruption of the thumb with only a nubbin remaining. The third and fifth fingers have constriction rings of the middle and distal phalanges, respectively. The fourth digit is amputated distally with a small fragment of amnion attached to the tip. *See Sources & Acknowledgments.*

vascular malformations may lead to disruption of distal structures, and urogenital malformations that cause oligohydramnios can cause fetal deformations. Thus a given constellation of birth defects in an individual may represent combinations of malformations, deformations, and disruptions.

Genetic, Genomic, and Environmental Causes of Malformations

Malformations have many causes (Fig. 14-4). Chromosome imbalance accounts for approximately 25%, of which autosomal trisomies for chromosomes 21, 18, and 13 (see Chapter 6) are some of the most common. The recent clinical application of genome-wide arrays in comparative genomic hybridization (CGH or array-CGH; see Chapter 5) has revealed small, de novo submicroscopic deletions and/or duplications, also known as copy number variants (CNVs), in as many as 10% of individuals with birth defects. An additional 20% are caused by mutations in single genes. Some malformations, such as **achondroplasia** or **Waardenburg syndrome,** are inherited as autosomal dominant traits. Many heterozygotes with birth defects, however, represent new mutations that are so severe that they are genetic lethals and are therefore often found to be isolated cases within families (see Chapter 7). Other malformation syndromes are inherited in an autosomal or X-linked recessive pattern, such as the **Smith-Lemli-Opitz syndrome** or the **Lowe syndrome,** respectively.

Another approximately 40% of major birth defects have no identifiable cause but recur in families of affected children with a greater frequency than would be expected on the basis of the population frequency and are thus considered to be multifactorial diseases (see Chapter 8). This category includes well-recognized birth defects such as **cleft lip with or without cleft palate,** and **congenital heart defects.**

The remaining 5% of birth defects are thought to result from exposure to certain environmental agents— drugs, infections, alcohol, chemicals, or radiation— or from maternal metabolic disorders such as poorly

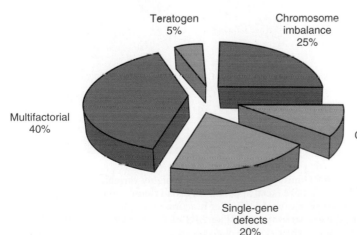

Figure 14-4 The relative contribution of single-gene defects, chromosome abnormalities, copy number variants, multifactorial traits, and teratogens to birth defects.

controlled maternal diabetes mellitus or maternal phenylketonuria (see Chapter 12). Such agents are called **teratogens** (derived, inelegantly, from the Greek word for monster plus *-gen,* meaning cause) because of their ability to cause malformations (discussed later in this chapter).

Pleiotropy: Syndromes and Sequences

A birth defect resulting from a single underlying causative agent may result in abnormalities of more than one organ system in different parts of the embryo or in multiple structures that arise at different times during intrauterine life, a phenomenon referred to as **pleiotropy.** The agent responsible for the malformation could be either a mutant gene or a teratogen. Pleiotropic birth defects come about in two different ways, depending on the mechanism by which the causative agent produces its effect. When the causative agent causes multiple abnormalities in parallel, the collection of abnormalities is referred to as a **syndrome.** If, however, a mutant gene or teratogen affects only a single organ system at one point in time, and it is the perturbation of that organ system that causes the rest of the constellation of pleiotropic defects to occur as secondary effects, the malformation is referred to as a **sequence.**

Pleiotropic Syndromes. The autosomal dominant **branchio-oto-renal dysplasia syndrome** exemplifies a pleiotropic syndrome. It has long been recognized that patients with branchial arch anomalies affecting

development of the ear and neck structures are at high risk for having renal anomalies. The branchio-oto-renal dysplasia syndrome, for example, consists of abnormal cochlear and external ear development, cysts and fistulas in the neck, renal dysplasia, and renal collecting duct malformations. The mechanism of this association is that a conserved set of genes and proteins are used by mammals to form both the ear and the kidney. The syndrome is caused by mutations in one such gene, *EYA1,* which encodes a protein phosphatase that functions in both ear and kidney development. Similarly, the **Rubinstein-Taybi syndrome,** caused by loss of function in a transcriptional coactivator, results in abnormalities in the transcription of many genes that depend on this coactivator being present in a transcription complex for normal expression (Fig. 14-5).

Sequences. In contrast, an example of a sequence is the U-shaped cleft palate and small mandible referred to as the **Robin sequence** (Fig. 14-6). This sequence comes about because a restriction of mandibular growth before the ninth week of gestation causes the tongue to lie more posteriorly than is normal, interfering with normal closure of the palatal shelves, thereby causing a cleft palate. The Robin sequence can be an isolated birth defect of unknown cause or can be due to extrinsic impingement on the developing mandible by a twin in utero. This phenotype can also be one of several features of a condition known as **Stickler syndrome,** in which mutations in the gene encoding a subunit of type II

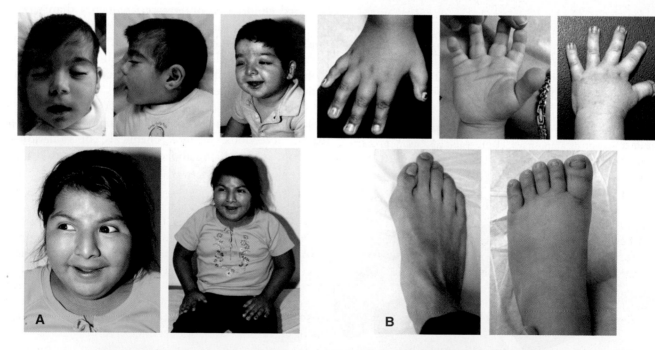

Figure 14-5 Physical characteristics of patients with Rubinstein-Taybi syndrome, a highly variable and pleiotropic syndrome of developmental delay, distinctive facial appearance, broad thumbs and large toes, and congenital heart defects. The syndrome is caused by loss-of-function mutations in one of two different but closely related transcriptional coactivators, *CBP* or *EP300.* **A,** Distinctive facial features. **B,** Appearance of hands and feet. *See Sources & Acknowledgments.*

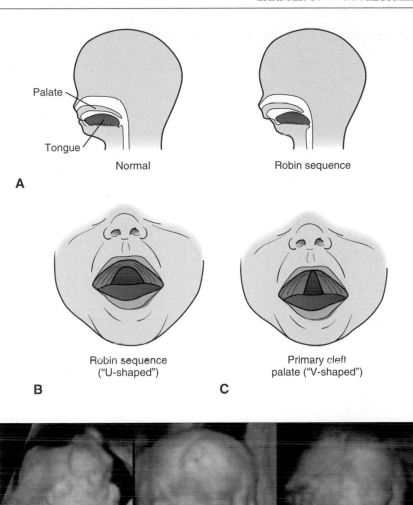

Figure 14-6 **A,** Hypoplasia of the mandible and resulting posterior displacement of the tongue lead to the Robin sequence, in which the tongue obstructs palatal closure. **B,** Posterior placement of the tongue in the Robin sequence causes a *deformation* of the palate during development, leading to the constellation of a small chin and a U-shaped cleft palate involving the soft palate and extending into the hard palate. **C,** In contrast, *primary* cleft palate resulting from failure of closure of maxillary ridges is a *malformation* that begins in the anterior region of the maxilla and extends posteriorly to involve first the hard palate and then the soft palate, and it is often V-shaped. **D,** The delay in jaw development can be observed by serial three-dimensional fetal scans, from as early as 17 weeks (*left*) to 20 weeks (*middle*) and 29 weeks (*right*). *See Sources & Acknowledgments.*

collagen result in an abnormally small mandible as well as other defects in stature, joints, and eyes. The Robin sequence in the Stickler syndrome is a sequence because the mutant collagen gene itself is not responsible for the failure of palatal closure; the cleft palate is secondary to the primary defect in jaw growth. Whatever the cause, a cleft palate due to the Robin sequence must be distinguished from a true primary cleft palate, which has other causes with differing prognoses and implications for the child and family. Knowledge of dysmorphology and developmental genetic principles is thus necessary to properly diagnose each condition and to recognize that different prognoses are associated with the different primary causes.

INTRODUCTION TO DEVELOPMENTAL BIOLOGY

The examples introduced briefly in the previous section serve to illustrate the principle that the clinical practice of medical genetics rests on a foundation of the basic science of developmental biology. For this reason, it behooves practitioners to have a working knowledge of some of the basic principles of developmental biology and to be familiar with the ways that abnormal function of genes and pathways affect development and, ultimately, their patients.

Developmental biology is concerned with a single, unifying question: *How can a single cell transform itself into a mature animal?* In humans, this transformation occurs each time a single fertilized egg develops into a human being with more than 10^{13} to 10^{14} cells, several hundred recognizably distinct cell types, and dozens of tissues. This process must occur in a reliable and predictable pattern and time frame.

Developmental biology has its roots in embryology, which was based on observing and surgically manipulating developing organisms. Early embryological studies, carried out in the 19th and early 20th centuries with readily accessible amphibian and avian embryos, determined that embryos developed from single cells and defined many of the fundamental processes of development. Much more recently, the application of molecular biology, genetics, and genomics to embryology has transformed the field by allowing scientists to study and manipulate development by a broad range of powerful biochemical and molecular techniques.

Development and Evolution

A critically important theme in developmental biology is its relationship to the study of evolution. Early in development, the embryos of many species look similar. As development progresses, the features shared between species are successively transformed into more specialized features that are, in turn, shared by successively fewer but more closely related species. A comparison of embryological characteristics among and within evolutionarily related organisms shows that developmental attributes (e.g., fingers) specific to certain groups of animals (e.g., primates) are built on a foundation of less specific attributes common to a larger group of animals (e.g., mammals), which are in turn related to structures seen in an even larger group of animals (e.g., the vertebrates). Structures in different organisms are termed **homologous** if they evolved from a structure present in a common ancestor (Fig. 14-7). In the case of the forelimb, the various ancestral lineages of the three species shown in Figure 14-7, tracing all the way back to their common predecessor, share a common attribute: a functional forelimb. The molecular developmental mechanism that created those limb

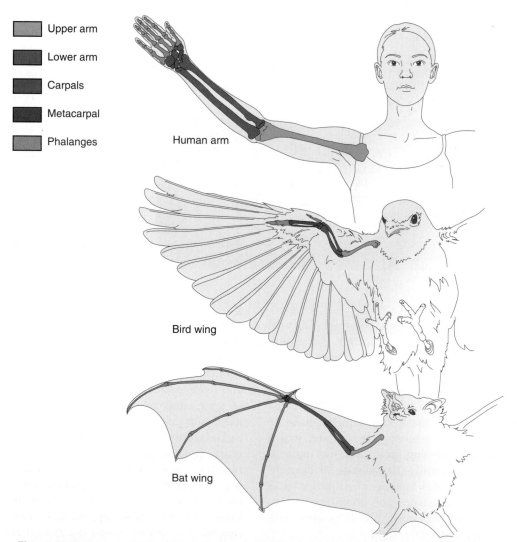

Upper arm

Lower arm

Carpals

Metacarpal

Phalanges

Human arm

Bird wing

Bat wing

Figure 14-7 Diagram of the upper limb of three species: human, bird, and bat. Despite the superficially dissimilar appearance of the human arm and hand, the avian wing, and the bat wing, the similarity in their underlying bone structure and functionality reveals the homology of the forelimbs of all three species. In contrast, the two superficially similar wings in the bird and bat are analogous, not homologous structures. Although both the bird and bat wings are used for flying, they are constructed quite differently and did not evolve from a winglike structure in a common ancestor. *See Sources & Acknowledgments.*

structures is shared across all three of the contemporary species.

Not all similarity is due to homology, however. Evolutionary studies also recognize the existence of **analogous** structures, those that appear similar but arose independently of one another, through different lineages that *cannot* be traced back to a common ancestor with that structure. The molecular pathways that generate analogous structures are unlikely to be evolutionarily conserved. In the example shown in Figure 14-7, the wing structures of the bat and the birds arose independently in evolution to facilitate the task of aerial movement. The evolutionary lineages of these two animals do not share a common ancestor with a primitive wing-like structure from which both bats and birds inherited wings. On the contrary, one can readily see that the birds developed posterior extensions from the limb to form a wing, whereas bats evolved wings through spreading the digits of their forelimbs and connecting them with syndactylous tissue. This situation is termed **convergent evolution.**

The evolutionary conservation of developmental processes is critically important to studies of human development because the vast majority of such research cannot (for important ethical reasons) be performed in humans (see Chapter 19). Thus, to understand a developmental observation, scientists use animal models to investigate normal and abnormal developmental processes. The ability to extend the results to humans is completely dependent on the evolutionary conservation of mechanisms of development and homologous structures.

GENES AND ENVIRONMENT IN DEVELOPMENT

Developmental Genetics

Development results from the action of genes interacting with cellular and environmental cues. The gene products involved include transcriptional regulators, diffusible factors that interact with cells and direct them toward specific developmental pathways, the receptors for such factors, structural proteins, intracellular signaling molecules, and many others. It is therefore not surprising that most of the numerous developmental disorders that occur in humans are caused by chromosomal, subchromosomal or gene mutations.

Even though the genome is clearly the primary source of information that controls and specifies human development, the role of genes in development is often mistakenly described as a "master blueprint." In reality, however, the genome does *not* resemble an architect's blueprint that specifies precisely how the materials are to be used, how they are to be assembled, and their final dimensions; it is not a literal description of the final form that all embryological and fetal structures will take. Rather, the genome specifies a set of interacting

proteins and noncoding RNAs (see Chapter 3) that set in motion the processes of growth, migration, differentiation, and apoptosis that ultimately result, with a high degree of probability, in the correct mature structures. Thus, for example, there are no genetic instructions directing that the phalanx of a digit adopt an hourglass shape or that the eye be spherical. These shapes arise as an implicit consequence of developmental processes, thereby generating correctly structured cells, tissues, and organs.

Probability

Although genes are the primary regulators of development, other processes must also play a role. That development is regulated but not determined by the genome is underscored by the important role that probability plays in normal development. For example, in the mouse, a mutation in the formin gene produces **renal aplasia** in only approximately 20% of mice who carry the mutation, even when such carriers are genetically identical. Given that inbred strains of mice are genetically identical throughout their genomes, the 20% penetrance of the formin mutation cannot be explained by different modifying gene variants in the mice affected with renal agenesis versus the mice who are unaffected. Instead, it appears likely that the formin mutation shifts the balance of some developmental process by increasing the probability that a threshold for causing renal aplasia is exceeded, much as we explored in Chapter 8 when discussing complex patterns of inheritance in humans. Thus carrying a formin mutation will not always lead to renal aplasia, but it sometimes will, and neither the rest of the genome nor nongenetic factors are responsible for development of the defect in only a minority of animals. Probabilistic processes provide a rich source of interindividual variation that can lead to a range of developmental outcomes, some normal and some not. Thus it is not the case in development that "nothing is left to chance."

Environmental Factors

As indicated earlier, the local environment in which a cell or tissue finds itself plays a central role in providing a normal developmental context. It is therefore not unexpected that drugs or other agents introduced from the environment can be teratogens, often because they interfere with intrinsic molecules that mediate the actions of genes. Identification of the mechanism of teratogenesis has obvious implications not only for clinical medicine and public health but also for basic science; understanding how teratogens cause birth defects can provide insight into the underlying developmental pathways that have been disturbed and result in a defect.

Because the molecular and cellular pathways used during development are often not employed in similar developmental processes after adulthood, teratogens that cause serious birth defects may have few or no side effects in adult patients. One important example of this concept is **fetal retinoid syndrome,** seen in fetuses of pregnant women who took the drug isotretinoin during pregnancy. Isotretinoin is an oral retinoid that is used systemically for the treatment of severe acne. It causes major birth defects when it is taken by a pregnant woman because it mimics the action of endogenous retinoic acid, a substance that in the developing embryo and fetus diffuses through tissues and interacts with cells, causing them to follow particular developmental pathways.

Different teratogens often cause very specific patterns of birth defects, the risk for which depends critically on the gestational age at the time of exposure, the vulnerability of different tissues to the teratogen, and the level of exposure during pregnancy. One of the best examples is **thalidomide syndrome.** Thalidomide, a sedative widely used in the 1950s, was later found to cause a high incidence of malformed limbs in fetuses exposed between 4 and 8 weeks of gestation because of its effect on the vasculature of the developing limb. Another example is the **fetal alcohol syndrome.** Alcohol causes a particular pattern of birth defects involving primarily the central nervous system because it is relatively more toxic to the developing brain and related craniofacial structures than to other tissues.

Some teratogens, such as x-rays, are also mutagens. A fundamental distinction between teratogens and mutagens is that mutagens cause damage by creating heritable alterations in genetic material, whereas teratogens act directly and transiently on developing embryonic tissue. Thus fetal exposure to a mutagen can cause an increased risk for birth defects or other diseases (e.g., cancer) throughout the life of the exposed individual and even in his or her offspring, whereas exposure to a teratogen increases the risk for birth defects for current but not for subsequent pregnancies.

BASIC CONCEPTS OF DEVELOPMENTAL BIOLOGY

Overview of Embryological Development

Developmental biology has its own set of core concepts and terminology that may be confusing or foreign to the student of genetics. We therefore provide a brief summary of a number of key concepts and terms used in this chapter (see Box on next page).

Cellular Processes during Development

During development, cells divide (**proliferate**), acquire novel functions or structures (**differentiate**), move within the embryo (**migrate**), and undergo programmed cell death (often through **apoptosis**). These four basic cellular processes act in various combinations and in different ways to allow **growth** and **morphogenesis** (literally, the "creation of form"), thereby creating an embryo of normal size and shape, containing organs of the

appropriate size, shape, and location, and consisting of tissues and cells with the correct architecture, structure, and function.

Although growth may seem too obvious to discuss, growth itself is carefully regulated in mammalian development, and unregulated growth is disastrous. The mere doubling (one extra round of cell division) of cell number (hyperplasia) or the doubling of cell size (hypertrophy) of an organism is likely to be fatal. Dysregulation of growth of segments of the body can cause severe deformity and dysfunction, such as in hemihyperplasia and other segmental overgrowth disorders (Fig. 14-8). Furthermore, the exquisite differential regulation of growth can change the shape of a tissue or an organ.

Morphogenesis is accomplished in the developing organism by the coordinated interplay of the mechanisms introduced in this section. In some contexts, morphogenesis is used as a general term to describe all of development, but this is formally incorrect because morphogenesis has to be coupled to the process of growth discussed here to generate a normally shaped and functioning tissue or organ.

Human Embryogenesis

This description of human development begins where Chapter 2 ends, with fertilization. After fertilization, the embryo undergoes a series of cell divisions without overall growth, termed *cleavage.* The single fertilized egg undergoes four divisions to yield the 16-cell morula by day 4 (Fig. 14-9). At day 5, the embryo transitions to become a **blastocyst,** in which cells that give rise to the placenta form a wall, inside of which the cells that will make the embryo itself aggregate to one side into what is referred to as the **inner cell mass.** This is the point at which the embryo acquires its first obvious manifestation of **polarity,** an axis of asymmetry that divides the inner cell mass (most of which goes on to form the mature organism) from the embryonic tissues that will go on to form the chorion, an extraembryonic tissue (e.g., placenta) (Fig. 14-10). The inner cell mass then separates again into the **epiblast,** which will make the embryo proper, and the **hypoblast,** which will form the amniotic membrane.

The embryo implants in the endometrial wall of the uterus in the interval between days 7 and 12 after fertilization. After implantation, gastrulation occurs, in which cells rearrange themselves into a structure consisting of three cellular compartments, termed the **germ layers,** comprising the ectoderm, mesoderm, and endoderm. The three germ layers give rise to different structures. The endodermal lineage forms the central visceral core of the organism. This includes the cells lining the main gut cavity, the airways of the respiratory system, and other similar structures. The mesodermal lineage gives rise to kidneys, heart, vasculature, and structural or supportive functions in the organism. Bone and

CORE CONCEPTS AND TERMINOLOGY IN HUMAN DEVELOPMENTAL BIOLOGY

Blastocyst: a stage in **embryogenesis** after the **morula,** in which cells on the outer surface of the morula secrete fluid and form a fluid-filled internal cavity within which is a separate group of cells, the **inner cell mass,** which will become the **fetus** itself (see Fig. 14-10).

Chimera: an embryo made up of two or more cell lines that differ in their genotype. Contrast with **mosaic.**

Chorion: membrane that develops from the outer cells of the **blastocyst** and goes on to form the placenta and the outer layer of the sac in which the **fetus** develops.

Determination: the stage in development in which cells are irreversibly committed to forming a particular tissue.

Dichorionic twins: monozygotic twins arising from splitting of the embryo into two parts, before formation of the blastocyst, so that two independent blastocysts develop.

Differentiation: the acquisition by a cell of novel characteristics specific for a particular cell type or tissue.

Ectoderm: the primary embryonic **germ layer** that gives rise to the nervous system and skin.

Embryo: the stage of a developing human organism between fertilization and 9 weeks of gestation, when separation into placental and embryonic tissues occurs.

Embryogenesis: the development of the **embryo.**

Embryonic stem cells: cells derived from the **inner cell mass** that under appropriate conditions can differentiate into all of the cell types and tissues of an **embryo** and form a complete, normal fetus.

Endoderm: the primary embryonic **germ layer** that gives rise to many of the visceral organs and lining of the gut.

Epiblast: a differentiated portion of the inner cell mass that gives rise to the **embryo** proper.

Fate: the ultimate destination for a cell that has traveled down a developmental pathway.

Fetus: the stage of the developing human between 9 weeks of gestation and birth.

Gastrulation: the stage of development just after implantation in which the cells of the **inner cell mass** rearrange themselves into the three **germ layers. Regulative development** ceases at gastrulation.

Germ cell: the cells that are the progenitors of the gametes. These cells are allocated early in development and undergo sex-specific differentiation.

Germ layers: three distinct layers of cells that arise in the inner cell mass, the **ectoderm, mesoderm,** and **endoderm,** which develop into distinctly different tissues in the embryo.

Hypoblast: the differentiated portion of the inner cell mass that contributes to fetal membranes (amnion).

Inner cell mass: a group of cells inside the **blastocyst** destined to become the **fetus.**

Mesoderm: the primary embryonic **germ layer** that gives rise to connective tissue, muscles, bones, vasculature, and the lymphatic and hematopoietic systems.

Monoamniotic twins: monozygotic twins resulting from cleavage of part of the inner cell mass (epiblast) but without cleavage of the part of the inner cell mass that forms the amniotic membrane (hypoblast).

Monochorionic twins: monozygotic twins resulting from cleavage of the inner cell mass without cleavage of the cells on the outside of the **blastocyst.**

Monozygotic twins: twins arising from a single fertilized egg, resulting from cleavage during embryogenesis in the interval between the first cell division of the zygote and gastrulation.

Morphogen: a substance produced by cells in a particular region of an **embryo** that diffuses from its point of origin through the tissues of the embryo to form a concentration gradient. Cells undergo **specification** and then **determination** to different **fates,** depending on the concentration of morphogen they experience.

Morphogenesis: the creation of various structures during **embryogenesis.**

Morula: a compact ball of 16 cells produced after four cell divisions of the **zygote.**

Mosaic: an individual who develops from a single fertilized egg but in whom mutation after conception results in cells with two or more genotypes. Contrast with **chimera.**

Mosaic development: a stage in development in which cells have already become committed to the point that removal of a portion of an embryo will not allow normal embryonic development.

Multipotent stem cell: a stem cell capable of self-renewal as well as of developing into many different types of cells in a tissue, but not an entire organism. These are often called adult stem cells or tissue progenitor cells.

Organogenesis: the creation of individual organs during **embryogenesis.**

Pluripotent cell: an early stem cell capable of self-renewal as well as of becoming any cell in any tissue, including the germ cells. **Embryonic stem cells** are pluripotent.

Progenitor cell: a cell that is traversing a developmental pathway on its way to becoming a fully differentiated cell.

Regulative development: a stage in development in which cells have not yet become determined so that the cells that remain after removal of a portion of an embryo can still form a complete organism.

Specification: a step along the path of differentiation in which cells acquire certain specialized attributes characteristic of a particular tissue but can still be influenced by external cues to develop into a different type of cell or tissue.

Stem cell: a cell that is capable both of generating another stem cell (self-renewal) and of differentiating into specialized cells within a tissue or an entire organism.

Zygote: the fertilized egg, the first step in **embryogenesis.**

---

muscle are nearly exclusively mesodermal and have the two main functions of structure (physical support) and providing the necessary physical and nutritive support of the hematopoietic system. The ectoderm gives rise to the central and peripheral nervous systems and the skin. During the complicated movements that occur in gastrulation, the embryo also establishes the major axes of the final body plan: anterior-posterior (cranial-caudal), dorsal-ventral (back-front), and left-right axes, which are discussed later.

The next major stages of development involve the initiation of the nervous system, establishment of the basic body plan, and then **organogenesis,** which occupies weeks 4 to 8. The position and basic structures of

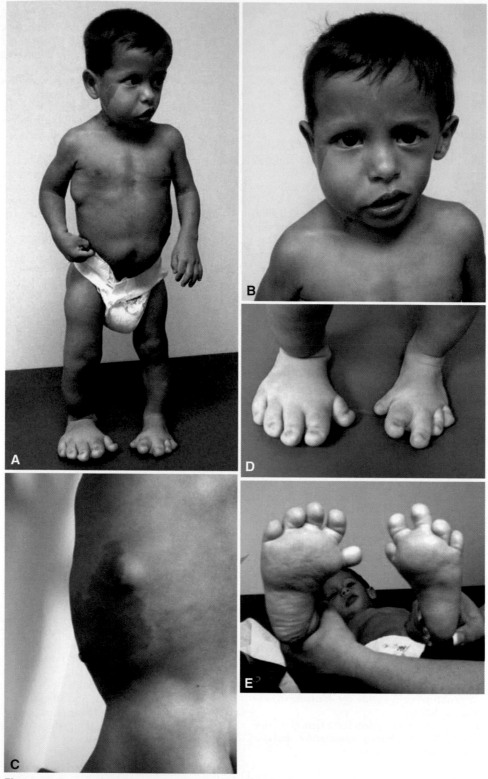

Figure 14-8 The clinical consequences of dysregulated growth in a child with Proteus syndrome, a congenital segmental overgrowth disorder affecting his face, abdomen, and right leg. Affected children are usually normal-appearing at birth but then in the first year begin to develop asymmetrical and disproportionate overgrowth of body parts. There are multiple malformations of the vascular system, including veins, capillaries and lymphatics; the osseous skeleton; and the connective tissue. The disorder is caused by somatic mosaicism for de novo activating mutations in *AKT1*, encoding a cell growth–promoting protein, which explains why the condition is always sporadic and occurs in an irregular pattern throughout the body in different affected individuals. *See Sources & Acknowledgments.*

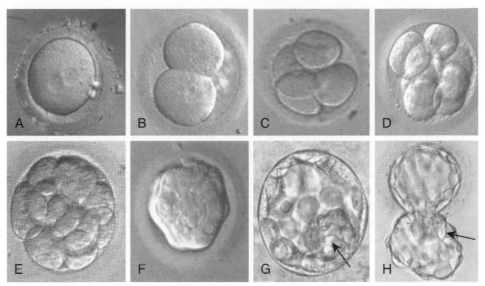

Figure 14-9 Human development begins with cleavage of the fertilized egg. **A,** The fertilized egg at day 0 with two pronuclei and the polar bodies. **B,** A two-cell embryo at day 1 after fertilization. **C,** A four-cell embryo at day 2. **D,** The eight-cell embryo at day 3. **E,** The 16-cell stage later in day 3, followed by the phenomenon of compaction, whereby the embryo is now termed a *morula* (**F,** day 4). **G,** T formation of the blastocyst at day 5, with the inner cell mass indicated by the *arrow.* Finally, the embryo (*arrow*) hatches from the zona pellucida (**H**). *See Sources & Acknowledgments.*

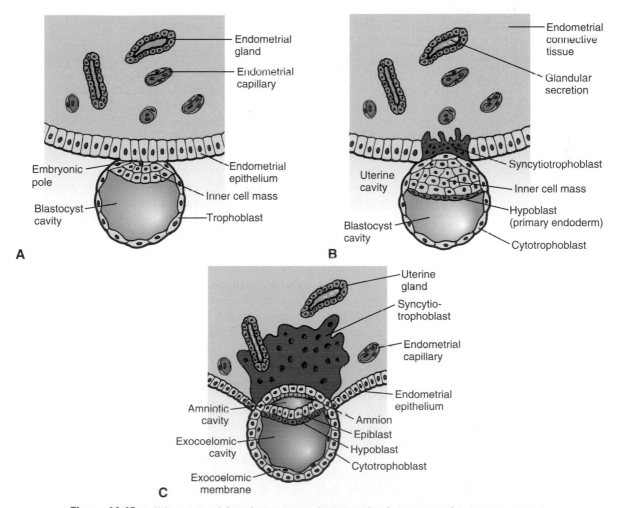

Figure 14-10 Cell lineage and fate during preimplantation development. Embryonic age is given in time after fertilization in humans: **A,** 6 days. **B,** 7 days. **C,** 8 days post fertilization. *See Sources & Acknowledgments.*

all of the organs are now established, and the cellular components necessary for their full development are now in place. It is during this phase of embryonic development that **neural tube defects** occur, as we explore next.

Neural Tube Defects

Neural tube defects (NTDs) are among the most common and devastating birth defects. **Anencephaly** and **spina bifida** are NTDs that frequently occur together in families and are considered to have a common pathogenesis. In anencephaly, the forebrain, overlying meninges, vault of the skull, and skin are all absent. Many infants with anencephaly are stillborn, and those born alive survive a few hours at most. Approximately two thirds of affected infants are female. In spina bifida, there is failure of fusion of the arches of the vertebrae, typically in the lumbar region. There are varying degrees of severity, ranging from spina bifida occulta, in which the defect is in the bony arch only, to spina bifida aperta, in which a bone defect is also associated with meningocele (protrusion of meninges) or meningomyelocele (protrusion of neural elements as well as meninges through the defect; see Fig. 17-3).

As a group, NTDs are a leading cause of stillbirth, death in early infancy, and handicap in surviving children. Their incidence at birth is variable, ranging from almost 1% in Ireland to 0.2% or less in the United States. The frequency also appears to vary with social factors and season of birth and oscillates widely over time (with a marked decrease in recent years; see later discussion).

A small proportion of NTDs have known specific causes, for example, amniotic bands (see Fig. 14-3), some single-gene defects with pleiotropic expression, some chromosomal disorders, and some teratogens. Most NTDs, however, are isolated defects of unknown cause.

Maternal Folic Acid Deficiency and Neural Tube Defects. NTDs were long believed to follow a multifactorial inheritance pattern determined by multiple genetic and environmental factors, as introduced generally in Chapter 8. It was therefore a stunning discovery to find that the single greatest factor in causing NTDs is a vitamin deficiency. The risk for NTDs was found to be inversely correlated with maternal serum folic acid levels during pregnancy, with a threshold of 200 µg/L, below which the risk for NTD becomes significant. Along with reduced blood folate levels, elevated homocysteine levels were also seen in the mothers of children with NTDs, suggesting that a biochemical abnormality was present at the step of recycling of tetrahydrofolate to methylate homocysteine to methionine (see Fig. 12-8). Folic acid levels are strongly influenced by dietary intake and can become depressed during pregnancy even with a typical intake of approximately 230 µg/day. The impact of folic acid deficiency is exacerbated by a genetic variant of the enzyme 5,10-methylenetetrahydrofolate reductase (MTHFR), caused by a common missense mutation that makes the enzyme less stable than normal. Instability of this enzyme hinders the recycling of tetrahydrofolate and interferes with the methylation of homocysteine to methionine.

The mutant allele is so common in many populations that between 5% and 15% of the population is homozygous for the variant. In studies of infants with NTDs and their mothers, it was found that mothers of infants with NTDs were twice as likely as controls to be homozygous for the mutant allele encoding the unstable enzyme. How this enzyme defect contributes to NTDs and whether the abnormality is a direct result of elevated homocysteine levels, depressed methionine levels, or some other metabolic derangement remain undefined.

Prevention of Neural Tube Defects. There are two approaches to preventing NTDs. The first is to educate women to supplement their diets with folic acid 1 month before conception and continuing for 2 months after conception during the period when the neural tube forms. Dietary supplementation with 400 to 800 µg of folic acid per day for women who plan their pregnancies has been shown to reduce the incidence of NTDs by more than 75%. Much active discussion is ongoing as to whether the entire food supply should be supplemented with folic acid as a public health measure to avoid the problem of women failing to supplement their diets individually during pregnancy.

The second approach is to apply prenatal screening for all pregnancies and offer prenatal diagnosis to high-risk pregnancies. Prenatal diagnosis of anencephaly and most cases of open spina bifida relies on detecting excessive levels of **alpha-fetoprotein** (AFP) and other fetal substances in the amniotic fluid and by **ultrasonographic scanning**, as we shall discuss further in Chapter 17. However, less than 5% of all patients with NTDs are born to women with previous affected children. For this reason, screening of all pregnant women for NTDs by measurements of AFP and other fetal substances in maternal serum is now widespread. Thus we can anticipate that a combination of preventive folic acid therapy and maternal AFP screening will provide major public health benefits by drastically reducing the incidence of NTDs.

Human Fetal Development

The embryonic phase of development occupies the first 2 months of pregnancy and is followed by the **fetal phase** of development, which is concerned primarily with the maturation and further differentiation of the components of the organs. For some organ systems, development does not cease at birth. For example, the brain undergoes substantial postnatal development, and limbs undergo epiphyseal growth and ultimately closure after puberty.

The Germ Cell: Transmitting Genetic Information

In addition to growth and differentiation of somatic tissues, the organism must also specify which cells will go on to become the gametes of the mature adult. The **germ cell compartment** serves this purpose. As described in Chapter 2, cells in the germ cell compartment become committed to undergoing gametogenesis and meiosis in order that the species can pass on its genetic complement and facilitate the recombination and random assortment of chromosomes. In addition, the sex-specific epigenetic imprint that certain genes require must be reset within the germ cell compartment (see Chapters 3, 6, and 7).

The Stem Cell: Maintaining Regenerative Capacity in Tissues

In addition to specifying the program of differentiation that is necessary for development, the organism must also set aside tissue-specific **stem cells** that can regenerate differentiated cells during adult life. The best-characterized example of these cells is in the hematopoietic system. Among the 10^{11} to 10^{12} nucleated hematopoietic

cells in the adult organism are approximately 10^4 to 10^5 cells that have the potential to generate any of the more specialized blood cells on a continuous basis during a lifetime. Hematopoietic stem cells can be transplanted to other humans and completely reconstitute the hematopoietic system (see Chapter 13). A system of interacting gene products maintains a properly sized pool of hematopoietic stem cells. These regulators permit a balance between the maintenance of stem cells through self-replication and the generation of committed precursor cells that can go on to develop into the various mature cells of the hematopoietic system (Fig. 14-11) (see Box).

Fate, Specification, and Determination

As an undifferentiated cell undergoes the process of differentiation, it moves through a series of discrete steps in which it manifests various distinct functions or attributes until it reaches its ultimate destination, referred to as its **fate** (e.g., when a precursor cell becomes an erythrocyte, a keratinocyte, or a cardiac myocyte). In the developing organism, these attributes not only vary across the recognizable cell types but also change over time. Early during differentiation, a cell undergoes **specification** when it acquires specific characteristics but can

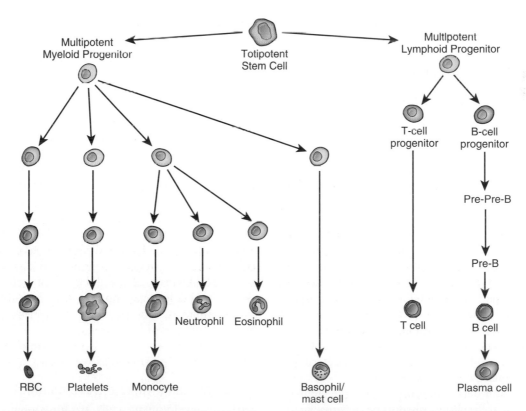

Figure 14-11 The development of blood cells is a continuous process that generates a full complement of cells from a single, totipotent hematopoietic stem cell. This hematopoietic stem cell is a committed stem cell that differentiated from a more primitive mesodermal stem cell. RBC, Red blood cell. *See Sources & Acknowledgments.*

EMBRYONIC STEM CELL TECHNOLOGY

Inner cell mass cells are believed to be capable of forming any tissue in the body. This is suspected of being true in humans (but has never been tested for obvious ethical reasons) but has been proved to be true in mice. The full developmental potential of inner cell mass cells is the basis of the experimental field of **embryonic stem cell** technology in mice, a technology that is crucial for generating animal models of human genetic disease (Fig. 14-12). In this

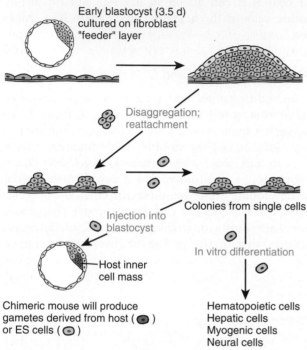

Early blastocyst (3.5 d) cultured on fibroblast "feeder" layer

Disaggregation; reattachment

Colonies from single cells

Injection into blastocyst

In vitro differentiation

Host inner cell mass

Chimeric mouse will produce gametes derived from host ( ) or ES cells ( )

Hematopoietic cells
Hepatic cells
Myogenic cells
Neural cells

Figure 14-12 Embryonic stem (ES) cells are derived directly from the inner cell mass, are euploid, and can contribute to the germline. Cultured ES cells differentiated in vitro can give rise to a variety of different cell types.

technique, mouse inner cell mass cells are grown in culture as embryonic stem cells and undergo genetic manipulation to introduce a given mutation into a specific gene. These cells are then injected into the inner cell mass of another early mouse embryo. The mutated cells are incorporated into the inner cell mass of the recipient embryo and contribute to many tissues of that embryo, forming a **chimera** (a single embryo made up of cells from two different sources). If the mutated cells contribute to the germline in a chimeric animal, the offspring of that animal can inherit the engineered mutations. The ability of the recipient embryo to tolerate the incorporation of these pluripotent, nonspecified cells, which then undergo specification and can contribute to any tissue in a living mouse, is the converse of regulative development, the ability of an embryo to tolerate removal of some cells.

Human stem cells (HSCs) made from unused fertilized embryos are the subject of intensive research as well as ethical controversy. Although the use of HSCs for cloning an entire human being is considered highly unethical and universally banned, current research is directed toward generating particular cell types from HSCs to provide cellular models of human genetic diseases or to repair damaged tissues and organs, a goal of **regenerative medicine** (see Chapter 13).

Induced pluripotent stem (iPS) cells are another source of early stem cells that can be cultured and differentiated in vitro into particular cell types. Human iPS cells are derived through reprogramming of readily available and ethically uncontroversial somatic cells, such as fibroblasts, to very early stem cells through the introduction of certain transcription factors into the cells (e.g., the transcription factors Oct4, Sox2, cMyc, and Klf4). This technology makes what were previously inaccessible tissues from patients with genetic disorders, such as cardiac myocytes from patients with cardiomyopathies, or central nervous system neurons from patients with neurodegenerative diseases available for research and, ultimately, perhaps tissue-based therapy using their own gene-corrected iPS cells. Shinya Yamanaka was awarded the 2012 Nobel Prize in Physiology or Medicine for his demonstration of the feasibility of creating iPS cells.

still be influenced by environmental cues (signaling molecules, positional information) to change its ultimate fate. These environmental clues are primarily derived from neighboring cells by direct cell-cell contact or by signals received at the cell surface from soluble substances, including positional information derived from where a cell sits in a gradient of various **morphogens.** Eventually a cell either irreversibly acquires attributes or has irreversibly been committed to acquire those attributes (referred to as **determination**). With the exception of the germ cell and stem cell compartments just described, all cells undergo specification and determination to their ultimate developmental fate.

Specification and determination involve the stepwise acquisition of a stable cellular phenotype of gene expression specific to the particular fate of each cell—nerve cells make synaptic proteins but do not make hemoglobin, whereas red blood cells do not make synaptic

proteins but must make hemoglobin. With the exception of lymphocyte precursor cells undergoing DNA rearrangements in the T-cell receptor or immunoglobulin genes (see Chapter 3), the particular gene expression profile responsible for the differentiated cellular phenotype does not result from permanent changes in DNA sequence. Instead, the regulation of gene expression depends on **epigenetic** changes, such as stable transcription complexes, modification of histones in chromatin, and methylation of DNA (see Chapter 3). The epigenetic control of gene expression is responsible for the loss of developmental **plasticity,** as we discuss next.

Regulative and Mosaic Development

Early in development, cells are functionally equivalent and subject to dynamic processes of specification, a phenomenon known as **regulative development**. In regulative development, removal or ablation of part of an embryo can be compensated for by the remaining similar cells. In contrast, later in development, each of the cells in some parts of the embryo has a distinct fate, and in each of those parts, the embryo only appears to be homogeneous. In this situation, known as **mosaic** development, loss of a portion of an embryo would lead to the failure of development of the final structures that those cells were fated to become. Thus the developmental plasticity of the embryo generally declines with time.

Regulative Development and Twinning

That early development is primarily regulative has been demonstrated by basic embryological experiments and confirmed by observations in clinical medicine. Identical (**monozygotic**) twins are the natural experimental evidence that early development is regulative. The most common form of identical twinning occurs in the second half of the first week of development, effectively splitting the inner cell mass into two halves, each of which develops into a normal fetus (Fig. 14-13). Were the embryo even partly regulated by mosaic development at this stage, the twins would develop only partially and consist of complementary parts. This is clearly not the case, because twins are generally completely normally developed and eventually attain normal size through prenatal and postnatal growth.

The various forms of monozygotic twinning demonstrate regulative development at several different stages. **Dichorionic twins** result from cleavage at the four-cell stage. **Monochorionic twins** result from a cleaved inner cell mass. **Monoamniotic twins** result from an even later cleavage, in this case within the bilayered embryo, which then forms two separate embryos but only one extraembryonic compartment that goes on to make the single amnion. All of these twinning events demonstrate that these cell populations can reprogram their development to form complete embryos from cells that, if cleavage had not occurred, would have contributed to only part of an embryo.

The successful application of the technique of **preimplantation diagnosis** (see Chapter 17) also illustrates that early human development is regulative. In this procedure, male and female gametes are harvested from the presumptive parents and fertilized in vitro (Fig. 14-14; see also Fig. 17-1). When these fertilized embryos have reached the eight-cell stage (at day 3), a biopsy microneedle is used to remove some of the cells of the developing blastocyst. The isolated cell with its clearly visible nucleus can then be examined using a variety of appropriate cytogenetic or genomic tests to ascertain if the embryo is suitable for implantation. Embryos composed of the remaining seven cells that are not affected by the disease can then be selected and implanted in the mother. The capacity of the embryo to recover from the biopsy of one of its eight cells is attributable to regulative development. Were those cells removed by biopsy fated to form a particular part or segment of the body (i.e., governed by mosaic development), one would predict that these parts of the body would be absent or defective in the mature individual. Instead, the embryo has compensatory mechanisms to replace those cells, which then undergo normal development as specified by their neighboring cells.

Mosaic Development

Embryonic development generally proceeds from more regulative to more mosaic development. Typical identical twinning early in development, as mentioned earlier,

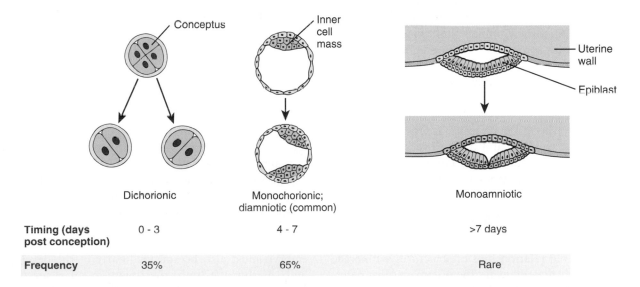

	Dichorionic	Monochorionic; diamniotic (common)	Monoamniotic
Timing (days post conception)	0 - 3	4 - 7	>7 days
Frequency	35%	65%	Rare

Figure 14-13 The arrangement of placental membranes in monozygotic twins depends on the timing of the twinning event. Dichorionic twins result from a complete splitting of the entire embryo, leading to duplication of all extraembryonic tissues. Monochorionic diamniotic twins are caused by division of the inner cell mass at the blastocyst stage. Monoamniotic twins are caused by division of the epiblast but not the hypoblast.

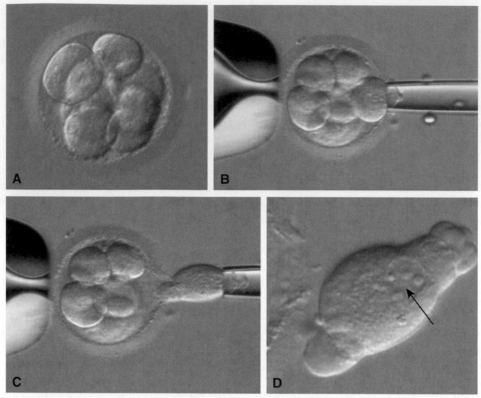

Figure 14-14 Blastomere biopsy of a human cleavage stage embryo. **A,** Eight-cell embryo, day 3 after fertilization. **B,** Embryo on holding pipette (*left*) with biopsy pipette (*right*) breaching the zona pellucida. **C,** Blastomere removal by suction. **D,** Blastomere removed by biopsy with a clearly visible single nucleus (indicated by *arrow*). *See Sources & Acknowledgments.*

is an illustration of regulative development. However, *later* embryo cleavage events result in the formation of **conjoined twins,** which are two fetuses that share body structures and organs because the cleavage occurred after the transition from regulative to mosaic development, too late to allow complete embryos.

Interestingly, in some adult nonhuman species, ablation of a specific tissue may not limit development. For example, the mature salamander can regenerate an entire tail when it is cut off, apparently retaining a population of cells that can reestablish the developmental program for the tail after trauma. One of the goals of research in developmental biology is to understand this process in other species and potentially harness it in practice for human regenerative medicine.

Axis Specification and Pattern Formation

A critical function of the developing organism is to specify the spatial relationships of structures within the embryo. In early development, the organism must determine the relative orientation of a number of body segments and organs, involving the establishment of three axes:

- The head-to-tail axis, which is termed the **cranial-caudal** or **anterior-posterior** axis, is established very

early in embryogenesis and is probably determined by the entry position of the sperm that fertilizes the egg. (It is referred to as the rostral-caudal axis later in development.)
- The **dorsal-ventral** axis is the second dimension, and here, too, a series of interacting proteins and signaling pathways are responsible for determining dorsal and ventral structures. The morphogen sonic hedgehog (discussed later) participates in setting up the axis of dorsal-ventral polarity along the spinal cord.
- Finally, a **left-right** axis must be established. The left-right axis is essential for proper heart development and positioning of viscera; for example, an abnormality in the X-linked gene *ZIC3*, involved in left-right axis determination, is associated with cardiac anomalies and **situs inversus,** in which some thoracic and abdominal viscera are on the wrong side of the chest and abdomen.

The three axes that must be specified in the whole embryo must also be specified early in the developing limb. Within the limb, the organism must specify the proximal-distal axis (shoulder to fingertip), the anterior-posterior axis (thumb to fifth finger), and the dorsal-ventral axis (dorsum to palm). On a cellular scale, individual cells also develop an axis of polarity, for example, the basal-apical axis of the proximal renal tubular cells or the axons and dendrites of a neuron. Thus, specifying axes in the whole embryo, in limbs, and in cells is a fundamental process in development.

Once an organismal axis is determined, the embryo then overlays a patterning program onto that axis. Conceptually, if axis formation can be considered as the drawing of a line through an undeveloped mass of cells and specifying which end is to be the head and which end the tail, then patterning is the division of the embryo into segments and the assignment to these segments of an identity, such as head, thorax, or abdomen. The *HOX* genes (discussed in the next section) have major roles in determining the different structures that develop along the anterior-posterior axis. The end result of these pattern specification programs is that cells or groups of cells are assigned an identity related primarily to their position within the organism. This identity is subsequently used by the cells as an instruction to specify how development should proceed.

Pattern Formation and the HOX Gene System

The **homeobox (*HOX*)** gene system, first described in the fruit fly *Drosophila melanogaster,* constitutes a paradigm in developmental biology. *HOX* genes are so named because the proteins they encode are transcription factors that contain a conserved DNA-binding motif called the homeodomain. The segment of the gene encoding the homeodomain is called a *homeobox,* thus giving the gene family its name, *HOX.*

Many species of animals have *HOX* genes, and the homeodomains encoded by these genes are similar; however, different species contain different numbers of *HOX* genes; for example, fruit flies contain 8 and humans nearly 40. The 40 human *HOX* genes are organized into four clusters on four different chromosomes. Strikingly, the order of the individual genes within the clusters is conserved across species. The human *HOX* gene clusters (Fig. 14-15) were generated by a series of gene duplication events, conceptually similar to those described in Chapter 11 for the evolution of the globin gene family. Initially, ancient events duplicated the original ancestral *HOX* gene in tandem along a single chromosome. Subsequent duplications of this single set of *HOX* genes and relocation of the new gene set to other locations in the genome resulted in four unlinked *HOX* gene clusters in humans (and other mammals) named *HOXA, HOXB, HOXC,* and *HOXD.*

Unique combinations of *HOX* gene expression in small groups of cells, located in particular regions of the

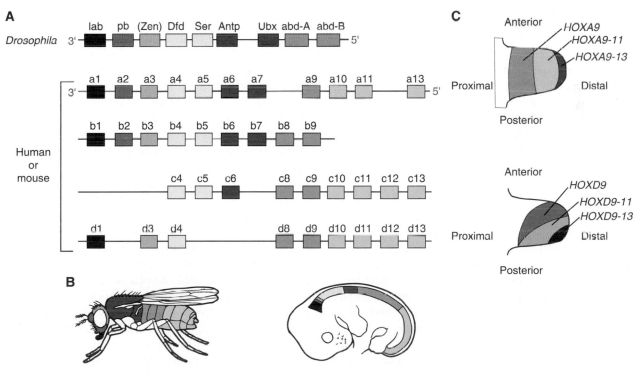

Figure 14-15 Action and arrangement of *HOX* genes. A, An ancestral *HOX* gene cluster in a common ancestor of vertebrates and invertebrates has been quadruplicated in mammals, and individual members of the ancestral cluster have been lost. **B,** The combination of *HOX* genes expressed in adjacent regions along the anteroposterior axis of developing embryos selects a unique developmental fate (as color-coded in the segments of the fly and human embryo). **C,** In the developing limbs, different combinations of *HOXA* and *HOXD* genes are expressed in adjacent zones that help specify developmental fate along the proximal-distal and anterior-posterior axes. *See Sources & Acknowledgments.*

embryo, help determine the developmental fate of those regions. Just as specific combinations of *HOX* genes from the single *HOX* gene cluster in the fly are expressed along the anterior-posterior axis of the body and regulate different patterns of gene expression and therefore different body structures (see Fig. 14-15), mammals use a number of *HOX* genes from different clusters to accomplish similar tasks. Early, in the whole embryo, HOX transcription factors specify the anterior-posterior axis: the *HOXA* and *HOXB* clusters, for example, act along the rostral-caudal axis to determine the identity of individual vertebrae and somites. Later in development, the *HOXA* and *HOXD* clusters determine regional identity along the axes of the developing limb.

One interesting aspect of *HOX* gene expression is that the order of the genes in a cluster parallels the position in the embryo in which that gene is expressed and the time in development when it is expressed (see Fig. 14-15). In other words, the position of a *HOX* gene in a cluster is collinear with both the timing of expression and the location of expression along the anterior-posterior axis in the embryo. For example, in the *HOXB* cluster, the genes expressed first and in the anterior portion of the embryo are at one end of the cluster; the order of the rest of the genes in the cluster parallels the order in which they are expressed, both by location along the anterior-posterior axis of the embryo and by timing of expression. Although this gene organization is distinctly unusual and is not a general feature of gene organization in the genome (see Chapter 3), a similar phenomenon is seen within another developmentally regulated human gene family, the globin gene clusters (see Chapter 11). In both cases, the association of spatial organization in the genome with temporal expression in development presumably reflects long-range regulatory elements in the genome that govern the epigenetic packaging and accessibility of different genes at different times in the embryo.

The *HOX* gene family thus illustrates several important principles of developmental biology and evolution:
- *First*, a group of genes functions together to accomplish similar general tasks at different times and places in the embryo.
- *Second*, homologous structures are generated by sets of homologous transcription factors derived from common evolutionary predecessors. For example, flies and mammals have a similar basic body plan (head anterior to the trunk, with limbs emanating from the trunk, cardiorespiratory organs anterior to digestive), and that body plan is specified by a set of genes that were passed down through common evolutionary predecessors.
- And *third*, although it is not usually the case with genes involved in development, the *HOX* genes show a remarkable genomic organization within a cluster that correlates with their function during development.

CELLULAR AND MOLECULAR MECHANISMS IN DEVELOPMENT

In this section, we review the basic cellular and molecular mechanisms that regulate development (see Box). We illustrate each mechanism with a human birth defect or disease that results from the failure of each of these normal mechanisms.

FUNDAMENTAL MECHANISMS OPERATING IN DEVELOPMENT

- Gene regulation by transcription factors
- Cell-cell signaling by direct contact and by morphogens
- Induction of cell shape and polarity
- Cell movement
- Programmed cell death

Gene Regulation by Transcription Factors

Transcription factors control development by controlling the expression of other genes, some of which are also transcription factors. Groups of transcription factors that function together are referred to as **transcriptional regulatory modules,** and the functional dissection of these modules is an important task of the developmental geneticist and, increasingly, of genome biologists. Some transcription factors activate target genes and others repress them. Still other transcription factors have both activator and repressor functions (so-called bifunctional transcription factors); noncoding RNAs such as microRNAs also interact with target sequences and can activate or repress gene expression. The recruitment of these various activators and repressors within chromatin can be guided by histone modifications such as acetylation, and the regulation of histone modifications is accomplished by histone acetyltransferases and deacetylases (see Chapter 3). These epigenetic changes to histones are marks that indicate whether a particular gene is likely to be active or inactive. Regulatory modules control development by causing different combinations of transcription factors to be expressed at different places and at different times to direct the spatiotemporal regulation of development. By directing differential gene expression across space and time, the binding of various transcriptional regulatory modules to transcriptional complexes is controlled by histone modifications and is a central element of the development of the embryo.

A transcriptional regulatory complex consists of a large number of general transcription factors joined with the specific transcription factors that are responsible for creating the selectivity of a transcriptional complex (Fig. 14-16). Most general transcription factors are found in thousands of transcriptional complexes

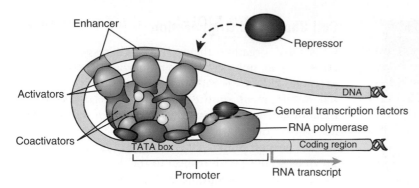

Figure 14-16 General transcription factors, shown in *blue*, and RNA polymerase bind to *cis*-acting sequences closely adjacent to the messenger RNA (mRNA) transcriptional start site; these *cis*-acting sequences are collectively referred to as the promoter. More distal enhancer or silencer elements bind specialized and tissue-specific transcription factors. Coactivator proteins facilitate a biochemical interaction between specialized and general transcription factors. *See Sources & Acknowledgments.*

throughout the genome, and, although each is essential, their roles in development are nonspecific. Specific transcription factors also participate in forming transcription factor complexes, mostly under the control of epigenetic marks of histone modifications, but only in specific cells or at specific times in development, thereby providing the regulation of gene expression that allows developmental processes to be exquisitely controlled.

The importance of transcription factors in normal development is illustrated by an unusual mutation of *HOXD13* that causes **synpolydactyly,** an incompletely dominant condition in which heterozygotes have interphalangeal webbing and extra digits in their hands and feet. Rare homozygotes have similar but more severe abnormalities and also have bone malformations of the hands, wrists, feet, and ankles (Fig. 14-17). The *HOXD13* mutation responsible for synpolydactyly is caused by expansion of a polyalanine tract in the aminoterminal domain of the protein; the normal protein contains 15 alanines, whereas the mutant protein contains 22 to 24 alanines. The polyalanine expansion that causes synpolydactyly is likely to act by a gain-of-function mechanism (see Chapter 11), as heterozygosity for a *HOXD13* loss-of-function mutation has only a mild effect on limb development, characterized by a rudimentary extra digit between the first and second metatarsals and between the fourth and fifth metatarsals of the feet. Regardless of the exact mechanism, this condition demonstrates that a general function for *HOX* genes is to determine regional identity along specific body axes during development.

Morphogens and Cell to Cell Signaling

One of the hallmarks of developmental processes is that cells must communicate with each other to develop proper spatial arrangements of tissues and cellular subtypes. This communication occurs through cell signaling mechanisms. These cell-cell communication systems are commonly composed of a cell surface receptor and the molecule, called a **ligand,** that binds to it. On ligand binding, receptors transmit their signals through intracellular signaling pathways. One of the common ligand-receptor pairs is the fibroblast growth factors and their

receptors. There are 23 recognized members of the fibroblast growth factor gene family in the human, and many of them are important in development. The fibroblast growth factors serve as ligands for tyrosine kinase receptors. Abnormalities in fibroblast growth factor receptors cause diseases such as **achondroplasia (Case 2)** (see Chapter 7) and certain syndromes that involve abnormalities of craniofacial development, referred to as **craniosynostoses** because they demonstrate premature fusion of cranial sutures in the skull.

One of the best examples of a developmental morphogen is **hedgehog,** originally discovered in *Drosophila* and named for its ability to alter the orientation of epidermal bristles. Diffusion of the hedgehog protein creates a gradient in which different concentrations of the protein cause surrounding cells to assume different fates. In humans, several genes closely related to *Drosophila* hedgehog also encode developmental morphogens; one example is the gene sonic hedgehog (*SHH*). Although the specific programs controlled by hedgehog in *Drosophila* are very different from those controlled by its mammalian counterparts, the underlying themes and molecular mechanisms are similar. For example, secretion of the SHH protein by the notochord and the floor plate of the developing neural tube generates a gradient that induces and organizes the different types of cells and tissues in the developing brain and spinal cord (Fig. 14-18A). SHH is also produced by a small group of cells in the limb bud to create what is known as the **zone of polarizing activity,** which is responsible for establishing the posterior side of the developing limb bud and the asymmetrical pattern of digits within individual limbs (see Fig. 14-18B).

Mutations that inactivate the *SHH* gene in humans cause birth defects that may be inherited as autosomal dominant traits, which demonstrates that a 50% reduction in gene expression is sufficient to produce an abnormal phenotype, presumably by altering the magnitude of the hedgehog protein gradient. Affected individuals usually exhibit **holoprosencephaly** (failure of the midface and forebrain to develop), leading to cleft lip and palate, hypotelorism (eyes that are closely spaced together), and absence of forebrain structures. On occasion, however, the clinical findings are mild or subtle such as, for

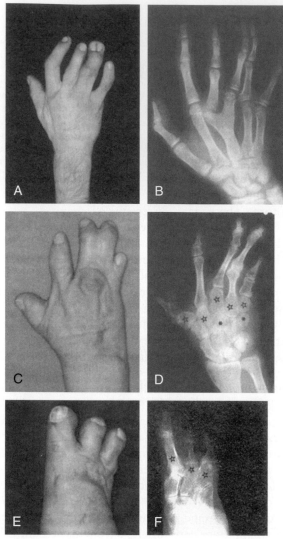

Figure 14-17 An unusual gain-of-function mutation in *HOXD13* creates an abnormal protein with a dominant negative effect. Photographs and radiographs show the synpolydactyly phenotype. **A** and **B,** Hand and radiograph of an individual heterozygous for a *HOXD13* mutation. Note the branching metacarpal III and the resulting extra digit IIIa. The syndactyly between digits has been partially corrected by surgical separation of III and IIIa-IV. **C** and **D,** Hand and radiograph of an individual homozygous for a *HOXD13* mutation. Note syndactyly of digits III, IV, and V and their single knuckle; the transformation of metacarpals I, II, III, and V to short carpal-like bones *(stars);* two additional carpal bones *(asterisks);* and short second phalanges. The radius, ulna, and proximal carpal bones appear normal. **E** and **F,** Foot and radiograph of the same homozygous individual. Note the relatively normal size of metatarsal I, the small size of metatarsal II, and the replacement of metatarsals III, IV, and V with a single tarsal-like bone *(stars). See Sources & Acknowledgments.*

example, a single central incisor or partial absence of the corpus callosum (Fig. 14-19). Because variable expressivity has been observed in members of the same family, it cannot be due to different mutations and instead must reflect the action of modifier genes at other loci, chance, environment, or some combination of all three.

Cell Shape and Organization

Cells must organize themselves with respect to their position and polarity in their microenvironment. For example, kidney epithelial cells must undergo differential development of the apical and basal aspects of their organelles to effect reabsorption of solutes. The acquisition of polarity by a cell can be viewed as the cellular version of axis determination (as discussed in a previous section) with respect to the development of the overall embryo. Under normal circumstances, each renal tubular cell elaborates on its cell surface a filamentous structure, known as a primary cilium. One hypothesis is that the primary cilium is designed to sense fluid flow in the developing kidney tubule and signal the cell to stop proliferating and to polarize. Another hypothesis is that the primary cilium is a sort of cellular antenna that concentrates signal transduction components to facilitate activation or repression of developmental pathways.

There is substantial evidence that the sonic hedgehog signal transduction pathway acts in this fashion. Adult **polycystic kidney disease** (Case 37) is caused by loss of function of one of two protein components of primary cilia, polycystin 1 or polycystin 2, so that the cells fail to sense fluid flow or to activate or repress signal transduction pathways properly. As a result, they continue to proliferate and do not undergo the appropriate developmental program of polarization, in which they stop dividing and display polarized expression of certain proteins on either the apical or basal aspect of the tubular epithelial cells (Fig. 14-20). The continued cell division leads to the formation of cysts, fluid-filled spaces lined by renal tubular cells.

Cell Migration

Programmed cell movement is critical in development, and nowhere is it more important than in the central nervous system. The central nervous system is developed from the neural tube, a cylinder of cells created during weeks 4 to 5 of embryogenesis. Initially, the neural tube is only a single cell layer thick, a pseudostratified columnar epithelium. Once sufficient neuroepithelial cells are produced by symmetrical division, these cells divide asymmetrically as neural stem cells. These neural stem cells stretch from the apical surface adjacent to the ventricle to the basal surface. The nucleus of these neural stem cells is adjacent to the apical surface in the ventricular cell layer situated adjacent to the ventricle, and the fiber of these cells stretches to the basal or pial surface as the so-called radial glial cells. These radial glia are one type of neural stem cells, which divide asymmetrically to generate new neural stem cells as well as committed neuronal precursors and secondary neural stem cells. These set up more basally located neural stem cells that can amplify the number of cells

A

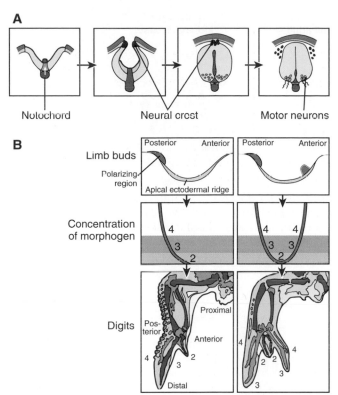

B

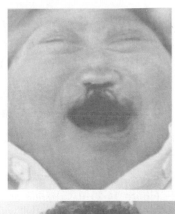

Figure 14-18 A, Transverse section of the developing neural tube. Sonic hedgehog protein released from the notochord diffuses upward to the ventral portion of the developing neural tube (*brown*); high concentrations immediately above the notochord induce the floor plate, whereas lower concentrations more laterally induce motor neurons. Ectoderm above (dorsal to) the neural tube releases bone morphogenetic proteins that help induce neural crest development at the dorsal edge of the closing neural tube (*dark purple*). **B,** Morphogenetic action of the sonic hedgehog (SHH) protein during limb bud formation. SHH is released from the zone of polarizing activity (labeled polarizing region in **B**) in the posterior limb bud to produce a gradient (shown with its highest levels as 4, declining to 2). Mutations or transplantation experiments that create an ectopic polarizing region in the anterior limb bud cause a duplication of posterior limb elements. *See Sources & Acknowledgments.*

Figure 14-19 Variable expressivity of an *SHH* mutation. The mother and her daughter carry the same missense mutation in *SHH,* but the daughter is severely affected with microcephaly, abnormal brain development, hypotelorism, and a cleft palate, whereas the only manifestation in the mother is a single central upper incisor. *See Sources & Acknowledgments.*

produced from a given radial glial progenitor. Postmitotic neuronal precursors then migrate outwards toward the pial surface along the radial glia. The central nervous system is built by waves of migration of these neuronal precursors. The neurons that populate the inner layers of the cortex migrate earlier in development, and each successive wave of neurons passes through the previously deposited, inner layers to form the next outer layer (Fig. 14-21).

Lissencephaly (literally, "smooth brain") is a severe abnormality of brain development causing profound intellectual disability. This developmental defect is one component of the **Miller-Dieker syndrome** (Case 32), which is caused by a contiguous gene deletion syndrome that involves one copy of the *LIS1* gene on chromosome 17. When there is loss of *LIS1* function, the progressive

waves of migration of cortical neurons do not occur in an organized fashion because of reduced speeds of migration. The result is a thickened, hypercellular cerebral cortex with undefined cellular layers and poorly developed gyri, thereby making the surface of the brain appear smooth.

In addition to the neuronal migrations described, another remarkable example of cell migration involves the neural crest, a population of cells that arises from the dorsolateral aspect of the developing neural tube (see Fig. 14-18A). Neural crest cells must migrate from their original location at the dorsal and lateral surface of the neural tube to remarkably distant sites, such as the ventral aspect of the face, the ear, the heart, the gut, and many other tissues, including the skin, where they differentiate into pigmented melanocytes.

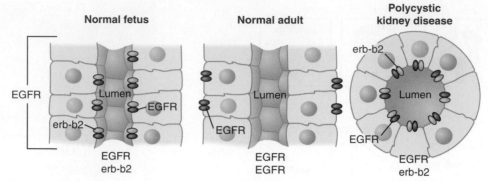

Figure 14-20 Polarization of epidermal growth factor receptor (EGFR) in epithelium from a normal fetus, a normal adult, and a patient with polycystic kidney disease. Fetal cells and epithelial cells from patients with polycystic kidney disease express a heterodimer of EGFR and erb-b2 at apical cell membranes. In normal adults, tubular epithelia express homodimeric complexes of EGFR at the basolateral membrane. *See Sources & Acknowledgments.*

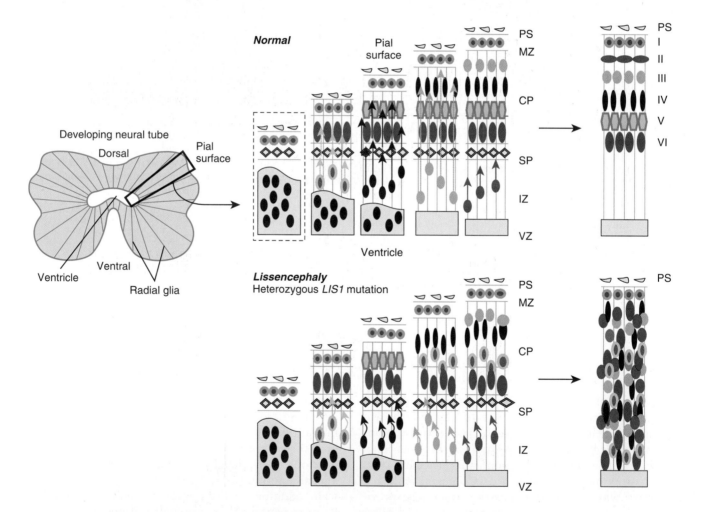

Figure 14-21 The role of neuronal migration in normal cortical development and the defective migration in individuals heterozygous for an *LIS1* mutation causing lissencephaly. *Top,* A radial slice is taken from a normal developing neural tube of the mouse, showing the progenitor cells at the ventricular zone (VZ). These cells divide, differentiate into postmitotic cells, and migrate radially along a scaffold made up of glia. The different shapes and colors represent the cells that migrate and form the various cortical layers: IZ, intermediate zone; SP, subplate; CP, cortical plate; MZ, marginal zone; PS, pial surface. The six distinguishable layers of the normal cortex (molecular, external granular, external pyramidal, internal granular, internal pyramidal, multiform) that occupy the region of the cortical plate are labeled I through VI. *Bottom,* Aberrant migration and failure of normal cortical development seen in lissencephaly. *See Sources & Acknowledgments.*

Population of the gut by neural crest progenitors gives rise to the autonomic innervation of the gut; failure of that migration leads to the aganglionic colon seen in **Hirschsprung disease (Case 22).** The genetics of Hirschsprung disease are complex (see Chapter 8), but a number of key signaling molecules have been implicated. One of the best characterized is the *RET* proto-oncogene. As discussed in Chapter 8, mutations in *RET* have been identified in approximately 50% of patients with Hirschsprung disease.

Another example of defects in neural crest development is the group of birth defects known as the **Waardenburg syndrome,** which includes defects in skin and hair pigmentation, coloration of the iris, and colon innervation (Fig. 14-22). This syndrome can be caused by mutations in at least four different transcription factors, each resulting in abnormalities in neural crest development.

Programmed Cell Death

Programmed cell death is a critical function in development and is necessary for the morphological development of many structures. It occurs wherever tissues need to be remodeled during morphogenesis, as during the separation of the individual digits, in perforation of the anal and choanal membranes, or in the establishment of communication between the uterus and vagina.

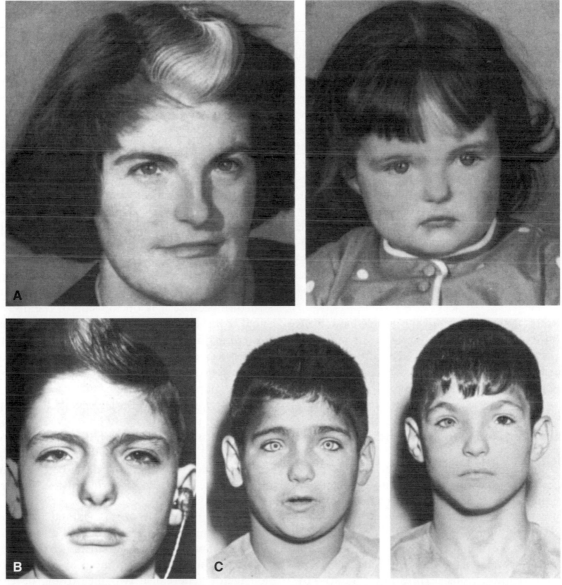

Figure 14-22 **Patients with type I Waardenburg syndrome. A,** Mother and daughter with white forelocks. **B,** A 10-year-old with congenital deafness and white forelock. **C,** Brothers, one of whom is deaf. There is no white forelock, but the boy on the right has heterochromatic irides. Mutations of *PAX3*, which encodes a transcription factor involved in neural crest development, cause type I Waardenburg syndrome. *See Sources & Acknowledgments.*

One major form of programmed cell death is **apoptosis**. Studies of mice with loss-of-function mutations in the *Foxp1* gene indicate that apoptosis is required for the remodeling of the tissues that form portions of the ventricular septum and cardiac outflow tract (**endocardial cushions**), to ensure the normal positioning of the origins of the aortic and pulmonary vessels. By eliminating certain cells, the relative position of the cushions is shifted into their correct location. It is also suspected that defects of apoptosis underlie some other forms of human congenital heart disease (see Chapter 8), such as the conotruncal heart defects of **DiGeorge syndrome** caused by deletion of the *TBX1* gene located in chromosome 22q11 (see Chapter 6). Apoptosis also occurs during development of the immune system to eliminate lymphocyte lineages that react to self, thereby preventing autoimmune disease.

INTERACTION OF DEVELOPMENTAL MECHANISMS IN EMBRYOGENESIS

Embryogenesis requires the coordination of multiple developmental processes in which proliferation, differentiation, migration, and apoptosis all play a part. For example, many processes must occur to convert a mass of mesoderm into a heart or a layer of neuroectoderm into a spinal cord. To understand how these processes interact and work together, developmental biologists typically study embryogenesis in a model organism, such as worms, flies, or mice. The general principles elucidated by these simpler, more easily manipulated systems can then be applied to understanding developmental processes in humans.

The Limb as a Model of Organogenesis

The vertebrate limb is a relatively simple and well-studied product of developmental processes. There is no genomic specification for a human arm to be approximately 1 m long, with one proximal bone, two bones in the forelimb, and 27 bones in the hand. Instead, the limb results from a series of regulated processes that specify development along three axes, the proximal-distal axis, the dorsal-ventral axis, and the anterior-posterior axis (Fig. 14-23).

Limbs begin as protrusions of proliferating cells, the **limb buds**, along the lateral edge of the mesoderm of the human embryo in the fourth week of development. The location of each limb bud along the anterior-posterior axis of the embryo (head-to-tail axis) is associated with the expression of a specific transcription factor at each location, Tbx4 for the hindlimbs and Tbx5 for the forelimbs, whose expression is induced by various combinations of fibroblast growth factor ligands. Thus the primarily proliferative process of limb bud outgrowth is activated by growth factors and transcription factors.

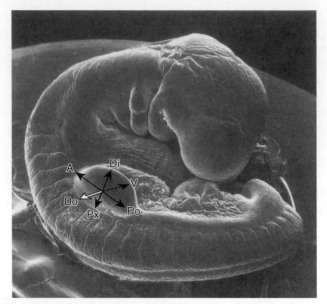

Figure 14-23 This scanning electron micrograph of a 4-week human embryo illustrates the early budding of the forelimb. Overlaid onto the bud are the three axes of limb specification: Do-V, dorsal-ventral (dorsal comes out of the plane of the photo, ventral goes into the plane of the photo); Px-Di, proximal-distal; and A-Po, anterior-posterior. *See Sources & Acknowledgments.*

The limb bud grows primarily in an outward, lateral expansion of the proximal-distal axis of the limb (see Fig. 14-18B). Whereas proximal-distal expansion of the limb is the most obvious process, the two other axes are established soon after the onset of limb bud outgrowth. The anterior-posterior axis is set up soon after limb bud outgrowth, with the thumb considered to be an anterior structure, because it is on the edge of the limb facing the upper body. The fifth finger is a posterior structure because it is on the side of the limb bud oriented toward the lower part of the body. During limb formation, the morphogen SHH is expressed in the posterior aspect of the developing limb bud, and its expression level forms a gradient that is primarily responsible for setting up the anterior-posterior axis in the developing limb (see Fig. 14-18B). Defects in anterior-posterior patterning in the limb cause excessive digit patterning, manifested as polydactyly, or failure of complete separation of developing digits, manifested as syndactyly. The dorsal-ventral axis is also established, resulting in a palm or sole on the ventral side of the hand and foot, respectively.

One can now begin to understand the mechanisms underlying birth defect syndromes by applying knowledge from molecular developmental biology to human disorders. For example, mutations in the *GLI3* transcription factor gene cause two pleiotropic developmental anomaly syndromes, the **Greig cephalopolysyndactyly syndrome (GCPS)** and the **Pallister-Hall syndrome** (see Fig. 14-1). These two syndromes comprise distinct combinations of limb, central nervous system, craniofacial,

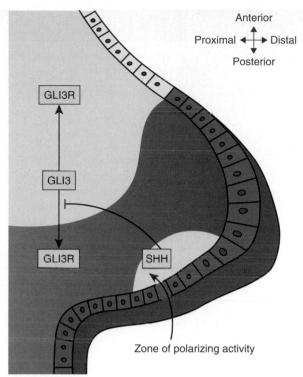

Figure 14-24 Schematic diagram of the anterior-posterior and proximal-distal axes of the limb bud and its molecular components. In this diagram, the anterior aspect is up and the distal aspect is to the right. SHH expression occurs in the zone of polarizing activity of the posterior limb bud, and SHH is activated by the *dHand* gene. SHH inhibits conversion of the GLI3 transcription factor to GLI3R in the posterior regions of the limb bud. However, SHH activity does not extend to anterior regions of the bud. The absence of SHH allows GLI3 to be converted to GLI3R (a transcriptional repressor) in the anterior limb bud. By this mechanism, the anterior-posterior axis of the limb bud is established with a gradient of GLI3 versus GLI3R. *See Sources & Acknowledgments.*

airway, and genitourinary anomalies that are caused by perturbed balance in the production of two variant forms of GLI3, referred to as GLI3 and GLI3R, as shown in Figure 14-24. GLI3 is part of the SHH signaling pathway. SHH signals, in part, through a cell surface receptor encoded by a gene called *PTCH1*, which is concentrated in the cilium of cells during development. Mutations in *PTCH1* cause the **nevoid basal cell carcinoma syndrome**. Also known as **Gorlin syndrome,** this syndrome comprises craniofacial anomalies and occasional polydactyly that are similar to those seen in GCPS, but in addition, Gorlin syndrome manifests

dental cysts and susceptibility to basal cell carcinoma. By considering Gorlin syndrome and GCPS, one can appreciate that the two disorders share phenotypic manifestations precisely because the genes that are mutated in the two disorders have overlapping effects in the same developmental genetic pathway. A third protein in the SHH signaling pathway, the CREB-binding protein, or CBP, is a transcriptional coactivator of the GLI3 transcription factor. Mutations in CBP cause the **Rubinstein-Taybi syndrome** (see Fig. 14-5), which also shares phenotypic manifestations with GCPS and Gorlin syndrome.

CONCLUDING COMMENTS

Many other examples of this phenomenon could be cited, but the key points to emphasize are that genes are the primary regulators of developmental processes, their protein products function in developmental genetic pathways, and these pathways are employed in related developmental processes in a number of organ systems. Understanding the molecular basis of gene function, how those functions are organized into modules, and how abnormalities in those modules cause and correlate with malformations and pleiotropic syndromes forms the basis of the modern clinical approach to human birth defects. The understanding of these developmental pathways in great detail may also provide an avenue in the future to devise therapies that target appropriate parts of these pathways.

GENERAL REFERENCES

Carlson BM: *Human embryology and developmental biology*, ed 5, Philadelphia, 2014, WB Saunders.
Dye FJ: *Dictionary of developmental biology and embryology*, ed 2, New York, 2012, Wiley-Blackwell.
Epstein CJ, Erickson RP, Wynshaw-Boris AJ, editors: *Inborn errors of development: the molecular basis of clinical disorders of morphogenesis*, ed 2, New York, 2008, Oxford University Press.
Gilbert SF: *Developmental biology*, ed 10, Sunderland, MA, 2013, Sinauer Associates.
Wolpert L, Tickle C: *Principles of development*, ed 4, New York, 2011, Oxford University Press.

REFERENCES SPECIFIC TO PARTICULAR TOPICS

Acimovic I, Vilotic A, Pesl M, et al: Human pluripotent stem cell-derived cardiomyocytes as research and therapeutic tools, *Biomed Res Int* 2014:512831, 2014.
Ross CA, Akimov S: Human induced pluripotent stem cells: potential for neurodegenerative diseases, *Hum Mol Genet* 23(R1):R17–R26, 2014.

PROBLEMS

1. What is the difference between regulative and mosaic development? What is the significance of these two stages of development for reproductive genetics and prenatal diagnosis?

2. Match the terms in the left-hand column with the terms that best fit in the right-hand column.

a. Erasure of imprinting during germ cell development	1. Totipotency
	2. Morphogen
b. Position-dependent development	3. Epigenetic regulation of gene expression
c. Regulative development	4. Monozygotic twinning
d. Embryonic stem cells	

3. Match the terms in the left-hand column with the terms that best fit in the right-hand column.

a. Amniotic band	1. U-shaped cleft palate
b. Polydactyly	2. Thalidomide
c. Inadequate amniotic fluid	3. *GLI3* mutation
d. Limb reduction	4. Disruption
e. Robin sequence	5. Deformation

4. What type of diploid cells would not be appropriate nucleus donors in an animal cloning experiment and why?

5. For discussion: Why do some mutations in transcription factors result in developmental defects even when they are present in the heterozygous state?

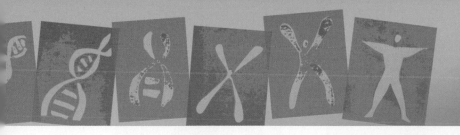

CHAPTER 15

Cancer Genetics and Genomics

Cancer is one of the most common and serious diseases seen in clinical medicine. There are 14 million new cases of cancer diagnosed each year and over 8 millions deaths from the disease worldwide. Based on the most recent statistics available, cancer treatment costs $80 billion per year in direct health care expenditures in the United States alone. Cancer is invariably fatal if it is not treated. Identification of persons at increased risk for cancer before its development is an important objective of genetics research. And for both those with an inherited predisposition to cancer as well those in the general population, early diagnosis of cancer and its early treatment are vital, and both are increasingly reliant on advances in genome sequencing and gene expression analysis.

NEOPLASIA

Cancer is the name used to describe the more virulent forms of **neoplasia**, a disease process characterized by uncontrolled cellular proliferation leading to a mass or tumor (**neoplasm**). The abnormal accumulation of cells in a neoplasm occurs because of an imbalance between the normal processes of cellular proliferation and cellular attrition. Cells proliferate as they pass through the cell cycle and undergo mitosis. Attrition, due to programmed cell death (see Chapter 14), removes cells from a tissue. For a neoplasm to be a cancer, however, it must also be **malignant,** which means that not only is its growth uncontrolled, it is also capable of invading neighboring tissues that surround the original site (the **primary** site) and can spread (**metastasize**) to more distant sites (Fig. 15-1). Tumors that do not invade or metastasize are not cancerous but are referred to as **benign** tumors, although their abnormal function, size or location may make them anything but benign to the patient.

Cancer is not a single disease but rather comes in many forms and degrees of malignancy. There are three main classes of cancer:

- **Sarcomas,** in which the tumor has arisen in mesenchymal tissue, such as bone, muscle, or connective tissue, or in nervous system tissue;
- **Carcinomas,** which originate in epithelial tissue, such as the cells lining the intestine, bronchi, or mammary ducts; and
- **Hematopoietic** and **lymphoid** malignant neoplasms, such as leukemia and lymphoma, which spread throughout the bone marrow, lymphatic system, and peripheral blood.

Within each of the major groups, tumors are classified by site, tissue type, histological appearance, degree of malignancy, chromosomal aneuploidy, and, increasingly, by which gene mutations and abnormalities in gene expression are found within the tumor.

In this chapter, we describe how genetic and genomic studies demonstrate that *cancer is fundamentally a genetic disease.* We describe the kinds of genes that have been implicated in initiating cancer and the mechanisms by which dysfunction of these genes can result in the disease. Second, we review a number of heritable cancer syndromes and demonstrate how insights gained into their pathogenesis have illuminated the basis of the much more common, sporadic forms of cancer. We also examine some of the special challenges that such heritable syndromes present for medical genetics and genetic counseling. Third, we illustrate ways in which genetics and genomics have changed both how we think about the causes of cancer and how we diagnose and treat the disease. Genomics—in particular the identification of mutations, altered epigenomic modifications, and abnormal gene expression in cancer cells—is vastly expanding our knowledge of why cancer develops and is truly changing cancer diagnosis and treatment.

GENETIC BASIS OF CANCER
Driver and Passenger Gene Mutations

The application to the study of cancer of powerful new sequencing technologies for genome sequencing (see Chapter 4) and RNA expression studies (see Chapter 3) has brought remarkable new clarity to our understanding of the origins of cancer. By analyzing many thousands of samples obtained from more than 30 types of human cancer, researchers are building **The Cancer Genome Atlas,** a public catalog of mutations, epigenomic modifications, and abnormal gene expression profiles found in a wide variety of cancers. Although the project is still under way, the results to date from these studies are striking. The number of mutations present in a tumor can vary from just a few to many tens of thousands. Most mutations found through sequencing

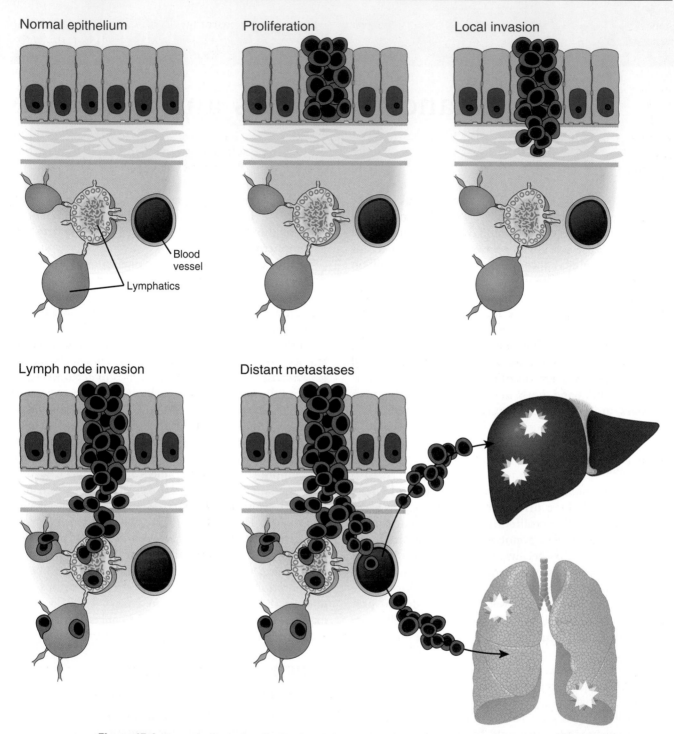

Normal epithelium

Proliferation

Local invasion

Blood vessel

Lymphatics

Lymph node invasion

Distant metastases

Figure 15-1 **General scheme for development of a carcinoma in an epithelial tissue such as colonic epithelium.** The diagram shows progression from normal epithelium to local proliferation, invasion across the lamina propria, spread to local lymph nodes, and final distant metastases to liver and lung.

of tumor tissue appear to be random, are not recurrent in particular cancer types, and probably occurred as the cancer developed, rather than directly causing the neoplasia to develop or progress. Such mutations are referred to as **"passenger" mutations.** However, a subset of a few hundred genes has been repeatedly found to be

mutated at high frequency in many samples of the same type of cancer or even in multiple different types of cancers, mutated in fact far too frequently to simply be passenger mutations. These genes are thus presumed to be involved in the development or progression of the cancer itself and are therefore referred to as **"driver"** genes, that is, they harbor mutations (so-called **driver gene mutations**) that are likely to be causing a cancer to develop or progress. Although many driver genes are specific to particular tumor types, some, such as those in the *TP53* gene encoding the p53 protein, are found in the vast majority of cancers of many different types. Although the most common driver genes are now known, it is likely that additional, less abundant driver genes will be identified as The Cancer Genome Atlas continues to grow.

Spectrum of Driver Gene Mutations

Many different genome alterations can act as driver gene mutations. In some cases, a single nucleotide change or small insertion or deletion can be a driver mutation. Large numbers of cell divisions are required to produce an adult organism of an estimated 10^{14} cells from a single-cell zygote. Given a frequency of 10^{-10} replication errors per base of DNA per cell division, and an estimated 10^{15} cell divisions during the lifetime of an adult, replication errors alone result in thousands of new single nucleotide or small insertion/deletion mutations in the genome in *every cell* of the organism. Some environmental agents, such as carcinogens in cigarette smoke or ultraviolet or X-irradiation, will increase the rate of mutations around the genome. If, by chance, mutations occur in critical driver genes in a particular cell, then the oncogenic process may be initiated.

Chromosome and subchromosomal mutations (see Chapters 4 and 5) can also serve as driver mutations. Particular translocations are sometimes highly specific for certain types of cancer and involve specific genes (e.g., the *BCR-ABL* translocation in **chronic myelogenous leukemia**) **(Case 10)**; in contrast, other cancers can show complex rearrangements in which chromosomes break into numerous pieces and rejoin, forming novel and complex combinations (a process known as "**chromosome shattering**"). Finally, large genomic alterations involving many kilobases of DNA can form the basis for loss of function or increased function of one or more driver genes. Large genomic alterations include deletions of a segment of a chromosome or multiplication of a chromosomal segment to produce regions with many copies of the same gene (**gene amplification**).

The Cellular Functions of Driver Genes

The nature of some driver gene mutations comes as no surprise: the mutations directly affect specific genes that regulate processes that are readily understood to be important in oncogenesis. These processes include cell-cycle regulation, cellular proliferation, differentiation and exit from the cell cycle, growth inhibition by cell-cell contacts, and programmed cell death (apoptosis). However, the effects of other driver gene mutations are not so readily understood and include genes that act more globally and indirectly affect the expression of many other genes. Included in this group are genes encoding products that maintain genome and DNA integrity or genes that affect gene expression, either at the level of transcription by epigenomic changes, at the post-transcriptional level through effects on messenger RNA (mRNA) translation or stability, or at the post-translational level through their effects on protein turnover (Table 15-1). Other driver genes affect translation, for example, genes that encode **noncoding RNAs** from which regulatory **microRNAs (miRNAs)** are derived (see Chapter 3). Many miRNAs have been found to be either greatly overexpressed or down-regulated in

TABLE 15-1 Classes of Driver Genes Mutated in Cancer

Genes with Specific Effects on Cellular Proliferation or Apoptosis	Genes with Global Effects on Genome or DNA Integrity or on Gene Expression
Cell-cycle regulation Cell-cycle checkpoint proteins Cellular proliferation signaling • Transcription factors • Receptor and membrane-bound tyrosine kinases • Growth factors • Intracellular serine-threonine kinases • PI3 kinases • G proteins and G protein–coupled receptors • mTOR signaling • Wnt/β-catenin signaling • Transcription factors Differentiation and lineage survival • Transcription factors protecting specific cell lineages • Genes involved in exit from cell cycle into G_0 Apoptosis	Genome integrity • Chromosome segregation • Genome and gene mutation • DNA repair • Telomere stability Gene expression: abnormal metabolites affecting activity of multiple genes/gene products Gene expression: epigenetic modifications of DNA/chromatin • DNA methylation and hydroxymethylation • Chromatin histone methylation, demethylation, and acetylation • Nucleosome remodeling • Chromatin accessibility and compaction (SWI/SNF complexes) Gene expression: post-transcriptional alterations • Aberrant mRNA splicing • MicroRNAs affecting mRNA stability and translation Gene expression: protein stability/turnover

mRNA, Messenger RNA; mTOR, mammalian target of rapamycin; PI3, phosphatidylinositol-3.

various tumors, sometimes strikingly so. Because each miRNA may regulate as many as 200 different gene targets, overexpression or underexpression of miRNAs may have widespread oncogenic effects because many driver genes will be dysregulated. Noncoding miRNAs that impact gene expression and contribute to oncogenesis are referred to as **oncomirs**.

Figure 15-2 is a diagram outlining how mutations in specific regulators of growth and in global guardians of DNA and genome integrity perturb normal homeostasis (see Fig. 15-2A), leading to a vicious cycle causing loss of cell cycle control, uncontrolled proliferation, interrupted differentiation, and defects in apoptosis (see Fig. 15-2B).

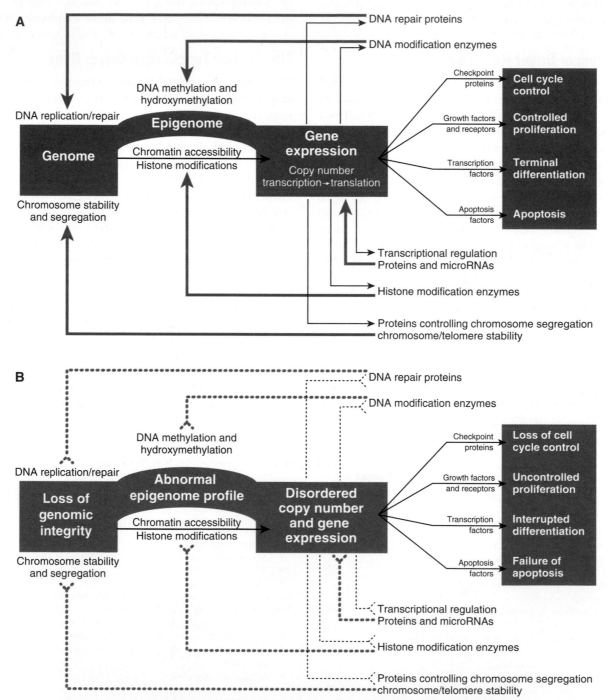

Figure 15-2 **A, Overview of normal genetic pathways controlling normal tissue homeostasis.** The information encoded in the genome (*black arrows*) results in normal gene expression, as modulated by the epigenomic state. Many genes provide negative feedback (*purple arrows*) to ensure normal homeostasis. **B, Perturbations in neoplasia.** Abnormalities in gene expression (*dotted black arrows*) lead to a vicious cycle of positive feedback (*brown dotted lines*) of progressively more disordered gene expression and genome integrity.

Activated Oncogenes and Tumor Suppressor Genes

Both classes of driver genes—those with specific effects on cellular proliferation or survival and those with global effects on genome or DNA integrity (see Table 15-1)—can be further subdivided into one of two functional categories depending on how, if mutated, they drive oncogenesis.

The first category includes **proto-oncogenes**. These are normal genes that, when mutated in very particular ways, become driver genes through alterations that lead to *excessive levels of activity*. Once mutated in this way, driver genes of this type are referred to as **activated oncogenes**. Only a single mutation at one allele can be sufficient for activation, and the mutations that activate a proto-oncogene can range from highly specific point mutations causing dysregulation or hyperactivity of a protein, to chromosome translocations that drive overexpression of a gene, to gene amplification events that create an overabundance of the encoded mRNA and protein product (Fig. 15-3).

The second, and more common, category of driver genes includes **tumor suppressor genes (TSGs)**, mutations in which cause a *loss of expression* of proteins necessary to control the development of cancers. To drive oncogenesis, loss of function of a TSG typically requires mutations at both alleles. There are many ways that a cell can lose the function of TSG alleles. Loss-of-function mechanisms can range from missense, nonsense, or frame-shift mutations to gene deletions or loss of a part or even an entire chromosome. Loss of function of TSGs can also result from epigenomic transcriptional silencing due to altered chromatin conformation or promoter methylation (see Chapter 3), or from translational silencing by miRNAs or disturbances in other components of the translational machinery (see Box).

Cellular Heterogeneity within Individual Tumors

The accumulation of driver gene mutations does not occur synchronously, in lockstep, in every cell of a tumor. To the contrary, cancer evolves along multiple lineages within a tumor, as chance mutational and epigenetic events in different cells activate proto-oncogenes and cripple the machinery for maintaining genome integrity, leading to more genetic changes in a vicious cycle of more mutations and worsening growth control. The lineages that experience an enhancement of growth, survival, invasion, and distant spread will come to predominate as the cancer evolves and progresses (see Box). In this way, the original clone of neoplastic cells evolves and gives rise to multiple sublineages, each carrying a set of mutations and epigenomic alterations that are different from but overlap with what is carried in other sublineages. The profile of mutations and epigenomic changes can differ between the primary and its metastases, between different metastases, and even between the cells of the original tumor or within a single metastasis. A paradigm for the development of cancer, as illustrated in Figure 15-4, provides a useful conceptual framework for considering the role of genomic and epigenomic changes in the evolution of cancer, a point we emphasize throughout this chapter. It is a general model that applies to all cancers.

Although the focus of this chapter is on genomic and epigenomic changes within the tumor, the surrounding normal tissue also plays an important role by providing the blood supply that nourishes the tumor, by permitting cancer cells to escape from the tumor and metastasize,

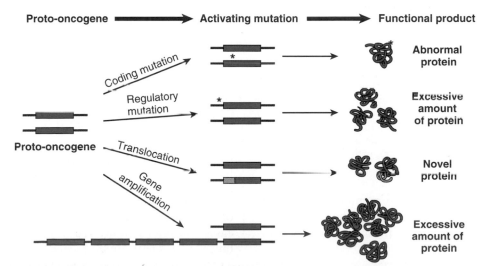

Figure 15-3 Different mutational mechanisms leading to proto-oncogene activation. These include a single point mutation leading to an amino acid change that alters protein function, mutations or translocations that increase expression of an oncogene, a chromosome translocation that produces a novel product with oncogenic properties, and gene amplification leading to excessive amounts of the gene product.

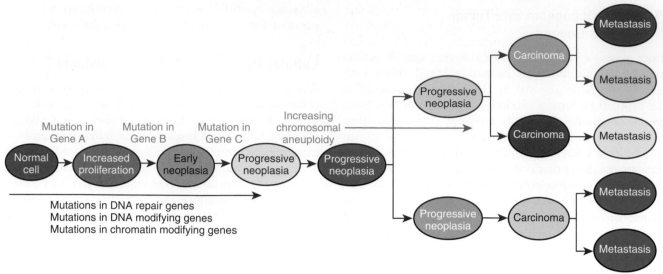

Figure 15-4 Stages in the evolution of cancer. Increasing degrees of abnormality are associated with sequential loss of tumor suppressor genes from several chromosomes and activation of proto-oncogenes, with or without a concomitant defect in DNA repair. Multiple lineages, carrying different mutations and epigenomic profiles, occur within the primary tumor itself, between the primary and metastases and between different metastases.

GENETIC BASIS OF CANCER

Regardless of whether a cancer occurs sporadically in an individual, as a result of **somatic mutation,** or repeatedly in many individuals in a family as a hereditary trait, cancer is a genetic disease.

- Genes in which mutations cause cancer are referred to as **driver genes**, and the cancer-causing mutations in these genes are **driver mutations**.
- Driver genes fall into two distinct categories: **activated oncogenes** and **tumor suppressor genes (TSGs).**
- An **activated oncogene** is a mutant allele of a **proto-oncogene,** a class of normal cellular protein-coding genes that promotes growth and survival of cells. Oncogenes facilitate malignant transformation by stimulating proliferation or inhibiting apoptosis. Oncogenes encode proteins such as the following:
  - Proteins in signaling pathways for cell proliferation
  - Transcription factors that control the expression of growth-promoting genes
  - Inhibitors of programmed cell death machinery
- A TSG is a gene in which loss of function through mutation or epigenomic silencing directly removes normal regulatory controls on cell growth or leads indirectly to such losses through an increased mutation rate or aberrant gene expression. TSGs encode proteins involved in many aspects of cellular function, including maintenance

of correct chromosome number and structure, DNA repair proteins, proteins involved in regulating the cell cycle, cellular proliferation, or contact inhibition, just to name a few examples.

- **Tumor initiation** can be caused by different types of genetic alterations. These include mutations such as the following:
  - Activating or gain-of-function mutations, including gene amplification, point mutations, and promoter mutations, that turn one allele of a proto-oncogene into an oncogene
  - Ectopic and heterochronic mutations (see Chapter 11) of proto-oncogenes
  - **Chromosome translocations** that cause misexpression of genes or create chimeric genes encoding proteins with novel functional properties
  - Loss of function of both alleles, or a dominant negative mutation of one allele, of TSGs
- **Tumor progression** occurs as a result of accumulating additional genetic damage, through mutations or epigenetic silencing, of driver genes that encode the machinery that repairs damaged DNA and maintains cytogenetic normality. A further consequence of genetic damage is altered expression of genes that promote vascularization and the spread of the tumor through local invasion and distant metastasis.

and by shielding the tumor from immune attack. Thus cancer is a complex process, both within the tumor and between the tumor and the normal tissues that surround it.

CANCER IN FAMILIES

Although essentially all individuals are at risk for some cancer at some point during their lives, many

forms of cancer have a higher incidence in relatives of patients than in the general population. In some cases, this increased incidence is due primarily to inheritance of a single mutant gene with high penetrance. These mutations result in **hereditary cancer syndromes** (see, for examples, Cases 7, 15, 29, 39, and 48) following mendelian patterns of inheritance that were presented in Chapter 7. Among these syndromes, we currently know of approximately 100 different genes in which deleterious mutations increase the risk for cancer many-fold higher than in the general population. There are also many dozens of additional genetic disorders that are not usually considered to be hereditary cancer syndromes and yet include some increased predisposition to cancer (Case 6) (for example, the ten- to twenty-fold increased lifetime risk for leukemia in Down syndrome [see Chapter 6]). These clear examples notwithstanding, it is important to emphasize that not all families with an apparently increased incidence of cancer can be explained by known mendelian or clearly recognized genetic disorders. These families likely represent the effects of both shared environment and one or more genetic variants that increase susceptibility and are therefore classified as multifactorial, with complex inheritance (see Chapter 8), as will be explored later in this chapter.

Although individuals with a hereditary cancer syndrome represent probably less than 5% of all patients with cancer, identification of a genetic basis for their disease has great importance both for clinical management of these families and for understanding cancer in general. First, the relatives of individuals with strong hereditary predispositions, which are most often due to mutations in a single gene, can be offered testing and counseling to provide appropriate reassurance or more intensive monitoring and therapy, depending on the results of testing. Second, as is the case with many common diseases, understanding the hereditary forms of the disease provides crucial insights into disease mechanisms that go far beyond the rare hereditary forms themselves. These general concepts are illustrated in the examples discussed in the sections that follow.

Activated Oncogenes in Hereditary Cancer Syndromes

Multiple Endocrine Adenomatosis, Type 2

The type A variant of **multiple endocrine adenomatosis, type 2** (MEN2) is an autosomal dominant disorder characterized by a high incidence of medullary carcinoma of the thyroid that is often but not always associated with pheochromocytoma, benign parathyroid adenomas, or both. Patients with the rarer type B variant, termed MEN2B, have, in addition to the tumors seen in patients with MEN2A, thickening of nerves and the development of benign neural tumors, known as **neuromas,** on the mucosal surface of the mouth and lips and along the gastrointestinal tract.

The mutations responsible for MEN2 are in the *RET* gene. Individuals who inherit an activating mutation in *RET* have a greater than 60% chance of developing a particular type of thyroid carcinoma (medullary), although more sensitive tests, such as blood tests for thyrocalcitonin or urinary catecholamines synthesized by pheochromocytomas, are abnormal in well above 90% of heterozygotes for MEN2.

RET encodes a cell-surface protein that contains an extracellular domain that can bind signaling molecules and a cytoplasmic tyrosine kinase domain. Tyrosine kinases are a class of enzymes that phosphorylate tyrosines in proteins. Tyrosine phosphorylation initiates a signaling cascade of changes in protein-protein and DNA-protein interactions and in the enzymatic activity of many proteins (Fig. 15-5). Normally, tyrosine kinase receptors must bind specific signaling molecules in order to undergo the conformational change that makes them enzymatically active and able to phosphorylate other cellular proteins. The mutations in *RET* that cause MEN2A increase its kinase activity even in the absence of its ligand (a state referred to as constitutive activation).

The *RET* gene is expressed in many tissues of the body and is required for normal embryonic development of autonomic ganglia and kidney. It is unclear why germline activating mutations in this proto-oncogene result in a particular cancer of distinct histological types restricted to specific tissues, whereas other tissues in which the oncogene is expressed do not develop tumors. Interestingly, *RET* is the same gene implicated in Hirschsprung disease (Case 22) (see Chapter 8), although those mutations are usually loss-of-function, not activating, mutations. There are, however, some families in which the *same* mutation in *RET* can act as an activated oncogene in some tissues (such as thyroid) and cause MEN2A, while not having sufficient function in other tissues, such as the developing enteric neurons of the gastrointestinal tract, resulting in Hirschsprung disease. Thus even the identical mutation can have different effects on different tissues.

The Two-Hit Theory of Tumor Suppressor Gene Inactivation in Cancer

As introduced earlier, whereas the proteins encoded by proto-oncogenes promote cancer when activated or overexpressed, mutations in TSGs contribute to malignancy by a different mechanism, the loss of function of both alleles of the gene. The products of many TSGs have now been isolated and characterized, some of which are presented in Table 15-2.

The existence of TSG mutations leading to cancer was proposed some five decades ago to explain why certain tumors can occur in either hereditary or sporadic forms (Fig. 15-6; see discussion later in this section). It was suggested that the hereditary form of the childhood cancer **retinoblastoma** (see next section) might be initiated when a cell in a person heterozygous for a *germline* mutation in the retinoblastoma TSG, required to prevent the development of the cancer, undergoes a *second, somatic* event that inactivates the other retinoblastoma gene allele. As a consequence of this second somatic event, the cell loses function of both alleles, giving rise to a tumor. In the sporadic form of retinoblastoma, both

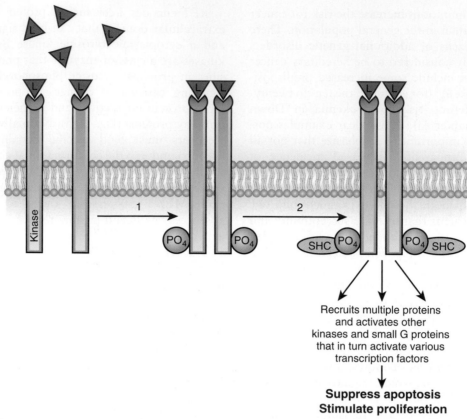

Figure 15-5 Schematic diagram of the function of the Ret receptor, the product of the *RET* proto-oncogene. Upon binding of a ligand (L), such as glial-derived growth factor or neurturin, to the extracellular domain, the protein dimerizes and activates its intracellular kinase domain to autophosphorylate specific tyrosine residues. These then bind the SHC adaptor protein, which sets off multiple cascades of complex protein interactions involving other serine-threonine and phosphatidylinositol kinases and small G proteins, which in turn activate other proteins, ultimately activating certain transcription factors that suppress apoptosis and stimulate cellular proliferation. Mutations in *RET* that result in type A variant of multiple endocrine adenomatosis, type 2 (MEN2A) cause inappropriate dimerization and activation of its own intrinsic kinase without ligand binding.

TABLE 15-2 Selected Tumor Suppressor Genes

Gene	Gene Product and Possible Function	Disorders in Which the Gene Is Affected	
		Familial	Sporadic
RB1	p110 Cell cycle regulation	Retinoblastoma	Retinoblastoma, small cell lung carcinomas, breast cancer
TP53	p53 Cell cycle regulation	Li-Fraumeni syndrome	Lung cancer, breast cancer, many others
APC	APC Multiple roles in regulating proliferation and cell adhesion	Familial adenomatous polyposis	Colorectal cancer
VHL	VHL Forms part of a cytoplasmic destruction complex with APC that normally inhibits induction of blood vessel growth when oxygen is present	von Hippel-Lindau syndrome	Clear cell renal carcinoma
BRCA1, BRCA2	BRCA1, BRCA2 Chromosome repair in response to double-stranded DNA breaks	Familial breast and ovarian cancer	Breast cancer, ovarian cancer
MLH1, MSH2	MLH1, MSH2 Repair nucleotide mismatches between strands of DNA	Lynch syndrome	Colorectal cancer

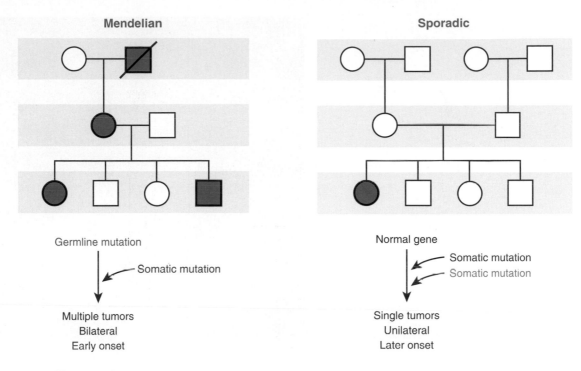

Figure 15-6 Comparison of mendelian and sporadic forms of cancers such as retinoblastoma and familial polyposis of the colon. See text for discussion.

alleles are also inactivated, but in this case, the inactivation results from two somatic events occurring in the same cell.

This so-called two-hit model is now widely accepted as the explanation for many hereditary cancers in addition to retinoblastoma, including **familial polyposis coli, familial breast cancer, neurofibromatosis type 1 (NF1), Lynch syndrome, and Li-Fraumeni syndrome.**

Tumor Suppressor Genes in Autosomal Dominant Cancer Syndromes

Retinoblastoma

Retinoblastoma is the prototype of diseases caused by mutation in a TSG and is a rare malignant tumor of the retina in infants, with an incidence of approximately 1 in 20,000 births (Fig. 15-7) (Case 39). Diagnosis of a retinoblastoma must usually be followed by removal of the affected eye, although smaller tumors, diagnosed at an early stage, can be treated by local therapy so that vision can be preserved.

Approximately 40% of cases of retinoblastoma are of the heritable form, in which the child (as just discussed and as represented generally by the family shown in Figure 15-6) inherits one mutant allele at the retinoblastoma locus (*RB1*) through the germline from either a heterozygous parent or, more rarely, from a parent with germline mosaicism for an *RB1* mutation (see Chapter 7). In these children, retinal cells, which like all the other cells of the body are already carrying one inherited defective *RB1* allele, suffer a somatic mutation

or other alteration in the remaining normal allele, leading to loss of both copies of the *RB1* gene and initiating development of a tumor in each of those cells (Fig. 15-8).

The disorder appears to be inherited as a dominant trait because the large number of primordial retinoblasts and their rapid rate of proliferation make it very likely that a somatic mutation will occur as a second hit in one or more of the more than 10^6 retinoblasts already carrying an inherited *RB1* mutation. Because the chance of a second hit is so great, it occurs frequently in more than one cell, and thus heterozygotes for the disorder often have tumors arising at multiple sites, such as **multifocal tumors** in one eye, in both eyes (**bilateral retinoblastoma**), or in both eyes, as well as in the pineal gland (referred to as "trilateral" retinoblastoma). It is worth emphasizing, however, that the occurrence of a second hit is a matter of chance and does not occur 100% of the time; the penetrance of retinoblastoma therefore, although greater than 90%, is not complete.

The other 60% of cases of retinoblastoma are sporadic; in these cases, *both RB1* alleles in a single retinal cell have been mutated or inactivated independently by chance, and the child does not carry an *RB1* mutation inherited through the germline. Because two hits in the same cell is a statistically rare event, there is usually only a single clonal tumor, and the retinoblastoma is found at one location (unifocal) in one eye only. Unilateral tumor is no guarantee that the child does not have the heritable form of retinoblastoma, however, because

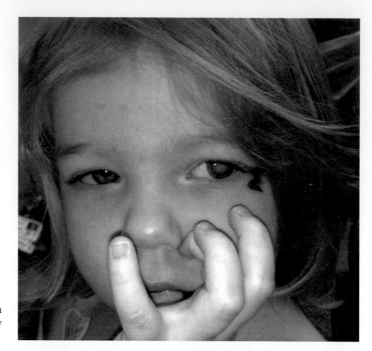

Figure 15-7 Retinoblastoma in a young girl, showing as a white reflex in the affected left eye when light reflects directly off the tumor surface. *See Sources & Acknowledgments.*

15% of patients with the heritable type develop a tumor in only one eye. Another difference between hereditary and sporadic tumors is that the average age at onset of the sporadic form is in early childhood, later than in infants with the heritable form (see Fig. 15-6), reflecting the longer time needed on average for two mutations, rather than one, to occur.

In a small percentage of patients with retinoblastoma, the mutation responsible is a cytogenetically detectable deletion or translocation of the portion of chromosome 13 that contains the *RB1* gene. Such chromosomal changes, if they also disrupt genes adjacent to *RB1,* may lead to dysmorphic features in addition to retinoblastoma.

Nature of the Second Hit. Typically, for retinoblastoma as well as for the other hereditary cancer syndromes, the first hit is an inherited mutation, that is, a change in the DNA sequence. The second hit, however, can be caused by a variety of genetic, epigenetic, or genomic mechanisms (see Fig. 15-8); although it is most often a somatic mutation, loss of function *without* mutation, such as occurs with epigenetic silencing (see Chapter 3), has also been observed in some cancer cells. Although a number of mechanisms have been documented, the common theme is loss of function of *RB1.* The *RB1* gene product, p110 Rb1, is a phosphoprotein that normally regulates entry of the cell into the S phase of the cell cycle (see Chapter 2). Thus loss of the *RB1* gene and/or absence of the normal *RB1* gene product (by whatever mechanism) deprives cells of an important checkpoint and allows uncontrolled proliferation (see Table 15-2).

Loss of Heterozygosity. In addition to mutations and epigenetic silencing, a novel genomic mechanism was uncovered when geneticists made an unusual but highly significant discovery when they compared DNA polymorphisms at the *RB1* locus in DNA from normal cells to those in the retinoblastoma tumor from the same patient. Individuals with retinoblastoma who were heterozygous at polymorphic loci flanking the *RB1* locus in normal tissues (see Fig. 15-8) had tumors that contained alleles from only one of their two chromosome 13 homologues, revealing a **loss of heterozygosity** (**LOH**) in tumor DNA in and around the *RB1* locus. Furthermore, in familial cases, the retained chromosome 13 markers were the ones inherited from the affected parent, that is, the chromosome with the abnormal *RB1* allele. Thus, in these cases, LOH represents the second hit of the remaining allele. LOH may occur by interstitial deletion, but there are other mechanisms as well, such as mitotic recombination or monosomy 13 due to nondisjunction (see Fig. 15-8).

LOH is the most common mutational mechanism by which the function of the remaining normal *RB1* allele is disrupted in heterozygotes, although each of the mechanisms shown in Figure 15-8 have been documented in different patients. LOH is a feature of tumors in a number of cancers, both heritable and sporadic, and is often considered evidence for the existence of a TSG in the region of LOH.

Familial Breast Cancer due to Mutations in *BRCA1* and *BRCA2*

Breast cancer is common. Among all cases of this disease, a small proportion (≈3% to 5%) appears to be

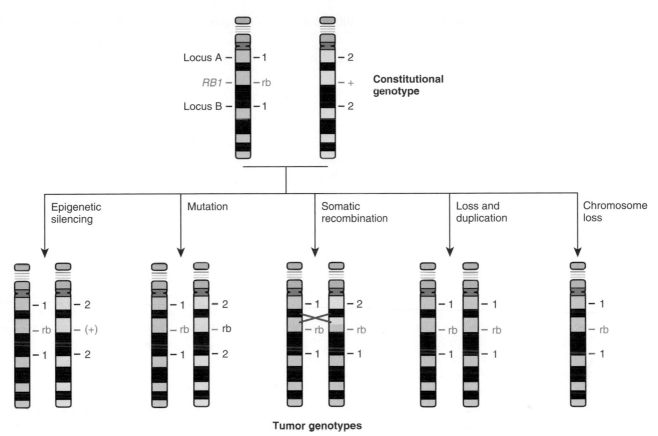

Figure 15-8 Chromosomal mechanisms that could lead to loss of heterozygosity for DNA markers at or near a tumor suppressor gene in an individual heterozygous for an inherited germline mutation. The figure depicts the events that constitute the "second hit" that leads to retinoblastoma with loss of heterozygosity (LOH). Local events such as mutation, gene conversion, or transcriptional silencing by promoter methylation, however, could cause loss of function of both *RB1* genes without producing LOH. 1, normal allele, rb, the mutant allele.

due to a highly penetrant dominantly inherited mendelian predisposition that increases the risk for female breast cancer fourfold to sevenfold over the 12% lifetime risk observed in the general female population. In these families, one often sees features characteristic of hereditary (as opposed to sporadic) cancer: multiple affected individuals in a family, earlier age at onset, frequent multifocal, bilateral disease or second independent primary breast tumor, and second primary cancers in other tissues such as ovary and prostate.

Although a number of genes in which mutations cause highly penetrant mendelian forms of breast cancer have been discovered from family studies, the two genes responsible for the majority of all hereditary breast cancers are *BRCA1* and *BRCA2* (Case 7). Together, these two TSGs account for approximately one half and one third, respectively, of autosomal dominant familial breast cancer. Numerous mutant alleles of both genes have now been catalogued. Mutations in *BRCA1* and *BRCA2* are also associated with a significant increase in the risk for ovarian and fallopian duct cancer in female heterozygotes. Moreover, mutations in *BRCA2*

and, to a lesser extent, *BRCA1*, also account for 10% to 20% of all male breast cancer and increase the risk for male breast cancer ten to sixtyfold over the 0.1% lifetime risk observed among males in the general population (Table 15-3).

The gene products of *BRCA1* and *BRCA2* are nuclear proteins contained within the same multiprotein complex. This complex has been implicated in the cellular response to double-stranded DNA breaks, such as occur normally during homologous recombination or abnormally as a result of damage to DNA. As might be expected for any TSG, tumor tissue from heterozygotes for *BRCA1* and *BRCA2* mutations frequently demonstrates LOH with loss of the normal allele.

Penetrance of **BRCA1** ***and*** **BRCA2** ***Mutations.*** Presymptomatic detection of women at risk for development of breast cancer as a result of any of these susceptibility genes relies on detecting clearly pathogenic mutations by gene sequencing. For the purposes of patient management and counseling, it would be helpful to know the lifetime risk for development of

TABLE 15-3 Lifetime Cancer Risks in Carriers of *BRCA1* or *BRCA2* Mutations Compared to the General Population

Cancer Type	General Population Risk	Cancer Risk When Mutation Present	
		BRCA1	**BRCA2**
Breast in females	12%	50%-80%	40%-70%
Second primary breast in females	3.5% within 5 yr	27% within 5 yr	12% within 5 yr
	Up to 11%		40%-50% at 20 yr
Ovarian	1%-2%	24%-40%	11%-18%
Male breast	0.1%	1%-2%	5%-10%
Prostate	15% (N. European origin)	<30%	<39%
	18% (African Americans)		
Pancreatic (both sexes)	0.50%	1%-3%	2%-7%

Data from Petrucelli N, Daly MB, Feldman GL: *BRCA1* and *BRCA2* hereditary breast and ovarian cancer. Updated September 26, 2013. In Pagon RA, Adam MP, Bird TD, et al, editors: GeneReviews[Internet], Seattle, University of Washington, Seattle, 1993-2014, http://www.ncbi.nlm.nih.gov/books/NBK1247/.

breast cancer in individuals, whether male or female, carrying particular mutations in the *BRCA1* and *BRCA2* genes, compared with the risk in the general male or female population (see Table 15-3). Initial studies showed a greater than 80% risk for breast cancer by the age of 70 years in women heterozygous for deleterious *BRCA1* mutations, with a somewhat lower estimate for *BRCA2* mutation carriers. These estimates relied on estimates of the risk for development of cancer in female relatives within families ascertained because breast cancer had already occurred many times in family members; that is, families in which the particular *BRCA1* or *BRCA2* mutation was highly penetrant.

When similar risk estimates were made from population-based studies, however, in which women carrying *BRCA1* and *BRCA2* mutations were *not* selected because they were members of families in which many cases of breast cancer had already developed, the risk estimates were lower and ranged from 40% to 50% by the age of 70 years. The discrepancy between the penetrance of mutant alleles in families with multiple occurrences of breast cancer and the penetrance seen in women identified by population screening and *not* by family history suggests that other genetic or environmental factors must play a role in the ultimate penetrance of *BRCA1* and *BRCA2* mutations in women heterozygous for these mutations.

In addition to mutations in *BRCA1* and *BRCA2*, mutations in other genes can also cause autosomal dominantly inherited breast cancer syndromes, albeit less commonly. These syndromes, which include the **Li-Fraumeni, hereditary diffuse gastric cancer, Peutz-Jeghers,** and **Cowden syndromes,** demonstrate lifetime breast cancer risks that approach those seen in carriers of *BRCA1* or *BRCA2* mutations, as well as risks for other forms of cancer such as sarcomas, brain tumors, and carcinomas of the stomach, thyroid, and small intestine.

Clinicians faced with a family with multiple affected individuals with breast cancer often look for distinguishing signs in the patient and in the family history to help guide the choice of which genes to analyze (see

Box). However, the rapid decline in the cost of gene or even genome-wide sequencing has allowed the development of **gene panels** in which a dozen or more candidate genes can be accurately and simultaneously tested for mutations, often at a cost that is equivalent or even less than what was charged previously to analyze just one or two genes.

Hereditary Colon Cancer

Colorectal cancer, a malignancy of the epithelial cells of the colon and rectum, is one of the most common forms of cancer. It affects approximately 1.3 million individuals worldwide per year (150,000 of whom are in the United States) and is responsible for approximately 10% to 15% of *all* cancer. Most cases are sporadic, but a small proportion of colon cancer cases are familial, among which are two autosomal dominant conditions: **familial adenomatous polyposis (FAP)** and **Lynch syndrome (LS),** along with their variants.

Familial Adenomatous Polyposis. FAP (Case 15) and its subvariant, **Gardner syndrome,** together have an incidence of approximately 1 per 10,000. In FAP heterozygotes, benign adenomatous polyps numbering in the many hundreds develop in the colon during the first two decades of life. In almost all cases, one or more of the polyps becomes malignant. Surgical removal of the colon (colectomy) prevents the development of malignancy. Because this disorder is inherited as an autosomal dominant trait, relatives of affected persons must be examined periodically by colonoscopy. FAP is caused by loss-of-function mutations in a TSG known as the *APC* gene (so-named because the condition used to be called adenomatous polyposis coli). Gardner syndrome is also due to mutations in *APC* and is therefore allelic to FAP. Patients with Gardner syndrome have, in addition to the adenomatous polyps with malignant transformation seen in FAP, other extracolonic anomalies, including osteomas of the jaw and desmoids, which are tumors arising in the muscle of the abdominal wall. Although the relatives of an individual affected with Gardner syndrome who also carry the same *APC* mutation tend

DIAGNOSTIC CRITERIA FOR HEREDITARY CANCER SYNDROMES

Li-Fraumeni Syndrome (LFS): *Chompret Criteria**

- Proband with tumor belonging to LFS tumor spectrum (e.g., soft tissue sarcoma, osteosarcoma, brain tumor, premenopausal breast cancer, adrenocortical carcinoma, leukemia, lung bronchoalveolar cancer) before age 46 years AND at least one first- or second-degree relative with LFS tumor (except breast cancer if proband has breast cancer) before age 56 years or with multiple tumors; OR
- Proband with multiple tumors (except multiple breast tumors), two of which belong to LFS tumor spectrum and first of which occurred before age 46 years; OR
- Patient with adrenocortical carcinoma or choroid plexus tumor, irrespective of family history

Hereditary Diffuse Gastric Cancer Syndrome

- Family history of diffuse gastric cancer with two or more cases of gastric cancer, with at least one diffuse gastric cancer diagnosed before age 50 years
- Family with multiple lobular breast cancer

Peutz-Jeghers Syndrome

- Peutz-Jeghers–type hamartomatous polyps in the small intestine as well as in the stomach, large bowel, and extraintestinal sites, including the renal pelvis, bronchus, gallbladder, nasal passages, urinary bladder, and ureters
- Pigmented macules on the face, around oral mucosa and the perianal region, most pronounced in childhood

Cowden Syndrome

- Early-onset breast cancer, particularly before age 40
- Macrocephaly, especially 63 cm or larger in males, 60 cm or larger in females
- Thyroid cancer, particularly follicular type, before age 50
- Goiter, Hashimoto thyroiditis
- Dysplastic gangliocytoma of the cerebellum (Lhermitte-Duclos disease)
- Intestinal hamartomas
- Esophageal glycogenic acanthosis
- Skin findings of tricholemmomas or penile freckling
- Papillomas of oral cavity

*From Tinat J, Bougeard G, Baert-Desurmont S, et al: 2009 Version of the Chompret criteria for Li Fraumeni syndrome, *J Clin Oncol* 27:e108, 2009.

---

to also show the extracolonic manifestations of Gardner syndrome, the same mutation in unrelated individuals has been found to cause only FAP in one individual and Gardner syndrome in another. Thus whether or not an individual has FAP or Gardner syndrome is not simply due to which mutation is present in the *APC* gene but is likely affected by genetic variation elsewhere in the genome.

Lynch Syndrome. Approximately 2% to 4% of cases of colon cancer are attributable to LS (Case 29). LS is characterized by autosomal dominant inheritance of colon cancer in association with a small number of adenomatous polyps that begin during early adulthood. The number of polyps is generally quite small, in contrast to the hundreds to thousands of adenomatous polyps seen with FAP. Nonetheless, the polyps in LS have high potential to undergo malignant transformation. Heterozygotes for the most commonly mutated LS gene have an approximately 80% lifetime risk for development of cancer of the colon; female heterozygotes have a somewhat smaller risk (approximately 70%) but also have an approximately 40% risk for endometrial cancer. There are also additional risks of 10% to 20% for cancer of the biliary or urinary tract and the ovary. Sebaceous gland tumors of the skin (**Muir-Torre syndrome**) may be the first presenting sign in LS; thus the presence of such tumors in a patient should raise suspicion of a possible hereditary colon cancer syndrome.

LS results from loss-of-function mutations in one of four distinct but related DNA repair genes (*MLH1, MSH2, MSH6,* and *PMS2)* that encode **mismatch repair** proteins. Although all four of these genes have been implicated in LS in different families, *MLH1* and *MSH2* are together responsible for the vast majority of LS, whereas the others have been found in only a few patients and are often associated with a lesser degree of mismatch repair deficiency and lower penetrance. Like the *BRCA1* and *BRCA2* genes, the LS mismatch repair genes are TSGs involved in maintaining the integrity of the genome. Unlike *BRCA1* and *BRCA2*, however, the LS genes are not involved in double-stranded DNA break repair. Instead, their role is to repair incorrect DNA base pairing (i.e., pairing other than A with T or C with G) that can arise during DNA replication.

At the cellular level, the most striking phenotype of cells lacking mismatch repair proteins is an enormous increase in both point mutations and mutations occurring during replication of simple DNA repeats, such as a segment containing a string of the same base, for example $(A)_n$, or a microsatellite, such as $(TG)_n$ (see Chapter 4). Microsatellites are believed to be particularly vulnerable to mismatch because slippage of the strand being synthesized on the template strand can occur more readily when a short tandem repeat is being synthesized. Such instability, referred to as the **microsatellite instability-positive (MSI+)** phenotype, occurs at two orders of magnitude higher frequency in cells lacking both copies of a mismatch repair gene. The MSI+ phenotype is easily seen in DNA as three, four, or even more alleles of a microsatellite polymorphism in a single individual's tumor DNA (Fig. 15-9). It is estimated that cells lacking both copies of a mismatch repair gene may carry 100,000 mutations within simple repeats throughout the genome.

Because of the increase in mutation rate in these classes of sequence, loss of function of mismatch repair

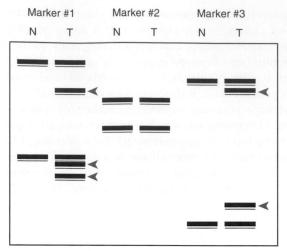

Figure 15-9 Gel electrophoresis of three different microsatellite polymorphic markers in normal (N) and tumor (T) samples from a patient with a mutation in *MSH2* and microsatellite instability. Although marker #2 shows no difference between normal and tumor tissues, genotyping at markers #1 and #3 reveals extra alleles (*blue arrows*), some smaller, some larger, than the alleles present in normal tissue.

genes will lead to somatic mutations in other driver genes. Two such driver genes have been isolated and characterized. The first is *APC*, whose normal function and role in FAP were described previously. The second is the gene *TGFBR2*, in which mutations also cause an autosomal dominant hereditary colon cancer syndrome. *TGFBR2* encodes transforming growth factor β receptor II, a serine-threonine kinase that inhibits intestinal cell division. *TGFBR2*, is particularly vulnerable to mutation when mismatch repair proteins are lost because it contains a stretch of 10 adenines encoding three lysines within its coding sequence; deletion of one or more of these As results in a frameshift and loss-of-function mutation. LS is an excellent example of how a gene, like *MLH1*, which has a global effect on mutation rate throughout the genome, can be a driver gene through its effect on other genes, such as *TGFBR2*, that are more specifically involved in driving the development of a cancer.

Mutations in Tumor Suppressor Genes Causing Autosomal Recessive Pediatric Cancer Syndromes

As expected from the important role that DNA replication and repair enzymes play in mutation surveillance and prevention, inherited defects that alter the function of repair enzymes can lead to a dramatic increase in the frequency of mutations of all types, including those that lead to cancer.

Mutations in the LS mismatch repair genes are frequent enough in the population for there to be rare individuals with *two* germline mutations in one of the

LS genes. Although much rarer than autosomal dominant forms of LS just discussed, this condition, known as **constitutional mismatch repair syndrome**, results in a markedly elevated risk for many cancers during childhood, including colorectal and small bowel cancer, as well as some cancers not associated with LS, such as leukemia in infancy and various types of brain tumors in childhood.

Several other well-known autosomal recessive disorders, including **xeroderma pigmentosum (Case 48)**, **ataxia-telangiectasia, Fanconi anemia,** and **Bloom syndrome,** are also due to loss of function of proteins required for normal DNA repair or replication. Patients with these rare conditions have a high frequency of chromosome and gene mutations and, as a result, a markedly increased risk for various types of cancer, particularly leukemia or, in the case of xeroderma pigmentosum, skin cancers in sun-exposed areas. Clinically, radiography must be used with extreme caution, if at all, in patients with ataxia-telangiectasia, Fanconi anemia, and Bloom syndrome, and exposure to sunlight must be avoided in patients with xeroderma pigmentosum.

Although these syndromes are rare autosomal recessive disorders, heterozygotes for these gene defects are much more common and appear to be at increased risk for malignant neoplasia. For example, Fanconi anemia, in which homozygotes have a number of congenital anomalies, bone marrow failure, leukemia, and squamous cell carcinoma of the head and neck, is a **chromosome instability syndrome** resulting from mutations of at least 18 different loci involved in DNA and chromosome repair. In the aggregate, Fanconi anemia has a population frequency of approximately 1 to 5 per million, which translates to a carrier frequency of approximately 1 to 2 per 500. One of these Fanconi anemia loci turns out to be the known hereditary cancer gene *BRCA2*. Others include *BRIP1*, *PALB2*, and *RAD51C* (discussed in the next section), which are known to increase susceptibility to breast cancer in heterozygotes. Similarly, female heterozygotes for certain ataxia-telangiectasia mutations have overall a twofold increased risk for breast cancer compared with controls and a fivefold higher risk for breast cancer before the age of 50 years. Thus heterozygotes for these chromosome instability syndromes constitute a sizeable pool of individuals at increased risk for cancer.

Testing for Germline Mutations Causing Hereditary Cancer

As introduced earlier, although some sporadic cancers will be truly sporadic and due entirely to somatic mutation(s), others likely reflect a predisposition to a specific cancer due to familial variants in one or more genes. This raises the possibility of using genetic testing or even whole-genome sequencing to screen for germline mutations that might inform risk estimates for members of the general population or for families with insufficient family history to implicate a hereditary cancer syndrome. Here we illustrate the issues involved in the case of two common neoplasias, breast cancer and colorectal cancer.

BRCA1 and *BRCA2* Testing

Identification of a germline mutation in *BRCA1* or *BRCA2* in a patient with breast cancer is of obvious importance for genetic counseling and cancer risk management for the patient's children, siblings, and other relatives, who may or may not be at increased risk. Such testing is, of course, also important for the patient's own management. For instance, in addition to removal of the cancer, a woman found to carry a *BRCA1* mutation might also choose to have a prophylactic mastectomy on the unaffected breast or a bilateral oophorectomy simultaneously to minimize the number of separate surgeries and anesthesia exposures. Finding a mutation in the proband or a first-degree relative would also allow mutation-specific testing in the rest of the family.

Importantly, however, the fraction of all female breast cancer patients whose disease is caused by a germline mutation in either the *BRCA1* or *BRCA2* gene is small, with estimates that vary between 1% and 3% in populations unselected for family history of breast or ovarian cancer, or for age at onset of the disease. Male breast cancer is 100 times less common than female breast cancer, but when it occurs, the frequency of germline mutations in hereditary breast cancer genes, particularly *BRCA2*, is 16%.

Until quite recently, the cost of mutation analysis in *BRCA1* and *BRCA2* was used to justify limiting gene sequencing to those patients most likely to be carrying a mutation, such as all male breast cancer patients and all women younger than 50 years with breast cancer, women with bilateral breast cancer, or women with first- and second-degree relatives with ovarian cancer or breast cancer. However, as the cost of sequencing falls, and large gene panels of breast cancer susceptibility genes, including *BRCA1* and *BRCA2*, can now be analyzed for less than it cost previously to sequence just *BRCA1* and *BRCA2* alone, the guidelines of just a few years ago will inevitably undergo reevaluation.

Colorectal Cancer Germline Mutation Testing

Only 4% of patients with colon cancer, not selected for a family history of cancer, carry a mutation in one of the four mismatch repair genes *MLH1, MSH2, MSH6,* and *PMS2* causing LS; an even smaller fraction contain *APC* mutations causing FAP. As with breast cancer, geneticists need to balance the cost and yield of sequencing hereditary colorectal cancer genes in every patient with colon cancer against the obvious importance of finding such a mutation for the patient and his or her family.

For LS, clinical factors such as the presence of multiple polyps, an early age at onset (before the age of 50 years), the location of the tumor in more proximal portions of the colon, the presence of a second tumor or history of colorectal cancer, a family history of colorectal or other cancers (particularly endometrial cancer), and cancer in relatives younger than 50 years of age, all boost the probability that a patient with colon cancer is carrying a mutation in a mismatch repair gene. Molecular studies of the tumor tissue, to look for evidence of the MSI+ phenotype (as discussed earlier in this chapter) or evidence of absent MSH2 and/or MSH6 protein by antibody staining in the tumor, also increase the probability that an individual patient with colorectal cancer carries a germline mismatch repair mutation. Unfortunately, loss of MLH1 protein staining in tumors due to promoter methylation is a frequent epigenetic finding in sporadic colon cancers and is therefore much less predictive of a germline LS mutation.

Combining clinical and molecular criteria allows the identification of a subset of all colorectal cancer patients in whom the probability of finding a mismatch repair mutation is much greater than 4%. These patients are clearly the most cost-effective group in which sequencing could be recommended. However, as with all such attempts at cost-effectiveness, limiting the number of patients studied to increase the yield of patients with positive sequencing inevitably results in missing a sizeable minority (20%) of patients with germline mismatch repair mutations. Again, the cost of mutation analysis must be reevaluated as the technology gets less expensive. More detailed discussions of gene testing will be presented in Chapter 18.

For FAP, the presence of hundreds of adenomatous polyps developing at an early age, multiple sebaceous adenomas, or the extracolonic signs of Gardner syndrome are sufficient to trigger germline testing for an *APC* mutation. There are, however, certain *APC* mutations that result in many fewer polyps and no extracolonic features (referred to as "attenuated FAP"). Attenuated FAP can be confused clinically with LS, but the tumors generally lack mismatch repair defects or microsatellite instability.

FAMILIAL OCCURRENCE OF CANCER

Cancer can also show increased incidence in families without fitting a clear-cut mendelian pattern. For example, it is estimated that as many as 20% of all breast cancer cases occurring in families that lack a clear, highly penetrant mendelian disorder nonetheless have a significant genetic contribution, as revealed by twin and family studies (see Chapter 8). The observed increase in cancer risk when relatives are affected may be due to mutations in a single gene but with penetrance that is sufficiently reduced to obscure any mendelian inheritance pattern. For example, in breast cancer, mutations in a gene such as *PALB2* can increase lifetime risk for breast cancer to approximately 25% by age 55 and approximately 40% by age 85. A lack of obvious breast cancer risk in men with *PALB2* mutations further obscures the inheritance pattern, although there is a significant increased risk for pancreatic cancer in men with these reduced penetrance alleles. Mutations in *BRIP1* and *RAD51C* have similar effects.

The bulk of familial cancer is, however, likely to be a complex disorder caused by both genetic and shared environmental factors (see Chapter 8). The degree of complex familial cancer risk can be assessed by epidemiological studies that compare how often the disease occurs in relatives versus the general population. The age-specific incidence of many forms of cancer in family members of probands is increased over the incidence of the same cancer in an age-matched cohort in the general population (Fig. 15-10). This increased risk has been observed in individuals whose first-degree relatives (parent or sibling) are affected by a wide variety of different cancers, with an even greater increase in incidence when an individual's parent and sibling are both affected.

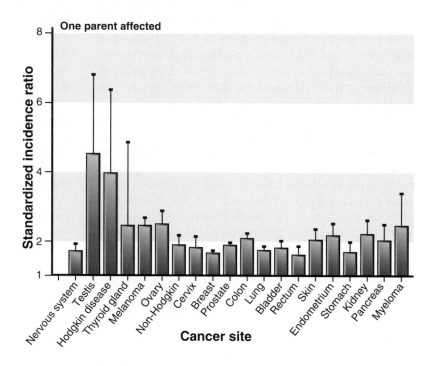

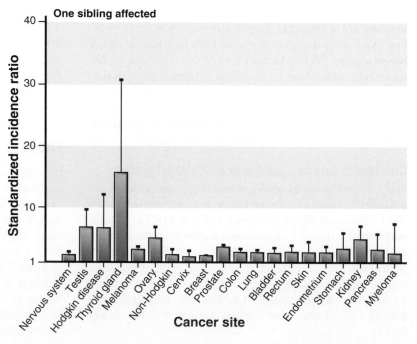

Figure 15-10 Standardized incidence ratios (SIRs) for cancers at various sites in first-degree relatives (child or sibling) of an affected person. A SIR is similar to the relative risk ratio (λ_r) that is based on *prevalence* of disease (as described in Chapter 8), except SIR is the ratio of the *incidence* of cases of cancer in relatives divided by the number expected from the incidence in an age-matched group in the general population. Error bars reflect 95% confidence limits on the SIRs. *See Sources & Acknowledgments.*

For example, population-based epidemiological studies have shown that approximately 5% of all individuals in North America and Western Europe will develop colorectal cancer in their lifetime, but the lifetime risk for colorectal cancer is increased twofold to threefold over the average population risk if one first-degree relative is affected.

In agreement with the likely complex inheritance of cancers, genome-wide association studies (see Chapter 10) have identified more than 150 mostly common variants associated with a variety of cancers. Prostate cancer, in particular, shows multiple associations with single nucleotide polymorphisms located in the intergenic or intronic regions of over a dozen loci. Unfortunately, odds ratios for most of these associations are all less than 2.0, and most are less than 1.3, therefore accounting for at most 20% of the observed familial risk for prostate cancer. Overall, then, although the role of inherited variants in the genome is clear, we cannot yet explain in detail the increased familial tendencies of most cancers. Whether common variants do not capture all of the risk or there are unrecognized environmental exposures in common between family members remain nonexclusive possibilities.

SPORADIC CANCER

Previously, we introduced the concept of activation of oncogenes by a variety of mutational mechanisms (see Fig. 15-3). Here, we explore these mechanisms and their effects in greater detail, particularly in the context of sporadic cancers.

Activation of Oncogenes by Point Mutation

Many mutated oncogenes were first identified by molecular studies of cell lines derived from sporadic cancers. One of the first activated oncogenes discovered was a mutant *RAS* gene derived from a bladder carcinoma cell line. *RAS* encodes one of a large family of small guanosine triphosphate (GTP)–binding proteins (so-called **G proteins**) that serve as molecular "on-off" switches to activate or inhibit downstream molecules. Remarkably, the activated oncogene and its normal counterpart proto-oncogene differed at only a single nucleotide. The alteration led to synthesis of an abnormal Ras protein that was able to signal continuously, thus stimulating cell division and changing it into a tumor. *RAS* point mutations are now known in many tumors, and the *RAS* genes have been shown experimentally to be the mutational target of known carcinogens, a finding that supports a role for mutated *RAS* genes in the development of many cancers.

To date, nearly 50 human proto-oncogenes have been identified as driver mutations in sporadic cancer. Only a few of these proto-oncogenes have also been found to be inherited in a hereditary cancer syndrome.

Activation of Oncogenes by Chromosome Translocation

As pointed out previously (see Fig. 15-3), oncogene activation is not always the result of a DNA mutation. In some instances, a proto-oncogene is activated by a subchromosomal mutation, typically a translocation. More than 40 oncogenic chromosome translocations have been described to date, primarily in sporadic leukemias and lymphomas but also in a few rare connective tissue sarcomas. Although originally detected only by cytogenetic analysis, such chromosome alterations can be detected now by whole-genome sequence analysis (see Fig. 5-7), even using cell-free DNA in plasma samples from cancer patients.

In some cases, translocation breakpoints lie within the introns of two genes, thereby fusing two genes into one abnormal gene that encodes a chimeric protein with novel oncogenic properties. The best-known example is the translocation between chromosomes 9 and 22, the so-called Philadelphia chromosome that is seen in **chronic myelogenous leukemia (CML)** (Fig. 15-11) (Case 10). The translocation moves the proto-oncogene *ABL1*, a tyrosine kinase, from its normal position on chromosome 9q to a gene of unknown function, *BCR*, on chromosome 22q. The translocation results in the synthesis of a novel, chimeric protein, **BCR-ABL1**, containing a portion of the normal Abl protein with

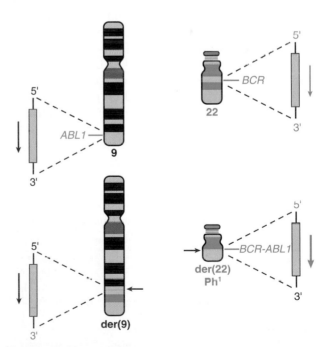

Figure 15-11 The **Philadelphia chromosome translocation, t(9;22)(q34;q11).** The Philadelphia chromosome (Ph[1]) is the derivative chromosome 22, which has exchanged part of its long arm for a segment of material from chromosome 9q that contains the *ABL1* oncogene. Formation of the chimeric *BCR-ABL1* gene on the Ph[1] chromosome is the critical genetic event in the development of chronic myelogenous leukemia.

increased tyrosine kinase activity. The enhanced tyrosine kinase activity of the novel protein encoded by the chimeric gene is the primary event causing the chronic leukemia. New, highly effective drug therapies for CML, such as imatinib, have been developed, based on inhibition of this tyrosine kinase activity.

In other cases, a translocation activates an oncogene by placing it downstream of a strong, constitutive promoter belonging to a different gene. Burkitt lymphoma is a B-cell tumor in which the *MYC* proto-oncogene is translocated from its normal chromosomal position at 8q24 to a position distal to the immunoglobulin heavy chain locus at 14q32 or the immunoglobulin light chain genes on chromosomes 22 and 2. The function of the Myc protein is still not entirely known, but it appears to be a transcription factor with powerful effects on the expression of a number of genes involved in cellular proliferation, as well as on telomerase expression (see later discussion). The translocation brings enhancer or other transcriptional activating sequences, normally associated with the immunoglobulin genes, near to the *MYC* gene (Table 15-4). These translocations allow unregulated *MYC* expression, resulting in uncontrolled cell division.

Telomerase as an Oncogene

Another type of oncogene is the gene-encoding **telomerase,** a reverse transcriptase that is required to synthesize the hexamer repeat, TTAGGG, a component of telomeres at the ends of chromosomes. Telomerase is needed because, during normal semiconservative replication of DNA (see Chapter 2), DNA polymerase can only add nucleotides to the 3′ end of DNA and cannot complete the synthesis of a growing strand all the way out to the very end of that strand on the chromosome arm; thus, in the absence of a specific mechanism to allow replication of telomeres, the end of each chromosome arm would shorten each and every cell division.

In human germline cells and embryonic cells, telomeres contain approximately 15 kb of the telomeric

repeat. As cells differentiate, telomerase activity declines in all somatic tissues; as telomerase function is lost, telomeres shorten, with a loss of approximately 35 bp of telomeric repeat DNA with each cell division. After hundreds of cell divisions, the chromosome ends become damaged, leading cells to stop dividing and to enter G_0 of the cell cycle; the cells will ultimately undergo apoptosis and die.

In contrast, in highly proliferative cells of tissues such as bone marrow, telomerase expression persists, allowing self-renewal. Similarly, telomerase persistence is observed in many tumors, which permits tumor cells to proliferate indefinitely. In some cases, telomerase activity results from chromosome or genome mutations that directly up-regulate the telomerase gene; in others, telomerase may be only one of many genes whose expression is altered by a transforming oncogene, such as *MYC*.

Loss of Tumor Suppressor Gene in Sporadic Cancer
TP53 and *RB1* in Sporadic Cancers

Although Li-Fraumeni syndrome, caused by a dominantly inherited germline mutation in the *TP53* gene, is a rare familial syndrome, somatic mutation causing a loss of function of both alleles of *TP53* is one of the most common genetic alterations seen in sporadic cancer (see Table 15-2). Mutations of *TP53*, deletion of the segment of chromosome 17p that includes *TP53*, or loss of the entire chromosome 17 is frequently and repeatedly seen in a wide range of sporadic cancers. These include breast, ovarian, bladder, cervical, esophageal, colorectal, skin, and lung carcinomas; glioblastoma of the brain; osteogenic sarcoma; and hepatocellular carcinoma.

The retinoblastoma gene *RB1* is also frequently mutated in many sporadic cancers, including breast cancer. For example, 13q14 LOH in human breast cancers is associated with loss of *RB1* mRNA in the tumor tissues. In still other cancers, the *RB1* gene is

TABLE 15-4 Characteristic Chromosome Translocations in Selected Human Malignant Neoplasms

Neoplasm	Chromosome Translocation	Percentage of Cases	Proto-oncogene Affected
Burkitt lymphoma	t(8;14)(q24;q32)	80	*MYC*
	t(8;22)(q24;q11)	15	
	t(2;8)(q11;q24)	5	
Chronic myelogenous leukemia	t(9;22)(q34;q11)	90-95	*BCR-ABL1*
Acute lymphocytic leukemia	t(9;22)(q34;q11)	10-15	*BCR-ABL1*
Acute lymphoblastic leukemia	t(1;19)(q23;p13)	3-6	*TCF3-PBX1*
Acute promyelocytic leukemia	t(15;17)(q22;q11)	≈95	*RARA-PML*
Chronic lymphocytic leukemia	t(11;14)(q13;q32)	10-30	*BCL1*
Follicular lymphoma	t(14;18)(q32;q21)	≈100	*BCL2*

Based on Croce CM: Role of chromosome translocations in human neoplasia, *Cell* 49:155-156, 1987; Park M, van de Woude GF: Oncogenes: genes associated with neoplastic disease. In Scriver CR, Beaudet AL, Sly WS, Valle D, editors: *The molecular and metabolic bases of inherited disease,* ed 6, New York, 1989, McGraw-Hill, pp 251-276; Nourse J, Mellentin JD, Galili N, et al: Chromosomal translocation t(1;19) results in synthesis of a homeobox fusion mRNA that codes for a potential chimeric transcription factor, *Cell* 60:535-545, 1990; and Borrow J, Goddard AD, Sheer D, Solomon E: Molecular analysis of acute promyelocytic leukemia breakpoint cluster region on chromosome 17, *Science* 249:1577-1580, 1990.

intact and its mRNA appears to be at or near normal levels, yet the RB1 protein is deficient. This anomaly has now been explained by the recognition that *RB1* can be down-regulated in association with overexpression of the oncomir *miR-106a*, which targets *RB1* mRNA and blocks its translation.

CYTOGENETIC CHANGES IN CANCER

Aneuploidy and Aneusomy

As introduced in Chapter 5, cytogenetic changes are hallmarks of cancer, whether sporadic or familial, particularly in later and more malignant or invasive stages of tumor development. Cytogenetic alterations suggest that a critical element of cancer progression includes defects in genes involved in maintaining chromosome stability and integrity and ensuring accurate mitotic segregation.

Initially, most of the cytogenetic studies of tumor progression were carried out in leukemias because the tumor cells were amenable to being cultured and karyotyped by standard methods. For example, when CML, with the 9;22 Philadelphia chromosome, evolves from the typically indolent chronic phase to a severe, life-threatening blast crisis, there may be several additional cytogenetic abnormalities, including numerical or structural changes, such as a second copy of the 9;22 translocation chromosome or an isochromosome for 17q. In advanced stages of other forms of leukemia, other translocations are common. In contrast, a vast array of chromosomal abnormalities are seen in most solid tumors. Cytogenetic abnormalities found repeatedly in a specific type of cancer are likely to be driver chromosome mutations involved in the initiation or progression of the malignant neoplasm. A current focus of cancer research is to develop a comprehensive cytogenetic and genomic definition of these abnormalities, many of which result in enhanced proto-oncogene expression or the loss of TSG alleles. Whole-genome sequencing is replacing cytogenetic analysis in many instances, because it provides a level of sensitivity and precision well beyond detection of cytologically visible genome changes.

Gene Amplification

In addition to translocations and other rearrangements, another cytogenetic aberration seen in many cancers is **gene amplification**, a phenomenon in which many additional copies of a segment of the genome are present in the cell (see Fig. 15-3). Gene amplification is common in many cancers, including neuroblastoma, squamous cell carcinoma of the head and neck, colorectal cancer, and malignant glioblastomas of the brain. Amplified segments of DNA are readily detected by comparative genome hybridization or whole-genome sequencing and appear as two types of cytogenetic change in routine chromosome analysis: **double minutes** (very small

accessory chromosomes) and **homogeneously staining regions** that do not band normally and contain multiple, amplified copies of a particular DNA segment. How and why double minutes and homogeneously staining regions develop are poorly understood, but amplified regions are known to include extra copies of proto-oncogenes such as the genes encoding Myc, Ras, and epithelial growth factor receptor, which stimulate cell growth, block apoptosis, or both. For example, amplification of the *MYCN* proto-oncogene encoding N-Myc is an important clinical indicator of prognosis in the childhood cancer **neuroblastoma**. *MYCN* is amplified more than 200-fold in 40% of advanced stages of neuroblastoma; despite aggressive treatment, only 30% of patients with advanced disease survive 3 years. In contrast, *MYCN* amplification is found in only 4% of early-stage neuroblastoma, and the 3-year survival is 90%. Amplification of genes encoding the targets of chemotherapeutic agents has also been implicated as a mechanism for the development of drug resistance in patients previously treated with chemotherapy.

APPLYING GENOMICS TO INDIVIDUALIZE CANCER THERAPY

Genomics is already having a major impact on diagnostic precision and optimization of therapy in cancer. In this section, we describe how one such approach, **gene expression profiling**, is used to guide diagnosis and treatment.

Gene Expression Profiling and Clustering to Create Signatures

Comparative hybridization techniques can be used to measure simultaneously the level of mRNA expression of some or all of the estimated 20,000 protein-coding genes in any human tissue sample. A measurement of mRNA expression in a sample of tissue constitutes a **gene expression profile** specific to that tissue. Figure 15-12 depicts a hypothetical, idealized situation of eight samples, four from each of two types of tumor, A and B, profiled for 100 different genes. The expression profile derived from expression arrays for this simple example is already substantial, consisting of 800 expression values. In a real expression profiling experiment, however, hundreds of samples may be analyzed for the expression of all human genes, producing a massive data set of millions of expression values. Organizing the data and analyzing them to extract key information are challenging problems that have inspired the development of sophisticated statistical and bioinformatic tools. Using such tools, one can organize the data to find groups of genes whose expression seems to correlate, that is, increase or decrease together, between and among the samples. Grouping genes by their patterns of expression across samples is termed **clustering**.

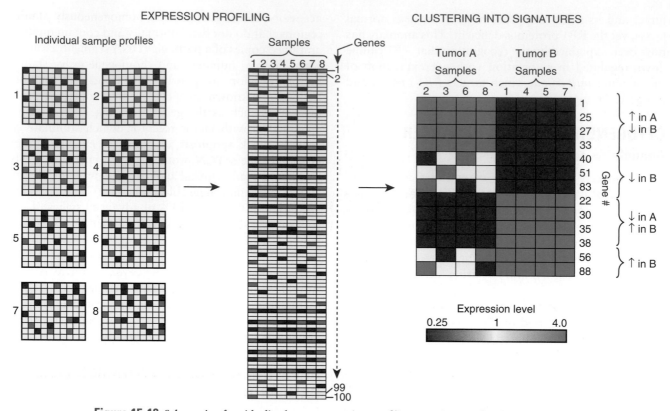

Figure 15-12 Schematic of an idealized gene expression profiling experiment of eight samples and 100 genes. *Left,* Individual arrays of gene sequences spotted on glass or silicon chips are used for comparative hybridization of eight different samples relative to a common standard. *Red* indicates decreased expression compared with control, *green* indicates increased expression, and *yellow* is unchanged expression. (In this schematic, *red, yellow,* and *green* represent decreased, equal or increased expression, whereas a real experiment would provide a continuous quantitative reading with shades of red and green.) *Center,* All 800 expression measurements are organized so that the relative expression for each gene, 1 through 100, is put in order vertically in a column under the number of each sample. *Right,* Clustering into signatures involves only those 13 genes that showed correlation across subsets of samples. Some genes have reciprocal (high versus low) expression in the two tumors; others show a correlated increase or decrease in one tumor and not the other.

Clusters of gene expression can then be tested to determine if any correlate with particular characteristics of the samples of interest. For example, profiling might indicate that a cluster of genes with a correlated expression profile is found more frequently in samples from tumor A than from tumor B, whereas another cluster of genes with a correlated expression profile is more frequent in samples derived from tumor B than from tumor A. Clusters of genes whose expression correlates with each other and with a particular set of samples constitute a so-called **expression signature** characteristic of those samples. In the hypothetical profiles in Figure 15-12, certain genes have a correlated expression that serves as a signature for tumor A; tumor B has a signature derived from the correlated expression of a different subset of these 100 genes.

Application of Gene Signatures

The application of gene expression profiles to characterize tumors is useful in a number of ways.

• First, it increases our ability to discriminate between different tumors in ways that complement the standard criteria applied by pathologists to characterize tumors, such as histological appearance, cytogenetic markers, and expression of specific marker proteins. Once distinguishing signatures for different tumor types (e.g., tumor A versus tumor B) are defined using known samples, the expression pattern of *unknown* tumor samples can then be compared with the expression signatures for tumor A and tumor B and classified as A-like, B-like, or neither, depending on how well their expression profiles match the signatures of A and B. Pathologists have used expression profiling to make difficult distinctions between tumors that require very different management approaches. These include distinguishing large B-cell lymphoma from Burkitt lymphoma, differentiating primary lung cancers from squamous cell carcinomas of the head and neck metastatic to lung, and identifying the tissue of origin of a cryptic primary tumor whose

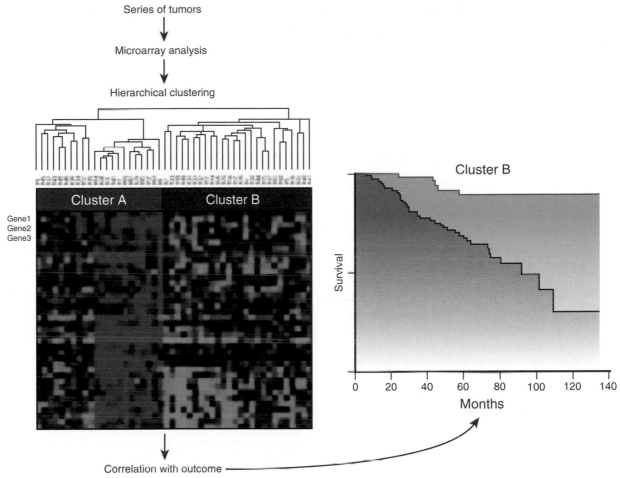

Figure 15-13 Expression patterns for a series of genes (along the vertical axis at *left*) for series of patient tumors, with the tumors arranged along the horizontal axis at *top* so that tumors with more similar expression patterns are grouped more closely together. The tumors appear to generally cluster into two groups, which are then correlated with long-term survival. *See Sources & Acknowledgments.*

metastasis gives too little information to allow its classification.

- Second, different signatures may be found to correlate with known clinical outcomes, such as prognosis, response to therapy, or any other outcome of interest. If validated, such signatures can be applied prospectively to help guide therapy in newly diagnosed patients.
- Finally, for basic research, clustering may reveal previously unsuspected connections of functional importance among genes involved in a disease process.

Gene Expression Profiling in Cancer Prognosis

Choosing the appropriate therapy for most cancers is difficult for patients and their physicians alike, because recurrence is common and difficult to predict. Better characterization of each patient's cancer as to recurrence risk and metastatic potential would clearly be beneficial for deciding between more or less aggressive courses of surgery and/or chemotherapy. For example, in breast

cancer, although presence of the estrogen and progesterone receptors, amplification of the human epidermal growth factor receptor 2 (*HER2*) oncogene, and absence of metastatic tumor in lymph nodes found on dissection of axillary lymphatics are strong predictors of better response to therapy and prognosis, they are still imprecise. Expression profiling (Fig. 15-13) is opening up a promising new avenue for clinical decision making in the management of breast cancer, as well as in other cancers, including lymphoma, prostate cancer, and metastatic adenocarcinomas of diverse tissue origins (lung, breast, colorectal, uterine, and ovarian).

Gene expression profiling of various sets of genes is clinically available for use in the management of breast, colon, and ovarian cancer; which genes and how many are included in the profile depends on the tumor type and vendor. Although the clinical utility and cost-effectiveness continue to be debated (see Chapter 18), there is a general consensus that combinations of clinical and gene expression data in patients newly diagnosed with cancer will provide better prospective estimates of

TABLE 15-5 Cancer Treatments Targeted to Specific Activated Driver Oncogenes

Tumor Type	Driver Gene and Mutation	Representative FDA-Approved Targeted Therapeutic	Mechanism of Action
Breast cancer	Amplified HER2	Trastuzumab	Anti-HER2 monoclonal antibody
Non–small cell lung cancer	Activated EGFR	Gefitinib	Tyrosine kinase inhibitor
Chronic myelogenous leukemia and gastrointestinal stromal tumor	Activated receptor tyrosine kinases Abl, KIT, and PDGF	Imatinib, nilotinib, and dasatinib	Tyrosine kinase inhibitor
Non–small cell lung cancer	Translocated ALK	Crizotinib	Tyrosine kinase inhibitor
Melanoma	Activated MEK	Trametinib	Serine-threonine kinase inhibitor
Melanoma	Activated BRAF kinase	Vemurafenib	Serine-threonine kinase inhibitor

ALK, Anaplastic lymphoma kinase; EGFR, epidermal growth factor receptor; FDA, U.S. Food and Drug Administration; HER2, human epidermal growth factor receptor 2; MEK, mitogen-activated extracellular signal-regulated kinase; PDGF, platelet-derived growth factor.

prognosis and improved guidance of therapy. It is hoped that by improving the accuracy of prognosis with tumor expression profiling, oncologists can choose to forgo more vigorous and expensive chemotherapies in patients who do not need and/or will not benefit from them.

The fact that the prognosis of practically every single patient could be associated with a particular combination of clinical features, genome sequence, and expression signatures underscores a crucial point about cancer: each person's cancer is a unique disorder. The genomic and gene expression heterogeneity among patients who all carry the same cancer diagnosis should not be surprising. *Every patient is unique in the genetic variants he or she carries, including those variants that will affect how the cancer develops and the body responds to it.* Moreover, the clonal evolution of a cancer implies that chance mutational and epigenetic events will likely occur in different and unique combinations in every patient's particular cancer.

Targeted Cancer Therapy

Until recently, most nonsurgical cancer treatment relied on cytotoxic agents, such as chemotherapeutic agents or radiation, designed to preferentially kill tumor cells while attempting to spare normal tissues. Despite tremendous successes in curing such diseases as childhood acute lymphocytic leukemia and Hodgkin lymphoma, most cancer patients in whom complete removal of the tumor with surgery is no longer possible receive remission, not cure, of their disease, usually at the cost of substantial toxicity from cytotoxic agents. The discovery of specific driver genes and their mutations in cancers has opened a new avenue for precisely targeted, less toxic treatments. Activated oncogenes are tempting targets for cancer therapy through direct blockade of their aberrant function. This can include blocking an activated cell surface receptor by monoclonal antibodies, or targeted inhibition of intracellular constitutive kinase activity with drugs designed to specifically inhibit their enzymatic activities.

The proof of principle for this approach was established with the development of imatinib, a highly effective inhibitor of a number of tyrosine kinases, including the ABL1 kinase in CML. Prolonged remissions of this disease have been seen, in some cases with apparently indefinite postponement of the transformation into a virulent acute leukemia (blast crisis) that so often meant the end of a CML patient's life. Additional kinase inhibitors have been developed to target other activated oncogene driver genes in a variety of tumor types (Table 15-5).

The initial results with targeted therapies, although very promising in some cases, have not led to permanent cures in most patients because tumors develop resistance to the targeted therapy. The outgrowth of resistant tumors is not surprising. First, as previously discussed, cancer cells are highly mutable, and their genomes undergo recurrent mutation. Even if only a small minority of cells acquire resistance through either mutation of the targeted oncogene itself, or through a compensatory mutation elsewhere, the tumor can progress even in the face of oncogene inhibition. Newer compounds that can overcome drug resistance are being developed and used in clinical trials. Ultimately, combination therapy that targets different driver genes may be required, based on the idea that a tumor cell is less likely to develop resistance in multiple unrelated pathways targeted by a combination of agents.

CANCER AND THE ENVIRONMENT

Although the theme of this chapter emphasizes the genetic basis of cancer, there is no contradiction in considering the role of environment in carcinogenesis. By environment, we include exposure to a wide variety of different types of agents—food, natural and artificial radiation, chemicals, even which viruses and bacteria are colonizing the gut. The risk for cancer shows significant variation among different populations and even within the same population in different environments. For example, gastric cancer is almost three times as common among Japanese in Japan as among Japanese living in Hawaii or Los Angeles.

In some cases, environmental agents act as mutagens that cause somatic mutations; the somatic mutations, in turn, are responsible for carcinogenesis. According to some estimates based chiefly on data from the aftermath of the atomic bombings of Hiroshima and Nagasaki, as much as 75% of the risk for cancer may be environmental in origin. In other cases, there appears to be a correlation between certain exposures and risk for cancer, such as the benefits of dietary fiber or low-dose aspirin therapy in lowering colon cancer risks. The nature of environmental agents that increase or reduce the risk for cancer, the assessment of the additional risk associated with exposure, and ways of protecting the population from such hazards are matters of strong public concern.

Radiation

Ionizing radiation is known to increase the risk for cancer. Everyone is exposed to some degree of ionizing radiation through background radiation (which varies greatly from place to place) and medical exposure. The risk is dependent on the age at exposure, being greatest for children younger than 10 years and for older adults.

Although there are still large areas of uncertainty about the magnitude of the effects of radiation (especially low-level radiation) on cancer risk, some information can be gleaned from events involving large-scale release of radiation into the environment. The data for survivors of the Hiroshima and Nagasaki atomic bombings, for example, show a long latency period, in the 5-year range for leukemia but up to 40 years for some tumors. In contrast, there has been little increase in cancer detectable among populations exposed to ionizing radiation by the more recent nuclear accident at Chernobyl, with the exception of a significant fivefold to sixfold increase in thyroid cancer among the most heavily exposed children living in Belarus. The increase in thyroid cancer is almost certainly caused by the radioactive iodine ^{131}I that was present in the nuclear material released from the damaged reactor and was taken up and concentrated within the thyroid gland.

Chemical Carcinogens

Interest in the carcinogenic effect of chemicals dates at least to the 18th century, when the high incidence of scrotal cancer in young chimney sweeps was noticed. Today, there is concern about many possible chemical carcinogens, especially tobacco, components of the diet, industrial carcinogens, and toxic wastes. Documentation of the risk of exposure is often difficult, but the level of concern is such that all clinicians should have a working knowledge of the subject and be able to distinguish between well-established facts and areas of uncertainty and debate.

The precise molecular mechanisms by which most chemical carcinogens cause cancer are still the subject of extensive research. One illustrative example of how a chemical carcinogen may contribute to the development of cancer is that of **hepatocellular carcinoma,** the fifth most common cancer worldwide. In many parts of the world, hepatocellular carcinoma occurs at increased frequency because of ingestion of aflatoxin B1, a potent carcinogen produced by a mold found on peanuts. Aflatoxin has been shown to mutate a particular base in the *TP53* gene, causing a G to T mutation in codon 249, thus converting an arginine codon to serine in the critically important p53 protein. This mutation is found in nearly half of all hepatocellular carcinomas in patients from parts of the world in which there is a high frequency of contamination of foodstuffs by aflatoxin, but it is not found in similar cancers in patients whose exposure to aflatoxin in food is low. The Arg249Ser mutation in p53 enhances hepatocyte growth and interferes with the growth control and apoptosis associated with wild-type p53; LOH of *TP53* in hepatocellular carcinoma is associated with a more malignant appearance of the cancer. Although aflatoxin B1 alone is capable of causing hepatocellular carcinoma, it also acts synergistically with chronic hepatitis B and C infections.

A more complicated situation occurs with an exposure to complex mixtures of chemicals, such as the many known or suspected carcinogens and mutagens found in cigarette smoke. The epidemiological evidence is overwhelming that cigarette smoke increases the risk for lung cancer and throat cancer, as well as other cancers. Cigarette smoke contains polycyclic hydrocarbons that are converted to highly reactive epoxides that cause mutations by directly damaging DNA. The relative importance of these substances and how they might interact in carcinogenesis are still being elucidated.

The case of cigarette smoking also raises another interesting issue. Why do only some cigarette smokers get lung cancer? The case of cancer and cigarette smoking provides an important example of the interaction between environmental and genetic factors to either enhance or prevent the carcinogenic effects of chemicals. The enzyme **aryl hydrocarbon hydroxylase (AHH)** is an inducible protein involved in the metabolism of polycyclic hydrocarbons, such as those found in cigarette smoke. AHH converts hydrocarbons into an epoxide form that is more easily excreted by the body but is also carcinogenic. AHH activity is encoded by members of the *CYP1* family of cytochrome P450 genes (see Chapter 18). The *CYP1A1* gene is inducible by cigarette smoke, but the inducibility is variable in the population because of different alleles at the *CYP1A1* locus. People who carry a "high-inducibility" allele, particularly those who are smokers, appear to be at an *increased risk for* lung cancer, with odds ratios of 4 to 5 compared to individuals without the cancer-susceptibility *CYP1A1* alleles. On the other hand, homozygotes for the recessive "low-inducibility" allele appear to be *less* likely to develop lung cancer, possibly because their AHH is less effective at converting the hydrocarbons to highly reactive carcinogens.

Similarly, individuals homozygous for alleles in the *CYP2D6* gene that reduce the activity of another cytochrome P450 enzyme appear to be more resistant to the potential carcinogenic effects of cigarette smoke or occupational lung carcinogens (e.g., asbestos or polycyclic aromatic hydrocarbons). Normal or ultrafast metabolizers, on the other hand, who carry alleles that increase the activity of the Cyp2D6 enzyme, have a four fold greater risk for lung cancer than do slow metabolizers. This risk increases to 18-fold among persons exposed routinely to lung carcinogens. A similar association has been reported for bladder cancer.

Although the precise genetic and biochemical basis for the apparent differences in cancer susceptibility within the normal population remains to be determined, these associations could have significant public health consequences and may point eventually to a way of identifying persons who are genetically at a higher risk for the development of cancer.

GENERAL REFERENCES

Garraway LA, Lander ES: Lessons from the cancer genome, *Cell* 153:17–37, 2013.
International Agency for Research on Cancer (IARC), World Health Organization: 2014. www.cruk.org/cancerstats.
Schneider L: *Counseling about cancer*, ed 3, New York, 2011, Wiley-Liss.

Shen H, Laird PW: Interplay between the cancer genome and epigenome, *Cell* 153:38–55, 2013.
Vogelstein B, Papadopoulos N, Velculescu VE, et al: Cancer genome landscapes, *Science* 339:1546–1558, 2013.

SPECIFIC REFERENCES

Chen P-S, Su J-L, Hung M-C: Dysregulation of microRNAs in cancer, *J Biomed Sci* 19:90, 2012.
Chin L, Anderson JN, Futreal PA: Cancer genomics, from discovery science to personalized medicine, *Nat Med* 17:297–303, 2011.
Di Leva G, Garofalo M, Croce CM: MicroRNAs in cancer, *Annu Rev Pathol Mech Dis* 9:287–314, 2014.
Kiplivaara O, Aaltonen LA: Diagnostic cancer genome sequencing and the contribution of germline variants, *Science* 339:1559–1562, 2013.
Lal A, Panos R, Marjanovic M, et al: A gene expression profile test to resolve head & neck squamous versus lung squamous cancers, *Diagn Pathol* 8:44, 2013.
Reis-Filho J, Pusztai L: Gene expression profiling in breast cancer: classification, prognostication, and prediction, *Lancet* 378:1812–1823, 2011.
Watson IR, Takahashi K, Futreal PA, et al: Emerging patterns of somatic mutations in cancer, *Nat Rev Genet* 14:703–718, 2013.
Wogan GN, Hecht SS, Felton JS, et al: Environmental and chemical carcinogenesis, *Semin Cancer Biol* 14:473–486, 2004.
Wong MW, Nordfors C, Mossman D, et al: BRIP1, PALB2, and RAD51C mutation analysis reveals their relative importance as genetic susceptibility factors for breast cancer, *Breast Cancer Res Treat* 127:853–859, 2011.

USEFUL WEBSITES

The Cancer Genome Atlas:
http://cancergenome.nih.gov/abouttcga/overview

PROBLEMS

1. A patient with retinoblastoma has a single tumor in one eye; the other eye is free of tumors. What steps would you take to try to determine whether this is sporadic or heritable retinoblastoma? What genetic counseling would you provide? What information should the parents have before a subsequent pregnancy?

2. Discuss possible reasons why colorectal cancer is an adult cancer, whereas retinoblastoma affects children.

3. Many tumor types are characterized by the presence of an isochromosome for the long arm of chromosome 17. Provide a possible explanation for this finding.

4. Many children with Fanconi anemia have limb defects. If an affected child requires surgery for the abnormal limb, what special considerations arise?

5. Wanda, whose sister has premenopausal bilateral breast cancer, has a greater risk for developing breast cancer herself than Wilma, whose sister has premenopausal breast cancer in only one breast. Both Wanda and Wilma, however, have a greater risk than does Winnie, who has a completely negative family history. Discuss the role of molecular testing in these women. What would their breast cancer risks be if a pathogenic *BRCA1* or *BRCA2* mutation were found in the affected relative? What if no mutations were found?

6. Propose a theory for why so few hereditary cancer syndromes, inherited as autosomal dominant diseases, are caused by activated oncogenes, whereas so many are caused by germline mutations in a tumor suppressor gene (TSG).

CHAPTER 16

Risk Assessment and Genetic Counseling

In this chapter, we present the fundamentals of the practice of genetic counseling as applied to families in which an individual is known or suspected to have a hereditary condition. Genetic counseling includes a discussion of the natural history of the disease as well as determination of the risk for disease in other family members based on the inheritance pattern, empirical risk figures, and medical testing, especially molecular genetic and genomic testing. Counseling includes a discussion of approaches available to mitigate or reduce the risk for heritable disease. Finally, the counselor carries out a careful assessment of the psychological and social impact of the diagnosis on the patient and family and works to help the family cope with the presence of a heritable condition.

FAMILY HISTORY IN RISK ASSESSMENT

Family history is clearly of great importance in diagnosis and risk assessment. Applying the known rules of mendelian inheritance, as introduced in Chapter 7, allows the geneticist to provide accurate evaluations of risk for disease in relatives of affected individuals. Family history is also important when a geneticist assesses the risk for complex disorders, as discussed in Chapter 8 and elsewhere in this book. Because a person's genes are shared with his or her relatives, family history provides the clinician with information on the impact that an individual's genetic makeup might have on one's health, using the medical history of relatives as an indicator of one's own genetic susceptibilities. Furthermore, family members often share environmental factors, such as diet and behavior, and thus relatives provide information about both shared genes and shared environmental factors that may interact to cause the common, genetically complex diseases. Having a first-degree relative with a common disease of adulthood—such as cardiovascular disease, cancer of the breast, cancer of the colon or prostate, type 2 diabetes, osteoporosis, or asthma—raises an individual's risk for the disease approximately twofold to threefold relative to the general population, a moderate increase compared with the average population risk (see Box). As discussed in Chapter 8, the more first-degree relatives one has with a complex trait and the earlier in life the disease occurs in a family member, the greater the load of susceptibility genes and environmental exposures likely to be present in the patient's family. Thus consideration of family history can lead to the designation of a patient as being at high risk for a particular disease on the basis of family history. For example, a male with three male first-degree relatives with prostate cancer has an 11-fold greater relative risk for development of the disease than does a man with no such family history.

FAMILY HISTORY IN RISK ASSESSMENT

High Risk

- Age at onset of a disease in a first-degree relative relatively early compared to the general population
- Two affected first-degree relatives
- One first-degree relative with late or unknown disease onset and an affected second-degree relative with premature disease from the same lineage
- Two second-degree maternal or paternal relatives with at least one having premature onset of disease
- Three or more affected maternal or paternal relatives
- Presence of a "moderate-risk" family history on both sides of the pedigree

Moderate Risk

- One first-degree relative with late or unknown onset of disease
- Two second-degree relatives from the same lineage with late or unknown disease onset

Average Risk

- No affected relatives
- Only one affected second-degree relative from one or both sides of the pedigree
- No known family history
- Adopted person with unknown family history

From Scheuner MT, Wang SJ, Raffel LJ, et al: Family history: a comprehensive genetic risk assessment method for the chronic conditions of adulthood, *Am J Med Genet* 71:315-324, 1997; quoted in Yoon PW, Scheuner MT, Peterson-Oehlke KL, et al: Can family history be used as a tool for public health and preventive medicine? *Genet Med* 4:304-310, 2002.

Determining that an individual is at increased risk on the basis of family history can have an impact on individual medical care. For example, two individuals with deep venous thrombosis—one with a family history of unexplained deep venous thrombosis in a relative younger than 50 years and another with no family history of any coagulation disorder—should receive different management with respect to testing for factor V Leiden or prothrombin 20210G>A and anticoagulation therapy (see Chapter 8). Similarly, having a first-degree relative with colon cancer is sufficient to trigger the initiation of colon cancer screening by colonoscopy at the age of 40 years, 10 years earlier than for the general population. This is because the cumulative incidence for development of the disease for someone 40 years old with a positive family history equals the risk for someone at the age of 50 years with no family history (see Fig. 18-1). The increase in risk is even more pronounced if two or more relatives have had the disease, an empirical observation that has driven standards of clinical care for screening in this condition.

Family history is admittedly an indirect method of assessing the contribution of an individual's own genetic variants to health and disease susceptibility. Direct detection of genetic risk factors and demonstrating that they are valid for guiding health care is a major challenge in applying genomics to medicine, as we will take up in Chapter 18.

GENETIC COUNSELING IN CLINICAL PRACTICE

Clinical genetics is concerned with the diagnosis and management of the medical, social, and psychological aspects of hereditary disease. As in all other areas of medicine, it is essential in clinical genetics to do the following:

- Make a correct diagnosis, which often involves laboratory testing, including genetic testing to find the mutations responsible.
- Help the affected person and family members understand and come to terms with the nature and consequences of the disorder.
- Provide appropriate treatment and management, including referrals to other specialist providers as needed.

Just as the unique feature of genetic disease is its tendency to recur within families, the unique aspect of genetic counseling is its focus on both the original patient and also on members of his or her family, both present and future. Genetic counselors have a responsibility to do the following:

- Work with the patient to inform other family members of their potential risk.
- Offer mutation or other testing to provide the most precise risk assessments possible for other family members.
- Explain what approaches are available to the patient and family members to modify these risks.

Finally, genetic counseling is not limited to the provision of information and identification of individuals at risk for disease; rather, it is a process of exploration and communication. Genetic counselors define and address the complex psychosocial issues associated with a genetic disorder in a family and provide psychologically oriented counseling to help individuals adapt and adjust to the impact and implications of the disorder in the family. For this reason, genetic counseling may be most effectively accomplished through periodic contact with the family as the medical or social issues become relevant to the lives of those involved (see Box earlier).

Genetic Counseling Providers

Clinical genetics is particularly time-consuming in comparison with other clinical fields because it requires extensive preparation and follow-up in addition to time for direct contact with patients. In many countries, genetic counseling is provided by physicians. However, in the United States, Canada, the United Kingdom, and a few other countries, genetic counseling services are often provided by **genetic counselors** or **nurse geneticists,** professionals specially trained in genetics *and* counseling, who serve as members of a health care team with physicians. Genetic counseling in the United States and Canada is a self-regulating health profession with its own board (the American and Canadian Boards of Genetic Counselors) for accreditation of training programs and certification of practitioners. Some states in the United States are also licensing genetic counselors. Nurses with genetics expertise are accredited through a separate credentialing commission.

Genetic counselors and nurse geneticists play an essential role in clinical genetics, participating in many aspects of the investigation and management of genetic problems. A genetic counselor is often the first point of contact that a patient has with clinical genetic services, provides genetic counseling directly to individuals, helps patients and families deal with the many psychological and social issues that arise during genetic counseling, and continues in a supportive role and as a source of information after the clinical investigation and formal counseling have been completed. Genetic counselors are also active in the field of genetic testing; they provide close liaison among the referring physicians, the diagnostic laboratories, and the families themselves. Their special expertise is invaluable to clinical laboratories because explaining and interpreting genetic testing to patients and referring physicians often requires a sophisticated knowledge of genetics and genomics, as well as excellent communication skills.

Common Indications for Genetic Counseling

Table 16-1 lists some of the most common situations that lead people to pursue genetic counseling. Individuals seeking genetic counseling (referred to as the

TABLE 16-1 Common Indications for Genetic Counseling

- Previous child with multiple congenital anomalies, intellectual disability, or an isolated birth defect such as neural tube defect or cleft lip and palate
- Personal history or family history of a hereditary condition, such as cystic fibrosis, fragile X syndrome, congenital heart defect, hereditary cancer, or diabetes
- Pregnancy at risk for a chromosomal or hereditary disorder
- Consanguinity
- Teratogen exposure, such as to occupational chemicals, medications, alcohol
- Repeated pregnancy loss or infertility
- Newly diagnosed abnormality or genetic condition
- Before undertaking genetic testing and after receiving results, particularly in testing for susceptibility to late-onset disorders, such as hereditary cancer syndromes or neurological disease
- As follow-up for a positive result of a newborn test, as with phenylketonuria; a heterozygote (preconception carrier) screening test, such as Tay-Sachs; or a positive first- or second-trimester maternal serum screen, a noninvasive prenatal screen by free fetal DNA analysis or abnormal fetal ultrasound examination results

TABLE 16-2 Genetic Counseling Case Management

• Collection of information
  Family history
    (questionnaire)
  Medical history
  Tests or additional
    assessments
• Assessment
  Physical examination
  Laboratory and radiological
    testing
  Validation or establishment
    of diagnosis—if possible
	• Counseling
	  Nature and consequence of
	    disorder
	  Recurrence risk
	  Availability of further or
	    future testing
	• Decision making
	  Referral to other specialists,
	    health agencies, support
	    groups
	• Continuing clinical
	    assessment, especially if no
	    diagnosis
	• Psychosocial support

consultands) may themselves be the probands in the family, or they may be the parents of an affected child or have relatives with a potential or known genetic condition. Genetic counseling is also an integral part of prenatal testing (see Chapter 17) and of genetic testing and screening programs (discussed in Chapter 18).

Established standards of medical care require that providers of genetic services obtain a history that includes family and ethnic information, inquire as to possible consanguinity, advise patients of the genetic risks to them and other family members, offer genetic testing or prenatal diagnosis when indicated, and outline the various treatment or management options for reducing the risk for disease. Although genetic counseling case management must be individualized for each patient's needs and situation, a generic approach can be summarized (Table 16-2). In general, patients are *not told* what decisions to make with regard to the various testing and management options but are instead provided with information and support in coming to a

decision that seems most appropriate for the patients, the consultands, and their families. This approach to counseling, referred to as **nondirective counseling**, has its origins in the setting of prenatal counseling, where the guiding principle is respect for an individual couple's autonomy, that is, their right to make reproductive choices free of coercion (see Chapter 19).

Managing the Risk for Recurrence in Families

Many families seek genetic counseling to ascertain the risk for heritable disease in their children and to learn what options are available to reduce the risk for recurrence of the particular genetic disorder in question. Genetic laboratory tests for carrier testing (karyotyping, biochemical analysis, or genome analysis) are frequently used to determine the actual risk to couples with a family history of a genetic disorder. Genetic counseling is recommended both before and after such testing, to assist consultands in making an informed decision to undergo testing, as well as to understand and to use the information gained through testing.

When family history or laboratory testing indicate an increased risk for a hereditary condition in a future pregnancy, prenatal diagnosis, described in Chapter 17, is one approach that can often be offered to families. Prenatal diagnosis is, however, by no means a universal solution to the risk for genetic problems in offspring. There are disorders for which prenatal diagnosis is not available and, for many parents, pregnancy termination is not an acceptable option, even if prenatal diagnosis is available. Preimplantation diagnosis by blastocyst or blastomere biopsy (see Chapter 17) avoids the problems of pregnancy termination but requires in vitro fertilization.

Other measures besides prenatal diagnosis are available for the management of recurrence and include the following:

- Genetic laboratory tests for carrier testing can sometimes reassure couples with a family history of a genetic disorder that they themselves are *not* at increased risk for having a child with a specific genetic disease. In other cases, such tests indicate that the couple *is* at increased risk. Genetic counseling is recommended both before and after such testing, to assist consultands in making an informed decision to undergo testing, as well as understanding and using the information gained through testing.
- If the parents plan to have no more children or no children at all, **contraception or sterilization** may be their choice, and they may need information about the possible procedures or an appropriate referral.
- **Adoption** is a possibility for parents who want a child or more children.
- **Artificial insemination** may be appropriate if the father has a gene for an autosomal dominant or X-linked defect or has a heritable chromosome defect,

but it is obviously not indicated if it is the mother who has such a defect. Artificial insemination is also useful if both parents are carriers of an autosomal recessive disorder. In vitro fertilization with a **donated egg** may be appropriate if the mother has an autosomal dominant defect or carries an X-linked disease. In either case, genetic counseling and appropriate genetic tests of the sperm or egg donor should be part of the process.

If the parents decide to terminate a pregnancy, provision of relevant information and support is an appropriate part of genetic counseling. Periodic follow-up through additional visits or by telephone is often arranged for a few months or more after a pregnancy termination.

Psychological Aspects

Patients and families dealing with a risk for a genetic disorder or coping with the illness itself are subject to varying degrees of emotional and social stress. Although this is also true of nongenetic disorders, the concern generated by knowledge that the condition might recur, the guilt or censure felt by some individuals, and the need for reproductive decisions can give rise to severe distress. Many persons have the strength to deal personally with such problems; they prefer receiving even bad news to remaining uninformed, and they make their own decisions on the basis of the most complete and accurate information they can obtain. Other persons require much more support and may need referral for psychotherapy. The psychological aspects of genetic counseling are beyond the scope of this book, but several texts cited in the General References at the end of this chapter give an introduction to this important field (see Box).

Genetic counselors often refer a patient and family with a genetic disorder or birth defect to family and patient **support groups**. These organizations, which can be focused either on a single disease or on a group of diseases, can help those concerned to share their experience, to learn how to deal with the day-to-day problems caused by the disorder, to hear of new developments in therapy or prevention, and to promote research into the condition. Many support groups have Internet sites and electronic chat rooms, through which patients and families give and receive information and advice, ask and answer questions, and obtain much needed emotional support. Similar disease-specific, self-help organizations are active in many nations around the world.

DETERMINING RECURRENCE RISKS

The estimation of recurrence risks is a central concern in genetic counseling. Ideally, it is based on knowledge of the genetic nature of the disorder in question and on

GENETIC COUNSELING AND RISK ASSESSMENT

The purpose of genetic counseling is to provide information and support to families at risk for having, or who already have, members with birth defects or genetic disorders. Genetic counseling helps the family or individual to do the following:

- Comprehend the medical facts, including the diagnosis, the probable course of the disorder, and the available management.
- Understand the way heredity contributes to the disorder and the risk for recurrence for themselves and other family members.
- Understand the options for dealing with the risk for recurrence.
- Identify those values, beliefs, goals, and relationships affected by the risk for or presence of hereditary disease.
- Choose the course of action that seems most appropriate to them in view of their risk, their family goals, and their ethical and religious standards.
- Make the best possible adjustment to the disorder or to the risk for recurrence of that disorder, or both, by providing supportive counseling to families and making referrals to appropriate specialists, social services, and family and patient support groups.

the pedigree of the particular family being counseled. The family member whose risk for a genetic disorder is to be determined is usually a relative of a proband, such as a sibling of an affected child or a living or future child of an affected adult. In some families, especially for some autosomal dominant and X-linked traits, it may also be necessary to estimate the risk for more remote relatives.

When a disorder is known to have single-gene inheritance, the recurrence risk for specific family members can usually be determined from basic mendelian principles (Fig. 16-1; also see Chapter 7). On the other hand, risk calculations may be less than straightforward if there is reduced penetrance or variability of expression, or if the disease is frequently the result of new mutation, as in many X-linked and autosomal dominant disorders. Laboratory tests that give equivocal results can add further complications. Under these circumstances, mendelian risk estimates can sometimes be modified by means of applying **conditional probability** to the pedigree (see later), which takes into account information about the family that may increase or decrease the underlying mendelian risk.

In contrast to single-gene disorders, the underlying mechanisms of inheritance for most chromosomal or genomic disorders and complex traits are unknown, and estimates of recurrence risk are based on previous experience (Fig. 16-2). This approach to risk assessment is valuable if there are reliable data on the frequency of recurrence of the disorder in families and if the

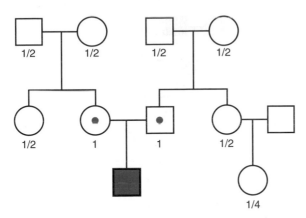

Figure 16-1 Pedigree of a family with an autosomal recessive condition. The probability of being a carrier is shown beneath each individual symbol in the pedigree.

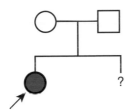

Figure 16-2 Empirical risk estimates in genetic counseling. A family with no other positive family history has one child affected with a disorder known to be multifactorial or chromosomal. What is the recurrence risk? If the child is affected with spina bifida, the empirical risk to a subsequent child is approximately 4%. If the child has Down syndrome, the empirical risk for recurrence would be approximately 1% if the karyotype is trisomy 21, but it might be substantially higher if one of the parents is a carrier of a Robertsonian translocation involving chromosome 21 (see Chapter 6).

phenotype is not heterogeneous. However, when a particular phenotype has an undetermined risk or can result from a variety of causes with different frequencies and with widely different risks, estimation of the recurrence risk is hazardous at best. In a later section, the estimation of recurrence risk in some typical clinical situations, both straightforward and more complicated, is considered.

Risk Estimation by Use of Mendel's Laws When Genotypes Are Fully Known

The simplest risk estimates apply to families in which the relevant genotypes of all family members are known or can be inferred. For example, if both members of a couple are known to be heterozygous carriers of an autosomal recessive condition because they have a child with the disorder or because of carrier testing, the risk (probability) is one in four with each pregnancy that the child will inherit two mutant alleles and inherit the disease (Fig. 16-3A). Even if the couple were to have six unaffected children subsequent to the affected child

(Fig. 16-3B), the risk in the eighth, ninth, or tenth pregnancy would still be one in four for each pregnancy (assuming there is no misattributed paternity for the first affected child).

Risk Estimation by Use of Conditional Probability When Alternative Genotypes Are Possible

In contrast to the simple case just described, situations arise in which the genotypes of the relevant individuals in the family are not *definitively* known; the risk for recurrence will be very different, depending on whether or not the consultand is a carrier of an abnormal allele of a disease gene. For example, the chance that a woman, who is known from her first marriage to be a carrier of cystic fibrosis (CF), might have an affected child depends on the chance that her husband by her second marriage is a carrier (Fig. 16-3C). The risk for the partner's being a carrier depends on his ethnic background (see Chapter 9). For the general non-Hispanic white population, this chance is approximately 1 in 22. Therefore the chance that a known carrier and her unrelated partner would have an affected first child is the product of these probabilities, or $\frac{1}{22} \times \frac{1}{4} = \frac{1}{88}$ (approximately 1.1%).

Of course, if the husband really were a carrier, the chance that the child of two carriers would be a homozygote or a compound heterozygote for mutant CF alleles is one in four. If the husband were not a carrier, then the chance of having an affected child is zero. Suppose, however, that one cannot test his carrier status

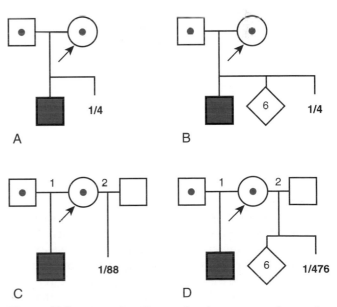

Figure 16-3 Series of pedigrees showing autosomal recessive inheritance with contrasting recurrence risks. A and B, The genotypes of the parents are known. C, The genotype of the consultand's second partner is inferred from the carrier frequency in the population. D, The inferred genotype is modified by additional pedigree information. *Arrows* indicate the consultand. Numbers indicate recurrence risk in the consultand's next pregnancy.

directly. A carrier risk of 1 in 22 is the best estimate one can make for individuals of his ethnic background and no family history of CF without direct carrier testing; in fact, however, a person either is a carrier or is not. The problem is that we do not know. In this situation, the more opportunities the male in Figure 16-3C (who *may or may not* be a carrier of a mutant gene) has to pass on the mutant gene and fails to do so, the less likely it would be that he is indeed a carrier. Thus, if the couple were to come for counseling already with six children, none of whom is affected (Fig. 16-3D), it would seem reasonable, intuitively, that the husband's chance of being a carrier should be less than the 1 in 22 risk that the childless male partner in Figure 16-3C was assigned on the basis of the population carrier frequency. In this situation, we apply conditional probability (also known as **Bayesian analysis,** based on Bayes's theorem on probability published in 1763), a method that takes advantage of *phenotypic* information in a pedigree to assess the relative probability of two or more alternative *genotypic* possibilities and to condition the risk on the basis of that information. In Figure 16-3D, the chance that the second husband is a carrier is actually 1 in 119, and the chance that this couple would have a child with CF is therefore 1 in 476, not 1 in 88, as calculated in Fig. 16-3C. Some examples of the use of Bayesian analysis for risk assessment in pedigrees are examined in the following section.

Conditional Probability

To illustrate the application of Bayesian analysis, consider the pedigrees shown in Figure 16-4. In Family A, the mother II-1 is an **obligate carrier** for the X-linked bleeding disorder hemophilia A because her father was affected. Her risk for transmitting the mutant factor VIII (*F8*) allele responsible for hemophilia A is 1 in 2, and the fact that she has already had four unaffected sons does *not* reduce this risk. Thus the risk that the consultand (III-5) is a carrier of a mutant *F8* allele is 1 in 2 because she is the daughter of a known carrier.

In Family B, however, the consultand's mother (individual II-2) may or may not be a carrier, depending on whether she has inherited a mutant *F8* allele from her mother, I-1. If III-5 were the only child of her mother, III-5's risk for being a carrier would be 1 in 4, calculated as ½ (her mother's risk for being a carrier) × ½ (her risk for inheriting the mutant allele from her mother). Short of testing III-5 directly for the mutant allele, we cannot tell whether she is a carrier. In this case, however, the fact that III-5 has four unaffected brothers is relevant because every time II-2 had a son, the chance that the son would be unaffected is only 1 in 2 if II-2 *were* a carrier, whereas it is a near certainty (probability = 1) that the son would be unaffected if II-2 were, in fact, *not* a carrier at all. With each son, II-2 has, in effect, tested her carrier status by placing herself at a 50% risk for having an affected son. To have four unaffected sons

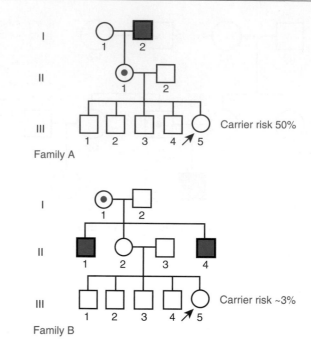

Figure 16-4 Modified risk estimates in genetic counseling. The consultands in the two families are at risk for having a son with hemophilia A. In Family A, the consultand's mother is an obligate heterozygote; in Family B, the consultand's mother may or may not be a carrier. Application of Bayesian analysis reduces the risk for being a carrier to only approximately 3% for the consultand in Family B but not the consultand in Family A. See text for derivation of the modified risk.

might suggest that maybe her mother is not a carrier. Bayesian analysis allows one to take this kind of indirect information into account in calculating whether II-2 is a carrier, thus modifying the consultand's risk for being a carrier. In fact, as we show in the next section, her carrier risk is far lower than 50%.

Identify the Possible Scenarios

To translate this intuition into actual risk calculation, we use a Bayesian probability calculation. First, we list *all* possible alternative genotypes that may be present in the relevant individuals in the pedigree (Fig. 16-5). In this case, there are three scenarios, each reflecting a different combination of alternative genotypes:

A. II-2 *is* a carrier, but the consultand *is not.*
B. II-2 and the consultand are *both* carriers.
C. II-2 is *not* a carrier, which implies that the consultand could not be one either because there is no mutant allele to inherit.

Why do we not consider the possibility that the consultand is a carrier even though II-2 is not? We do not list this scenario because it would require that *two* mutations in the same gene occur independently in the same family, one inherited by the probands and one new mutation in the consultand, a scenario so vanishingly unlikely that it can be dismissed out of hand.

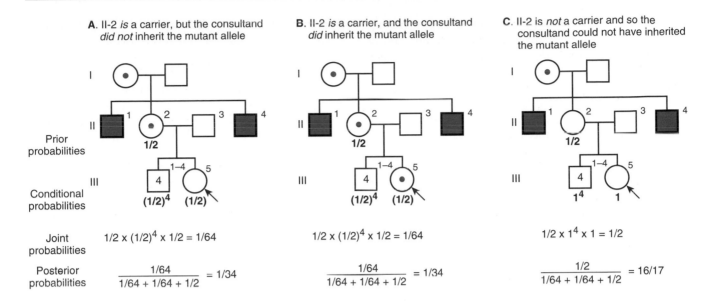

Figure 16-5 Conditional probability used to estimate carrier risk for a consultand in a family with hemophilia in which the prior probability of the carrier state is determined by mendelian inheritance from a known carrier at the top of the pedigree. These risk estimates, based on genetic principles, can be further modified by considering information obtained from family history, carrier detection testing, or molecular genetic methods for direct detection of the mutation in the affected boy, with use of Bayesian calculations. A to C, The three mutually exclusive situations that could explain the pedigree.

First, we draw the three possible scenarios as pedigrees (as in Fig. 16-5) and write down the probability of individual II-2's being a carrier or not. This is referred to as her **prior probability** because it depends simply on her risk for carrying a mutant allele inherited from her known carrier mother, I-1, and it has not been modified ("conditioned") at all by her own reproductive history.

Next, we write down the probabilities that individuals III-1 through III-4 would be unaffected under each scenario. These probabilities are different, depending on whether II-2 is a carrier or not. If she *is* a carrier (situations A and B), then the chance that individuals III-1 through III-4 would all be unaffected is the chance that each did not inherit II-2's mutant *F8* allele, which is 1 in 2 for each of her sons or $(\frac{1}{2})^4$ for all four. In situation C, however, II-2 is *not* a carrier, so the chance that her four sons would all be unaffected is 1 because II-2 does not have a mutant *F8* to pass on to any of them. These are called **conditional probabilities** because they are probabilities affected by the condition of whether II-2 is a carrier.

Similarly, we can write down the probability that the consultand (III-5) is a carrier. In A, she did not inherit the mutant allele from her carrier mother, with a probability of 1 in 2. In B, she did inherit the mutant allele (probability = $\frac{1}{2}$). In C, her mother is not a carrier, and so III-5 has essentially a 100% chance of not being a carrier. Multiply the prior and conditional probabilities together to form the **joint probabilities** for each situation, A, B, and C.

Finally, we determine what *fraction* of the total joint probability is represented by any scenario of interest; this is called the **posterior probability** of each of the three situations. Because III-5 is the consultand and wants to know her risk for being a carrier, we need the posterior probability of situation B, which is:

$$\frac{\frac{1}{64}}{\frac{1}{64} + \frac{1}{64} + \frac{1}{2}} = \frac{1}{34} - \approx 3\%$$

If we wish to know the chance that II-2 is a carrier, we add the posterior probabilities of the two situations in which she is a carrier, A and B, to get a carrier risk of 1 in 17, or approximately 6%.

If III-5 were also to have unaffected sons, her carrier risk could also be modified downward by a Bayesian calculation. However, if II-2 were to have an affected child, then she would have proved herself a carrier, and III-5's risk would thus become 1 in 2. Similarly, if III-5 were to have an affected child, then she must be a carrier, and Bayesian analysis would no longer be necessary.

Bayesian analysis may seem to some like mere statistical maneuvering. However, the analysis allows genetic counselors to quantify what seemed to be intuitively likely from inspection of the pedigree: the fact that the consultand had four unaffected brothers provides support for the hypothesis that her mother is not a carrier. The analysis having been performed, the final risk that III-5 is a carrier can be used in genetic counseling. The risk that her first child will have hemophilia A is $\frac{1}{34} \times \frac{1}{4}$, or less than 1%. This risk is appreciably below the prior probability estimated without taking into account the genetic evidence provided by her brothers.

Conditional Probability in X-Linked Lethal Disorders

Because any severe X-linked disorder is manifested in the hemizygous male, an isolated case (no family history) of such a disorder may represent either a new gene mutation (in which case the mother is not a carrier) or inheritance of a mutant allele from his unaffected carrier mother; we do not consider the small but real chance of gonadal mosaicism for the mutation in the mother (see Chapter 7). Estimation of the recurrence risk depends on knowing the chance that she could be a carrier. Bayesian analysis can be used to estimate carrier risks in X-linked lethal diseases such as Duchenne muscular dystrophy (DMD) and severe ornithine transcarbamylase deficiency.

Consider the family at risk for DMD shown in Figure 16-6. The consultand, III-2, wants to know her risk for being a carrier. There are three possible scenarios, each with dramatically different risk estimates for the family:

A. III-1's condition may be the result of a new mutation. In this case, his sister and maternal aunt are not at significant risk for being a carrier.

B. His mother, II-1, is a carrier, but her condition is the result of a new mutation. In this case, his sister (III-2) has a 1 in 2 risk for being a carrier, but his maternal aunt is not at risk for being a carrier because his grandmother, I-1, is not a carrier.

C. His mother is a carrier who inherited a mutant allele from her carrier mother (I-1). In this case, all of the

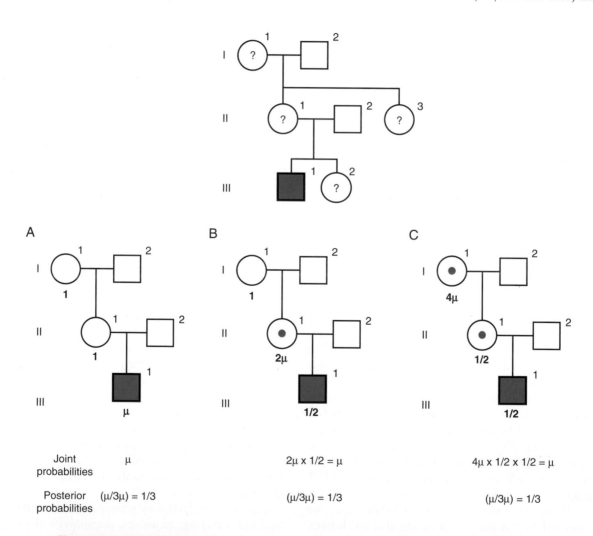

Figure 16-6 Conditional probability used to determine carrier risks for females in a family with an X-linked genetic lethal disorder in which the prior probability of being a carrier has to be calculated by assuming that the carrier frequency is not changing from generation to generation, and that the mutation rates are the same in males and females. *Top,* Pedigree of a family with an X-linked genetic lethal disorder. *Bottom,* The three mutually exclusive situations that could explain the pedigree. **A,** The proband is a new mutation. **B,** The mother of the proband is a new mutation. **C,** The mother of the proband inherited the mutation from her carrier mother, the grandmother of the proband.

female relatives have either a 1 in 2 or a 1 in 4 risk for being carriers.

How can we use conditional probability to determine the carrier risks for the female relatives of III-1 in this pedigree? If we proceed as we did previously with the hemophilia family in Figure 16-4, what do we use as the prior probability that individual I-1 is a carrier? We do not have pedigree information, as we did in the hemophilia pedigree, from which to calculate these prior probabilities. We can, however, use some simple assumptions that the frequency of the disease is unchanging and the new mutation rate is equal in males and females to estimate the prior probability (see Box).

PRIOR PROBABILITY THAT A FEMALE IN THE POPULATION IS A CARRIER OF AN X-LINKED LETHAL DISORDER

Suppose H is the population frequency of female carriers of an X-linked lethal disorder. Assume H is constant from generation to generation.

Suppose the mutation rate at this X-linked locus in any one gamete = μ. Assume μ is the same in males and females. Mutation rate μ is a small number, in the range of 10^{-4} to 10^{-6} (see Chapter 4).

Then, there are three mutually exclusive ways that any female could be a carrier:
1. She inherits a mutant allele from a carrier mother = $\frac{1}{2} \times H$.
   or
2. She receives a newly mutant allele on the X she receives from her mother = μ.
   or
3. She receives a newly mutant allele on the X she receives from her father = μ.

The chance a female is a carrier is the sum of the chance that she inherited a preexisting mutation and the chance that she received a new mutation from her mother or from her father.

$$H = (\tfrac{1}{2} \times H) + \mu + \mu = H/2 + 2\mu$$

Solving for H, you get the chance that a random female in the population is a carrier of a particular X-linked disorder = 4μ. Note that half of this 4μ, 2μ, is the probability she is a carrier by inheritance, and the other 2μ is the probability that she is a carrier by new mutation.

The chance a random female in the population is *not* a carrier is $1 - 4\mu \cong 1$ (because μ is a very small number).

Now we can use this value 4μ from the Box as the prior probability that a woman is a carrier of an X-linked lethal disorder (see Fig. 16-6). For the purpose of calculating the chance that II-1 is a carrier, we ignore the female relatives II-3 and III-2 because there is nothing about them, such as phenotype, laboratory testing, or reproductive history, that conditions whether II-1 is a carrier.

- A. III-1 is a new mutation with probability μ. His mother and grandmother are both noncarriers, each

of which has a probability of $1 - 4\mu \cong 1$. The joint probability is $\mu \times 1 \times 1 = \mu$.
- B. I-1 is a noncarrier, and so II-1 must be the product of a maternal or paternal new mutation and not a carrier by inheritance because we are specifying in scenario B that I-1 is *not* a carrier. The chance that a female will be a carrier by new mutation only is $\mu + \mu = 2\mu$ (and *not* 4μ). The joint probability is therefore $2\mu \times \frac{1}{2} = \mu$.
- C. Individuals I-1 and II-1 are both carriers. As explained in the Box, the chance that I-1 is a carrier has a prior probability of 4μ. For II-1 to be a carrier, she must have inherited the mutant allele from her mother, which has probability 1 in 2. In addition, the chance that II-1 has passed the mutant allele on to her affected son is also 1 in 2. The joint probability is therefore $4\mu \times \frac{1}{2} \times \frac{1}{2} = \mu$.

The posterior probabilities are now easy to calculate as $\mu/(\mu + \mu + \mu) = \frac{1}{3}$ each for scenarios A, B, and C. The affected boy has a 1 in 3 chance of being affected because of a new mutation (situation A), whereas his mother II-1 is a carrier in both B and C and therefore has a $\frac{1}{3} + \frac{1}{3} = \frac{2}{3}$ chance of being a carrier. The grandmother, I-1, is a carrier only in C, and so her chance of being a carrier is 1 in 3.

With these risk figures for the core individuals in the pedigree, we can then calculate the carrier risks for the female relatives II-3 and III-2. III-2's risk for being a carrier is $\frac{1}{2} \times$ [the chance II-1 is a carrier] = $\frac{1}{2} \times \frac{2}{3} = \frac{1}{3}$. The risk that II-3 is a carrier is $\frac{1}{2} \times$ [the chance I-1 is a carrier] = $\frac{1}{2} \times \frac{1}{3} = \frac{1}{6}$. In all of these calculations, for the sake of simplicity, we are ignoring the small but very real possibility of germline mosaicism (see Chapter 7). In a real genetic counseling situation, however, the possibility of mosaicism cannot be ignored.

Disorders with Incomplete Penetrance

To estimate the recurrence risk for disorders with incomplete penetrance, the probability that an apparently unaffected person actually carries the mutant gene in question must be considered. Figure 16-7 shows a pedigree of **split hand deformity**, an autosomal dominant abnormality with incomplete penetrance discussed in Chapter 7. An estimate of penetrance can be made from a single pedigree if it is large enough, or from a review of published pedigrees; we use 70% in our example. That means that a heterozygote for a mutation that causes split hand deformity has a 30% chance of *not* showing the phenotype. The pedigree shows several people who must carry the mutant gene but do not express it (i.e., in whom the defect is not penetrant), I-1 or I-2 (assuming no somatic or germline mosaicism) and II-3. The other unaffected family members may or may not carry the mutant gene.

If III-4, the daughter of a known affected heterozygote, is the consultand, she either may have escaped inheriting the mutant allele from her affected mother or

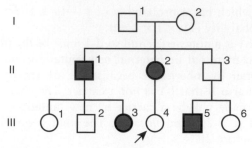

Figure 16-7 Pedigree of family with split hand deformity and lack of penetrance in some individuals.

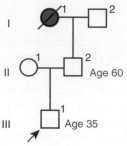

Figure 16-9 **Age-modified risks for genetic counseling in dominant Parkinson disease.** That the consultand's father is asymptomatic at the age of 60 years reduces the consultand's final risk for carrying the gene to approximately 12.5%. That the consultand himself is asymptomatic reduces the risk only slightly, because most patients carrying the mutant allele for this disorder will be asymptomatic at the age of 35 years.

did inherit it but is not expressing the phenotype because penetrance is incomplete in this disorder. There are two possibilities (Fig. 16-8). In A, III-4 is not a carrier with prior probability of 1 in 2. If she does not carry the mutant allele, she will not have the phenotype, so the joint probability for A is 1 in 2. In B, III-4 is a carrier, also with prior probability 1 in 2. Here, we must apply the conditional probability that she is a carrier but does not show the phenotype, which has probability of 1 − penetrance = 1 − 0.7 = 0.3, so the joint probability for B is $\frac{1}{2} \times 0.3 = 0.15$. The posterior probability that III-4 is a carrier without expressing the phenotype is therefore $\frac{3}{13} = \approx 23\%$.

Disorders with Late Age at Onset

Many autosomal dominant conditions characteristically show a late age at onset, beyond the age of reproduction. Thus it is not uncommon in genetic counseling to ask whether a person of reproductive age who is at risk for a particular autosomal dominant disorder carries the gene. One example of such a disorder is a rare, familial

form of **Parkinson disease (PD)** inherited as an autosomal dominant condition.

Consider the dominant PD pedigree in Figure 16-9 in which the consultand, an asymptomatic 35-year-old man, wishes to know his risk for PD. His prior risk for having inherited the *PD* gene from his affected grandmother is 1 in 4. Considering that perhaps only 5% of persons with this rare form of PD show symptoms at his age, he would not be expected to show signs of the disease *even if he had inherited the mutant allele*. The more significant aspect of the pedigree, however, is that the consultand's father (II-2) is asymptomatic at the age of 60 years, an age by which perhaps two thirds of persons with this form of PD show symptoms and one third do not.

As shown in Figure 16-10, there are three possibilities:
A. His father did not inherit the mutant allele, so the consultand is not at risk.
B. His father inherited the mutant allele and is asymptomatic at the age of 60 years, but the consultand did not.
C. His father inherited the mutant allele and is asymptomatic. The consultand inherited it from his father and is asymptomatic at the age of 35 years.

The father's chance of carrying the mutant allele (situations B and C) is 25%; the consultand's chance of having the mutant allele (situation C only) is 12%. Providing these recurrence risks in genetic counseling requires careful follow-up. If, for example, the consultand's father were to develop symptoms of PD, the risks would change dramatically.

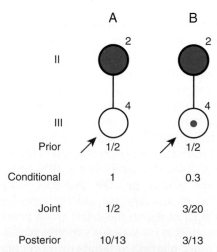

Figure 16-8 **Conditional probability calculation for the risk for the carrier state in the consultand in Figure 16-7.** There are two possibilities: either she is not a carrier (**A**) or she is a carrier (**B**). Her failure to demonstrate the phenotype lowers her carrier risk from the prior probability of 1 in 2 (50%) to 3 in 13 (23%).

	A	B
Prior	1/2	1/2
Conditional	1	0.3
Joint	1/2	3/20
Posterior	10/13	3/13

EMPIRICAL RECURRENCE RISKS
Counseling for Complex Disorders

Genetic counselors deal with many conditions that are not single-gene disorders. Instead, counselors may be

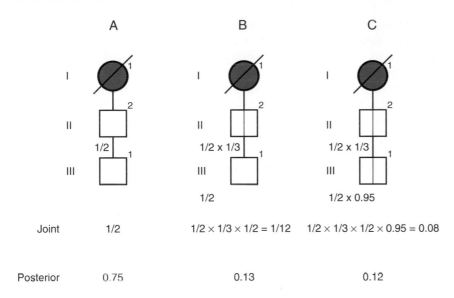

	A	B	C
Joint	1/2	$1/2 \times 1/3 \times 1/2 = 1/12$	$1/2 \times 1/3 \times 1/2 \times 0.95 = 0.08$
Posterior	0.75	0.13	0.12

Figure 16-10 Three scenarios pertaining to the Parkinson disease pedigree in Figure 16-9. Individual II-2 is a nonpenetrant carrier (*vertical line* inside the symbol) in scenarios B and C. Individual III-1 is a nonpenetrant carrier in scenario C.

called on to provide risk estimates for complex trait disorders with a strong genetic component and familial clustering, such as cleft lip and palate, congenital heart disease, meningomyelocele, psychiatric illness, and coronary artery disease (see Chapter 8). In these situations, the risk for recurrence in first-degree relatives of affected individuals may be increased over the background incidence of the disease in the population. For the vast majority of these disorders, however, we do not know the relevant underlying genetic variants or how they interact with each other or with the environment to cause disease.

As the information gained through the Human Genome Project is applied to the problem of diseases with complex inheritance, physicians and genetic counselors and other health professionals in the years ahead will have more of the information they need to provide accurate molecular diagnosis and risk assessment and to develop rational preventive and therapeutic measures. In the meantime, however, geneticists must rely on **empirically derived risk figures** to give patients and their relatives some answers to their questions about disease risk and how to manage that risk. Recurrence risks are estimated empirically by studying as many families with the disorder as possible and observing how frequently the disorder recurs. The observed frequency of a recurrence is taken as an **empirical recurrence risk.** With time, research should make empirical recurrence risks obsolete, replacing them with individualized assessments of risk based on knowledge of a person's genotype and environmental exposures.

Another area in which empirical recurrence risks must be applied is for chromosomal abnormalities (see Chapter 6). When one member of a couple is carrying a chromosomal or genome abnormality, such as a balanced translocation or a chromosomal inversion, the risk for a liveborn, chromosomally unbalanced child

depends on a number of factors. These include the following:

- Whether the couple was ascertained through a previous liveborn, chromosomally abnormal child, in which case a viable offspring with the chromosome abnormality is clearly possible, or the ascertainment was through chromosome or genome studies for infertility or recurrent miscarriage
- The chromosomes involved, which region of the chromosome was affected, and the size of the regions that could be potentially trisomic or monosomic in the fetus
- Whether the mother or father is the carrier of the balanced translocation or inversion

These factors must all be considered when empirical recurrence risks are determined for a couple in which one member is carrying a balanced translocation or a seemingly "normal" genomic copy number variant.

Empirical recurrence risks are also applied when both parents are chromosomally normal but have a child with, for example, trisomy 21. In this case, the age of the mother plays a major role in that, in a young woman younger than 30 years, recurrence risk for trisomy 21 is approximately 5 per 1000 and the risk for any chromosome abnormality is approximately 10 per 1000 as opposed to the population risk of approximately 1.6 per 1000 live births. Over age 30, however, the age-specific risk becomes the dominant factor, and the fact of a previously affected child with trisomy 21 plays much less of a role in determining recurrence risk.

Genetic counselors must use caution in applying empirical risk figures to a particular family. First, empirical estimates are an average over what is undoubtedly a group of heterogeneous disorders with different mechanisms of inheritance. In any one family, the real recurrence risk may actually be higher or lower than the average. Second, empirical risk estimates use history to make predictions about future occurrences; if the underlying biological causes are changing through time, data from the past may not be accurate for the future.

For example, neural tube defects (myelomeningocele and anencephaly) occur in approximately 3.3 per 1000 live births in the U.S. white population. If, however, a couple has a child with a neural tube defect, the risk in the next pregnancy has been shown to be 40 per 1000 (13 times higher; see Table 8-9). The risks remained elevated compared with the general population risk for more distantly related individuals; a second-degree relative (e.g., a nephew or niece) of an individual with a neural tube defect was found to have a 1.7% chance of a similar birth defect. Thus, as we saw in Chapter 8, neural tube defects manifest many of the features typical of multifactorial inheritance. However, these empirical recurrence risks were calculated before widespread folic acid supplementation. With folate supplementation before conception and during early pregnancy, these recurrence risk figures have fallen dramatically (see Chapter 8). This is not because the allelic variants in the families have changed, but rather because a critical environmental factor has changed.

Finally, it is important to emphasize that empirical figures are derived from a particular population, and so the data from one ethnic group, socioeconomic class, or geographical location may not be accurate for an individual from a different background. Nonetheless, such figures are useful when patients ask genetic counselors to give a best estimate for recurrence risk for disorders with complex inheritance.

Genetic Counseling for Consanguinity

Consanguineous couples sometimes request genetic counseling before they have children because an increased risk for birth defects in their offspring is widely appreciated. In the absence of a family history for a known autosomal recessive condition, we use empirical risk figures for the offspring of consanguineous couples, based on population surveys of birth defects in children born to first-cousin couples compared with nonconsanguineous couples (Table 16-3).

TABLE 16-3 Incidence of Birth Defects in Children Born to Nonconsanguineous and First-Cousin Couples

	Incidence of First Birth Defect in Sibship (per 1000)	Incidence of Recurrence of Any Birth Defect in Subsequent Children in Sibship (per 1000)
First-cousin marriage	36	68
Nonconsanguineous marriage	15	30

Data from Stoltenberg C, Magnus P, Skrondal A, Lie RT: Consanguinity and recurrence risk of birth defects: a population-based study, *Am J Med Genet* 82:424-428, 1999.

These results provide empirical risk figures in the counseling of first cousins. Although the relative risk for abnormal offspring is higher for related than for unrelated parents, it is still quite low: approximately double in the offspring of first cousins, compared with baseline risk figures for any abnormality of 15 to 20 per 1000 for any child, regardless of consanguinity. This increased risk is not exclusively for single-gene autosomal recessive diseases but includes the entire spectrum of single-gene and complex trait disorders. However, any couple, consanguineous or not, who has a child with a birth defect is at greater risk for having another child with a birth defect in a subsequent pregnancy.

These risk estimates for consanguinity may be slightly inflated given they are derived from communities in which first-cousin marriages are widespread and encouraged. These are societies in which the degree of relationship (coefficient of inbreeding) between two first cousins may actually be greater than the theoretical $\frac{1}{16}$ due to multiple other lines of relatedness (see Chapter 9). Furthermore, these same societies may also limit marriages to individuals from the same clan, leading to substantial population stratification, which also increases the rate of autosomal recessive disease beyond what might be expected based on mutant allele frequency alone (see Chapter 9).

MOLECULAR AND GENOME-BASED DIAGNOSTICS

Advances over the past few years in mutation detection have provided major improvements in risk assessment, carrier detection, and prenatal diagnosis, in many cases allowing determination of the presence or absence of particular mutations with essentially 100% accuracy. Laboratory testing for direct detection of disease-causing mutations is now available for more than 3000 genes involved in well over 4000 genetic conditions. With our expanding knowledge of the genes involved in hereditary disease and the rapidly falling cost of DNA sequencing, direct detection of mutations in a patient's or family member's genomic DNA to make a molecular diagnosis has become standard of care for many conditions. DNA samples for analysis are available from such readily accessible tissues as a buccal scraping or blood sample, but also from tissues obtained by more invasive testing, such as chorionic villus sampling or amniocentesis (see Chapter 17).

Mutation detection is most commonly performed using one of two different techniques, depending on the nature of the mutations in question. Comprehensive sequencing of polymerase chain reaction (PCR) products made from the coding regions and splice sites immediately adjacent to coding exons is effective when the mutation is a single nucleotide variant or small insertion or deletion. However, when the mutation is a large deletion involving one or more exons, attempts to sequence PCR products made from primers that fall into the deleted region is highly problematic. The sequencing will simply fail if the deletion is in an X-linked gene in a male or, even worse, can be misleading because it will yield only the sequence from the other copy of the gene on the homologous autosome. Duplications are even more challenging because they may yield a perfectly normal sequence unless the primers used for amplification happen to straddle the junction of a duplicated segment. For deletions and duplications, a variety of other methods are available that detect deletions or duplications by providing a quantitative measure of the copy number of the deleted or duplicated region.

For most genetic conditions, the majority of pathogenic mutations are single nucleotide or small insertion/deletion mutations that are well detected by sequencing. One major exception is DMD, in which point mutations or small insertions or deletions account for only approximately 34% of mutations, whereas large deletions and insertions account for 60% and 6%, respectively, of the mutations in patients with DMD. In a patient with DMD, one might start with measuring the copy number of segments of DNA across the entire gene to look for deletion or duplication and, if normal, consider sequencing.

Gene Panels and "Clinical Whole Exomes"

For many hereditary disorders (including hereditary retinal degeneration, deafness, hereditary breast and ovarian cancer, congenital myopathy, mitochondrial disorders, familial thoracic aortic aneurysm syndrome, and hypertrophic or dilated cardiomyopathies), there is substantial locus heterogeneity, that is, a large number of genes are known to be mutated in different families with these disorders. When faced with an individual patient with one of these highly heterogeneous disorders in whom the particular gene and mutations responsible for the disorder are *not* known, recent advances in DNA sequencing make it possible to analyze large panels of dozens to well over 100 genes simultaneously and cost-effectively for mutations in every gene in which mutations have been seen previously to cause the disorder.

In disorders for which even a large panel of relevant genes cannot be formulated for a particular phenotypically defined disorder, diagnosis might still be possible by analyzing the coding exons of every gene (i.e., by whole-exome sequencing) or by sequencing the entire genome in a search for disease-causing mutations (see Chapter 4). For example, two reported series of so-called clinical whole exome testing, one from the United States and one from Canada, showed substantial success. In a 2013 study from the United States, 250 patients with primarily undiagnosed neurological disorders underwent whole-exome sequencing and 62 (≈25%) received a diagnosis. Interestingly, among the patients receiving a diagnosis, four were likely to have had two disorders at the same time, which made a clinical diagnosis very difficult because the patients' phenotype did not match any single known disorder. In another study in 2014 by the Canadian FORGE Consortium, approximately 1300 patients representing 264 disorders known or suspected of being hereditary, but for which the genes involved were unknown, underwent whole-exome sequencing. Mutations highly likely to explain the disorders were found in 60%; at least half of the genes had not been previously known to be involved in human disease. Of great interest in both studies was that a large number of patients carried de novo disease-causing mutations in genes not previously suspected of causing disorders. These mutations, because they are de novo, are extremely difficult to find by standard gene discovery methods as described in Chapter 10, such as linkage or association, and therefore pose particular challenges for genetic counseling and risk assessment.

Variant Interpretation and "Variants of Unknown Significance"

The use of large gene panels and, even more so, whole-exome or whole-genome sequencing raises special issues for sequence interpretation and risk assessment. As the number of genes being studied increases, the number of differences between an individual's sequence and that of an arbitrary reference sequence also increases; consequently many previously undescribed variants will be found whose pathogenetic significance is unknown. These are referred to as "variants of unknown significance" (VUSs). This is particularly the case for missense mutations that result in the substitution of one amino acid for another in the encoded protein.

Interpreting variants is a challenging and demanding area for all professional geneticists engaged in providing molecular diagnostic services. The American College of Medical Genetics and Genomics has recommended that variants be assigned into one of five categories, ranging from definitely pathogenic to definitely benign (see Box). Only those variants with a high probability of being disease-causing are communicated to the medical provider and patient. It is a matter of debate whether a record of all VUSs should be retained by the testing laboratory and attached to a patient's record, thereby remaining available for updating as new information becomes available to allow reclassification as either benign or pathogenic. Thus risk assessment and genetic counseling in this context are ongoing and iterative processes, continually evaluating newly available information and communicating this to medical providers and patients as appropriate.

ASSESSING THE CLINICAL SIGNIFICANCE OF A GENE VARIANT

The American College of Genetics and Genomics recommends that all variants detected during gene sequencing (whether from targeted, whole-exome, or whole-genome sequencing) be classified on a five-level scale, spanning pathogenic, likely pathogenic, of uncertain significance, likely benign, and benign variants. Specialists in molecular diagnostics, human genomics, and bioinformatics have developed a series of criteria for assessing where a mutation sits among these five categories. In the vast majority of cases, none of these criteria is absolutely definitive but must be considered together to provide an overall assessment of how likely any variant is to be pathogenic. These criteria include the following:

- **Population frequency**—If a variant has been seen frequently in a sizeable fraction of normal individuals (>2% of the population), it is considered less likely to be disease causing. Being frequent is, however, no guarantee a variant is benign because autosomal recessive conditions or disorders with low penetrance may be due to a disease-causing variant that may be surprisingly common among unaffected individuals because most carriers will be asymptomatic. Conversely, the vast majority of variants (>98%) found when sequencing a large gene panel or in a whole-exome or whole-genome sequence are rare (occur in 1% of the population or less), so being rare is no guarantee it is disease causing!

- **In silico assessment**—There are many software tools designed to evaluate how likely a missense variant is to be pathogenic by determining if the amino acid at that position is highly conserved or not in orthologous proteins in other species and how likely it is that a particular amino acid substitution would be tolerated. Such tools are less than precise and are generally never used by themselves for categorizing variants for clinical use. They are, however, improving with time and are playing a role in variant assessment. A comparable set of bioinformatic tools is being developed to assess the pathogenicity of other types of variants, such as potential splice site variants or even noncoding sequence variants.

- **Functional data**—If a particular variant has been shown to affect in vitro biochemical activity, a function in cultured cells, or the health of a model organism, then it is less likely to be benign. However, it remains possible that a particular variant will appear benign by these criteria and still be disease-causing in humans because of a prolonged human life span, environmental triggers, or compensatory genes in the model organism not present in humans.

- **Segregation data**—If a particular variant has been seen to be coinherited with a disease in one or more families, or, conversely, does not track with a disease in the family under investigation, then it is more or less likely to be pathogenic. Of course, when only a few individuals are affected, the variant and disease may appear to track by random chance; the number of times a variant and disease must be coinherited to be considered not by chance alone is not firmly fixed but is generally accepted to be at least 5, if not 10. Finding affected individuals in the family who do not carry the variant would be strong evidence *against* the variant being pathogenic, but finding unaffected individuals who do carry the variant is less persuasive if the disorder is known to have reduced penetrance.

- **De novo mutation**—The appearance of a severe disorder in a child along with a new mutation in a coding exon that neither parent carries (de novo mutation) is additional evidence the variant is likely to be pathogenic. However, between 1 and 2 new mutations occur in the coding regions of genes in every child (see Chapter 4), and so the fact that a mutation is de novo is not definitive for the mutation being pathogenic.

- **Variant characterization**—A variant may be a synonymous change, a missense mutation, a nonsense mutation, a frameshift with a premature termination downstream, or a highly conserved splice site mutation. The impact on the function of the gene can be inferred but, once again, is not definitive. For example, a synonymous change that does not change an amino acid codon might be thought to be benign but may have deleterious effects on normal splicing and be pathogenic (see examples in Chapter 12). Conversely, premature termination or frameshift mutations might be considered to be always deleterious and disease causing. However, such mutations occurring at the far 3′ end of a gene may result in a truncated protein that is still quite capable of functioning and therefore be a benign change.

- **Prior occurrence**—A variant that has been seen before multiple times in affected patients, as recorded in collections of variants found in patients with a similar disorder, is important additional evidence for the variant being pathogenic. Even if a missense variant is novel, that is, has never been described before, it is more likely to be pathogenic if it occurs at the same position in the protein where other known pathogenic missense mutations have occurred.

Another important aspect of how to use molecular and genome-based diagnostic testing in families is the selection of the best person(s) to test. If the consultand is also the affected proband, then molecular testing is appropriate. If, however, the consultand is an unaffected, at-risk individual, with an affected relative serving as the indication for having genetic counseling, it is best to test the affected person rather than the consultand, if logistically possible. This is because a negative mutation test in the consultand is a so-called **uninformative negative**; that is, we do not know if the test was negative because (1) the gene or mutation responsible for disease in the proband was not covered by the test, or (2) the consultand in fact did not inherit a variant that we could have detected had we found the disease-causing variant in the affected proband in the family. Once the mutation or mutations responsible for a particular disorder are found in the proband, then the other members of the family no longer need comprehensive gene sequencing. The DNA of family members can be assessed with less expensive testing only for the presence or absence of the specific mutations already found in the family. If a family member tests negative under these circumstances, the test is a "true" negative that eliminates any elevated risk due to his or her having an affected relative.

GENERAL REFERENCES

Buckingham L: *Molecular diagnostics: fundamentals, methods and clinical applications*, ed 2, Philadelphia, 2011, F.A. Davis and Co.

Gardner RJM, Sutherland GR, Shaffer LG: *Chromosome abnormalities and genetic counseling*, ed 4, Oxford, 2011, Oxford University Press.

Harper PS: *Practical genetic counseling*, ed 7, London, 2010, Hodder Arnold.

Uhlmann WR, Schuette JL, Yashar B: *A guide to genetic counseling*, New York, 2009, Wiley-Blackwell.

Young ID: *Introduction to risk calculation in genetic counseling*, ed 3, New York, 2007, Oxford University Press.

REFERENCES FOR SPECIFIC TOPICS

Beaulieu CL, Majewski J, Schwartzentruber J, et al: FORGE Canada Consortium: Outcomes of a 2-year national rare-disease gene-discovery project, *Am J Hum Genet* 94:809–817, 2014.

Biesecker LG, Green RC: Diagnostic clinical genome and exome sequencing, *N Engl J Med* 370:2418–2425, 2014.

Brock JA, Allen VM, Keiser K, et al: Family history screening: use of the three generation pedigree in clinical practice, *J Obstet Gynaecol Can* 32:663–672, 2010.

Guttmacher AE, Collins FS, Carmona RH: The family history—more important than ever, *N Engl J Med* 351:2333–2336, 2004.

Richards CS, Bale S, Bellissimo DB, et al: ACMG recommendations for standards for interpretation and reporting of sequence variations: Revisions 2007, *Genet Med* 10:294–300, 2008.

Sheridan E, Wright J, Small N, et al: Risk factors for congenital anomaly in a multiethnic birth cohort: an analysis of the Born in Bradford study, *Lancet* 382:1350–1359, 2013.

Yang Y, Muzny DM, Reid JG, et al: Clinical whole-exome sequencing for the diagnosis of mendelian disorders, *N Engl J Med* 369:1502–1511, 2013.

Zhang VW, Wang J: Determination of the clinical significance of an unclassified variant, *Methods Mol Biol* 837:337–348, 2012.

PROBLEMS

1. You are consulted by a couple, Dorothy and Steven, who tell the following story. Dorothy's maternal grandfather, Bruce, had congenital stationary night blindness, which also affected Bruce's maternal uncle, Arthur; the family history appears to fit an X-linked inheritance pattern. (There is also an autosomal dominant form.) Whether Bruce's mother was affected is unknown. Dorothy and Steven have three unaffected children: a daughter, Elsie, and two sons, Zack and Peter. Elsie is planning to have children in the near future. Dorothy wonders whether she should warn Elsie about the risk that she might be a carrier of a serious eye disorder. Sketch the pedigree, and answer the following.
   a. What is the chance that Elsie is heterozygous?
   b. An ophthalmologist traces the family history in further detail and finds evidence that in this pedigree, the disorder is not X-linked but autosomal dominant. There is no evidence that Dorothy's mother Rosemary was affected. On this basis, what is the chance that Elsie is heterozygous?

2. A deceased boy, Nathan, was the only member of his family with Duchenne muscular dystrophy (DMD). He is survived by two sisters, Norma (who has a daughter, Olive) and Nancy (who has a daughter, Odette). His mother, Molly, has two sisters, Maud and Martha. Martha has two unaffected sons and two daughters, Nora and Nellie. Maud has one daughter, Naomi. No carrier tests are available because the mutation in the affected boy remains unknown.
   a. Sketch the pedigree, and calculate the posterior risks for all these females, using information provided in this chapter.
   b. Suppose prenatal diagnosis by DNA analysis is available only to women with more than a 2% risk that a pregnancy will result in a son with DMD. Which of these women would not qualify?

3. In a village in Wales in 1984, 13 boys were born in succession before a girl was born. What is the probability of 13 successive male births? What is the probability of 13 successive births of a single sex? What is the probability that after 13 male births, the 14th child will be a boy?

4. Let H be the population frequency of carriers of hemophilia A. The incidence of hemophilia A in males (I) equals the chance that a maternal $F8$ gene has a new mutation (μ) from a noncarrier mother *plus* the chance it was inherited as a preexisting mutation from a carrier mother ($\frac{1}{2} \times H$). Adding these two terms gives $I = \mu + (\frac{1}{2} \times H)$. H is the chance a carrier inherits the mutation from a surviving, reproducing affected father ($I \times f$) (where f is the fitness of hemophilia) *plus* the chance of a new paternal mutation (μ) *plus* the chance of a new maternal mutation (μ) *plus* the chance of inheriting it from a carrier mother ($\frac{1}{2} \times H$). Adding these four terms gives $H = (I \times f) + \mu + \mu + (\frac{1}{2})H$.
   a. If hemophilia A has a fitness (f) of approximately 0.70, that is, hemophiliacs have approximately 70% as many offspring as do controls, then what is the incidence of affected males? of carrier females? (Answer in terms of multiples of the mutation rate.) If a woman has a son with an isolated case of hemophilia A, what is the risk that she is a carrier? What is the chance that her next son will be affected?
   b. For DMD, $f = 0$. What is the population frequency of affected males? Of carrier females?
   c. Color blindness is thought to have normal fitness ($f = 1$). What is the incidence of carrier females if the frequency of color blind males is 8%?

5. Ira and Margie each have a sibling affected with cystic fibrosis.
   a. What are their prior risks for being carriers?
   b. What is the risk for their having an affected child in their first pregnancy?
   c. They have had three unaffected children and now wish to know their risk for having an affected child. Using Bayesian analysis to take into consideration

that they have already had three unaffected children, calculate the chance that their next child will be affected.

6. A 30-year-old woman with myotonic dystrophy comes in for genetic counseling. Her son, aged 14 years, shows no symptoms, but she wishes to know whether he will be affected with this autosomal dominant condition later in life. Approximately half of individuals carrying the mutant gene are asymptomatic before the age of 14 years. What is the risk that the son will eventually develop myotonic dystrophy? Should you test the child for the expanded repeat in the gene for myotonic dystrophy?

7. A couple arrives in your clinic with their 7-month-old son, who has been moderately developmentally delayed from birth. The couple is contemplating having additional children, and you are asked whether this could be a genetic disorder.
   a. Is this possible, and if so, what pattern or patterns of inheritance would fit this story?
   b. On taking a detailed family history, you learn that both parents' families were originally from the same small village in northern Italy. How might this fact alter your assessment of the case?
   c. You next learn that the mother has two sisters and five brothers. Both sisters have developmentally delayed children. How might this alter your assessment of the case?

8. You are addressing a Neurofibromatosis Association parents' meeting. A severely affected woman, 32 years old, comments that she is not at risk for passing on the disorder because her parents are not affected, and her neurofibromatosis therefore is due to a new mutation. Comment.

9. The figure shows the family from Figure 16-6, but with additional information that the consultand III-2 has two unaffected sons. There are now seven possible scenarios to explain this pedigree. List the scenarios, and use them to calculate the carrier risk for individual III-2.

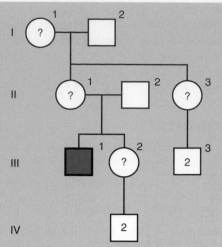

The family from Figure 16-6 but now with additional information consisting of unaffected males that must be used to modify the carrier risks for females in the pedigree.

10. An alternative approach to calculating the carrier risk for III-2 (refer to pedigree in problem 9) is to break the pedigree apart and do the calculations stepwise, a method referred to as the **dummy consultand method**. Instead of calculating the joint probabilities of all seven scenarios to determine the posterior probability that III-2 is a carrier, one ignores III-2 and her two children for the moment, makes individual II-1 serve as a dummy consultand, and calculates II-1's risk for being a carrier *without using any conditional information provided by III-2*. Then, with the carrier risk for II-1 in hand, determine the prior probability that III-2 is a carrier and then condition that risk by use of the fact that she has two unaffected male children. How does the carrier risk for III-2 calculated by the dummy consultand method compare with the risk calculated by the comprehensive method in Table 16-3? How about the carrier risk for II-1? How does the risk calculated by the dummy consultand method compare with the risk calculated by the comprehensive method in Table 16-3?

CHAPTER 17

Prenatal Diagnosis and Screening

The goal of prenatal diagnosis and screening is to inform pregnant women and couples about the risk for birth defects or genetic disorders in their fetus and to provide them with informed choices on how to manage that risk. Some couples known to be at an elevated risk for having a child with a specific birth defect choose to forego having children. Prenatal diagnosis allows them to undertake a pregnancy with the knowledge that the presence or absence of the disorder in the fetus can be confirmed by testing. Many women or couples at risk for having a child with a severe genetic disorder decide to undertake a pregnancy and have been able to have healthy children because of the availability of prenatal diagnosis and the option of terminating an affected pregnancy if necessary. In some, prenatal testing can reassure and reduce anxiety, especially among high-risk groups. For still others, prenatal diagnosis allows physicians to plan prenatal treatment of a fetus with a genetic disorder or birth defect. If prenatal treatment is not possible, diagnosis during pregnancy can alert parents and physicians to arrange for appropriate management for the impending birth of an affected child in terms of psychological preparation of the family, pregnancy and delivery management, and postnatal care.

Prenatal diagnosis is the term traditionally applied to testing a fetus already known to be at an elevated risk for a genetic disorder to determine if the fetus is affected or not with the disorder in question. The elevated risk is usually recognized because of the birth of a previous child with the disease, a family history of the disorder, a positive parental carrier test, or when a **prenatal screening** test (discussed later in this chapter) indicates an increased risk. Prenatal diagnosis often, but not always, requires an **invasive procedure** such as **chorionic villus sampling (CVS)** or **amniocentesis** (both discussed later in this chapter) to acquire fetal cells or amniotic fluid for analysis. Prenatal diagnosis is meant to be as definitive as possible, giving a "yes or no" answer as to whether the fetus is affected with a particular disorder.

Prenatal screening, on the other hand, has traditionally referred to testing for certain common birth defects such as chromosomal aneuploidies, neural tube defects, and other structural anomalies in pregnancies *not* known to be at an increased risk for a birth defect or genetic disorder. Screening tests were developed because common birth defects most often occur in pregnancies not known to be at any increased risk and therefore the parents would not have been offered prenatal diagnosis. Screening tests are typically **noninvasive,** based on obtaining a maternal blood sample or on imaging, usually by ultrasonography or magnetic resonance imaging (MRI). Screening tests are typically designed to be inexpensive and sufficiently low risk to make them suitable for screening all pregnant women in a population regardless of their risk.

The ultimate goal of prenatal diagnosis is to inform couples about the risk for particular birth defects or genetic disorders in their offspring and to provide them with informed choices on how to manage that risk. In contrast, the goal of prenatal screening is to identify pregnancies for which prenatal diagnostic testing should be offered. Screening tests do not give a "yes or no" diagnostic answer about whether an abnormality is present. Rather, the risk for a birth defect derived from screening falls along a continuum relative to the background risk for an age-matched control group. The cut off for what is considered a positive screen is carefully set to balance sensitivity and specificity (i.e., false-negative and false-positive rates). Screening tests generally allow higher false-negative rates than would be acceptable for a diagnostic test to keep false-positive rates to a reasonable level, generally below 5%.

Traditionally therefore the distinction between prenatal diagnostic testing and prenatal screening has been made based on:
- Whether or not the pregnancy was known to be at risk for a particular disorder
- Whether the goal of the testing was a definitive diagnosis of a particular disorder or an assessment of risk relative to the background population risk
- Whether the test was invasive or noninvasive

Now, however, because of improvements in the safety of invasive procedures and advances in technology, the need to distinguish between diagnosis and screening is becoming much less clear. CVS or amniocentesis coupled with chromosomal microarray analysis (CMA) (see Chapter 5) is now being offered to every pregnant woman as a screening test, not only for the common chromosomal aneuploidies but also for other genomic imbalances, regardless of risk assessment based on personal or family history or noninvasive screening test results. Prenatal diagnosis is expanding beyond testing for specific disorders for which the fetus is known to be at risk to include any copy number abnormalities detectable by CMA and may, in the near future, include whole-genome sequence analysis of the fetus.

The purpose of this chapter is to discuss these various approaches to screening and diagnosis and to review the methodologies and indications as currently being used in this very rapidly changing field. The reader is cautioned, however, that because of technological advances in the methods available for assessing the fetus and the fetal genome, standards of care in prenatal screening and diagnosis are in flux.

METHODS OF PRENATAL DIAGNOSIS

Invasive Testing

Amniocentesis

Invasive testing utilizes CVS or amniocentesis to obtain fetal tissues. Amniocentesis refers to the procedure of inserting a needle into the amniotic sac and removing a sample of amniotic fluid transabdominally (Fig. 17-1A). The amniotic fluid contains cells of fetal origin that can be cultured for diagnostic tests. Before amniocentesis, ultrasonographic scanning is routinely used to assess fetal viability, gestational age (by determining various biometric parameters such as head circumference, abdominal circumference, and femur length), the number of fetuses, volume of amniotic fluid, normality of fetal anatomical structures, and position of the fetus and placenta to allow the optimal position for needle insertion. Amniocentesis is performed on an outpatient basis typically between the 16th and 20th week after the first day of the last menstrual period.

In addition to fetal chromosome and genome analysis, the concentration of **alpha-fetoprotein (AFP)** can be assayed in amniotic fluid to detect open **neural tube defects (NTDs)** (see Chapters 8 and 14). AFP is a fetal glycoprotein produced mainly in the liver, secreted into the fetal circulation, and excreted through the kidneys into the amniotic fluid. AFP enters the maternal bloodstream through the placenta, amniotic membranes, and maternal-fetal circulation. It can therefore be assayed either in amniotic fluid (amniotic fluid AFP [AFAFP]) or in maternal serum (maternal serum AFP [MSAFP]). Both assays are extremely useful for assessing the risk for an open NTD but also for other reasons (see later discussion).

AFP concentration is measured by an immunoassay, a relatively simple and inexpensive method that can be applied to all amniotic fluid samples, regardless of the specific indication for the amniocentesis. To interpret an AFAFP, one compares the level to the normal range for each gestational age. If the AFAFP level is elevated (relative to the normal range for that particular gestational age), one must look for an open NTD as well as for

TABLE 17-1 Causes of Elevated Amniotic Fluid Alpha-Fetoprotein Other Than Neural Tube Defect

- Fetal blood contamination
- Fetal death
- Twin pregnancy
- Fetal abnormalities, including ventral wall defects (omphalocele or gastroschisis) and at least one form of congenital nephrosis, as well as other rare problems
- Other unexplained variation in the normal AFP concentration of amniotic fluid
- False-positive elevation due to overestimation of gestational age

Note: Some of these causes of an elevated amniotic fluid AFP level can be confirmed or ruled out by ultrasonographic examination.
AFP, Alpha-fetoprotein.

causes other than an open NTD. Factors potentially leading to abnormally high concentrations of AFP in amniotic fluid are shown in Table 17-1. When the AFAFP assay is used in conjunction with ultrasonographic scanning at 18 to 19 weeks' gestation, approximately 99% of fetuses with open spina bifida and virtually all fetuses with anencephaly can be identified.

If amniocentesis is performed for any reason, both the concentration of AFP in the amniotic fluid and a chromosome analysis of amniotic fluid cells are determined to screen for open NTDs and chromosomal and other genomic abnormalities, respectively. Other tests are performed only for specific indications.

Complications. The major complication associated with midtrimester amniocentesis at 16 to 20 weeks of gestation is a 1 in 300 to 1 in 500 risk for inducing miscarriage over the baseline risk of pregnancy loss of approximately 1% to 2% for any pregnancy at this stage of gestation. Other complications are rare, including leakage of amniotic fluid, infection, and injury to the fetus by needle puncture. Early amniocentesis performed between 10 and 14 weeks is no longer recommended because of an increased risk for amniotic fluid leakage, a threefold increased risk for spontaneous abortion, and an approximately sixfold to sevenfold increased risk for talipes equinovarus (clubfeet), over the 0.1% to 0.3% population risk. Early amniocentesis has now been replaced by chorionic villus sampling (see next section).

Chorionic Villus Sampling

CVS involves the biopsy of tissue from the villi of the chorion transcervically or transabdominally, generally between the 10th and 13th weeks of pregnancy (see Fig. 17-1B). Chorionic villi are derived from the trophoblast, the extraembryonic part of the blastocyst (Fig. 17-2), and are a ready source of fetal tissue for biopsy. As with amniocentesis, ultrasonographic scanning is used before CVS to determine the best approach for sampling.

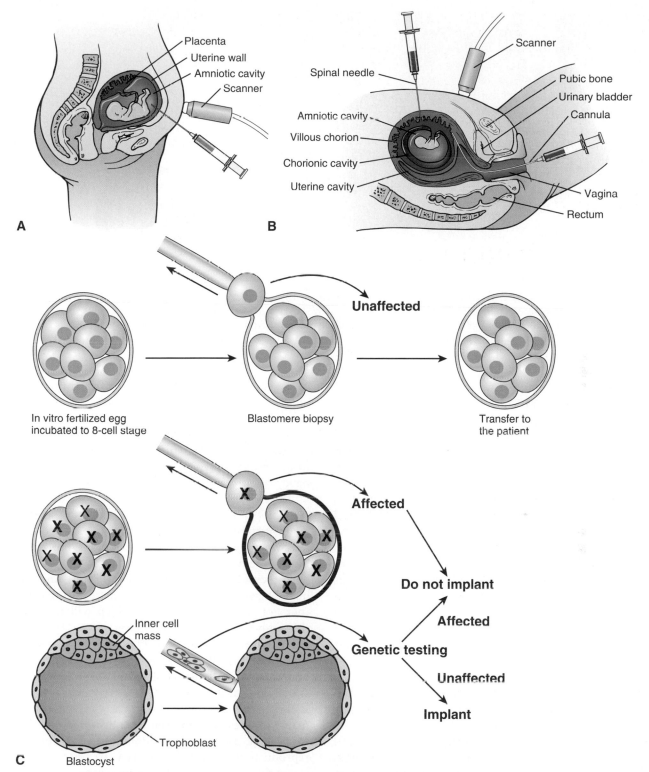

Figure 17-1 A, Amniocentesis. A needle is inserted transabdominally into the amniotic cavity, and a sample of amniotic fluid (usually approximately 20 mL) is withdrawn by syringe for diagnostic studies (e.g., chromosome studies, enzyme measurements, or DNA analysis). Ultrasonography is routinely performed before or during the procedure. **B,** Chorionic villus sampling (CVS). Two alternative approaches are drawn: transcervical (by means of a flexible cannula) and transabdominal (with a spinal needle). In both approaches, success and safety depend on use of ultrasound imaging (scanner). **C,** Preimplantation genetic diagnosis (PGD). Eggs are removed and used for in vitro fertilization. For blastomere biopsy, the fertilized embryos are incubated for 3 days, to the 8- to 16-cell stage, and a single blastomere is removed and undergoes genetic testing for a chromosomal abnormality or mendelian disorder. In this example, the embryo is affected ("X") and after testing would not be implanted. In the blastocyst biopsy, approximately five trophectoderm cells (which will go to make the placenta and not the embryo proper) are removed and tested. Only those embryos that are unaffected will be implanted in the patient's uterus to establish a pregnancy.

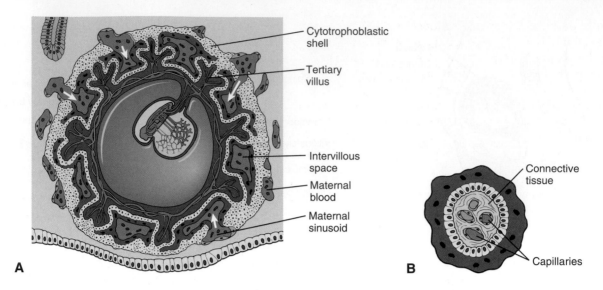

Figure 17-2 Development of the tertiary chorionic villi and placenta. **A,** Cross section of an implanted embryo and placenta at approximately 21 days. **B,** Cross section of a tertiary villus showing establishment of circulation in mesenchymal core, cytotrophoblast, and syncytiotrophoblast. *See Sources & Acknowledgments.*

The major advantage of CVS compared with midtrimester amniocentesis is that CVS allows the results to be available at an early stage of pregnancy, thus reducing the period of uncertainty and allowing termination, if it is elected, to be performed in the first trimester. However, unlike after amniocentesis, AFAFP cannot be assayed at this stage. Evaluation for a possible open NTD thus must be done by other methods, including MSAFP screening, amniocentesis for AFAFP, and ultrasonography.

The success of chromosome analysis by karyotype or CMA is the same as with amniocentesis (i.e., more than 99%). However, approximately 1% of CVS samplings yield ambiguous results because of chromosomal mosaicism (including true mosaicism and pseudomosaicism; see later); in these situations, follow-up with amniocentesis is recommended to establish whether the fetus has a chromosomal abnormality.

Complications. In prenatal diagnostic centers experienced in performing CVS, the rate of fetal loss is only slightly increased over the baseline risk of 2% to 5% in any pregnancy of 7 to 12 weeks of gestation and approximates the 1 in 300 to 1 in 500 risk seen with amniocentesis. Although there were initial reports of an increase in the frequency of birth defects, particularly limb reduction defects, after CVS, this increase has not been confirmed in large series of CVS procedures performed after 10 weeks of gestation by experienced physicians.

Preimplantation Genetic Diagnosis

Preimplantation genetic diagnosis (PGD) refers to testing during **in vitro fertilization (IVF)** to select embryos free

of a specific genetic condition before transfer to the uterus (see Fig. 17-1C). This technology was developed in an effort to offer an alternative option to abortion for those couples at significant risk for a specific genetic disorder or aneuploidy in their offspring, allowing them to undertake a pregnancy even when opposed to pregnancy termination.

The two most common approaches are single **blastomere** biopsy and **blastocyst** biopsy. In blastomere biopsy, a single cell is removed from the embryo 3 days after IVF when there are 8 to 16 cells present. For blastocyst biopsy, the fertilized egg is cultured for 5 to 6 days until a blastocyst has developed (see Fig. 17-1C), and approximately five cells are moved from the trophectoderm (but not the inner cell mass, which will develop into the embryo itself; see Chapter 14). Diagnosis by polymerase chain reaction (PCR) has been undertaken for a number of single-gene disorders; chromosome abnormalities can also be detected using fluorescence in situ hybridization or CMA (see Chapters 4 and 5). Embryos that are found *not* to carry the genetic abnormality in question can then be transferred and allowed to implant, as is routinely done after IVF for assisted reproduction. Affected embryos are discarded. Data currently available on this technology suggest that there are no detrimental effects to embryos that have undergone biopsy.

Although PGD by blastomere biopsy has been performed many thousands of times worldwide, it is not without controversy. First, molecular analysis of a single cell is technically challenging; accuracy varies, with false-positive rates around 6% and false-negative rates around 1%, significantly higher than with analysis of specimens obtained by CVS or amniocentesis. The more recently developed blastocyst biopsy method provides more cellular material, with an apparently greater accuracy, but extensive studies are still ongoing. Second, although PGD was developed to avoid the ethical, religious, and psychological difficulties with pregnancy terminations, it still raises ethical concerns for those who consider the practice of discarding affected embryos as akin to abortion.

Noninvasive Prenatal Diagnosis

Prenatal Diagnosis of Anomalies by Ultrasonography

High-resolution, real-time scanning is widely used for general assessment of fetal age, multiple pregnancies, and fetal viability. Long-term follow-up assessments have failed to provide any evidence that ultrasonography is harmful to the fetus or the mother. The equipment and techniques used by ultrasonographers now allow the detection of many malformations by routine ultrasonography (Figs. 17-3 and 17-4). Once a malformation

has been detected or is suspected on routine ultrasound examination, a detailed ultrasound study in three and even four dimensions (three dimensions over time, as with fetal echocardiography) may be indicated. With improvements in ultrasound resolution, an increasing number of structural fetal anomalies can be detected in the late first trimester (Table 17-2; see Fig. 17-3).

A number of fetal abnormalities detectable by ultrasound examination are associated with chromosomal aneuploidy, including trisomy 21, trisomy 18, trisomy 13, 45,X and many other abnormal karyotypes (Table 17-3). These abnormalities may also occur as isolated findings in a chromosomally normal fetus. Table 17-3 compares the prevalence of fetal chromosome defects in fetuses when one of these common ultrasound examination abnormalities is present as an isolated finding versus when it is one of multiple abnormalities. The likelihood of a chromosomally abnormal fetus increases dramatically when a fetal abnormality detected by ultrasound examination is only one of many abnormalities.

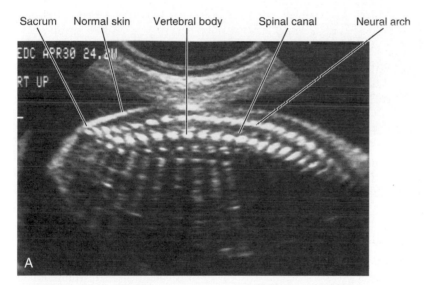

Sacrum   Normal skin   Vertebral body   Spinal canal   Neural arch

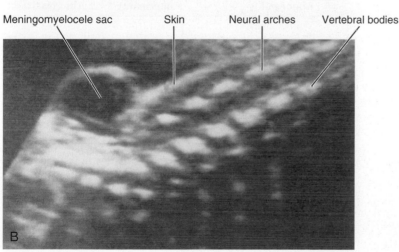

Meningomyelocele sac   Skin   Neural arches   Vertebral bodies

Figure 17-3 Ultrasonograms of spinal canal and neural tube. **A,** Normal fetus at 24 weeks of gestation; longitudinal midline view, with the sacrum to the *left,* thoracic spine to the *right.* Note the two parallel rows of white echoes that represent the neural arches. Also shown are echoes of the vertebral bodies and the overlying intact skin. **B,** Fetus with a neural tube defect, clearly showing the meningomyelocele sac protruding through the skin. *See Sources & Acknowledgments.*

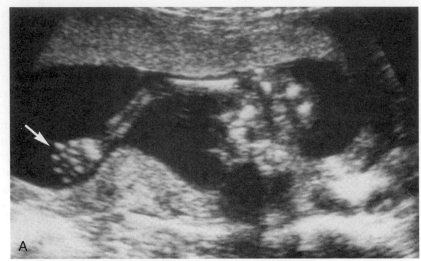

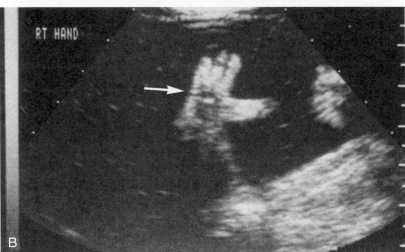

Figure 17-4 Ultrasonograms of hands (*arrows*). A, Normal fetus. B, Fetus with Holt-Oram syndrome, an autosomal dominant defect with congenital heart defects (often an atrial septal defect) and variable limb abnormalities caused by mutations in the *TBX5* transcription factor gene. Note that there are only three obvious fingers and a thumb. The thumb is abnormal in shape (large and thick) and in position. *See Sources & Acknowledgments.*

TABLE 17-2 Examples of Fetal Anomalies That Can Be Diagnosed or Ruled Out by Prenatal Diagnostic Ultrasonography

Single-Gene Disorders

- Holoprosencephaly
- Infantile polycystic kidney disease
- Meckel-Gruber syndrome (an autosomal recessive disorder with encephalocele, polydactyly, and polycystic kidneys)
- Fryns syndrome (an autosomal recessive disorder with abnormalities of the face, diaphragm, limbs, genitourinary tract, and central nervous system)

Disorders Usually Thought of as Multifactorial

- Cleft lip and other facial malformations
- Clubfoot
- Congenital heart defects
- Neural tube defects

Anomalies That May Indicate a Syndrome

- Abnormal genitalia
- Cystic hygroma
- Polydactyly
- Omphalocele
- Radial ray defects

TABLE 17-3 Prevalence of Chromosome Defects in Fetuses with Selected Isolated and Multiple Sonographically Detected Abnormalities

Abnormality	Percent of Fetuses with Abnormal Karyotype	
	If Isolated Abnormality	If Multiple Abnormalities
Ventriculomegaly	2	17
Choroid plexus cysts	≪1	48
Cystic hygroma	52	71
Nuchal edema	19	45
Diaphragmatic hernia	2	49
Heart defects	16	66
Duodenal atresia	38	64
Exomphalos	8	46
Renal abnormalities	3	24

Modified from Snijders RJM, Nicolaides KH: *Ultrasound markers for fetal chromosomal defects*, New York, 1996, Parthenon.

The finding of a normal fetus can be cautiously reassuring, whereas the identification of a fetus with an abnormality allows the couple the option of either appropriate pregnancy and delivery management or pregnancy termination. Consultation with a clinical genetics unit or perinatal unit should be initiated for counseling and further investigation should multiple congenital anomalies be found by ultrasonography or MRI.

Prenatal Ultrasonography for Diagnosis of Single-Gene Disorders

In some single-gene disorders for which DNA testing is possible but a blood or tissue sample is unavailable for DNA or biochemical studies, diagnostic ultrasonography can be useful for prenatal diagnosis. For example, Figure 17-4B shows an abnormal fetal hand detected by ultrasound examination in a pregnancy at 50% risk for **Holt-Oram syndrome,** an autosomal dominant disorder characterized by congenital heart disease in association with hand anomalies.

Ultrasonography can also be useful when the risk for a genetic disorder is uncertain and no definitive DNA-based testing is available.

Prenatal Ultrasonography for Diagnosis of Multifactorial Disorders

A number of isolated abnormalities that may recur in families and that are believed to have multifactorial inheritance can also be identified by ultrasonography (see Table 17-2), including neural tube malformations (see Fig. 17-3). Fetal echocardiography is also available at many centers for a detailed assessment of pregnancies at risk for a congenital heart defect (Table 17-4).

TABLE 17-4 Some Examples of Indications for Fetal Echocardiography*

Maternal Indications (% Risk for Congenital Heart Defect)
- Insulin-dependent diabetes mellitus (3%-5%)
- Phenylketonuria (15%)
- Teratogen exposure
- Thalidomide (10% if 20-36 days post conception)
- Phenytoin (2%-3%)
- Alcohol (25% with fetal alcohol syndrome)
- Maternal congenital heart disease (5%-10% for most lesions)

Fetal Indications
- Abnormal general fetal ultrasound examination results
- Arrhythmia
- Chromosome abnormalities
- Nuchal thickening
- Nonimmune hydrops fetalis

Familial Indications
- Mendelian syndromes
- Paternal congenital heart disease (2%-5%)
- Previously affected child with congenital heart lesion (2%-4%, higher for certain lesions)

*This list is not comprehensive, and indications vary between centers.

Determination of Fetal Sex

Ultrasound examination can be used to determine fetal sex as early as 13 weeks' gestation. This determination may be an important prelude or adjunct in the prenatal diagnosis of certain X-linked recessive disorders (e.g., hemophilia) for those women identified to be at increased risk. A couple may decide not to proceed with invasive testing if a female (and therefore likely unaffected) fetus is identified by ultrasound examination.

INDICATIONS FOR PRENATAL DIAGNOSIS BY INVASIVE TESTING

There are a number of well-accepted indications for prenatal testing by invasive procedures (see Box). Because of the increased incidence of certain trisomies with increasing age of the mother, the most common indication for invasive prenatal diagnosis is to test for **Down syndrome** (trisomy 21) and the two other, more severe autosomal trisomies, trisomy 13 and trisomy 18 (see Chapter 6). For this reason, prenatal diagnosis was most often used in the past in the setting of **advanced maternal age.** Current clinical guidelines, however, do not support using maternal age as the sole indicator for invasive testing for aneuploidies and, instead, recommend risk assessment be made by one or more of the noninvasive screening methods described later in this chapter.

In addition to fetal chromosome abnormalities, there are over 2000 genetic disorders for which genetic testing is available. Prenatal testing by amniocentesis or CVS can be offered with genetic counseling to couples known to be at risk for any of these disorders, but whether or not a couple considers the fetus to be at significant risk and the condition sufficiently burdensome to justify an invasive procedure and possible pregnancy termination is a personal, individual decision each couple must make for itself.

The traditional clinical approach to invasive prenatal diagnosis is to offer these procedures only in pregnancies for which the fetus has an increased risk for a specific condition, as indicated by family history, a positive screening test result, or other well-defined risk factors (but not maternal age alone). Reserving invasive testing for pregnancies with a documented increased risk for aneuploidy is supported by 2011 Practice Guidelines from the Society of Obstetricians and Gynaecologists of Canada and the International Society for Prenatal Diagnosis. In contrast, the American College of Obstetricians and Gynecologists (ACOG) has recommended that amniocentesis or CVS be made available to *all* women regardless of age and without a prior screening test indicating increased risk.

It is important to stress that invasive prenatal diagnosis cannot be used to rule out *all* possible fetal abnormalities. It is limited to determining whether the fetus

PRINCIPAL INDICATIONS FOR PRENATAL DIAGNOSIS BY INVASIVE TESTING

- Previous child with de novo chromosomal aneuploidy or other genomic imbalance

  Although the parents of a child with chromosomal aneuploidy may have normal chromosomes themselves, in some situations there may still be an increased risk for a chromosomal abnormality in a subsequent child. For example, if a woman at 30 years of age has a child with Down syndrome, her recurrence risk for *any* chromosomal abnormality is approximately 1 per 100, compared with the age-related population risk of approximately 1 per 390. Parental mosaicism is one possible explanation of the increased risk, but in the majority of cases, the mechanism of the increase in risk is unknown.

- Presence of structural chromosomal or genome abnormality in one of the parents

  Here, the risk for a chromosome abnormality in a child varies according to the type of abnormality and sometimes the parent of origin. The greatest risk, 100% for Down syndrome, occurs only if either parent has a 21q21q Robertsonian translocation (see Chapter 6).

- Family history of a genetic disorder that may be diagnosed or ruled out by biochemical or DNA analysis

  Most of the disorders in this group are caused by single-gene defects with 25% or 50% recurrence risks. Cases in which the parents have been diagnosed as carriers after a population screening test, rather than after the birth of an affected child, are also in this category. Mitochondrial disorders pose special challenges for prenatal diagnosis.

- Family history of an X-linked disorder for which there is no specific prenatal diagnostic test

When there is no alternative method, the parents of a boy affected with an X-linked disorder may use fetal sex determination to help them decide whether to continue or to terminate a subsequent pregnancy because the recurrence risk may be as high as 25%. For X-linked disorders, such as Duchenne muscular dystrophy and hemophilia A and B, however, for which prenatal diagnosis by DNA analysis is available, the fetal sex is first determined and DNA analysis is then performed if the fetus is male. In either of the situations mentioned, **preimplantation genetic diagnosis** (see text) may be an option for allowing the transfer to the uterus of only those embryos determined to be unaffected for the disorder in question.

- Risk for a neural tube defect (NTD)

  First-degree relatives (and second-degree relatives at some centers) of patients with NTDs are eligible for amniocentesis because of an increased risk for having a child with an NTD; many open NTDs, however, can now be detected by other noninvasive tests, as described in this chapter.

- Increased risk as determined by maternal serum screening, ultrasound examination, and noninvasive prenatal screening test of cell-free DNA

  Genetic assessment and further testing are recommended when fetal abnormalities are suspected on the basis of routine screening by maternal serum screening and fetal ultrasound examination.

- The pregnant woman or couple wishes invasive testing

  Although limited at one time to a pregnant woman with no increased risk other than advanced maternal age, some current professional guidelines call for invasive testing to be offered to all couples.

---

has (or probably has) a specific condition detectable with the diagnostic testing method being used.

PRENATAL SCREENING

Prenatal screening has traditionally relied on both ultrasonography and measuring various proteins and hormones (referred to as *analytes*) whose levels in maternal serum are altered when a fetus is affected by a trisomy or an NTD. More recently, the field of prenatal screening and obstetrical genetics has taken a great leap forward with the discovery that maternal serum contains not only useful analytes but also cell-free DNA, of which a certain fraction is fetal in origin. Sequencing of this cell-free DNA using advanced technologies, as discussed later in this chapter, has made noninvasive screening for trisomies more sensitive and accurate compared to traditional analyte screening.

Screening for Neural Tube Defects

The AFAFP test described earlier is indicated for pregnancies that are undergoing amniocentesis due to a known high risk for an open NTD. However, because an estimated 95% of infants with NTDs are born into families with *no* known history of this malformation, a relatively simple screening test, such as the noninvasive MSAFP test, constitutes an important tool for prenatal diagnosis, prevention, and management.

When the fetus has an open NTD, the concentration of AFP in maternal serum is likely to be higher than normal, just as we saw previously in amniotic fluid. This observation is the basis for the use of MSAFP measurement at 16 weeks as a screening test for open NTDs. There is considerable overlap between the normal range of MSAFP and the range of concentrations found when the fetus has an open NTD (Fig. 17-5). Although an elevated MSAFP concentration is by no means specific to a pregnancy with an open NTD, many of the other causes of elevated MSAFP concentration can be distinguished from open NTDs by fetal ultrasonography (Table 17-5).

MSAFP is also not perfectly sensitive, because its assessment depends on statistically defined cutoff values. If an elevated concentration is defined as two multiples of the median value in pregnancies without any abnormality that could raise the AFP concentration, one can estimate that 20% of fetuses with open NTDs remain undetected. However, lowering the cutoff to improve sensitivity would be at the expense of reduced specificity, thereby increasing the false-positive rate.

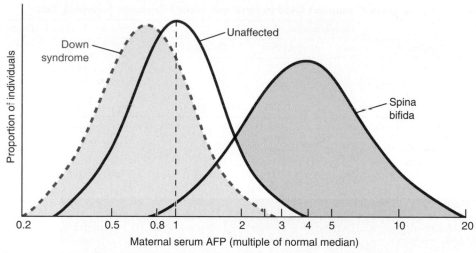

Figure 17-5 Maternal serum alpha-fetoprotein (AFP) concentration, expressed as multiples of the median, in normal fetuses, fetuses with open neural tube defects, and fetuses with Down syndrome. *See Sources & Acknowledgments.*

TABLE 17-5 Causes of Elevated Maternal Serum Alpha-Fetoprotein Concentration

Gestational age older than calculated	Sacrococcygeal teratomas
Spina bifida	Renal anomalies
Anencephaly	Urinary obstruction
Congenital skin defects	Polycystic kidney
Pilonidal cysts	Absent kidney
Abdominal wall defects	Congenital nephrosis
Gastrointestinal defects	Osteogenesis imperfecta
Obstruction	Low birth weight
Liver necrosis	Oligohydramnios
Cloacal exstrophy	Multiple gestation
Cystic hygroma	Decreased maternal weight

From Cunningham FG, MacDonald PC, Gant NF, et al: *Williams obstetrics,* ed 20, Stamford, CT, 1997, Appleton & Lange, p 972.

The combined use of the MSAFP assay with detailed diagnostic ultrasonography (see later discussion) approaches the accuracy of AFAFP assay and ultrasonography for the detection of open NTDs. Thus first-degree, second-degree, or more remote relatives of patients with NTDs may have an MSAFP assay (at 16 weeks) followed by detailed ultrasound examination (at 18 weeks) rather than undergoing amniocentesis.

Screening for Down Syndrome and Other Aneuploidies

More than 70% of all children with major autosomal trisomies are born to women who lack known risk factors, including advanced maternal age (see Fig. 6-1). A solution to this problem was first suggested by the unexpected finding that MSAFP concentration (measured, as just discussed, during the second trimester as a screening test for NTD) was depressed in many pregnancies later discovered to have an autosomal trisomy, particularly trisomies 18 and 21. MSAFP concentration alone has far too much overlap between unaffected pregnancies and Down syndrome pregnancies to be a useful screening tool on its own (see Fig. 17-5). However, a battery of maternal serum protein analytes has now been developed that in combination with specific ultrasound measurements has the necessary sensitivity and specificity to be useful for screening. These batteries of tests are now recommended for noninvasive screening, although not for definitive diagnosis, during the first and second trimesters of all pregnancies regardless of maternal age.

First-Trimester Screening

First-trimester screening is ideally performed between 11 and 13 weeks of gestation and relies on measuring the level of certain analytes in maternal serum in combination with a highly targeted ultrasonographic examination. The analytes used are **pregnancy-associated plasma protein A (PAPP-A)** and the hormone **human chorionic gonadotropin (hCG),** either as total hCG or as its free β subunit. PAPP-A is depressed below the normal range in all trisomies; hCG (or free β-hCG) is elevated in trisomy 21 but depressed in the other trisomies (Table 17-6). Analyte measurements are combined with ultrasonographic measurement of **nuchal translucency (NT),** defined as the thickness of the echo-free space between the skin and the soft tissue overlying the dorsal aspect of the cervical spine caused by subcutaneous edema of the fetal neck. An increase in NT is commonly seen in trisomies 21, 13, and 18 and in 45,X fetuses (Fig. 17-6). NT varies with age of the fetus and thus must be determined with reference to gestational age.

TABLE 17-6 Elevation and Depression of Parameters Used in First- and Second-Trimester Screening Tests

	First-Trimester Screen			Second-Trimester Screen			
	Nuchal Translucency	PAPP-A	Free β-hCG	uE₃	AFP	hCG	Inhibin A
Trisomy 21	↑	↓	↑	↓	↓	↑	↑
Trisomy 18	↑	↓	↓	↓	↓	↓	—
Trisomy 13	↑	↓	↓	↓	↓	↓	—
Neural tube defect	—	—	—	—	↑↑	—	—

AFP, Alpha-fetoprotein; β-hCG, human chorionic gonadotropin β subunit; PAPP-A, pregnancy-associated plasma protein A; uE₃, unconjugated estriol.

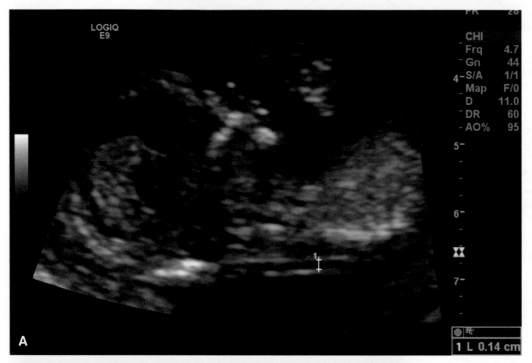

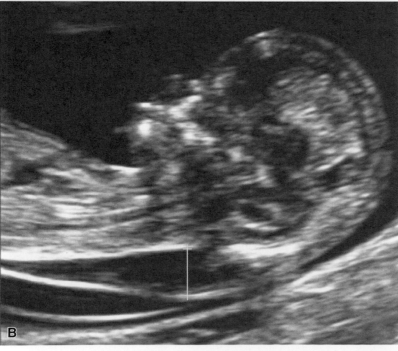

Figure 17-6 Nuchal translucency measurements at 11 weeks of gestation. Nuchal translucency is a dark, echo-free zone beneath the skin in an ultrasonographic "sagittal section" through the fetus and is marked by two "+" signs connected by a *yellow line*. The average nuchal translucency is 1.2 mm at 11 weeks of gestation (95th percentile up to 2 mm) and 1.5 mm at 14 weeks of gestation (95th percentile up to 2.6 mm). **A,** Nuchal translucency of 1.2 mm in a normal 11-week fetus, the average for a normal fetus at this gestational age. **B,** Increased nuchal translucency of 5.9 mm, which is nearly 20 standard deviations above the mean and associated with a greatly increased risk for Down syndrome. *See Sources & Acknowledgments.*

Second-Trimester Screening

Second-trimester screening is usually accomplished by measuring hCG in combination with three other analytes: MSAFP, **unconjugated estriol**, and **inhibin A**. This battery of tests is referred to as a **quadruple screen**. All of these substances are depressed below the normal range in all trisomies with the exception of hCG, which is elevated in trisomy 21 but depressed in the other trisomies, and inhibin A, which is elevated in trisomy 21 but not significantly affected in the other trisomies (see Table 17-6). Levels of these analytes can be affected by a number of factors, including race, smoking, IVF pregnancy, and maternal diabetes, and laboratories generally adjust for these variables. Extremely low levels of unconjugated estriol may be indicative of rare genetic conditions such as steroid sulfatase deficiency or the Smith-Lemli-Opitz syndrome.

Noninvasive Prenatal Screening by Analysis of Cell-Free Fetal DNA

The field of prenatal screening and obstetrical genetics is being revolutionized by the joining together of two major advances in the field of genomics, one biological and the other technological, to produce a new prenatal screening technology known as **noninvasive prenatal screening** (**NIPS**) (also sometimes referred to as noninvasive prenatal testing, NIPT). The biological discovery is that after 7 weeks of gestation, the serum of a pregnant woman contains fetal DNA that is not contained in the nucleus of a cell but is floating freely in the maternal circulation. Approximately 2% to 10% of the cell-free DNA in maternal blood is derived from the placental trophoblasts and is therefore fetal in origin. This **cell-free fetal DNA**, although mixed with DNA of maternal origin, provides a sample of the fetal genome that is available for analysis without the need for an invasive procedure. The technological breakthrough is the development of high-throughput sequencing methods that allow the sequencing of millions of individual DNA molecules in a mixture.

NIPS makes highly accurate, noninvasive screening of pregnancies for the common autosomal and sex chromosome aneuploidies possible, with sensitivities and specificities approaching 99% for trisomy 21. Cell-free fetal DNA in maternal serum has also been used to genotype the fetus at the Rh locus (see Chapter 9) and to determine fetal sex. Further refinements in the analysis of cell-free DNA will make noninvasive testing for many other genetic disorders, including many single-gene disorders, available for clinical care in the future.

Sequencing cell-free DNA in maternal serum has been implemented for aneuploidy detection in a number of different ways by different providers; an example designed to illustrate the concept is given here. Total cell-free DNA is subjected to next-generation sequencing, and millions of molecules of DNA are each mapped to its particular chromosome of origin (Fig. 17-7). The number of molecules that map to each chromosome is determined, without knowing which of the fragments is fetal and which maternal. Because chromosome 21 constitutes approximately 1.5% of total DNA in the genome, approximately 1.5% of total fragments should be assigned to chromosome 21 if the fetus and mother have two normal copies of chromosome 21. If, however, the fetus has trisomy 21, more sequences than expected will map to chromosome 21—a small but significant increase relative to the number of sequences that map to an appropriate reference chromosome or to the full set of chromosomes *not* including chromosome 21. A similar calculation can be used for the other common autosomal trisomies and for sex chromosome aneuploidies as well.

Although NIPS provides a substantial improvement in sensitivity and specificity for fetal trisomies, particularly trisomy 21, and sex chromosome aneuploidies, it remains a screening test, not a diagnostic test. NIPS can also be used to detect Y chromosome sequences in maternal serum for the purposes of determining fetal sex; the test has false-positive and false-negative rates in the 1% to 2% range.

Integrated Screening Strategies

For standard first-trimester and second-trimester screening by ultrasonography and maternal serum analytes, a cutoff, chosen to keep false positives at 5%, results in sensitivities of first- and second-trimester screening, as shown in Table 17-7. Based on these parameters, a strategy was developed for combining the results of first-trimester and second-trimester testing to increase the ability to detect pregnancies with autosomal trisomies, particularly trisomy 21 (Fig. 17-8). These strategies have the advantage of giving couples found to be at significantly increased risk on the basis of first-trimester testing alone the choice of early invasive testing by CVS, rather than having to wait for second-trimester screening and use amniocentesis. The most common strategy, however, is to combine the risk as determined from first- and second-trimester screening tests in a sequential manner (see Fig. 17-8). In this stepwise sequential strategy, couples are identified as "screen positive" for Down syndrome once an ultrasound examination has confirmed fetal age and the estimated risk is found to be elevated. A couple showing increased risk by serum analyte screening can then be offered either NIPS or prenatal chromosome analysis (see Fig. 17-8). Without NIPS, this strategy can detect up to 95% of all Down syndrome cases with an approximately 5% false-positive rate. If NIPS is added, the sensitivity for trisomy 21 rises to greater than 99% with a less than

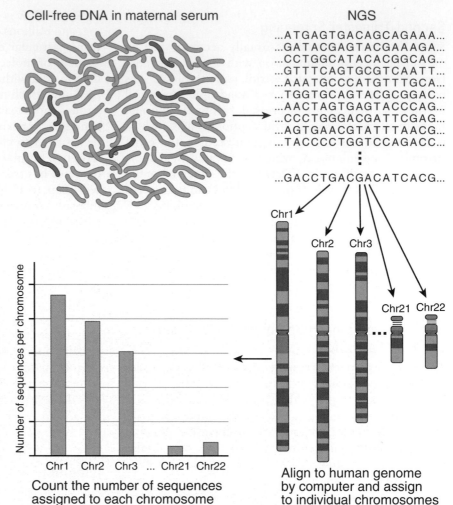

Cell-free DNA in maternal serum

NGS

...ATGAGTGACAGCAGAAA...
...GATACGAGTACGAAAGA...
...CCTGGCATACACGGCAG...
...GTTTCAGTGCGTCAATT...
...AAATGCCCATGTTTGCA...
...TGGTGCAGTACGCGGAC...
...AACTAGTGAGTACCCAG...
...CCCTGGGACGATTCGAG...
...AGTGAACGTATTTAACG...
...TACCCCTGGTCCAGACC...

...GACCTGACGACATCACG...

Chr1 Chr2 Chr3 Chr21 Chr22

Align to human genome
by computer and assign
to individual chromosomes

Number of sequences per chromosome

Chr1 Chr2 Chr3 ... Chr21 Chr22

Count the number of sequences
assigned to each chromosome

Figure 17-7 Schematic diagram of noninvasive prenatal screening (NIPS) for trisomies by next-generation sequencing of cell-free DNA in maternal serum. Fetal component of maternal serum cell-free DNA shown in *red*, maternal contribution in *blue*. Millions of molecules of DNA are sequenced and assigned to each chromosome by computerized alignment against the human genome. Highly accurate measurements of small but significant increases in the fraction of molecules assigned to chromosome 13, 18, 21 or X compared to a reference indicate increased risk for trisomy of each of these chromosomes.

TABLE 17-7 Sensitivity and False-Positive Rates for Trisomy 21 for Various Prenatal Screening Methods

Screening Test	Sensitivity	False Positive Rate (1 Specificity)
First-trimester triple screen	≈85%	5%
Second-trimester quadruple screen	≈81%	5%
Combined first and second trimester	≈95%	5%
Noninvasive prenatal screening	>99%	<1%

Modified from Malone FD, Canick JA, Ball RH, et al: First-trimester or second-trimester screening, or both, for Down's syndrome, *N Engl J Med* 353:2001-2011, 2005; and Bianchi DW, Parker RL, Wentworth J, et al: DNA sequencing versus standard prenatal aneuploidy screening, *N Engl J Med* 370:799-808, 2014.

1% false-positive rate. Sensitivity for other trisomies is in the 90% to 95% range, but still with a remarkably low false-positive rate of less than 1%. Although NIPS is relatively new and more data are needed, initial measurements of sensitivity and false-positive rate for NIPS

appear to provide improved screening parameters compared to currently available standard serum analyte screening. If this remarkable sensitivity and specificity are borne out, *it is anticipated that NIPS may replace serum analyte screening for aneuploidies*; MSAFP screening would remain, however, for NTDs.

As with any screening test in medicine, it is critical for couples to be informed that screening for birth defects with measurement of maternal serum analytes, ultrasound scanning, and NIPS is a screening tool and not a definitive diagnostic test. They must also be counseled that screening tests will not reliably detect chromosome abnormalities other than the common trisomies and sex chromosome aneuploidies, mosaicism, or single-gene defects. Furthermore, only the second-trimester quadruple screen, which includes MSAFP, is helpful for detecting an open NTD in the fetus. Finally, women whose screening test result is considered to be "negative" must also be counseled that their risk for having a child with Down syndrome or another aneuploidy or NTD, although greatly reduced, is not zero.

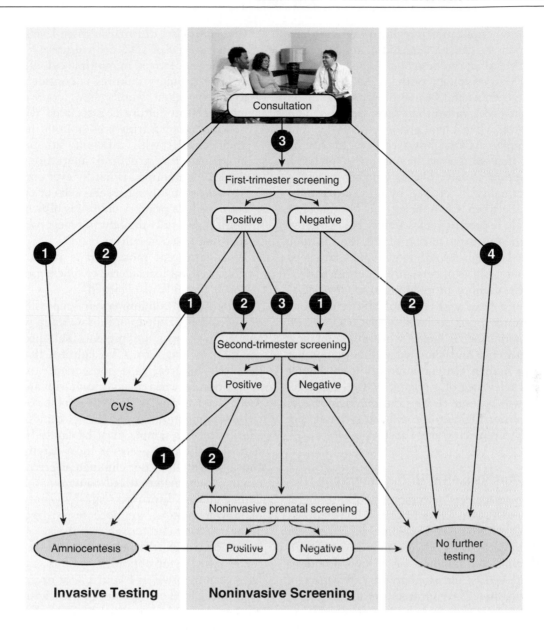

Figure 17-8 Prenatal screening decision tree used by physicians and patients to decide on which screening and diagnostic modalities to use to decide whether invasive testing is indicated. A number of different options shown as different alternative pathways are available at various points in the tree.

LABORATORY STUDIES

Cytogenetics in Prenatal Diagnosis

Either amniocentesis or CVS can provide fetal cells for karyotyping as well as for biochemical or DNA analysis. PGD, being a single-cell technique, has been used only for a limited number of DNA analyses and cannot be used for biochemical studies. Preparation and analysis of chromosomes from cultured amniotic fluid cells or cultured chorionic villi require 7 to 10 days, although chorionic villi can also be used for karyotyping after short-term incubation. Although short-term incubation using rapid metaphase analysis of villous cytotrophoblast tissue provides a result more quickly, it yields relatively poorer quality preparations, in which the banding resolution is inadequate for detailed analysis, as well as a higher rate of mosaicism. Some laboratories use both techniques, but if only one is used, long-term culture of the cells of the mesenchymal core is the technique of choice at present.

Fluorescence in situ hybridization (see Chapters 4 and 5) makes it possible to screen interphase nuclei in fetal cells for the common aneuploidies of chromosomes 13, 18, 21, X, and Y immediately after amniocentesis or CVS. This approach for prenatal cytogenetic assessment requires 1 to 2 days and can be used when rapid aneuploidy testing is indicated.

CMA (see Chapter 5) is replacing karyotyping for prenatal diagnosis under certain circumstances. Copy number variants (CNVs), including chromosome mutations, such as duplications, triplication, and deletions (see Chapter 4), can be detected at much higher resolution by CMA than can be accomplished even with high-resolution karyotyping. Both ACOG and the Society of Obstetricians and Gynaecologists of Canada have advised that CMA, rather than karyotyping, should be the first-line test when a fetal abnormality is detected by ultrasonography. ACOG, however, goes further by recommending that *all* women having invasive testing be *given the option* to have CMA, whether or not a structural abnormality is detected by ultrasonography. The difference between CMA being the first-line test when an anomaly is present versus simply being offered as an option when invasive testing of any kind is done reflects the fact that high-resolution genome analysis by CMA detects many CNVs of currently uncertain clinical significance. The number of false-positive CMA tests will be lower in a fetus with an anomaly than in one without because the prior probability that a CNV will be of clinical significance is higher when an anomaly is present. As experience and knowledge of copy number variation in the human genome improves (see Chapter 4), the medical relevance of a greater and greater fraction of CNVs will become clearer and the incidence of variants of uncertain significance will fall to levels that will result in CMA replacing fetal karyotyping for nearly all indications.

Chromosome Analysis after Ultrasonography

Because some birth defects detectable by ultrasonography are associated with chromosome abnormalities, CMA of amniotic fluid cells, chorionic villus cells, or (much more rarely) fetal blood cells obtained by insertion of a needle into an umbilical vessel (**cordocentesis**) may be indicated after ultrasonographic detection of such an abnormality. Chromosome abnormalities are more frequently found when multiple rather than isolated malformations are detected (see Table 17-3). The karyotypes most often seen in fetuses ascertained by abnormal ultrasonographic findings are the common autosomal trisomies (21, 18, and 13), 45,X (Turner syndrome), and unbalanced structural abnormalities. The presence of a cystic hygroma can indicate a 45,X karyotype, but it can also occur in Down syndrome and trisomy 18, as well as in fetuses with normal karyotypes. Thus complete chromosome assessment is indicated.

Problems in Prenatal Chromosome Analysis

Mosaicism. Mosaicism refers to the presence of two or more cell lines in an individual or tissue sample (see Chapter 7). Because invasive prenatal techniques, particularly CVS, sample extraembryonic tissues of the placenta, and not the fetus itself, mosaicism found in cultured fetal cells may be difficult to interpret. The prenatal geneticist must determine if the fetus itself is truly mosaic and understand the clinical significance of any apparent mosaicism.

Cytogeneticists distinguish three levels of mosaicism in amniotic fluid or CVS cell cultures:

1. Mosaicism detected in multiple colonies from several *different* primary cultures is considered **true mosaicism**. Postnatal studies have confirmed that true mosaicism in culture is associated with a high risk that mosaicism is truly present in the fetus. The probability varies with different situations, however; mosaicism for structural aberrations of chromosomes, for example, is hardly ever confirmed.

2. Mosaicism involving several cells or colonies of cells from a *single* primary culture is difficult to interpret, but it is generally thought to reflect **pseudomosaicism** that has arisen in culture.

3. When apparent mosaicism is restricted to only a single cell, it is considered to reflect pseudomosaicism and is typically disregarded.

Maternal cell contamination is a possible explanation of some cases of apparent mosaicism in which both XX and XY cell lines are present. This problem is more common in long-term CVS cultures than in amniotic fluid cell cultures, as a consequence of the intimate association between the chorionic villi and the maternal tissue (see Fig. 17-2). To minimize the risk for maternal cell contamination, any maternal decidua present in a chorionic villus sample must be carefully dissected and removed, although even the most careful dissection of chorionic villi does not eliminate every cell of maternal origin. When maternal cell contamination is suspected and cannot be disproved (e.g., by genotyping with use of polymorphisms), amniocentesis is recommended to allow a second chromosome analysis.

In CVS studies, discrepancies between the karyotypes found in the cytotrophoblast, villous stroma, and fetus have been reported in 1% to 2% of pregnancies studied at 10 to 11 weeks of gestation. Mosaicism is sometimes present in the placenta but absent in the fetus, a situation termed **confined placental mosaicism** (Fig. 17-9). On occasion, a liveborn infant or fetus with nonmosaic trisomy 13 or trisomy 18 has been reported in a pregnancy with placental mosaicism with both the trisomic cell line and a normal cell line. This finding suggests that when the zygote is trisomic, a normal placental cell lineage, established by postzygotic loss of the additional chromosome in a progenitor cell of the cytotrophoblast, can improve the probability of intrauterine survival of a trisomic fetus.

Confined placental mosaicism for any chromosome (but particularly for trisomy 7, 11, 14, or 15) raises the additional concern that the fetal diploidy may have actually arisen by **trisomy rescue**. This term refers to the loss of an extra chromosome postzygotically, an event that presumably allows fetal viability. If the fetus has retained two copies of a chromosome from the same parent, however, the result is **uniparental disomy** (see Chapter 5). Because some genes on the chromosome mentioned are imprinted, uniparental disomy must be

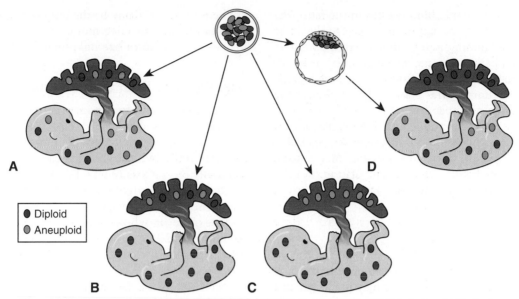

Figure 17-9 The different types of mosaicism that may be detected by prenatal diagnosis. **A,** Generalized mosaicism affecting both the fetus and placenta. **B,** Confined placental mosaicism with normal *(red)* and abnormal *(green)* cell lineages present. **C,** Confined placental mosaicism with only an abnormal cell lineage present. **D,** Mosaicism confined to the embryo. *See Sources & Acknowledgments.*

excluded; two maternal copies of chromosome 15, for example, cause Prader-Willi syndrome, and two paternal copies are associated with Angelman syndrome (see Chapter 5).

CMA can detect some, but not all, cases of mosaicism. Because CMA uses pooled DNA from tissues or cultured cells and does not examine individual cells the way karyotyping does, it is less sensitive for the detection of mosaicism. Mosaicism in which 10% of the cells are aneuploid is difficult to detect as a copy number change by CMA, whereas 10% mosaicism will be detected with greater than 99% probability when 50 cells are examined by karyotype, as is typically done for the assessment of possible mosaicism. CMA is even less sensitive for detecting mosaicism for a copy number variation of only a segment of a chromosome unless it constitutes more than 20% to 25% of the cells under study.

Confirmation and interpretation of apparent mosaicism are among the most difficult challenges in genetic counseling for prenatal diagnosis because, at present, clinical outcome information on the numerous possible types and extents of mosaicism is limited. Further studies (e.g., amniocentesis that follows CVS, cordocentesis that follows amniocentesis) as well as the medical literature may provide some guidance, but the interpretation sometimes still remains uncertain. Ultrasonographic scanning may provide some reassurance if normal growth is observed and if no congenital anomalies can be demonstrated.

Parents should be counseled in advance of the possibility that mosaicism may be found and that the interpretation of mosaicism may be uncertain. After birth,

an effort should be made to verify any abnormal chromosome findings suspected on the basis of prenatal diagnosis. In the case of termination, verification should be sought by analysis of fetal tissues. Confirmation of mosaicism, or lack thereof, may prove helpful with respect to medical management as well as for genetic counseling of the specific couple and other family members.

Culture Failure. If couples are to have an opportunity to consider termination of a pregnancy when an abnormality is found in the fetus, they should be provided with the information at the earliest possible time. Because prenatal diagnosis is always a race against time, the rate of culture failure can be a concern; fortunately, this rate is low. When a CVS culture fails to grow, there is time to repeat the chromosome study with amniocentesis. If an amniotic fluid cell culture fails, either repeated amniocentesis or cordocentesis can be offered, depending on fetal age.

Unexpected Adverse Findings. On occasion, prenatal chromosome analysis performed primarily to rule out aneuploidy reveals some other unusual chromosome finding, for example, a rare chromosomal rearrangement or a marker chromosome (see Chapter 5). In such a case, because the significance of the finding in the fetus cannot be assessed until the parental karyotypes are known, both parents should be karyotyped to determine whether the finding seen in the fetus is de novo or inherited. Unbalanced or de novo structural rearrangements may cause serious fetal abnormalities (see Chapter 6). If one parent is found to be a carrier of a structural rearrangement seen in unbalanced form in the fetus, the consequences for the fetus may be serious. On the other hand, if the same finding is seen in a normal parent, it is likely to be a benign change without untoward consequences. Potential exceptions to this guideline include the possibility of uniparental disomy in a region of the genome that contains imprinted genes. In this situation, an inherited balanced rearrangement may cause serious fetal abnormalities. This possibility can be excluded if there has been a previous transmission of the same balanced rearrangement from a parent of origin of the same sex as the transmitting parent in the current pregnancy.

Biochemical Assays for Metabolic Diseases

Although any disorder for which the genetic basis is known and the responsible mutation(s) identified can be diagnosed prenatally by DNA analysis, more than 100 metabolic disorders can also be diagnosed by biochemical analysis of chorionic villus tissue or cultured amniotic fluid cells; a few rare conditions can even be identified directly by assay of a substance in amniotic fluid. Most metabolic disorders are rare in the general population but have a high recurrence risk (usually 25% within sibships, because most are recessive conditions). Because each condition is rare, the experience of the laboratory performing the prenatal diagnostic testing is important; thus referral to specialized centers is preferable. Whenever possible, biochemical assay on *direct* chorionic villus tissue (as opposed to cultured tissue) is preferred to avoid misinterpretation of results due to the expansion in culture of contaminating maternal cells. Access to a cultured cell line from a proband in the family is highly advisable so that the laboratory can confirm its ability to detect the biochemical abnormality in the proband before the assay is attempted in CVS or amniotic fluid cells from the pregnancy at risk.

Of course, many metabolic disorders cannot be diagnosed by enzyme assay of chorionic villus tissue or cultured amniotic fluid cells because the enzyme is not expressed in those tissues or a reliable in vitro biochemical assay has not been developed. For these disorders, DNA sequencing to look for the pathogenic mutations responsible can be performed. Nonetheless, biochemical tests have one significant advantage over DNA analysis in some cases: whereas DNA analysis by direct detection of a mutation is accurate only for that mutation and not for other alleles at the locus, biochemical testing can detect abnormalities caused by any mutant allele that has a significant effect on the function of the protein. This advantage is particularly significant for disorders characterized by a high degree of allelic heterogeneity, genes in which mutant alleles occur in regions of the gene that are not routinely sequenced, or by a high proportion of new mutations. In addition, biochemical testing may be the only option for prenatal diagnosis if the causative mutations in the family have not been identified or are unknown.

Fetal DNA and Fetal Genome Analysis

As the specific basis for an increasing number of inherited disorders is determined (see Chapter 12), many conditions (some of which were not previously detectable prenatally by other means) can now be diagnosed prenatally by analysis of fetal DNA. Any technique used for direct mutation screening can be used for prenatal diagnosis, from allele-specific or gene-specific tests to whole-exome or whole-genome sequencing. As of the beginning of 2015, genetic testing registries report the clinical availability of DNA-based prenatal testing for more than 5000 genetic disorders caused by mutations in over 3500 genes. The degree of certainty of the diagnosis approaches 100% when direct detection of a mutation is possible, but testing will fail if the disorder in the patient is due to a mutation different from the one that is being sought.

Numerous diseases cannot yet be diagnosed prenatally, but every month additional disorders are added to the list of conditions for which prenatal diagnosis is possible either by biochemical testing or by DNA analysis. One of the contributions of medical geneticists to medical practice in general is keeping up with these rapid changes and serving as a central source of information about the current status of prenatal testing.

Prenatal diagnosis by DNA analysis may not be predictive of the exact clinical presentation in an affected pregnancy in the case of disorders characterized by variable expressivity. For example, in neurofibromatosis type 1 (Case 34), a specific mutation may lead to a severe clinical manifestation in one family member and a mild manifestation in another. **Mitochondrial disorders** (see Chapters 7 and 12) that result from mutations in mitochondrial DNA are particularly challenging for prenatal counseling because the mutations are almost always heteroplasmic, and it is difficult to predict the fraction of defective mitochondrial genomes any one fetus will inherit. Although there is uncertainty concerning the degree of heteroplasmy that will be passed on from mother to fetus, DNA analysis of samples from the fetus obtained by CVS or amniocentesis is likely to reflect the overall degree of heteroplasmy in the fetus and therefore should be a reliable indicator of the burden of pathogenic mitochondrial mutations in the fetus.

Although **whole-exome** or **whole-genome sequencing of fetal DNA** is not yet part of routine care, it is technically feasible now, and discussions are ongoing concerning whether whole-genome analysis by exome or genome sequencing of fetal DNA could serve as a prenatal screening test (see earlier discussion of NIPS). The ethical concerns posed by whole-genome analysis of fetuses are substantial. These include presymptomatic diagnosis of adult disorders, particularly those for which no treatments are known, stigmatization, damage to the parent-child relationship, and the impact of having to provide counseling for massive amounts of currently uninterpretable information arising from the discovery of **variants of uncertain significance**. This will be an area that bears watching closely in the years ahead, with increasingly important ethical and policy implications for the practice of fetal medicine and prenatal genetics.

GENETIC COUNSELING FOR PRENATAL DIAGNOSIS AND SCREENING

The majority of genetic counselors practice in the setting of a prenatal diagnosis program. The professional staff of a prenatal diagnosis program (physician, nurse, and genetic counselor) must obtain an accurate family history and determine whether other previously unsuspected genetic problems should also be considered on the basis of family history or ethnic background.

Ethnic background, even in the absence of a positive family history, may indicate the need for carrier tests in the parents in advance of prenatal diagnostic testing. For example, in a couple referred for any reason, one must discuss carrier testing for autosomal recessive disorders with increased frequency in various ethnic groups. Such disorders include thalassemia in individuals of Mediterranean or Asian background, sickle cell anemia in Africans or African Americans, and various disorders in the fetus of an Ashkenazi Jewish couple. However, because it is becoming increasingly difficult to assign a single ethnicity to each patient, the use of universal carrier screening panels, in which patients are tested for a large array of genetic disorders irrespective of apparent or stated ethnicity, are becoming more and more common.

The complexities posed by the availability of different tests (including the distinction between screening tests and diagnostic tests), the many different and distinctive indications for testing in different families, the subtleties of interpretation of test results, and the personal, ethical, religious, and social issues that enter into reproductive decision making all make the provision of prenatal diagnosis services a challenging arena for counselors. Parents considering prenatal diagnosis for any reason need information that will allow them to understand their situation and to give or withhold consent for the procedure. Genetic counseling of candidates for prenatal diagnosis usually deals with the following:

- The risk that the fetus will be affected
- The nature and probable consequences of the specific problem
- The risks and limitations of the procedures to be used
- The time required before a report can be issued
- The possible need for a repeated procedure in the event of a failed attempt

In addition, the couple must be advised that a result may be difficult to interpret, further tests and consultation may be required, and even then the results may not necessarily be definitive.

Elective Pregnancy Termination

In most cases, the findings in prenatal diagnosis are normal, and parents to be are reassured that their baby will be unaffected by the condition in question. Unfortunately, in a small proportion of cases, the fetus is found to have a serious genetic defect. Because effective prenatal therapy is not available for most disorders (see Chapter 13), the parents may then choose to terminate the pregnancy. Few issues today are as hotly debated as elective abortion, but despite legal restrictions in some jurisdictions, elective abortion is widely used. Among all elective abortions, those performed because of prenatal diagnosis of an abnormality in a fetus account for only a very small proportion. Without a means of legal termination of pregnancy, prenatal diagnosis would not have developed into the accepted medical procedure that it has become.

Some pregnant women who would not consider termination nevertheless request prenatal diagnosis to reduce anxiety or to prepare for the birth of a child with a genetic disorder. This information may be used for psychological preparation as well as for management of the delivery and of the newborn infant.

At the level of public health, prenatal diagnosis combined with elective termination has led to a major decline in the incidence in certain population groups of a few serious disorders, such as β-thalassemia (see Chapter 11) and Tay-Sachs disease (see Chapter 12). Similar data for the effects of prenatal screening, diagnosis, and elective termination on the birth incidence of Down syndrome in the United States are conflicting, however. Estimates range from a 24% *increase* to a 15% *decrease* in the numbers of babies born with Down syndrome over the 15- to 20-year time period ending in 2005. These data must be viewed against an estimated 34% *increase* in affected pregnancies that was expected due to a rise in average maternal age. The frequency with which couples carrying a Down syndrome pregnancy terminate a pregnancy also varies tremendously between different societies. For example, whereas approximately two thirds of couples in the United States choose to terminate a Down syndrome pregnancy, nearly 90% of couples in the United Kingdom choose to terminate such pregnancies.

Impact of Prenatal Diagnosis

It must be stressed, however, that the principal advantage of prenatal diagnosis is not to the population but to the immediate family. Parents at risk for having a child with a serious abnormality can undertake pregnancies that they may otherwise not have risked, with the knowledge that they can learn early in a pregnancy whether the fetus has the abnormality and can make an informed decision about whether or not to continue the pregnancy.

Although the great majority of prenatal diagnoses end in reassurance, options available to parents in the event of an abnormality—of which termination of pregnancy is only one—should be discussed. *Above all, the parents must understand that in undertaking prenatal diagnosis, they are under no implied obligation to terminate a pregnancy in the event that an abnormality is detected.* The objective of prenatal diagnosis is to determine whether the fetus is affected or unaffected with the disorder in question. Diagnosis of an affected fetus may, at the least, allow the parents to prepare emotionally and medically for the management of a newborn with a disorder.

In closing, the reader is again cautioned that because of technological advances in the methods available for assessing the fetus and the fetal genome, and because of ongoing discussions of social and ethical norms and governmental policies concerning prenatal diagnosis in different cultures and countries around the globe, standards of care in prenatal screening and diagnosis will continue to be subject to modification and refinement.

GENERAL REFERENCES

Gardner RJM, Sutherland GR, Shaffer LG: *Chromosome abnormalities and genetic counseling*, ed 4, New York, 2011, Oxford University Press.

Milunsky A, Milunsky J: *Genetic disorders and the fetus: diagnosis, prevention, and treatment*, ed 6, Chichester, West Sussex, England, 2010, Wiley-Blackwell.

SPECIFIC REFERENCES

American College of Obstetricians and Gynecologists Committee on Genetics: The use of chromosomal microarray analysis in prenatal diagnosis, *Obstet Gynecol* 122:1374–1377, 2009.

American College of Obstetricians and Gynecologists Committee on Genetics: Noninvasive prenatal testing for fetal aneuploidy, *Obstet Gynecol* 120:1532–1534, 2012.

Bianchi D: From prenatal genomic diagnosis to fetal personalized medicine: progress and challenges, *Nat Med* 18:1041–1051, 2012.

Bianchi DW, Parker RL, Wentworth J, et al: DNA sequencing versus standard prenatal aneuploidy screening, *N Engl J Med* 370(9):799–808, 2014.

Bodurtha J, Strauss JF: Genomics and perinatal care, *N Engl J Med* 366:64–73, 2012.

Chitayat D, Langlois S, Wilson RD, et al: Prenatal screening for fetal aneuploidy in singleton pregnancies, *J Obstet Gynaecol Can* 33:736–750, 2011.

Dugoff L: Application of genomic technology in prenatal diagnosis, *N Engl J Med* 367:2249–2251, 2012.

Duncan A, Langlois S, SOGC Genetics Committee, et al: Use of array genomic hybridization technology in prenatal diagnosis in Canada, *J Obstet Gynaecol Can* 33:1256–1259, 2011.

Fan HC, Gu W, Wang J, et al: Non-invasive prenatal measurement of the fetal genome, *Nature* 487:320–324, 2012.

Gregg A, Gross SJ, Best RG, et al: ACMG statement on noninvasive prenatal screening for fetal aneuploidy, *Genet Med* 15:395–398, 2013.

Malone FD, Canick JA, Ball RH, et al: First-trimester and second-trimester screening, or both, for Down's Syndrome, *N Engl J Med* 353:2001–2011, 2005.

McArthur SJ, Leigh D, Marshall JT, et al: Blastocyst trophectoderm biopsy and preimplantation genetic diagnosis for familial monogenic disorders and chromosomal translocations, *Prenat Diagn* 28:434–442, 2008.

Norwitz ER, Levy B: Noninvasive prenatal testing: the future is now, *Rev Obstet Gynecol* 6:48–62, 2013.

Talkowski ME, Ordulu Z, Pillalamarri V, et al: Clinical diagnosis by whole-genome sequencing of a prenatal sample, *N Engl J Med* 367:2226–2232, 2012.

Wapner RJ, Martin CL, Levy B, et al: Chromosomal microarray versus karyotyping for prenatal diagnosis, *N Engl J Med* 367:2175–2184, 2012.

Yurkiewicz IR, Korf BR, Lehmann LS: Prenatal whole-genome sequencing – is the quest to know a fetus' future ethical? *N Engl J Med* 370:195–197, 2014.

PROBLEMS

1. Match the term in the top section with the appropriate comment in the bottom section.
   a. Rh immune globulin
   b. 10th week of pregnancy
   c. Cordocentesis
   d. Mosaicism
   e. 16th week of pregnancy
   f. Alpha-fetoprotein in maternal serum
   g. Aneuploidy
   h. Cystic hygroma
   i. Chorionic villi
   j. Amniotic fluid

   _____ method of obtaining fetal blood for karyotyping
   _____ usual time at which amniocentesis is performed
   _____ increased level when fetus has neural tube defect
   _____ contains fetal cells viable in culture
   _____ major cytogenetic problem in prenatal diagnosis
   _____ ultrasonographic diagnosis indicates possible Turner syndrome
   _____ risk increases with maternal age
   _____ usual time at which chorionic villus sampling (CVS) is performed
   _____ derived from extraembryonic tissue
   _____ used to prevent immunization of Rh-negative women

2. A couple has a child with Down syndrome, who has a 21q21q translocation inherited from the mother. Could prenatal diagnosis be helpful in the couple's next pregnancy? Explain.

3. Cultured cells from a chorionic villus sample show two cell lines, 46,XX and 46,XY. Does this necessarily mean the fetus is abnormal? Explain.

4. What two main types of information about a fetus can be indicated (although not proved) by assay of alpha-fetoprotein, human chorionic gonadotropin, and unconjugated estriol in maternal serum during the second trimester?

5. A couple has had a first-trimester spontaneous abortion in their first pregnancy and requests counseling.
   a. What proportion of all pregnancies abort in the first trimester?
   b. What is the most common genetic abnormality found in such cases?
   c. Assuming that there are no other indications, should this couple be offered prenatal diagnosis for their next pregnancy?

6. A young woman consults a geneticist during her first pregnancy. Her brother was previously diagnosed with Duchenne muscular dystrophy and had since died. He was the only affected person in her family. The woman had been tested biochemically and found to have elevated creatine kinase levels, indicating she is a carrier of the disease.

   Unfortunately, no DNA analysis had been conducted on the woman's brother to determine what type of mutation in the *DMD* gene he had. The woman was investigated by molecular analysis and found to be heterozygous (*A1/A2*) for a microsatellite marker closely linked to the *DMD* gene. No relatives except the parents of the woman were available for analysis.
   a. Can the phase of the mutation in the woman be determined from analysis of the available individuals?
   b. Can this information be used to diagnose her pregnancy?
   c. What other molecular analysis could be performed on the fetus?

7. Discuss the relative advantages and disadvantages of the following diagnostic procedures, and cite types of disorders for which they are indicated or not indicated: amniocentesis, CVS, first-semester maternal serum screening, second trimester screening, noninvasive screening of cell-free fetal DNA (noninvasive prenatal screening [NIPS]).

8. Suppose the frequency of Down syndrome is 1 in 600 in pregnancies in women younger than 35 years. Consider the following two strategies for prenatal detection of the disorder:
   - All pregnant women younger than 35 years are offered CVS or amniocentesis.
   - All pregnant women undergo a sequential screening strategy, as follows: All participate in first-trimester screening with pregnancy-associated plasma protein A (PAPP-A), human chorionic gonadotropin (hCG), and nuchal translucency. Sensitivity is 84% with a false-positive rate of 5%. Those who score positive are offered CVS, and all use it. Those who score negative are screened during the second trimester with a quadruple maternal serum screening, which has 81% sensitivity and a 5% false-positive rate. Those who score positive are offered amniocentesis, and all use it.

   Assuming that a population of 600,000 women younger than 35 years are pregnant:
   a. How many CVS procedures or amniocenteses are done overall, given these two strategies?
   b. What fraction of the total expected number of affected fetuses is detected under the two strategies? What fraction is missed?
   c. How many CVS or amniocentesis procedures would need to be done to detect one fetus with Down syndrome under these two strategies?

CHAPTER 18

Application of Genomics to Medicine and Personalized Health Care

The last several chapters have been dedicated to introducing various aspects of the applications of modern genomics to the practice of medicine. In Chapter 15, we described powerful new genomic technologies, such as identifying the mutations present in a tumor and profiling its pattern of RNA expression, that are currently being used for determining prognosis and choosing appropriate targeted therapies for individual cancer patients. In Chapter 16, we discussed how modern genomic approaches are expanding our capabilities in risk assessment and genetic counseling for patients and families dealing with heritable disease. Chapter 17 focused on prenatal genetics and the advances in prenatal diagnosis made possible by genomics.

Finally, in this chapter, we explore other applications of genomics to individualized health care: screening asymptomatic individuals for risk or susceptibility to disease in them or their family members and applying that knowledge to improve health care. We will first describe population screening and present one of the best-established and highly successful forms of genetic screening, the detection of abnormalities in newborns at high risk for preventable illness. We then present some of the basic concepts and applications of **pharmacogenomics** and how knowledge of individual variation affecting drug therapy can be used to improve therapeutic efficacy and reduce adverse events. Finally, we discuss screening of patients for genetic susceptibility based on their genome sequence and review some of the concepts and methods of genetic epidemiology commonly used to evaluate screening for susceptibility genotypes.

GENETIC SCREENING IN POPULATIONS

Genetic screening is a population-based method for identifying persons with increased susceptibility for a genetic disease. Screening at the population level is not to be confused with testing for affected persons or carriers within families already identified because of family history, as we explored in Chapter 16. Although family history is a very useful tool (Fig. 18-1), no one except

an identical twin has all of the same gene variants that another family member has. Family history is therefore only an indirect means of assessing the contribution that an individual's own combination of genetic variants might make to disease. Family history is also an insensitive indicator of susceptibility because it depends on overt disease actually occurring in the relatives of the individual patient.

The challenge going forward is to screen populations, independent of family history and independent of clinical status, for variants relevant to health and disease and to apply this information to make risk assessments that can be used to improve the health care of an individual patient and his or her family.

The objective of population screening is to examine all members of a designated population, regardless of family history. Applying this information requires that we demonstrate that genetic risk factors are valid indicators of actual risk in an individual patient and, if they are valid, how useful such information is in guiding health care. Genetic screening is an important public health activity that will become more significant as more and better screening tests become available for determining genetic susceptibilities for disease.

Newborn Screening

The best-known population screening efforts in genetics are the government-supported or government-mandated programs that identify presymptomatic infants with diseases for which early treatment can prevent or at least ameliorate the consequences (Table 18-1). For newborn screening, the presence of disease is generally not assessed by determining the genotype directly. Instead, in most instances, asymptomatic newborns are screened for abnormalities in the level of various substances in the blood. Abnormalities in these metabolites trigger further evaluation to either confirm or rule out the presence of a disorder. Exceptions to this paradigm of using a biochemical measurement to detect a disease-causing genotype are screening programs for abnormalities in

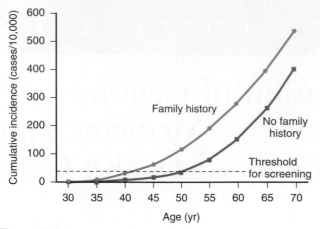

Figure 18-1 Cumulative incidence (per 10,000) of colon cancer versus age in individuals with and without a family history of the disease.

TABLE 18-1 Some Conditions for Which Newborn Screening Has Been Implemented

Condition	Frequency (per 100,000 newborns)*
Congenital hearing loss	200
Sickle cell disease	47
Hypothyroidism	28
Phenylketonuria	3
Congenital adrenal hyperplasia	2
Severe combined immunodeficiency	2
Galactosemia	2
Maple syrup urine disease	≤1
Homocystinuria	≤1
Biotinidase deficiency	≤1

*Approximate values in the United States.

hearing, in which the phenotype itself is the target of screening and intervention (see later).

Many of the general issues concerning genetic screening are highlighted by newborn screening programs. A determination of the appropriateness of newborn screening for any particular condition is based on a standard set of criteria involving clinical validity and clinical utility (see Box). The design of newborn screening tests includes keeping false-negative rates low so that true-positive cases are not missed, without making the test so nonspecific as to drive the false-positive rate unacceptably high. False-positive results cause unnecessary anxiety to the parents and also increase the costs, because more unaffected infants have to be recalled for retesting; at the other extreme, false-negative results vitiate the purpose of having a screening program. The criterion that the public health system infrastructure be capable of handling the care of affected newborns identified by screening is often underemphasized in discussions of the clinical utility of screening, but must also be considered in deciding whether to institute screening for any given condition.

The prototype condition that satisfies all of these criteria is **phenylketonuria** (see Chapter 12). For decades, finding elevated levels of phenylalanine in a spot of blood on filter paper obtained soon after birth has been the mainstay of neonatal screening for phenylketonuria and other forms of hyperphenylalaninemia in the United States, all the provinces of Canada, and nearly all developed countries. A positive screen result, followed by definitive confirmation of the diagnosis, led to the institution of dietary phenylalanine restriction early in infancy, thereby preventing irreversible intellectual disability.

GENERAL CRITERIA FOR AN EFFECTIVE NEWBORN SCREENING PROGRAM

Analytic Validity

• A rapid and economic laboratory test is available that detects the appropriate metabolite.

Clinical Validity

• The laboratory test is highly sensitive (no false-negatives) and reasonably specific (few false-positives). Positive predictive value is high.

Clinical Utility

• Treatment is available.
• Early institution of treatment, before symptoms become manifest, reduces or prevents severe illness.
• Routine observation and physical examination will not reveal the disorder in the newborn—a test is required.
• The condition is frequent and serious enough to justify the expense of screening; that is, screening is cost-effective.
• The public health system infrastructure is in place to inform the newborn's parents and physicians of the results of the screening test, to confirm the test results, and to institute effective treatment and counseling.

Two other conditions that are widely targeted for newborn screening are **congenital deafness** and **congenital hypothyroidism**. Newborn screening for hearing loss is mandated in 37 states in the United States and three provinces in Canada. Approximately half of all congenital deafness is due to single-gene defects (Case 13). Infants found to have hearing impairments by newborn screening receive intervention with sign language, cochlear implants, and other communication aids early in life, thereby improving their long-term language skills and intellectual abilities beyond those seen if the impairment is discovered later in childhood. Screening for congenital hypothyroidism, a disorder whose genetic basis is known in only 10% to 15% of cases but is easily treatable, is universal in the United States and Canada and is also routine in many other countries. Thyroid hormone replacement therapy started early in infancy completely prevents the severe and irreversible intellectual disability caused by congenital hypothyroidism. Thus both hypothyroidism and deafness easily fulfill the criteria for newborn screening.

A number of other disorders, such as galactosemia, sickle cell disease (Case 42), biotinidase deficiency (see Chapter 12), severe combined immunodeficiency, and congenital adrenal hyperplasia (see Chapter 6), are part of neonatal screening programs in many or most states, but not all. Which disorders should be the target of newborn screening varies from state to state in the United States. However, many states have instituted screening for a group of 32 conditions, following the recommendations of a panel convened by the Secretary of the Department of Health and Human Services.

Standards for newborn screening differ widely across the globe. Which disorders should be the target of newborn screening varies from province to province in Canada without a national consensus. As of 2014, the United Kingdom's national program to screen newborns across all jurisdictions included just five disorders, with the exception of Northern Ireland, which already tests for seven disorders; the United Kingdom is considering adding three additional disorders.

Tandem Mass Spectroscopy

For many years, most newborn screening was performed by a test specific for each individual condition. For example, phenylketonuria screening was based on a microbial or a chemical assay that tested for elevated phenylalanine level (see previous section). This situation has changed dramatically with the application of the technology of **tandem mass spectrometry (TMS)**. Not only can a neonatal blood spot be examined accurately and rapidly for an elevation of phenylalanine, with fewer false positives than with the older testing methods, but TMS analysis can simultaneously detect a few dozen other biochemical disorders as well. Some of these, such as homocystinuria (see Chapter 12) or maple syrup urine disease, were already being screened for by individual tests (Table 18-2). TMS, however, does not replace the disease-specific testing methods for other disorders currently included in newborn screening,

TABLE 18-2 Disorders Detectable by Tandem Mass Spectrometry

A. Amino Acid Disorders
- Classical phenylketonuria (PKU)
- Variant PKU
- Guanosine triphosphate cyclohydrolase 1 (GTPCH) deficiency (biopterin deficiency)
- 6-pyruvoyl-tetrahydropterin synthase (PTPS) deficiency (biopterin deficiency)
- Dihydropteridine reductase (DHPR) deficiency (biopterin deficiency)
- Pterin-4α-carbinolamine dehydratase (PCD) deficiency (biopterin deficiency)
- Argininemia/arginase deficiency
- Argininosuccinic acid lyase deficiency (ASAL deficiency)
- Citrullinemia, type I/argininosuccinic acid synthetase deficiency (ASAS deficiency)
- Citrullinemia, type II (citrin deficiency)
- Gyrate atrophy of the choroid and retina
- Homocitrullinuria, hyperornithinemia, hyperammonemia (HHH)
- Homocystinuria/cystathionine beta-synthase deficiency (CBS deficiency)
- Methionine adenosyltransferase deficiency (MAT deficiency)
- Maple syrup urine disease—(MSUD)
- Prolinemia
- Tyrosinemia, types I, II, III, and transient
- Ornithine transcarbamylase deficiency (OTC deficiency)
- Remethylation defects (MTHFR, MTR, MTRR, Cbl D v1, Cbl G deficiencies)

B. Organic Acid Disorders
- 2-methyl-3-hydroxybutyryl-CoA dehydrogenase deficiency
- 2-methylbutyryl-CoA dehydrogenase deficiency
- 3-hydroxy-3-methylglutaryl-CoA lyase deficiency (HMG CoA lyase deficiency)
- 3-methylcrotonyl-CoA carboxylase deficiency (3MCC deficiency)
- 3-methylglutaconic aciduria (MGA), type I (3-methylglutaconyl-CoA hydratase deficiency)
- Beta-ketothiolase deficiency (BKT)
- Ethylmalonic encephalopathy (EE)
- Glutaricacidemia type-1 (GA-1)
- Isobutyryl-CoA dehydrogenase deficiency
- Isovalericacidemia (IVA)
- Malonicaciduria
- Methylmalonicacidemia, mut −
- Methylmalonicacidemia, mut 0
- Methylmalonicacidemia (Cbl A, B)
- Methylmalonicacidemia (Cbl C, D)
- Multiple carboxylase deficiency (MCD)
- Propionicacidemia (PA)

C. Fatty Acid Oxidation Disorders
- Carnitine transporter deficiency
- Carnitine-acylcarnitine translocase deficiency (CAT deficiency)
- Carnitine palmitoyltransferase deficiency-type 1 (CPT-1 deficiency)
- Carnitine palmitoyltransferase deficiency-type 2 (CPT-2 deficiency)
- Long chain hydroxyacyl CoA dehydrogenase deficiency (LCHAD deficiency)
- Medium chain acyl-CoA dehydrogenase deficiency (MCAD deficiency)
- Medium/short chain L-3-hydroxy acyl-CoA dehydrogenase deficiency (M/SCHAD deficiency)
- Multiple acyl-CoA dehydrogenase deficiency (MAD deficiency)/glutaric acidemia type-2 (GA-2)
- Short chain acyl-CoA dehydrogenase deficiency (SCAD deficiency)
- Trifunctional protein deficiency (TFP deficiency)
- Very long chain acyl-CoA dehydrogenase deficiency (VLCAD deficiency)
- Forminimoglutamic acid (FIGLU) disorder

Cbl, cobalamin; MTHFR, methylene tetrahydrofolate reductase; MTR, 5-methyltetrahydrofolate-homocysteine methyltransferase; MTRR, methionine synthase reductase.

Modified from California Newborn Screening Program, January 2012, http://www.cdph.ca.gov/programs/nbs/Documents/NBS-DisordersDetectable011312.pdf.

such as galactosemia, biotinidase deficiency, congenital adrenal hyperplasia, and sickle cell disease.

TMS also provides a reliable method for newborn screening for some disorders that fit the criteria for screening but had no reliable newborn screening program in place. For example, **medium-chain acyl-CoA dehydrogenase (MCAD) deficiency** is a disorder of fatty acid oxidation that is usually asymptomatic but manifests clinically when the patient becomes catabolic. Detection of MCAD deficiency at birth can be lifesaving because affected infants and children are at very high risk for life-threatening hypoglycemia in early childhood during the catabolic stress caused by an intercurrent illness, such as a viral infection, and nearly 25% of children with undiagnosed MCAD deficiency will die with their first episode of hypoglycemia. The metabolic derangement can be successfully managed if it is treated promptly. In MCAD deficiency, alerting parents and physicians to the risk for metabolic decompensation is the primary goal of screening, because the children are healthy between attacks and do not require daily management other than avoidance of prolonged fasting.

In addition to providing a rapid test for many disorders for which newborn screening either is already being done or can easily be justified, TMS also identifies infants with inborn errors, such as methylmalonic acidemia, that have *not* generally been the targets of newborn screening because of their rarity and difficulty of providing definitive therapy that will prevent the progressive neurological impairment. TMS can also identify abnormal metabolites whose significance for health are uncertain. For example, **short-chain acyl-CoA dehydrogenase (SCAD) deficiency**, another disorder of fatty acid oxidation, is most often asymptomatic, although a few patients may have difficulties with episodic hypoglycemia. Thus a positive TMS screen result is not particularly predictive of developing symptomatic SCAD later in life. Although TMS can identify many metabolic disorders, does the benefit of detecting disorders such as SCAD deficiency outweigh the negative impact of raising parental concern unnecessarily for most newborns whose test result is positive but who will never be symptomatic? Thus not every disorder detected by TMS fits the criteria for newborn screening. Some public health experts argue, therefore that only those metabolites of proven clinical utility should be reported to parents and physicians.

PHARMACOGENOMICS

One area of medicine that is receiving a lot of attention for potential application of genomics to personalized medical care is **pharmacogenomics,** the study of the many differences between individuals in how they respond to drugs because of allelic variation in genes affecting drug metabolism, efficacy, and toxicity. Drug treatment failures and adverse drug reactions occur in more than 2 million patients each year in the United States alone, resulting in ongoing morbidity and an estimated 100,000 excess deaths. The development of a genetic profile that predicts efficacy, toxicity, or an adverse drug reaction is likely to have immediate benefit in allowing physicians to choose a drug for which the patient will benefit without risk for an adverse event, or to decide on a dosage that ensures adequate therapy and minimizes complications.

The U.S. Food and Drug Administration has recognized the importance of pharmacogenetic variation in individual response to drug treatment by including pharmacogenetic information on the labels that come with a broad range of pharmaceuticals (Table 18-3). As with all other aspects of personalized medicine, however, the cost-effectiveness of such testing must be proved if it is to become part of accepted medical care.

There are two ways that genetic variation affects drug therapy. The first is the effect of variation on **pharmacokinetics,** that is, the rate at which the body absorbs, transports, metabolizes, or excretes drugs or their metabolites. The second is the variation affecting **pharmacodynamics,** that is, differences in the way the body responds to a drug. Thus, most broadly, pharmacogenetics encompasses any genetically determined variation in "what the body does to the drug" and in "what the drug does to the body," whereas pharmacogenomics refers to the sum total of all relevant genetic variation affecting drug therapy.

TABLE 18-3 Gene-Drug Combinations for Which There Is Pharmacogenetic Information in Their U.S. Food and Drug Administration Package Inserts*

Gene	Drug(s)
CYP2C19	Clopidogrel, voriconazole, omeprazole, pantoprazole, esomeprazole, diazepam, nelfinavir, rabeprazole
CYP2C9	Celecoxib, warfarin
CYP2D6	Atomoxetine, venlafaxine, risperidone, tiotropium bromide inhalation, tamoxifen, timolol maleate, fluoxetine, cevimeline, tolterodine, terbinafine, tramadol and acetaminophen, clozapine, aripiprazole, metoprolol, propranolol, carvedilol, propafenone, thioridazine, protriptyline, tetrabenazine, codeine
DPYD	Capecitabine, fluorouracil
G6PD	Rasburicase, dapsone, primaquine, chloroquine
HLA-B*1502	Carbamazepine
HLA-B*5701	Abacavir **(Case 1)**
NAT	Rifampin, isoniazid, and pyrazinamide; isosorbide dinitrate and hydralazine hydrochloride
TPMT	Azathioprine, thioguanine, mercaptopurine
UGT1A1	Irinotecan, nilotinib
VKORC1	Warfarin

*Constitutional variants only; chemotherapy whose usage is affected by somatic mutations are not included.

Variation in Pharmacokinetic Response
Variation in Drug Metabolism: Cytochrome P-450

The human cytochrome P-450 proteins are a large family of 56 different functional enzymes, each encoded by a different *CYP* gene. The cytochromes P-450 are grouped into 20 families according to amino acid sequence homology. Three of these families—*CYP1, CYP2,* and *CYP3*—contain enzymes that are promiscuous in the substrates they will act on and that participate in metabolizing a wide array of substances from outside the body (**xenobiotics**), including drugs. Six cytochrome P-450 genes (*CYP1A1, CYP1A2, CYP2C9, CYP2C19, CYP2D6,* and *CYP3A4*) are especially important because the enzymes they encode are responsible for the metabolism of more than 90% of all commonly used drugs (Fig. 18-2).

For many drugs, the action of a cytochrome P-450 is to begin the process of detoxification through a series of reactions that render the drug less active and easier to excrete. Some drugs, however, are themselves inactive **prodrugs** whose conversion into an active metabolite by a cytochrome P-450 is required for the drug to have any therapeutic effect.

Many of the *CYP* genes important for drug metabolism (including *CYP1A2, CYP2C9, CYP2C19, CYP2D6,* and *CYP3A4*) are highly polymorphic, with alleles that result in absent, decreased, or increased enzyme activity, thereby affecting the rate at which many drugs are metabolized, with real functional consequences for how individuals respond to drug therapy (see Table 18-3). As one example, *CYP2D6,* the primary cytochrome in the metabolism of more than 70 different drugs, has dozens of reduced, absent, or increased activity alleles, leading to normal, poor, or ultrafast metabolism (see Table on metabolizer phenotypes later). Missense mutations decrease the activity of this cytochrome; alleles with no activity are caused by splicing or frameshift mutations. In contrast, the *CYP2D6*1XN* allele is actually a series of copy number variation alleles in which the *CYP2D* gene is present in three, four, or more copies on one chromosome. Predictably, these copy number polymorphisms produce high levels of the enzyme. There are dozens more alleles that do not affect the function of the protein and are therefore considered to be wild-type. Various combinations of these four classes of alleles produce quantitative differences in metabolizing activity, resulting in three main phenotypes: **normal** (also called "extensive") metabolizers, **poor** metabolizers, and **ultrafast** metabolizers (Fig. 18-3).

Depending on whether a drug is itself an active compound or is a prodrug that requires activation by a cytochrome P-450 enzyme to have its pharmacological effect, poor metabolizers may either accumulate toxic levels of the drug or fail to have therapeutic efficacy because of poor activation of a prodrug. In contrast, ultrafast metabolizers are at risk for being undertreated by a drug with doses inadequate to maintain blood levels in the therapeutic range, or they may suffer

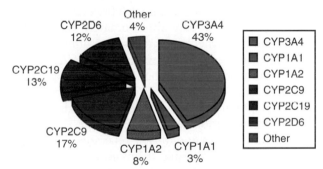

Figure 18-2 Contribution of individual cytochrome P-450 enzymes to drug metabolism.

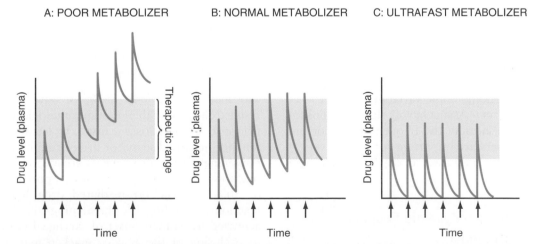

Figure 18-3 Serum drug levels after repeated doses of a drug (*arrows*) in three individuals with different phenotypic profiles for drug metabolism. **A,** Poor metabolizer accumulates drug to toxic levels. **B,** Normal (extensive) metabolizer reaches steady-state levels within the therapeutic range. **C,** Ultrafast metabolizer fails to maintain serum levels within the therapeutic range.

Metabolizer Phenotypes Arising from Various Combinations of *CYP2D6* Alleles

		Allele on One Chromosome			
		Wild-type	Reduced	Absent	Increased
Allele on other chromosome	Wild-type	Normal			
	Reduced	Normal	Poor		
	Absent	Normal	Poor	Poor	
	Increased	Ultrafast	—	—	—

TABLE 18-4 Frequency of Poor *CYP2D6* and *CYP2C19* Metabolizers in Various Population Groups

	Population Frequency of Poor Metabolizers (%)	
Ethnic Origin of Population	*CYP2D6*	*CYP2C19*
Sub-Saharan Africa	3.4	4.0
Native American	0	2
Asian	0.5	15.7
White	7.2	2.9
Middle Eastern/North Africa	1.5	2.0
Pacific Islander	0	13.6

Data from Burroughs VJ, Maxey RW, Levy RA: Racial and ethnic differences in response to medicines: towards individualized pharmaceutical treatment, *J Natl Med Assoc* 94(Suppl):1-26, 2002.

overdose due to too rapid conversion of a prodrug to its active metabolite. For example, codeine is a weak narcotic drug that exerts most of its analgesic effect on conversion to morphine, a bioactive metabolite with a 10-fold higher potency. This conversion is carried out by the CYP2D6 enzyme. Poor metabolizers, quite common in some populations, carrying loss-of-function alleles at the *CYP2D6* locus fail to convert codeine to morphine and thereby receive little therapeutic benefit; in contrast, ultrafast metabolizers can become rapidly intoxicated with low doses of codeine. A number of children have died from codeine overdoses due to having an ultrafast metabolizer phenotype.

As with many forms of genetic variation (see Chapter 9), the frequency of many of the alleles in the cytochromes P-450 differs among different populations (Table 18-4). For example, a slow metabolizing phenotype for *CYP2D6* that is present in 1 in 14 whites is rare in Asia and nearly absent in Native Americans and Pacific Islanders. Similarly, slow metabolizing alleles at *CYP2C19* show striking ethnic variability, with 1 in 33 whites but nearly 1 in 6 Asians having slow metabolism. These ethnic differences in the frequency of poor and ultrafast metabolizers are important for the delivery of personalized genetic medicine to ethnically heterogeneous populations.

Variation in Pharmacodynamic Response
Malignant Hyperthermia

Malignant hyperthermia is a rare autosomal dominant condition in which there may be a dramatic adverse response to the administration of many commonly used inhalational anesthetics (e.g., halothane) and depolarizing muscle relaxants (e.g., succinylcholine). Soon after induction of anesthesia, the patients develop life-threatening fever, sustained muscle contraction, and attendant hypercatabolism. The fundamental physiological abnormality in the disease is an elevation of the level of ionized calcium in the sarcoplasm of muscle. This increase leads to muscle rigidity, elevation of body temperature, rapid breakdown of muscle (rhabdomyolysis), and other abnormalities. The condition is an important if not a common cause of death during anesthesia. The incidence is 1 in 50,000 adults undergoing anesthesia but for unknown reasons is 10-fold higher in children.

Malignant hyperthermia is most frequently associated with mutations in a gene called *RYR1,* encoding an intracellular calcium ion channel. However, mutations in *RYR1* account for only approximately half of cases of malignant hyperthermia. At least five other loci have now been identified, one of which is the *CACNL1A3* gene, which encodes the α_1 subunit of a dihydropyridine-sensitive calcium channel. Precisely why the abnormalities in calcium handling in muscle found with *RYR1* or *CACNL1A3* mutations make the muscle sensitive to inhalation anesthetics and muscle relaxants and precipitate malignant hyperthermia is unknown.

The need for special precautions when at-risk persons require anesthesia is obvious. Cooling blankets, muscle relaxants, and cardiac antiarrhythmics may all be used to prevent or reduce the severity of the response if an unsuspected attack occurs, and alternative anesthetics can be given to patients at risk.

Adverse Drug Reactions

The majority (75% to 80%) of adverse drug events result from predictable, nonimmunological drug toxicities such as overdoses caused by medication errors, renal or hepatic disease, or drug-drug interactions. The remaining adverse drug events are mostly unpredictable reactions to the drugs; of these, approximately 25% to 50% are true IgE-mediated drug hypersensitivity reactions, including life-threatening anaphylaxis characterized by sudden onset of laryngeal edema, leading to occlusion of the airway, marked hypotension, and cardiac arrhythmias.

The remaining 50 to 75% of adverse drug reactions are genetically determined nonallergic immune reactions, manifesting as widespread damage to skin and mucous membranes, referred to as **Stevens-Johnson syndrome (SJS)** and, in its more serious extreme form, **toxic epidermal necrolysis (TEN)**. Although rare, TEN is a very serious adverse drug reaction that causes denuding of large areas of skin and carries a mortality rate of 30% to 40%. There is a strong correlation between particular drugs and certain human leukocyte antigen (HLA) alleles in the major histocompatibility complex (see Chapters 4 and 8) that results in SJS and TEN. For example, individuals who take the retroviral drug abacavir (Case 1) and carry the *HLA-B*5701* allele have a 50% risk for SJS or TEN, whereas those without it never develop this skin reaction in response to the drug. Because approximately 5% of Europeans carry the *HLA-B*5701* allele, the risk for a severe drug reaction in abacavir-treated patients from this ethnic background is especially significant. The allele is less frequent in Asian populations ($\approx$1%) and even less frequent in Africans (<1%). HLA typing is therefore standard of care for any patient for whom one is contemplating beginning abacavir. A similar situation exists with the use of the antiseizure medication carbamazepine and *HLA-B*1502*, which is present in 10% to 20% of certain Chinese populations (see Table 18-3).

PHARMACOGENOMICS AS A COMPLEX TRAIT

The examples of pharmacogenomics provided in this chapter primarily involve variation at single genes and its effect on drug treatment. In truth, most drug response is a complex trait. A drug may have its effect directly or through more active metabolites, each of which may then be metabolized by different pathways and exert its effects on various targets. Thus variants at more than one locus may interact, synergistically or antagonistically, either to potentiate or to reduce the effectiveness of a drug or to increase its toxic side effects. A comprehensive **pharmacogenomic profile** that takes into account multiple genetic variants as well as environmental effects, including other drugs, for their aggregate impact on the outcome of drug therapy is necessary before we can have highly precise and predictive information to guide drug therapy. If a pharmacogenomic profile is sufficiently predictive of drug response, it could be used to predict the probable efficacy or side effects of the medication in an individual before the drug is administered, and to identify those patients who should be treated more aggressively and monitored to be sure that the drug achieves therapeutic levels. The ultimate goal is that patients receive the best drug at the right dose and avoid potentially dangerous side effects. We expect pharmacogenomics to become increasingly more important in the delivery of personalized, precision medicine in the years ahead.

SCREENING FOR GENETIC SUSCEPTIBILITY TO DISEASE
Genetic Epidemiology

Epidemiological studies of risk factors for disease rely on population studies that measure disease prevalence or incidence and determine whether certain risk factors (e.g., genetic, environmental, social) are present in individuals with versus without disease. **Genetic epidemiology** is concerned with how genotypes and environmental factors interact to increase or decrease susceptibility to disease. Epidemiological studies generally follow one of three different strategies: case-control, cross-sectional, and cohort design (see Box).

STRATEGIES USED IN GENETIC EPIDEMIOLOGY

- **Case-control:** Individuals with and without the disease are selected, and the genotypes and environmental exposures of individuals in the two groups are determined and compared.
- **Cross-sectional:** A random sample of the population is selected and divided into those with and without the disease, and their genotypes and environmental exposures are determined and compared.
- **Cohort:** A sample of the population is selected and observed for some time to ascertain who does or does not develop disease, and their genotypes and environmental exposures are determined and compared. The cohort may be selected at random or may be targeted to individuals who share a genotype or an environmental exposure.

Cohort and cross-sectional studies not only capture information on the relative risk conferred by different genotypes but, if they are random population samples, also provide information on the prevalence of the disease and the frequency of the various genotypes under study. A randomly selected cohort study, in particular, is the most accurate and complete approach in that phenotypes that take time to appear have a better chance of being detected and scored; they are, however, more expensive and time-consuming. Cross-sectional studies, on the other hand, suffer from underestimation of the frequency of the disease. First, if the disease is rapidly fatal, many of the patients with disease and carrying a risk factor will be missed. Second, if the disease shows age-dependent penetrance, some patients carrying a risk factor will actually not be scored as having the disease. Case-control studies, on the other hand, allow researchers to efficiently target individuals, particularly with relatively rare phenotypes that would require very large sample sizes in a cross-sectional or cohort study. However, unless a study is based on complete ascertainment of individuals with a disease (e.g., in a population register or surveillance program) or uses a random sampling scheme, a case-control study cannot capture information on the population prevalence of the disease.

Disease Association

A genetic **disease association** is the relationship in a population between a susceptibility or protective genotype and a disease phenotype (see Chapter 10). The susceptibility or protective genotype can be an allele (in either a heterozygote or a homozygote), a genotype at one locus, a haplotype containing alleles at neighboring loci, or even combinations of genotypes at multiple unlinked loci. Whether a disease association between genotype and phenotype is statistically significant can be determined from standard statistical tests, such as the chi-square test, whereas how strongly associated the genotype and phenotype are is given by the odds ratio or relative risk, as discussed in Chapter 10. The relationship between some of these concepts is best demonstrated by means of a 2 × 2 table.

Clinical Validity and Utility

Finding the genetic contributions to health and disease is of obvious importance for research into underlying disease etiology and pathogenesis, as well as for identifying potential targets for intervention and therapy. In medical practice, however, whether to screen individuals for increased susceptibilities to illness depends on the **clinical validity** and **clinical utility** of the test. That is, how predictive of disease is a positive test, and how useful is it to have this information?

Determination of the Predictive Value of a Test

Genotype	Disease		Total
	Affected	Unaffected	
Susceptibility genotype present	a*	b	a + b
Susceptibility genotype absent	c	d	c + d
Total	a + c	b + d	a + b + c + d = N

Frequency of the susceptibility genotype = (a + b)/N

Disease prevalence = (a + c)/N (with random sampling or a complete population survey)

Relative Risk Ratio:

$$= \frac{a/(a + b)}{c/(c + d)}$$

$$RRR = \frac{\text{Disease prevalence in carriers of susceptibility genotype}}{\text{Disease prevalence in noncarriers of susceptibility genotype}}$$

Sensitivity: Fraction of individuals with disease who have the susceptibility genotype = a/(a + c)

Specificity: Fraction without disease who do not have the susceptibility genotype = d/(b + d)

Positive predictive value: Proportion of individuals with the susceptibility genotype who have or will develop a particular disease = a/(a + b)

Negative predictive value: Proportion of individuals without the susceptibility genotype who do not have or will not develop a particular disease = d/(c + d)

*The values of a, b, c, and d are derived from a random sample of the population, divided into those with and without the susceptibility genotype, and then examined for the disease (with or without longitudinal follow-up, depending on whether it is a cross-sectional or cohort study) (see later).

Clinical Validity

Clinical validity is the extent to which a test result is predictive for disease. Clinical validity is captured by the two concepts of **positive predictive value** and **negative predictive value**. The positive predictive value is the frequency with which a group of individuals who test positive have or will develop the disease. For mendelian disorders, the positive predictive value of a genotype is the penetrance. Conversely, the negative predictive value is the frequency with which a group of individuals who test negative are free of disease and remain so. When faced with an individual patient, the practitioner of personalized genetic medicine needs to know more than just whether there is an association and its magnitude (i.e., relative risk or odds ratio). It is important to know clinical validity (i.e., how well the test predicts the presence or absence of disease).

Susceptibility Testing Based on Genotype

The positive predictive value of a genotype that confers susceptibility to a particular disease depends on the relative risk for disease conferred by one genotype over another and on the prevalence of the disease. Figure 18-4 provides the positive predictive value for genotype frequencies ranging from 0.5% (rare) to 50% (common), which confer a relative risk that varies from low (twofold) to high (100-fold), when the prevalence of the disease ranges from relatively rare (0.1%) to more common (5%). As the figure shows, the value of the test as a predictor of disease increases substantially when one is dealing with a common disorder due to a relatively rare susceptibility genotype that confers a high relative risk, compared with the risk for individuals who do not carry the genotype. The converse is also clear; testing for a common genotype that confers a modest relative risk is of limited value as a predictor of disease.

We will illustrate the use of the 2 × 2 table in assessing the role of susceptibility alleles in a common disorder, **colorectal cancer**. Shown in the following Box are data from a population-based study of colorectal cancer risk conferred by a polymorphic variant in the *APC* gene (see Chapter 15) (Case 15) that changes isoleucine to lysine at position 1307 of the protein (Ile1307Lys). This variant has an allele frequency of approximately 3.1% among Ashkenazi Jews, which means that approximately 1 in 17 individuals is a heterozygote (and 1 in 1000 are homozygous) for the allele. The prevalence of colon cancer among Ashkenazi Jews is 1%. The Ile1307Lys variant, common enough to be heterozygous in approximately 1 in 17 Ashkenazi Jews, confers a 2.4-fold increased risk for colon cancer relative to individuals without the allele. However, the small positive predictive value (≈2%) means that an individual who tests positive for this allele has only a 2% chance of developing colorectal cancer. If this had been a cohort study that allowed complete ascertainment of everyone in whom colorectal cancer was going to develop, the penetrance would, in effect, be only 2%.

THE Ile1307Lys ALLELE OF THE *APC* GENE AND COLON CANCER

| Allele | Colon Cancer | | |
	Affected	Unaffected	Total
Lys1307	7	310	317
Ile1307	38	4142	4180
Total	45	4452	4497

- Relative Risk Ratio = RRR

$$= \frac{\text{Disease prevalence in allele carriers}}{\text{Disease prevalence in noncarriers}}$$

$$= \frac{7/317}{38/4180} = 2.4$$

- **Sensitivity:** Fraction of individuals with colon cancer who have the Lys1307 allele = 7/45 = 16%
- **Specificity:** Fraction without colon cancer who do not have the Lys1307 allele = 4142/4452 = 93%
- **Positive predictive value:** Fraction of individuals with the Lys1307 allele who develop colon cancer = 7/317 = 2%
- **Negative predictive value:** Fraction of individuals without the Lys1307 allele who do not develop colon cancer = 99%

Data from Woodage T, King SM, Wacholder S, et al: The APCl1307K allele and cancer risk in a community based study of Ashkenazi Jews. *Nat Genet* 20:62-65, 1998.

Clinical Utility

The clinical utility of a test is more difficult to assess than clinical validity, because it has different meanings for different people. In its narrowest sense, the clinical utility of a test is that the result is medically **actionable,** that is, the result will change what medical care an individual receives and, as a consequence, will improve the outcome of care, both medically and economically. At the other end of the spectrum is clinical utility broadly defined as *any* piece of information an individual patient might be interested in having, for any reason, including simply for the sake of knowing.

In a patient who tests positive for the *APC* Ile1307Lys allele, how does a positive predictive value of 2% translate into clinical utility for medical practice? One critical factor is a public health economic one: can the screening be shown to be **cost-effective**? Is the expense of the testing outweighed by improving health outcomes while reducing health care costs, disability, and loss of earning power? In the example of screening for the *APC* Ile1307Lys allele in Ashkenazi Jews, more frequent screening or the use of different approaches to screening for colon cancer may be effective. Screening methods (occult stool blood testing versus fecal DNA testing, or sigmoidoscopy versus full colonoscopy) differ in expense, sensitivity, specificity, and potential for hazard, and so deciding which regimen to follow has important implications for the patient's health and health care costs.

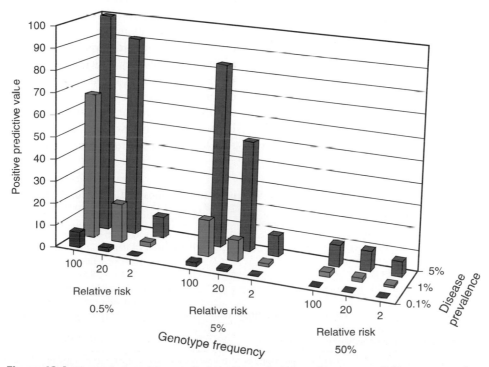

Figure 18-4 Theoretical positive predictive value calculations for a susceptibility genotype for a disease, over a range of genotype frequencies, disease prevalences, and relative risks for disease conferred by the genotype.

Even with demonstrable clinical validity and actionable clinical utility, demonstrating that testing improves health care is not always straightforward. For example, 1 in 200 to 1 in 250 white individuals are homozygous for a Cys282Tyr mutation in the *HFE* gene associated with **hereditary hemochromatosis**, a disorder characterized by iron overload that can silently lead to extensive liver damage and cirrhosis (Case 20). A simple intervention—regular phlebotomy to reduce total body iron stores—can prevent hepatic cirrhosis. The susceptibility genotype is common, and 60% to 80% of Cys282Tyr homozygotes show biochemical evidence of increased body iron stores, which suggests that screening would be a reasonable and cost-effective measure to identify asymptomatic individuals who should undergo further testing and, if indicated, the institution of regular phlebotomy. However, most Cys282Tyr homozygotes (>90% to 95%) remain *clinically* asymptomatic, leading to the argument that the positive predictive value of *HFE* gene testing for liver disease in hereditary hemochromatosis is too low to justify population screening. Nonetheless, some of these largely asymptomatic patients *do* have signs of clinically occult fibrosis and cirrhosis on liver biopsy, indicating that the Cys282Tyr homozygote may actually be at a higher risk for liver disease than previously thought. Thus some would argue for population screening to identify individuals in whom regular prophylactic phlebotomy should be instituted. The clinical utility of such population screening remains controversial and will require additional research to determine the natural history of the disease and whether the silent fibrosis and cirrhosis seen on liver biopsy represent the early stages of what will be a progressive illness.

APOE testing in **Alzheimer disease (AD)** (see Chapter 12) (Case 4) is another example of the role of a careful assessment of clinical validity and clinical utility in applying genetic testing to personalized medicine. As discussed in Chapter 8, heterozygotes for the ε4 allele of the *APOE* gene are at a two- to threefold increased risk for development of AD compared with individuals without an *APOE* ε4 allele. *APOE* ε4/ε4 homozygotes are at a eightfold increased risk. An analysis of both the clinical validity and clinical utility of *APOE* testing, including calculation of the positive predictive value for asymptomatic and symptomatic individuals, is shown later (Table 18-5).

As can be seen from these positive predictive values for asymptomatic people in the age bracket 65 to 74 years, the presence of a single ε4 allele is not a strong predictor of whether AD will develop, despite the threefold increased risk for the disease conferred by the ε4 allele compared with those without an ε4 allele. Thus AD will not develop in the majority of individuals heterozygous for an ε4 allele identified through *APOE* testing as being at increased risk. Even with two ε4 alleles, which occurs in approximately 1.5% of the population and is associated with a 8-fold increased risk relative to genotypes without ε4 alleles, there is still less than a one in four chance of developing AD. *APOE* testing for the ε4 allele, is therefore not recommended in *asymptomatic* individuals but is being used by some practitioners in the evaluation of individuals with symptoms and signs of dementia.

The utility of testing asymptomatic individuals at their *APOE* locus to assess risk for AD is also controversial. First, knowing that one is at increased risk for AD through *APOE* testing does not lead to any preventive or therapeutic options. Thus, under a strict definition of clinical utility—that is, the result is actionable and leads to changes in medical management—there would appear to be little value in *APOE* testing for AD risk.

There may be, however, positive and negative outcomes of testing that are psychological or economic in nature and more difficult to assess than the purely clinical factors. For example, testing positive for a susceptibility genotype could empower patients with knowledge of their risks as they make important lifestyle decisions. On the other hand, it has been suggested that knowing of an increased risk through *APOE* testing might cause significant emotional and psychological distress. However, careful studies of the impact of receiving *APOE* genotype information have shown little harm in

TABLE 18-5 Clinical Validity and Utility of *APOE* Population Screening and Diagnostic Testing for Alzheimer Disease

	Population Screening	Diagnostic Testing
Clinical validity	Asymptomatic individuals aged 65-74 yr Population prevalence of AD = 3% PPV given ε2/ε4 or ε3/ε4 = 6% PPV given ε4/ε4 = 23%	Individuals aged 65-74 yr with symptoms of dementia Proportion of dementia patients with AD = ≈60% PPV given ε2/ε4 or ε3/ε4 = ≈75% PPV given ε4/ε4 = ≈98%
Clinical utility	No intervention possible to prevent disease	Increases suspicion that another, potentially treatable cause of dementia may be present
	Psychological distress for most people with ε4 alleles who are not likely to develop AD False reassurance for those without ε4 alleles	Reduces unnecessary testing

Positive predictive value (PPV) calculations are based on a population prevalence of Alzheimer disease (AD) of approximately 3% in individuals aged 65 to 74 years, an allele frequency for the ε4 allele in whites of 10% to 15%, a relative risk of approximately 3 for one ε4 allele, and a relative risk of approximately 20 for two ε4 alleles.

appropriately counseled individuals with a family history of AD who wished to know if they were at increased risk.

Finally, patients who test negative for the ε4 alleles could be falsely reassured that they are at no increased risk for the disorder, despite having a positive family history or other risk factors for dementia. Balancing all of these considerations, *APOE* testing is still not recommended in asymptomatic individuals even in light of such a strong genotype-disease association, because of the low positive predictive value and lack of clinical utility, rather than because such information is clearly harmful.

As in all of medicine, the benefits and costs for each component of personalized genetic medicine need to be clearly demonstrated but also continually reassessed. The requirement for constant reevaluation is obvious; imagine how the recommendations for *APOE* testing, despite its low positive predictive value, might change if a low-risk and inexpensive medical intervention were discovered that could prevent or significantly delay the onset of dementia.

Heterozygote Screening

In contrast to screening for genetic disease in newborns or for genetic susceptibility in patients, screening for carriers of mendelian disorders has, as its main purpose, the identification of individuals who are themselves healthy but are at substantial (25% or higher) risk for having children with a severe autosomal recessive or X-linked illness. The principles of heterozygote screening are shown in the accompanying Box.

CRITERIA FOR HETEROZYGOTE SCREENING PROGRAMS

- High frequency of carriers, at least in a specific population
- Availability of an inexpensive and dependable test with very low false-negative and false-positive rates
- Access to genetic counseling for couples identified as heterozygotes
- Availability of prenatal diagnosis
- Acceptance and voluntary participation by the population targeted for screening

To provide a sufficient yield of carriers, current heterozygote screening programs have typically focused on particular ethnic groups in which the frequency of mutant alleles is high. In contrast to newborn screening, as discussed previously in this chapter, heterozygote screening is voluntary and focuses on individuals who identify themselves as being members of a particular high-risk ethnic group. Heterozygote screening has been used extensively for a battery of disorders for which carrier frequency is relatively high: Tay-Sachs disease (Case 43) (the prototype of carrier screening) (see Chapter 12), Gaucher disease, and Canavan disease in the Ashkenazi Jewish population; sickle cell disease (Case 42) in the African American population of North America; and β-thalassemia (Case 44) in high-incidence areas, especially in Cyprus and Sardinia or in extended consanguineous families from Pakistan (see Chapter 11).

Carrier screening for cystic fibrosis (Case 12) has become standard of care for couples contemplating a pregnancy. As discussed in Chapter 12, more than 1000 different disease-causing mutations have been described in the *CFTR* gene. Although the vast majority of disease-causing mutations in the *CFTR* gene can be readily detected with greater than 99% sensitivity when the entire gene is sequenced, sequencing the entire gene in every couple seeking preconception carrier testing is expensive if carried out on a population-wide basis, particularly in individuals with low prior probability of carrying a mutation (Table 18-6). Instead, panels of specific mutations have been designed to detect just the most common mutations in various ethnic groups using a relatively inexpensive platform. The sizes of these panels range from one proposed by the American College of Medical Genetics and Genomics, which contains the most common 23 mutations found in ethnic groups with the highest frequency of the disease, such as non-Hispanic whites, to more extensive panels of more than 60 different mutations that include mutations found more commonly in populations with lower frequencies of disease, such as Africans or Asians (see Table 18-6). Because these allele-specific methods are designed to detect only the most common mutations, their sensitivity is less than 100%, ranging from 88% to 90% in non-Hispanic whites, and from 64% to 72%

TABLE 18-6 Cystic Fibrosis Carrier Frequencies by Ethnic Group before and after Negative Carrier Test with Standard and Expanded Allele-Specific Panel

Ethnic Group	Incidence of Cystic Fibrosis	Carrier Probability without Testing	Carrier Probability if Negative for ACMG Panel	Carrier Probability if Negative for Expanded Panel
White	1 in 3,200	1/25	1/214	1/266
African American	1 in 15,300	1/65	1/183	1/236
Hispanic American	1 in 9,500	1/46	1/162	1/282
Asian American	1 in 32,100	1/90	1/176	1/198
Ashkenazi Jewish	1 in 3,300	1/25	1/417	1/610

ACMG, American College of Medical Genetics and Genomics.

in African Americans. One may anticipate that as the cost of comprehensive sequencing falls, allele-specific methods with less than 100% sensitivity may be superseded, but, for the near future, the cost-effectiveness of allele-specific methods remains a reasonable argument for their continued use in the appropriate setting.

As the cost of mutation detection using allele-specific detection methods has fallen, it is becoming much less compelling that carrier screening needs to be restricted to a small number of mutant alleles common in certain ethnic groups in genes that are known to be associated with disease. It is possible now to obtain expanded carrier screening beyond disorders common to particular ethnic groups, such as cystic fibrosis, sickle cell trait, or thalassemia, to include carrier status for more than 100 additional autosomal recessive and X-linked disorders. With the use of sequencing instead of allele-specific detection methods, there is no longer any limit on which genes and which mutant alleles in these genes can theoretically be detected. Rare mutant alleles in genes associated with known disease will be found, thereby raising the sensitivity of carrier detection methods. Sequencing, however, also has the ability to uncover variants, particularly missense changes, of unknown pathogenicity in disease genes as well as in genes whose role in the disease is unknown (see Chapter 16). Unless great care is taken in assessing the clinical validity of rare variants detected by sequencing, the frequency of false-positive carrier test results will increase.

The impact of carrier screening in lowering the incidence of a genetic disease can be dramatic. Carrier screening for Tay-Sachs disease in the Ashkenazi Jewish population has been carried out since 1969. Screening followed by prenatal diagnosis, when indicated, has already lowered the incidence of Tay-Sachs disease by 65% to 85% in this ethnic group. In contrast, attempts to screen for carriers of sickle cell disease in the U.S. African American community have been less effective and have had little impact on the incidence of the disease so far. The success of carrier screening programs for Tay-Sachs disease, as well as the relative failure for sickle cell anemia, underscores the importance of community consultation, community engagement, and the availability of genetic counseling and prenatal diagnosis as critical requirements for an effective program.

PERSONALIZED GENOMIC MEDICINE

More than a century ago, the British physician-scientist Archibald Garrod proposed the concept of **chemical individuality**, in which each of us differs in our health status and susceptibility to various illnesses because of our individual genetic makeup. Indeed, in 1902, he wrote:

> ...the factors which confer upon us our predisposition and immunities from disease are inherent in our very

> chemical structure, and even in the molecular groupings which went to the making of the chromosomes from which we sprang.

The goal of **personalized genomic medicine** is to use knowledge of an individual's genetic variants relevant to maintaining health or treating illness as a routine part of medical care.

Now, more than a hundred years after Garrod's visionary pronouncement, in the era of human genomics, we have the means to assess an individual's genotype at all relevant loci by whole-genome sequencing (WGS) or, less comprehensively, by whole-exome sequencing (WES) to characterize the genetic underpinnings of each person's unique "chemical individuality." In addition to genomic approaches to prenatal screening of the fetus for aneuploidy by maternal cell-free DNA, as described in Chapter 17, WGS and WES are being studied for analyzing fetal DNA obtained by invasive procedures, newborn screening, screening asymptomatic adults for increased predisposition to various diseases, identifying couples that are heterozygotes for autosomal recessive or X-linked diseases that could affect their children before conception, and for finding pharmacogenetic variants relevant to drug therapy.

The National Health Service of the United Kingdom is preparing to sequence the genomes of 100,000 people by 2017, with the eventual aim of having the sequence of every individual in the country in a database to use for developing personalized prevention and treatment. Hospitals, pharmaceutical companies, and the U.S. Department of Veterans Affairs are also beginning large-scale sequencing of hundreds of thousands of individuals. Although these efforts are focusing initially on mining the data for genetic variants that contribute to disease or for finding novel drug targets, they are also proposing to study how to use the genomic information to design personalized prevention and treatment strategies.

The application of WGS and WES to personalized medicine is not without controversy, however. One issue is cost. Although sequencing per se is many orders of magnitude less expensive now than when the original Human Genome Project was being carried out, the interpretation of such sequences remains very time consuming and expensive. Despite the time and effort put into interpretation, we are still unable to assign any clinical significance to the vast majority of all variants found through sequencing. There is widespread concern that individuals and their health care providers, when confronted with variants of uncertain significance (see Chapter 16), will seek additional expensive and unnecessary testing, with all the attendant expense and potential for complications that result from any medical test. There is the additional concern that even when a variant is known to be pathogenic and shown to be highly penetrant in families with multiple affected individuals, the actual penetrance when the variant is found through population screening in individuals with a negative family history may be much less.

Personalized genomic medicine is only one component of **precision medicine,** which, in its broadest sense, requires health care providers to merge genomic information with other kinds of information, such as physiological or biochemical measures, developmental history, environmental exposures, and social experiences. The ultimate goal is to provide more precise diagnosis, counseling, preventive intervention, management, and therapy. This effort has begun, but a lot of work remains before personalized genomic medicine becomes a part of mainstream medicine.

GENERAL REFERENCES

Feero WG, Guttmacher AE, Collins FS: Genomic medicine—an updated primer, *N Engl J Med* 362:2001–2011, 2010.

Ginsburg G, Willard HF, editors: *Genomic and personalized medicine,* ed 2 (vols 1 & 2), New York, 2012, Elsevier. 1305 pp.

Kitzmiller JP, Groen DK, Phelps MA, et al: Pharmacogenomic testing: relevance in medical practice, *Cleve Clin J Med* 78:243–257, 2011.

Schrodi SJ, Mukherjee S, Shan Y, et al: Genetic-based prediction of disease traits: prediction is very difficult, especially about the future, *Frontiers Genet* 5:1–18, 2014.

REFERENCES FOR SPECIFIC TOPICS

Amstutz U, Carleton BC: Pharmacogenetic testing: time for clinical guidelines, *Pharmacol Therapeutics* 89:924–927, 2011.

Bennett MJ: Newborn screening for metabolic diseases: saving children's lives and improving outcomes, *Clin Biochem* 47(9):693–694, 2014.

Dorschner MO, Amendola LM, Turner EH, et al: Actionable, pathogenic incidental findings in 1,000 participants' exomes, *Am J Hum Genet* 93:631–640, 2013.

Ferrell PB, McLeod HL: Carbamazepine, *HLA-B*1502* and risk of Stevens-Johnson syndrome and toxic epidermal necrolysis: US FDA recommendations, *Pharmacogenomics* 9:1543–1546, 2008.

Green RC, Roberts JS, Cupples LA, et al: Disclosure of APOE genotype for risk of Alzheimer's disease, *N Engl J Med* 361:245–254, 2009.

Kohane IS, Hsing M, Kong SW: Taxonomizing, sizing, and overcoming the incidentalome, *Genet Med* 14:399–404, 2012.

Mallal S, Phillips E, Carosi G, et al: HLA-B*5701 screening for hypersensitivity to abacavir, *N Engl J Med* 358:568–579, 2008.

McCarthy JJ, McLeod HL, Ginsburg GS: Genomic medicine: a decade of successes, challenges and opportunities, *Sci Transl Med* 5:189sr4, 2013.

Topol EJ: Individualized medicine from prewomb to tomb, *Cell* 157:241–253, 2014.

Urban TJ, Goldstein DB: Pharmacogenetics at 50: genomic personalization comes of age, *Sci Transl Med* 6:220ps1, 2014.

PROBLEMS

1. In a population sample of 1,000,000 Europeans, idiopathic cerebral vein thrombosis (iCVT) occurred in 18, consistent with an expected rate of 1 to 2 per 100,000. All the women were tested for factor V Leiden (FVL). Assuming an allele frequency of 2.5% for FVL, how many homozygotes and how many heterozygotes for FVL would you expect in this sample of 1,000,000 people, assuming Hardy-Weinberg equilibrium?

   Among the affected individuals, two were heterozygotes for FVL and one was homozygous for FVL. Set up a 3 × 2 table for the association of the homozygous FVL genotype, the heterozygous FVL genotype, and the wild-type genotype for iCVT.

   What is the relative risk for iCVT in a FVL heterozygote versus the wild-type genotype? What is the risk in a FVL homozygote versus wild-type? What is the sensitivity of testing positive for either one or two FVL alleles for iCVT? Finally, what is the positive predictive value of being homozygous for FVL? heterozygous?

2. In a population sample of 100,000 European women taking oral contraceptives, deep venous thrombosis (DVT) of the lower extremities occurred in 100, consistent with an expected rate of 1 per 1,000. Assuming an allele frequency of 2.5% for factor V Leiden (FVL), how many homozygotes and how many heterozygotes for FVL would you expect in this sample of 100,000 women, assuming Hardy-Weinberg equilibrium?

   Among the affected individuals, 58 were heterozygotes for FVL and three were homozygous for FVL. Set up a 3 × 2 table for the association of the homozygous FVL genotype, the heterozygous FVL genotype, and the wild-type genotype for DVT of the lower extremity.

   What is the relative risk for DVT in a FVL heterozygote using oral contraceptives versus women with the wild-type genotype taking oral contraceptives? What is the risk in a FVL homozygote versus wild-type? What is the sensitivity of testing positive for either one or two FVL alleles for DVT while taking oral contraceptives? Finally, what is the positive predictive value for DVT of being homozygous for FVL while taking oral contraceptives? Heterozygous?

3. What steps should be taken when a phenylketonuria (PKU) screening test comes back positive?

4. Newborn screening for sickle cell disease can be performed by hemoglobin electrophoresis, which separates hemoglobin A and S, thereby identifying individuals who are heterozygotes as well as those who are homozygotes for the sickle cell mutation. What potential benefits might accrue from such testing? What harms?

5. Toxic epidermal necrolysis (TEN) and the Stevens-Johnson syndrome (SJS) are two related, life-threatening skin reactions that occur in approximately 1 per 100,000 individuals in China, most commonly as a result of exposure to the antiepileptic drug carbamazepine. These conditions carry a significant mortality rate of 30% to 35% (TEN) and 5% to 15% (SJS). It was observed that individuals who suffered this severe immunological reaction carried a particular major histocompatibility complex class 1 allele, *HLA-B*1502,* as do 8.6% of the Chinese population. In a retrospective cohort study of 145 patients who received carbamazepine therapy, 44 developed either

TEN or SJS. Of these, all 44 carried the *HLA-B*1502* allele, whereas only three of the patients who received the drug without incident were *HLA-B*1502* positive. What is the sensitivity, specificity, and positive predictive value of this allele for TEN or SJS in patients receiving carbamazepine?

6. In 1997, a young female college student died suddenly of cardiac arrhythmia after being startled by a fire alarm in her college dormitory in the middle of the night. She had recently been prescribed an oral antihistamine, terfenadine, for hay fever by a physician at the school. Her parents reported that she would take her medications every morning with breakfast, which consisted of grapefruit juice, toast, and caffeinated coffee. Her only other medication was oral itraconazole, which she was given by a dermatologist in her home town to treat a stubborn toenail fungus that she considered unsightly. Terfenadine was removed from the U.S. market in 1998.

Do a literature search on sudden cardiac death associated with terfenadine, relating possibly genetic and environmental factors that might have interacted to cause this young woman's death.

CHAPTER 19

Ethical and Social Issues in Genetics and Genomics

Human genetics and genomics are having a major impact in all areas of medicine and across all age-groups, and their importance will only grow as knowledge increases and the power and reach of sequencing technology improves. Yet, no single area of medical practice raises as many challenging ethical, social, and policy issues in so many areas of medicine and across so broad a spectrum of age-groups, including the fetus, neonates, children, prospective parents, and adults.

There are many categories of information that genetics and genomics deals with, ranging from ancestry and personal heritage to diagnosis of treatable or untreatable disease to explanations for familial traits to concerns about what has been or might be passed on to the next generation. Some of these were introduced in previous chapters; others are presented in this chapter. But, as we shall see, they all pose ethical, legal, social, personal, and policy challenges. And if that is true today, it will become only more commonplace in the years and decades ahead, as genome sequences (and the data-rich landscape of genomic and medical information) become available for millions—and eventually hundreds of millions—of individuals worldwide.

The ethical and social issues raised because of new information and capabilities in human genetics and genomics are especially relevant to decisions in the area of reproduction (see Chapter 17) because of the absence of a societal consensus on the religious and ethical concerns about abortion and assisted reproductive technologies. The damaging legacy of eugenics (discussed later in this chapter) hangs over discussions of reproductive genetics, now especially timely in light of the ability to evaluate the sequence of fetal genomes. Finally, privacy concerns also loom large because genetic and genomic information, in the absence of any other demographical information, may still render an individual and his or her personal sensitive health information uniquely identifiable. Yet, we share DNA variation with our family members and indeed with all of mankind, and thus privacy concerns need to be balanced against the benefits that could be derived from making personal genetic information available to other family members and to society at large.

In this chapter, we will review some of the most challenging ethical and societal issues that arise from the application of genetics and genomics to medicine. These relate to prenatal diagnosis, presymptomatic testing, the duty to inform family members of genetic conditions in the family, and the policy challenges arising from the discovery of genetic variants that confer increased risk for disease that are found incidental to diagnostic testing for another indication.

PRINCIPLES OF BIOMEDICAL ETHICS

Four cardinal principles are frequently considered in any discussion of ethical issues in medicine:
- **Respect for individual autonomy,** safeguarding an individual's rights to control his or her medical care and medical information, free of coercion
- **Beneficence,** doing good
- **avoidance of maleficence,** "first of all, do no harm" (from the Latin, *primum non nocere*)
- **Justice,** ensuring that all individuals are treated equally and fairly

Complex ethical issues arise when these principles are perceived to be in conflict with one another. The role of ethicists working at the interface between society and medical genetics is to weigh and balance conflicting demands, each of which has a claim to legitimacy based on one or more of these cardinal principles.

ETHICAL DILEMMAS ARISING IN MEDICAL GENETICS

In this section, we focus our discussion on some of the ethical dilemmas arising in medical genetics, dilemmas that will only become more difficult and complex as genetics and genomics research expands our knowledge (Table 19-1). The list of issues discussed here is by no means exhaustive, nor are the issues necessarily independent of one other.

Ethical Dilemmas in Genetic Testing
Prenatal Genetic Testing

Geneticists are frequently asked to help couples use prenatal diagnosis or assisted reproductive technology to avoid having offspring with a serious hereditary disorder. For some hereditary disorders, prenatal diagnosis remains controversial, particularly when the diagnosis leads to a decision to terminate the pregnancy for a disease that causes various kinds of physical or intellectual disabilities but is not fatal in infancy. Prenatal diagnosis is equally controversial for adult-onset disorders, particularly ones that may be managed or treated. A debate is ongoing in the community of persons who have a physical or intellectual disability and deaf patients and their families (to name only a few examples) about whether prenatal diagnosis and abortion for these disorders are ethically justified. The dilemma lies in attempting to balance, on the one hand, respect for the autonomy of parents' reproductive decision making about the kind of family they wish to have versus, on the other hand, an assessment of whether aborting a fetus affected with a disability compatible with life is fair to the fetus or to the broader community of persons with a disability or people with hearing impairment.

The dilemma also arises when a couple makes a request for prenatal diagnosis in a pregnancy that is at risk for what most people would not consider a disease or disability at all. Particularly troubling is prenatal diagnosis for selection of sex for reasons other than reducing the risk for sex-limited or X-linked disease. Many genetics professionals are concerned that couples are using assisted reproductive technologies, such as in vitro fertilization and blastomere biopsy, or prenatal sex determination by ultrasonography and abortion, to balance the sexes of the children in their family or to avoid having children of one or the other sex for social and economic reasons prevalent in their societies. There are already clear signs of a falling ratio of female to male infants from 0.95 to less than 0.85 in certain areas of the world where male children are more highly prized.

Other areas of ethical debate include seeking prenatal diagnosis to avoid recurrence of a disorder associated with a mild or cosmetic defect or for putative genetic enhancement, such as genetic variants affecting muscle physiology and therefore athletic prowess. Other examples are the use of prenatal diagnosis and possible pregnancy termination for what is considered by society to be a normal phenotype, such as hearing or typical stature, in a family in which both parents are deaf or have achondroplasia and consider their phenotypes to be important components of their family identity, not disabilities. Such dilemmas have so far been more theoretical than real. Although surveys of couples with deafness or achondroplasia show that the couples are concerned about having children who are not deaf or do not have achondroplasia, the vast majority would not actually use prenatal diagnosis and abortion to avoid having children who do not share their conditions.

In the future, particular alleles and genes that contribute to complex traits, such as intelligence, personality, stature, and other physical characteristics, will likely be identified. Will such nonmedical criteria be viewed as a justifiable basis for prenatal diagnosis? Some might argue that parents are already expending tremendous effort and resources on improving the *environmental* factors that contribute to healthy, successful children. They might therefore ask why they should not try to improve the *genetic* factors as well. Others consider prenatal selection for particular desirable genes a dehumanizing step that treats children simply as commodities fashioned for their parents' benefit. Once again, the ethical dilemma is in attempting to balance respect for the autonomy of parents' reproductive decision making with an assessment of whether it is just or beneficial to terminate a pregnancy when a fetus has a strictly cosmetic problem or carries what are perceived to be undesirable alleles or is even of the "wrong" sex. Does a health professional have, on the one hand, a responsibility and, on the other hand, the right to decide for a couple when a disorder is not serious enough to warrant prenatal diagnosis and abortion or assisted reproduction?

There is little consensus among geneticists as to *where* or even *whether* one can draw the line in deciding what constitutes a trait serious enough to warrant prenatal testing.

Genetic Testing for Predisposition to Disease

Another area of medical genetics and genomics in which ethical dilemmas frequently arise is genetic testing of asymptomatic individuals for diseases that may have an onset in life later than the age at which the molecular testing is to be performed. The ethical principles of respect for individual autonomy and beneficence are central to testing in this context. At one end of the spectrum is testing for late-onset, highly penetrant neurological disorders, such as Huntington disease (see Chapter 12) (Case 24). For such diseases, individuals carrying a mutant allele may be asymptomatic but will almost certainly develop a devastating illness later in life for which there is currently little or no treatment. For these asymptomatic individuals, is knowledge of the test result more beneficial than harmful, or vice versa? There is no simple answer. Studies demonstrate that some individuals at risk for Huntington disease choose not to undergo testing and would rather not know their risk, whereas others choose to undergo testing. Those who choose testing and test positive have been shown to sometimes have a transient period of depression, but with few suffering severe depression, and many report positive benefits in terms of the knowledge provided to make life decisions about marriage and choice of career. Those who choose testing and are found *not* to carry the trinucleotide expansion allele report positive benefits of relief but can also experience negative emotional responses due to guilt for no longer being at risk for a disease that either affects or threatens to affect many of their close relatives. In any case, the decision to undergo testing is a highly personal one that must be made only after thorough review of the issues with a genetics professional.

The balance for or against testing of unaffected, at-risk individuals shifts when testing indicates a predisposition to a disease for which intervention and early treatment are available. For example, in autosomal dominant hereditary breast cancer, individuals carrying various mutations in *BRCA1* or *BRCA2* have a 50% to 90% chance of developing breast or ovarian cancer (see Chapter 15) (Case 7). Identification of heterozygous carriers would be of benefit because individuals at risk could choose to undergo more frequent surveillance or have preventive surgery, such as mastectomy, oophorectomy, or both, recognizing that these measures can reduce but not completely eliminate the increased risk for cancer. What if surveillance and preventive measures were more definitive, as they are in familial adenomatous polyposis, for which prophylactic colectomy is a proven preventive measure (see Chapter 15) and (Case 15)? Upon testing for any predisposing gene mutation(s), individuals incur the risk for serious psychological distress, stigmatization in their social lives, and discrimination in insurance and employment (see later). How are respect for a patient's autonomy, the physician's duty not to cause harm, and the physician's desire to prevent illness to be balanced in these different situations?

Geneticists would all agree that the decision to be tested or not to be tested is not one made in a vacuum. The patient must make an informed decision using all available information concerning the risk for and severity of the disease, the effectiveness of preventive and therapeutic measures, and the potential harm that could arise from testing.

Genetic Testing of Asymptomatic Children

Additional ethical complexity arises when genetic testing involves minor children (younger than 18 years), particularly children too young to even give assent, because now the basic principles of bioethics need to be considered in the case of both the child and the parents. There are several reasons why parents may wish to have their children tested for a disease predisposition. Testing asymptomatic children for alleles that predispose to disease can be beneficial, even lifesaving, if interventions that decrease morbidity or increase longevity are available. One example is testing the asymptomatic sibling of a child with medium-chain acyl-CoA dehydrogenase deficiency (see Chapter 18) and (Case 31).

However, some have argued that even in situations where there are currently no clear medical interventions that might benefit the child, it is the parents' duty to inform and prepare their children for the future possibility of development of a serious illness. The parents may also seek this information for their own family planning or to avoid what some parents consider the corrosive effects of keeping important information about their children from them. Testing children, however, carries the same risks for serious psychological damage, stigmatization, and certain kinds of insurance discrimination as does testing adults (see later). Children's autonomy—their ability to make decisions for themselves about their own genetic constitution—must also now be balanced with the desire of parents to obtain and use such information.

A different but related issue arises in testing children for the carrier state of a disease that poses no threat to their health but places them at risk for having affected children. Once again, the debate centers on the balance between respect for children's autonomy in regard to their own procreation and the desire on the part of well-meaning parents to educate and prepare children for the difficult decisions and risks that lie ahead once they reach childbearing age.

Most bioethicists believe (and the American College of Medical Genetics and Genomics [ACMG] agrees) that, unless there is a clear benefit to the medical care of the child, genetic testing of asymptomatic children for adult-onset disease or for a carrier state should be done only when the child is sufficiently old and mature, as in late adolescence or on reaching adulthood, to decide for himself or herself whether to seek such testing.

Incidental and Secondary Findings from Whole-Exome and Whole-Genome Sequencing

Another area of controversy has arisen in patients who have given consent for whole-exome or whole-genome sequencing (WES/WGS) to find a genetic basis for their undiagnosed diseases (see Chapters 10 and 18). Laboratories searching the exomes or genomes of such patients usually develop a primary candidate gene list based on the phenotype of the patient. The laboratory considers deleterious mutations in these genes as their *primary findings*, that is, the results that are actively being sought as the primary target of the testing. In the process of analyzing a whole exome or genome, however, deleterious mutations may be discovered *incidentally* in genes known to be associated with diseases unrelated to the phenotype for which the sequencing test was originally conducted (see Chapter 16). If the mutations uncovered as incidental findings cause serious diseases that can be ameliorated or prevented, then is there benefit of drawing up a list of genes that every laboratory doing WES/WGS variants would deliberately analyze in every patient, even though they are not relevant to the primary goal of finding the genetic cause for a patient's unexplained disease? Mutations in this list of genes would be *secondary findings* that would be sought regardless of whether the patient wishes to know these results, because his or her providers deem the benefit of knowing is so compelling for the patient's health that it outweighs the requirement of patient autonomy, to be able to choose what kind of information he or she wants to know.

The ACMG made an initial attempt to draw up a list of secondary findings that a laboratory should seek. The current list includes 56 genes, most of which are involved in serious hereditary cancer and cardiovascular syndromes that are (1) life threatening, (2) not readily diagnosable before the onset of symptoms, and (3) preventable or treatable. The secondary finding gene list is subject to ongoing refinement and will presumably grow over time. Furthermore, whether a given gene mutation should always be a secondary finding that must be sought is also undergoing reevaluation. The current ACMG recommendation is that patients should be provided with appropriate counseling and then given the opportunity to agree or to refuse to have such secondary findings looked for and reported.

Ethical Dilemmas in Newborn Screening

Although newborn screening programs (see Chapter 18) are generally accepted as one of the great triumphs of modern genetics in improving public health, questions about newborn screening still arise. First, should parents be asked to provide active consent or can they simply be offered the opportunity to "opt out" of the program. Second, who has access to samples and data, and how can we make sure that samples, such as DNA, are not used for purposes other than the screening tests for which they were collected and for which consent was given (or at least, not withheld)? In the United States, these questions came to a head in the area of newborn screening in the state of Texas when a group of parents of children sued the state because blood spots obtained through an "opt-out" process for newborn screening

had been diverted to the Department of Defense and private companies and used for purposes other than newborn screening, without parental consent. Texas agreed to destroy their collection of more than 5 million blood spots. In doing so, the state lost samples that could have been used for legitimate purposes, such as developing new newborn screening tests and for quality control of current testing efforts.

PRIVACY OF GENETIC INFORMATION

Legal protections for genetic information are not uniform across the globe or even within different jurisdictions in the same countries. In the United States, the primary set of regulations governing the privacy of health information, including genetic information, is the Privacy Rule of the **Health Insurance Portability and Accountability Act (HIPAA)**. The HIPAA rule sets criminal and civil penalties for disclosing such information without authorization to others, including other providers, except under a defined set of special circumstances. Genetic information, however, receives special attention because it has implications for other family members.

Privacy Issues for Family Members in a Family History

Patients are free to provide their physicians with a complete family medical history or communicate with their physicians about conditions that run in the family. The HIPAA Privacy Rule does not prevent individuals from gathering medical information about their family members or from deciding to share this information with their health care providers.

This information becomes part of the individual's medical record and is treated as "protected health information" about the individual but is *not* protected health information for the family members included in the medical history. In other words, only patients, and not their family members, may exercise their rights under the HIPAA Privacy Rule to their own family history information in the same fashion as any other information in their medical records, including the ability to elect to control disclosure to others.

Duty to Warn and Permission to Warn

A patient's desire to have his or her medical information kept confidential is one facet of the concept of patient autonomy, in which patients have the right to make their own decisions about how their individual medical information is used and communicated to others. Genetics, however, more than any other branch of medical practice, is concerned with both the patient and the family. A serious ethical and legal dilemma can arise in the practice of genetic medicine when a patient's insistence that his or her medical information be kept strictly private restrains the geneticist from letting other family members know about their risk for a condition, even when such information could be beneficial to their own health and the health of their children (see Box). In this situation, is the genetics practitioner obligated to respect the patient's autonomy by keeping information confidential, or is the practitioner permitted or, more forcefully, does the practitioner have a duty to inform other family members and/or their providers? Is there a duty to warn? If so, is informing the patient that he or she should share the information with relatives sufficient to discharge the practitioner's duty?

Judges have ruled in a number of court cases in the United States on whether or not a health care practitioner is permitted or is even required to override a patient's wishes for confidentiality. The precedent-setting case was not one involving genetics. In the 1976 State Supreme Court case in California, *Tarasoff v the Regents of the University of California,* judges ruled that a psychiatrist who failed to warn law enforcement that his client had declared an intention to kill a young woman was liable in her death. The judges declared that this situation is no different from one in which physicians have a duty to protect the contacts of a patient with a

contagious disease by warning them that the patient has the disease, even against the express wishes of the patient. In the realm of genetics, a duty to warn was mandated in a case in New Jersey, *Safer v Estate of Pack* (1996), in which a panel of three judges concluded that a physician had a duty to warn the daughter of a man with familial adenomatous polyposis of her risk for colon cancer. The judges wrote that "there is no essential difference between the type of genetic threat at issue here and the menace of infection, contagion, or a threat of physical harm." They added that the duty to warn relatives is not automatically fulfilled by telling the patient that the disease is hereditary and that relatives need to be informed.

Guidelines from international health organizations, individual national health policy groups, and professional medical organizations are not unanimous on this issue. Furthermore, in the United States, the inconsistent case law from state courts must also be considered with respect to legislative and regulatory mandates, particularly the HIPAA Privacy Rule.

Contrary to widespread belief, the HIPAA Privacy Rule permits a physician to disclose protected health information about a patient to another health care

---

DUTY TO WARN: PATIENT AUTONOMY AND PRIVACY VERSUS PREVENTING HARM TO OTHERS

A woman first presents with an autosomal dominant disorder at the age of 40 years, undergoes testing, and is found to carry a particular mutation in a gene known to be involved in this disorder. She is planning to discuss the results with her teenage daughter but insists that her younger adult half-siblings (from her father's second marriage after her mother's and father's divorce) not be told that they might be at risk for this disorder and that testing is available. How does a practitioner reconcile the obligation to respect the patient's right to privacy with a desire not to cause her relatives harm by failing to inform them of their risk?

There are many questions to answer in determining whether "a serious threat to another person's health or safety" exists to justify unauthorized disclosure of risk to a relative.

Clinical Questions

- What is the penetrance of the disorder, and is it age dependent? How serious is the disorder? Can it be debilitating or life-threatening? How variable is the expressivity? Are there interventions that can reduce the risk for disease or prevent it altogether? Is this a condition that will be identified by routine medical care, once it is symptomatic, in time for institution of preventive or therapeutic measures?
- The risk to half-siblings of the patient is either 50% or negligible, depending on which parent passed the mutant allele to the patient. What does the family history reveal, if anything, about the parent in common between the patient and her half-siblings? Is the patient's mother still alive and available for testing?

Counseling Questions

- Was the patient informed at the time of testing that the results might have implications for other family members? Did she understand in advance that she might be asked to warn her relatives?
- What are the reasons for withholding the information? Are there unresolved issues, such as resentment, feelings of abandonment, or emotional estrangement, that are sources of psychological pain that could be addressed for her own benefit as well as to help the patient clarify her decision making?
- Are the other family members already aware of the possibility of this hereditary disease, and have they made an informed choice not to seek testing themselves? Would the practitioner's warning be seen as an unwarranted intrusion of psychologically damaging information, or would their risk come as a complete surprise?

Legal and Practical Questions

- Does the practitioner have the information and resources required to contact all the half-siblings without the cooperation of the patient?
- Could the practitioner have reached an understanding, or even a formal agreement, with the patient in advance of testing that she would help in informing her siblings? Would asking for such an agreement be seen as coercive and lead to the patient's depriving herself of the testing she needs for herself and her children?
- What constitutes adequate discharge of the practitioner's duty to warn? Is it sufficient to provide a form letter for the patient to show to relatives that discloses the absolute minimal amount of information needed to inform them of a potential risk?

provider who is treating a family member of the physician's patient without the individual's authorization, unless the patient has explicitly chosen to impose additional restrictions on the use or disclosure of his or her protected health information. For example, an individual who has obtained a genetic test may request that the health care provider not disclose the test results. If the health care provider agrees to the restriction, the HIPAA rule prevents disclosing such information without authorization to providers treating other family members who are seeking to identify their own genetic health risks. However, the health care provider should discuss such restrictions with the patient in advance of doing the test and is not obligated to agree to the requested restriction.

Although the genetics practitioner is most knowledgeable about the clinical aspects of the disease, the relevance of the family history, and the family risk assessment, the many legal and ethical controversies surrounding HIPAA and the duty to warn suggest that consultation with legal and bioethics experts is advisable should a conflict arise over the release of a patient's medical information.

Use of Genetic Information by Employers and Insurers

The fourth major ethical principle is justice—the requirement that everyone be able to benefit equally from progress in medical genetics. Justice is a major concern in the area of the use of genetic information in employment and health insurance. Whether healthy individuals could be denied employment or health insurance because they carry a genetic predisposition to disease was not a settled issue in the United States until passage of the landmark **Genetic Information Nondiscrimination Act (GINA)** of 2008. Under this act, private employers with 15 or more employees are prohibited from deliberately seeking or using genetic information, including family history, to make an employment decision because genetic information was not considered to be relevant to an individual's current ability to work. Similarly, GINA prohibits most group health insurers from denying insurance or adjusting group premiums based on the genetic information of members of the group.

Outside of the United States, however, equivalent GINA laws are not in place. For some countries with national health systems and with private health insurance that is not risk-rated, genetic discrimination in health insurance may not be an issue. However, for most other countries (and in the area of employment in all other countries), there is widespread agreement that genetic discrimination should not be permitted, but legislation banning the practice remains to be enacted.

Significantly, GINA does not apply in the area of life, disability, and long-term care insurance. Insurers that sell such products insist that they must have access to all pertinent genetic information about an individual that the individual himself or herself has when making a decision to purchase one of these policies. Life insurance companies calculate their premiums on the basis of actuarial tables of age-specific survival averaged over the population; premiums will not cover losses if individuals with private knowledge that they are at higher risk for disease conceal this information and buy extra life or long-term disability insurance, a practice referred to as **adverse selection**. If adverse selection were widespread, the premiums for the entire population would have to increase so that in essence, the entire population would be subsidizing the increased coverage for a minority. Adverse selection is likely to be a real phenomenon in some circumstances; in one study of asymptomatic individuals tested for the *APOE* ε4 allele, those who chose to know that they tested positive were found to be nearly six times more likely to purchase extra long-term care insurance than those who did not choose to know their *APOE* genotype. Knowledge that one carried an *APOE* ε4 allele did not, however, affect life, health, or disability insurance purchases.

At present, there is little evidence that life insurance companies have actually engaged in discriminatory underwriting practices on the basis of genetic testing. Nevertheless, the fear of such discrimination, and the negative impact that discrimination would have on people obtaining clinical testing for their own health benefit as well as on their willingness to participate in genetic research, has led to proposals to ban the use of genetic information in life insurance. In the United Kingdom, for example, life insurance companies have voluntarily agreed to an extended moratorium on the use of genetic information in most life underwriting, except when large policies are involved or in the case of Huntington disease, for which disclosure of a positive test result by the patient is required.

There must be, however, a clear distinction between what are already phenotypic manifestations of a disease, such as hypertension, hypercholesterolemia, and diabetes mellitus, and what are predisposing alleles, such as *BRCA1* mutations (see Chapter 15) and *APOE* ε4 alleles (see Chapters 8 and 18), that may never result in overt disease in the individual who carries such an allele.

EUGENIC AND DYSGENIC EFFECTS OF MEDICAL GENETICS
The Problem of Eugenics

The term **eugenics**, introduced by Darwin's cousin Francis Galton in 1883, refers to the improvement of a population by selection of only its "best" specimens for breeding. Plant and animal breeders have followed this practice since ancient times. In the late 19th century, Galton and others began to promote the idea of using selective breeding to improve the human species, thereby initiating the so-called eugenics movement, which was widely advocated for the next half-century. The so-called ideal qualities that the eugenics movement sought to promote through the encouragement of certain kinds of human breeding were more often than not defined by social, ethnic, and economic prejudices and fed by anti-immigrant and racist sentiments in society. What we now would consider a lack of education was described then as familial "feeble-mindedness"; what we now would call rural poverty was considered by eugenicists to be hereditary "shiftlessness." The scientific difficulties in determining whether traits or characteristics are heritable and to what extent heredity contributes to a trait were badly overestimated because most human traits, even those with some genetic component, are complex in their inheritance pattern and are influenced strongly by environmental factors. Thus, by the middle of the last century, many scientists began to appreciate the theoretical and ethical difficulties associated with eugenics programs.

Eugenics is commonly thought to have been largely discredited when it was resurrected and used in Nazi Germany as a justification for mass murder. However, it should be pointed out that in North America and Europe, involuntary sterilization of institutionalized individuals deemed to be mentally incompetent or disabled was carried out under laws passed in the early part of the 20th century in support of eugenics and was continued for many years after the Nazi regime was destroyed.

Genetic Counseling and Eugenics

Genetic counseling, with the aim of helping patients and their families manage the pain and suffering caused by genetic disease, should not be confounded with the eugenic goal of reducing the incidence of genetic disease or the frequency of alleles considered deleterious in the population. Helping patients and families come to free and informed decisions, particularly concerning reproduction, without coercion, forms the basis for the concept of **nondirective counseling** (see Chapter 16). Nondirectiveness asserts that individual autonomy is paramount and must not to be subordinated to reducing the burden of genetic disease on society or to a theoretical goal of "improving the gene pool," a totalitarian concept that echoes the Nazi doctrine of racial hygiene. Some, however, have argued that true nondirective counseling is a myth, often acclaimed but not easy to accomplish, because of the personal attitudes and values the counselor brings to the counseling session.

Nonetheless, despite the difficulties in attaining the ideal of nondirective counseling, the ethical principles of respect for autonomy, beneficence, avoidance of maleficence, and justice remain at the heart of all genetic counseling practice, particularly in the realm of individual reproductive decision making.

The Problem of Dysgenics

The opposite of eugenics is **dysgenics,** a deterioration in the health and well-being of a population by practices that allow the accumulation of deleterious alleles. In this regard, the long-term impact of activities in medical genetics that can affect gene frequencies and the incidence of genetic disease may be difficult to predict.

In the case of some single-gene defects, medical treatment can have a dysgenic effect by reducing selection against a particular genotype, thereby allowing the frequency of harmful genes and consequently of disease to increase. The effect of relaxed selection is likely to be more striking for autosomal dominant and X-linked disorders than for autosomal recessive disorders, in which the majority of mutant alleles are in silent heterozygous carriers.

For example, if successful treatment of Duchenne muscular dystrophy were to be achieved, the incidence of the disease would rise sharply because the *DMD* genes of the affected males would then be transmitted to all their daughters. The effect of this transmission would be to greatly increase the frequency of carriers in the population. In contrast, if all persons affected with cystic fibrosis could survive and reproduce at a normal rate, the incidence of the disease would rise from 1 in 2000 to only approximately 1 in 1550 over 200 years. Common genetic disorders with complex inheritance, discussed in Chapter 8, could theoretically also become more common if selection were removed, although it is likely that as with autosomal recessive diseases, most of the many susceptibility alleles are distributed among unaffected individuals. Consequently, reproduction by affected persons would have little effect on susceptibility allele frequencies.

As prenatal diagnosis (see Chapter 17) becomes widespread, increasing numbers of pregnancies in which the fetus has inherited a genetic defect may be terminated. The effect on the overall incidence of disease is quite variable. In a disorder such as Huntington disease, prenatal diagnosis and pregnancy termination would have a large effect on the incidence of the responsible gene. For most other severe X-linked or autosomal dominant disorders, some reduction might occur, but the disease will continue to recur owing to new mutations. In the case of autosomal recessive conditions, the effect on the frequency of the mutant allele, and consequently of the disease, of aborting all homozygous affected pregnancies would be small because most of these alleles are carried silently by heterozygotes.

One theoretical concern is the extent to which pregnancy termination for genetic reasons is followed by **reproductive compensation**—that is, by the birth of additional, unaffected children, many of whom are carriers of the deleterious gene. Some families with X-linked disorders have chosen to terminate pregnancies in which the fetus was male, but of course, daughters in such families, although unaffected, may be carriers. Thus reproductive compensation has the potential long-term consequence of increasing the frequency of the genetic disorder that led to the loss of an affected child.

GENETICS IN MEDICINE

The 20th century will be remembered as the era that began with the rediscovery of Mendel's laws of inheritance and their application to human biology and medicine, continued with the discovery of the role of DNA in heredity, and culminated in the completion of the Human Genome Project. At the beginning of the 21st century, the human species has, for the first time:

- A complete representative sequence of its own DNA
- A comprehensive, albeit likely incomplete, inventory of its genes
- A vigorous ongoing effort to identify and characterize mutations and polymorphic variants in DNA sequence and copy number
- A rapidly expanding knowledge base in which various diseases and disease predispositions will be attributable to such variation
- Powerful new sequencing technologies that allow sequencing of an exome or genome at a tiny fraction of the cost of the first human genome sequence

With such knowledge comes powerful capabilities as well as great responsibilities. Ultimately, **genetics in medicine** is not about knowledge for its own sake, but for the sake of sustaining wellness, improving health, relieving suffering, and enhancing human dignity. The challenge confronting us all, both future health professionals and members of society at large, is to make sure that the advances in human genetics and genomics knowledge and technology are used responsibly, fairly, and humanely.

GENERAL REFERENCES

Beauchamp TL, Childress JF: *Principles of biomedical ethics*, ed 5, New York, 2001, Oxford University Press.
Kevles D: *In the name of eugenics: genetics and the uses of human heredity*, Cambridge, Mass, 1995, Harvard University Press.

REFERENCES FOR SPECIFIC TOPICS

Biesecker LG: Incidental variants are critical for genomics, *Am J Hum Genet* 92:648–651, 2013.
Elger B, Michaud K, Mangin P: When information can save lives: the duty to warn relatives about sudden cardiac death and environmental risks, *Hastings Center Report* 40:39–45, 2010.
HIPAA regulations on family history. http://www.hhs.gov/ocr/privacy/hipaa/faq/family_medical_history_information/index.html.
MacEwen JE, Boyer JT, Sun KY: Evolving approaches to the ethical management of genomic data, *Trends Genet* 29:375–382, 2013.
McGuire AL, Joffe S, Koenig BA, et al: Point-counterpoint. Ethics and genomic incidental findings, *Science* 340:1047–1048, 2013.
Offit K, Thom P: Ethicolegal aspects of cancer genetics, *Cancer Treat Res* 155:1–14, 2010.
Visscher PM, Gibson G: What if we had whole-genome sequence data for millions of individuals? *Genome Med* 5:80, 2013.
Yurkiewicz IR, Korf BR, Lehmann LS: Prenatal whole-genome sequencing—is the quest to know a fetus's future ethical? *N Engl J Med* 370:195–197, 2014.

PROBLEMS

1. A couple with two children is referred for genetic counseling because their younger son, a 12-year-old boy, has a movement disorder for which testing for juvenile Huntington disease **(Case 24)** is being considered. What are the ethical considerations for the family in testing?

2. A research project screened more than 40,000 consecutive, unselected births for the number of X chromosomes and the presence of a Y chromosome and correlated the sex chromosome karyotype with the sex assigned by visual inspection in the newborn nursery. The purpose of the project was to observe infants with sex chromosome abnormalities (see Chapter 6) prospectively for developmental difficulties. What are the ethical considerations in carrying out this project?

3. In the case described in the Box in the section on duty to warn, consider what might be your course of action if you were the genetic counselor and the disease in question were the following: hereditary breast and ovarian cancer due to *BRCA1* mutations (see Chapter 15) **(Case 7)**; malignant hyperthermia due to *RYR1* (ryanodine receptor) mutations (see Chapter 18); early-onset, familial Alzheimer disease due to a *PSEN1* (presenilin 1) mutation (see Chapter 12) **(Case 4)**; neurofibromatosis due to *NF1* mutations (see Chapter 7) **(Case 34)**; or type 2 diabetes mellitus **(Case 35)**.

4. Draw up a list of a dozen genes and disorders that you believe should be analyzed as secondary findings during a whole-exome or whole-genome sequence for undiagnosed diseases. Explain how and why you chose each of these dozen genes and conditions.

Clinical Case Studies Illustrating Genetic Principles

These 48 clinical vignettes illustrate genetic and genomic principles in the practice of medicine. Each vignette is followed by a brief explanation or description of the disease and its etiology, pathophysiology, phenotype, management, and inheritance risk. These explanations and descriptions are based on current knowledge and understanding; therefore, like most things in medicine and science, they are subject to refinement and change as our knowledge and understanding evolve. The description of each case uses standard medical terminology; student readers therefore may need to consult a medical dictionary for explanations. Each vignette is also followed by a few questions that are intended to initiate discussion of some basic genetic or clinical principles illustrated by the case.

The cases are not intended to be definitive or comprehensive or to set a standard of care; rather, they are simply illustrations of the application of genetic and genomic principles to the practice of medicine. Although the cases are loosely based on clinical experience, all individuals and medical details presented are fictitious.

Ada Hamosh, MD, MPH
Roderick R. McInnes, MD, PhD
Robert L. Nussbaum, MD
Huntington F. Willard, PhD
(With the assistance of Emily C. Lisi, MS CGC and Nara Sobreira, MD)

CASE PRESENTATIONS

ABACAVIR-INDUCED STEVENS-JOHNSON SYNDROME/ TOXIC EPIDERMAL NECROLYSIS (Genetically Determined Immunological Adverse Drug Reaction)

Autosomal Dominant

PRINCIPLES

- Pharmacogenetic test that has been widely adopted as standard of care
- Significant positive and negative predictive values
- Ethnic differences in the frequency of the predisposing allele

MAJOR PHENOTYPIC FEATURES

- Widespread red/purple patches on the skin and mucosal membranes (eye, mouth, genitalia) 10 to 14 days after beginning antiretroviral treatment with abacavir.
- Skin sloughing of greater than 30% of body surface area is referred to as toxic epidermal necrolysis; a similar rash but with sloughing of less than 10% of body surface area is referred to as Stevens-Johnson syndrome.

HISTORY AND PHYSICAL FINDINGS

P.R., a 37-year-old German man, was admitted to the hospital in 2001 with shortness of breath and confusion and found to have both *Pneumocystis carinii* pneumonia and *Toxoplasma gondii* encephalitis, opportunistic infections that occur commonly in the setting of newly diagnosed human immunodeficiency virus (HIV)-1 acquired immunodeficiency syndrome (AIDS). His CD4 cell count was 2/mm³ and HIV-1 viral load was 120,000 copies/mL. Treatment with trimethoprim-sulfamethoxazole was started, and he was started on antiretroviral therapy (ART) that included the nucleoside analogue reverse transcriptase inhibitor abacavir. His encephalitis and pneumonia cleared, and he was discharged from the hospital on oral antiparasitic treatment.

Two weeks after beginning ART, P.R. presented with a nonfebrile, generalized macular rash involving his palms and mouth. His blood pressure was 130/60 mm Hg, temperature was 37.1°C, pulse was 88 beats/min, he was breathing 15 breaths/min, and oxygen saturation was 96% on room air. He had a disseminated cutaneous eruption of discrete dark red macules on 90% of the body surface area, a detachment of 5% of the epidermis, genital ulcerations, erosive stomatitis, and conjunctival lesions with hyperemia but without keratitis or corneal erosions. The application of minor pressure to the skin resulting in sloughing of the skin (Nikolsky sign).

Skin biopsy was compatible with Stevens-Johnson syndrome. Because of previous reports of cutaneous hypersensitivity reactions with abacavir treatment, the drug was stopped, and he was transferred to a burn unit, monitored for further skin sloughing, and treated with supportive care. The epidermis began to heal over the next week, and the skin lesions resolved completely within 3 weeks. His ART was changed to a combination of protease inhibitors and different nucleoside analogue reverse transcriptase inhibitors without recurrence of the skin reaction. His viral load decreased to undetectable, and the CD4 count returned to normal.

One year later, when the increased susceptibility to SJS with abacavir therapy was shown to depend on human leukocyte antigen (HLA) genotype, he had HLA typing and was found to carry the SJS-abacavir susceptibility allele HLA-B*5701.

BACKGROUND

Adverse drug reactions are defined as harmful reactions caused by normal use of a drug at correct doses. The majority (75% to 80%) of all adverse drug reactions are caused by predictable, nonimmunological effects, some of which are due to genetically determined pharmacokinetic or pharmacodynamic differences between individuals. The remaining 20% to 25% of adverse drug events are caused by largely unpredictable effects that may or may not be immune-mediated. Immune-mediated reactions account for 5% to 10% of all drug reactions and represent true drug hypersensitivity, with immunoglobulin E–mediated drug allergies with hives or laryngeal swelling, falling into this category. A different kind of skin reaction, a generalized maculopapular rash, is also common with certain medications, including sulfa drug antibiotics.

One particularly dangerous adverse drug reaction is T-cell mediated damage to skin and mucous membranes, referred to as Stevens-Johnson syndrome (SJS), and its more serious extreme manifestation, toxic epidermal necrolysis (TEN) (Fig. C-1). Both SJS and TEN are characterized by malaise and fever, followed by rapid appearance of red/purple patches on the skin, which progress to sloughing of the skin, similar to what is seen with a thermal burn. Mucosal membranes (eye, mouth, genitalia) are frequently affected. In SJS, skin sloughing involves less than 10% of body surface area, whereas TEN involves sloughing of greater than 30% of the body surface area.

Histological features in the skin in drug-induced SJS/TEN patients include epidermal necrosis, in some cases extending through the full thickness of the epidermis as seen with thermal burn, individual keratinocyte apoptosis, subepidermal bullae, and dense dermal infiltrates with lymphocytes, as well as a substantial number of eosinophils or neutrophils.

The mortality rate in SJS/TEN ranges from 10% to 30%. Although SJS and TEN represent only a small fraction of all immune-mediated drug reactions, they are particularly severe and can be life threatening.

Pathogenesis

SJS/TEN is mediated by cytotoxic T cells. Molecular immunological studies have elucidated why T-cell–mediated hypersensitivity occurs in individuals with the HLA-B*5701 allele treated with abacavir. In HLA-B*5701–expressing cells cultured in the presence of abacavir, up to 25% of the peptides present in the antigen-presenting groove of their class I cell-surface antigen-presenting molecules are novel self-peptides that are not seen in the absence of abacavir. Abacavir appears

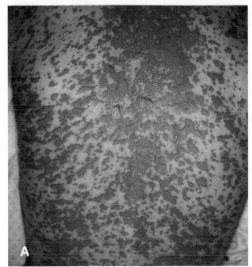

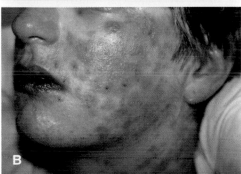

Figure C-1 A, Numerous coalescing dusky lesions with flaccid bullae and multiple sites of epidermal detachment involving 10% to 30% of skin surface. This extent of epidermal detachment is in the "zone of overlap" between Stevens-Johnson syndrome and toxic epidermal necrolysis. **B,** Stevens-Johnson syndrome, showing involvement of lips and mucous membranes of the mouth. *See Sources & Acknowledgments.*

to interact specifically with segments of the HLA-B*5701 peptide-binding groove, altering its binding properties. This alteration allows HLA-B*5701 to present novel peptides that happen to have a much higher cross-reactivity to self, including skin antigens. Drugs precipitate over 50% of SJS cases and up to 95% of TEN cases.

Management

Discontinuation of the offending drug and transfer to a burn unit for supportive care are the mainstays of treatment. Other therapies such as systemic corticosteroids and intravenous immunoglobulin have been suggested but have not to date been proven to be either beneficial or harmful.

Prevention

The 50% positive predictive value for SJS or TEN in HLA-B*5701–positive individuals treated with abacavir and the nearly 100% negative predictive value for SJS or TEN in individuals treated with abacavir who *lack* HLA-B*5701 led the Infectious Diseases Society of America and other international health services to require, as a standard of care, that only individuals who lack the HLA-B*5701 allele be given abacavir therapy. The varying frequency of the allele in different populations, however, and the availability of other nucleoside analogue reverse transcriptase inhibitors that lack the potential to cause SJS/TEN have led to a lively debate as to whether it is cost-effective to carry out HLA-B typing before starting abacavir treatment in everybody, or whether testing should be considered only in individuals from ethnic backgrounds with the highest frequencies of the HLA-B*5701 allele. Nonetheless, the combination of a 50% positive predictive value, a very high negative predictive value, and the life-threatening nature of SJS/TEN make screening a reasonable choice in *all* patients for whom abacavir treatment is being considered, regardless of ethnic background.

INHERITANCE RISK

As with all HLA alleles (see Chapter 8), inheritance is autosomal codominant. Studies of large cohorts of patients treated with abacavir have demonstrated that approximately 50% of patients carrying an HLA-B*5701 allele will develop SJS or TEN, whereas none of the patients without this antigen will develop these conditions.

The frequency of the HLA-B*5701 allele (and therefore the risk for abacavir-induced SJS and TEN) differs greatly among various ethnic groups (see Table).

Population	Frequency of the HLA-B*5701 Allele (%)
White	8-10
African American	2.5
Chinese	0-2
South Indian	5-20
Thai	4-10

Similar associations between SJS or TEN and other HLA alleles have been seen with the antiepileptic drug carbamazepine (HLA-B*1502), the uric acid–lowering drug allopurinol (HLA-B*5801) used for gout, and other commonly used medications.

QUESTIONS FOR SMALL GROUP DISCUSSION

1. Suggest a mechanism by which SJS/TEN might arise in individuals with different HLA-B alleles when exposed to different drugs.
2. Why might there be different frequencies of various HLA-B alleles in different ethnic groups?

REFERENCES

Downey A, Jackson C, Harun N, et al: Toxic epidermal necrolysis: review of pathogenesis and management, *J Am Acad Dermatol* 66:995–1003, 2012.

Mallal S, Phillips E, Carosi G, et al: HLA-B*5701 screening for hypersensitivity to abacavir, *N Engl J Med* 358:568–579, 2008.

Martin MA, Kroetz DL: Abacavir pharmacogenetics—from initial reports to standard of care, *Pharmacotherapy* 33:765–775, 2013.

Mockenhaupt M, Viboud C, Dunant A, et al: Stevens-Johnson syndrome and toxic epidermal necrolysis: assessment of medication risks with emphasis on recently marketed drugs: the EuroSCAR-study, *J Invest Dermatol* 128:35–44, 2008.

ACHONDROPLASIA (*FGFR3* Mutation, MIM 100800)

Autosomal Dominant

PRINCIPLES

- Gain-of-function mutations
- Advanced paternal age
- De novo mutation

MAJOR PHENOTYPIC FEATURES

- Age at onset: Prenatal
- Rhizomelic short stature
- Megalencephaly
- Spinal cord compression

HISTORY AND PHYSICAL FINDINGS

P.S., a 30-year-old healthy woman, was 27 weeks pregnant with her first child. A fetal ultrasound examination at 26 weeks' gestation identified a female fetus with macrocephaly and rhizomelia (shortening of proximal segments of extremities). P.S.'s spouse was 45 years of age and healthy; he had three healthy children from a previous relationship. Neither parent has a family history of skeletal dysplasia, birth defects, or genetic disorders. The obstetrician explained to the parents that their fetus had the features of achondroplasia. The infant girl was delivered at 38 weeks' gestation by cesarean section. She had the physical and radiographic features of achondroplasia, including frontal bossing, megalencephaly, midface hypoplasia, lumbar kyphosis, limited elbow extension, rhizomelia, trident hands, brachydactyly, and hypotonia. Consistent with her physical features, DNA testing identified an 1138G>A mutation leading to a glycine to arginine substitution at codon 380 (Gly380Arg) in the fibroblast growth factor receptor 3 gene (*FGFR3*).

BACKGROUND

Disease Etiology and Incidence

Achondroplasia (MIM 100800), the most common cause of human dwarfism, is an autosomal dominant disorder caused by specific mutations in *FGFR3*; two mutations, 1138G>A (≈98%) and 1138G>C (1% to 2%), account for more than 99% of cases of achondroplasia, and both result in the Gly380Arg substitution. Achondroplasia has an incidence of 1 in 15,000 to 1 in 40,000 live births and affects all ethnic groups.

Pathogenesis

FGFR3 is a transmembrane tyrosine kinase receptor that binds fibroblast growth factors. Binding of fibroblast growth factors to the extracellular domain of FGFR3 activates the intracellular tyrosine kinase domain of the receptor and initiates a signaling cascade. In endochondral bone, FGFR3 activation inhibits proliferation of chondrocytes within the growth plate and thus helps coordinate the growth and differentiation of chondrocytes with the growth and differentiation of bone progenitor cells.

The *FGFR3* mutations associated with achondroplasia are gain-of-function mutations that cause ligand-independent activation of FGFR3. Such constitutive activation of FGFR3 inappropriately inhibits chondrocyte proliferation within the growth plate and consequently leads to shortening of the long bones as well as to abnormal differentiation of other bones.

Guanine at position 1138 in the *FGFR3* gene is one of the most mutable nucleotides identified in *any* human gene. Mutation of this nucleotide accounts for nearly 100% of achondroplasia; more than 80% of patients have a de novo mutation. Such de novo mutations occur exclusively in the father's germline and increase in frequency with advanced paternal age (>35 years) (see Chapter 7).

Phenotype and Natural History

Patients with achondroplasia present at birth with rhizomelic shortening of the arms and legs, relatively long and narrow trunk, trident configuration of the hands, and macrocephaly with midface hypoplasia and prominent forehead. They have a birth length that is usually slightly less than normal, although occasionally within the low-normal range; their length or height falls progressively farther from the normal range as they grow.

In general, patients have normal intelligence, although most have delayed motor development. Their delayed motor development arises from a combination of hypotonia, hyperextensible joints (although the elbows have limited extension and rotation), mechanical difficulty balancing their large heads, and, less commonly, foramen magnum stenosis with brainstem compression.

Abnormal growth of the skull and facial bones results in midface hypoplasia, a small cranial base, and small cranial foramina. The midface hypoplasia causes dental crowding, obstructive apnea, and otitis media. Narrowing of the jugular foramina is believed to increase intracranial venous pressure and thereby to cause hydrocephalus. Narrowing of the foramen magnum causes compression of the brainstem at the craniocervical junction in approximately 10% of patients and results in an increased frequency of hypotonia, quadriparesis, failure to thrive, central apnea, and sudden death. Between 3% and 7% of patients die unexpectedly during their first year of life because of brainstem compression (central apnea) or obstructive apnea. Other medical complications include obesity, hypertension, lumbar spinal stenosis that worsens with age, and genu varum.

Management

Suspected on the basis of clinical features, the diagnosis of achondroplasia is usually confirmed by radiographic findings. DNA testing for *FGFR3* mutations can be helpful in ambiguous cases but is usually not necessary for the diagnosis to be made.

Throughout life, management should focus on the anticipation and treatment of the complications of achondroplasia. During infancy and early childhood, patients must be monitored for chronic otitis media, hydrocephalus, brainstem compression, and obstructive apnea and treated as necessary. Treatment of patients with brainstem compression by decompression of the craniocervical junction usually results in marked improvement of neurological function. During later childhood and through early adulthood, patients must be monitored for symptomatic spinal stenosis, symptomatic genu varum, obesity, hypertension, dental complications, and chronic otitis media and treated as necessary. Treatment of the

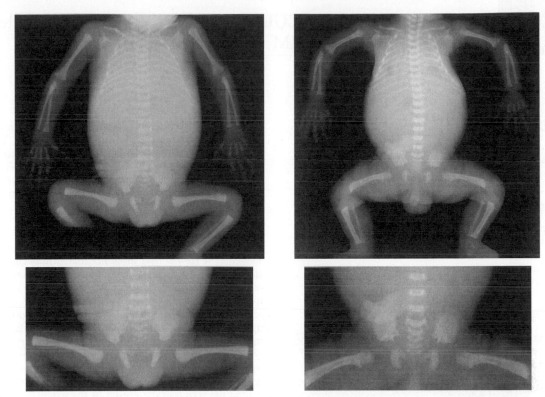

Figure C-2 Radiographs of a normal 34-week fetus (*left*) and a 34-week fetus with achondroplasia (*right*). Comparison of the upper frames shows rhizomelia and trident positioning of the fingers in the fetus with achondroplasia. Comparison of the lower frames illustrates the caudal narrowing of the interpedicular distance in the fetus with achondroplasia versus the interpedicular widening in the normal fetus. Also, the fetus with achondroplasia has small iliac wings shaped like an elephant's ear and narrowing of the sacrosciatic notch. *See Sources & Acknowledgments.*

spinal stenosis usually requires surgical decompression and stabilization of the spine. Obesity is difficult to prevent and control and often complicates the management of obstructive apnea and joint and spine problems.

Patients should avoid activities in which there is risk for injury to the craniocervical junction, such as collision sports, use of a trampoline, diving from diving boards, vaulting in gymnastics, and hanging upside down from the knees or feet on playground equipment.

Both growth hormone therapy and surgical lengthening of the lower legs have been promoted for treatment of the short stature. Both therapies remain controversial.

In addition to management of their medical problems, patients often need help with social adjustment both because of the psychological impact of their appearance and short stature and because of their physical handicaps. Support groups often assist by providing interaction with similarly affected peers and social awareness programs.

INHERITANCE RISK

For unaffected parents with a child affected with achondroplasia, the risk for recurrence in their future children is low but probably higher than for the general population because mosaicism involving the germline, although extremely rare in achondroplasia, has been documented. For relationships in which one partner is affected with achondroplasia, the risk for recurrence in each child is 50% because achondroplasia is an autosomal dominant disorder with full penetrance. For relationships in which both partners are affected, each child has a 50% risk for having achondroplasia, a 25% risk for having lethal homozygous achondroplasia, and a 25% chance of being of average stature. Cesarean section is required for a pregnancy in which an unaffected baby is carried by a mother with achondroplasia.

Prenatal diagnosis before 20 weeks of gestation is available only by molecular testing of fetal DNA, although the diagnosis can be made late in pregnancy by analysis of a fetal skeletal radiograph (Fig. C-2). The features of achondroplasia cannot be detected by prenatal ultrasonography before 24 weeks' gestation, whereas the more severe thanatophoric dysplasia type 2 (homozygous achondroplasia) can be detected earlier.

QUESTIONS FOR SMALL GROUP DISCUSSION

1. Name other disorders that increase in frequency with increasing paternal age. What types of mutations are associated with these disorders?
2. Discuss possible reasons that the *FGFR3* mutations 1138G>A and 1138G>C arise exclusively during spermatogenesis.
3. Marfan syndrome, Huntington disease, and achondroplasia arise as a result of dominant gain-of-function mutations. Compare and contrast the pathological mechanisms of these gain-of-function mutations.
4. In addition to achondroplasia, gain-of-function mutations in *FGFR3* are associated with hypochondroplasia and thanatophoric dysplasia. Explain how phenotypic severity of these three disorders correlates with the level of constitutive FGFR3 tyrosine kinase activity.

REFERENCES

Pauli RM: Achondroplasia. Available from: http://www.ncbi.nlm.nih.gov/books/NBK1152/.

Wright MJ, Irving MD: Clinical management of achondroplasia, *Arch Dis Child* 97:129–134, 2012.

AGE-RELATED MACULAR DEGENERATION (Complement Factor H Variants, MIM 603075)

Multifactorial

PRINCIPLES

- Complex inheritance
- Predisposing and resistance alleles, at several loci
- Gene-environment (smoking) interaction

MAJOR PHENOTYPIC FEATURES

- Age at onset: >50 years
- Gradual loss of central vision
- Drusen in the macula
- Changes in the retinal pigment epithelium
- Neovascularization (in "wet" form)

HISTORY AND PHYSICAL EXAMINATION

C.D., a 57-year-old woman, presents to her ophthalmologist for routine eye examination. She has not been evaluated in 5 years. She reports no change in visual acuity but has noticed that it takes her longer to adapt to changes in light level. Her mother was blind from age-related macular degeneration by her 70s. C.D. smokes a pack of cigarettes per day. On retinal examination, she has many drusen, yellow deposits found beneath the retinal pigment epithelium. A few are large and soft. She is told that she has early features of age-related macular degeneration, causing a loss of central vision that may progress to complete blindness over time. Although there is no specific treatment for this disorder, smoking cessation and oral administration of antioxidants (vitamins C and E and beta-carotene) and zinc are recommended as steps she can take to slow the progression of disease.

BACKGROUND

Disease Etiology and Incidence

Age-related macular degeneration (AMD, MIM 603075) is a progressive degenerative disease of the macula, the region of the retina responsible for central vision, which is critical for fine vision (e.g., reading). It is one of the most common forms of blindness in older adults. Early signs occur in 30% of all individuals older than 75 years; approximately one quarter of these individuals have severe disease with significant visual loss. AMD is rarely found in individuals younger than 55 years. Approximately 50% of the population-attributable genetic risk is due to a polymorphic variant, Tyr402His, in the complement factor H gene (CFH). In contrast, polymorphic variants in two other genes in the alternative complement pathway, factor B (CFB) and complement component 2 (C2), significantly reduce the risk for AMD (see Chapter 10).

In addition to the polymorphisms in the three complement factor genes, mutations at other loci have been implicated in a small percentage of patients with AMD, and they are classified as ARMD1 to ARMD12, depending on the susceptibility gene. In 7 of 402 patients with AMD, different heterozygous missense mutations were identified in the FBLN5 gene encoding fibulin 5, a component of the extracellular matrix involved in the assembly of elastin fibers. All patients had small circular drusen and retinal detachments. AMD was also seen among relatives of patients with Stargardt disease, an early-onset recessive form of macular degeneration seen in individuals homozygous for mutations in the ABCA4 gene. The affected relatives were heterozygous for ABCA4 mutations. Other ARMD genes include FBLN6, ERCC6, RAXL1, HTRA1, ARMS2, C3, TLR4, CST3, and CX3CR1. Mutations at each of these loci account for only a small proportion of the large number of individuals with AMD.

Pathogenesis

The pathobiology of AMD is characterized by inflammation. The current view is that inflammatory insults characteristic of aging have a greater impact in the retina of individuals predisposed to AMD because of reduced activity of the alternative complement pathway in limiting the inflammatory response. The inflammation damages the photoreceptors of the macula, causing retinal atrophy. AMD is further divided into "dry" (atrophic) and "wet" (neovascular or exudative) types. Early AMD is usually dry. Dry AMD is characterized by large soft drusen, the clinical and pathological hallmark of AMD. Drusen are localized deposits of extracellular material behind the retina in the region of the macula. Although small "hard" drusen, which are small granular deposits commonly found in normal retinas, are not associated with macular degeneration, large soft drusen are strongly linked with AMD and are harbingers of retinal damage. As AMD progresses, there is thinning and loss of retinal tissue in focal or patchy areas. In approximately 10% of patients, retinal pigment epithelium remodeling occurs at the site of large, soft drusen. There is invasion of the subretinal space by new blood vessels (neovascularization) that grow in from the choroid. These vessels are fragile, break, and bleed in the retina, resulting in wet AMD.

Drusen contain complement factors, including complement factor H (CFH). Given that CFH is a negative regulator of the alternative complement cascade and that the Tyr402His variant is less capable of inhibiting complement activation, Tyr402His appears to be a functional variant that predisposes to AMD. Importantly, the CFH variants confer increased risk for both the wet and dry forms, suggesting that these two manifestations of the disease have a common basis.

The Leu9His and Arg32Gln variants in factor B and the Glu318Asp and intron 10 variants of complement component 2 reduce the risk for AMD substantially (odds ratios of 0.45 and 0.36, respectively). The mechanism by which the variants in the factor B and complement component 2 genes *decrease* the risk for AMD is not yet known but is also likely to occur through their effect on complement activation.

Although it is clear that environmental factors contribute to AMD, the only nongenetic risk factor identified to date is smoking. Interestingly, smoking significantly decreases serum levels of CFH. The reason for the epidemic of AMD in developed countries is unknown.

Phenotype and Natural History

AMD leads to changes in the central retina that are readily apparent by ophthalmoscopy (Fig. C-3). Patients complain of loss of central vision, making reading and driving difficult or impossible. Visual loss is generally slowly progressive in dry AMD. In contrast, the bleeding from neovascularization

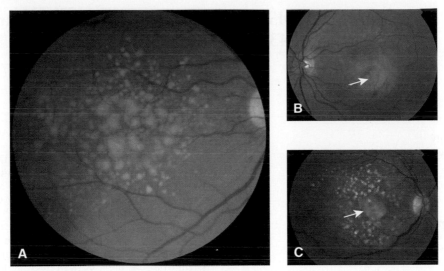

Figure C-3 A, Funduscopic image of numerous large, soft drusen in and around the region of the fovea (dryage-related macular degeneration). **B,** Neovascularization and scarring in the region of the fovea (*arrow*). **C,** Area of thinning and loss of retinal tissue at the fovea ("geographical atrophy"; *arrow*), which tends to protect against neovascularization. *See Sources & Acknowledgments.*

can lead to retinal detachment or bleeding under the retina and cause rapid vision loss. Peripheral vision is usually preserved.

Management

There is no specific treatment for the dry type of AMD. Smoking cessation is strongly indicated. Large clinical trials have suggested that for individuals with extensive intermediate-sized drusen or one large drusen, the use of antioxidants (vitamins A and E, beta-carotene) and zinc may slow progression of disease. Beta-carotene should probably not be used by smokers because some studies suggest it increases the risk for lung cancer and coronary heart disease.

For wet-type AMD, thermal laser photocoagulation, photodynamic therapy, and intravitreous injection of a vascular endothelial growth factor inhibitor (pegaptanib) may slow the rate of visual loss.

INHERITANCE RISK

The role of both genetic and environmental influences is demonstrated by twin studies showing concordance in monozygotic twins of 37%, far below the 100% expected for a purely genetic trait but still significantly greater than the 19% concordance in dizygotic twins, indicating there is a prominent genetic contribution to the disorder. First-degree relatives of patients are at a 4.2-fold greater risk for disease compared with the general population. Thus AMD falls into the category of a genetically complex disease. Despite ample evidence for familial aggregation in AMD, most affected individuals are

not in families in which there is a clear mendelian pattern of inheritance.

QUESTIONS FOR SMALL GROUP DISCUSSION

1. How could mutations in a complement factor account for a disease limited to the eye?
2. Suggest other types of proteins that could be implicated in AMD.
3. Discuss possible reasons that *ABCR* mutations account for such a small proportion of AMD if they are the main cause of Stargardt disease.
4. How would antibodies against vascular endothelial growth factor help in wet-type AMD? Suggest other diseases for which this treatment might be effective alone or in conjunction with other therapies.

REFERENCES

Arroyo JG: Age-related macular degeneration. Available at: http://uptodate.com.

Fritsche IG, Fariss RN, Stambolian D, et al: Age-related macular degeneration: genetics and biology coming together, *Ann Rev Genomics Hum Genet* 15:5.1–5.21, 2014.

Holz FG, Schmitz-Valkenberg S, Heckenstein M: Recent developments in the treatment of age-related macular degeneration, *J Clin Invest* 124:1430–1438, 2014.

Kourlas H, Schiller DS: Pegaptanib sodium for the treatment of neovascular age-related macular degeneration: a review, *Clin Ther* 28:36–44, 2006.

Ratnapriya R, Chew EY: Age-related degeneration—clinical review and genetics update, *Clin Genet* 84:160–166, 2013.

ALZHEIMER DISEASE (Cerebral Neuronal Dysfunction and Death, MIM 104300)

Multifactorial or Autosomal Dominant

PRINCIPLES

- Variable expressivity
- Genetic heterogeneity
- Gene dosage
- Toxic gain of function
- Risk modifier

MAJOR PHENOTYPIC FEATURES

- Age at onset: Middle to late adulthood
- Dementia
- β-Amyloid plaques
- Neurofibrillary tangles
- Amyloid angiopathy

HISTORY AND PHYSICAL FINDINGS

L.W. was an older woman with dementia. Eight years before her death, she and her family noticed a deficit in her short-term memory. Initially they ascribed this to the forgetfulness of "old age"; her cognitive decline continued, however, and progressively interfered with her ability to drive, shop, and look after herself. L.W. did not have findings suggestive of thyroid disease, vitamin deficiency, brain tumor, drug intoxication, chronic infection, depression, or strokes; magnetic resonance imaging of her brain showed diffuse cortical atrophy. L.W.'s brother, father, and two other paternal relatives had died of dementia in their 70s. A neurologist explained to L.W. and her family that normal aging is not associated with dramatic declines in memory or judgment and that declining cognition with behavioral disturbance and impaired daily functioning suggested a clinical diagnosis of familial dementia, possibly Alzheimer disease. The suspicion of Alzheimer disease was supported by her apolipoprotein E genotype: APOE ε4/ε4. L.W.'s condition deteriorated rapidly during the next year, and she died in hospice at 82 years of age. Her autopsy confirmed the diagnosis of Alzheimer disease.

BACKGROUND

Disease Etiology and Incidence

Approximately 10% of persons older than 70 years have dementia, and approximately half of them have Alzheimer disease (AD, MIM 104300). AD is a panethnic, genetically heterogeneous disease; less than 5% of patients have early-onset familial disease, 15% to 25% have late-onset familial disease, and 75% have sporadic disease. Approximately 10% of familial AD exhibits autosomal dominant inheritance; the remainder exhibits multifactorial inheritance.

Current evidence suggests that defects of β-amyloid precursor protein metabolism cause the neuronal dysfunction and death observed with AD. Consistent with this hypothesis, mutations associated with early-onset autosomal dominant AD have been identified in the β-amyloid precursor protein gene (APP), the presenilin 1 gene (PSEN1), and the presenilin 2 gene (PSEN2) (see Chapters 8 and 12). The prevalence of mutations in these genes varies widely, depending on the inclusion criteria of the study; 20% to 70% of patients with

early-onset autosomal dominant AD have mutations in PSEN1, 1% to 2% have mutations in APP, and less than 5% have mutations in PSEN2.

No mendelian causes of late-onset AD have been identified; however, both familial AD and sporadic late-onset AD are strongly associated with allele ε4 at the apolipoprotein E gene (APOE; see Chapter 8). The frequency of ε4 is 12% to 15% in normal controls compared with 35% in all patients with AD and 45% in patients with a family history of dementia.

There is evidence for at least a dozen additional AD loci in the genome. Evidence also suggests that mitochondrial DNA polymorphisms may be risk factors in Alzheimer disease. Finally, there have been associations between AD and various polymorphisms in many other genes.

Pathogenesis

As discussed in Chapter 12, β-amyloid precursor protein (βAPP) undergoes endoproteolytic cleavage to produce peptides with neurotrophic and neuroprotective activities. Cleavage of βAPP within the endosomal-lysosomal compartment produces a carboxyl-terminal peptide of 40 amino acids ($A\beta_{40}$); the function of $A\beta_{40}$ is unknown. In contrast, cleavage of APP within the endoplasmic reticulum or cis-Golgi produces a carboxyl-terminal peptide of 42 or 43 amino acids ($A\beta_{42/43}$). $A\beta_{42/43}$ readily aggregates and is neurotoxic in vitro and possibly in vivo. Patients with AD have a significant increase in $A\beta_{42/43}$ aggregates within their brains. Mutations in APP, PSEN1, and PSEN2 increase the relative or absolute production of $A\beta_{42/43}$. Approximately 1% of all cases of AD occur in patients with Down syndrome, who overexpress βAPP (because the gene for βAPP is on chromosome 21) and thus $A\beta_{42/43}$. The role of APOE ε4 is clear, but the mechanism is uncertain.

AD is a central neurodegenerative disorder, especially of cholinergic neurons of the hippocampus, neocortical association area, and other limbic structures. Neuropathological changes include cortical atrophy, extracellular neuritic plaques, intraneuronal neurofibrillary tangles (Fig. C-4), and amyloid deposits in the walls of cerebral arteries. The neuritic plaques (see Fig. C-4) contain many different proteins, including $A\beta_{42/43}$ and apolipoprotein E. The neurofibrillary tangles are composed predominantly of hyperphosphorylated tau protein; tau helps maintain neuronal integrity, axonal transport, and axonal polarity by promoting the assembly and stability of microtubules.

Phenotype and Natural History

AD is characterized by a progressive loss of cognitive function, including recent memory, abstract reasoning, concentration, language, visual perception, and visual-spatial function. Beginning with a subtle failure of memory, AD is often attributed initially to benign "forgetfulness." Some patients perceive their cognitive decline and become frustrated and anxious, whereas others are unaware. Eventually patients are unable to work, and they require supervision. Social etiquette and superficial conversation are often retained surprisingly well. Ultimately, most patients develop rigidity, mutism, and incontinence and are bedridden. Other symptoms associated with AD include agitation, social withdrawal, hallucinations, seizures,

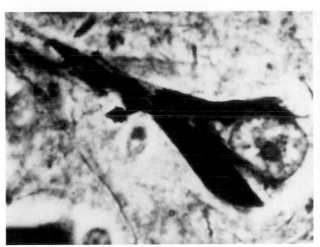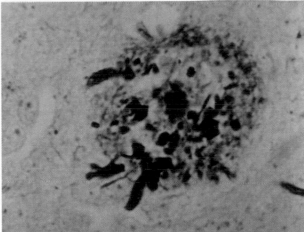

Figure C-4 A neurofibrillary tangle *(left)* and a neuritic plaque *(right)* observed on histopathological examination of the brain of an individual with Alzheimer disease. *See Sources & Acknowledgments.*

myoclonus, and parkinsonian features. Death usually results from malnutrition, infection, or heart disease.

Aside from the age at onset, early-onset AD and late-onset AD are clinically indistinguishable. Mutations in *PSEN1* are fully penetrant and usually cause rapidly progressive disease, with a mean onset at 45 years. Mutations in *APP* are fully penetrant and cause a rate of AD progression similar to that of late-onset AD; the age at onset ranges from the 40s to early 60s. Mutations in *PSEN2* may not be fully penetrant and usually cause slowly progressive disease with onset ranging from 40 to 75 years. In contrast to early-onset AD, late-onset AD develops after 60 to 65 years of age; the duration of disease is usually 8 to 10 years, although the range is 2 to 25 years. For both late-onset AD and AD secondary to *APP* mutations, the *APOE* allele ε4 is a dose-dependent modifier of onset; that is, the age at onset varies inversely with the number of copies of the ε4 allele (see Chapter 8).

Management

Except for patients in families segregating an AD-associated mutation, patients with dementia can be definitively diagnosed with AD only by autopsy; however, with rigorous adherence to diagnostic criteria, a clinical suspicion of AD is confirmed by neuropathological examination 80% to 90% of the time. The accuracy of the clinical suspicion increases to 97% if the patient is homozygous for the *APOE* ε4 allele.

Because no curative therapies are available for AD, treatment is focused on the amelioration of associated behavioral and neurological problems. Approximately 10% to 20% of patients have a modest decrease in the rate of cognitive decline if they are treated early in the disease course with agents that increase cholinergic activity.

INHERITANCE RISK

Old age, family history, female sex, and Down syndrome are the most important risk factors for AD. In Western populations, the empirical lifetime risk for AD is 5%. If patients have a first-degree relative in whom AD developed after 65 years, they have a threefold to sixfold increase in their risk for AD. If patients have a sibling in whom AD developed before 70

years and an affected parent, their risk is increased sevenfold to ninefold. *APOE* testing may be used as an adjunct diagnostic test in individuals seeking evaluation for signs and symptoms suggestive of dementia but is generally not used for predictive testing for AD in asymptomatic patients (see Chapter 18).

Patients with Down syndrome have an increased risk for AD. After the age of 40 years, nearly all patients with Down syndrome have neuropathological findings of AD, and approximately 50% manifest cognitive decline.

For families segregating autosomal dominant AD, each person has a 50% risk for inheriting an AD-causing mutation. With the exception of some *PSEN2* mutations, full penetrance and relatively consistent age at onset within a family facilitate genetic counseling. Currently, clinical DNA testing is available for *APP*, *PSEN1*, and *PSEN2*, as well as several other genes; DNA testing should be offered only in the context of genetic counseling.

QUESTIONS FOR SMALL GROUP DISCUSSION

1. Why is the *APOE* genotype not useful for predicting AD in asymptomatic individuals?
2. Why is AD usually a neuropathological diagnosis? What is the differential diagnosis for AD?
3. Mutation of *MAPT*, the gene encoding tau protein, causes frontotemporal dementia; however, *MAPT* mutations have not been detected in AD. Compare and contrast the proposed mechanisms by which abnormalities of tau cause dementia in AD and frontotemporal dementia.
4. Approximately 30% to 50% of the population risk for AD is attributed to genetic factors. What environmental factors are proposed for the remaining risk? What are the difficulties with conclusively identifying environmental factors as risks?

REFERENCES

Bird TD: Alzheimer disease overview. Available from: http://www.ncbi.nlm.nih.gov/books/NBK1161/.

Karch CM, Cruchaga C, Goate AM: Alzheimer's disease genetics: from the bench to the clinic, *Neuron* 83:11–26, 2014.

AUTISM/16p11.2 DELETION SYNDROME (Susceptibility to Autism Spectrum Disorders, MIM 611913)

Autosomal Dominant or De Novo

PRINCIPLES

- New technology adding to diagnostic yield
- Copy number variant (benign and pathogenic)
- Variant of uncertain significance
- Gene dosage effect
- Susceptibility loci
- Incomplete penetrance

MAJOR PHENOTYPIC FEATURES

- Age at onset: Birth or first 6 months of life
- Intellectual disability to normal intelligence
- Impaired social and communication skills or frank autism spectrum disorder
- Minor dysmorphic features

HISTORY AND PHYSICAL FINDINGS

M.L., a 3-year-old boy, was referred to a medical genetics clinic to identify the cause of his speech delay. Pregnancy and birth were uneventful. He walked around 14 months of age, and spoke his first words at 30 months. At 3 years of age, he had five words. His parents felt that he understood more than he could communicate, although his receptive language was also delayed. M.L. had no medical concerns, and his family history was noncontributory. A physical examination revealed minor dysmorphic features, including simple, low-set ears, a single transverse palmar crease on the left hand, and bilateral 2/3/4 toe syndactyly. His parents described him as a "loner"; he preferred to play alone rather than with his siblings or peers. Concerning behaviors included becoming very agitated with loud noises or irritating textures such as his shirt tag and throwing tantrums when his routine was changed. He was interested only in cars but preferred to play with their wheels or place them in groups rather than racing them. In the meantime, the geneticist ordered a chromosome microarray and fragile X DNA studies, due to his developmental delay with autistic features and mild dysmorphic features. The fragile X DNA test was normal. However, the single nucleotide polymorphism array revealed two copy number variants: a 550-kb deletion at 16p11.2 (thought to be pathogenic) and a 526-kb duplication at 21q22.12 (a variant of uncertain significance). Parental studies showed M.L.'s mother had the 21q duplication, but the 16p11.2 deletion was de novo. The family was counseled that the 16p11.2 deletion was likely the cause of M.L.'s autistic features and delays, and the 21q22.12 duplication was likely a benign variant.

BACKGROUND

Disease Etiology and Incidence

16p11.2 microdeletion syndrome (MIM 611913) is an autosomal dominant condition caused by an approximately 550-kb contiguous gene deletion on chromosome 16p11.2 (Fig. C-5). This recurring microdeletion contains 25 annotated genes. As a newly described condition, the prevalence of 16p11.2 microdeletion syndrome is still being determined. About 1% of individuals tested by array comparative genome hybridization (CGH) for autism spectrum disorder (ASD) have the common 16p11.2 microdeletion, and 0.1% of people tested for developmental delay or a psychiatric condition carry it while only 0.03% of people in the general population carry the microdeletion. Most microdeletions at 16p11.2 are de novo, but some are inherited from symptomatic parents or from healthy, cognitively normal parents. Therefore incomplete penetrance is evident in this condition.

Pathogenesis

16p11.2 microdeletion is one of many microdeletion/microduplications that recur due to low-copy repeat sequences (LCRs) with high sequence homology flanking the deleted or duplicated DNA (see Chapter 6). During replication, the DNA misaligns on these LCRs, causing nonallelic homologous recombination (NAHR) and consequent deletion or duplication of the DNA between the LCRs. It is unclear which of the 25 known genes in the interval leads to ASD and other phenotypic manifestations of the condition. Sequencing of many of these genes in individuals with autism has revealed mutations in several genes, but further studies are needed to validate these results.

Phenotype and Natural History

16p11.2 microdeletion syndrome is characterized by susceptibility to developmental delay/intellectual disability and/or ASD. Typically the delays present in children with 16p11.2 microdeletion are more pronounced in speech/language skills and socialization rather than motor functioning. Expressive language is usually more affected than receptive language. Features of ASD occur more frequently in this population than the general population, but the percentage of affected individuals who have a diagnosis of ASD is controversial and is certainly not 100%. Individuals with 16p11.2 microdeletion are more likely to be overweight or obese, particularly in adolescence and adulthood, perhaps due to haploinsufficiency of *SH2B1* and/or other genes. Seizures are somewhat more common in this population than the general population. Some individuals with this deletion have been found to have aortic valve abnormalities; a majority of individuals do not have heart malformations. Minor dysmorphic features may be present, but no specific features are characteristic of this disorder. Cognitively normal parents of children with 16p11.2 microdeletion syndrome have, however, been found to have the same microdeletion present in the child; thus intellectual disability and ASD features are not universal in this condition.

The reciprocal 16p11.2 microduplication carries a 14.5-fold increased risk for schizophrenia over the general population. This duplication has also been found in individuals with developmental delay/intellectual disability, ASD, and bipolar disorder. However, the 16p11.2 microduplication has been found in healthy controls and is more likely to be inherited

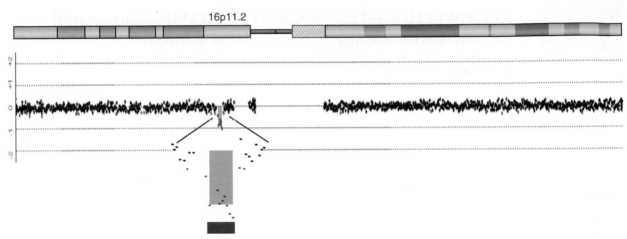

Figure C-5 **Chromosomal microarray analysis of a 16p11.2 deletion in a patient with autism spectrum disorder.** Chromosome 16 ideogram with probe coverage (*dots*) along the length of the chromosome. The log₂ ratio scale is shown on the left; probes with a normal ratio are shown in *black*, whereas probes with a ratio suggestive of either a loss or gain are shown in *green* and *red*, respectively. The deleted region is highlighted (*pink*) in the expanded region of the figure below. The *red bar* corresponds to the deleted region (≈600 kb), which is flanked by paired segmental duplications that mediate the deletion. *See Sources & Acknowledgments.*

from a healthy parent than the microdeletion. Thus the duplication probably increases susceptibility to delays or psychiatric disorders with low penetrance.

Array CGH is a powerful tool that has identified the etiology of developmental delay/intellectual disability, developmental disorders such as ASD, and/or multiple congenital anomalies in up to 20% of individuals tested. In general, the technology has changed the way that medical geneticists practice (see Chapters 5 and 6). However, uncertainty regarding results is an ever-present dilemma; variants of uncertain significance (VUSs; see Chapter 16) abound. Several recommendations have arisen to help determine the pathogenicity of results. The size and dosage effect of the CNV is important; loss of genomic material and large variations are more detrimental than gains and small changes, in general. However, small CNVs in a gene-rich area can cause phenotypic manifestations, whereas large CNVs in a gene-poor region may not. Parents of a child with a VUS should have array or FISH testing to determine if a CNV is inherited or de novo; an inherited VUS from a phenotypically normal parent is historically considered less likely to be pathogenic. However, as with 16p11.2 microdeletion and microduplication syndromes, incomplete penetrance can exist with many CNVs; therefore an inherited VUS cannot be ruled benign based only on this information.

Because of the potential for ambiguous results, providing genetic counseling to a family regarding the possible implications of testing both before and after array CGH testing is beneficial.

Management

Because of the higher prevalence of developmental delay/intellectual disability and ASD features in individuals with 16p11.2 microdeletion, referral to a developmental pediatrician or clinical psychologist is recommended for developmental assessment and placement in appropriate early intervention services, such as physical, occupational, and speech therapies. Social, behavioral, and educational interventions are also available for children with ASDs. An echocardiogram and/or electrocardiogram should be considered to look for aortic valve or other structural heart anomalies, and referral to a pediatric neurologist should be made if there is suspicion of seizure activity. Weight management and nutritional support should be provided because of the increased risk for obesity.

INHERITANCE RISK

16p11.2 deletion is usually de novo but can be inherited from a parent. When de novo, the recurrence risk for the parents is less than 5%, taking into account the risk for gonadal mosaicism. If one parent also carries the deletion, recurrence risk for the deletion is 50% for each subsequent pregnancy. Therefore, in order to provide appropriate genetic counseling, it is crucial to perform parental studies when a 16p11.2 abnormality is diagnosed in a child. However, due to incomplete penetrance, a child who inherits the deletion may not be affected with the same features as his or her sibling and may exhibit normal intelligence and behavior. Alternatively, an affected child may have more significant intellectual disability, autistic features, and/or health concerns.

QUESTIONS FOR SMALL GROUP DISCUSSION

1. Name other recurring microdeletion/microduplication syndromes caused by LCRs. What might be the impact of array CGH in detecting new recurrent syndromes?

2. In performing array CGH testing and whole-exome sequencing, what are some results that may give arise to ethical dilemmas? How would you counsel patients with these types of results, before and after the testing is ordered?

3. Deletions of a particular genomic region are typically more severe than duplications of the same region. In what situations would a duplication create a greater health risk than a deletion?

4. Why was a karyotype not ordered for this patient? Is there ever an indication for a karyotype? If so, what is it/are they?

REFERENCES

McCarthy S, Makarov V, Kirov G, et al: Microduplications of 16p11.2 are associated with schizophrenia, *Nat Genet* 41:1223–1227, 2009.

Miller DT, Nasir R, Sobeih MM, et al: 16p11.2 Microdeletion. Available from: http://www.ncbi.nlm.nih.gov/books/NBK11167/.

Simons VIP Consortium: Simons Variation in Individuals Project (Simons VIP): a genetics-first approach to studying autism spectrum and related neurodevelopmental disorders, *Neuron* 73:1063–1067, 2012.

Unique, the Rare Chromosomal Disorder Support Group. Available from: http://www.rarechromo.org.

Weiss LA, Shen Y, Korn JM, et al: Association between microdeletion and microduplication at 16p11.2 and autism, *N Engl J Med* 358:667–675, 2008.

BECKWITH-WIEDEMANN SYNDROME (Uniparental Disomy and Imprinting Defect, MIM 130650)

Chromosomal with Imprinting Defect

PRINCIPLES

- Multiple pathogenic mechanisms
- Imprinting
- Uniparental disomy
- Assisted reproductive technology

MAJOR PHENOTYPIC FEATURES

- Age at onset: Prenatal
- Prenatal and postnatal overgrowth
- Macroglossia
- Omphalocele
- Visceromegaly
- Embryonal tumor in childhood
- Hemihyperplasia
- Renal abnormalities
- Adrenocortical cytomegaly
- Neonatal hypoglycemia

HISTORY AND PHYSICAL FINDINGS

A.B., a 27-year-old gravida 1/para 0 woman, presented to a prenatal diagnostic center for level II ultrasonography and genetic counseling after a routine ultrasound examination revealed a male fetus, large for gestational age with possible omphalocele. The pregnancy, the first for each of his parents, was undertaken without assisted reproductive technology. After confirmation by level II ultrasonography, the family was counseled that the fetus had a number of abnormalities most consistent with Beckwith-Wiedemann syndrome, although other birth defects were also possible. The couple decided not to undergo amniocentesis. The baby, B.B., was delivered by cesarean section at 37 weeks with a birth weight of 9 pounds, 2 ounces and a notably large placenta. Omphalocele was noted, as were macroglossia and vertical ear lobe creases.

A genetics consultant made a clinical diagnosis of Beckwith-Wiedemann syndrome. When hypoglycemia developed, B.B. was placed in the newborn intensive care unit and was treated with intravenous administration of glucose for 1 week; the hypoglycemia resolved spontaneously. The findings on cardiac evaluation were normal, and the omphalocele was surgically repaired without difficulty. Methylation studies of the *KCNQOT1* gene confirmed an imprinting defect at 11p15 consistent with the diagnosis of Beckwith-Wiedemann syndrome. Abdominal ultrasound examination to screen for Wilms tumor was recommended every 3 months until B.B. was 8 years old, and measurement of serum alpha-fetoprotein level was recommended every 6 weeks as a screen for hepatoblastoma for the first 3 years of life. At a follow-up visit, the family was counseled that in view of their negative family history and normal parental karyotypes, the imprinting defect was consistent with sporadic Beckwith-Wiedemann syndrome, and the recurrence risk was low.

BACKGROUND

Disease Etiology and Incidence

Beckwith-Wiedemann syndrome (BWS, MIM 130650) is a panethnic syndrome that is usually sporadic but may rarely be inherited as an autosomal dominant. BWS affects approximately 1 in 13,700 live births.

BWS results from an imbalance in the expression of imprinted genes in the p15 region of chromosome 11. These genes include *KCNQOT1* and *H19*, noncoding RNAs (see Chapter 3), and *CDKN1C* and *IGF2*, which do encode proteins. Normally, *IGF2* and *KCNQOT1* are imprinted and expressed from the paternal allele only while *CDKN1C* and *H19* are expressed from the maternal allele only. *IGF2* encodes an insulin-like growth factor that *promotes* growth; in contrast, *CDKN1C* encodes a cell cycle suppressor that *constrains* cell division and growth. Transcription of *H19* and *KCNQOT1* RNA suppresses expression of the maternal copy of *IGF2* and the paternal copy of *CDKN1C*, respectively.

Unbalanced expression of 11p15 imprinted genes can occur through a number of mechanisms. Mutations in the maternal *CDKN1C* allele are found in 5% to 10% of sporadic cases and in 40% of families with autosomal dominant BWS. The majority of patients with BWS, however, have loss of expression of the maternal *CDKN1C* allele because of abnormal imprinting, not mutation. In 10% to 20% of individuals with BWS, loss of maternal *CDKN1C* expression and increased *IGF2* expression are caused by paternal isodisomy of 11p15. Because the somatic recombination leading to segmental uniparental disomy occurs after conception, individuals with segmental uniparental disomy are mosaic and may require testing of tissues other than blood to reveal the isodisomy. A few are BWS patients have a detectable chromosomal abnormality, such as maternal translocation, inversion of chromosome 11, or duplication of paternal chromosome 11p15. Rare microdeletions in *KCNQOT1* or *H19* that disrupt imprinting have also been found in BWS.

Pathogenesis

During gamete formation and early embryonic development, a different pattern of DNA methylation is established within the *KCNQOT1* and *H19* genes between males and females. Abnormal imprinting in BWS is most easily detected by analysis of DNA methylation at specific CpG islands in the *KCNQOT1* and *H19* genes. In 60% of patients with BWS, there is *hypo*methylation of the maternal *KCNQOT1*. In another 2% to 7% of patients, *hyper*methylation of the maternal *H19* gene decreases its expression, resulting in excess *IGF2* expression. Inappropriate *IGF2* expression from both parental alleles may explain some of the overgrowth seen in BWS. Similarly, loss of expression of the maternal copy of *CDKN1C* removes a constraint on fetal growth.

Figure C-6 Characteristic macroglossia in a 4-month-old male infant with Beckwith-Wiedemann syndrome. The diagnosis was made soon after birth on the basis of the clinical findings of macrosomia, macroglossia, omphalocele, a subtle ear crease on the right, and neonatal hypoglycemia. Organomegaly was absent. Karyotype was normal, and molecular studies showed hypomethylation of the *KCNQOT1* gene. *See Sources & Acknowledgments.*

Phenotype and Natural History

BWS is associated with prenatal and postnatal overgrowth. Up to 50% of affected individuals are premature and large for gestational age at birth. The placentas are particularly large, and pregnancies are frequently complicated by polyhydramnios. Additional complications in infants with BWS include omphalocele, macroglossia (Fig. C-6), neonatal hypoglycemia, and cardiomyopathy, all of which contribute to a 20% mortality rate. Neonatal hypoglycemia is typically mild and transient, but some cases of more severe hypoglycemia have been documented. Renal malformations and elevated urinary calcium level with nephrocalcinosis and nephrolithiasis are present in almost half of BWS patients. Hyperplasia of various body segments or of selected organs may be present at birth and may become more or less evident over time. Development is typically normal in individuals with BWS unless they have an unbalanced chromosome abnormality.

Children with BWS have an increased risk for development of embryonal tumors, particularly Wilms tumor and hepatoblastoma. The overall risk for neoplasia in children with BWS is approximately 7.5%; the risk is much lower after 8 years of age.

Management

Management of BWS involves treatment of presenting symptoms, such as omphalocele repair and management of hypoglycemia. Special feeding techniques or speech therapy may be required due to the macroglossia. Surgical intervention may be necessary for abdominal wall defects, leg length discrepancies, and renal malformations. If hypercalciuria is present, medical therapy may be instituted to reduce calcium excretion. Periodic screening for embryonal tumors is essential because these are fast-growing and dangerous neoplasias. The current recommendations for monitoring for tumors are an abdominal ultrasound examination every 3 months for the first 8 years of life and measurement of serum alpha-fetoprotein level for hepatoblastoma every 6 weeks for the first few years of life. In addition, an annual renal ultrasound examination for affected individuals between age 8 years and midadolescence is recommended to identify those with nephrocalcinosis or medullary sponge kidney disease.

RECURRENCE RISK

The recurrence risk for siblings and offspring of children with BWS varies greatly with the molecular basis of their condition.

Prenatal screening for pregnancies not previously known to be at increased risk for BWS by ultrasound examination and maternal serum alpha-fetoprotein assay may lead to the consideration of chromosome analysis and/or molecular genetic testing. Specific prenatal testing is possible by chromosome analysis for families with an inherited chromosome abnormality or by molecular genetic testing for families in whom the molecular mechanism of BWS has been defined.

Increased Risk for Beckwith-Wiedemann Syndrome with Assisted Reproductive Technologies

Assisted reproductive technologies (ARTs), such as in vitro fertilization (IVF) and intracytoplasmic sperm injection, have become commonplace, accounting now for 1% to 2% of all births in many countries. Retrospective studies demonstrated that ART had been used 10 to 20 times more frequently in pregnancies that resulted in infants with BWS compared with controls. The risk for BWS after IVF is estimated to be 1 in 4000, which is threefold higher than in the general population.

The reason for the increased incidence of imprinting defects with ART is unknown. The incidence of Prader-Willi syndrome (Case 38), a defect in paternal imprinting, has *not* been shown to be increased with IVF, whereas the frequency of Angelman syndrome, a maternal imprinting defect, *is* increased with IVF, suggesting a specific relationship between ART and maternal imprinting. Because the paternal imprint takes place well before IVF, whereas maternal imprinting takes place much closer to the time of fertilization, a role for IVF itself in predisposing to imprinting defects merits serious study.

QUESTIONS FOR SMALL GROUP DISCUSSION

1. Discuss possible reasons for embryonal tumors in BWS. Why would these decline in frequency with age?
2. Discuss reasons why imprinted genes frequently affect fetal size. Name another condition associated with uniparental disomy for another chromosome.
3. Besides imprinting defects, discuss other genetic disorders that may cause infertility and yet can be passed on by means of ART.
4. In addition to mutations in the genes implicated in BWS, discuss how a mutation in the imprinting locus control region could cause BWS.

REFERENCES

Jacob KJ, Robinson WP, Lefebvre L: Beckwith-Wiedemann and Silver-Russell syndromes: opposite developmental imbalances in imprinted regulators of placental function and embryonic growth, *Clin Genet* 84.326–334, 2013.

Shuman C, Beckwith JB, Smith AC, et al: Beckwith-Wiedemann syndrome. Available from: http://www.ncbi.nlm.nih.gov/books/NBK1394/.

Uyar A, Seli E: The impact of assisted reproductive technologies on genomic imprinting and imprinting disorders, *Curr Opin Obstet Gynecol* 26:210–221, 2014.

HEREDITARY BREAST AND OVARIAN CANCER
(*BRCA1* and *BRCA2* Mutations)
Autosomal Dominant

PRINCIPLES

- Tumor-suppressor gene
- Multistep carcinogenesis
- Somatic mutation
- Incomplete penetrance and variable expressivity
- Founder effect

MAJOR PHENOTYPIC FEATURES

- Age at onset: Adulthood
- Breast cancer
- Ovarian cancer
- Prostate cancer
- Multiple primary cancers

HISTORY AND PHYSICAL FINDINGS

S.M., a 27-year-old previously healthy woman, was referred to the cancer genetics clinic by her gynecologist after being diagnosed with breast cancer. She was concerned about her children's risk for development of cancer and about her risk for development of ovarian cancer. Her mother, two maternal aunts, and maternal grandfather had breast cancer; her mother had also had ovarian cancer (Fig. C-7). The genetic counselor explained that the family history of breast cancer was indicative of an inherited predisposition and calculated that the proband's risk for carrying a mutation in the breast cancer susceptibility gene *BRCA1* or *BRCA2* was well above the threshold for considering gene sequencing. On the basis of the ensuing discussion of prognosis and recurrence risks, S.M. chose to pursue DNA sequencing of *BRCA1* and *BRCA2*. This testing showed that she had a premature termination mutation in one *BRCA2* allele that had been previously seen in other patients with early-onset breast cancer. During the discussion of the results, S.M. inquired whether her 6- and 7-year-old girls should be tested. The genetic counselor explained that because the mutations posed little risk in childhood, the decision to have genetic testing was better left until the children were mature enough to decide on the utility of such testing, and S.M. agreed.

Five adult relatives elected to have predictive testing, and four (including one male) were found to be carriers of the mutation; one of these four, a female, pursued prophylactic bilateral mastectomy. The risk for cancers at other sites was also discussed with all mutation carriers.

BACKGROUND
Disease Etiology and Incidence

Mutations of major cancer predisposition genes account for 3% to 10% of cases of breast cancer and have an estimated overall prevalence of 1 in 300 to 1 in 800. Two of these genes are *BRCA1* and *BRCA2*. In the general North American population, the prevalence of *BRCA1* mutations is between 1 in 500 and 1 in 1000; the prevalence of *BRCA2* mutations is approximately twice as high. There are, however, marked differences in ethnic distribution of deleterious mutations among families with two or more cases of breast or ovarian cancer. Mutations of *BRCA1* or *BRCA2* account for approximately 70% to 80% of *familial* breast cancer cases but only a small fraction of breast cancer overall (see Chapter 15).

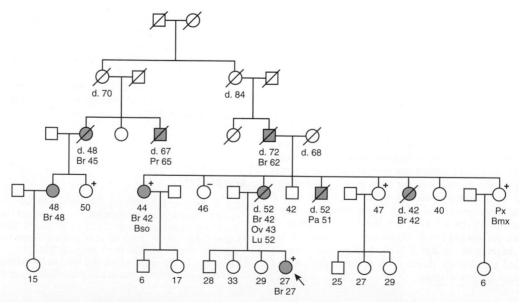

Figure C-7 Family segregating a *BRCA2* C3590G mutation. The proband, S.M., is indicated by an *arrow*. *Blue* symbols indicate a diagnosis of cancer. Ages are shown directly below the symbol. A *plus sign* identifies carriers of the *BRCA2* mutation, and a *minus sign* identifies noncarriers as determined by DNA sequencing. Cancer diagnoses are followed by the age at diagnosis. Cancer abbreviations: Br, breast; Lu, lung; Ov, ovarian; Pa, pancreatic; Pr, prostate. Other abbreviations: Bso, bilateral salpingo-oophorectomy; d., age at death; Px Bmx, prophylactic bilateral mastectomy. *See Sources & Acknowledgments.*

Pathogenesis

BRCA1 and *BRCA2* encode ubiquitously expressed nuclear proteins that are believed to maintain genomic integrity by regulating DNA repair, transcriptional transactivation, and the cell cycle.

Despite the ubiquitous expression of *BRCA1* and *BRCA2*, mutation of these genes predisposes predominantly to breast and ovarian neoplasias. Loss of *BRCA1* or *BRCA2* function probably permits the accumulation of other mutations that are directly responsible for neoplasia. Consistent with this hypothesis, breast and ovarian carcinomas from patients with mutations of *BRCA1* or *BRCA2* have chromosomal instability and frequent mutations in other tumor-suppressor genes.

Tumor formation in carriers of *BRCA1* or *BRCA2* germline mutations follows the two-hit hypothesis; that is, both alleles of either *BRCA1* or *BRCA2* lose function in tumor cells (see Chapter 15). Somatic loss of function by the second allele can occur by a variety of mechanisms, including loss of heterozygosity, intragenic mutation, or promoter hypermethylation. Because of the high frequency with which the second allele of *BRCA1* or *BRCA2* loses function, families segregating a germline *BRCA1* or *BRCA2* mutation exhibit autosomal dominant inheritance of neoplasia.

The population prevalence of individual *BRCA1* or *BRCA2* germline mutations varies widely and often suggests a founder effect. In Iceland, the *BRCA2* 999del5 mutation occurs on a specific haplotype and has a prevalence of 0.6% in that population. Among Ashkenazi Jews, the *BRCA1* 185delAG and 5382insC mutations and the *BRCA2* 6174delT mutation also occur on specific haplotypes and have prevalences of 1%, 0.4%, and 1.2%, respectively.

Phenotype and Natural History

Patients with *BRCA1* or *BRCA2* germline mutations have an increased risk for several cancers (see Table). In addition to the increased risk for ovarian and female breast cancer, *BRCA1* mutations confer an increased risk for prostate cancer, melanoma, and possibly for colon cancer. Similarly, in addition to ovarian and female breast cancer, germline *BRCA2* mutations increase the risk for prostate, pancreatic, bile duct, gallbladder, melanoma, and male breast cancers.

Among female carriers of a *BRCA1* or *BRCA2* germline mutation, the overall penetrance of breast cancer, ovarian cancer, or both is estimated to be approximately 50% to 80% for *BRCA1* mutations but lower for *BRCA2* mutations (40% for breast cancer and 10% for ovarian cancer). Approximately two thirds of families with a history of breast and ovarian cancer segregate a *BRCA1* mutation, whereas approximately two thirds of families with a history of male *and* female breast cancer segregate a *BRCA2* mutation.

Management

Current recommendations for women with a germline *BRCA1* or *BRCA2* mutation include frequent breast and ovarian examinations as well as imaging studies. Management of at-risk males includes frequent prostate and breast examinations and laboratory tests for evidence of prostate cancer. In families with known germline mutations, molecular analysis can focus surveillance or prophylaxis on members carrying a mutation. Total bilateral mastectomy may reduce the risk for breast cancer by more than 90%, although the risk is not abolished because some breast tissue often remains. Similarly, bilateral salpingo-oophorectomy may reduce the risk for ovarian cancer by more than 90%.

Cumulative Risk (%) by Age 70 Years

	Female		Male	
	Breast Cancer	Ovarian Cancer	Breast Cancer	Prostate Cancer
General population	8-10	1.5	<0.1	10
BRCA1 mutation carriers	40-87	16-63	?	25
BRCA2 mutation carriers	28-84	27	6-14	20

INHERITANCE RISK

Female sex, age, and family history are the most important risk factors for breast cancer. In Western populations, the cumulative female breast cancer incidence is 1 in 200 at 40 years, 1 in 50 at 50 years, and 1 in 10 by 70 years. If patients have a first-degree relative in whom breast cancer developed after 55 years, they have a 1.6 relative risk for breast cancer, whereas the relative risk increases to 2.3 if the breast cancer developed in the family member before 55 years and to 3.8 if it developed before 45 years. If the first-degree relative had bilateral breast cancer, the relative risk is 6.4.

Children of a patient with a *BRCA1* or *BRCA2* germline mutation have a 50% risk for inheriting that mutation. Because of incomplete penetrance and variable expressivity, the development and onset of cancer cannot be precisely predicted.

QUESTIONS FOR SMALL GROUP DISCUSSION

1. At what age and under what conditions might testing of an at-risk child be appropriate?
2. What is the risk for development of prostate cancer in a son if a parent carries a *BRCA1* germline mutation? A *BRCA2* germline mutation?
3. Currently, sequencing of the coding region of *BRCA1* detects only 60% to 70% of mutations in families with linkage to the gene. What mutations would sequencing miss? How should a report of "no mutation detected by sequencing" be interpreted and counseled? How would testing of an affected family member clarify the testing results?

REFERENCES

King M-C: The race to clone BRCA1, *Science* 343:1462–1465, 2014.

Lynch HT, Snyder C, Casey MJ: Hereditary ovarian and breast cancer: what have we learned? *Ann Oncol* 24(Suppl 8):83–95, 2013.

Mavaddat N, Peock S, Frost D, et al: Cancer risks for BRCA1 and BRCA2 mutation carriers, *J Natl Cancer Inst* 105:812–822, 2013.

Metcalfe KA, Kim-Sing C, Ghadirian P, et al: Health care provider recommendations for reducing cancer risks among women with a *BRCA1* or *BRCA2* mutation, *Clin Genet* 85:21–30, 2014.

CHARCOT-MARIE-TOOTH DISEASE TYPE 1A
(*PMP22* Mutation or Duplication, MIM 118220)
Autosomal Dominant

PRINCIPLES

- Genetic heterogeneity
- Gene dosage
- Recombination between repeated DNA sequences

MAJOR PHENOTYPIC FEATURES

- Age at onset: Childhood to adulthood
- Progressive distal weakness
- Distal muscle wasting
- Hyporeflexia

HISTORY AND PHYSICAL FINDINGS

During the past few years, J.T., an 18-year-old woman, had noticed a progressive decline in her strength, endurance, and ability to run and walk. She also complained of frequent leg cramps exacerbated by cold and recent difficulty stepping over objects and climbing stairs. She did not recollect a precedent illness or give a history suggestive of an inflammatory process, such as myalgia, fever, or night sweats. No other family members had similar problems or a neuromuscular disorder. On examination, J.T. was thin and had atrophy of her lower legs, mild weakness of ankle extension and flexion, absent ankle reflexes, reduced patellar reflexes, footdrop as she walked, and enlarged peroneal nerves. She had difficulty walking on her toes and could not walk on her heels. The findings from her examination were otherwise normal. As part of her evaluation, the neurologist requested several studies, including nerve conduction velocities (NCVs). J.T.'s NCVs were abnormal; her median NCV was 25 m/sec (normal, >43 m/sec). Results of a subsequent nerve biopsy showed segmental demyelination, myelin sheath hypertrophy (redundant wrappings of Schwann cells around nerve fibers), and no evidence of inflammation. The neurologist explained that these results were strongly suggestive of a demyelinating neuropathy such as type 1 Charcot-Marie-Tooth disease, also known as hereditary motor and sensory neuropathy type 1. Explaining that the most common cause of type 1 Charcot-Marie-Tooth disease is a duplication of the peripheral myelin protein 22 gene (*PMP22*), the neurologist requested testing for this duplication. This test confirmed that J.T. had a duplicated *PMP22* allele and type 1A Charcot-Marie-Tooth disease.

BACKGROUND
Disease Etiology and Incidence

The Charcot-Marie-Tooth (CMT) disorders are a genetically heterogeneous group of hereditary neuropathies characterized by chronic motor and sensory polyneuropathy. CMT has been subdivided according to patterns of inheritance, neuropathological changes, and clinical features. By definition, type 1 CMT (CMT1) is an autosomal dominant demyelinating neuropathy; it has a prevalence of approximately 15 in 100,000 and is also genetically heterogeneous. CMT1A, which represents 70% to 80% of CMT1, is caused by increased dosage of PMP22 secondary to duplication of the *PMP22* gene on chromosome 17. De novo duplications account for 20% to 33% of CMT1A cases; of these, more than 90% arise during male meiosis.

Pathogenesis

PMP22 is an integral membrane glycoprotein. Within the peripheral nervous system, PMP22 is found in compact but not in noncompact myelin. The function of PMP22 has not been fully elucidated, but evidence suggests that it plays a key role in myelin compaction.

Dominant negative mutations within *PMP22* or increased dosage of PMP22 can each cause this peripheral polyneuropathy. Increased dosage of PMP22 arises by tandem duplication of a 1.5-Mb region in 17p11.2 flanked by repeated DNA sequences that are approximately 98% identical. Misalignment of these flanking repeat elements during meiosis can lead to unequal crossing over and formation of one chromatid with a duplication of the 1.5-Mb region and another with the reciprocal deletion. (The reciprocal deletion causes the disease hereditary neuropathy with pressure palsies [HNPP].) An individual inheriting a chromosome with the duplication will have three copies of a normal *PMP22* gene and thus overexpress PMP22 (see Chapter 6).

Overexpression of PMP22 or expression of dominant negative forms of PMP22 results in an inability to form and to maintain compact myelin. Nerve biopsy specimens from severely affected infants show a diffuse paucity of myelin, and nerve biopsy specimens from more mildly affected patients show segmental demyelination and myelin sheath hypertrophy. The mechanism by which PMP22 overexpression causes this pathological process remains unclear.

The muscle weakness and atrophy observed in CMT1 result from muscle denervation secondary to axonal degeneration. Longitudinal studies of patients have shown an age-dependent reduction in the nerve fiber density that correlates with the development of disease symptoms. In addition, evidence in murine models suggests that myelin is necessary for maintenance of the axonal cytoskeleton. The mechanism by which demyelination alters the axonal cytoskeleton and affects axonal degeneration has not been completely elucidated.

Phenotype and Natural History

CMT1A has nearly full penetrance, although the severity, onset, and progression of CMT1 vary markedly within and among families. Many affected individuals do not seek medical attention, either because their symptoms are not noticeable or because their symptoms are accommodated easily. On the other hand, others have severe disease that is manifested in infancy or in childhood.

Symptoms of CMT1A usually develop in the first two decades of life; onset after 30 years of age is rare. Typically symptoms begin with an insidious onset of slowly progressive weakness and atrophy of the distal leg muscles and mild sensory impairment (Fig. C-8). The weakness of the feet and legs leads to abnormalities of gait, a dropped foot, and eventually foot deformities (pes cavus and hammer toes) and loss of balance; it rarely causes patients to lose their ability to walk. Weakness of the intrinsic hand muscles usually occurs late in

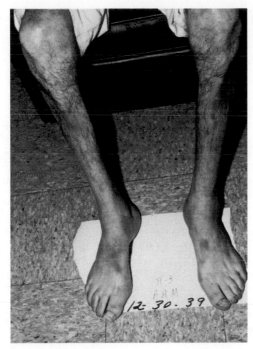

Figure C-8 Distal leg muscle wasting in an older man with the *PMP22* duplication. *See Sources & Acknowledgments.*

the disease course and, in severe cases, causes claw hand deformities because of imbalance between flexor and extensor muscle strength. Other associated findings include decreased or absent reflexes, upper extremity ataxia and tremor, scoliosis, and palpably enlarged superficial nerves. On occasion, the phrenic and autonomic nerves are also involved.

In electrophysiological studies, the hallmark of CMT1A is uniform slowing of NCVs in all nerves and nerve segments as a result of demyelination. The full reduction in NCVs is usually present by 2 to 5 years of age, although clinically apparent symptoms may not be manifested for many years.

Management

Although the diagnosis of CMT1 is suspected because of clinical, electrophysiological, and pathological features, a definitive diagnosis often depends on detection of a mutation. Inflammatory peripheral neuropathies are frequently difficult to distinguish from CMT1 and HNPP and, before the advent

of molecular testing many patients with inherited neuropathies were treated with immunosuppressants and experienced the associated morbidity without improvement of their neuropathy.

Treatment focuses on symptomatic management because curative therapies are currently unavailable for CMT1. Paralleling disease progression, therapy generally follows three stages: strengthening and stretching exercises to maintain gait and function, use of orthotics and special adaptive splints, and orthopedic surgery. Further deterioration may require use of ambulatory supports such as canes and walkers or, in rare, severely affected patients, a wheelchair. All patients should be counseled to avoid exposure to neurotoxic medications and chemicals.

INHERITANCE RISK

Because the *PMP22* duplication and most *PMP22* point mutations are autosomal dominant and fully penetrant, each child of an affected parent has a 50% chance for development of CMT1A. The variable expressivity of the *PMP22* duplication and *PMP22* mutations, however, makes prediction of disease severity impossible.

QUESTIONS FOR SMALL GROUP DISCUSSION

1. Genomic deletions and duplications frequently arise by recombination between repetitive sequences within the human genome (see Chapter 6). Name three disorders caused by deletion after presumed recombination between repetitive sequences. Which of these deletions are associated with a reciprocal duplication? What does the identification of a reciprocal duplication suggest about the mechanism of recombination? What does the absence of a reciprocal duplication suggest?

2. In general, genomic duplications are associated with less severe disease than genomic deletions. Duplication of a *PMP22* allele, however, usually causes more severe disease than deletion of a *PMP22* allele does. Discuss possible reasons for this.

3. Name two other diseases that are caused by a gene dosage effect.

REFERENCES

Bird TD: Charcot-Marie-Tooth neuropathy type 1. Available from: http://www.ncbi.nlm.nih.gov/books/NBK1205/.

Harel T, Lupski JR: Charcot-Marie-Tooth disease and pathways to molecular based therapies, *Clin Genet* 86:422–431, 2014.

PRINCIPLES

- Pleiotropy
- Haploinsufficiency
- Association versus syndrome

MAJOR PHENOTYPIC FEATURES

- *C*oloboma of the iris, retina, optic disc, or optic nerve
- *H*eart defects
- *A*tresia of the choanae
- *R*etardation of growth and development
- *G*enital abnormalities
- *E*ar anomalies
- Facial palsy
- Cleft lip
- Tracheoesophageal fistula

HISTORY AND PHYSICAL FINDINGS

Baby girl E.L. was the product of a full-term pregnancy to a 34-year-old gravida 1, para 1 mother after an uncomplicated pregnancy. At birth, it was noted that E.L.'s right ear was cupped and posteriorly rotated. Because of feeding difficulties, she was placed in the neonatal intensive care unit. Placement of a nasogastric tube was attempted but was unsuccessful in the right naris, demonstrating unilateral choanal atresia. A geneticist determined that she might have the CHARGE syndrome. Further evaluation included echocardiography, which revealed a small atrial septal defect, and ophthalmological examination demonstrating a retinal coloboma in the left eye. The atrial septal defect was repaired surgically without complications. She failed the newborn hearing screen and was subsequently diagnosed with mild to moderate sensorineural hearing loss. Testing for mutations in the gene associated with CHARGE syndrome, *CHD7*, demonstrated a 5418C>G heterozygous mutation in exon 26 that results in a premature termination codon (Tyr1806Ter). Mutation analyses in E.L.'s parents were negative, indicating that a de novo mutation had occurred in E.L. Consequently, the family was advised that the recurrence risk in future pregnancies was low but still possible due to parental germline mosaicism. At 1 year of age, E.L. was moderately delayed in gross motor skills and had speech delay. Her height and weight were at the 5th percentile, and head circumference was at the 10th percentile. Yearly follow-up was planned.

BACKGROUND

Disease Etiology and Incidence

CHARGE syndrome (MIM 214800) is an autosomal dominant condition with multiple congenital malformations caused by mutations in the *CHD7* gene in the majority of individuals tested. Estimated birth prevalence of the condition is 1 in 3000 to 1 in 12,000. However, the advent of genetic testing may reveal *CHD7* mutations in atypical cases, leading to recognition of a higher incidence.

Pathogenesis

The *CHD7* gene, located at 8q12, is a member of the superfamily of chromodomain helicase DNA-binding (CHD) genes. The proteins in this family are predicted to affect chromatin structure and gene expression in early embryonic development. The *CHD7* gene is expressed ubiquitously in many fetal and adult tissues, including the eye, cochlea, brain, central nervous system, stomach, intestine, skeleton, heart, kidney, lung, and liver. Over 500 heterozygous nonsense and missense mutations in the *CHD7* gene, as well as deletions in the 8q12 region encompassing *CHD7*, have been demonstrated in patients with CHARGE syndrome, indicating that haploinsufficiency for the gene causes the disease. Most mutations are novel, although a few hot spots for mutations in the gene exist. Some patients with CHARGE syndrome have no identifiable mutation in *CHD7*, suggesting that mutations in other loci may sometimes underlie the condition.

Phenotype and Natural History

The acronym CHARGE (*c*oloboma, *h*eart defects, *a*tresia of the choanae, *r*etardation of growth and development, *g*enital abnormalities, *e*ar anomalies), encompassing the most common features of the condition, was coined by dysmorphologists as a descriptive name for an association of abnormalities of unknown etiology and pathogenesis seen together more often than would be expected by chance. With the discovery of *CHD7* mutations in CHARGE, the condition is now considered to be a dysmorphic syndrome, a characteristic pattern of causally related anomalies (see Chapter 14). The current major diagnostic criteria are ocular coloboma (affecting the iris, retina, choroid, or disc with or without microphthalmia), choanal atresia (unilateral or bilateral; stenosis or atresia), cranial nerve anomalies (with unilateral or bilateral facial palsy, sensorineural deafness, or swallowing problems), and characteristic ear anomalies (external ear lop or cup-shaped ear, middle ear ossicular malformations, mixed deafness, and cochlear defects). A number of other abnormalities are found less often, such as cleft lip or palate, congenital heart defect, growth deficiency, and tracheoesophageal fistula or esophageal atresia. CHARGE syndrome is diagnosable if three or four major criteria or two major and three minor criteria are found (Fig. C-9).

Perinatal or early infant mortality (before 6 months of age) is seen in approximately half of affected patients and appears to be most highly correlated with the most severe congenital anomalies, including bilateral posterior choanal atresia and congenital heart defects. Gastroesophageal reflux is a significant cause of morbidity and mortality. Feeding problems are also common; as many as 50% of adolescent and adult patients require gastrostomy tube placement. Delayed puberty is found in the majority of patients with CHARGE syndrome. Developmental delay or mental retardation can range from mild to severe in the majority of individuals, and behavioral abnormalities (including hyperactivity, sleep disturbances, and obsessive-compulsive behavior) are frequent. As *CHD7* mutation testing delineates more individuals with CHARGE, the features of the condition may become better defined and the phenotypic spectrum widened.

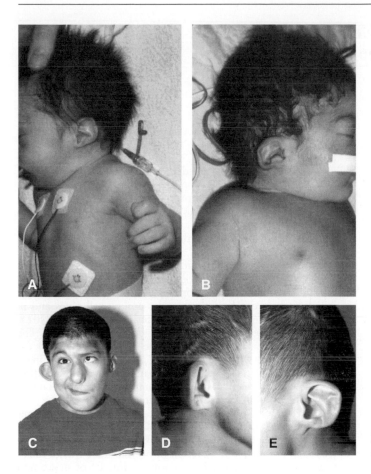

Figure C-9 Ear and eye anomalies in patients with CHARGE syndrome. *See Sources & Acknowledgments.*

Management

If CHARGE syndrome is suspected, thorough evaluation is warranted for possible choanal atresia or stenosis (unilateral), congenital heart defect, central nervous system abnormalities, renal anomalies, hearing loss, and feeding difficulties. Management consists of surgical correction of malformations and supportive care. Developmental evaluation is an important component of follow-up. With the availability of testing for *CHD7* mutations, a molecular diagnosis can be made in at least 50% of patients. *SEMA3E* mutations are another rare cause of the syndrome.

INHERITANCE RISK

Almost all cases of CHARGE syndrome are due to new dominant mutations, with the majority of mutations occurring in the paternal germline. Recurrence risk is therefore low for future offspring. There has been one known reported instance of monozygotic twins having CHARGE, as well as one family with two affected siblings (male and female). The latter situation suggests that germline mosaicism may be present with this condition. If a mutation in *CHD7* is found in an affected individual and both parents test negative for the mutation, the recurrence risk for future offspring would be less than 5%.

An affected individual has a 50% recurrence risk to his or her offspring.

QUESTIONS FOR SMALL GROUP DISCUSSION

1. Explain the difference between an association and a syndrome. Give an example of a common association.
2. By what mechanism could haploinsufficiency for a chromodomain protein cause the pleiotropic effects of CHARGE syndrome?
3. Why would you counsel the parents of a child with a proven de novo mutation in *CHD7* of a 5% recurrence risk? Would the risk change if their next child were affected?

REFERENCES

Hus P, Ma A, Wilson M, et al: CHARGE syndrome: a review, *J Paediatr Child Health* 50:504–511, 2014.

Janssen N, Bergman JE, Swertz MA, et al: Mutation update on the *CHD7* gene involved in CHARGE syndrome, *Hum Mutat* 33:1149–1160, 2012.

Lalani SR, Hefner MA, Belmont JW, et al: CHARGE syndrome. Available from: http://www.ncbi.nlm.nih.gov/books/NBK1117/.

Pauli S, von Velsen N, Burfeind P, et al: *CHD7* mutations causing CHARGE syndrome are predominantly of paternal origin, *Clin Genet* 8:234–239, 2012.

CASE 10

CHRONIC MYELOGENOUS LEUKEMIA
(*BCR-ABL1* Oncogene)

Somatic Mutation

PRINCIPLES

- Chromosomal abnormality
- Oncogene activation
- Fusion protein
- Multihit hypothesis
- Therapy targeted to an oncogene

MAJOR PHENOTYPIC FEATURES

- Age at onset: Middle to late adulthood
- Leukocytosis
- Splenomegaly
- Fatigue and malaise

HISTORY AND PHYSICAL FINDINGS

E.S., a 45-year-old woman, presented to her family physician for her annual checkup. She had been in good health and had no specific complaints. On examination, she had a palpable spleen tip but no other abnormal findings. Results of her complete blood count unexpectedly showed an elevated white blood cell count of 31×10^9/L and a platelet count of 650×10^9/L. The peripheral smear revealed basophilia and immature granulocytes. Her physician referred her to the oncology department for further evaluation. Her bone marrow was found to be hypercellular with an increased number of myeloid and megakaryocytic cells and an increased ratio of myeloid to erythroid cells. Cytogenetic analysis of her marrow identified several myeloid cells with a Philadelphia chromosome, der(22) t(9;22)(q34;q11.2). Her oncologist explained that she had chronic myelogenous leukemia, which, although indolent now, had a substantial risk for becoming a life-threatening leukemia in the next few years. She was also advised that although the only potentially curative therapy currently available is allogeneic bone marrow transplantation, newly developed drug therapy targeting the function of the oncogene in chronic myelogenous leukemia is able to induce or to maintain long-lasting remissions.

BACKGROUND

Disease Etiology and Incidence

Chronic myelogenous leukemia (CML, MIM 608232) is a clonal expansion of transformed hematopoietic progenitor cells that increases circulating myeloid cells. Transformation of progenitor cells occurs by expression of the *BCR-ABL1* oncogene. CML accounts for 15% of adult leukemia and has an incidence of 1 to 2 per 100,000; the age-adjusted incidence is higher in men than in women (1.3 to 1.7 versus 1.0; see Chapter 15).

Pathogenesis

Approximately 95% of patients with CML have a Philadelphia chromosome; the remainder have complex or variant translocations (see Chapter 15). The Abelson proto-oncogene (*ABL1*), which encodes a nonreceptor tyrosine kinase, resides on 9q34, and the breakpoint cluster region gene (*BCR*), which encodes a phosphoprotein, resides on 22q11. During the

formation of the Philadelphia chromosome, the *ABL1* gene is disrupted in intron 1 and the *BCR* gene in one of three breakpoint cluster regions; the *BCR* and *ABL1* gene fragments are joined head to tail on the derivative chromosome 22 (Fig. C-10). The *BCR-ABL1* fusion gene on the derivative chromosome 22 generates a fusion protein that varies in size according to the length of the BCR peptide attached to the amino terminus.

To date, the normal functions of ABL1 and BCR have not been clearly defined. ABL1 has been conserved fairly well throughout metazoan evolution. It is found in both the nucleus and cytoplasm and as a myristolated product associated with the inner cytoplasmic membrane. The relative abundance of ABL1 in these compartments varies among cell types and in response to stimuli. ABL1 participates in the cell cycle, stress responses, integrin signaling, and neural development. The functional domains of BCR include a coiled-coil motif for polymerization with other proteins, a serine-threonine kinase domain, a GDP-GTP exchange domain involved in regulation of Ras family members, and a guanosine triphosphatase–activating domain for regulating Rac and Rho GTPases.

Expression of ABL1 does not result in cellular transformation, whereas expression of the BCR-ABL1 fusion protein does. Transgenic mice expressing BCR-ABL1 develop acute leukemia at birth, and infection of normal mice with a retrovirus expressing BCR-ABL1 causes a variety of acute and chronic leukemias, depending on the genetic background. In contrast to ABL1, BCR-ABL1 has constitutive tyrosine kinase activity and is confined to the cytoplasm, where it avidly binds actin microfilaments. BCR-ABL1 phosphorylates

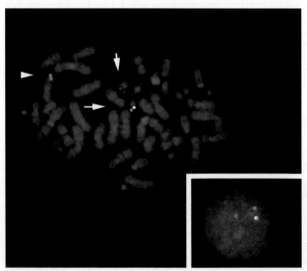

Figure C-10 **FISH analysis in metaphase and interphase (*inset*) cells for the detection of the t(9;22)(q34;q11.2) in CML.** The DNA is counterstained with DAPI. The probe is a mixture of DNA probes specific for the *BCR* gene (*red*) at 22q11.2 and for the *ABL1* gene (*green*) at 9q34. In cells with the t(9;22), a green signal is observed on the normal chromosome 9 (*arrowhead*) and a red signal on the normal chromosome 22 (*short arrow*). As a result of the translocation of *ABL1* to the der(22) chromosome, a yellow fusion signal (*long arrow*) is observed from the presence of both green and red signals together on the Philadelphia chromosome. *See Sources & Acknowledgments.*

several cytoplasmic substrates and thereby activates signaling cascades that control growth and differentiation and possibly adhesion of hematopoietic cells. Unregulated activation of these signaling pathways results in unregulated proliferation of the hematopoietic stem cell, release of immature cells from the marrow, and ultimately CML.

As CML progresses, it becomes increasingly aggressive. During this evolution, tumor cells of 50% to 80% of patients acquire additional chromosomal changes (trisomy 8, i(17q), or trisomy 19), another Philadelphia chromosome, or both. In addition to the cytogenetic changes, tumor-suppressor genes and proto-oncogenes are also frequently mutated in the progression of CML.

Phenotype and Natural History

CML is a biphasic or triphasic disease. The initial or chronic stage is characterized by an insidious onset with subsequent development of fatigue, malaise, weight loss, and minimal to moderate splenic enlargement. Over time, CML typically evolves to an accelerated phase and then to a blast crisis, although some patients progress directly from the chronic phase to the blast crisis. CML progression includes development of additional chromosomal abnormalities within tumor cells, progressive leukocytosis, anemia, thrombocytosis or thrombocytopenia, increasing splenomegaly, fever, and bone lesions. Blast crisis is an acute leukemia in which the blasts can be myeloid, lymphoid, erythroid, or undifferentiated. The accelerated phase is intermediate between the chronic phase and blast crisis.

Approximately 85% of patients are diagnosed in the chronic phase. Depending on the study, the median age at diagnosis ranges from 45 to 65 years, although all ages can be affected. Untreated, the rate of progression from the chronic phase to blast crisis is approximately 5% to 10% during the first 2 years and then 20% per year subsequently. Because blast crisis is rapidly fatal, demise parallels progression to blast crisis.

Management

Recognition of the molecular basis of CML led to the development of a specific BCR-ABL1 tyrosine kinase inhibitor, imatinib mesylate (Gleevec). This drug is now the first line of treatment for CML. More than 85% of patients have a clear cytogenetic response after imatinib therapy, with disappearance of the t(9;22) in cells obtained by bone marrow aspirates. Cytogenetic response corresponds to a large reduction in CML disease burden to levels below 10^9 to 10^{10} leukemic cells. Few patients (<5%), however, show no evidence of the BCR-ABL1 fusion gene by polymerase chain reaction analysis, indicating that even in remission, most patients have a residual leukemia burden of at least 10^6 to 10^7 cells. Of patients with complete hematological and cytogenetic remission, more than 95% remained in control for more than 3.5 years. Patients in blast crisis also respond with improved 12-month survival of 32%, but relapses are common. In these patients, imatinib resistance is frequent (60% to 90%), in association with point mutations that render the ABL1 kinase resistant to the drug or, less commonly, with BCR-ABL1 gene amplification.

Although allogeneic bone marrow transplantation (BMT) is the only known curative therapy, the success of imatinib mesylate has limited the population of patients to whom BMT is offered to those with the highest success rate (patients younger than 40 years with an HLA-matched sibling donor, in whom BMT success is quoted at 80%) and to those in blast crisis. The success of BMT depends on the stage of CML, the age and health of the patient, the bone marrow donor (related versus unrelated), the preparative regimen, the development of graft-versus-host disease, and the post-transplantation treatment. Much of the long-term success of BMT depends on a graft-versus-leukemia effect, that is, a graft-versus-host response directed against the leukemic cells. After BMT, patients are monitored frequently for relapse by reverse transcriptase polymerase chain reaction to detect BCR-ABL1 transcripts and treated as necessary. If BMT fails, patients often respond to infusion of BMT donor-derived T cells, consistent with a graft-versus-leukemia mechanism of action of BMT for CML.

Patients in blast crisis are usually treated with imatinib mesylate, cytotoxic agents and, if possible, BMT. Unfortunately, only 30% of patients have a related or unrelated HLA-matched bone marrow donor. The outcome of these therapies for blast crisis remains poor.

INHERITANCE RISK

Because CML arises from a somatic mutation that is not found in the germline, the risk for a patient's passing the disease to his or her children is zero.

QUESTIONS FOR SMALL GROUP DISCUSSION

1. What is the multihit hypothesis? How does it apply to neoplasia?
2. Discuss two additional mechanisms of proto-oncogene activation in human cancer.
3. Neoplasias graphically illustrate the effects of the accumulation of somatic mutations; however, other less dramatic diseases arise, at least in part, through the accumulation of somatic mutations. Discuss the effect of somatic mutations on aging.
4. Many somatic mutations and cytogenetic rearrangements are never detected because the cells containing them do not have a selective advantage. What advantage does the Philadelphia chromosome confer?
5. Name other cancers caused by fusion genes resulting in oncogene activation. Which others have been successfully targeted?

REFERENCES

Druker BJ. Translation of the Philadelphia chromosome into therapy for CML, *Blood* 112:4808–4817, 2008.

Jabbour E, Cortes J, Ravandi F, et al: Targeted therapies in hematology and their impact on patient care: chronic and acute myeloid leukemia, *Semin Hematol* 50:271–283, 2013.

Krause DS, Van Etten RA: Tyrosine kinases as targets for cancer therapy, *N Engl J Med* 353:172–187, 2005.

CROHN DISEASE (Increased Risk from *NOD2* Mutations)

Multifactorial Inheritance

PRINCIPLES

- Multifactorial inheritance
- Autoimmune disease
- Ethnic predilection

MAJOR PHENOTYPIC FEATURES

- Episodic abdominal pain, cramping, and diarrhea
- Occasional hematochezia (blood in the stool)
- May involve any segment of the intestinal tract
- Transmural ulceration and granulomas of the gastrointestinal tract
- Fistulas
- Patchy involvement usually of the terminal ileum and ascending colon
- Extraintestinal manifestations including inflammation of the joints, eyes, and skin

HISTORY AND PHYSICAL FINDINGS

P.L. is a 14-year-old white male brought to the emergency department by his mother for severe right lower quadrant pain and nausea without vomiting or fever. His history revealed intermittent nonbloody diarrhea for 1 year, no significant constipation, 1-hour postprandial lower quadrant abdominal pain relieved by defecation, and nocturnal abdominal pain that awakens him from sleep. The patient's developmental history was normal except that his growth was noted to have dropped from the 50th-75th percentile to the 25th percentile during the past 2 years. Family history was significant in that a first cousin on the paternal side had Crohn disease. Physical examination revealed peritoneal signs, hyperactive bowel sounds, and diffuse lower abdominal pain to palpation without palpable masses or organomegaly. A stool guaiac test was trace positive. Peripheral blood count revealed only a slightly elevated white blood cell count and a slight microcytic hypochromic anemia. Urinalysis and abdominal plain films were unremarkable. A computed tomographic scan showed mucosal inflammation extending from the distal ileum into the ascending colon. Upper endoscopy and colonoscopy with biopsy were performed, revealing transmural ulceration of the distal ileum with moderate to severe ulceration of the ileocecal junction, consistent with Crohn disease.

Subsequent genetic testing identified a Gly908Arg mutation on one allele of the *NOD2* (*CARD15*) gene, confirming the diagnosis of Crohn disease.

BACKGROUND

Disease Etiology and Incidence

Inflammatory bowel disease (IBD) is a chronic inflammatory disease of the gastrointestinal tract that primarily affects adolescents and young adults. The disease is divided into two major categories, Crohn disease (CD) and ulcerative colitis, each of which occurs with approximately equal frequency in the population. IBD affects 1 in 500 to 1 in 1000 individuals, with a twofold to fourfold increased prevalence in individuals of Ashkenazi Jewish backgrounds compared with non–Ashkenazi Jewish whites. Both disorders show substantial familial clustering and increased concordance rates in monozygotic twins, but they do not follow a mendelian inheritance pattern and are therefore classified as multifactorial. Three different common variants in the *NOD2* gene (also known as *CARD15*) have been found to significantly increase the risk for development of CD (but not of ulcerative colitis) with an additive effect; heterozygotes have a 1.5-fold to fourfold increased risk, whereas homozygotes or compound heterozygotes have a 15- to 40-fold increased risk. The absolute risk among homozygotes or compound heterozygotes therefore approaches 1% to 2%.

Pathogenesis

Because of inflammation in the intestinal tract, IBD was long thought to be an autoimmune disease. Case-control studies in whites revealed three single nucleotide polymorphisms (SNPs) with strong evidence for association with the disease; all three were found to be in the coding exons of the *NOD2* gene and to cause either amino acid substitutions (Gly908Arg and Arg702Trp) or premature termination of the protein (3020insC), an intracellular pattern recognition receptor named for its *n*ucleotide-binding *o*ligomerization *d*omain. Additional studies in several independent cohorts of CD patients have confirmed that these variants are strongly associated with CD.

The NOD2 protein binds to gram-negative bacterial cell walls and participates in the inflammatory response to bacteria by activating the nuclear factor κB (NF-κB) transcription factor in mononuclear leukocytes. The three variants all reduce the ability of the NOD2 protein to activate NF-κB, suggesting that the variants in this gene alter the ability of monocytes in the intestinal wall to respond to resident bacteria, thereby predisposing to an abnormal, inflammatory response. Thus *NOD2* variants are likely to be the alleles actually responsible for increased susceptibility to CD at this IBD locus, which is designated IBD1.

The *NOD2* variants are clearly neither necessary nor sufficient to cause CD. They are not necessary because, although half of all white patients with CD have one or two copies of a *NOD2* variant, half do not. Estimates are that the *NOD2* variants account, at most, for 20% of the genetic contribution to IBD in whites. Furthermore, the particular variants associated with disease risk in Europe are not found in Asian or African populations, and CD in these populations shows no association with *NOD2*. The variants are also not sufficient to cause the disease. The *NOD2* variants are common in Europe; 20% of the population are heterozygous for these alleles yet show no signs of IBD. Even in the highest-risk genotype, those who are homozygotes or compound heterozygotes for the *NOD2* variants, penetrance is less than 10%. The low penetrance points strongly to other genetic or environmental factors that act on genotypic susceptibility at the *NOD2* locus. The obvious connection between an IBD and structural protein variants in the NOD2 protein, a modulator of the innate antibacterial inflammatory response, is a strong clue that the intestinal microenvironment may be an important environmental factor contributing to pathogenesis.

Phenotype and Natural History

Presenting in adolescence or young adulthood, CD most often affects segments of the gastrointestinal tract, such as the

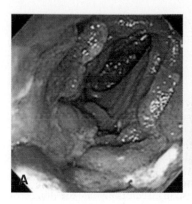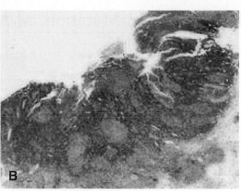

Figure C-11 **A,** Endoscopic appearance of ileitis in a patient with Crohn disease. **B,** Multiple granulomas in the wall of the small intestine in a patient with Crohn disease. *See Sources & Acknowledgments.*

terminal small intestine (ileum) and portions of the ascending colon, but it can occur anywhere within the digestive tract, with granulomatous inflammation (Fig. C-11) that penetrates the wall of the intestine and produces narrowing and scarring. Onset is usually insidious with a history of nocturnal abdominal pain, diarrhea, and gradual weight loss.

Fistulous tracks and intraabdominal abscesses can occur and may be life-threatening. Hospitalization is frequent, and surgery for abscesses may be necessary in CD. Symptoms outside the gastrointestinal tract in CD may include arthritis of the spine and joints, inflammation of the eyes (uveitis), skin involvement (erythema nodosum and pyoderma gangrenosum), primary sclerosing cholangitis, and hypercoagulability. There is also an increased risk for adenocarcinoma of the intestine in long-standing CD, although the risk is not as great as the substantial risk in ulcerative colitis.

Management

Currently there is no cure for IBD. The goals of treatment include induction of remission, maintenance of remission, minimization of side effects of treatment, and improvement of the quality of life. Five main categories of drugs are used alone or in combination to treat CD flare-ups: antiinflammatory medications, corticosteroids, antibiotics, immune modulators, and mixed inflammatory-immune modulators. All of the antiinflammatory medications are derivatives of mesalamine, and the choice of which antiinflammatory medication to use is based on side effect profile and location of disease within the intestine. During the acute phase of the disease, corticosteroids are the mainstay of therapy. These medications, combined with dietary modification, are used to decrease the severity of the disease and to prevent flare-ups. Because fiber is poorly digested, its intake should be reduced in patients with CD. As a result of chronic inflammation and scarring, malnutrition is common. Folate, iron, calcium, and vitamin B_{12} commonly need to be supplemented. Surgery to remove diseased bowel, to drain abscesses, and to close fistulous tracks is often necessary.

INHERITANCE RISK

The empirical risk for development of IBD is approximately 1% to 8% in a sibling of an IBD patient and falls to 0.1% to 0.2% in second-degree relatives, findings not compatible with classic autosomal recessive or dominant inheritance. However, this sibling recurrence is still high compared with the risk in the general population (the relative risk ratio, λ_s, for siblings is between 10 and 30) (see Chapter 8). In one large twin registry, monozygotic twins showed a concordance rate for CD of 44%; dizygotic twins were concordant only 4% of the time. Concordance in ulcerative colitis was only 6% in monozygotic twins but still much higher than in dizygotic twins, in whom no concordant twins were observed. Thus the genetic epidemiological data all strongly support classification of IBD as a disorder with a strong genetic contribution but with complex inheritance.

QUESTIONS FOR SMALL GROUP DISCUSSION

1. Discuss possible environmental factors that play a role in CD.
2. How could variation in innate immunity interact with these environmental factors?
3. How should a family member of a patient with CD who is found to have one of the *NOD2* variants be counseled? Should the testing be done? Why or why not?

REFERENCES

Baumgart DC, Sandborn WJ: Crohn's disease, *Lancet* 380:1590–1605, 2012.
Lees CW, Barrett JC, Parkes M, et al: New IBD genetics: common pathways with other diseases, *Gut* 60:1739–1753, 2011.
Parkes M. The genetics universe of Crohn's disease and ulcerative colitis, *Dig Dis* 30(Suppl 1):78–81, 2012.
Van Limbergen J, Wilson DC, Satsangi J: The genetics of Crohn's disease, *Ann Rev Genomics Hum Genet* 10.89–116, 2009.

PRINCIPLES

- Ethnic variation in mutation frequency
- Variable expressivity
- Tissue-specific expression of mutations
- Genetic modifiers
- Environmental modifiers

MAJOR PHENOTYPIC FEATURES

- Age at onset: Neonatal to adulthood
- Progressive pulmonary disease
- Exocrine pancreatic insufficiency
- Obstructive azoospermia
- Elevated sweat chloride concentration
- Growth failure
- Meconium ileus

HISTORY AND PHYSICAL FINDINGS

J.B., a 2-year-old boy, was referred to the pediatric clinic for evaluation of poor growth. During infancy, J.B. had diarrhea and colic that resolved when an elemental formula was substituted for his standard formula. As table foods were added to his diet, he developed malodorous stools containing undigested food particles. During his second year, J.B. grew poorly, developed a chronic cough, and had frequent upper respiratory infections. No one else in the family had poor growth, feeding disorders, or pulmonary illnesses. On physical examination, J.B.'s weight and height plotted less than the 3rd percentile and his head circumference at the 10th percentile. He had a severe diaper rash, diffuse rhonchi, and mild clubbing of his digits. The physical examination was otherwise normal. After briefly discussing a few possible causes of J.B.'s illness, the pediatrician requested several tests, including a test for sweat chloride concentration by pilocarpine iontophoresis; the sweat chloride level was 75 mmol/L (normal, <40 mmol/L; indeterminate, 40 to 60 mmol/L), a level consistent with cystic fibrosis. On the basis of this result and the clinical course, the pediatrician diagnosed J.B.'s condition as cystic fibrosis. J.B. and his parents were referred to the cystic fibrosis clinic for further counseling, mutation testing, and treatment. (Note: J.B. was born in 2008 prior to implementation of cystic fibrosis newborn screening in his home state.)

BACKGROUND

Disease Etiology and Incidence

Cystic fibrosis (CF, MIM 219700) is an autosomal recessive disorder of epithelial ion transport caused by mutations in the CF transmembrane conductance regulator gene (*CFTR*) (see Chapter 12). Although CF has been observed in all races, it is predominantly a disease of those of northern European ancestry. The live birth incidence of CF ranges from 1 in 313 among the Hutterites of southern Alberta, Canada, to 1 in 90,000 among the Asian population of Hawaii. Among all U.S. whites, the incidence is 1 in 3200.

Pathogenesis

CFTR is an anion channel that conducts chloride and bicarbonate. It is regulated by ATP and by phosphorylation by cAMP-dependent protein kinase. CFTR facilitates the maintenance of hydration of airway secretions through the transport of chloride and inhibition of sodium uptake (see Chapter 12). Dysfunction of CFTR can affect many different organs, particularly those that secrete mucus, including the upper and lower respiratory tracts, pancreas, biliary system, male genitalia, intestine, and sweat glands.

The dehydrated and viscous secretions in the lungs of patients with CF interfere with mucociliary clearance, inhibit the function of naturally occurring antimicrobial peptides, provide a medium for growth of pathogenic organisms, and obstruct airflow. Within the first months of life, these secretions and the bacteria colonizing them initiate an inflammatory reaction. The release of inflammatory cytokines, host antibacterial enzymes, and bacterial enzymes damages the bronchioles. Recurrent cycles of infection, inflammation, and tissue destruction decrease the amount of functional lung tissue and eventually lead to respiratory failure (Fig. C-12).

Loss of CFTR chloride transport into the pancreatic duct impairs the hydration of secretions and leads to the retention of exocrine enzymes in the pancreas. Damage from these retained enzymes eventually causes fibrosis of the pancreas.

CFTR also regulates the uptake of sodium and chloride from sweat as it moves through the sweat duct. In the absence of functional CFTR, the sweat has an increased sodium chloride content, and this is the basis of the historical "salty baby syndrome" and the diagnostic sweat chloride test.

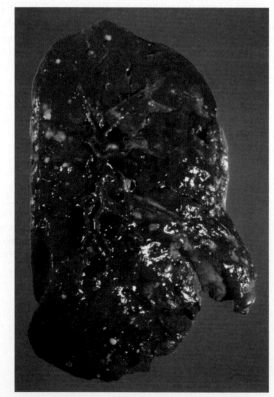

Figure C-12 **A median cross section of a lung from a patient with cystic fibrosis (CF).** Note the mucous plugs and purulent secretions within the airways. *See Sources & Acknowledgments.*

Phenotype and Natural History

CF classically manifests in early childhood, although approximately 4% of patients are diagnosed in adulthood; 15% to 20% of patients present at birth with meconium ileus, and the remainder present with chronic respiratory complaints (rhinitis, sinusitis, obstructive lung disease) poor growth, or both later in life. The poor growth results from a combination of increased caloric expenditure because of chronic lung infections and malnutrition from pancreatic exocrine insufficiency. Five percent to 15% of patients with CF do not develop pancreatic insufficiency. More than 95% of male patients with CF are azoospermic because of congenital bilateral absence of the vas deferens. The progression of lung disease is the chief determinant of morbidity and mortality. Most patients die of respiratory failure and right ventricular failure secondary to the destruction of lung parenchyma and high pulmonary vascular resistance (cor pulmonale); the current average life expectancy is over 38 years in North America and many other regions of the world.

In addition to CF, mutations within *CFTR* have been associated with a spectrum of diseases, including obstructive azoospermia, idiopathic pancreatitis, disseminated bronchiectasis, allergic bronchopulmonary aspergillosis, atypical sinopulmonary disease, and asthma. Some of these disorders are associated with mutations within a single *CFTR* allele; others, like CF, are observed only when mutations are present in both *CFTR* alleles. A direct causative role for mutant *CFTR* alleles has been established for some but not all of these disorders.

A correlation between particular *CFTR* mutant alleles and disease severity exists only for pancreatic insufficiency. Secondary mutations or polymorphisms within a *CFTR* allele may alter the efficiency of splicing or protein maturation and thereby extend the spectrum of disease associated with some mutations. In addition, some mutations in *CFTR* cause disease manifestations only in certain tissues; for example, some mutations affecting the efficiency of splicing have a greater effect on wolffian duct derivatives than in other tissues because of a tissue-specific need for full-length transcript and protein. Environmental factors, such as exposure to cigarette smoke, markedly worsen the severity of lung disease among patients with CF.

Management

Because nearly 2000 different mutations and variants have been described across the *CFTR* gene, the diagnosis of CF is usually based on clinical criteria and sweat chloride concentration. Sweat chloride concentrations are normal in 1% to 2% of patients with CF; in these patients, however, an abnormal nasal transepithelial potential difference measurement is usually diagnostic of CF.

Currently there are no curative treatments of CF, although improved symptomatic management has increased the average longevity from early childhood to between 30 and 40 years. The objectives of medical therapy for CF are clearance of pulmonary secretions, control of pulmonary infection, pancreatic enzyme replacement, adequate nutrition, and prevention of intestinal obstruction. Although medical therapy slows the progression of pulmonary disease, the only effective treatment of respiratory failure in CF is lung transplantation. Pancreatic enzyme replacement and supplementation of fat-soluble vitamins treat the malabsorption effectively; because of increased caloric needs and anorexia, however, many patients also require caloric supplements. New drugs are being developed that either correct or enhance the function of CFTR proteins carrying certain mutations. Most patients also require extensive counseling to deal with the psychological effects of having a chronic fatal disease.

Newborn screening for CF has been implemented in all 50 U.S. states and in most Canadian provinces because detection in the newborn period, prevents the malnutrition seen in clinically undiagnosed pancreatic-insufficient patients. Long-term effects on survival and pulmonary disease progression are unclear.

INHERITANCE RISK

A couple's empirical risk for having a child affected with CF varies greatly, depending on the frequency of CF in their ethnic groups. For North Americans who do not have a family history of CF and are of northern European ancestry, the empirical risk for each to be a carrier is approximately 1 in 29, and such a couple's risk for having an affected child is therefore 1 in 3200. For couples who already have a child affected with CF, the risk for future children to have CF is 1 in 4. In 1997, a National Institutes of Health consensus conference recommended offering CF carrier testing to all pregnant women and couples considering a pregnancy in the United States. The American College of Obstetrics and Gynecology adopted those recommendations.

Prenatal diagnosis is based on identification of proven disease-causing *CFTR* mutations in DNA from fetal tissue, such as chorionic villi or amniocytes. Effective identification of affected fetuses usually requires that the mutations responsible for CF in a family have already been identified.

QUESTIONS FOR SMALL GROUP DISCUSSION

1. Newborn screening for CF can be performed by testing immunoreactive trypsinogen (IRT) alone or by IRT followed by mutation screening. Discuss the risks and benefits of adding *CFTR* mutation screening to a newborn screening panel.

2. The most common CF mutation is ΔF508; it accounts for approximately 70% of all mutant *CFTR* alleles worldwide. For a couple of northern European origin, what is their risk for having an affected child if each tests negative for ΔF508? If one tests positive and the other tests negative for ΔF508?

3. What constitutes disease—a mutation in a gene or the phenotype caused by that mutation? Does detection of a mutation in the *CFTR* gene of patients with congenital bilateral absence of the vas deferens mean they have CF?

REFERENCES

Barrett PM, Alagely A, Topol EJ: Cystic fibrosis in an era of genomically guided therapy, *Hum Mol Genet* 21(R1):R66–R71, 2012.

Boyle MP, de Boeck K: A new era in the treatment of cystic fibrosis: correction of the underlying CFTR defect, *Lancet Respir Med* 1:158–163, 2013.

Cystic Fibrosis Mutation Database. Available at: http://www.genet.sickkids.on.ca/cftr/.

Ferec C, Cutting GR. Assessing the disease-liability of mutations in CFTR, *Cold Spring Harb Perspect Med* 2:a009480, 2012.

Milla CE: Cystic fibrosis in the era of genomic medicine, *Curr Opin Pediatr* 25:323–328, 2013.

Tsui LC, Dorfman R: The cystic fibrosis gene: a molecular genetic perspective, *Cold Spring Harb Perspect Med* 3:a009472, 2013.

DEAFNESS (NONSYNDROMIC) (*GJB2* Mutation, MIM 220290)

Autosomal Dominant and Recessive

PRINCIPLES

- Allelic heterogeneity with both dominant and recessive inheritance patterns
- Newborn screening
- Cultural sensitivity in counseling

MAJOR PHENOTYPIC FEATURES

- Congenital deafness in the recessive form
- Progressive childhood deafness in the dominant form

HISTORY AND PHYSICAL FINDINGS

R.K. and J.K. are a couple referred to the genetics clinic by their ears, nose, and throat specialist because their 6-week-old daughter, B.K., was diagnosed with congenital hearing loss. The child was initially identified by routine neonatal hearing testing (evoked otoacoustic emissions testing) and then underwent formal auditory brainstem response (ABR) testing, which demonstrated moderate hearing impairment.

Both of B.K.'s parents are of European ancestry. Neither parent has a personal or family history of hearing difficulties in childhood, although the father thought that his aunt might have had some hearing difficulties in her old age. B.K. was the product of a full-term, uncomplicated pregnancy.

On examination, B.K. was nondysmorphic. There was no evidence of craniofacial malformation affecting the pinnae or external auditory canals. Tympanic membranes were visible and normal. Ophthalmoscope examination was limited because of the patient's age, but no abnormalities were seen. There was no goiter. Skin was normal.

Laboratory testing revealed a hearing loss of 60 dB bilaterally in the middle- and high-frequency ranges (500 to 2000 Hz and >2000 Hz). Electrocardiography results were normal. Computed tomography scans of petrous bone and cochlea were normal, without malformation or dilatation of the canals.

DNA from B.K. was examined for mutations in the *GJB2* gene. She was found to be homozygous for the common frameshift mutation 35delG in the *GJB2* gene.

BACKGROUND

Disease Etiology and Incidence

Approximately 1 in 500 to 1000 neonates has clinically significant congenital hearing impairment, which arises either from defects of the conductive apparatus in the middle ear or from neurological defects. It is estimated that approximately one third to one half of congenital deafness has a genetic etiology. Of the hereditary forms, approximately three quarters are nonsyndromic, characterized by deafness alone; one quarter is syndromic, that is, associated with other manifestations.

Among inherited forms of nonsyndromic deafness, mutations of *GJB2* are among the more common causes, although mutations in several dozen other genes can also lead to nonsyndromic deafness. *GJB2* mutations cause DFNB1 (MIM 220290), which accounts for half of congenital nonsyndromic autosomal recessive deafness, as well as DFNA3 (MIM

601544), a rare form of childhood-onset, progressive, autosomal dominant deafness. The mutation 35delG accounts for approximately two thirds of identified autosomal recessive *GJB2* mutations in white populations but not in other ethnic groups. Among the Chinese, for example, a different mutation—235delC—is the predominant mutation in *GJB2* causing DFNB1.

Pathogenesis

The *GJB2* gene encodes connexin 26, one of a family of proteins that form gap junctions. Gap junctions create pores between cells, allowing exchange of ions and passage of electrical currents between cells. Connexin 26 is highly expressed in the cochlea, the inner ear organ that transduces sound waves to electrical impulses. The failure to form functional gap junctions results in loss of cochlear function but does not affect the vestibular system or auditory nerve.

Phenotype and Natural History

Autosomal recessive deafness due to *GJB2* mutations is congenital and may be mild to profound (Fig. C-13). Cognitive deficits are *not* a component of the disorder if the hearing impairment is detected early and the child is referred for proper management to allow the development of spoken or sign language.

Autosomal dominant deafness due to *GJB2* mutations also occurs. It has an early childhood onset and is associated with progressive, moderate to severe, high-frequency sensorineural hearing loss. Like the autosomal recessive disease, it also is not associated with cognitive deficits.

Management

The diagnosis of congenital deafness is usually made through newborn screening. Newborn screening is carried out either

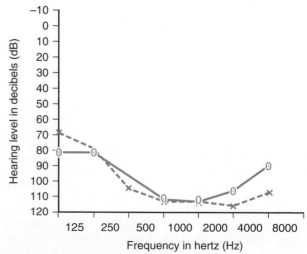

Figure C-13 Profound hearing loss in a child homozygous for mutations in the *GJB2* gene. *X* and *O* represent left and right ear, respectively. Normal hearing level is 0 to 20 dB throughout the frequency range. *See Sources & Acknowledgments.*

by measuring otoacoustic emissions, which are sounds caused by internal vibrations from within a normal cochlea, or by automated ABR, which detects electrical signals in the brain generated in response to sound. With the introduction of universal newborn screening, the average age at diagnosis has fallen to 3 to 6 months, allowing early intervention with hearing aids and other forms of therapy. Infants in whom therapy is initiated before 6 months of age show improvement in language development compared with infants identified at an older age.

As soon as deafness is identified, the child needs to be referred for early intervention, regardless of the cause of the deafness. By consulting with professionals such as audiologists, cochlear implant teams, otolaryngologists, and speech pathologists about the benefits and the drawbacks of different options, parents can be helped to choose those that seem best for their families. Age-appropriate, intensive language therapy with sign language and spoken language with hearing assistance with hearing aids can be instituted as early as possible. Parents can be offered the option of an early cochlear implant, a device that bypasses the dysfunctioning cochlea. Use of cochlear implants before 3 years of age is associated with better oral speech and language outcomes than those in patients receiving an implant later in childhood.

During the newborn period, clinically distinguishing between some forms of syndromic deafness and nonsyndromic deafness is difficult because some syndromic features, such as the goiter in Pendred syndrome or the retinitis pigmentosa in any of the Usher syndromes, may have an onset late in childhood or adolescence. However, a definitive diagnosis is often important for prognosis, management, and counseling; therefore a careful family history and DNA analysis for mutations in the *GJB2* gene as well as in other genes are key to such a diagnosis. Importantly, distinguishing among nonsyndromic forms of deafness is often critical for selecting proper therapy.

INHERITANCE RISK

The form of severe congenital deafness caused by loss-of-function mutations in *GJB2* (DFNB1) is inherited in a typical autosomal recessive manner. Unaffected parents are both carriers of one normal and one altered gene. Two carrier parents have one chance in four with each pregnancy of having a child with congenital deafness. Prenatal diagnosis by direct detection of the mutation in DNA is available.

Among families segregating nonsyndromic progressive deafness with childhood onset due to *GJB2* mutations (DFNA3), inheritance is autosomal dominant, and the risk for an affected parent to have a deaf child is one in two for each pregnancy.

QUESTIONS FOR SMALL GROUP DISCUSSION

1. Why might certain missense mutations in *GJB2* cause *dominant,* progressive hearing loss, whereas another mutation (frameshift) results in *recessive,* nonprogressive hearing loss?
2. What special considerations and concerns might arise in providing genetic counseling to a deaf couple about the risk for their having a child with hearing loss? What is meant by the term *deaf culture?*
3. Mutation testing detects only 95% of the *GJB2* mutations among white families known to have autosomal recessive deafness secondary to *GJB2* defects. Also, many sequence variations have been detected in the *GJB2* gene. If a couple with a congenitally deaf child presented to you and mutation analysis detected a *GJB2* sequence variation, not previously associated with disease, in only one parent, how would you counsel them regarding recurrence risk and genetic etiology? Would your counseling be different if the sequence variation had been previously associated with disease, and what would constitute significant association? Would your counseling be different if the child had early childhood onset of progressive deafness?
4. Why might a child with a cochlear implant learn sign language in addition to spoken language?
5. Because mutations in many different genes can underlie recessive forms of nonsyndromic deafness, discuss various approaches to molecular diagnosis of the gene responsible in any given case: *GJB2* testing, testing a panel of known genes, whole-exome sequencing, or whole-genome sequencing.

REFERENCES

Duman D, Tekin M: Autosomal recessive nonsyndromic deafness genes: a review, *Front Biosci* 17:2213–2236, 2012.

Shearer AE, Smith RJ: Genetics: advances in genetic testing for deafness, *Curr Opin Pediatr* 24:679–686, 2012.

Smith RJH, Shearer AE, Hildebrand MS, et al: Deafness and hereditary hearing loss overview. Available from: http://www.ncbi.nlm.nih.gov/books/NBK1434/.

Smith RJH, Van Camp G: Nonsyndromic hearing loss and deafness, DFNB1. Available from: http://www.ncbi.nlm.nih.gov/books/NBK1272/.

Vona B, Muller T, Nanda I, et al: Targeted next-generation sequencing of deafness genes in hearing-impaired individuals uncovers informative mutations, *Genet Med* 16:945–953, 2014.

DUCHENNE MUSCULAR DYSTROPHY
(Dystrophin [*DMD*] Mutation, MIM 310200)

X-Linked

PRINCIPLES

- High frequency of new mutations
- Allelic heterogeneity
- Manifesting carriers
- Phenotypic variability

MAJOR PHENOTYPIC FEATURES

- Age at onset: Childhood
- Muscle weakness
- Calf pseudohypertrophy
- Mild intellectual compromise
- Elevated serum creatine kinase level

HISTORY AND PHYSICAL FINDINGS

A.Y., a 6-year-old boy, was referred for mild developmental delay. He had difficulty climbing stairs, running, and participating in vigorous physical activities; he had decreased strength and endurance. His parents, two brothers, and one sister were all healthy; no other family members were similarly affected. On examination, he had difficulty jumping onto the examination table, a Gowers sign (a sequence of maneuvers for rising from the floor; Fig. C-14), proximal weakness, a waddling gait, tight heel cords, and apparently enlarged calf muscles. His serum creatine kinase level was 50-fold higher than normal. Because the history, physical examination findings, and elevated creatine kinase level strongly suggested a myopathy, A.Y. was referred to the neurogenetics clinic for further evaluation. Results of his muscle biopsy showed marked variation of muscle fiber size, fiber necrosis, fat and connective tissue proliferation, and no staining for dystrophin. On the basis of these results, A.Y. was given a provisional diagnosis of Duchenne muscular dystrophy, and he was tested for deletions of the dystrophin gene; he was found to have a deletion of exons 45 through 48. Subsequent testing showed his mother to be a carrier. The family was therefore counseled that the risk for affected sons was 50%, the risk for affected daughters was low but dependent on skewing of X inactivation, and the risk for carrier daughters was 50%. Because her carrier status placed her at a high risk for cardiac complications, the mother was referred for a cardiac evaluation.

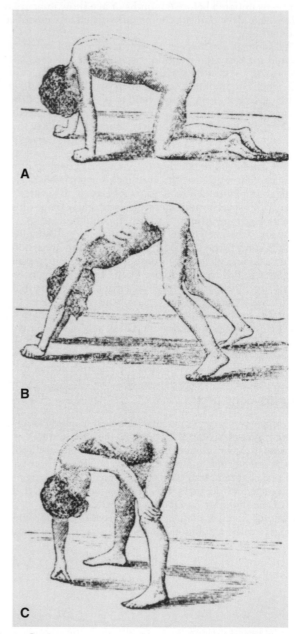

Figure C-14 Drawing of a boy with Duchenne muscular dystrophy rising from the ground, illustrating the Gowers maneuver. *See Sources & Acknowledgments.*

BACKGROUND

Disease Etiology and Incidence

Duchenne muscular dystrophy (DMD, MIM 310200) is a panethnic, X-linked progressive myopathy caused by mutations within the *DMD* gene. It has an incidence of approximately 1 in 3500 male births.

Pathogenesis

DMD encodes dystrophin, an intracellular protein that is expressed predominantly in smooth, skeletal, and cardiac muscle as well as in some brain neurons (see Chapter 12). In skeletal muscle, dystrophin is part of a large complex of sarcolemma-associated proteins that confers stability to the sarcolemma (see Fig. 12-20).

DMD mutations that cause DMD include large deletions (60% to 65%), large duplications (5% to 10%), and small deletions, insertions, or nucleotide changes (25% to 30%). Most large deletions occur in one of two hot spots. Nucleotide changes occur throughout the gene, predominantly at CpG dinucleotides. De novo mutations arise with comparable frequency during oogenesis and spermatogenesis; most of the de novo large deletions arise during oogenesis, whereas most of the de novo nucleotide changes arise during spermatogenesis.

Mutations causing a dystrophin null phenotype effect more severe muscle disease than mutant *DMD* alleles expressing partially functional dystrophin. A consistent genotype-phenotype correlation has not been defined for the intellectual impairment.

Phenotype and Natural History

Males

DMD is a progressive myopathy resulting in muscle degeneration and weakness. Beginning with the hip girdle muscles and neck flexors, the muscle weakness progressively involves the shoulder girdle and distal limb and trunk muscles. Although occasionally manifesting in the newborn period with hypotonia or failure to thrive, male patients usually present between the ages of 3 and 5 years with gait abnormalities. By 5 years of age, the majority of patients use a Gowers maneuver and have calf pseudohypertrophy, that is, enlargement of the calf through replacement of muscle by fat and connective tissue. By 12 years of age, most patients are confined to a wheelchair and have or are developing contractures and scoliosis. Most die of impaired pulmonary function and pneumonia; the median age at death is 18 years.

Nearly 95% of patients with DMD have some cardiac compromise (dilated cardiomyopathy, electrocardiographic abnormalities, or both), and 84% have demonstrable cardiac involvement at autopsy. Chronic heart failure develops in nearly 50% of patients. Rarely, cardiac failure is the presenting complaint for patients with DMD. Although dystrophin is also present in smooth muscle, smooth muscle complications are rare. These complications include gastric dilatation, ileus, and bladder paralysis.

Patients with DMD have an average IQ approximately 1 standard deviation below the mean, and nearly one third have some degree of intellectual disability. The basis of this impairment has not been established.

Females

The age at onset and the severity of DMD in females depend on the degree of skewing of X inactivation (see Chapter 6). If the X chromosome carrying the mutant *DMD* allele is active in most cells, females develop signs of DMD; if the X chromosome carrying the normal *DMD* allele is predominantly active, females have few or no symptoms of DMD. Regardless of whether they have clinical symptoms of skeletal muscle weakness, most carrier females have cardiac abnormalities, such as dilated cardiomyopathy, left ventricle dilatation, and electrocardiographic changes.

Management

The diagnosis of DMD is based on family history and either DNA analysis or muscle biopsy to test for immunoreactivity for dystrophin.

Currently there are no curative treatments of DMD, although improved symptomatic management has increased the average longevity from late childhood to early adulthood. The objectives of therapy are slowing of disease progression, maintenance of mobility, prevention and correction of contractures and scoliosis, weight control, and optimization of pulmonary and cardiac function. Glucocorticoid therapy can slow the progression of DMD for several years. Several experimental therapies, including gene transfer, are under investigation. Most patients also require extensive counseling to deal with the psychological effects of having a chronic fatal disease.

INHERITANCE RISK

One third of mothers who have a single affected son will not themselves be carriers of a mutation in the *DMD* gene (see Chapter 16). Determination of the carrier state in females in the approximately 30% to 35% of DMD families with a point mutation or small indel has proved difficult in the past because of the large number of exons in the dystrophin gene. Advances in DNA sequencing, however, have made targeted exome sequencing much more effective. Counseling of recurrence risk must take into account the high rate of germline mosaicism (currently estimated to be 14%).

If a mother is a carrier, each son has a 50% risk for DMD and each daughter has a 50% risk for inheriting the *DMD* mutation. Reflecting the random nature of X chromosome inactivation, daughters inheriting the *DMD* mutation have a low risk for DMD; however, for reasons not fully understood, their risk for cardiac abnormalities may be as high as 50% to 60%. If a mother is apparently not a carrier by DNA testing, she still has an approximately 7% risk for having a boy with DMD due to germline mosaicism (see Chapter 7). Counseling and possibly prenatal diagnosis are indicated for these mothers.

QUESTIONS FOR SMALL GROUP DISCUSSION

1. Why is DMD considered a genetic lethal condition? What features define a condition as being genetically lethal?
2. Discuss what mechanisms may cause a gender bias in different types of mutation. Name several diseases other than DMD in which this occurs. In particular, discuss the mechanism and high frequency of mutations at CpG dinucleotides during spermatogenesis.
3. How is the rate of germline mosaicism determined for a disease? Name several other diseases with a high rate of germline mosaicism.
4. Contrast the phenotype of Becker muscular dystrophy with DMD. What is the postulated basis for the milder phenotype of Becker muscular dystrophy?

REFERENCES

Darras BT, Miller DT, Urion DK: Dystrophinopathies. Available from: http://www.ncbi.nlm.nih.gov/books/NBK1119/.

Fairclough RJ, Wood MJ, Davies KE: Therapy for Duchenne muscular dystrophy: renewed optimism from genetic approaches, *Nat Rev Genet* 14:373–378, 2013.

Shieh PB: Muscular dystrophies and other genetic myopathies, *Neurol Clin* 31:1009–1029, 2013.

FAMILIAL ADENOMATOUS POLYPOSIS (*APC* Mutation, MIM 175100)

Autosomal Dominant

PRINCIPLES

- Tumor-suppressor gene
- Multistep carcinogenesis
- Somatic mutation
- Cytogenetic and genomic instability
- Variable expressivity

MAJOR PHENOTYPIC FEATURES

- Age at onset: Adolescence through mid-adulthood
- Colorectal adenomatous polyps
- Colorectal cancer
- Multiple primary cancers

HISTORY AND PHYSICAL FINDINGS

R.P., a 35-year-old man, was referred to the cancer genetics clinic by his oncologist. He had just undergone a total colectomy; the colonic mucosa had more than 2000 polyps and pathological changes consistent with adenomatous polyposis coli. In addition to his abdominal scars and colostomy, he had retinal pigment abnormalities consistent with congenital hypertrophy of the retinal pigment epithelium. Several of his relatives had died of cancer. He did not have a medical or family history of other health problems. On the basis of the medical history and suggestive family history, the geneticist counseled R.P. that he most likely had familial adenomatous polyposis. The geneticist explained the surveillance protocol for R.P.'s children and the possibility of using molecular testing to identify those children at risk for familial adenomatous polyposis. Because R.P. did not have contact with his family and family studies were therefore not possible, R.P. elected to proceed with direct screening of the adenomatous polyposis coli gene (*APC*); he had a nonsense mutation in exon 15 of one *APC* allele.

BACKGROUND

Disease Etiology and Incidence

At least 50% of individuals in Western populations develop a colorectal tumor, including benign polyps, by the age of 70 years, and approximately 10% of these individuals eventually develop colorectal carcinoma. Approximately 15% of colorectal cancer is familial, including familial adenomatous polyposis (FAP, MIM 175100) and hereditary nonpolyposis colorectal cancer. FAP is an autosomal dominant cancer predisposition syndrome caused by inherited mutations in the *APC* gene. It has a prevalence of 2 to 3 per 100,000 and accounts for less than 1% of colon cancers. Somatic *APC* mutations also occur in more than 80% of sporadic colorectal tumors (see Chapter 15).

Pathogenesis

The APC protein directly or indirectly regulates transcription, cell adhesion, the microtubular cytoskeleton, cell migration, crypt fission, apoptosis, and cell proliferation. It forms complexes with several different proteins, including β-catenin.

Both alleles of *APC* must be inactivated for adenoma formation. The high frequency of somatic loss of function in the second *APC* allele defines FAP as an autosomal dominant condition. As described in Chapter 15, this somatic loss of function can occur by a variety of mechanisms, including loss of heterozygosity, intragenic mutation, transcriptional inactivation, and, rarely, dominant negative effects of the inherited mutant allele. More than 95% of intragenic *APC* mutations cause truncation of the APC protein. Loss of functional APC usually results in high levels of free cytosolic β-catenin; free β-catenin migrates to the nucleus, binds to T-cell factor 4, and inappropriately activates gene expression. Consistent with this mechanism, mutations of the β-catenin gene have been identified in some colorectal carcinomas without *APC* mutations.

Although loss of functional APC causes affected cells to form dysplastic foci within intestinal crypts, these cells are not cancerous and must acquire other somatic mutations to progress to cancer (see Chapter 15). This progression is characterized by cytogenetic instability resulting in the loss of large chromosomal segments and, consequently, loss of heterozygosity. Specific genetic alterations implicated in this progression include activation of the *KRAS* or *NRAS* oncogenes, inactivation of a tumor-suppressor gene on 18q, inactivation of the *TP53* gene, and alterations in methylation leading to transcriptional silencing of tumor-suppressor genes. As cells accumulate mutations, they become increasingly neoplastic and eventually form invasive and metastatic carcinomas.

Phenotype and Natural History

FAP is characterized by hundreds to thousands of colonic adenomatous polyps (Fig. C-15). It is diagnosed clinically by the presence of either more than 100 colorectal adenomatous polyps or between 10 and 100 polyps in an individual with a relative with FAP. Adenomatous polyps usually appear between 7 and 40 years of age and rapidly increase in number. If untreated, 7% of patients develop colorectal cancer by 21 years of age, 87% by 45 years, and 93% by 50 years.

Although nonpenetrance is very rare, patients with germline mutations of *APC* do not necessarily develop adenomas or colorectal cancer; they are only predisposed. The rate-limiting step in adenoma formation is somatic mutation of the wild-type *APC* allele. Progression of an adenoma to carcinoma requires the accumulation of other genetic alterations. Patients with FAP are at much greater risk than the general population for development of colorectal carcinoma for two reasons. First, although the average time to progress from adenoma to carcinoma is approximately 23 years, these patients develop adenomas earlier in life and are less likely to die of other causes before the development of carcinoma. Second, although less than 1% of adenomas progress to carcinoma, patients have tens to thousands of adenomas, each with the potential to transform to carcinoma. Thus the likelihood that at least one adenoma will progress to become an adenocarcinoma is a near certainty.

The penetrance and expressivity of *APC* mutations depend on the particular *APC* mutation, genetic background, and environment. Mutations in different regions of the gene are variously associated with Gardner syndrome (an association of colonic adenomatous polyposis, osteomas, and soft tissue

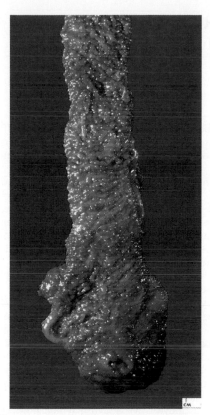

Figure C-15 The mucosa of an ascending colon resected from a patient with familial adenomatous polyposis. Note the enormous number of polyps. *See Sources & Acknowledgments.*

tumors), congenital hypertrophy of the retinal pigment epithelium, attenuated adenomatous polyposis coli, or Turcot syndrome (colon cancer and central nervous system tumors, usually medulloblastoma). Modifier genes in the human genome may cause patients with identical germline mutations to have dissimilar clinical features. Many studies of sporadic colorectal tumorigenesis identify an enhanced risk for individuals consuming diets high in animal fat; therefore, given the common mechanism of tumorigenesis, diet is likely to play a role in FAP as well.

Management

Early recognition of FAP is necessary for effective intervention, that is, prevention of colorectal cancer. After the development of polyps, definitive treatment is total colectomy with ileoanal pull-through. Recommended surveillance for patients at risk for FAP is colonoscopy every 1 to 2 years beginning at 10 to 12 years of age. To focus this surveillance, molecular testing is recommended to identify at-risk family members.

INHERITANCE RISK

The empirical lifetime risk for colorectal cancer among Western populations is 5% to 6%. This risk is markedly modified by family history. Patients who have a sibling with adenomatous polyps but no family history of colorectal cancer have a 1.78 relative risk; the relative risk increases to 2.59 if a sibling developed adenomas before the age of 60 years. Patients with a first-degree relative with colorectal cancer have a 1.72 relative risk; this relative risk increases to 2.75 if two or more first-degree relatives had colorectal cancer. If an affected first-degree relative developed colorectal cancer before 44 years of age, the relative risk increases to more than 5.

In contrast to these figures for all colorectal cancer, a patient with FAP or an *APC* germline mutation has a 50% risk for having a child affected with FAP in each pregnancy. The absence of a family history of FAP does not preclude the diagnosis of FAP in a parent because approximately 20% to 30% of patients have a new germline *APC* mutation. Prenatal diagnosis is available by linkage analysis or by testing for the mutation if the mutation in the parent has been defined. Because of intrafamilial variation in expressivity, the severity, time at onset, and associated features cannot be predicted.

Germline *APC* mutations are not detected in between 10% and 30% of individuals with a clinical phenotype of typical FAP and in 90% of individuals with "attenuated" FAP (FAP phenotype, except there are fewer than 100 adenomas). Among these patients, 10% are germline homozygotes or compound heterozygotes for a mutation in the DNA repair gene *MYH*; another 10% carry one mutant *MYH* allele in their germline. Heterozygosity for a mutant *MYH* allele increases the risk for colon cancer threefold; having both alleles mutant increases risk 50-fold. A patient with FAP and no *APC* mutation should be investigated for *MYH* mutations, particularly if there is a family history suggestive of autosomal recessive inheritance (see FAP2, MIM 608456).

QUESTIONS FOR SMALL GROUP DISCUSSION

1. Name additional disorders that demonstrate autosomal dominant inheritance but are recessive at the cellular level. Why do these diseases exhibit autosomal dominant inheritance if two mutations are required for expression of the disease?
2. Discuss some other mendelian disorders that have modeled or provided insights into more common diseases, including at least one for cancer and one for dementia.
3. What does the association of attenuated adenomatous polyposis coli with early truncations of APC suggest about the biochemical basis of attenuated adenomatous polyposis coli compared with classic FAP?

REFERENCES

Jasperson KW, Burt RW: APC-associated polyposis conditions. Available from: http://www.ncbi.nlm.nih.gov/books/NBK1345/.

Jenkins MA, Croitoru ME, Monga N, et al: Risk of colorectal cancer in monoallelic and biallelic carriers of MYH mutations: a population-based case-family study, *Cancer Epidemiol Biomarkers Prev* 15:312–314, 2006.

Kerr SE, Thomas CB, Thibodeau SN, et al: APC germline mutations in individuals being evaluated for familial adenomatous polyposis: a review of the Mayo Clinic experience with 1591 consecutive tests, *J Mol Diagn* 15:31–43, 2013.

FAMILIAL HYPERCHOLESTEROLEMIA (Low-Density Lipoprotein Receptor [*LDLR*] Mutation, MIM 143890)

Autosomal Dominant

PRINCIPLES

- Environmental modifiers
- Founder effects
- Gene dosage
- Genetic modifiers

MAJOR PHENOTYPIC FEATURES

- Age at onset: Heterozygote—early to middle adulthood; homozygote—childhood
- Hypercholesterolemia
- Atherosclerosis
- Xanthomas
- Arcus corneae

HISTORY AND PHYSICAL FINDINGS

L.L., a previously healthy 45-year-old French Canadian poet, was admitted for a myocardial infarction. He had a small xanthoma on his right Achilles tendon. His brother also had coronary artery disease (CAD); his mother, maternal grandmother, and two maternal uncles had died of CAD. In addition to his family history and sex, his risk factors for CAD and atherosclerosis included an elevated level of low-density lipoprotein (LDL) cholesterol, mild obesity, physical inactivity, and cigarette smoking. On the basis of family history, L.L. was believed to have an autosomal dominant form of hypercholesterolemia. Molecular analysis revealed that he was heterozygous for a deletion of the 5′ end of the LDL receptor gene (*LDLR*), a mutation found in 59% of French Canadians with familial hypercholesterolemia. Screening of his children revealed that two of the three children had elevated LDL cholesterol levels. The cardiologist explained to L.L. that in addition to drug therapy, effective treatment of his CAD required dietary and lifestyle changes, such as a diet low in saturated fat and low in cholesterol, increased physical activity, weight loss, and smoking cessation. L.L. was not compliant with treatment and died a year later of a myocardial infarction.

BACKGROUND

Disease Etiology and Incidence

Familial hypercholesterolemia (FH, MIM 143890) is an autosomal dominant disorder of cholesterol and lipid metabolism caused by mutations in *LDLR* (see Chapter 12). FH occurs among all races and has a prevalence of 1 in 500 in most white populations. It accounts for somewhat less than 5% of patients with hypercholesterolemia.

Pathogenesis

The LDL receptor, a transmembrane glycoprotein predominantly expressed in the liver and adrenal cortex, plays a key role in cholesterol homeostasis. It binds apolipoprotein B-100, the sole protein of LDL, and apolipoprotein E, a protein found on very-low-density lipoproteins, intermediate-density lipoproteins, chylomicron remnants, and some high-density lipoproteins. Hepatic LDL receptors clear approximately 50%
of intermediate-density lipoproteins and 66% to 80% of LDL from the circulation by endocytosis; poorly understood LDL receptor–independent pathways clear the remainder of the LDL.

Mutations associated with FH occur throughout *LDLR*; 2% to 10% are large insertions, deletions, or rearrangements mediated by recombination between *Alu* repeats within *LDLR*. Some mutations appear to be dominant negative. Most mutations are private mutations, although some populations—such as Lebanese, French Canadians, South African Indians, South African Ashkenazi Jews, and Afrikaners—have common mutations and a high prevalence of disease because of founder effects.

Homozygous or heterozygous mutations of *LDLR* decrease the efficiency of intermediate-density lipoprotein and LDL endocytosis and cause accumulation of plasma LDL by increasing production of LDL from intermediate-density lipoproteins and decreasing hepatic clearance of LDL. The elevated plasma LDL levels cause atherosclerosis by increasing the clearance of LDL through LDL receptor–independent pathways, such as endocytosis of oxidized LDL by macrophages and histiocytes. Monocytes, which infiltrate the arterial intima and endocytose oxidized LDL, form foam cells and release cytokines that cause proliferation of smooth muscle cells of the arterial media. Initially, the smooth muscle cells produce sufficient collagen and matrix proteins to form a fibrous cap over the foam cells; because foam cells continue to endocytose oxidized LDL, however, they eventually rupture through the fibrous cap into the arterial lumen and trigger the formation of thrombi, a common cause of strokes and myocardial infarction.

Environment, sex, and genetic background modify the effect of LDL receptor mutations on LDL plasma levels and thereby the occurrence of atherosclerosis. Diet is the major environmental modifier of LDL plasma levels; for example, most Tunisian FH heterozygotes have LDL levels in the normal North American range and rarely develop cardiovascular disease and xanthomas. Similarly, Chinese FH heterozygotes living in China rarely have xanthomas and cardiovascular disease, whereas Chinese FH heterozygotes living in Western societies have clinical manifestations similar to those of white FH heterozygotes. Dietary cholesterol suppresses the synthesis of LDL receptors, thereby raising plasma LDL levels; this effect of dietary cholesterol is potentiated by saturated fatty acids, such as palmitate from dairy products, and ameliorated by unsaturated fatty acids, such as oleate and linoleate. Because a similar diet does not elevate LDL levels equally among patients, other environmental and genetic factors must also influence LDL metabolism. A few families with FH segregate a different dominant locus that reduces plasma LDL, providing evidence for a genetic modifier. Other forms of FH include type B hypercholesterolemia (MIM 144010), caused by ligand-defective apolipoprotein B-100, and autosomal dominant hypercholesterolemia (MIM 603776), due to *PCSK9* mutations. In subjects with the *LDLR* mutation IVS14+1G-A, the phenotype can be altered by a single nucleotide polymorphism (SNP) in *APOA2*, a SNP in *EPHX2*, or a SNP in *GHR*. A SNP in the promoter region of the G-substrate gene (*GSBS*) correlates with elevated plasma total cholesterol levels. A SNP in intron 17 of *ITIH4* was

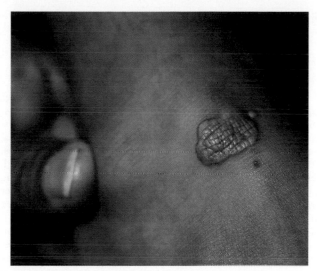

Figure C-16 An Achilles tendon xanthoma from a patient with familial hypercholesterolemia. *See Sources & Acknowledgments.*

Age- and Sex-Specific Rates (%) of CAD and Death in Familial Hypercholesterolemia Heterozygotes

Age	Males		Females	
	CAD	Death	CAD	Death
30	5	—	0	—
40	20-24	—	0-3	0
50	45-51	25	12-20	2
60	75-85	50	45-57	15
70	100	80	75	30

CAD, Coronary artery disease.
From Rader DJ, Hobbs HH: Disorders of lipoprotein metabolism. In Kasper DL, Braunwald E, Fauci AS, et al, editors: *Harrison's principles of internal medicine,* ed 16, New York, 2004, McGraw-Hill.

associated with hypercholesterolemia susceptibility in a Japanese population.

Phenotype and Natural History

Hypercholesterolemia, the earliest finding in FH, usually manifests at birth and is the only clinical finding through the first decade in heterozygous patients; at all ages, the plasma cholesterol concentration is greater than the 95th percentile in more than 95% of patients. Arcus corneae and tendon xanthomas begin to appear by the end of the second decade and by death, 80% of FH heterozygotes have xanthomas (Fig. C-16). Nearly 40% of adult patients have recurrent nonprogressive polyarthritis and tenosynovitis. As tabulated, the development of CAD among FH heterozygotes depends on age and gender. In general, the untreated cholesterol level is greater than 300 mg/dL.

Homozygous FH presents in the first decade with tendon xanthomas and arcus corneae. Without aggressive treatment, homozygous FH is usually lethal by the age of 30 years. The untreated cholesterol concentration is between 600 and 1000 mg/dL.

Management

Elevated plasma LDL cholesterol and a family history of hypercholesterolemia, xanthomas, or premature CAD strongly suggest a diagnosis of FH. Confirmation of the diagnosis requires quantification of LDL receptor function in the patient's skin fibroblasts or identification of the *LDLR* mutation. In most populations, the plethora of *LDLR* mutations precludes direct DNA analysis unless a particular mutation is strongly suspected. The absence of DNA confirmation does not interfere with management of FH patients, however, because a definitive molecular diagnosis of FH does not provide prognostic or therapeutic information beyond that already derived from the family history and determination of plasma LDL cholesterol level.

Regardless of whether they have FH, all patients with elevated LDL cholesterol levels require aggressive normalization of the LDL cholesterol concentration to reduce their risk for CAD; rigorous normalization of the LDL cholesterol concentration can prevent and reverse atherosclerosis. In FH heterozygotes, rigorous adherence to a low-fat, high-carbohydrate

diet usually produces a 10% to 20% reduction in LDL cholesterol, but most patients also require treatment with one or a combination of three classes of drugs: bile acid sequestrants, 3-hydroxy-3-methylglutaryl coenzyme A reductase inhibitors, and nicotinic acid (see Chapter 13). Current recommendations are initiation of drug therapy at 10 years of age for patients with an LDL cholesterol level of more than 190 mg/dL and a negative family history for premature CAD, and at 10 years of age for patients with an LDL cholesterol level of more than 160 mg/dL and a positive family history for premature CAD. Among FH homozygotes, LDL apheresis can reduce plasma cholesterol levels by as much as 70%. The therapeutic effectiveness of apheresis is increased when it is combined with aggressive statin and nicotinic acid therapy. Liver transplantation has also been used on rare occasions.

INHERITANCE RISK

Because FH is an autosomal dominant disorder, each child of an affected parent has a 50% chance of inheriting the mutant *LDLR* allele. Untreated FH heterozygotes have a 100% risk for development of CAD by the age of 70 years if male and a 75% risk if female (see Table). Current medical therapy markedly reduces this risk by normalizing plasma cholesterol concentration.

QUESTIONS FOR SMALL GROUP DISCUSSION

1. What insights does FH provide into the more common polygenic causes of atherosclerosis and CAD?
2. Familial defective apolipoprotein B-100 is a genocopy of FH. Why?
3. Vegetable oils are hydrogenated to make some margarines. What effect would eating margarine have on LDL receptor expression compared with vegetable oil consumption?
4. Discuss genetic susceptibility to infection and potential heterozygote advantage in the context of the role of the LDL receptor in hepatitis C infection.

REFERENCES

Rader DJ, Hobbs HH: Disorders of lipoprotein metabolism. In Longo D, Fauci AS, Kasper DL, et al, editors: *Harrison's principles of internal medicine,* ed 18, New York, 2012, McGraw-Hill.

Sniderman AD, Tsimikas S, Fazio S: The severe hypercholesterolemia phenotype: clinical diagnosis, management, and emerging therapies, *J Am Coll Cardiol* 63:1935–1947, 2014.

Varghese MJ: Familial hypercholesterolemia: a review, *Ann Pediatr Cardiol* 7:107–117, 2014.

Youngblom E, Knowles JW: Familial hypercholesterolemia. Available from: http://www.ncbi.nlm.nih.gov/books/NBK174884/.

FRAGILE X SYNDROME (*FMR1* Mutation, MIM 300624)
X-Linked

PRINCIPLES

- Triplet repeat expansion
- Somatic mosaicism
- Sex-specific anticipation
- DNA methylation
- Haplotype effect

MAJOR PHENOTYPIC FEATURES

- Age at onset: Childhood
- Intellectual disability
- Dysmorphic facies
- Male postpubertal macroorchidism

HISTORY AND PHYSICAL FINDINGS

R.L., a 6-year-old boy, was referred to the developmental pediatrics clinic for evaluation of intellectual disability and hyperactivity. He had failed kindergarten because he was disruptive, was unable to attend to tasks, and had poor speech and motor skills. His development was delayed, but he had not lost developmental milestones: he sat by 10 to 11 months, walked by 20 months, and spoke two or three clear words by 24 months. He had otherwise been in good health. His mother and maternal aunt had mild childhood learning disabilities, and a maternal uncle was intellectually disabled. The findings from his physical examination were normal except for hyperactivity. The physician recommended several tests, including a chromosomal microarray, thyroid function studies, and DNA analysis for fragile X syndrome. Diagnostic analysis of the *FMR1* gene was consistent with fragile X syndrome.

BACKGROUND
Disease Etiology and Incidence

Fragile X syndrome (MIM 300624) is an X-linked disorder of intellectual disability that is caused by mutations in the *FMR1* gene on Xq27.3 (see Chapter 12). It has an estimated prevalence of 16 to 25 per 100,000 in the general male population and half that in the general female population. The disorder accounts for 3% to 6% of intellectual disability among boys with a positive family history of cognitive deficits and no birth defects.

Pathogenesis

The *FMR1* gene product, FMRP, is expressed in many cell types but most abundantly in neurons. The FMRP protein may chaperone a subclass of mRNAs from the nucleus to the translational machinery.

More than 99% of *FMR1* mutations are expansions of a $(CGG)_n$ repeat sequence in the 5′ untranslated region of the gene (see Chapter 12). In normal alleles of *FMR1*, the number of CGG repeats ranges from 6 to approximately 50. In disease-causing alleles or full mutations, the number of repeats is more than 200. Alleles with more than 200 CGG repeats usually have hypermethylation of the CGG repeat sequence and the adjacent *FMR1* promoter (Fig. C-17). Hypermethylation epigenetically inactivates the *FMR1* promoter, causing a loss of FMRP expression.

FMR1 full mutations arise from premutation alleles (approximately 59 to 200 CGG repeats) with maternal transmission of a mutant *FMR1* allele but not with paternal transmission; in fact, premutations often shorten with paternal transmission. Full mutations do not arise from normal alleles. Because the length of an unstable CGG repeat increases each generation when it is transmitted by a female, increasing numbers of affected offspring are usually observed in later generations of an affected family; this phenomenon is referred to as genetic anticipation (see Chapter 7).

The risk for premutation expansion to a full mutation increases as the repeat length of the premutation increases (see Fig. 7-23). Not all premutations, however, are equally

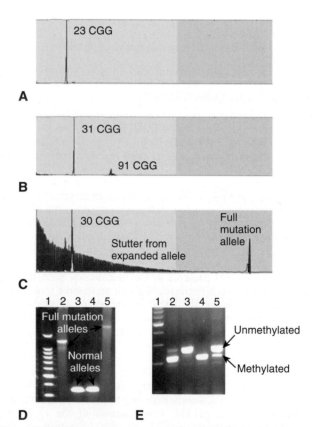

Figure C-17 Polymerase chain reaction (PCR) analysis of *FMR1* CGG repeat number in a normal male (**A**), a premutation female (**B**), and a full mutation female (**C**). The number of CGG repeats is on the x-axis, and fluorescence intensity is on the Y-axis. Normal and premutation ranges are boxed in *gray*; full mutation range is boxed in *pink* with characteristic stutter from the repeat-targeted primer in the *gray box*. **D,** Agarose gel separation of expanded alleles after PCR with *FMR1*-specific primers. *Lane 1,* size markers; *Lane 2,* full mutation male mosaic for approximately 280 and approximately 350 CGG alleles; *Lanes 3* and *4,* normal males; *Lane 5,* full mutation male. **E,** Methylation-sensitive PCR determines methylation status of alleles near the premutation/full mutation boundary in males; positions of methylated and unmethylated alleles are indicated. *Lane 1,* size markers; *Lane 2,* abnormal male with mosaic methylation with approximately 140, approximately 350, and approximately 770 CGG alleles; *Lane 3,* male with expanded unmethylated repeat in the premutation range; *Lane 4,* affected male with complete methylation; *Lane 5,* male with mosaic methylation. Note that methylation status in females can only be determined by Southern blot (see Fig. 7-22). *See Sources & Acknowledgments.*

predisposed to expand. Although premutations are relatively common, progression to a full mutation has been observed only on a limited number of haplotypes; that is, there is a haplotype predisposition to expansion. This haplotype predisposition may relate partly to the presence of a few AGG triplets embedded within the string of CGG repeats; these AGG triplets appear to inhibit expansion of the string of CGG repeats, and their absence in some haplotypes therefore may predispose to expansion.

Phenotype and Natural History

Fragile X syndrome causes moderate intellectual disability in affected males and mild intellectual deficits in affected females. Most affected individuals also have behavioral abnormalities, including hyperactivity, hand flapping or biting, temper tantrums, poor eye contact, and autistic features. The physical features of males vary in relation to puberty such that before puberty, they have somewhat large heads but few other distinctive features; after puberty, they frequently have more distinctive features (long face with prominent jaw and forehead, large ears, and macro-orchidism). Because these clinical findings are not unique to fragile X syndrome, the diagnosis depends on molecular detection of mutations. Patients with fragile X syndrome have a normal life span.

Nearly all males and 40% to 50% of females who inherit a full mutation will have fragile X syndrome. The severity of the phenotype depends on repeat length mosaicism and repeat methylation (see Fig. C-17). Because full mutations are mitotically unstable, some patients have a mixture of cells with repeat lengths ranging from premutation to full mutation (repeat length mosaicism). All males with repeat length mosaicism are affected but often have higher mental function than those with a full mutation in every cell; females with repeat length mosaicism are normal to fully affected. Similarly, some patients have a mixture of cells, with and without methylation of the CGG repeat (repeat methylation mosaicism). All males with methylation mosaicism are affected but often have higher mental function than those with a hypermethylation in every cell; females with methylation mosaicism are normal to fully affected. Very rarely, patients have a full mutation that is unmethylated in all cells; whether male or female, these patients vary from normal to fully affected. In addition, in females, the phenotype is dependent on the degree of skewing of X chromosome inactivation (see Chapter 6).

Female carriers of premutations (but not full mutations) are at a 20% risk for premature ovarian failure. Male premutation carriers are at risk for the fragile X associated tremor/ataxia syndrome (FXTAS). FXTAS manifests as late-onset, progressive cerebellar ataxia and intention tremor. Affected individuals may also have loss of short-term memory, executive function, and cognition as well as parkinsonism, peripheral neuropathy, lower limb proximal muscle weakness, and autonomic dysfunction. Penetrance of FXTAS is age-dependent, manifesting in 17% in the sixth decade, in 38% in the seventh decade, in 47% in the eighth decade, and in three fourths of those older than 80 years. FXTAS may manifest in some female premutation carriers.

Management

No curative treatments are currently available for fragile X syndrome. Therapy focuses on educational intervention and pharmacological management of the behavioral problems.

INHERITANCE RISK

The risk that a woman with a premutation will have an affected child is determined by the size of the premutation, the sex of the fetus, and the family history. Empirically, the risk to a premutation carrier of having an affected child can be as high as 50% for each male child and 25% for each female child but depends on the size of the premutation. On the basis of analysis of a relatively small number of carrier mothers, the recurrence risk appears to decline as the premutation decreases from 100 to 59 repeats. Prenatal testing is available by use of fetal DNA derived from chorionic villi or amniocytes.

QUESTIONS FOR SMALL GROUP DISCUSSION

1. Discuss haplotype bias in disease; that is, the effect of haplotype on mutation development (fragile X syndrome), disease severity (sickle cell disease), or predisposition to disease (autoimmune diseases).
2. Fragile X syndrome, myotonic dystrophy, Friedreich ataxia, Huntington disease, and several other disorders are caused by expansion of repeat sequences. Contrast the mechanisms or proposed mechanisms by which expansion of the repeat causes disease for each of these disorders. Why do some of these disorders show anticipation, whereas others do not?
3. The sex bias in transmission of *FMR1* mutations is believed to arise because FMRP expression is necessary for production of viable sperm. Compare the sex bias in transmitting fragile X syndrome and Huntington disease. Discuss mechanisms that could explain biases in the transmitting sex for various diseases.
4. What family history and diagnostic information are necessary before prenatal diagnosis is undertaken for fragile X syndrome?
5. How would you counsel a pregnant woman carrying a 46,XY fetus with 60 repeats? A 46,XX fetus with 60 repeats? A 46,XX fetus with more than 300 repeats?

REFERENCES

Besterman AD, Wilke SA, Milligan TE, et al: Towards an understanding of neuropsychiatric manifestations in fragile X premutation carriers, *Future Neurol* 9:227–239, 2014.

Hagerman R, Hagerman P: Advances in clinical and molecular understanding of the FMR1 premutation and fragile X-associated tremor/ataxia syndrome, *Lancet Neurol* 12:786–798, 2013.

Saul RA, Tarleton JC: *FMR1*-related disorders. Available from: http://www.ncbi.nlm.nih.gov/books/NBK1384/.

Tassone F: Newborn screening for fragile X syndrome, *JAMA Neurol* 71:355–359, 2014.

TYPE I (NON-NEURONOPATHIC) GAUCHER DISEASE
(*GBA1* Mutation, MIM 230800)

Autosomal Recessive

PRINCIPLES

- Variable expression
- Asymptomatic homozygotes

MAJOR PHENOTYPIC FEATURES

- Age at onset: Childhood or early adulthood
- Hepatosplenomegaly
- Anemia
- Thrombocytopenia
- Bone pain
- Short stature

HISTORY AND PHYSICAL FINDINGS

An 8-year-old Ashkenazi Jewish girl presented to clinic with easy bleeding and bruising, excessive fatigue, short stature, and enlargement of the belly. Abdominal ultrasound examination showed enlarged liver and spleen; complete blood count showed pancytopenia, and skeletal survey showed Erlenmeyer flask deformity Her parents were healthy and had another 6-year-old healthy child. Neither parent had a family history of bone anomalies, blood disease, or liver and spleen disease. Consistent with her clinical history and physical features, she had decreased β-glucocerebrosidase activity in leukocytes. DNA testing identified an Asn370Ser homozygous mutation in *GBA1*.

BACKGROUND

Disease Etiology and Incidence

Type 1 (non-neuronopathic) Gaucher disease (MIM 230800) is the most prevalent lysosomal storage disorder as well as the most common Gaucher disease phenotype, accounting for more than 90% of all Gaucher disease patients. It is an autosomal recessive disorder caused by mutations in *GBA1* gene causing β-glucocerebrosidase deficiency. Type 1 Gaucher disease has a prevalence worldwide of 1 in 50,000 to 1 in 100,000, but it is as high as approximately 1 in 480 to 1280 in individuals of Ashkenazi heritage.

Pathogenesis

The defect in Gaucher disease is an inherited deficiency of the lysosomal enzyme acid β-glucosidase (glucocerebrosidase), which results in the accumulation of glucocerebroside within lysosomes of macrophages. Systemic accumulation of these glycolipid-lipid engorged cells (known as Gaucher cells) results in variable combinations of splenomegaly with associated abdominal discomfort; anemia associated with chronic fatigue; bleeding due to thrombocytopenia and/or Gaucher disease–related coagulopathy; hepatomegaly; abnormal results for tests of liver function; and a diverse pattern of bone disease. Increased susceptibility to infections may result from impaired neutrophil function and neutropenia. Rarely, the lungs, lymphatic system, skin, eyes, kidneys, and the heart are involved and, in the rare neuronopathic forms, neurodegenerative disease results. Gaucher disease is traditionally classified into three broad phenotypic categories: type 1 (non-neuronopathic disease); type 2 (MIM 230900), a fulminant neuronopathic disease that is fatal during infancy; and type 3 (MIM 231000), chronic neuronopathic disease, which usually results in death in childhood or early adult life. More than 350 *GBA1* mutations have been identified to date. The *GBA1* mutation spectrum varies widely according to ethnic group, and homozygosity for the *Asn370Ser* mutation is the most common genotype in the Ashkenazim, in whom it accounts for 70% of all disease alleles. The most common disease allele of *GBA1* worldwide is the Leu444Pro mutation, which occurs in the sequence of the closely linked pseudogene; it is believed that gene conversion leads to the mutation in the active gene. The most frequent genotype of type 1 Gaucher disease in populations of European descent is Asn370Ser/Leu444Pro. This genotype generally leads to more severe disease compared with Asn370Ser homozygosity.

Phenotype and Natural History

The broadest phenotypic spectrum in Gaucher disease with respect to age of onset, rate of progression, and organs affected occurs in type 1 Gaucher disease. Symptoms may present in early childhood, and asymptomatic grandparents may be identified by their symptomatic grandchildren. Patients are often identified by splenomegaly in childhood or early adulthood. Hepatomegaly is usually not as severe as splenomegaly. Thrombocytopenia and anemia are easily observed, and skeletal findings such as osteopenia, osteonecrosis, bone pain, short stature, scoliosis, and multiple fractures are also noted. B-cell lymphoma is an atypical manifestation, but multiple myeloma may occur as a late complication.

Management

A bone marrow or liver biopsy is sometimes done to diagnose Gaucher disease by revealing typical Gaucher cells, macrophages filled with lipid material, but the gold standard is to confirm deficient β-glucocerebrosidase activity in leukocytes. This noninvasive test is both sensitive and specific. Genetic tests can be used as an effective tool for diagnosis as well, but molecular analysis of *GBA1* is complicated by the presence of a highly homologous pseudogene that harbors several mutations, which, if present in the active gene, would lead to Gaucher disease. A negative screen for common *GBA1* mutations does not exclude Gaucher disease. Thus, sequencing of the entire coding region of *GBA1* is recommended in patients strongly suspected of Gaucher disease when a screen for common mutations is negative. Mutation analysis of the *GBA1* may provide some prognostic information, although there is considerable variation of disease severity among patients with identical *GBA1* genotypes. Increased activity of biomarkers, such as acid phosphatase, angiotensin-converting enzyme, chitotriosidase, and ferritin, can also be used to identify disease activity.

Since 1993, when recombinant β-glucocerebrosidase became available, enzyme replacement treatment (ERT) has remarkably improved the clinical outcome of Gaucher disease patients (Fig. C-18). Particularly, hepatosplenomegaly and hematological abnormalities show notable improvement under ERT in patients with both the non-neuronopathic and chronic

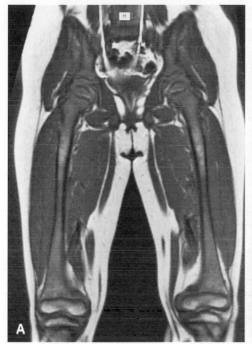
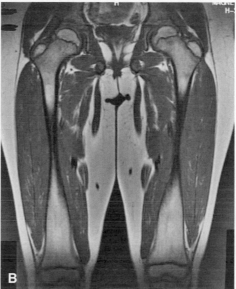

Figure C-18 A, T1-weighted coronal image of the pelvis and femurs of a 5-year-old child with Gaucher disease before treatment with enzyme therapy. There is classic Erlenmeyer flask deformity of the distal femurs and abnormal marrow distribution with intermediate signal in the proximal and distal femoral epiphyses and diaphyses. **B**, T1-weighted coronal image of the femurs approximately 1 year after initiating enzyme therapy for treatment of Gaucher disease. The marrow distribution shows increased signal, indicating it has returned to normal. Erlenmeyer flask deformities of the femora remain, however. *See Sources & Acknowledgments.*

neuronopathic types. ERT is currently the standard treatment for non-neuronopathic Gaucher disease, and notably, levels of biomarkers (including chitotriosidase, acid phosphatase, and angiotensin-converting enzyme) decrease with ERT. However, because the recombinant enzyme cannot cross the blood-brain barrier, it cannot prevent neurological deterioration in patients with neuronopathic Gaucher disease. Other variants of macrophage-targeted enzyme replacement therapies for this disorder are undergoing clinical trials: velaglucerase, a human fibroblast–derived enzyme, was recently approved for treatment of type 1 Gaucher disease and taliglucerase, a plant-derived enzyme is in clinical trials.

Substrate reduction therapy (SRT; miglustat [Zavesca]) is approved for patients with mild Gaucher disease who are unable to receive ERT. SRT with N-butyldeoxynojirimycin (miglustat), a small iminosugar molecule, reversibly inhibits glucosylceramide synthase, the ceramide-specific glucosyltransferase that catalyzes the first committed step in glycosphingolipid synthesis, and in this way reduces intracellular storage of glucosylceramide. Recent data confirmed miglustat efficacy in the long-term maintenance therapy of type 1 Gaucher disease with a positive impact of miglustat on both bone marrow and bone tissue. A more specific and potent inhibitor of glucosylceramide synthesis, eliglustat tartrate, is currently in Phase III trials, having shown efficacy and safety in Phase II trials.

INHERITANCE RISK

For unaffected parents with a child affected with type 1 Gaucher, the risk for recurrence in their future children is 25%, and knowledge of the *GBA1* mutation in a proband facilitates family screening for genetic counseling purposes because heterozygote carriers cannot be reliably identified by enzyme assays. Penetrance, however, can be very variable.

QUESTIONS FOR SMALL GROUP DISCUSSION

1. Name and discuss other disorders for which enzyme replacement therapy has been used.
2. How do the mutations in *GBA1* affect mRNA and protein production?
3. The reason for the a high rate of asymptomatic homozygotes for the *N370S* mutation in *GBA1* is unknown. What possible explanations might there be for this finding?
4. How has chemical chaperone therapy been used in Gaucher disease?

REFERENCES

Desnick RJ, Schuchman EH: Enzyme replacement therapy for lysosomal diseases: lessons from 20 years of experience and remaining challenges, *Annu Rev Genomics Hum Genet* 13:307–335, 2012.

Mignot C, Gelot A, De Villemeur TB: Gaucher disease, *Handb Clin Neurol* 113:1709–1715, 2013.

Pastores GM, Hughes DA: Gaucher disease. Available from: http://www.ncbi.nlm.nih.gov/books/NBK1269/.

Thomas AS, Mehta A, Hughes DA: Gaucher disease: haematological presentations and complications, *Br J Haematol* 165:427–440, 2014.

GLUCOSE-6-PHOSPHATE DEHYDROGENASE DEFICIENCY (*G6PD* Mutation, MIM 305900)

X-Linked

PRINCIPLES

- Heterozygote advantage
- Pharmacogenetics

MAJOR PHENOTYPIC FEATURES

- Age at onset: Neonatal
- Hemolytic anemia
- Neonatal jaundice

HISTORY AND PHYSICAL FINDINGS

L.M., a previously healthy 5-year-old boy, presented to the emergency department febrile, pale, tachycardic, tachypneic, and minimally responsive; his physical examination was otherwise normal. The morning before presentation, he had been in good health, but during the afternoon, he had abdominal pain, headache, and fever; by late evening, he was tachypneic and incoherent. He had not ingested any medications or known toxins, and results of a urine toxicology screen were negative. Results of other laboratory tests showed massive nonimmune intravascular hemolysis and hemoglobinuria. After resuscitation, L.M. was admitted to the hospital; the hemolysis resolved without further intervention. L.M. was of Greek ethnicity; his parents were unaware of a family history of hemolysis, although his mother had some cousins in Europe with a "blood problem." Further inquiry revealed that the morning before admission, L.M. had been eating fava beans from the garden while his mother was working in the yard. The physician explained to the parents that L.M. probably was deficient for glucose-6-phosphate dehydrogenase (G6PD) and that because of this, he had become ill after eating fava beans. Subsequent measurement of L.M.'s erythrocyte G6PD activity confirmed that he had G6PD deficiency. The parents were counseled concerning L.M.'s risk for acute hemolysis after exposure to certain drugs and toxins and given a list of compounds that L.M. should avoid.

BACKGROUND

Disease Etiology and Incidence

G6PD deficiency (MIM 305900), a hereditary predisposition to hemolysis, is an X-linked disorder of antioxidant homeostasis that is caused by mutations in the *G6PD* gene. In areas in which malaria is endemic, G6PD deficiency has a prevalence of 5% to 25%; in nonendemic areas, it has a prevalence of less than 0.5% (Fig. C-19). Like sickle cell disease, G6PD deficiency appears to have reached a substantial frequency in some areas because it confers some resistance to malaria and thus a survival advantage to individuals heterozygous for G6PD deficiency (see Chapter 9).

Pathogenesis

G6PD is the first enzyme in the hexose monophosphate shunt, a pathway critical for generating nicotinamide adenine dinucleotide phosphate (NADPH). NADPH is required for the regeneration of reduced glutathione. Within erythrocytes, reduced glutathione is used for the detoxification of oxidants produced by the interaction of hemoglobin and oxygen and by exogenous factors such as drugs, infection, and metabolic acidosis.

Most G6PD deficiency arises because mutations in the X-linked *G6PD* gene decrease the catalytic activity or the stability of G6PD, or both. When G6PD activity is sufficiently depleted or deficient, insufficient NADPH is available to regenerate reduced glutathione during times of oxidative stress. This results in the oxidation and aggregation of intracellular proteins (Heinz bodies) (see Fig. 11-8) and the formation of rigid erythrocytes that readily hemolyze.

With the more common *G6PD* alleles, which cause the protein to be unstable, deficiency of G6PD within erythrocytes worsens as erythrocytes age. Because erythrocytes do not have nuclei, new G6PD mRNA cannot be synthesized; thus erythrocytes are unable to replace G6PD as it is degraded. During exposure to an oxidative stress episode, therefore, hemolysis begins with the oldest erythrocytes and progressively involves younger erythrocytes, depending on the severity of the oxidative stress.

Phenotype and Natural History

As an X-linked disorder, G6PD deficiency predominantly and most severely affects males. Rare symptomatic females have a skewing of X chromosome inactivation such that the X chromosome carrying the mutant *G6PD* allele is the active X chromosome in erythrocyte precursors (see Chapter 6).

The severity of G6PD deficiency depends not only on sex, but also on the specific *G6PD* mutation. In general, the mutation common in the Mediterranean basin (i.e., G6PD B⁻ or Mediterranean) tends to be more severe than those mutations common in Africa (i.e., G6PD A⁻ variants) (see Fig. C-19). In erythrocytes of patients with the Mediterranean variant, G6PD activity decreases to insufficient levels 5 to 10 days after erythrocytes appear in the circulation, whereas in the erythrocytes of patients with the G6PD A⁻ variants, G6PD activity decreases to insufficient levels 50 to 60 days after erythrocytes appear in the circulation. Therefore most erythrocytes are susceptible to hemolysis in patients with severe forms of G6PD deficiency, such as G6PD Mediterranean, but only 20% to 30% are susceptible in patients with G6PD A⁻ variants.

G6PD deficiency most commonly manifests as either neonatal jaundice or acute hemolytic anemia. The peak incidence of neonatal jaundice occurs during days 2 and 3 of life. The severity of the jaundice ranges from subclinical to levels compatible with kernicterus; the associated anemia is rarely severe. Episodes of acute hemolytic anemia usually begin within hours of an oxidative stress and end when G6PD-deficient erythrocytes have hemolyzed; therefore, the severity of the anemia associated with these acute hemolytic episodes is proportionate to the deficiency of G6PD and the oxidative stress. Viral and bacterial infections are the most common triggers, but many drugs and toxins can also precipitate hemolysis. The disorder called favism results from hemolysis secondary to the ingestion of fava beans by patients with more severe forms of G6PD deficiency, such as G6PD Mediterranean; fava beans contain β-glycosides, naturally occurring oxidants.

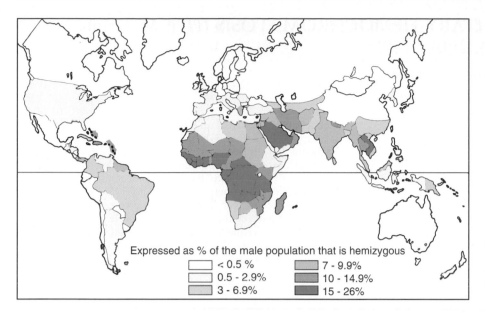

Figure C-19 World distribution of G6PD deficiency. The frequencies of G6PD-deficient males in the various countries are also the allele frequencies because the gene is X-linked. *See Sources & Acknowledgments.*

Expressed as % of the male population that is hemizygous

< 0.5 %	7 - 9.9%
0.5 - 2.9%	10 - 14.9%
3 - 6.9%	15 - 26%

In addition to neonatal jaundice and acute hemolytic anemia, G6PD deficiency rarely causes congenital or chronic nonspherocytic hemolytic anemia. Patients with chronic nonspherocytic hemolytic anemia generally have a profound deficiency of G6PD that causes chronic anemia and an increased susceptibility to infection. The susceptibility to infection arises because the NADPH supply within granulocytes is inadequate to sustain the oxidative burst necessary for killing phagocytosed bacteria.

Management

G6PD deficiency should be suspected in patients of African, Mediterranean, or Asian ancestry who present with either an acute hemolytic episode or neonatal jaundice. G6PD deficiency is diagnosed by measurement of G6PD activity in erythrocytes; this activity should be measured only when the patient has had neither a recent transfusion nor a recent hemolytic episode. (Because G6PD deficiency occurs primarily in older erythrocytes, measurement of G6PD activity in the predominantly young erythrocytes present during or immediately after a hemolytic episode often gives a false-negative result.)

The key to management of G6PD deficiency is prevention of hemolysis by prompt treatment of infections and avoidance of oxidant drugs (e.g., sulfonamides, sulfones, nitrofurans) and toxins (e.g., naphthalene). Although most patients with a hemolytic episode will not require medical intervention, those with severe anemia and hemolysis may require resuscitation and erythrocyte transfusions. Patients presenting with neonatal jaundice respond well to the same therapies as for other infants with neonatal jaundice (hydration, light therapy, and exchange transfusions).

INHERITANCE RISK

Each son of a mother carrying a *G6PD* mutation has a 50% chance of being affected, and each daughter has a 50% chance of being a carrier. Each daughter of an affected father will be a carrier, but each son will be unaffected because an affected father does not contribute an X chromosome to his sons. The risk that carrier daughters will have clinically significant symptoms is low because sufficient skewing of X chromosome inactivation is relatively uncommon.

QUESTIONS FOR SMALL GROUP DISCUSSION

1. The consumption of fava beans and the occurrence of G6PD deficiency are coincident in many areas. What evolutionary advantage might the consumption of fava beans give populations with G6PD deficiency?
2. Several hundred different mutations have been described that cause G6PD deficiency. Presumably, all of these mutations have persisted because of selection. Discuss heterozygote advantage in the context of G6PD deficiency.
3. What is pharmacogenetics? How does G6PD deficiency illustrate the principles of pharmacogenetics?

REFERENCES

Dunn IIF: The triumph of good over evil: protection by the sickle gene against malaria, *Blood* 121:20–25, 2013.

Howes RE, Battle KE, Satyagraha AW, et al: G6PD deficiency: global distribution, genetic variants and primaquine therapy, *Adv Parasitol* 81:133–201, 2013.

Luzzatto L, Seneca E: G6PD deficiency: a classic example of pharmacogenetics with on-going clinical implications, *Br J Haematol* 164:469–480, 2014.

HEREDITARY HEMOCHROMATOSIS (*HFE* Mutation, MIM 235200)

Autosomal Recessive

PRINCIPLES

- Incomplete penetrance and variable expressivity
- Sex differences in penetrance
- Population screening versus at-risk testing
- Molecular versus biochemical testing

MAJOR PHENOTYPIC FEATURES

- Age at onset: 40 to 60 years in males; after menopause in females
- Fatigue, impotence, hyperpigmentation (bronzing), diabetes, cirrhosis, cardiomyopathy
- Elevated serum transferrin iron saturation
- Elevated serum ferritin level

HISTORY AND PHYSICAL FINDINGS

S.F. was a 30-year-old healthy white man referred to the genetics clinic for counseling because his 55-year-old father had just been diagnosed with cirrhosis due to hereditary hemochromatosis. History and physical examination findings were normal. His transferrin iron saturation was 48% (normal, 20% to 50%). His serum ferritin level was normal (<300 ng/mL), and liver transaminase activities were normal. Given that S.F. was an obligate carrier for the condition and his mother had an 11% population risk for being a carrier of mutations in the hereditary hemochromatosis gene *(HFE)*, his prior risk for having inherited two mutant *HFE* alleles was 5.5%. S.F. chose to have his *HFE* gene examined for the two common hemochromatosis variants. Molecular testing revealed that he was homozygous for the Cys282Tyr mutation, putting him at risk for development of hemochromatosis. He was referred to his primary care provider to follow serum ferritin levels every 3 months and to institute therapy as needed.

BACKGROUND

Disease Etiology and Incidence

Hereditary hemochromatosis (MIM 235200) is a disease of iron overload that occurs in some individuals with homozygous or compound heterozygous mutations in the *HFE* gene. Most patients (90% to 95%) with hereditary hemochromatosis are homozygous for a Cys282Tyr mutation; the remaining 5% to 10% of affected individuals are compound heterozygotes for the Cys282Tyr and another mutation, His63Asp. Homozygosity for His63Asp does not lead to clinical hemochromatosis unless there is an additional cause of iron overload. The carrier rate in the white population of North America is approximately 11% for Cys282Tyr and approximately 27% for His63Asp, which means that approximately 1 in 330 individuals will be Cys282Tyr homozygotes and an additional 1 in 135 will be compound heterozygotes for *HFE* disease-causing mutations. The frequency of these mutations is far lower in Asians, Africans, and Native Americans.

The penetrance of clinical hereditary hemochromatosis has been difficult to determine; estimates vary from 10% to 70%, depending on whether hereditary hemochromatosis is defined as organ damage due to pathological iron overload or by biochemical evidence of an elevation of ferritin and transferrin saturation. In a family-based study, for example, between 75% and 90% of homozygous relatives of affected individuals were asymptomatic. Population-based studies have provided estimates of penetrance based on biochemical evidence of hereditary hemochromatosis of 25% to 50%, but penetrance may be higher if liver biopsies are performed to find occult cirrhosis. Whatever the penetrance, it is clear that males are affected more frequently than females and that Cys282Tyr/His63Asp compound heterozygotes are at much lower risk for hereditary hemochromatosis than Cys282Tyr homozygotes. Although the exact value of the penetrance in Cys282Tyr homozygotes remains to be definitively determined, penetrance is clearly incomplete.

At least four additional iron overload disorders labeled hemochromatosis have been identified on the basis of clinical, biochemical, and genetic characteristics. Juvenile hemochromatosis, or hemochromatosis type 2 (*HFE2*), is autosomal recessive and is divided into two forms: HFE2A, caused by mutation in *HJV*, and HFE2B, caused by mutation in *HAMP*. Hemochromatosis type 3, an autosomal recessive disorder, is caused by mutation in *TFR2*. Hemochromatosis type 4, an autosomal dominant disorder, is caused by mutation in *SLC40A1*.

Pathogenesis

Hereditary hemochromatosis is a disorder of iron overload. Body stores of iron are determined largely by dietary iron absorption from enterocytes of the small intestine and release of endogenous iron from macrophages that phagocytose red blood cells. Iron release from enterocytes and macrophages is regulated by a circulating iron response hormone, hepcidin, which is synthesized in the liver and released to block further iron absorption when iron supplies are adequate. Mutant *HFE* interferes with hepcidin signaling, which results in the stimulation of enterocytes and macrophages to release iron. The body therefore continues to absorb and recycle iron, despite an iron-overloaded condition.

Ultimately, a small proportion of individuals with two mutations in the *HFE* gene will develop symptomatic iron overload. Early symptoms include fatigue, arthralgia, decreased libido, and abdominal pain. An additional presentation is the finding of elevated transferrin iron saturation or ferritin on routine screening. Late findings of iron overload include hepatomegaly, cirrhosis (Fig. C-20), hepatocellular carcinoma, diabetes mellitus, cardiomyopathy, hypogonadism, arthritis, and increased skin pigmentation (bronzing). Males develop symptoms between the ages of 40 and 60 years. Women, who are reported to develop symptoms at 10% to 50% the rate of men, do not develop symptoms until after menopause. Prognosis is excellent in patients diagnosed and treated before the development of cirrhosis. Patients diagnosed with cirrhosis and treated effectively with phlebotomy still have a 10% to 30% risk for liver cancer years later.

Management

The diagnosis of clinical *HFE*-associated hereditary hemochromatosis in individuals with clinical findings consistent with *HFE*-associated hereditary hemochromatosis and the

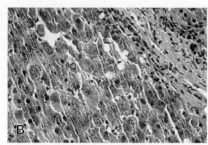
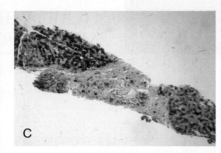

Figure C-20 **Liver of patient with hereditary hemochromatosis showing iron deposition and cirrhosis. A,** Low-power view showing area of fibrosis (*arrow;* hematoxylin and eosin stain). **B,** Higher power view showing iron deposition (*brown* pigment seen within hepatocytes) next to an area of fibrosis (hematoxylin and eosin stain). **C,** Perl's stain in which iron stains *dark blue.* Heavy staining in hepatocytes flanks an area of fibrosis with much less iron deposition. *See Sources & Acknowledgments.*

diagnosis of biochemical *HFE*-associated hereditary hemochromatosis is typically based on finding elevated transferrin-iron saturation of 45% or higher and serum ferritin concentration above the upper limit of normal (i.e., >300 ng/mL in men and >200 ng/mL in women), as well as two *HFE*-associated hereditary hemochromatosis–causing mutations on confirmatory *HFE* gene testing. Although serum ferritin concentration may increase progressively over time in untreated individuals with *HFE*-associated hereditary hemochromatosis, it is not specific for *HFE*-associated hereditary hemochromatosis and therefore cannot be used alone for identification of individuals with *HFE*-associated hereditary hemochromatosis.

Individuals with an at-risk genotype are monitored with serum ferritin levels annually. If the level is higher than 50 ng/mL, phlebotomy to remove a unit of blood is recommended to maintain normal levels. Phlebotomy is repeated until a normal ferritin concentration is achieved. Failure to achieve a normal ferritin concentration within 3 months of starting phlebotomy is a poor prognostic sign. Once the ferritin concentration is below 50 ng/mL, maintenance phlebotomy is performed every 3 to 4 months for men and every 6 to 12 months for women. Symptomatic patients with initial ferritin concentrations of more than 1000 ng/mL should undergo liver biopsy to determine if cirrhosis is present. Patients found to have biochemical abnormalities should undergo phlebotomy weekly until the hematocrit is 75% of the baseline and the ferritin concentration is below 50 ng/mL.

INHERITANCE RISK

Hereditary hemochromatosis is an autosomal recessive disorder with reduced penetrance. The sibs of an affected individual have a 25% chance of having two mutations. The child of an affected individual will be a carrier and has a 5% risk for having two mutations if both parents are white. Because of the apparently low penetrance of this disease, universal population screening for *HFE* mutations is not indicated. However, because of the prevalence of the disorder, the uncertainty as to the true penetrance, and the availability of easy and effective treatment, one-time screening of serum transferrin iron saturation and ferritin concentrations in adult white, non-Hispanic men of northern European descent may be justified.

QUESTIONS FOR SMALL GROUP DISCUSSION

1. Why do women have a much lower incidence of clinical hemochromatosis?
2. Besides phlebotomy, what dietary interventions would be indicated to prevent iron overload?
3. Discuss the possible reasons for the high prevalence of the Cys282Tyr mutation among whites.

REFERENCES

Barton JC: Hemochromatosis and iron overload: from bench to clinic, *Am J Med Sci* 346:403–412, 2013.

Kanwar P, Kowdley KV: Diagnosis and treatment of hereditary hemochromatosis: an update, *Expert Rev Gastroenterol Hepatol* 7:517–530, 2013.

Kowdley KV, Bennett RL, Motulsky AG: *HFE*-associated hereditary hemochromatosis. Available from: http://www.ncbi.nlm.nih.gov/books/NBK1440/.

HEMOPHILIA (*F8* or *F9* Mutation, MIM 307600 and MIM 306900)

X-Linked

PRINCIPLES

- Intrachromosomal recombination
- Transposable element insertion
- Variable expressivity
- Protein replacement therapy

MAJOR PHENOTYPIC FEATURES

- Age at onset: Infancy to adulthood
- Bleeding diathesis
- Hemarthroses
- Hematomas

HISTORY AND PHYSICAL FINDINGS

S.T., a healthy 38-year-old woman, scheduled an appointment for counseling regarding her risk for having a child with hemophilia. She had a maternal uncle who had died in childhood from hemophilia and a brother who had had bleeding problems as a child. Her brother's bleeding problems had resolved during adolescence. No other family members had bleeding disorders. The geneticist explained to S.T. that her family history was suggestive of an X-linked abnormality of coagulation such as hemophilia A or B and that her brother's improvement was particularly suggestive of the hemophilia B variant factor IX Leyden. To confirm the diagnosis of hemophilia, the geneticist told S.T. that her brother should be evaluated first because identification of an isolated carrier is difficult. S.T. talked to her brother, and he agreed to an evaluation. Review of his records showed that he had been diagnosed with factor IX deficiency as a child but now had nearly normal plasma levels of factor IX. DNA mutation analysis confirmed that he had a mutation in the *F9* gene promoter, consistent with factor IX Leyden. Subsequent testing of S.T. showed that she did not carry the mutation identified in her brother.

BACKGROUND

Disease Etiology and Incidence

Hemophilia A (MIM 307600) and hemophilia B (MIM 306900) are X-linked disorders of coagulation caused by mutations in the *F8* and *F9* genes, respectively. Mutations of *F8* cause deficiency or dysfunction of clotting factor VIII; mutations of *F9* cause deficiency or dysfunction of clotting factor IX.

Hemophilia is a panethnic disorder without racial predilection. Hemophilia A has an incidence of 1 in 5000 to 10,000 newborn males. Hemophilia B is far more rare, with an incidence of 1 in 100,000.

Pathogenesis

The coagulation system maintains the integrity of the vasculature through a delicate balance of clot formation and inhibition. The proteases and protein cofactors composing the clotting cascade are present in the circulation as inactive precursors and must be sequentially activated at the site of injury to form a fibrin clot. Timely and efficient formation of a clot requires exponential activation or amplification of the protease cascade. Clotting factors VIII and IX, which complex together, are key to this amplification; they activate clotting factor X, and active factor X, in turn, activates more factor IX and factor VIII (see Figure 8-8). Factor IX functions as a protease and factor VIII as a cofactor. Deficiency or dysfunction of either factor IX or factor VIII causes hemophilia.

Mutations of *F8* include deletions, insertions, inversions, and point mutations. The most common mutation is an inversion deleting the carboxyl terminus of factor VIII; it accounts for 25% of all hemophilia A and for 40% to 50% of severe hemophilia A. This inversion results from an intrachromosomal recombination between sequences in intron 22 of *F8* and homologous sequences telomeric to *F8*. Another intriguing class of mutation involves retrotransposition of L1 repeats into the gene. For all *F8* mutations, the residual enzymatic activity of the factor VIII–factor IX complex correlates with the severity of clinical disease (see Table).

Many different *F9* mutations have been identified in patients with hemophilia B; but in contrast to the frequent partial inversion of *F8* in hemophilia A, a common *F9* mutation has not been identified for hemophilia B. Factor IX Leyden is an unusual *F9* variant caused by point mutations in the *F9* promoter; it is associated with very low levels of factor IX and severe hemophilia during childhood, but spontaneous resolution of hemophilia occurs at puberty as factor IX levels nearly normalize. For each *F9* mutation, the residual enzymatic activity of the factor VIII–factor IX complex correlates with the severity of clinical disease (see Table).

Phenotype and Natural History

Hemophilia is classically a male disease, although rarely females can be affected because of skewed X chromosome inactivation. Clinically, hemophilia A and hemophilia B are indistinguishable. Both are characterized by bleeding into soft tissues, muscles, and weight-bearing joints (Fig. C-21). Bleeding occurs within hours to days after trauma and often continues for days or weeks. Those with severe disease are usually diagnosed as newborns because of excessive cephalohematomas or prolonged bleeding from umbilical or circumcision wounds. Patients with moderate disease often do not develop hematomas or hemarthroses until they begin to crawl or walk and therefore escape diagnosis until that time. Patients with mild disease frequently present in adolescence or adulthood with hemarthroses or prolonged bleeding after surgery or trauma.

Hemophilia A and hemophilia B are diagnosed and distinguished by measurement of factor VIII and IX activity levels. For both hemophilia A and hemophilia B, the level of factor VIII or IX activity predicts the clinical severity.

Clinical Classification and Clotting Factor Levels

Classification	% Activity (Factor VIII or IX)
Severe	<1
Moderate	1-5
Mild	5-25

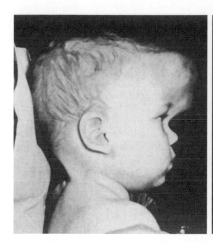
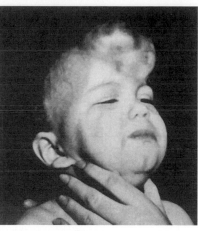

Figure C-21 **Subcutaneous hematoma of the forehead in a young boy with hemophilia.** The photograph was taken 4 days after a minor contusion. The appearance of the forehead returned to normal in 6 months. *See Sources & Acknowledgments.*

Management

The diagnosis of hemophilia A is established by identifying low factor VIII clotting activity in the presence of a normal von Willebrand factor level. Molecular genetic testing of *F8*, the gene encoding factor VIII, identifies disease-causing mutations in as many as 98% of individuals with hemophilia A. The diagnosis of hemophilia B is established by identifying low factor IX clotting activity. Molecular genetic testing of *F9*, the gene encoding factor IX, identifies disease-causing mutations in more than 99% of individuals with hemophilia B. Both tests are available clinically.

Although current gene therapy trials show promise, no curative treatments are available for hemophilia A and hemophilia B except for liver transplantation (see Chapter 13). Currently the standard of care is intravenous replacement of the deficient factor. Factor replacement therapy has increased life expectancy from an average of 1.4 years in the early 1900s to approximately 65 years today.

INHERITANCE RISK

If a woman has a family history of hemophilia, her carrier status can be determined by linkage analysis or by identification of the *F8* or *F9* mutation segregating in the family. Routine mutation identification used to be available only for the common *F8* inversion, but advances in DNA sequencing have made targeted exome sequencing much more effective. Carrier detection by enzyme assay is difficult and not widely available.

If a mother is a carrier, each son has a 50% risk for hemophilia, and each daughter has a 50% risk for inheriting the *F8* or *F9* mutation. Reflecting the low frequency of clinically significant skewing of X chromosome inactivation, daughters inheriting an *F8* or *F9* mutation have a low risk for hemophilia.

If a mother has a son with hemophilia but no other affected relatives, her a priori risk for being a carrier depends on the type of mutation. Point mutations and the common *F8* inversions almost always arise in male meiosis; as a result, 98% of mothers of a male with one of these mutations are carriers due

to a new mutation in their father (the affected male's maternal grandfather). In contrast, deletion mutations usually arise during female meiosis. If there is no knowledge of the mutation type, then approximately one third of patients are assumed to have a new mutation in *F8* or *F9*. Through the application of Bayes's theorem, this risk can be modified by considering the number of unaffected sons in the family (see Chapter 16).

QUESTIONS FOR SMALL GROUP DISCUSSION

1. What other diseases are caused by recombination between repeated genome sequences? Compare and contrast the recombination mechanism observed with hemophilia A with that observed with Smith-Magenis syndrome and with familial hypercholesterolemia.
2. One of the more unusual mutations in *F8* is insertion of an L1 element into exon 14. What are transposable elements? How do transposable elements move within a genome? Name another disease caused by movement of transposable elements.
3. In patients with hemophilia B due to factor IX Leyden, why does the deficiency of factor IX resolve during puberty?
4. Compare and contrast protein replacement for hemophilia to that for Gaucher disease. Approximately 10% of patients with hemophilia develop a clinically significant antibody titer against factor VIII or IX. Why? Is there a genetic predisposition to development of antibodies against the replacement factors? How could this immune reaction be circumvented? Would gene therapy be helpful for patients with antibodies?
5. Discuss current approaches to gene therapy in hemophilia B.

REFERENCES

Konkle BA, Josephson NC, Nakaya Fletcher S: Hemophilia A. Available from: http://www.ncbi.nlm.nih.gov/books/NBK1404/.

Konkle BA, Josephson NC, Nakaya Fletcher S: Hemophilia B. Available from: http://www.ncbi.nlm.nih.gov/books/NBK1495/.

Santagostino E, Fasulo MR: Hemophilia A and hemophilia B: different types of diseases? *Semin Thromb Hemost* 39:697–701, 2013.

HIRSCHSPRUNG DISEASE (Neurocristopathy, MIM 142623)

Autosomal Dominant, Autosomal Recessive, or Multigenic

PRINCIPLES

- Genetic heterogeneity
- Incomplete penetrance and variable expressivity
- Genetic modifiers
- Sex-dependent penetrance

MAJOR PHENOTYPIC FEATURES

- Age at onset: Neonatal to adulthood
- Constipation
- Abdominal distention
- Enterocolitis

HISTORY AND PHYSICAL FINDINGS

S.L. and P.L. were referred to the genetics clinic to discuss their risk for having another child with Hirschsprung disease; their daughter had been born with long-segment Hirschsprung disease and was doing well after surgical removal of the aganglionic segment of colon. On examination and by history, the daughter did not have signs or symptoms of other diseases. The mother knew of an uncle and a brother who had died in infancy of bowel obstruction. The genetic counselor explained that in contrast to short-segment Hirschsprung disease, long-segment disease usually segregates as an autosomal dominant trait with incomplete penetrance and is often caused by mutations in the *RET* (*re*arranged during *transfection*) gene, which encodes a cell surface tyrosine kinase receptor. Subsequent testing showed that the affected daughter and the mother were heterozygous for a loss of function mutation in *RET*.

BACKGROUND

Disease Etiology and Incidence

Hirschsprung disease (HSCR, MIM 142623) is the congenital absence of parasympathetic ganglion cells in the submucosal and myenteric plexuses along a variable length of the intestine (Fig. C-22); aganglionosis extending from the internal anal sphincter to proximal of the splenic flexure is classified as long-segment disease, whereas aganglionosis with a proximal limit distal to the splenic flexure is classified as short-segment disease. Approximately 70% of HSCR occurs as an isolated trait, 12% in conjunction with a recognized chromosomal abnormality, and 18% in conjunction with multiple congenital anomalies (Waardenburg-Shah syndrome, Mowat-Wilson syndrome, Goldberg-Shprintzen megacolon syndrome, and congenital central hypoventilation syndrome).

Isolated or nonsyndromic HSCR is a panethnic, incompletely penetrant, sex-biased disorder with intrafamilial and interfamilial variation in expressivity; it has an incidence from 1.8 per 10,000 live births among Europeans to 2.8 per 10,000 live births among Asians. Long-segment disease, including total colonic aganglionosis, generally segregates as an autosomal dominant low-penetrance disorder; short-segment disease usually exhibits autosomal recessive or multigenic inheritance.

Pathogenesis

The enteric nervous system forms predominantly from vagal neural crest cells that migrate craniocaudally during the 5th to 12th weeks of gestation. Some enteric neurons also migrate cranially from the sacral neural crest; however, correct migration and differentiation of these cells depend on the presence of vagal neural crest cells.

HSCR arises from premature arrest of craniocaudal migration of vagal neural crest cells in the hindgut and thus is characterized by the absence of parasympathetic ganglion cells in the submucosal and myenteric plexuses of the affected intestine. The genes implicated in HSCR include *RET, EDNRB, EDN3, GDNF,* and *NRTN*. How mutations in these genes cause premature arrest of the craniocaudal migration of vagal neural crest cells remains undefined. Regardless of the mechanism, the absence of ganglion cells causes loss of peristalsis and thereby intestinal obstruction.

RET is the major susceptibility gene for isolated HSCR. Nearly all families with more than one affected patient demonstrate linkage to the *RET* locus. However, mutations in the *RET* coding sequence can be identified in only approximately 50% of patients with familial HSCR and in 15% to 35% of patients with sporadic HSCR. In addition, within families segregating mutant *RET* alleles, penetrance is only 65% in males and 45% in females. A common noncoding variant within a conserved, enhancer-like sequence in intron 1 of *RET* has been shown to be associated with HSCR and accounts for incomplete penetrance and sex differences (see Chapter 8). In addition, the variant is far more frequent in Asians than in whites, explaining the population differences.

Phenotype and Natural History

Patients with HSCR usually present early in life with impaired intestinal motility; however, 10% to 15% of patients are not identified until after a year of life. Approximately 50% to 90% of patients fail to pass meconium within the first 48 hours of life. After the newborn period, patients can present with constipation (68%), abdominal distention (64%), emesis (37%), or occasionally diarrhea; 40% of these patients have a history of delayed passage of meconium.

Untreated, HSCR is generally fatal. Failure to pass stool sequentially causes dilatation of the proximal bowel, increased intraluminal pressure, decreased blood flow, deterioration of the mucosal barrier, bacterial invasion, and enterocolitis. Recognition of HSCR before the onset of enterocolitis is essential to reducing morbidity and mortality.

HSCR frequently occurs as part of a syndrome or complex of malformations. As a neurocristopathy, HSCR is part of a continuum of diseases involving tissues of neural crest origin such as peripheral neurons, Schwann cells, melanocytes, conotruncal cardiac tissue, and endocrine and paraendocrine tissues. An illustration of this continuum is Waardenburg syndrome type IV, which is characterized by HSCR, deafness, and the absence of epidermal melanocytes.

Management

The diagnosis of HSCR requires histopathological demonstration of the absence of enteric ganglion cells in the distal rectum (see Fig. C-22C). Biopsy specimens for such testing are usually

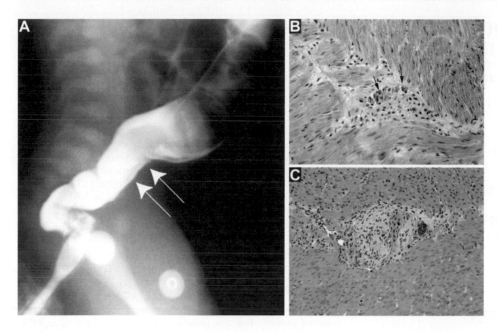

Figure C-22 A, Barium enema study of a 3-month-old child with Down syndrome with a history of severe constipation. Note the narrowing of the distal colon, with the transition from dilated to narrowed colon demarcated by *arrows*; subsequent mucosal biopsy showed an absence of myenteric ganglion cells consistent with Hirschsprung disease. **B,** Normal myenteric ganglion. Myenteric ganglion cells (*arrows*) are located normally in the plexus between the longitudinal and circular layers of the muscularis propria. **C,** Aganglionic distal bowel of Hirschsprung disease. *See Sources & Acknowledgments.*

Sex-Dependent Recurrence Risk in Siblings of a Proband with HSCR

Sex of Proband	Sex of Sibling	Proband Phenotype	Recurrence Risk (%)
Male	Male	Long-segment HSCR	17
		Short-segment HSCR	5
	Female	Long-segment HSCR	13
		Short-segment HSCR	1
Female	Male	Long-segment HSCR	33
		Short-segment HSCR	5
	Female	Long-segment HSCR	9
		Short-segment HSCR	3

HSCR, Hirschsprung disease.
From Parisi M: Hirschsprung disease overview. Available from: http://www.ncbi.nlm.nih.gov/books/NBK1439/.

obtained by suction biopsies of the rectal mucosa and submucosa.

Definitive treatment of HSCR involves removal or bypassing of the aganglionic segment of bowel. The surgical procedure also usually involves the anastomosis of innervated bowel to the anal sphincter rather than a permanent colostomy. The prognosis for surgically treated patients is generally good, and most patients achieve fecal continence; however, a few patients have postoperative problems including enterocolitis, strictures, prolapse, perianal abscesses, and incontinence.

INHERITANCE RISK

Nonsyndromic HSCR has a 4:1 predominance in males versus females as well as variable expressivity and incomplete penetrance. The empirical recurrence risk for HSCR in siblings is dependent on the sex of the proband, the length of aganglionosis in the proband, and the sex of the sibling (see Table).

Prenatal counseling is complicated by the incomplete penetrance and variable expressivity. Even if a mutation has been identified in a family, generally neither the prediction of short- or long-segment HSCR nor the prediction of nonsyndromic or syndromic HSCR is possible. Moreover, prenatal diagnosis is currently further complicated by the poor availability of molecular testing.

QUESTIONS FOR SMALL GROUP DISCUSSION

1. Mutations in the *RET* gene also cause multiple endocrine neoplasia; how do these mutations generally differ from those observed in HSCR disease? On occasion, the same mutation can cause both HSCR and multiple endocrine neoplasia; discuss possible explanations for this.
2. Discuss how stochastic, genetic, and environmental factors can cause incomplete penetrance, and give examples of each.
3. Haddad syndrome (congenital central hypoventilation and HSCR) has also been associated with mutations of *RET, GDNF,* and *EDN3.* Describe the developmental relationship and pathology of HSCR and congenital central hypoventilation.
4. Mutations of the transcription factor *SOX10* cause Waardenburg syndrome type IV plus dysmyelination of the central and peripheral nervous system. Mutations of the endothelin pathway cause HSCR and Waardenburg syndrome type IV without dysmyelination. Mutations of *RET* and its ligands cause HSCR but not Waardenburg syndrome type IV or dysmyelination. Discuss what these observations say about the relationship between these three pathways and their regulation of neural crest cells.
5. Compare and contrast the various forms of multigenic inheritance, that is, additive, multiplicative, mixed multiplicative, and epistatic inheritance.

REFERENCES

Emison E, McCallion AS, Cashuk CS, et al: A common sex dependent mutation in a RET enhancer underlies Hirschsprung disease risk, *Nature* 434:857–863, 2005.
Langer JC: Hirschsprung disease, *Curr Opin Pediatr* 25:368–374, 2013.
Parisi MA: Hirschsprung disease overview. Available from: http://www.ncbi.nlm.nih.gov/books/NBK1439/.

HOLOPROSENCEPHALY (NONSYNDROMIC FORM)
(Sonic Hedgehog (*SHH*) Mutation, MIM 236100)
Autosomal Dominant

PRINCIPLES

- Developmental regulatory gene
- Genetic heterogeneity
- Position-effect mutations
- Incomplete penetrance and variable expressivity

MAJOR PHENOTYPIC FEATURES

- Age at onset: Prenatal
- Ventral forebrain maldevelopment
- Facial dysmorphism
- Developmental delay

HISTORY AND PHYSICAL FINDINGS

Dr. D., a 37-year-old physicist, presented to the genetics clinic with his wife because their first child died at birth of holoprosencephaly. The pregnancy had been uncomplicated, and the child had a normal karyotype. Neither he nor his wife reported any major medical problems. Dr. D. had been adopted as a child and did not know the history of his biological family; his wife's family history was not suggestive of any genetic disorders. Careful examination of Dr. D. and his wife showed that he had an absent superior labial frenulum and slight hypotelorism but no other dysmorphic findings. His physician explained to him that the holoprosencephaly in his child and his absent superior labial frenulum and slight hypotelorism were suggestive of autosomal dominant holoprosencephaly. Subsequent molecular testing confirmed that Dr. D. had a mutation in the sonic hedgehog gene (*SHH*).

BACKGROUND

Disease Etiology and Incidence

Holoprosencephaly (HPE, MIM 236100) has a birth incidence of 1 in 10,000 to 1 in 12,000 and is the most common human congenital brain defect. It affects twice as many girls as boys.

HPE results from a variety of causes, including chromosomal and single-gene disorders, environmental factors such as maternal diabetes, and possibly maternal exposure to cholesterol-lowering agents (statins). The disorder occurs both in isolation and as a feature of various syndromes, such as Smith-Lemli-Opitz syndrome. Nonsyndromic familial HPE, when inherited, is predominantly autosomal dominant, although both autosomal recessive and X-linked inheritance have been reported. Approximately 25% to 50% of all HPE is associated with a chromosomal abnormality; the nonrandom distribution of chromosomal abnormalities predicts at least 12 different HPE loci, including 7q36, 13q32, 2p21, 18p11.3, and 21q22.3.

SHH, the first gene identified with mutations causing HPE, maps to 7q36. *SHH* mutations account for approximately 30% to 40% of familial nonsyndromic autosomal dominant HPE but for less than 5% of nonsyndromic HPE overall. Other genes implicated in autosomal dominant nonsyndromic HPE are *ZIC2*, accounting for 5%; *SIX3* and *TGIF*, each accounting for 1.3%; *PTCH1, CDON, GLI2, FOXH1*, and *NODAL, HPE6*, and *HPE8* are rare causes.

Pathogenesis

SHH is a secreted signaling protein required for developmental patterning in both mammals and insects (see Chapter 14).

Human *SHH* mutations are loss-of-function mutations. Some of the cytogenetic abnormalities affecting *SHH* expression are translocations that occur 15 to 256 kb 5′ to the coding region of *SHH*. These translocations are referred to as position-effect mutations because they do not mutate the coding sequence but disrupt distant regulatory elements, chromatin structure, or both, and thereby alter *SHH* expression.

Phenotype and Natural History

The prosencephalic malformations of HPE follow a continuum of severity but are usually subdivided into alobar HPE (no evidence of an interhemispheric fissure), semilobar HPE (posterior interhemispheric fissure only), and lobar HPE (ventricular separation and almost complete cortical separation) (Fig. C-23). Among HPE patients with a normal karyotype, 63% have alobar HPE, 28% have semilobar HPE, and 9% have lobar HPE. Other commonly associated central nervous system malformations include undivided thalami, dysgenesis of the corpus callosum, hypoplastic olfactory bulbs, hypoplastic optic bulbs and tracts, and pituitary dysgenesis.

The spectrum of facial dysmorphism in HPE extends from cyclopia to normal and usually reflects the severity of the central nervous system malformations. Dysmorphic features associated with, but not diagnostic of, HPE include microcephaly or macrocephaly, anophthalmia or microphthalmia, hypotelorism or hypertelorism, dysmorphic nose, palatal anomalies, bifid uvula, a single central incisor, and absence of a superior labial frenulum.

Delayed development occurs in nearly all patients with HPE. The severity of delay correlates with the severity of central nervous system malformation; that is, patients with normal brain imaging usually have normal intelligence. In addition to delayed development, patients frequently have seizures, brainstem dysfunction, and sleep dysregulation.

Among HPE patients without chromosomal abnormalities, survival varies inversely with the severity of the facial phenotype. Patients with cyclopia or ethmocephaly usually do not survive a week; approximately 50% of patients with alobar HPE die before 4 to 5 months of age, and 80% before a year. Approximately 50% of patients with isolated semilobar or lobar HPE survive the first year.

Management

Patients with HPE require an expeditious evaluation within the first few days of life. Treatment is symptomatic and supportive. Aside from the medical concerns of the patient, a major part of management includes counseling and supporting the parents, as well as defining the cause of HPE.

INHERITANCE RISK

Etiologically, HPE is extremely heterogeneous, and the recurrence risk in a family is dependent on identification of the underlying cause. Diabetic mothers have a 1% risk for having

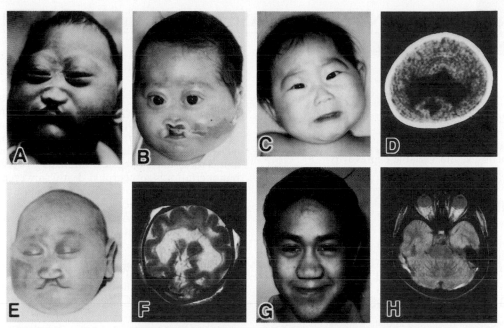

Figure C-23 **Holoprosencephaly (HPE) in patients with *SHH* mutations. A,** Microcephaly, absence of nasal bones, midline cleft palate, and semilobar HPE. **B,** Semilobar HPE, premaxillary agenesis, and midline cleft lip. **C** and **D,** Mild facial findings with severe semilobar HPE on magnetic resonance imaging. **E** and **F,** Microcephaly, prominent optic globes, premaxillary agenesis, and cleft lip, with semilobar HPE on magnetic resonance imaging. **G** and **H,** Microcephaly, ocular hypotelorism, flat nose without palpable cartilage, midface and philtrum hypoplasia, normal intelligence, and normal brain on magnetic resonance imaging. All patients have *SHH* mutations. Patients **A** and **B** also have mutations of *TGIF*, and patient **C** also has a mutation in *ZIC2*. *TGIF* mutations indirectly decrease SHH expression. *See Sources & Acknowledgments.*

a child with HPE. For parents of a patient with a cytogenetic anomaly, the recurrence risk depends on whether one of them has a cytogenetic abnormality that gave rise to the abnormality in the patient. For parents of patients with syndromic HPE, the recurrence risk depends on the recurrence risk for that syndrome. In the absence of a family history of HPE or a cytogenetic or syndromic cause of HPE, parents and siblings must be examined closely for microforms, subtle features associated with HPE such as an absent frenulum or a single central upper incisor. For parents with a negative family history, no identifiable causes of HPE, and no microforms suggestive of autosomal dominant HPE, the empirical recurrence risk is approximately 4% to 5%. In some cases, digenic inheritance may explain the reduced penetrance of some *SHH* mutations.

Although autosomal recessive and X-linked HPE have been reported, most families with an established mode of inheritance exhibit autosomal dominant inheritance. The penetrance of autosomal dominant HPE is approximately 70%. Among obligate carriers of autosomal dominant HPE, the risk for having a child affected with severe HPE is 16% to 21% and with a microform, 13% to 14%. The phenotype of the carrier does not affect the risk for having an affected child, nor does it predict the severity if the child is affected.

Molecular testing for certain of the HPE mutations is currently available as a clinical service. Severe HPE can be detected by prenatal ultrasound examination at 16 to 18 weeks of gestation.

QUESTIONS FOR SMALL GROUP DISCUSSION

1. What factors might explain the variable expressivity and penetrance of *SHH* mutations among siblings?
2. Discuss genetic disorders with a sex bias and the mechanisms underlying the sex bias. As examples, consider Rett syndrome to illustrate embryonic sex-biased lethality, pyloric stenosis to illustrate a sex bias in disease frequency, and coronary heart disease in familial hypercholesterolemia to illustrate a sex bias in disease severity.
3. Considering the many loci associated with HPE, discuss why mutations in different genes give rise to identical phenotypes.
4. Considering that *GLI3* is in the signal transduction cascade of SHH, discuss why *GLI3* loss-of-function mutations do not give rise to the same phenotype as *SHH* loss-of-function mutations.
5. Discuss the role of cholesterol in brain morphogenesis.

REFERENCES

Kauvar EF, Muenke M: Holoprosencephaly: recommendations for diagnosis and management, *Curr Opin Pediatr* 22:687–695, 2010.
Solomon BD, Gropman A, Muenke M: Holoprosencephaly overview. Available from: http://www.ncbi.nlm.nih.gov/books/NBK1530/.

PRINCIPLES

- Triplet repeat expansion
- Novel property mutation
- Sex-specific anticipation
- Reduced penetrance and variable expressivity
- Presymptomatic counseling

MAJOR PHENOTYPIC FEATURES

- Age at onset: Late childhood to late adulthood
- Movement abnormalities
- Cognitive abnormalities
- Psychiatric abnormalities

HISTORY AND PHYSICAL FINDINGS

M.P., a 45-year-old man, presented initially with declining memory and concentration. As his intellectual function deteriorated during the ensuing year, he developed involuntary movements of his fingers and toes as well as facial grimacing and pouting. He was aware of his condition and became depressed. He had been previously healthy and did not have a history of any similarly affected relatives; his parents had died in their 40s in an automobile accident. M.P. had one healthy daughter. After an extensive evaluation, the neurologist diagnosed M.P.'s condition as Huntington disease. The diagnosis of Huntington disease was confirmed by a DNA analysis showing 43 CAG repeats in one of his *HD* alleles (normal, <26). Subsequent presymptomatic testing of M.P.'s daughter showed that she had also inherited the mutant *HD* allele (Fig. C-24). Both received extensive counseling.

BACKGROUND

Disease Etiology and Incidence

Huntington disease (HD, MIM 143100) is a panethnic, autosomal dominant, progressive neurodegenerative disorder that is caused by mutations in the *HD* gene (see Chapter 12). The prevalence of HD ranges from 3 to 7 per 100,000 among western Europeans to 0.1 to 0.38 per 100,000 among Japanese. This variation in prevalence reflects the variation in distribution of *HD* alleles and haplotypes that predispose to mutation.

Pathogenesis

The *HD* gene product, huntingtin, is ubiquitously expressed. The function of huntingtin remains unknown.

Disease-causing mutations in *HD* usually result from an expansion of a polyglutamine-encoding CAG repeat sequence in exon 1; normal *HD* alleles have 10 to 26 CAG repeats, whereas mutant alleles have more than 36 repeats (see Chapter 12). Approximately 3% of patients develop HD as the result of a new CAG repeat expansion; 97% inherit a mutant *HD* allele from an affected parent. New mutant *HD* genes arise from expansion of a premutation (27 to 35 CAG repeats) to a full mutation. When a patient inherits a full mutation from a parent carrying a permutation, that parent is nearly always the father.

Expansion of the huntingtin polyglutamine tract appears to confer a deleterious novel property and to be both necessary

and sufficient for the induction of an HD-like phenotype. In addition to the diffuse, severe atrophy of the neostriatum that is the hallmark of HD, expression of mutant huntingtin causes neuronal dysfunction, generalized brain atrophy, changes in neurotransmitter levels, and accumulation of neuronal nuclear and cytoplasmic aggregates. Ultimately, expression of mutant huntingtin leads to neuronal death; however, it is likely that clinical symptoms and neuronal dysfunction precede the development of intracellular aggregates and neuronal death. The mechanism by which expression of this expanded polyglutamine tract causes HD remains unclear.

Phenotype and Natural History

The patient's age at disease onset is inversely proportional to the number of *HD* CAG repeats. Patients with adult-onset

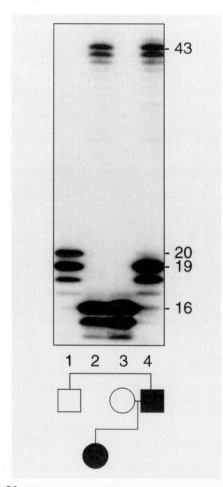

Figure C-24 Segregation of an *HD* gene mutation in a family with Huntington disease and a Southern blot analysis of polymerase chain reaction products derived from amplification of the CAG repeats in exon 1 of the *HD* gene. Each allele generates a full-length fragment as well as two or more shorter fragments because of difficulties with the polymerase chain reaction (PCR) across a triplet repeat. Notice that the affected father and daughter both have an allele with a full mutation (43 CAG repeats) and a normal allele (19 and 16 repeats, respectively). The daughter's unaffected mother and her unaffected paternal uncle have *HD* alleles with a normal number of CAG repeats. *See Sources & Acknowledgments.*

disease usually have 40 to 55 repeats; those with juvenile-onset disease usually have more than 60 repeats (see Fig. 7-20). Patients with 36 to 39 *HD* CAG repeats exhibit reduced penetrance; that is, they may or may not develop HD in their lifetime. Apart from the relationship to the age at onset, the number of repeats does not correlate with other features of HD.

Instability and expansion of the CAG repeats within mutant *HD* alleles often results in anticipation, that is, progressively earlier ages at onset with succeeding generations. Once the number of CAG repeats is 36 or more, the CAG repeat length generally expands during paternal transmission; expansions during maternal transmission are less frequent and shorter than are expansions during paternal transmission. Because the age of onset is inversely correlated with CAG repeat length, individuals with a juvenile onset have a massive expansion of the CAG repeat; it turns out that approximately 80% of such juvenile patients inherit the massively expanded HD allele from their father who is already carrying a full mutation.

Approximately one third of patients present with psychiatric abnormalities; two thirds present with a combination of cognitive and motor disturbances. The mean age at presentation is 35 to 44 years; approximately one quarter of patients develop HD after the age of 50 years, however, and one tenth before the age of 20 years. The median survival after diagnosis is 15 to 18 years, and the mean age at death is approximately 55 years.

HD is characterized by progressive motor, cognitive, and psychiatric abnormalities. The motor disturbances involve both voluntary and involuntary movement. Initially these movements interfere little with daily activities but generally become incapacitating as HD progresses. Chorea, which is present in more than 90% of patients, is the most common involuntary movement; it is characterized by nonrepetitive, nonperiodic jerks that cannot be suppressed voluntarily. Cognitive abnormalities begin early in the disease course and affect all aspects of cognition; language is usually affected later than are other cognitive functions. Behavioral disturbances, which usually develop later in the disease course, include social disinhibition, aggression, outbursts, apathy, sexual deviation, and increased appetite. The psychiatric manifestations, which can develop at any time in the disease course, include personality changes, affective psychosis, and schizophrenia.

In the end stages of HD, patients usually develop such severe motor impairments that they are fully dependent on others. They also experience weight loss, sleep disturbances, incontinence, and mutism. Their behavioral disturbances decrease as the disease advances.

Management

Currently no curative treatments are available for HD. Therapy focuses on supportive care as well as pharmacological management of the behavioral and neurological problems.

INHERITANCE RISK

Each child of a parent with HD has a 50% risk for inheriting a mutant *HD* allele. Except for those alleles with incomplete penetrance (36 to 39 CAG repeats), all children inheriting a mutant *HD* allele will develop HD if they have a normal life span.

Children of fathers carrying a premutation have an empirical risk of approximately 3% for inheriting an *HD* allele in which the premutation has expanded to a full mutation. Not all males carrying a premutation, however, are equally likely to transmit a full mutation.

Presymptomatic testing and prenatal testing are available through analysis of the number of CAG repeats within exon 1 of the *HD* gene. Presymptomatic testing and prenatal testing are forms of predictive testing (see Chapter 16) and are best interpreted after confirmation of a CAG expansion in an affected family member. Recommendations have been made regarding presymptomatic genetic testing for untreatable conditions such as Huntington disease, including the need for neurological and psychological evaluation before testing and the need for psychological support from family members or friends. Additionally, the patient is required to be 18 years of age or older and competent to make an informed decision regarding his or her desire to have presymptomatic test results. The implications of such results are obviously life altering.

QUESTIONS FOR SMALL GROUP DISCUSSION

1. Patients with heterozygous and homozygous mutations of *HD* have similar clinical expression of HD, whereas individuals with deletion of one *HD* allele on chromosome 4p have a normal phenotype. How can this be explained?
2. Some studies suggest that a father with a premutation and an affected child has a higher risk of transmitting a full mutation than a father with a premutation and no affected children. Discuss possible mechanisms for this predisposition to transmit *HD* mutations.
3. Expansion of *HD* premutations to full mutations occurs through the male germline, whereas expansion of *FMR1* (fragile X syndrome) premutations to full mutations occurs through the female germline. Discuss possible mechanisms for sex biases in disease transmission.
4. By international consensus, asymptomatic at-risk children are not tested for *HD* mutations because testing removes the child's choice to know or not know, testing results open the child to familial and social stigmatization, and testing results could affect educational and career decisions. When would it be appropriate to test asymptomatic at-risk children? What advances in medicine are necessary to make testing of all asymptomatic at-risk children acceptable? (Consider the reasoning underlying newborn screening.)

REFERENCES

Bordelon YM: Clinical neurogenetics: Huntington disease, *Neurol Clin* 31:1085–1094, 2013.
Warby SC, Graham RK, Hayden MR: Huntington disease. Available from: http://www.ncbi.nlm.nih.gov/books/NBK1305/.

HYPERTROPHIC CARDIOMYOPATHY (Cardiac Sarcomere Gene Mutations, MIM 192600)

Autosomal Dominant

PRINCIPLES

- Locus heterogeneity
- Age-related penetrance
- Variable expressivity

MAJOR PHENOTYPIC FEATURES

- Age at onset: Adolescence and early adulthood (age 12 to 21 years)
- Left ventricular hypertrophy
- Myocardial crypts or scarring
- Elongated mitral leaflets
- Diastolic dysfunction
- Heart failure
- Sudden death

HISTORY AND PHYSICAL FINDINGS

A 30-year-old healthy man presents to clinic with dyspnea, palpitation, and chest pain. His father has congestive heart failure, and his brother had sudden cardiac death at 18 years of age while practicing football. The cardiologist explained to the patient the possibility of hypertrophic cardiomyopathy running in his family. Cardiac examination showed double apical impulse and laterally displaced, split second heart sound, fourth heart sound present, jugular venous pulse, and double carotid arterial pulse. Echocardiogram showed asymmetrical septal hypertrophy with no structural anomalies, diagnostic of hypertrophic cardiomyopathy. Consistent with his clinical history, physical features, and family history, DNA testing identified an Arg403Gln mutation in *MYH7*.

BACKGROUND

Disease Etiology and Incidence

Hypertrophic cardiomyopathy (HCM, MIM 192600), the most common monogenic cardiovascular disease, is an autosomal dominant disorder caused by mutations in approximately 20 genes encoding proteins of the cardiac sarcomere. Of those patients with positive genetic tests, approximately 70% are found to have mutations in the two most common genes, *MYH7* and *MYBPC3*, whereas other genes, including those encoding troponin T, troponin I, *Tropomyosin 1*, and alpha-actin each account for a small proportion of patients (1% to 5%).

Pathogenesis

Over 1500 mutations have been reported in genes encoding thick and thin myofilament proteins of the sarcomere or contiguous Z disk. Mutations in several additional sarcomere (or calcium-handling) genes have been proposed, but with less evidence supporting pathogenicity.

Approximately 60% of adult and pediatric patients with a family history of HCM will have a sarcomere mutation identified. In contrast, only approximately 30% of patients without a family history will have positive results, often due to sporadic or de novo mutations (65% of the probands) that may, however be passed on to the next generation. Approximately

3% to 4% of males with HCM will have unrecognized Fabry disease, a lysosomal storage disorder caused by mutations in the α galactosidase A gene.

Phenotype and Natural History

HCM is characterized by left ventricular hypertrophy (LVH) (Fig. C-25) in the absence of predisposing cardiac conditions (e.g., aortic stenosis) or cardiovascular conditions (e.g., longstanding hypertension). The clinical manifestations of HCM range from asymptomatic to progressive heart failure to sudden cardiac death and vary from individual to individual even within the same family. Common symptoms include

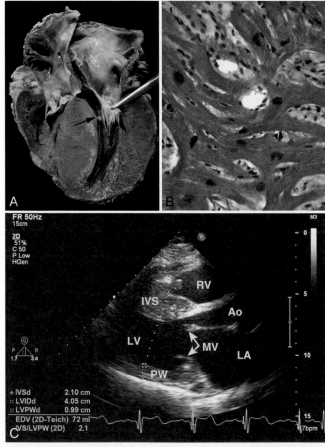

Figure C-25 Hypertrophic cardiomyopathy with asymmetric septal hypertrophy. A, The septal muscle bulges into the left ventricular outflow tract, and the left atrium is enlarged. The anterior mitral leaflet has been reflected away from the septum to reveal a fibrous endocardial plaque (*arrow*) (see text). **B,** Histological appearance demonstrating myocyte disarray, extreme hypertrophy, and exaggerated myocyte branching, as well as the characteristic interstitial fibrosis (collagen is *blue* in this Masson trichrome stain). **C,** Echocardiographic appearance of hypertrophic cardiomyopathy. Parasternal long-axis view from a patient with hypertrophic cardiomyopathy demonstrating asymmetrical septal hypertrophy. The interventricular septum (IVS) measures 2.1 cm (normal 0.6 to 1.0 cm), the posterior wall measures 0.99 cm. Ao, Aorta; LA, left atrium; LV, left ventricle; MV, mitral valve; PW, posterior wall; RV, right ventricle. *See Sources & Acknowledgments.*

shortness of breath (particularly with exertion), chest pain, palpitations, orthostasis, presyncope, and syncope. Most often the LVH of HCM becomes apparent during adolescence or young adulthood, although it may also develop late in life, in infancy, or in childhood.

Management

Before identification of the genes responsible for HCM, diagnosis of HCM could only be made through integration of examination, electrocardiogram, echocardiogram, and invasive angiographic/hemodynamic studies, disproportionately identifying patients with left ventricular outflow obstruction. Genetic testing for HCM is now available and can provide important insights into family management by definitively identifying at-risk relatives (i.e., those who have inherited the family's pathogenic mutation). Mutations found are considered pathogenic based on the following criteria: (1) cosegregation with the HCM phenotype in family members; (2) previously reported or identified as a cause of HCM; (3) absent from unrelated and ethnic-matched normal controls; (4) important alteration in protein structure and function; and (5) amino acid sequence change in a region of the protein otherwise highly conserved through evolution. However, even with the use of these criteria, a considerable number of variants are classified as VUSs (see Chapter 16), making the test results ambiguous.

Because diagnosis and sudden cardiac death risk are both linked to the presence of LVH and because the penetrance of LVH is age-dependent, clinical evaluation must continue longitudinally. Screening is annual during adolescence and early adulthood when LVH most commonly emerges. Early childhood screening is appropriate if there is a family history of early-onset disease or other concerns. During adulthood, screening is recommended approximately every 5 years or in response to clinical changes because LVH can develop late in life. The first-line screening consists of clinical testing with cardiac imaging and electrocardiography to identify phenotype-positive relatives. Genetic testing can also be used to identify family members at risk for developing disease who do not have LVH.

No treatments to prevent disease development or to reverse established manifestations currently exist. The treatment of manifestations includes medical management of diastolic dysfunction, medical or surgical management of ventricular outflow obstruction, restoration and maintenance of sinus rhythm in those with atrial fibrillation, implantable cardioverter-defibrillator in survivors of cardiac arrest and those at high risk for cardiac arrest, medical treatment for heart failure, and consideration for cardiac transplantation when necessary. The prevention of secondary complications includes consideration of anticoagulation in those with persistent or paroxysmal atrial fibrillation to reduce the risk for thromboembolism; consideration of antibiotic prophylaxis when necessary; and during the pregnancy of a woman with HCM, care by an experienced cardiologist and obstetrician trained in high-risk obstetrics. Patients should avoid competitive endurance training, burst activities (e.g., sprinting), intense isometric exercise (e.g., heavy weight lifting), dehydration, hypovolemia (i.e., use diuretics with caution), and medications that decrease afterload (e.g., angiotensin-converting enzyme [ACE] inhibitors, angiotensin receptor blockers, and other direct vasodilators). But the consensus recommendations from the 36th Bethesda Conference do not exclude individuals carrying pathogenic mutations but without manifestations of the disease from sports.

INHERITANCE RISK

HCM follows autosomal dominant inheritance with variable expressivity and incomplete, age-related penetrance; each first-degree relative of an affected patient has a 50% chance of carrying the mutation and potentially developing HCM. Alternatively, sporadic cases may be due to de novo mutations in the proband but absent from the parents.

Disease prevention based on genetic testing is currently available in the form of assisted reproduction using preimplantation genetic diagnosis (PGD) or prenatal diagnosis with pregnancy termination in the event of an affected fetus (see Chapter 17).

QUESTIONS FOR SMALL GROUP DISCUSSION

1. Name other disorders that show age-related penetrance. What types of mutations are associated with these disorders?
2. Discuss possible reasons for locus heterogeneity in HCM.
3. What are the criteria to classify a variant as benign?
4. When is genetic testing indicated in a proband with suspected HCM?

REFERENCES

Cirino AL, Ho C: Familial hypertrophic cardiomyopathy overview. Available from: http://www.ncbi.nlm.nih.gov/books/NBK1768/.

Ho CY: Genetic considerations in hypertrophic cardiomyopathy, *Prog Cardio Dis* 54:456–460, 2012.

Maron BJ, Maron MS, Semsarian C: Genetics of hypertrophic cardiomyopathy after 20 years: clinical perspectives, *J Am Coll Cardiol* 60:705–715, 2012.

Maron BJ, Zipes DP: 36th Bethesda Conference: eligibility recommendations for competitive athletes with cardiovascular abnormalities, *J Am Coll Cardiol* 45:1312–1375, 2005.

INSULIN-DEPENDENT (TYPE 1) DIABETES MELLITUS
(Autoimmune Destruction of Islet β Cells, MIM 222100)

Multifactorial

PRINCIPLES

- Polygenic disease
- Environmental trigger
- Susceptibility allele
- Protective allele

MAJOR PHENOTYPIC FEATURES

- Age at onset: Childhood through adulthood
- Polyuria, polydipsia, polyphagia
- Hyperglycemia
- Ketosis
- Wasting

HISTORY AND PHYSICAL FINDINGS

F.C., a 45-year-old father with late-onset diabetes mellitus, was referred to the genetics clinic for counseling regarding his children's risk for diabetes. F.C. developed glucose intolerance (inability to maintain normal blood glucose levels after ingestion of sugar) at the age of 39 years and fasting hyperglycemia at 45 years. He did not have a history of other medical or surgical problems. The findings from his physical examination were normal except for moderate abdominal obesity; his body mass index [weight in kilograms/(height in meters)²] was 31.3, with the excess adiposity distributed preferentially around his waist. He had five children by two different partners; a child from each relationship had developed insulin-dependent (type 1) diabetes mellitus (IDDM) before 10 years of age. His sister developed IDDM as a child and died during adolescence from diabetic ketoacidosis. The geneticist explained that given his family history, F.C. might have a late-onset form of IDDM and that his current, non–insulin-dependent diabetes mellitus was probably an antecedent to the development of IDDM. After discussing the possible causes of and prognostic factors for the development of IDDM, F.C. elected to enroll himself and his children, who were all minors, in a research protocol studying the prevention of IDDM. As part of that study, he and his children were tested for anti-islet antibodies. Both he and an unaffected daughter had a high titer of anti-islet antibodies; the daughter also had an abnormal glucose tolerance test result but not fasting hyperglycemia. As part of the study protocol, F.C. and his daughter were prescribed low-dose insulin injections.

BACKGROUND

Disease Etiology and Incidence

IDDM (now typically called type 1 diabetes, MIM 222100) is usually caused by autoimmune destruction of islet β cells in the pancreas; this autoimmune reaction is triggered by an unknown mechanism. The destruction of islet β cells causes insulin deficiency and thereby dysregulation of anabolism and catabolism, resulting in metabolic changes similar to those observed in starvation (Fig. C-26). Among North American white individuals, IDDM is the second most common chronic disease of childhood, increasing in prevalence from 1 in 2500 at 5 years of age to 1 in 300 at 18 years of age.

Pathogenesis

IDDM usually results from a genetic susceptibility and subsequent environmental insult (see Chapter 8) and only very rarely from an environmental insult or a genetic mutation alone. Although approximately 90% of IDDM occurs in patients without a family history of diabetes, observations supporting a genetic predisposition include differences in concordance between monozygotic (33% to 50%) and dizygotic twins (1% to 14%), familial clustering, and differences in prevalence among different populations (see Chapter 8). More than a dozen different genetic susceptibility loci have been reported in humans, although few have been identified consistently and reproducibly. One of the few confirmed loci is the HLA locus that may account for as much as 30% to 60% of the genetic susceptibility. Approximately 95% of white patients express a DR3 or DR4 allele, or both, compared with 50% of controls; this association apparently arises not because DR3 and DR4 are susceptibility alleles, but because of linkage disequilibrium between DR and DQ. The DQβ1*0201 allele, which segregates with DR3, and DQβ1*0302, which segregates with DR4, appear to be the primary susceptibility alleles. In contrast, DQβ1*602, which segregates with DR2, appears to be a protective allele; that is, it negates the effect of a susceptibility allele when both are present. Both of the DQβ1 susceptibility alleles have a neutral amino acid at position 57, a site within the putative antigen-binding cleft, whereas protective or neutral DQβ1 alleles have an aspartic acid at position 57. This substitution of an uncharged amino acid for aspartic acid is predicted to change the specificity of antigen binding to the DQ molecule.

Evidence supporting an environmental component to the induction of IDDM in genetically susceptible individuals includes a concordance of less than 50% among monozygotic twins, a seasonal variation in incidence, and an increased incidence of diabetes among children with congenital rubella.

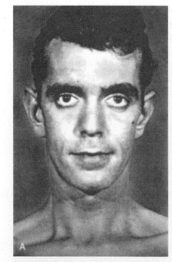

Figure C-26 **A 28-year-old man with insulin-dependent diabetes mellitus. A,** Photograph after 3 weeks of polydipsia and polyuria. **B,** Photograph after 5-kg weight gain with 10 days of insulin replacement. *See Sources & Acknowledgments.*

Proposed environmental triggers include viral infections and early exposure to bovine albumin. Exposure to viruses and bovine albumin could cause autoimmune destruction of β cells by molecular mimicry, that is, sharing of antigenic determinants between β-cell proteins and the virus or bovine albumin. Approximately 80% to 90% of newly diagnosed patients with IDDM have anti-islet cell antibodies. These autoantibodies recognize cytoplasmic and cell surface determinants such as glutamic acid decarboxylase, carboxypeptidase H, ganglioside antigens, islet cell antigen 69 (ICA69), and a protein tyrosine phosphatase. Glutamic acid decarboxylase and ICA69, respectively, share epitopes with coxsackievirus B4 and bovine serum albumin.

Thus, IDDM appears to be an autoimmune disease, although the precise role of islet cell autoantibodies remains uncertain. Additional evidence for an autoimmune mechanism in IDDM includes an increased prevalence of other autoimmune diseases, mononuclear cell infiltrates of islets, and recurrent β-cell destruction after transplantation from a monozygotic twin. However, two lines of evidence suggest that progression to IDDM involves more than the development of autoantibodies. First, less than 1% of the general population develops diabetes although 10% have islet autoantibodies; and second, first-degree relatives and schoolchildren have remission rates of 10% to 78% for islet cell antibodies.

Phenotype and Natural History

Loss of insulin reserve occurs during a few to many years. The earliest sign of abnormality is the development of islet autoantibodies when blood glucose concentrations, glucose tolerance (ability to maintain normal blood glucose levels after ingestion of sugar), and insulin responses to glucose are normal. This period is followed by a phase of decreased glucose tolerance but normal fasting blood glucose concentration. With continued loss of β cells, fasting hyperglycemia eventually develops, but sufficient insulin is still produced to prevent ketosis; during this period, patients have non–insulin-dependent diabetes mellitus. Eventually, insulin production falls below a critical threshold, and patients become dependent on exogenous insulin supplements and have a propensity to ketoacidosis. Younger patients generally progress through these phases more rapidly than do older patients.

Although the acute complications of diabetes can be controlled by administration of exogenous insulin, the loss of endogenous insulin production causes many problems, including atherosclerosis, peripheral neuropathy, renal disease, cataracts, and retinopathy. Approximately 50% of patients eventually develop renal failure. The occurrence and severity of these complications are related to the genetic background and degree of metabolic control. Rigorous control of blood glucose levels reduces the risk for complications by 35% to 75%.

Management

Although pancreatic or islet transplantation can cure IDDM, the paucity of tissue for transplantation and complications of immunosuppression limit this therapy. Management of most patients emphasizes intensive control of blood glucose levels by injection of exogenous insulin.

The development of islet autoantibodies several years before the onset of IDDM has led to the development of studies to predict and prevent IDDM. The administration of insulin or nicotinamide appears to delay the development of IDDM in some patients.

INHERITANCE RISK

The risk for IDDM in the general population is approximately 1 in 300. With one affected sibling, the risk increases to 1 in 14 (1 in 6 if HLA identical, 1 in 20 if HLA haploidentical). The risk increases to 1 in 6 with a second affected first-degree relative in addition to an affected sibling and to 1 in 3 with an affected monozygotic twin. Children of an affected mother have a 1 in 50 to 1 in 33 risk for development of IDDM, whereas children of an affected father have a 1 in 25 to 1 in 16 risk. This paternity-related increased risk appears to be limited to fathers with an HLA DR4 allele.

QUESTIONS FOR SMALL GROUP DISCUSSION

1. Discuss the difficulties of identifying the genetic components of polygenic diseases.
2. How might HLA susceptibility alleles affect susceptibility and protective alleles affect protection?
3. Discuss the underlying mechanisms for prevention of IDDM by exogenous insulin injections.
4. Compare risk counseling for fathers and mothers with IDDM. Discuss the teratogenic risks and mechanisms of maternal diabetes.

REFERENCES

Alemzadeh R, Ali O: Diabetes mellitus. In Kliegman RM, Stanton BF, St. Geme JW, et al, editors: *Nelson textbook of pediatrics*, ed 19, Philadelphia, 2011, WB Saunders, pp 1968–1997.

Bluestone JA, Herold K, Eisenbarth G: Genetics, pathogenesis and clinical interventions in type 1 diabetes, *Nature* 464:1293–1300, 2010.

Chiang JL, Kirkman MS, Laffel LM, et al: Type 1 diabetes through the life span: a position statement of the American Diabetes Association, *Diabetes Care* 37:2034–2054, 2014.

INTRAUTERINE GROWTH RESTRICTION
(Abnormal Fetal Karyotype)
Spontaneous Chromosomal Deletion

PRINCIPLES

- Prenatal diagnosis
- Ultrasound screening
- Interstitial deletion
- Cytogenetic and genome analysis
- Genetic counseling
- Pregnancy management options

MAJOR PHENOTYPIC FEATURES

- Age at onset: Prenatal
- Intrauterine growth restriction
- Increased nuchal fold
- Dysmorphic facies

HISTORY AND PHYSICAL FINDINGS

A.G. is a 26-year-old gravida 2, para 1 white woman referred for ultrasonography for detailed examination of fetal anatomy. A.G. denied any medication, drug, or alcohol exposure in the pregnancy, and both parents were in good health. The biometric parameters from the fetal anatomy study suggested a 17.5-week fetus. On the basis of first-trimester ultrasound dating and the date of the patient's last menstrual period, however, the fetus should have been at approximately 21 weeks of gestation. This discrepancy suggested symmetrical fetal intrauterine growth restriction. Further evaluation also revealed increased nuchal fold measurements of 6.1 to 7.3 mm. The couple was counseled regarding the increased risk for fetal aneuploidy, and amniocentesis was elected. The chromosome results revealed an interstitial chromosome 4p deletion, with karyotype 46,XX,del(4)(p15.1p15.32). Parental chromosomes were normal. After extensive genetic counseling, the couple decided to terminate the pregnancy. The autopsy revealed a 19-week fetus by size (22.5 weeks by dates) with bilateral epicanthal folds, low-set and posteriorly rotated ears, prominent nasal bridge, and micrognathia. Redundant posterior nuchal skin was also noted.

BACKGROUND
Disease Etiology and Incidence

Intrauterine growth restriction (IUGR) is diagnosed when a fetus or neonate is less than 10th percentile for weight (<2500 g for a neonate born at term in the United States) (Fig. C-27). A newborn with IUGR should be distinguished from a newborn who is small for gestational age (SGA) who is also below the 10th percentile in size but is small for physiological reasons, such as the size of the parents. Approximately 7% of pregnancies result in a fetus who is SGA, of which approximately 1 in 8 is truly IUGR.

IUGR may result from uteroplacental insufficiency, exposure to drugs or alcohol, congenital infections, or intrinsic genetic limitations of growth potential. Fetuses with growth restriction due to nutritional compromise tend to have less retardation of head growth compared to the rest of the body. Several chromosomal disorders are associated with IUGR, and a finding of early or symmetrical IUGR increases the

likelihood that a fetus is affected with a chromosomal abnormality such as trisomy 18, triploidy, or maternal uniparental disomy for chromosome 7 or 14. Nuchal fold measurements of more than 3 mm in the first trimester (11 to 14 weeks) and of 6 mm or more in the second trimester are considered increased and are associated with a greater risk for Down syndrome. Approximately one in seven fetuses with a second-trimester nuchal thickening will have Down syndrome. The ultrasound findings in A.G.'s fetus increased the suspicion of aneuploidy and led to the identification of the small interstitial deletion in 4p, which is the likely explanation for the fetal abnormalities.

The etiology and incidence of such a rare deletion are not entirely understood, especially in light of the normal parental chromosomes. Most de novo deletions are considered to originate at meiosis, but they may also arise during mitosis, before meiosis in gametogenesis, so that a parent is a gonadal mosaic. Gonadal mosaicism cannot be ruled out with any certainty by fibroblast or lymphoblast testing of the parents; consequently, prenatal testing should be offered in future pregnancies.

Pathogenesis

The deletion breakpoints on the short arm of chromosome 4 in 46,XX,del(4)(p15.1p15.32) flank a 14.5-Mb segment of DNA. Analysis of the human genome sequence in this region indicates that 47 known protein-coding genes exist within this

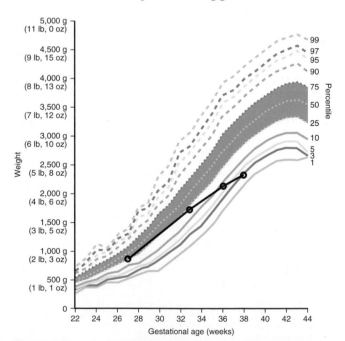

Figure C-27 Intrauterine growth curve for a fetus with trisomy 18 (*black line*), superimposed on a standard intrauterine and postnatal growth chart averaged for both sexes over the U.S. population (shown in *blue*). The aneuploid fetus's growth curve begins at the 30th percentile at 27 weeks of gestation but then cuts across percentile lines, as shown, culminating in birth at 38 weeks with fetal weight just below the third percentile. Fetal weight during pregnancy is estimated by a formula that combines ultrasound measurements of the distance between the parietal bones of the fetal skull (biparietal diameter), head circumference, abdominal circumference, and femur length. *See Sources & Acknowledgments.*

deleted region; haploinsufficiency for one or more of these genes is the likely cause of the phenotype of this fetus. Chromosome microarray testing allows more precise definition of breakpoints in deletions or duplications than does prenatal karyotype, thus determining which genes are involved in the deleted area. This may add more precise prognostic information if the involvement of critical genes is in question.

Phenotype and Natural History

All pregnancies regardless of family, medical, or pregnancy history are at an approximate 3% to 5% risk for developmental disabilities or a birth defect in the infant. Although this couple was not at increased risk, the routine second-trimester ultrasound findings increased the suspicion of fetal aneuploidy. The finding of an interstitial deletion is likely to explain the ultrasound findings. Although this exact deletion had not been reported previously, many deletions of the short arm of chromosome 4 have been associated with birth defects. For example, Wolf-Hirschhorn syndrome is due to a microdeletion of 4p, resulting in both severe intellectual disabilities and physical anomalies. FISH analysis in this fetus revealed that the sequences for the Wolf-Hirschhorn critical region at 4p16.3 were present on both copies of chromosome 4 and that the deletion in this case was more proximal, in band p15. In this case, as with any substantial loss or gain of material on an autosome not previously reported in other patients, the outcome is likely to involve both physical and neurological impairment, the severity of which cannot be predicted.

Management

No curative treatments are available for chromosome abnormalities. The overriding question for many couples regarding the outcome for their unborn child is whether the fetus is at risk for intellectual disability or a significant birth defect. In light of the already present ultrasound anomalies and the identified chromosomal abnormality, this fetus will have sequelae, the extent of which is not predictable. In such cases, the couple is counseled in detail about the limited information and the inability to conclude with any certainty about the expected outcome of the pregnancy. The options include continuation of the pregnancy with expectant management, with or without giving the neonate up for adoption, or termination of pregnancy.

Follow-up ultrasound evaluations can assess fetal growth and development. Long-term progressive IUGR alone suggests a poor prognosis for the fetus. By the late second trimester, the majority of cardiac lesions that would require intervention at birth can usually be identified. Consultation with neonatologists and maternal-fetal medicine specialists can provide information regarding what to expect at delivery and the types of postnatal evaluations that should be considered. There may be advantages to arranging for delivery in a tertiary facility that provides specialized neonatal intensive care and surgery.

Termination of the pregnancy is currently legal in the United States, but not always available. In the second trimester of pregnancy, this procedure can be performed by either dilation and evacuation or induction of labor (prostaglandin induction). The former is usually not performed in pregnancies of more than 24 weeks' gestation. Prostaglandin induction provides the couple with the option of an autopsy, but with a known serious chromosome anomaly, the information from an autopsy provides no additional information that would affect recurrence risk or prenatal testing options in a future pregnancy. The emotional and physical benefits and disadvantages of the two procedures should be outlined in detail before the patient's decision to terminate, should that option be chosen (see Chapter 17). In the majority of states in the U.S., pregnancy terminations are not covered by private health insurance, even when the indication is for a severe birth defect diagnosed prenatally. The costs can be in excess of thousands of dollars, and the financial burden of this procedure may affect decision making in some individuals.

Finally, the parents can be offered the option of giving the neonate up for adoption if they decide that termination is either not an option or unaffordable, or because the anomalies were identified too late in the pregnancy to allow termination.

INHERITANCE RISK

De novo deletions have a low recurrence risk, due to the chance of undetectable gonadal mosaicism in either parent. Prenatal testing, such as chorionic villus sampling or amniocentesis, is available for future pregnancies, although the risk for miscarriage from these procedures may be comparable to the actual empirical risk for a recurrence.

QUESTIONS FOR SMALL GROUP DISCUSSION

1. What is the difference between the terms *small for gestational age* (SGA) and *intrauterine growth restriction* (IUGR)?
2. What would be the advantages and disadvantages of performing amniocentesis for karyotype at 24 weeks of gestation in a pregnancy thought to have IUGR, even if the societal regulations and family situation preclude a pregnancy termination if the amniocentesis demonstrates a chromosomal abnormality?

REFERENCES

Bianchi D, Crombleholme T, D'Alton M, et al: *Fetology: diagnosis and management of the fetal patient*, ed 2, New York, 2010, McGraw Hill.
Gardner RJM, Sutherland GR, Shaffer LG: *Chromosome abnormalities and genetic counseling*, ed 4, Oxford, England, 2012, Oxford University Press.
South ST, Corson VL, McMichael JL, et al: Prenatal detection of an interstitial deletion in 4p15 in a fetus with an increased nuchal skin fold measurement, *Fetal Diagn Ther* 20:58–63, 2005.

LONG QT SYNDROME (Cardiac Ion Channel Gene Mutations; MIM 192500)

Autosomal Dominant or Recessive

PRINCIPLES

- Locus heterogeneity
- Incomplete penetrance
- Genetic susceptibility to medications

MAJOR PHENOTYPIC FINDINGS

- QTc prolongation (>470 msec in males, >480 msec in females)
- Tachyarrhythmias (torsades de pointes)
- Syncopal episodes
- Sudden death

HISTORY AND PHYSICAL FINDINGS

A.B. is a 30-year-old woman with long QT (LQT) syndrome who presents to the genetics clinic with her husband because they are contemplating a pregnancy. The couple wants to know the recurrence risk for this condition in their children and the genetic testing and prenatal diagnosis options that might be available to them. She is also concerned about potential risks to her own health in carrying a pregnancy. The patient was diagnosed with the LQT syndrome in her early 20s when she was evaluated after the sudden death of her 15-year-old brother. Overall, she is a healthy individual with normal hearing, no dysmorphic features, and an otherwise negative review of systems. She has never had any fainting episodes. Subsequently, electrocardiographic findings confirmed the diagnosis of the syndrome in A.B., and a paternal aunt but not in her father, who had a normal QTc interval. Molecular testing revealed a missense mutation in *KCNH2*, one that had been previously seen in other families with Romano-Ward syndrome, type LQT2. A.B. was initially prescribed β-blockade medication, which she is continuing, but her cardiologists decided that the less than total efficacy of β-blockers in LQT2 and the previous lethal event in her brother justified the use of an implantable cardioverter-defibrillators in A.B. and her affected relatives. A.B. is the first person in her family to pursue genetic counseling for the LQT syndrome.

BACKGROUND

Disease Etiology and Incidence

The LQT syndromes are a heterogeneous, panethnic group of disorders referred to as channelopathies because they are caused by defects in cardiac ion channels. The overall prevalence of LQT disorders is approximately 1 in 5000 to 7000 individuals.

The genetics underlying LQT syndromes is complex. First, there is locus heterogeneity. Mutations in at least five known cardiac ion channel genes (*KCNQ1*, *KCNH2*, *SCN5A*, *KCNE1*, and *KCNE2*) are responsible for most cases of LQT; mutations in additional genes are known, but much rarer. Second, different mutant alleles in the same locus can result in two distinct LQT syndromes with two different inheritance patterns, the Romano-Ward syndrome and the autosomal recessive Jervell and Lange-Nielsen syndrome (MIM 220400).

Pathogenesis

LQT syndrome is caused by repolarization defects in cardiac cells. Repolarization is a controlled process that requires a balance between inward currents of sodium and calcium and outward currents of potassium. Imbalances cause the action potential of cells to increase or decrease in duration, causing elongation or shortening, respectively, of the QT interval on electrocardiography. Most cases of LQT syndrome are caused by loss-of-function mutations in genes that encode subunits or regulatory proteins for potassium channels (genes whose names begin with *KCN*). These mutations decrease the outward, repolarization current, thereby prolonging the action potential of the cell and lowering the threshold for another depolarization. In other LQT syndrome patients, gain-of-function mutations in a sodium channel gene, *SCN5A*, lead to an increased influx of sodium, resulting in similar shifting of action potential and repolarization effects.

Phenotype and Natural History

The LQT syndromes are characterized by elongated QT interval and T-wave abnormalities on electrocardiography (Fig. C-28), including tachyarrhythmia and torsades de pointes, a ventricular tachycardia characterized by a change in amplitude and twisting of the QRS complex. Torsades de pointes is associated with a prolonged QT interval and typically stops

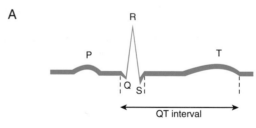

Rate-correction formula (Bazett's):

$$QTc \text{ (msec)} = \frac{QT \text{ (msec)}}{\sqrt{RR \text{ (sec)}}}$$

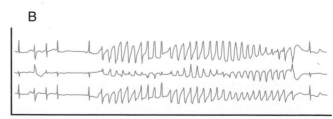

Figure C-28 A, Measurement of the QT interval from the electrocardiogram. The diagram depicts the normal electrocardiogram with the P wave representing atrial activation, the QRS complex representing ventricular activation and the start of ventricular contraction, and the T wave representing ventricular repolarization. Owing to heart rate sensitivity of the QT interval, this parameter is corrected (normalized) to heart rate (as reflected by the beat-to-beat RR interval), yielding the QTc. QT and QTc can both be expressed in milliseconds or seconds. **B,** Arrhythmia onset in long QT syndrome. Three simultaneous (and distinct) electrocardiographic channel recordings in a patient with QT prolongation and runs of continuously varying polymorphic ventricular tachycardia (torsades de pointes). Torsades de pointes may resolve spontaneously or progress to ventricular fibrillation and cardiac arrest. *See Sources & Acknowledgments.*

spontaneously but may persist and worsen to ventricular fibrillation.

In the most common LQT syndrome, Romano-Ward, syncope due to cardiac arrhythmia is the most frequent symptom; if undiagnosed or left untreated, it recurs and can be fatal in 10% to 15% of cases. However, between 30% and 50% of individuals with the syndrome never show syncopal symptoms. Cardiac episodes are most frequent from the preteen years through the 20s, with the risk decreasing over time. Episodes may occur at any age when triggered by QT-prolonging medications (see list at http://www.qtdrugs.org). Nonpharmacological triggers for cardiac events in the Romano-Ward syndrome differ on the basis of the gene responsible. LQT1 triggers are typically adrenergic stimuli, including exercise and sudden emotion. Individuals with LTQ2 are at risk with exercise and at rest and with auditory stimuli, such as alarm clocks and phones. LQT3 individuals have episodes with slower heart rates during rest periods and sleep. In addition, 40% of LQT1 cases are symptomatic before 10 years of age; in 10% of LTQ2 and rarely in LQT3 do symptoms occur before 10 years of age. There are at least 10 genes associated with LQT syndromes, of which two—*KCNQ1* and *KCNH2*—account for over 80% of cases.

The LQT syndrome exhibits reduced penetrance in terms of both electrocardiographic abnormalities and syncopal episodes. As many as 30% of affected individuals can have QT intervals that overlap with the normal range. Variable expression of the disorder can occur within and between families. Due to reduced penetrance, exercise electrocardiography is often used for diagnosis of at-risk family members but is not 100% sensitive.

LQT syndromes may be accompanied by other findings on physical examination. For example, Jervell and Lange-Nielsen syndrome (MIM 220400) is characterized by congenital, profound sensorineural hearing loss together with LQT syndrome. It is an autosomal recessive disorder caused by particular mutations within the same two genes (*KCNQ1* and *KCNE1*) implicated in the autosomal dominant Romano-Ward syndrome. Heterozygous relatives of Jervell and Lange-Nielsen syndrome patients are not deaf but have a 25% risk for LQT syndrome.

Management

Treatment of the LQT syndrome is aimed at prevention of syncopal episodes and cardiac arrest. Optimal treatment is influenced by identification of the gene responsible in a given case. For instance, β-blocker therapy before the onset of symptoms is most effective in LQT1 and, to a somewhat lesser extent, in LQT2, but its efficacy in LQT3 is reduced. β-Blockade therapy must be monitored closely for age-related dose adjustment, and it is imperative that doses are not missed. Pacemakers may be necessary for individuals with bradycardia; access to external defibrillators may be appropriate. Implantable cardioverter-defibrillators may be needed in individuals with LQT3 or in other individuals with the LQT syndrome in whom β-blocker therapy is problematic, such as in patients with asthma, depression, or diabetes and those with a history or family history of cardiac arrest. Medications such as the antidepressant amitriptyline, over-the-counter cold medications such as phenylephrine and diphenhydramine, or antifungal drugs, including fluconazole and ketoconazole, should be avoided because of their effect on prolonging the QT interval or causing increased sympathetic tone. Activities and sports likely to be associated with intense physical activity, emotion, or stress should also be avoided.

INHERITANCE RISK

Individuals with the Romano-Ward syndrome have a 50% chance of having a child with the inherited gene mutations. Most individuals have an affected (although perhaps asymptomatic) parent, because the rate of de novo mutations is low. A detailed family history and careful cardiac evaluation of family members are extremely important and could be lifesaving. The recurrence risk in siblings of patients with Jervell and Lange-Nielsen syndrome is 25%, as expected with an autosomal recessive condition. The penetrance of LQT alone, without deafness, is 25% in heterozygous carriers in Jervell and Lange-Nielsen syndrome families.

QUESTIONS FOR SMALL GROUP DISCUSSION

1. Some genetic syndromes rely on clinical evaluation, even with the availability of molecular testing, for diagnosis. In the case of LQT, how would you proceed with a patient thought to have LQT on family history? Why?
2. Discuss the ethics of testing minors in this condition.
3. You have just diagnosed a child with Jervell and Lange-Nielsen syndrome. What do you counsel the family in regard to recurrence risk and management for other family members?

REFERENCES

Alders M, Mannens MMAM: Romano-Ward syndrome. Available from: http://www.ncbi.nlm.nih.gov/books/NBK1129/.

Guidicessi JR, Ackerman MJ: Genotype- and phenotype-guided management of congenital long QT syndrome, *Curr Probl Cardiol* 38:417–455, 2013.

Martin CA, Huang CL, Matthews GD: The role of ion channelopathies in sudden cardiac death: implications for clinical practice, *Ann Med* 45:364–374, 2013.

Tranebjaerg L, Samson RA, Green GE: Jervell and Lange-Nielsen syndrome. Available from: http://www.ncbi.nlm.nih.gov/books/NBK1405/.

LYNCH SYNDROME (DNA Mismatch Repair Gene Mutations, MIM 120435)

Autosomal Dominant

PRINCIPLES

- Tumor susceptibility genes
- Multistep carcinogenesis
- Somatic mutation
- Microsatellite instability
- Variable expressivity and incomplete penetrance

MAJOR PHENOTYPIC FEATURES

- Age at onset: Middle adulthood
- Colorectal cancer
- Multiple primary cancers

HISTORY AND PHYSICAL FINDINGS

P.P., a 38-year-old banker and mother of three children, was referred to the cancer genetics clinic by her physician for counseling regarding her family history of cancer. Her father, brother, nephew, niece, and paternal uncle all developed colorectal cancer while her paternal grandmother had been diagnosed in her 40's with uterine cancer. P.P. did not have a history of medical or surgical problems. The findings from her physical examination were normal. The geneticist explained to P.P. that her family history was suggestive of Lynch syndrome (also known as hereditary nonpolyposis colon cancer, HNPCC), and that the most efficient and effective way to determine the genetic cause of Lynch syndrome in her family was through molecular testing of a living affected family member. After some discussion with her niece, the only surviving affected family member, P.P. and her niece returned to the clinic for testing. Testing of an archived tumor sample from the niece's resected colon identified microsatellite instability (MSI); subsequent sequencing of DNA from a blood sample obtained from the niece revealed a germline mutation in *MLH1*. P.P. did not carry the mutation; therefore the geneticist counseled that her risk and her children's risk for development of cancer were similar to that of the general population. Her unaffected brother was found to carry the mutation and continued to have an annual screening colonoscopy.

BACKGROUND

Disease Etiology and Incidence

At least 50% of individuals in Western populations develop a colorectal tumor by the age of 70 years, and approximately 10% of these individuals eventually develop colorectal cancer. Lynch syndrome (MIM 120435) is a genetically heterogeneous autosomal dominant cancer predisposition syndrome that is often caused by mutations in DNA mismatch repair genes. Lynch syndrome has a prevalence of 2 to 5 per 1000 and accounts for approximately 3% to 8% of colorectal cancer.

Pathogenesis

In most colorectal cancers, including in familial adenomatous polyposis, the tumor karyotype becomes progressively more aneuploid (see Chapter 15). Approximately 15% of colorectal cancers do not have such chromosomal instability but have insertion or deletion mutations in repetitive sequences (MSI).

Microsatellite instability occurs in 85% to 90% of Lynch syndrome tumors. Consistent with this observation, approximately 70% of Lynch syndrome families with carcinomas exhibiting MSI have germline mutations in one of four DNA mismatch repair genes: *MSH2, MSH6, MLH1,* or *PMS2.*

DNA mismatch repair reduces DNA replication errors by 1000-fold. Errors of DNA synthesis cause mispairing and deform the DNA double helix. A complex of mismatch repair proteins recruits other enzymes to excise the segment of newly synthesized mismatched DNA and resynthesize it.

As is typical for tumor suppressor genes, both alleles of a DNA mismatch repair gene must lose function to cause MSI. This somatic loss of function can occur by loss of heterozygosity, intragenic mutation, or promoter hypermethylation.

In Lynch syndrome, a number of microsatellite loci mutate during the progression from adenoma to carcinoma. Inactivation of genes containing microsatellite sequences could play key roles in tumor progression. For example, MSI induces frameshift mutations in the transforming growth factor receptor II gene (*TGFBR2*). Mutations within *TGFBR2* cause the loss of TGFβRII expression, which reduces the ability of TGFβ to inhibit the growth of colonic epithelial cells. *TGFBR2* mutations occur in early Lynch syndrome lesions and may contribute to the growth of adenomas. Lynch syndrome also results from epigenetic silencing of *MSH2* caused by deletion of 3′ exons of *EPCAM* and intergenic regions directly upstream of *MSH2.*

MLH1 and *MSH2* germline mutations account for approximately 90% of mutations in Lynch syndrome families. *MSH6* mutations account for an additional 7% to 10%, whereas *PMS2* mutations are found in fewer than 5% of cases.

Phenotype and Natural History

Although patients with Lynch syndrome develop polyps similar in number to those of the general population, they develop them at younger ages. Their median age at diagnosis with a colorectal adenocarcinoma is younger than 50 years, that is, 10 to 15 years younger than the general population (Fig. C-29). Patients with Lynch syndrome and a defined *MLH1* or *MSH2* germline mutation have an 80% lifetime risk for development of colorectal cancer; the penetrance of *MSH6* and *PMS2* mutations is much lower. Sixty percent to 70% of adenomas and carcinomas in Lynch syndrome occur between the splenic flexure and ileocecal junction. By way of contrast, most sporadic colorectal cancers (and cancer in familial adenomatous polyposis) (Case 15) occur in the descending colon and sigmoid. Carcinomas in Lynch syndrome are less likely to have chromosome instability and aneuploidy, and behave less aggressively than sporadic colon cancer. For this reason, patients with Lynch syndrome have a better age- and stage-adjusted prognosis than do patients with familial adenomatous polyposis or colorectal tumors with chromosome instability.

In addition to colorectal cancer, Lynch-associated cancers include cancer of the stomach, small bowel, pancreas, kidney, endometrium, and ovaries; cancers of the lung and breast are not associated (see Fig. C-29). Patients with Lynch syndrome and a defined germline mutation have a more than 90% lifetime risk for development of colorectal cancer, one of these associated cancers, or both.

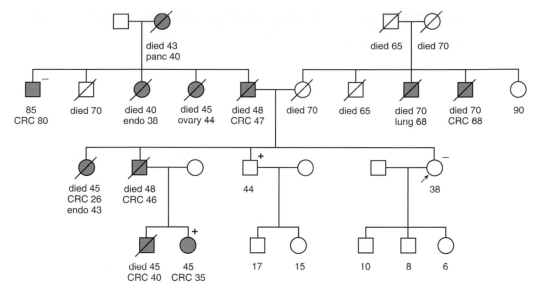

Figure C-29 **Family segregating an _MLH1_ mutation.** Note the frequent occurrence of colorectal cancer as well as other Lynch syndrome-associated cancers, such as endometrial cancer, pancreatic cancer, and ovarian cancer. Note that one family member had cancers of the colorectum and endometrium and that another had sporadic colon cancer (tested negative for family mutation). The consultand is indicated by an _arrow_. _Shaded symbols_ indicate a diagnosis of cancer. Ages are shown directly below each symbol. A _plus sign_ identifies carriers of the _MLH1_ mutation, and a _minus sign_ identifies noncarriers. Cancer diagnoses are followed by the age at diagnosis. CRC, Colorectal cancer; endo, endometrial cancer; lung, lung cancer; ovary, ovarian cancer; panc, pancreatic cancer. _See Sources & Acknowledgments._

Management

Lynch syndrome patients do not have distinguishing physical features. The minimal criteria for considering Lynch syndrome are the occurrence of colorectal cancer or another Lynch syndrome-associated tumor in three members of a family, at least two of whom are first-degree relatives, across two or more generations, and the development of colorectal cancer in at least one affected individual before the age of 50 years. In patients without a family history but with early-onset colorectal cancer, genetic testing for Lynch syndrome is ideally performed in a stepwise manner: evaluation of tumor tissue for MSI through molecular MSI testing and/or immunohistochemistry of the four mismatch repair proteins. The presence of MSI in the tumor alone is not sufficient to diagnosis Lynch syndrome because 10% to 15% of sporadic colorectal cancers exhibit MSI due to somatic methylation of the _MLH1_ promoter. Immunohistochemistry testing helps identify the mismatch repair gene that most likely harbors a germline mutation.

Early recognition of Lynch syndrome is necessary for effective intervention; surveillance colonoscopy of the proximal colon beginning at the age of 25 years increases life span expectancy by 13.5 years, and prophylactic surgical removal of the colon at the age of 25 years increases life span expectancy by more than 15 years. Surveillance endometrial biopsies and abdominal ultrasound scans for at-risk women have not proved to be effective preventive measures for the uterine or ovarian cancer seen in this condition. In families with known germline mutations, identification of the DNA mismatch repair gene mutation can focus surveillance on those patients carrying the mutation, but in Lynch syndrome families without an identified germline mutation, the absence of a mutation does not negate the need for frequent surveillance.

INHERITANCE RISK

The empirical Western general population risk for the development of colorectal cancer is 5% to 6%. This risk is markedly modified by family history. Patients with a first-degree relative with colorectal cancer have a 1.7 relative risk; this relative risk increases to 2.75 if two or more first-degree relatives had colorectal cancer. If an affected first-degree relative developed colorectal cancer before 44 years of age, the relative risk increases to more than 5.

In contrast, a patient with a DNA mismatch repair gene germline mutation has a 50% risk for having a child carrying a germline mutation. Each child carrying such a mutation has a lifetime cancer risk of up to 90%, assuming the 80% penetrance of _MLH1_ or _MSH2_ mutations is responsible for a cancer risk over and above the background risk in the general population for colon cancer and the other cancers of the types associated with Lynch syndrome. Prenatal diagnosis is highly controversial and not routine but is theoretically possible if the germline mutation has been identified in the parent. Because of incomplete penetrance and variation in expressivity, the severity and onset of Lynch syndrome and the occurrence of associated cancers cannot be predicted.

QUESTIONS FOR SMALL GROUP DISCUSSION

1. Compare the mechanisms of tumorigenesis in disorders of nucleotide excision repair, chromosomal instability, and microsatellite instability.
2. How should a patient with a family history of Lynch syndrome be counseled if testing for DNA mismatch repair gene mutations is positive? Negative?
3. Discuss the ethics of testing of minors for Lynch syndrome.

REFERENCES

Brenner H, Kloor M, Pox CP: Colorectal cancer, _Lancet_ 383:1490–1502, 2014.

Kohlmann W, Gruber SB: Lynch syndrome. 2004 [Updated 2014]. In Pagon RA, Bird TD, Dolan CR, et al, editors: _GeneReview._ Available from: http://www.ncbi .nlm.nih.gov/books/NBK1211/.

Matloff J, Lucas A, Polydorides AD, et al: Molecular tumor testing for Lynch syndrome in patients with colorectal cancer, _J Natl Compr Canc Netw_ 11:1380–1385, 2013.

MARFAN SYNDROME (*FBN1* Mutation, MIM 154700)

Autosomal Dominant

PRINCIPLES

- Dominant negative mutations
- Variable expressivity

MAJOR PHENOTYPIC FEATURES

- Age at onset: Early childhood
- Disproportionately tall stature
- Skeletal anomalies
- Ectopia lentis
- Mitral valve prolapse
- Aortic dilatation and rupture
- Spontaneous pneumothorax
- Lumbosacral dural ectasia

HISTORY AND PHYSICAL FINDINGS

J.L., a healthy 16-year-old high school basketball star, was referred to the genetics clinic for evaluation for Marfan syndrome. His physique was similar to that of his father. His father, a tall, thin man, had died during a morning jog; no other family members had a history of skeletal abnormalities, sudden death, vision loss, or congenital anomalies. On physical examination, J.L. had an asthenic habitus with a high arched palate, mild pectus carinatum, arachnodactyly, arm span–height ratio of 1.1, diastolic murmur, and stretch marks on his shoulders and thighs. He was referred for echocardiography, which showed dilatation of the aortic root with aortic regurgitation. An ophthalmological examination showed bilateral iridodonesis and slight superior displacement of the lenses. On the basis of his physical examination and testing results, the geneticist explained to J.L. and his mother that he had Marfan syndrome.

BACKGROUND

Disease Etiology and Incidence

Marfan syndrome (MIM 154700) is a panethnic, autosomal dominant, connective tissue disorder that results from mutations in the fibrillin 1 gene (*FBN1*, MIM 134797). This syndrome has an incidence of approximately 1 in 5000. Approximately 25% to 35% of patients have de novo mutations. Mutations leading to Marfan syndrome are scattered across the gene, and each mutation is usually unique to a family.

Pathogenesis

FBN1 encodes fibrillin 1, an extracellular matrix glycoprotein with wide distribution. Fibrillin 1 polymerizes to form microfibrils in both elastic and nonelastic tissues, such as the aortic adventitia, ciliary zonules, and skin.

Mutations affect fibrillin 1 synthesis, processing, secretion, polymerization, or stability. Studies of fibrillin 1 deposition and cell culture expression assays have generally suggested a dominant negative pathogenesis; that is, production of mutant fibrillin 1 inhibits formation of normal microfibrils by normal fibrillin 1 or stimulates inappropriate proteolysis of extracellular microfibrils. More recent evidence in mouse models of Marfan syndrome suggests that half-normal amounts of normal fibrillin 1 are insufficient to initiate effective microfibrillar

assembly. Thus haploinsufficiency may also contribute to disease pathogenesis.

In addition to Marfan syndrome, mutations in *FBN1* can cause other syndromes, including neonatal Marfan syndrome, isolated skeletal features, autosomal dominant ectopia lentis, and the MASS phenotype (marfanoid signs, including *m*itral valve prolapse or *m*yopia, borderline and nonprogressive *a*ortic enlargement, and nonspecific *s*keletal and *s*kin findings). In general, the phenotypes are fairly consistent within a family, although the severity of the phenotype may vary considerably. To date, clear genotype-phenotype correlations have not emerged. The intrafamilial and interfamilial variability suggests that environmental and epigenetic factors play a significant role in determining the phenotype.

Recent evidence in mouse models suggests that fibrillin 1 is not simply a structural protein, and that Marfan syndrome is not the result of structural weakness of the tissues. Rather, fibrillin 1 microfibrils normally bind and reduce the concentration and activity of growth factors in the TGFβ superfamily. Loss of fibrillin 1 leads to an increase in the local abundance of unbound TGFβ and in local activation of TGFβ signaling. This increased signaling contributes significantly to the pathogenesis, as shown by the rescue by TGFβ antagonists of the pulmonary and vascular changes seen in fibrillin 1-deficient mice (see Chap. 13).

PHENOTYPE AND NATURAL HISTORY

Marfan syndrome is a multisystem disorder with skeletal, ocular, cardiovascular, pulmonary, skin, and dural abnormalities. The skeletal abnormalities include disproportionate tall stature (arm span–height ratio > 1.05; upper to lower segment ratio < 0.85 in adults), arachnodactyly, pectus deformities, scoliosis, joint laxity, and narrow palate. Ocular abnormalities associated with Marfan syndrome are ectopia lentis (Fig. C-30), flat corneas, and increased globe length causing axial

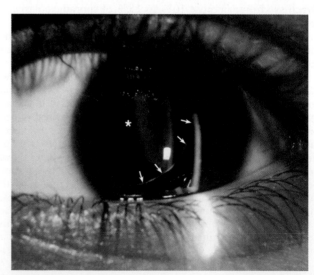

Figure C-30 Ectopia lentis. Slit-lamp view of the left eye of a patient with Marfan syndrome. The *asterisk* indicates the center of the lens that is displaced superior nasally; normally, the lens is in the center of the pupil. The *arrows* indicate the edge of the lens that is abnormally visible in the pupil. *See Sources & Acknowledgments.*

myopia. The cardiovascular abnormalities include mitral valve prolapse, aortic regurgitation, and dilatation and dissection of the ascending aorta. Striae atrophicae and recurrent herniae are common, and lumbosacral dural ectasia may occur. Spontaneous pneumothorax and apical blebs can be seen in this disorder.

Many features of Marfan syndrome develop with age. Skeletal anomalies such as anterior chest deformity and scoliosis worsen with bone growth. Subluxation of the lens is often present in early childhood but can progress through adolescence. Retinal detachment, glaucoma, and cataracts show increased frequency in Marfan syndrome. Cardiovascular complications manifest at any age and progress throughout life.

The major causes of premature death in patients with Marfan syndrome are heart failure from valve regurgitation and aortic dissection and rupture (see Fig. 13-6). As surgical and medical management of the aortic dilatation have improved, so has survival. Between 1972 and 1993, the median age of mortality in Marfan patients rose from 49 to 74 years for women and from 41 to 70 years for men.

Management

Marfan syndrome is a clinical diagnosis based upon the recognition of characteristic features in the ocular, skeletal, and cardiovascular systems and the integument. Aortic root dilatation and ectopia lentis carry disproportionate weight in the diagnostic criteria, given their relative specificity for this disorder. Although molecular confirmation of an *FBN1* mutation is not a requirement for diagnosis, it can play a pivotal role in children with emerging clinical manifestations or in atypically mild presentations of disease. *FBN1* gene sequencing lacks full specificity or sensitivity for Marfan syndrome and therefore cannot substitute for a comprehensive clinical evaluation. It can be of particular importance, however, for prenatal and presymptomatic diagnosis and in the discrimination of Marfan syndrome from other entities in the differential diagnosis, some of which require different treatment protocols.

There is presently no cure for Marfan syndrome; treatment therefore focuses on prevention and symptomatic management. Ophthalmological management includes frequent examinations, correction of the myopia and, often, lens replacement. Orthopedic management includes bracing or surgery for scoliosis. Pectus deformity repair is largely cosmetic. Physical therapy or orthotics can compensate for joint instability. Cardiovascular management includes a combination of medical and surgical therapy. Medical therapy attempts to prevent or to slow progression of aortic dilatation by reducing heart rate, blood pressure, and ventricular ejection force historically with β-adrenergic blockers. Recent work using mouse models of Marfan syndrome documents remarkable protection from aortic aneurysm and dissection with the use of angiotensin receptor blockers such as losartan, which work through a combination of reduction of hemodynamic stress and antagonism of the transforming growth β signaling cascade. Clinical trials of losartan in Marfan syndrome are on-going. Cardiovascular protection is also achieved through restriction of participation in contact sports, competitive sports, and isometric exercise. Prophylactic replacement of the aortic root is recommended when aortic dilatation or aortic regurgitation becomes sufficiently severe. Most patients now receive a valve-sparing aortic root replacement that eliminates the need for chronic anticoagulation.

Pregnancy can precipitate progressive aortic enlargement or dissection. The aortic dissections are believed to be secondary to the hormonal, blood volume, and cardiac output changes associated with pregnancy and parturition. Current evidence suggests that there is a high risk for dissection in pregnancy if the aortic root measures more than 4 cm at conception. Women can elect to undergo valve-sparing aortic replacement before pregnancy.

INHERITANCE RISK

Patients with Marfan syndrome have a 50% risk for having a child affected with Marfan syndrome. In families segregating Marfan syndrome, at-risk individuals can be identified by detecting the mutation (if known in the family) or by clinical evaluation. Prenatal diagnosis is available for families in which the *FBN1* mutation has been identified.

QUESTIONS FOR SMALL GROUP DISCUSSION

1. Homocystinuria has many overlapping features with Marfan syndrome. How can these two disorders be distinguished by medical history? by physical examination? by biochemical testing?
2. What are dominant negative mutations? What are gain-of-function mutations? Contrast the two. Why are dominant negative mutations common in connective tissue disorders?
3. If one wished to design a curative treatment for a disorder caused by dominant negative mutations, what must the therapy accomplish at a molecular level? How is this different from treatment of a disease caused by loss-of-function mutations?

REFERENCES

Bolar N, Van Laer L, Loeys BL: Marfan syndrome: from gene to therapy, *Curr Opin Pediatr* 24:498–504, 2012.

Cook JR, Ramirez F: Clinical, diagnostic, and therapeutic aspects of the Marfan syndrome, *Adv Exp Med Biol* 802:77–94, 2014.

Dietz HC: Marfan syndrome. 2001 [Updated 2014]. Available from: http://www.ncbi.nlm.nih.gov/books/NBK1335/.

Lacro RV, Guey LT, Dietz HC, et al: Characteristics of children and young adults with Marfan syndrome and aortic root dilation in a randomized trial comparing atenolol and losartan therapy, *Am Heart J* 165:828–835, 2013.

Yim ES: Aortic root disease in athletes: aortic root dilation, anomalous coronary artery, bicuspid aortic valve, and Marfan's syndrome, *Sports Med* 43:721–732, 2013.

MEDIUM-CHAIN ACYL-COA DEHYDROGENASE DEFICIENCY (*ACADM* Mutation, MIM 201450)

Autosomal Recessive

PRINCIPLES

- Loss-of-function mutations
- Newborn screening
- Early prevention

MAJOR PHENOTYPIC FEATURES

- Age at onset: Between 3 and 24 months
- Hypoketotic hypoglycemia
- Vomiting
- Lethargy
- Hepatic encephalopathy

HISTORY AND PHYSICAL FINDINGS

A.N., a 6-month-old previously healthy girl, presents to the emergency department with vomiting and lethargy. Parents are healthy first cousins and have a healthy 2-year-old boy. A.N. was born in a country without newborn screening. When in the hospital, the patient had a seizure and physical examination showed hepatomegaly. Blood glucose was 32 mg/dL and ketone bodies were absent. Electrospray ionization tandem mass spectrometry (TMS) showed elevations of C10:1, C8, and C10 acylcarnitines, diagnostic for medium-chain acyl-CoA dehydrogenase deficiency (Fig. C-31). Consistent with her clinical history, physical features, and TMS results, DNA testing identified a homozygous Lys304Glu mutation in the *ACADM* gene, which encodes medium-chain acyl-CoA dehydrogenase. Her asymptomatic brother was tested and found to also be homozygous for the *Lys304Glu* mutation.

BACKGROUND

Disease Etiology and Incidence

Disorders of fatty acid oxidation (FAODs) are a group of frequent inborn errors of metabolism with an estimated combined incidence of 1 in 9000. Medium-chain acyl-CoA dehydrogenase (MCAD) deficiency (MIM 201450), the most common defect in the mitochondrial fatty acid oxidation pathway, is an autosomal recessive disorder caused by mutations in the *ACADM* gene, which encodes the MCAD enzyme. In white populations, the frequency of MCAD deficiency is about the same as that for phenylketonuria, approximately 1 in 14,000. However, the incidence of MCAD deficiency is lower, 1 in 23,000, among Asians and African-Americans.

Pathogenesis

MCAD deficiency is caused by homozygous or compound heterozygous mutations in *ACADM*. The point mutation c.985A>G, which causes an amino acid change from lysine to glutamate at residue 304 (Lys304Glu) of the mature MCAD protein, is found in approximately 70% of mutant alleles of clinically ascertained patients, but neonatal screening shows over 90 different loss-of-function mutations to date.

MCAD is one of the enzymes involved in mitochondrial fatty acid β-oxidation, which fuels hepatic ketogenesis, a major source of energy once hepatic glycogen stores become depleted during prolonged fasting and periods of increased energy demand. The disorder is associated with hypoketotic hypoglycemia and characteristic accumulations of dicarboxylic acids, medium-chain acylglycines, and acylcarnitines in plasma and urine.

Phenotype and Natural History

Children with MCAD deficiency are normal at birth and appear healthy but typically become ill during infancy when, for example, an intercurrent viral illness causes increased metabolic stress and decreased calorie intake. Although the disorder ordinarily presents between ages 3 and 24 months, a later presentation, even in adulthood, is possible. The combination of increased energy demand and reduced caloric supply precipitates acute symptoms of vomiting, drowsiness, or lethargy. Seizures may occur. Hepatomegaly and liver disease are often present during an acute episode, which can quickly progress to coma and death. At first presentation, 25% to 50% of the patients die during a metabolic crisis. The prognosis is excellent once the diagnosis is established and frequent feedings are instituted to avoid any prolonged period of fasting.

Biochemical studies reveal hypoketotic hypoglycemia. Elevation of octanoylcarnitine (C8:0) in peripheral blood by flow injection electrospray ionization TMS is considered to be diagnostic for MCAD deficiency (see Fig. C-31).

Management

Metabolic decompensation can be fatal in patients in whom the diagnosis is not suspected. However, in patients in whom the diagnosis is suspected, decompensation can be prevented by avoiding fasting. If, however, a patient has increased energy demand and reduced oral intake because of an intercurrent illness or surgery, decompensation can be prevented or treated by the administration of intravenous glucose (10% dextrose plus electrolytes at 1.5 to 2 times maintenance) and carnitine (to promote efficient excretion of dicarboxylic acids). Prognosis is excellent when the diagnosis of MCAD deficiency is established and proper therapeutic measures are taken.

Given the frequency of the condition and the improved clinical outcome achieved by presymptomatic diagnosis and initiation of treatment, it became clear that FAODs, including MCAD deficiency, belonged on the list of disorders appropriate for newborn screening. As a result, acylcarnitine analysis by TMS of dried blood spots was added to many newborn screening programs starting in the mid-1990s. False-negative cases have been reported for MCAD deficiency, and are possible in all FAODs, because acylcarnitine profiles may be abnormal at birth but then give a false-negative normal result once the infant beings regular feeding. For this reason, enzyme measurement of MCAD activity in leukocytes or lymphocytes using phenylpropionyl-CoA as a substrate or molecular analysis of the *ACADM* gene, should be performed following a first positive newborn screen. Newborn screening for MCAD deficiency has been very successful because it has resulted in a 74% reduction in severe metabolic decompensation and/or death in these patients.

The mainstay in the treatment of MCAD deficiency is avoidance of fasting. In symptomatic patients, the most

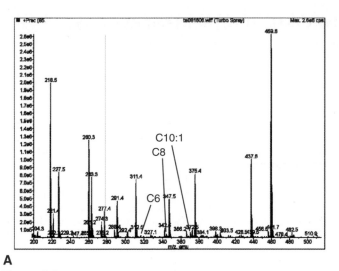
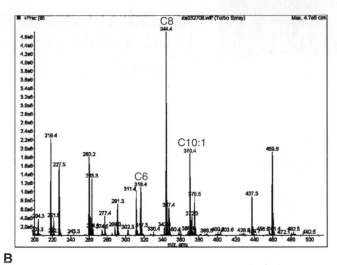

A   **B**

Figure C-31 Plasma acylcarnitine profiles obtained by flow injection electrospray ionization tandem mass spectrometry of butylated compounds. The peak heights, measured on the y-axis, indicate the amounts of various acylcarnitines containing either 6 carbons (C6), 8 carbons (C8), or 10 carbons with one unsaturated bond (C10:1), identified by their specific mass-to-charge ratio (m/z) measured in atomic mass units (amu) along the x-axis. **A,** Normal individual: the C6, C8, and C10:1 peaks are barely detectable. **B,** Medium-chain acyl-CoA dehydrogenase (MCAD) deficiency patient: the C6, C8, and C10:1 peaks are markedly elevated, particularly the elevation of C8, which is characteristic of MCAD deficiency. *See Sources & Acknowledgments.*

important aspect of treatment is reversal of catabolism and sustained anabolism by provision of simple carbohydrates by mouth (e.g., glucose tablets or sweetened, nondiet beverages) or intravenously if the patient is unable or unlikely to maintain or achieve anabolism through oral intake of food and fluids.

INHERITANCE RISK

MCAD deficiency is inherited in an autosomal recessive manner. At conception, each sib of an affected individual has a 25% chance of being affected, a 50% chance of being an asymptomatic carrier, and a 25% chance of being unaffected and not a carrier. Given that a clear genotype-phenotype correlation does not exist for MCAD deficiency and that individuals may remain asymptomatic until late adulthood, apparently unaffected sibs of an affected child should be tested for MCAD deficiency.

Prenatal diagnosis for pregnancies at risk for MCAD deficiency and other FAODs is possible by analysis of DNA extracted from fetal cells obtained by amniocentesis or chorionic villus sampling (CVS) (see Chapter 17). Both disease-causing alleles of an affected family member must be identified before prenatal testing can be performed. Prenatal diagnosis for pregnancies at increased risk is also possible by assay of MCAD enzymatic activity in CVS or amniocyte cultures. Amniocyte cultures can also be used for analysis of fatty acid oxidation as is done in fibroblast cultures. Preimplantation

genetic diagnosis may be available for couples in whom the disease-causing mutations have been identified. Prenatal diagnosis, however, carries its own inherent risks and offers no advantage to timely postnatal measurement of plasma acylcarnitine and urine acylglycine levels. Prompt postnatal testing and consultation with a biochemical geneticist are indicated.

QUESTIONS FOR SMALL GROUP DISCUSSION

1. What other FAODs are included in newborn screening programs?
2. What are the criteria for the inclusion of a disease in newborn screening programs?
3. Can individuals heterozygous for a mutation in *ACADM* be identified by newborn screening?
4. What are the false-positive and false-negative rates for MCAD deficiency by newborn screening?

REFERENCES

Lindner M, Hoffmann GF, Matern D: Newborn screening for disorders of fatty acid oxidation: experience and recommendations from an expert meeting, *J Inherit Metab Dis* 33:521–526, 2010.

Matern D, Rinaldo P: Medium-chain acyl-coenzyme A dehydrogenase deficiency. Available from: http://www.ncbi.nlm.nih.gov/books/NBK1424/.

Tein I: Disorders of fatty acid oxidation, *Handb Clin Neurol* 113:1675–1688, 2013.

MILLER-DIEKER SYNDROME (17p13.3 Heterozygous Deletion, MIM 247200)

Chromosomal Deletion

PRINCIPLES

- Microdeletion syndrome
- Contiguous gene disorder/genomic disorder
- Haploinsufficiency

MAJOR PHENOTYPIC FEATURES

- Age at onset: Prenatal
- Lissencephaly type 1 or type 2
- Facial dysmorphism
- Severe global intellectual disability
- Seizures
- Early death

HISTORY AND PHYSICAL FINDINGS

B.B., a 5-day-old boy born at 38 weeks of gestation, was admitted to the neonatal intensive care unit because of marked hypotonia and feeding difficulties. He was the product of an uncomplicated pregnancy, with a fetal ultrasound examination at 18 weeks of gestation. B.B. was born by spontaneous vaginal delivery; his Apgar scores were 8 at 1 minute and 9 at 5 minutes. He did not have a family history of genetic, neurological, or congenital disorders. On physical examination, B.B. had hypotonia and mild dysmorphic facial features, including bitemporal narrowing, depressed nasal bridge, small nose with anteverted nares, and micrognathia. The findings from the examination were otherwise normal. His evaluation had included normal serum electrolyte values, normal metabolic screen results, and normal study results for congenital infections. Brain ultrasound scan showed a hypoplastic corpus callosum, mild ventricular dilatation, and a smooth cortex. In addition to those studies, the genetics consultation team recommended magnetic resonance imaging (MRI) of the brain and a chromosomal microarray. MRI showed a thickened cerebral cortex, complete cerebral agyria, multiple cerebral heterotopias, hypoplastic corpus callosum, normal cerebellum, and normal brainstem. A chromosomal microarray revealed a 1.2-Mb deletion on 17p13.3, including the LIS1 gene. On the basis of these results, the geneticist explained to the parents that B.B. had Miller-Dieker syndrome. The parents declined further measures other than those to keep the baby comfortable, and B.B. died at 2 months of age.

BACKGROUND

Disease Etiology and Incidence

Miller-Dieker syndrome (MDS, MIM 247200) is a contiguous gene deletion syndrome caused by heterozygous deletion of 17p13.3; the mechanism underlying recurrent deletion of 17p13.3 has not yet been elucidated, but it may (like other microdeletion syndromes; see Chapter 6) involve recombination between low-copy repeated DNA sequences. MDS is a rare disorder, occurring in possibly 40 per 1 million births in all populations.

Pathogenesis

More than 50 genes have been mapped within the MDS deletion region in 17p13.3, but only the LIS1 gene (MIM 601545) has been associated with a specific phenotypic feature of MDS; heterozygosity for a LIS1 mutation causes lissencephaly. LIS1 encodes the brain isoform of the noncatalytic β subunit of platelet-activating factor acetylhydrolase (PAFAH). PAFAH hydrolyzes platelet-activating factor, an inhibitor of neuronal migration. PAFAH also binds to and stabilizes microtubules; preliminary observations suggest that PAFAH may play a role in the microtubule reorganization required for neuronal migration.

Haploinsufficiency of LIS1 alone, however, does not cause the other dysmorphic features associated with MDS. Mutations within LIS1 cause isolated lissencephaly sequence (ILS) (MIM 607432), that is, lissencephaly without other dysmorphism. Because all patients with MDS have dysmorphic facial features, this dysmorphism must be caused by haploinsufficiency of one or more different genes in the common MDS deletion interval.

PHENOTYPE AND NATURAL HISTORY

The features of MDS include brain dysgenesis, hypotonia, failure to thrive, and facial dysmorphism. The brain dysgenesis is characterized by lissencephaly type 1 (complete agyria) or type 2 (widespread agyria with a few sulci at the frontal or occipital poles), a cerebral cortex with four instead of six layers, gray matter heterotopias, and attenuated white matter (see Chapter 14). Some patients also have heart malformation and omphalocele.

A patient's facial features and an MRI finding of lissencephaly often suggest a diagnosis of MDS (Fig. C-32). Confirmation of the diagnosis, however, requires detection of a 17p13.3 deletion by chromosome analysis, FISH with a LIS1-specific probe, or chromosomal microarray. Approximately 60% of patients have a cytogenetically visible deletion of the MDS critical region. Normal FISH or microarray study results do not exclude the diagnosis of MDS; some patients have a partial gene deletion. Patients with ILS may have a mutation in LIS1, and males may have a mutation in DCX (an X-linked gene also associated with ILS).

Patients with MDS feed and grow poorly. Smiling, brief visual fixation, and nonspecific motor responses are the only developmental skills most patients acquire. In addition to intellectual disability, patients usually suffer from opisthotonos, spasticity, and seizures. Nearly all patients die by 2 years of age.

Management

MDS is incurable; therefore treatment focuses on the management of symptoms and palliative care. Nearly all patients require pharmacological management of their seizures. Also, many patients receive nasogastric or gastrostomy tube feedings because of poor feeding and repeated aspiration.

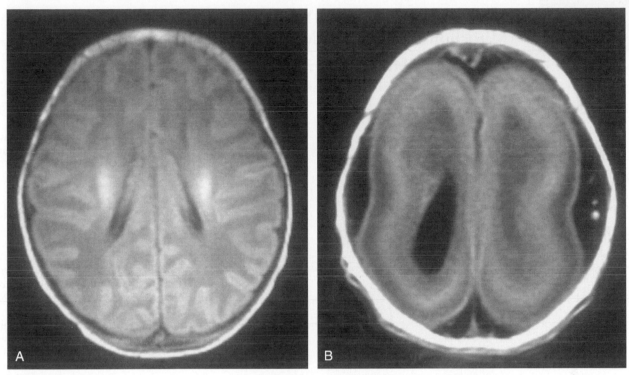

Figure C-32 Brain magnetic resonance images of infants without lissencephaly (**A**) and with Miller-Dieker syndrome (**B**). Note the smooth cerebral surface, the thickened cerebral cortex, and the classic "figure-8" appearance of the brain of the patient with Miller-Dieker syndrome. *See Sources & Acknowledgments.*

INHERITANCE RISK

Eighty percent of patients have a de novo microdeletion of 17p13.3, and 20% inherit the deletion from a parent who carries a balanced chromosomal rearrangement. Because of the frequency with which the deletion is inherited from a parent with a balanced translocation, karyotype analysis and FISH for *LIS1* should be performed in both parents. A parent with a balanced translocation involving 17p13.3 has approximately a 25% risk for having an abnormal liveborn child (MDS or dup17p) and approximately a 20% risk for pregnancy loss. In contrast, if a patient has MDS as a result of a de novo deletion, the parents have a low risk for recurrence of MDS in future children (but not zero, due to the possibility of gonadal mosaicism).

Although the brain malformations of MDS result from incomplete migration of neurons to the cerebral cortex during the third and fourth months of gestation, lissencephaly is not detected by fetal MRI or ultrasonography until late in gestation. Prenatal diagnosis of MDS requires detection of a 17p13.3 deletion in fetal chorionic villi or amniocytes.

QUESTIONS FOR SMALL GROUP DISCUSSION

1. Rubenstein-Taybi syndrome is caused either by deletion of 16p13.3 or by mutation of the *CREBBP* transcription factor. Compare and contrast the relationship of *CREBBP* and Rubenstein-Taybi syndrome with the relationship of *LIS1* and MDS. Why is MDS a contiguous gene deletion syndrome, whereas Rubenstein-Taybi syndrome is not?

2. Mutations of either *LIS1* on chromosome 17 or *DCX* on the X chromosome account for approximately 75% of ILS cases. What features of the family history and brain MRI scan can be used to focus testing on *DCX* as opposed to *LIS1*?

3. At 30 weeks of gestation, a woman has a fetal ultrasound examination showing fetal lissencephaly. The pregnancy was otherwise uncomplicated, and fetal ultrasound findings earlier in gestation had been normal. What counseling and evaluation are indicated? Discuss your counseling approach if she and her spouse wish to terminate the pregnancy at 32 weeks of gestation.

REFERENCES

Dobyns WB, Das S: LIS1-Associated lissencephaly/subcortical band heterotopia. Available from: http://www.ncbi.nlm.nih.gov/books/NBK5189/.

Hsieh DT, Jennesson MM, Thiele EA, et al: Brain and spinal manifestations of Miller-Dieker syndrome, *Neurol Clin Pract* 3:82–83, 2013.

Wynshaw-Boris A: Lissencephaly and LIS1: insights into molecular mechanisms of neuronal migration and development, *Clin Genet* 72:296–304, 2007.

MYOCLONIC EPILEPSY WITH RAGGED-RED FIBERS
(Mitochondrial tRNA^{lys} Mutation, MIM 545000)

Matrilineal, Mitochondrial

PRINCIPLES

- Mitochondrial DNA mutations
- Replicative segregation
- Expression threshold
- High mutation rate
- Accumulation of mutations with age
- Heteroplasmy

MAJOR PHENOTYPIC FEATURES

- Age at onset: Childhood through adulthood
- Myopathy
- Dementia
- Myoclonic seizures
- Ataxia
- Deafness

HISTORY AND PHYSICAL FINDINGS

R.S., a 15-year-old boy, was referred to the neurogenetics clinic for myoclonic epilepsy; his electroencephalogram was characterized by bursts of slow wave and spike complexes. Before the seizures developed, he had been well and developing normally. His family history was remarkable for a maternal uncle who had died of an undiagnosed myopathic disorder at 53 years; a maternal aunt with progressive dementia who had presented with ataxia at 37 years; and an 80-year-old maternal grandmother with deafness, diabetes, and renal dysfunction. On examination, R.S. had generalized muscle wasting and weakness, myoclonus, and ataxia. Initial evaluation detected sensorineural hearing loss, slowed nerve conduction velocities, and mildly elevated blood and cerebrospinal fluid lactate levels. Results of a subsequent muscle biopsy identified abnormal mitochondria, deficient staining for cytochrome oxidase, and ragged-red fibers—muscle fibers with subsarcolemmal mitochondria that stained red with Gomori trichrome stain. Molecular testing of a skeletal muscle biopsy specimen for mutations within the mitochondrial genome (mtDNA) identified a heteroplasmic mutation (8344G>A, tRNA^{lys} gene), a mutation known to be associated with myoclonic epilepsy with ragged-red fibers (MERRF), in 80% of the mtDNA from muscle. Subsequent testing of blood samples from R.S.'s mother, aunt, and grandmother confirmed that they also were heteroplasmic for this mutation. A review of the autopsy of the deceased uncle identified ragged-red fibers in some muscle groups. The physician counseled the family members (R.S.'s sibs and his mother's sibs) that they were either manifesting or nonmanifesting carriers of a deleterious mtDNA mutation compromising oxidative phosphorylation. No other family members chose to be tested for the mutation.

BACKGROUND

Disease Etiology and Incidence

MERRF (MIM 545000) is a rare panethnic disorder caused by mutations within the mtDNA in the tRNA^{lys} gene. More than 90% of patients have one of three mutations within this gene: 8344G>A accounts for 80% and 8356T>C and 8363G>A together account for an additional 10% (see Fig. 12-26). The disease is inherited maternally because mitochondria are inherited almost exclusively from the mother. Patients with MERRF are nearly always heteroplasmic for the mutant mitochondria (see Chapters 7 and 12).

Pathogenesis

Mitochondria generate energy for cellular processes by producing adenosine triphosphate (ATP) through oxidative phosphorylation. Five enzyme complexes, I to V, compose the oxidative phosphorylation pathway. Except for complex II, each complex has some components encoded within the mtDNA and some in the nuclear genome. The mtDNA encodes 13 of the polypeptides in the oxidative phosphorylation complexes as well as two ribosomal RNAs and 22 transfer RNAs (tRNAs) (see Fig. 12-26).

In MERRF, the activities of complexes I and IV are usually most severely reduced. The tRNA^{lys} mutations associated with MERRF reduce the amount of charged tRNA^{lys} in the mitochondria by 50% to 60% and thereby decrease the efficiency of translation so that at each lysine codon, there is a 26% chance of termination. Because complexes I and IV have the most components synthesized within the mitochondria, they are most severely affected.

Because each mitochondrion contains multiple mtDNAs and each cell contains multiple mitochondria, a cell can contain normal mtDNAs or abnormal mtDNAs in varying proportions; therefore expression of the MERRF phenotype in any cell, organ, or individual ultimately depends on the overall reduction in oxidative phosphorylation capacity. The threshold for expression of a deleterious phenotype depends on the balance between oxidative supply and demand. This threshold varies with age and among individuals, organ systems, and tissues.

The threshold for expression of the MERRF phenotype in an individual tissue heteroplasmic for a tRNA^{lys} can be exceeded either by an accumulation of mutations in the normal mtDNA or by increasing the proportion of mutant mtDNAs. Compared to nuclear DNA, mtDNA has a 10-fold higher mutation rate; this may result from exposure to a high concentration of oxygen free radicals from oxidative phosphorylation, a lack of protective histones, and ineffective DNA repair. Because mtDNA has no introns, random mutations usually affect coding sequences. Consistent with this increased mutation rate, mitochondrial efficiency declines gradually throughout adulthood, and as reserve oxidative phosphorylation activity declines, expression of defects in the oxidative phosphorylation pathway becomes increasingly likely.

Increases in the proportion of mutant mtDNA can occur by a combination of inheritance, preferential replication of mutant mtDNA, and selection. First, the children of heteroplasmic mothers have widely varying proportions of mtDNA genotypes because of replicative segregation, that is, random partitioning of mitochondria during expansion of the oogonial population, particularly because of the mitochondrial "genetic bottleneck" that occurs during oogenesis. Second, as heteroplasmic cells in an individual undergo mitosis, the proportion of mtDNA genotypes in daughter cells changes from that of the parent cell by replicative segregation. Third, because changes in the proportion of mtDNA genotypes affect the cellular phenotype, the mtDNA is subject to strong selective

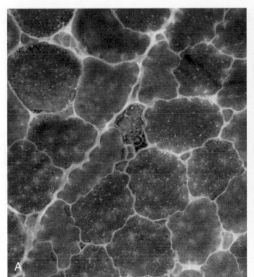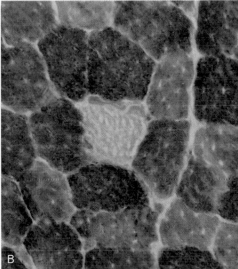

Figure C-33 **Quadriceps muscle histology. A,** Modified Gomori trichrome stain illustrating ragged-red fiber (×525). **B,** Cytochrome oxidase stain illustrating absence of cytochrome oxidase in an affected muscle fiber, consistent with a mitochondrial DNA defect (×525). *See Sources & Acknowledgments.*

pressures; the selective pressures vary among tissues and result in different mtDNA populations in different tissues of the same person. Thus, both intercellular and intergenerational mtDNA transmission follow the principles of population genetics.

Phenotype and Natural History

The classic MERRF phenotype includes myoclonic epilepsy and mitochondrial myopathy with ragged-red fibers (Fig. C-33). Other associated findings include abnormal brainstem evoked responses, sensorineural hearing loss, ataxia, renal dysfunction, diabetes, cardiomyopathy, and dementia. Onset of symptoms can be in childhood or adult life, and the course can be slowly progressive or rapidly downhill.

Because mtDNA genetics follows quantitative and stochastic principles, clinical features of affected relatives vary in pattern and severity and do not have an easily defined clinical course. The absence of ragged-red fibers in a muscle biopsy specimen does not exclude MERRF. Within pedigrees, phenotypes generally correlate well with the severity of the oxidative phosphorylation deficit, but correlation with the percentage of mutant mtDNA in skeletal muscle requires adjustment for age. In one pedigree, a young adult with 5% normal mtDNA in skeletal muscle, had a severe clinical and biochemical phenotype; other young adults with 15% normal mtDNA had normal phenotypes; and an older adult with 16% normal mtDNA had a severe phenotype. This expression pattern demonstrates that symptoms accumulate progressively as oxidative phosphorylation capacity drops below organ expression thresholds and that age-related declines in oxidative phosphorylation play a critical role in the appearance and progression of symptoms.

Management

Treatment is symptomatic and palliative. No specific therapies are currently available. Most patients are given coenzyme Q10 and L-carnitine supplements to optimize the activity of the oxidative phosphorylation complexes.

INHERITANCE RISK

The risk to children of affected males is essentially zero because, with only one known exception, children do not inherit paternal mtDNA. The risk to children of affected or unaffected females with a MERRF mutation cannot be estimated accurately by prenatal testing because the critical parameters defining disease in the child (replicative segregation, tissue selection, and somatic mtDNA mutations) cannot be predicted in advance.

Similarly, molecular testing of blood samples from at-risk family members is complicated by two general problems. First, because of replicative segregation and tissue selection, the mutation may not be detectable in blood; therefore a negative result does not exclude a family member as a carrier of the mtDNA mutation. Second, because of replicative segregation, a positive result predicts neither the proportion of mutant mtDNA in other tissues nor the expected severity of disease.

QUESTIONS FOR SMALL GROUP DISCUSSION

1. How does a mutant mtDNA molecule, arising de novo in a cell with hundreds of normal molecules, become such a significant fraction of the total that energy-generating capacity is compromised and symptoms develop?

2. How could mitochondrial mutations affecting oxidative phosphorylation accelerate the mutation rate of mtDNA?

3. How would mitochondrial mutations affecting oxidative phosphorylation accelerate aging?

4. In the fetus, oxygen tension is low and most energy is derived from glycolysis. How could this observation affect the prenatal expression of deleterious oxidative phosphorylation mutations?

REFERENCES

Abbott JA, Francklyn CS, Robey-Bond SM: Transfer RNA and human disease, *Front Genet* 5:158, 2014.

DiMauro S, Hirano M: MERRF. Available from: http://www.ncbi.nlm.nih.gov/books/NBK1520/.

Suzuki T, Nagao A, Suzuki T: Human mitochondrial tRNAs: biogenesis, function, structural aspects and diseases, *Ann Rev Genet* 45:299–329, 2011.

PRINCIPLES

- Variable expressivity
- Extreme pleiotropy
- Tumor-suppressor gene
- Loss-of-function mutations
- Allelic heterogeneity
- De novo mutations

MAJOR PHENOTYPIC FEATURES

- Age at onset: Prenatal to late childhood
- Café au lait spots
- Axillary and inguinal freckling
- Cutaneous neurofibromas
- Lisch nodules (iris hamartomas)
- Plexiform neurofibromas
- Optic glioma
- Specific osseous lesions

HISTORY AND PHYSICAL FINDINGS

L.M. was a 2-year-old girl referred because of five café au lait spots, of which three measured larger than 5 mm in diameter. She had no axillary or inguinal freckling, no osseous malformations, and no neurofibromas. Physical examination of her parents revealed no stigmata of neurofibromatosis. The consulting geneticist informed the parents and referring pediatrician that L.M. did not meet the clinical criteria for neurofibromatosis type 1.

L.M. returned to the genetics clinic at 5 years of age. She now had Lisch nodules in both eyes and 12 café au lait spots, 8 of which measured at least 5 mm in diameter. She also had axillary freckling bilaterally. She was given the diagnosis of neurofibromatosis 1; her parents were told that she likely had a de novo mutation and the recurrence risk was therefore low, but that gonadal mosaicism could not be excluded.

L.M.'s parents declined both molecular testing in L.M. and prenatal testing during their next pregnancy.

BACKGROUND

Disease Etiology and Incidence

Neurofibromatosis 1 (NF1, MIM 162200) is a panethnic autosomal dominant condition with symptoms most frequently expressed in the skin, eye, skeleton, and nervous system. NF1 results from mutations in the neurofibromin gene (*NF1*). The disease has an incidence of 1 in 3500 persons, making it one of the most common autosomal dominant genetic conditions. Approximately half of patients have de novo mutations; the mutation rate for the *NF1* gene is one of the highest known for any human gene, at approximately 1 mutation per 10,000 live births. Approximately 80% of the de novo mutations are paternal in origin, but there is no evidence for a paternal age effect increasing the mutation rate (see Chapter 4).

Pathogenesis

NF1 is a large gene (350 kb and 60 exons) that encodes neurofibromin, a protein widely expressed in almost all tissues but most abundantly in the brain, spinal cord, and peripheral nervous system. Neurofibromin is thought to regulate several intracellular processes, including the activation of Ras GTPase, thereby controlling cellular proliferation and acting as a tumor suppressor.

More than 500 mutations in the *NF1* gene have been identified; most are unique to an individual family. The clinical manifestations result from a loss of function of the gene product; 80% of the mutations cause protein truncation. A disease-causing mutation can be identified for more than 95% of individuals diagnosed with NF1.

NF1 is characterized by extreme clinical variability, both between and within families. This variability is probably caused by a combination of genetic, nongenetic, and stochastic factors. No clear genotype-phenotype correlations have been recognized, although large deletions are more common in NF1 patients with neurodevelopmental difficulties.

Phenotype and Natural History

NF1 is a multisystem disorder with neurological, musculoskeletal, ophthalmological, and skin abnormalities and a predisposition to neoplasia (Fig. C-34). A diagnosis of NF1 can be made if an individual meets two or more of the following criteria: six or more café au lait spots measuring at least 5 mm in diameter (if prepubertal) or 15 mm in diameter (if postpubertal); two or more neurofibromas of any type, or one plexiform neurofibroma; axillary or inguinal freckling; optic glioma; two or more Lisch nodules; a distinctive osseous phenotype (sphenoid dysplasia and thinning of the long bone cortex, with or without pseudarthrosis); or a first-degree relative with NF1.

Nearly all individuals with NF1 but no family history will meet clinical criteria by the age of 8 years. Children who have inherited NF1 can usually be identified clinically within the first year of life because they require only one other feature of the disease to be present.

Although penetrance is essentially complete, manifestations are extremely variable. Multiple café au lait spots are present in nearly all affected individuals, with freckling seen in 90% of cases. Many individuals with NF1 have only cutaneous manifestations of the disease and iris Lisch nodules. Numerous neurofibromas are usually present in adults. Plexiform neurofibromas are less common. Ocular manifestations include optic gliomas (which may lead to blindness) and iris Lisch nodules. The most serious bone complications are scoliosis, vertebral dysplasia, pseudarthrosis, and overgrowth. Stenosis of pulmonic, renal, and cerebral vessels and hypertension are also frequent. The most common neoplasms for children with NF1 (other than neurofibromas) are optic nerve gliomas, brain tumors, and malignant myeloid disorders. Approximately half of all children with NF1 will have a learning disability or attention deficits, which can persist into adulthood. Pheochromocytoma can also occur.

Individuals with features of NF1 limited to one region of the body, and who have unaffected parents, may be diagnosed with segmental (or regional) NF1. Segmental NF1 may

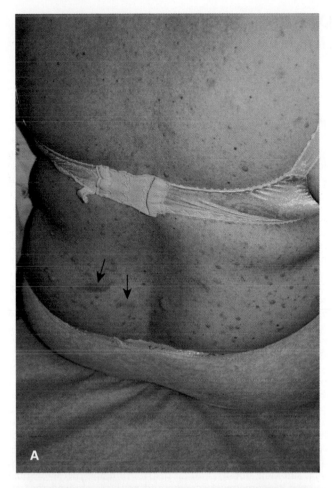

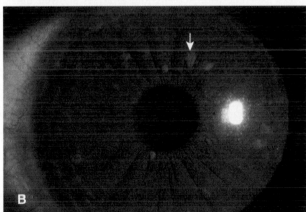

Figure C-34 A, Cutaneous manifestations of neurofibromatosis 1, including hundreds of small to medium-sized reddish papular neurofibromas and two large café au lait spots (*arrows*). **B,** Iris showing numerous Lisch nodules (one typical nodule is indicated by the *arrow*). *See Sources & Acknowledgments.*

represent an unusual distribution of clinical features by chance or somatic mosaicism for an *NF1* gene mutation.

Management

NF1 is a clinical diagnosis. Identification of mutations is not done routinely because of the size of the gene and the extreme allelic heterogeneity but is clinically available and be useful for patients in whom the diagnosis is less obvious.

No curative treatments are available, and therefore treatment focuses on symptomatic management. Ongoing surveillance in an individual with NF1 should include an annual physical examination conducted by someone familiar with NF1, annual ophthalmological evaluation in childhood (less frequent as an adult), regular developmental assessments in childhood, and regular blood pressure measurements.

The deformities caused by NF1 are the most distressing disease manifestation. Discrete cutaneous and subcutaneous neurofibromas can be surgically removed if they are disfiguring or inconveniently located. Plexiform neurofibromas causing disfigurement or impingement can also be surgically managed. However, surgical intervention for these neoplasms can be problematic because they are often intimately involved with nerves and have a tendency to grow back at the site of removal.

INHERITANCE RISK

Individuals with NF1 have a 50% risk for having a child affected with NF1, although the features may be different in an affected child. Prenatal diagnosis is available for those families in whom a causative *NF1* gene mutation has been identified. Although prenatal diagnosis is accurate, it will not provide much prognostic information because of the extreme phenotypic variability of the disease. Parents of an affected child who themselves show no signs of the disease are still at some small elevated recurrence risk in the next pregnancy because of the possibility of gonadal mosaicism, which has been documented with NF1.

QUESTIONS FOR SMALL GROUP DISCUSSION

1. Why is there such clinical variability in NF1? What factors could be influencing this phenotype?
2. Why is a positive family history of NF1 one of the major diagnostic criteria for this condition and not for other autosomal dominant conditions?
3. Review the major points of discussion with a family that desires prenatal testing for NF1 based on a known mutation in one of the parents.
4. How would a treatment of NF1 need to be targeted at the molecular level to specifically address the loss of function seen with this condition? How is that different from a disease caused by a dominant negative mutation?

REFERENCES

Friedman JM: Neurofibromatosis 1. Available from: http://www.ncbi.nlm.nih.gov/books/NBK1109/.

Hirbe AC, Gutmann DH: Neurofibromatosis type 1: a multidisciplinary approach to care, *Lancet Neurol* 13:834–843, 2014.

Pasmant E, Vidaud M, Vidaud D, et al: Neurofibromatosis type 1: from genotype to phenotype, *J Med Genet* 49:483–489, 2012.

NON–INSULIN-DEPENDENT (TYPE 2) DIABETES MELLITUS (Insulin Deficiency and Resistance, MIM 125853)

Multifactorial

PRINCIPLES

- Polygenic disease
- Environmental modifiers

MAJOR PHENOTYPIC FEATURES

- Age at onset: Childhood through adulthood
- Hyperglycemia
- Relative insulin deficiency
- Insulin resistance
- Obesity
- Acanthosis nigricans

HISTORY AND PHYSICAL FINDINGS

M.P. is a 38-year-old healthy male member of the Pima Indian tribe who requested information on his risk for development of non–insulin-dependent (type 2) diabetes mellitus (NIDDM or T2D). Both of his parents had had T2D; his father died at 60 years from a myocardial infarction and his mother at 55 years from renal failure. His paternal grandparents and one older sister also had T2D, but he and his four younger siblings did not have the disease. The findings from M.P.'s physical examination were normal except for mild obesity; he had a normal fasting blood glucose level but an elevated blood insulin level and abnormally high blood glucose levels after an oral glucose challenge. These results were consistent with early manifestations of a metabolic state likely to lead to T2D. His physician advised M.P. to change his lifestyle so that he would lose weight and increase his physical activity. M.P. sharply reduced his dietary fat consumption, began commuting to work by bicycle, and jogged three times per week; his weight decreased 10 kg, and his glucose tolerance and blood insulin level normalized.

BACKGROUND

Disease Etiology and Incidence

Diabetes mellitus (DM) is a heterogeneous disease composed of type 1 (also referred to as insulin-dependent DM, IDDM, or T1D) (Case 26) and Type 2 (also referred to as non–insulin-dependent DM, NIDDM, or T2D) diabetes mellitus (see Table). NIDDM/T2D (MIM 125853) accounts for 80% to 90% of all diabetes mellitus and has a prevalence of 6% to 7% among adults in the United States. For as yet unknown reasons, there is a strikingly increased prevalence of the disease among Native Americans from the Pima tribe in Arizona, in whom the prevalence of T2D is nearly 50% by the age of 35 to 40 years. Approximately 5% to 10% of patients with T2D have maturity-onset diabetes of the young (MODY, MIM 606391); 5% to 10% have a rare genetic disorder; and the remaining 70% to 85% have "typical T2D," a form of type 2 diabetes mellitus characterized by relative insulin deficiency and resistance. Despite significant efforts to identify genes that influence the risk for T2D (see Chapter 10), the molecular and genetic bases of typical T2D remain poorly defined.

Pathogenesis

T2D results from a derangement of insulin secretion and resistance to insulin action. Normally, basal insulin secretion follows a rhythmic pattern interrupted by responses to glucose loads. In patients with T2D, the rhythmic basal release of insulin is markedly deranged, responses to glucose loads are inadequate, and basal insulin levels are elevated, although low relative to the hyperglycemia of these patients.

Persistent hyperglycemia and hyperinsulinemia develop before T2D and initiate a cycle leading to T2D. The persistent hyperglycemia desensitizes the islet β-cell such that less insulin is released for a given blood glucose level. Similarly, the chronic elevated basal levels of insulin down-regulate insulin receptors and thereby increase insulin resistance. Furthermore, as sensitivity to insulin declines, glucagon is unopposed and its secretion increases; as a consequence of excessive glucagon, glucose release by the liver increases, worsening the hyperglycemia. Ultimately, this cycle leads to T2D.

Typical T2D results from a combination of genetic susceptibility and environmental factors. Observations supporting a genetic predisposition include differences in concordance between monozygotic and dizygotic twins, familial clustering, and differences in prevalence among populations. Whereas human inheritance patterns suggest complex inheritance, identification of the relevant genes in humans, although made difficult by the effects of age, gender, ethnicity, physical fitness, diet, smoking, obesity, and fat distribution, has met with some success. Genome-wide screens and analyses showed that an allele of a short tandem repeat polymorphism in the intron of the gene for a transcription factor, *TCF7L2*, is significantly associated with T2D in the Icelandic population. Heterozygotes (38% of the population) and homozygotes (7% of the population) have an increased relative risk for T2D of

Comparison of Type 1 and Type 2 Diabetes Mellitus

Characteristic	Type 1 (IDDM)	Type 2 (NIDDM)
Sex	Female = male	Female > male
Age at onset	Childhood and adolescence	Adolescence through adulthood
Ethnic predominance	Whites	African Americans, Mexican Americans, Native Americans
Concordance		
Monozygotic twins	33%-50%	69%-90%
Dizygotic twins	1%-14%	24%-40%
Family history	Uncommon	Common
Autoimmunity	Common	Uncommon
Body habitus	Normal to wasted	Obese
Acanthosis nigricans	Uncommon	Common
Plasma insulin	Low to absent	Normal to high
Plasma glucagon	High, suppressible	High, resistant
Acute complication	Ketoacidosis	Hyperosmolar coma
Insulin therapy	Responsive	Resistant or responsive
Oral hypoglycemic therapy	Unresponsive	Responsive

approximately 1.5-fold and 2.5-fold, respectively, over non-carriers. The increased risk due to the *TCF7L2* variant has been replicated in other population cohorts, including in the United States. The risk for T2D attributable to this allele is 21%. *TCF7L2* encodes a transcription factor involved in the expression of the hormone glucagon, which raises the blood glucose concentration and therefore works to oppose the action of insulin in lowering blood glucose.

Screens of Finnish and Mexican American groups have identified another predisposition variant, a Pro12Ala mutation in the *PPARG* gene, apparently specific to those populations and accounting for up to 25% of the population-attributable risk for T2D in these populations. The more common proline allele has a frequency of 85% and causes a modest increase in risk (1.25-fold) for diabetes. PPARG is a member of the nuclear hormone receptor family and is important in the regulation of adipocyte function and differentiation.

Evidence for an environmental component includes a concordance of less than 100% in monozygotic twins; differences in prevalence in genetically similar populations; and associations with lifestyle, diet, obesity, pregnancy, and stress. The body of experimental evidence suggests that although genetic susceptibility is a prerequisite for T2D, clinical expression of T2D is likely to be strongly influenced by environmental factors.

Phenotype and Natural History

T2D usually affects obese individuals in middle age or beyond, although an increasing number of children and younger individuals are affected as more become obese and sedentary. Depending on the apparent severity of the genetic susceptibility, some T2D patients are only mildly obese or not obese at all.

T2D has an insidious onset and is diagnosed usually by an elevated glucose level on routine examination. In contrast to patients with T1D, patients with T2D usually do not develop ketoacidosis. In general, the development of T2D is divided into three clinical phases. First, the plasma glucose concentration remains normal despite elevated blood levels of insulin, indicating that the target tissues for insulin action appear to be relatively resistant to the effects of the hormone. Second, postprandial hyperglycemia develops despite elevated insulin concentrations. Third, declining insulin secretion causes fasting hyperglycemia and overt diabetes.

In addition to hyperglycemia, the metabolic dysregulation resulting from islet β-cell dysfunction and insulin resistance causes atherosclerosis, peripheral neuropathy, renal disease, cataracts, and retinopathy (Fig. C-35). One in six patients with T2D will develop end-stage renal disease or will require a lower extremity amputation for severe vascular disease; one in five will become legally blind from retinopathy. The development of these complications is related to the genetic background and degree of metabolic control. Chronic hyperglycemia can be monitored by measurements of the percentage of hemoglobin that has become modified by glycosylation, referred to as hemoglobin A1c (HbA1c). Rigorous control of blood glucose levels, as determined by HbA1c levels as close to normal as possible (<7%), reduces the risk for complications by 35% to 75% and can extend the average life expectancy, which now averages 17 years after diagnosis, by a few years.

Management

Weight loss, increased physical activity, and dietary changes help many patients with T2D by markedly improving insulin sensitivity and control. Unfortunately, many patients are unable or unwilling to change their lifestyle sufficiently to accomplish this control and require treatment with oral

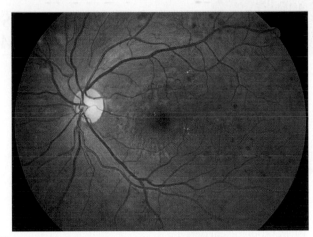

Figure C-35 Nonproliferative diabetic retinopathy in a patient with type 2 diabetes. Note the multiple "dot and blot" hemorrhages, the scattered "bread crumb" patches of intraretinal exudate- and, superonasally, a few cotton-wool patches. *See Sources & Acknowledgments.*

hypoglycemic agents, such as sulfonylureas and biguanides. A third class of agent, thiazolidinediones, reduces insulin resistance by binding to PPARG. A fourth category of medication, α-glucosidase inhibitors, which act to slow intestinal absorption of glucose, can also be used. Each of these classes of drugs has been approved as monotherapy for T2D. As they fail with progression of disease, an agent from another class can be added. Oral hypoglycemics are not as effective as weight loss, increased physical activity, and dietary changes for achieving glycemic control. To achieve glycemic control and possibly reduce the risk for diabetic complications, some patients require treatment with exogenous insulin; however, insulin therapy accentuates insulin resistance by increasing hyperinsulinemia and obesity.

INHERITANCE RISK

The population risk for T2D is highly dependent on the population under consideration; in most populations, this risk is 1% to 5%, although it is 6% to 7% in the United States. If a patient has one affected sibling, the risk increases to 10%; an affected sibling and another first-degree relative, the risk is 20%; and an affected monozygotic twin, the risk is 50% to 100%. In addition, because some forms of T2D are antecedents to T1D (**Case 26**), children of parents with T2D have an empirical risk of 1 in 10 for development of T1D.

QUESTIONS FOR SMALL GROUP DISCUSSION

1. How could civil engineering have a major impact on the treatment of patients with T2D?
2. What counseling should members, including children, of T2D families be given?
3. What factors are contributing to the rising prevalence of T2D?

REFERENCES

Bonnefond A, Froguel P, Vaxillaire M: The emerging genetics of type 2 diabetes, *Trends Mol Med* 16:407–416, 2010.

Diabetes Genetics Replication and Meta-analysis Consortium, et al: Genome-wide trans-ancestry meta-analysis provides insight into the genetic architecture of type 2 diabetes susceptibility, *Nat Genet* 46:234–244, 2014.

Thomsen SK, Gloyn AL: The pancreatic β cell: recent insights from human genetics, *Trends Endocrinol Metab* S1043–S2760, 2014.

ORNITHINE TRANSCARBAMYLASE DEFICIENCY
(OTC Mutation, MIM 311250)

X-Linked

PRINCIPLES

- Inborn error of metabolism
- X chromosome inactivation
- Manifesting heterozygotes
- Asymptomatic carriers
- Germline mutation rate much greater in spermatogenesis than in oogenesis

MAJOR PHENOTYPIC FEATURES

- Age at onset: Hemizygous male with null mutation—neonatal; heterozygous female—with severe intercurrent illness, postpartum, or never
- Hyperammonemia
- Coma

HISTORY AND PHYSICAL FINDINGS

J.S. is a 4-day-old male infant brought to the emergency department because he could not be aroused. The parents reported a history of 24 hours of decreased intake, vomiting, and increasing lethargy. He was the 3-kg, full-term product of an uncomplicated gestation born to a healthy 26-year-old primiparous woman. Physical examination showed a comatose, hyperpneic, nondysmorphic male newborn. Initial laboratory evaluation revealed a blood NH_3 concentration of 900 µM (normal in a newborn is <75) and elevated venous pH of 7.48, with a normal bicarbonate concentration and anion gap. A urea cycle disorder was suspected, so plasma amino acid levels were determined on an emergency basis. Glutamine was elevated at 1700 µM (normal, <700 µM), and citrulline was undetectable (normal is 7 to 34 µM) (Fig. C-36). Analysis of urine for organic acids was normal; urinary orotic acid was extremely elevated. Elevated urine orotic acid levels with low citrulline indicates a diagnosis of ornithine transcarbamylase deficiency pending confirmation by mutation analysis.

Further questioning of J.S.'s mother revealed that she had a lifelong aversion to protein and a brother who died in the first week of life of unknown causes. J.S. was started on intravenous sodium benzoate and sodium phenylacetate (Ammonul) and arginine hydrochloride supplementation. The child was transported by air to a tertiary care center equipped for neonatal hemodialysis. On arrival, his plasma NH_3 level had dropped to 700 µM. The parents were counseled about the high risk for brain damage from this degree of hyperammonemia. They elected to proceed with hemodialysis, which was well tolerated and resulted in decline of the blood NH_3 to less than 200 µM after 4 hours. The child was maintained on Ammonul and high calories from intravenous dextrose and intralipids until the NH_3 level was normal, at which point he was slowly started on a protein-restricted diet and monitored for hyperammonemia, especially during intercurrent illnesses. His prognosis remains guarded.

BACKGROUND
Disease Etiology and Incidence

Ornithine transcarbamylase (OTC) deficiency (MIM 311250) is a panethnic X-linked disorder of urea cycle metabolism caused by mutations of the gene encoding ornithine transcarbamylase (*OTC*). It has an incidence of 1 in 30,000 males. The exact incidence of manifesting females is unknown.

Pathogenesis

Ornithine transcarbamylase is an enzyme in the urea cycle (see Fig. C-36). The urea cycle is the mechanism by which waste nitrogen is detoxified and excreted. Complete deficiency of any enzyme within the cycle (except arginase) leads to severe hyperammonemia in the neonatal period. For patients with urea cycle defects, arginine becomes an essential amino acid (see Fig. C-36). In utero, excess nitrogen is metabolized by the mother. Postnatally, however, accumulation of waste nitrogen in the extremely catabolic period after birth leads to elevation of glutamine and alanine, the body's natural pools for nitrogen, and ultimately to elevated levels of NH_3 ion. Plasma NH_3 levels above 200 micromolar may cause brain damage; the degree of brain damage is related to how high the concentrations of NH_3 and glutamine in the blood rise and how long the elevations persist. Thus early detection and treatment are critical to outcome.

Males are hemizygous for the *OTC* gene and are therefore more severely affected by mutations in this gene. Because *OTC* undergoes random X chromosome inactivation (see Chapter 6), females are mosaic for expression of the mutation and can demonstrate a wide range of enzyme function and clinical severity. Female heterozygotes can be completely asymptomatic and able to eat as much protein as they wish. Alternatively, if their loss of OTC activity is more significant, they may find themselves avoiding dietary protein and subject to recurrent, symptomatic hyperammonemia.

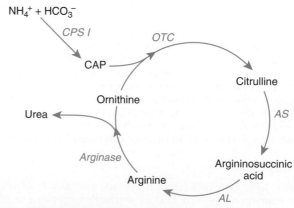

$$NH_4^+ + HCO_3^-$$

Figure C-36 **The urea cycle.** AL, Argininosuccinate lyase; AS, argininosuccinate synthetase; CAP, carbamoyl phosphate; CPS I, carbamoyl phosphate synthetase I; OTC, ornithine transcarbamylase.

Phenotype and Natural History

Males with complete OTC deficiency are born normal but begin vomiting, become lethargic, and eventually lapse into coma between 48 and 72 hours of age. Because they have been vomiting, they are usually dehydrated as well. Untreated males with null mutations usually die in the first week of life. Even if the patient with OTC deficiency is promptly and successfully treated in the neonatal period, the risk remains high for recurrent bouts of hyperammonemia, particularly during intercurrent illnesses, because complete control of severe OTC deficiency is difficult, even with dietary protein restriction and medications that divert the NH_3 to nontoxic pathways (see Chapter 13). With each episode of hyperammonemia, the patient may suffer brain damage or die in a matter of only a few hours after the onset of metabolic decompensation.

Girls (or boys with partial OTC deficiency) are usually asymptomatic in the neonatal period but may develop hyperammonemia during intercurrent febrile illnesses, such as influenza, or with excessive dietary protein intake. Other catabolic stresses, such as surgery, pregnancy, or long bone fracture, may also precipitate hyperammonemia. Like affected males, symptomatic females are at risk for brain damage and intellectual disability, but these serious complications can usually be prevented by anticipating them and instituting aggressive interventions to prevent catabolism.

OTC deficiency and carbamoyl phosphate synthetase deficiency (see Fig. C-36) cannot be detected by newborn screening. Abnormal metabolites that occur in other enzyme deficiencies within the urea cycle, however, can be detected by tandem mass spectrometry of serum amino acids (see Chapter 18).

Management

Plasma NH_3 concentration should be measured in any sick neonate. For most urea cycle defects, the pattern of abnormalities on quantitative amino acid determination is diagnostic. To distinguish between OTC deficiency and carbamoyl phosphate synthetase deficiency, both of which are characterized by very low or absent citrulline, it is necessary to measure urine orotic acid, which is elevated in OTC deficiency. Determination of urine organic acids is also important to rule out an organic aciduria, which can also present with hyperammonemia in the newborn period. Molecular testing is available to confirm the diagnosis.

Acutely hyperammonemic patients should be treated with a four-pronged approach: (1) 10% dextrose at twice the maintenance rate to provide calories in the form of sugar for gluconeogenesis and thereby reduce catabolism of endogenous proteins, and elimination of dietary protein intake; (2) intravenous Ammonul, a solution of sodium benzoate and sodium phenylacetate, both of which provide diversion therapy by driving the excretion of nitrogen independently of the urea cycle (see Chapter 13); (3) intravenous arginine hydrochloride to provide adequate amounts of arginine, an essential amino acid, and to drive any residual enzyme activity by ensuring adequate substrate to the urea cycle; and (4) if a patient does not respond to the initial bolus of these medications, hemodialysis.

Chronic management entails careful control of dietary calories and protein as well as oral phenylbutyrate. Maintenance of a high carbohydrate intake spares endogenous protein from being catabolized for gluconeogenesis; dietary protein restriction reduces the load of NH_3 requiring detoxification through the urea cycle. Phenylbutyrate is readily converted to phenylacetate, which promotes non–urea cycle dependent nitrogen excretion. The family must be carefully trained to look for early signs of hyperammonemia, such as irritability, vomiting, and sleepiness, so that the patient can be promptly brought to the hospital for intravenous treatment.

Because of the great difficulty in achieving metabolic control and the substantial risk for brain damage or death within hours of the onset of metabolic decompensation, liver transplantation to provide a functioning urea cycle should be presented as an option as soon as a patient has grown sufficiently (>10 kg) to tolerate the procedure.

INHERITANCE RISK

OTC deficiency is inherited as an X-linked trait. Because OTC deficiency is nearly always a genetic lethal disorder, approximately 67% of the mothers of affected infants would be expected to be carriers, as discussed in Chapters 7 and 16. Surprisingly, studies in families with OTC deficiency indicate that 90% of the mothers of affected infants are carriers. The reason for this discrepancy between the theoretical and actual carrier rates is that the underlying assumption of equal male and female mutation rates used for the theoretical calculation is incorrect. In fact, mutations in the *OTC* gene are much more frequent (≈50-fold) in the male germline than in the female germline. Most of the mothers of an isolated boy with OTC deficiency are carriers as a result of a new mutation inherited on the X chromosome they received from their fathers.

In a woman who is a carrier of a mutant OTC deficiency allele, her sons who receive the mutant allele will be affected and her daughters will be carriers who may or may not be symptomatic, depending on random X inactivation in the liver. Males with partial OTC deficiency who reproduce will have all carrier daughters and no affected sons. When the mutation in the family is known, prenatal testing by examination of the gene is available. Prenatal diagnosis by assay of the OTC enzyme is not practical because the enzyme is not expressed in chorionic villi or amniotic fluid cells.

QUESTIONS FOR SMALL GROUP DISCUSSION

1. Discuss the Lyon hypothesis and explain the variability of disease manifestations in females.
2. Why is arginine an essential amino acid in this disorder? Arginine is ordinarily not an essential amino acid in humans.
3. What organic acidurias cause hyperammonemia?
4. What are some of the reasons both for and against performing a liver transplant for OTC deficiency? Is a liver transplant for OTC deficiency more or less helpful than for other inborn errors of metabolism?

REFERENCE

Lichter-Konecki U, Caldovic L, Morizono H, et al: Ornithine transcarbamylase deficiency. Available from: http://www.ncbi.nlm.nih.gov/books/NBK154378/.

POLYCYSTIC KIDNEY DISEASE (*PKD1* and *PKD2* Mutations, MIM 173900 and MIM 613095)

Autosomal Dominant

PRINCIPLES

- Variable expressivity
- Genetic heterogeneity
- Two-hit hypothesis

MAJOR PHENOTYPIC FEATURES

- Age at onset: Childhood through adulthood
- Progressive renal failure
- Renal and hepatic cysts
- Intracranial saccular aneurysms
- Mitral valve prolapse
- Colonic diverticula

HISTORY AND PHYSICAL FINDINGS

Four months ago, P.J., a 35-year-old man with a history of mitral valve prolapse, developed intermittent flank pain. He eventually presented to his local emergency department with severe pain and hematuria. A renal ultrasound scan showed nephrolithiasis and polycystic kidneys consistent with polycystic kidney disease. The findings from his physical examination were normal except for a systolic murmur consistent with mitral valve prolapse, mild hypertension, and a slight elevation of serum creatinine concentration. His father and his sister had died of ruptured intracranial aneurysms, and P.J.'s son had died at 1 year of age from polycystic kidney disease. At the time of his son's death, the physicians had suggested that P.J. and his wife should be evaluated to see whether either of them had polycystic kidney disease; however, the parents elected not to pursue this evaluation because of guilt and grief about their son's death. P.J. was admitted for management of his nephrolithiasis. During this admission, the nephrologists told P.J. that he had autosomal dominant polycystic kidney disease.

BACKGROUND

Disease Etiology and Incidence

Autosomal dominant polycystic kidney disease (ADPKD1, MIM 173900) is genetically heterogeneous. Approximately 85% of patients have ADPKD1 caused by mutations in the *PKD1* gene; most of the remainder have ADPKD2 (MIM 613095) due to mutations of *PKD2*. A few families have not shown linkage to either of these loci, suggesting that there is at least one additional, as yet unidentified locus.

ADPKD is one of the most common genetic disorders and has a prevalence of 1 in 300 to 1 in 1000 in all ethnic groups studied. In the United States, it accounts for 8% to 10% of end-stage renal disease.

Pathogenesis

PKD1 encodes polycystin 1, a transmembrane receptor–like protein of unknown function. *PKD2* encodes polycystin 2, an integral membrane protein with homology to the voltage-activated sodium and calcium α_1 channels. Polycystin 1 and polycystin 2 interact as part of a heteromultimeric complex.

Cyst formation in ADPKD appears to follow a "two-hit" mechanism similar to that observed with tumor suppressor genes and neoplasia (see Chapter 15); that is, both alleles of either *PDK1* or *PDK2* must lose function for cysts to form. Polycystin-1 and -2 likely contribute to fluid-flow sensation by the primary cilium in renal epithelium within the same mechanotransduction pathway. The mechanism by which loss of either polycystin 1 or polycystin 2 function causes cyst formation has not been defined but includes mislocalization of cell surface proteins of developing renal tubular cells (see Chapter 14).

Phenotype and Natural History

ADPKD may manifest at any age, but symptoms or signs most frequently appear in the third or fourth decade. Patients present with urinary tract infections, hematuria, urinary tract obstruction (clots or nephrolithiasis), nocturia, hemorrhage into a renal cyst, or complaints of flank pain from the mass effect of the enlarged kidneys (Fig. C-37). Hypertension affects 20% to 30% of children and nearly 75% of adults with ADPKD. The hypertension is a secondary effect of intrarenal ischemia and activation of the renin-angiotensin system. Nearly half of patients have end-stage renal disease by 60 years of age. Hypertension, recurrent urinary tract infections, male sex, and early age of clinical onset are most predictive of early renal failure. Approximately 43% of patients presenting with ADPKD before or shortly after birth die of renal failure within the first year of life; end-stage renal disease, hypertension, or both develop in the survivors by 30 years of age.

ADPKD exhibits both interfamilial and intrafamilial variation in the age at onset and severity. Part of the interfamilial variation is secondary to locus heterogeneity because patients with ADPKD2 have milder disease than patients with ADPKD1. Intrafamilial variation appears to result from a

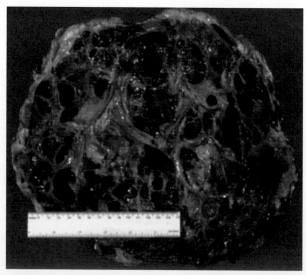

Figure C-37 Cross section of a kidney from a patient with autosomal dominant polycystic kidney disease showing large cysts and widespread destruction of the normal renal parenchyma. *See Sources & Acknowledgments.*

combination of environment and genetic background because the variability is more marked between generations than among siblings.

In addition to renal cysts, patients with ADPKD develop hepatic, pancreatic, ovarian, and splenic cysts as well as intracranial aneurysms, mitral valve prolapse, and colonic diverticula. Hepatic cysts are common in both ADPKD1 and ADPKD2, whereas pancreatic cysts are generally observed with ADPKD1. Intracranial saccular aneurysms develop in 5% to 10% of patients with ADPKD; however, not all patients have an equal risk for development of aneurysms because they exhibit familial clustering. Patients with ADPKD have an increased risk for aortic and tricuspid valve insufficiency, and approximately 25% develop mitral valve prolapse. Colonic diverticula are the most common extrarenal abnormality; the diverticula associated with ADPKD are more likely to perforate than those observed in the general population.

Management

In general, ADPKD is diagnosed by family history and renal ultrasound examination. The detection of renal cysts by ultrasound examination increases with age such that 80% to 90% of patients have detectable cysts by 20 years, and nearly 100% have them by 30 years. If necessary for prenatal diagnosis or identification of a related kidney donor, the diagnosis can be confirmed by mutation detection in most families.

The management and treatment of patients with ADPKD focus on slowing the progression of renal disease and minimizing its symptoms. Hypertension and urinary tract infections are treated aggressively to preserve renal function. Pain from the mass effect of the enlarged kidneys is managed by drainage and sclerosis of the cysts.

INHERITANCE RISK

Approximately 90% of patients have a family history of ADPKD; only 10% of cases result from de novo mutations of either *PDK1* or *PDK2*. Parents with ADPKD have a 50% risk

of having an affected child in each pregnancy. If the parents have had a child with onset of disease in utero, the risk for having another severely affected child is approximately 25%. In general, however, the severity of disease cannot be predicted because of variable expressivity. For families in whom the mutation is known, prenatal diagnostic testing is available via amniocentesis or CVS.

Siblings and parents of patients with ADPKD also have an increased risk for disease. Renal ultrasonography is the recommended method for the screening of family members.

QUESTIONS FOR SMALL GROUP DISCUSSION

1. Compare the molecular mechanism of cyst development in ADPKD with the development of neurofibromas in neurofibromatosis type 1.
2. Many mendelian diseases have variable expressivity that might be accounted for by modifier loci. How would one identify such loci?
3. Why is ADPKD occasionally associated with tuberous sclerosis? How might this illustrate a contiguous gene deletion syndrome?
4. How can ADPKD be distinguished from autosomal recessive polycystic kidney disease?
5. Linkage analysis of families segregating ADPKD requires the participation of family members in addition to the patient. What should be done if individuals crucial to the study do not want to participate?

REFERENCES

Chang MY, Ong AC: New treatments for autosomal dominant polycystic kidney disease, *Br J Clin Pharmacol* 76:524–535, 2013.

Eccles MR, Stayner CA: Polycystic kidney disease—where gene dosage counts, *F1000Prime Rep* 6:24, 2014.

Harris PC, Torres VE: Polycystic kidney disease, autosomal dominant. Available from: http://www.ncbi.nlm.nih.gov/books/NBK1246/.

PRADER-WILLI SYNDROME (Absence of Paternally Derived 15q11-q13, MIM 176270)

Chromosomal Deletion, Uniparental Disomy

PRINCIPLES

- Imprinting
- Uniparental disomy
- Microdeletion
- Recombination between repeated DNA sequences

MAJOR PHENOTYPIC FEATURES

- Age at onset: Infancy
- Infantile feeding difficulties
- Childhood hyperphagia and obesity
- Hypotonia
- Cognitive impairment
- Short stature
- Dysmorphism

HISTORY AND PHYSICAL FINDINGS

J.T. was born at 38 weeks' gestation after an uncomplicated pregnancy and delivery. She was the second child of nonconsanguineous parents. Shortly after birth, her parents and the nurses noticed that she was hypotonic and feeding poorly. Her parents and older sister were in good health; she did not have a family history of neuromuscular, developmental, genetic, or feeding disorders. Review of the medical record did not reveal a history of overt seizures, hypoxic insults, infection, cardiac abnormalities, or blood glucose or electrolyte abnormalities. On examination, J.T. did not have respiratory distress or dysmorphism; her weight and length were appropriate for gestational age; she was severely hypotonic with lethargy, weak cry, decreased reflexes, and a poor suck. Subsequent evaluation included testing for congenital infections and congenital hypothyroidism; measurements of blood ammonium, plasma amino acids, and urine organic acids; chromosomal microarray; and methylation testing for the Prader-Willi/Angelman region on 15q11-13 (see Chapter 6). The results of the methylation testing showed an abnormal methylation pattern consistent with Prader-Willi syndrome (one hypermethylated copy of *SNRPN*), and the chromosomal microarray revealed a deletion on chromosome 15q11-q13. The geneticist explained to the parents that J.T. had Prader-Willi syndrome. After much discussion and thought, J.T.'s parents decided that they were unable to care for a disabled child and gave her up for adoption.

BACKGROUND

Disease Etiology and Incidence

Prader-Willi syndrome (PWS, MIM 176270) is a panethnic developmental disorder caused by loss of expression of genes on paternally derived chromosome 15q11-q13. Loss of paternally expressed genes can arise by several mechanisms; approximately 70% of patients have a deletion of 15q11-q13, 25% have maternal uniparental disomy, less than 5% have mutations within the imprinting control element, and less than 1% have another chromosomal abnormality (see Chapter 6). PWS has an incidence of 1 in 10,000 to 1 in 15,000 live births.

Pathogenesis

Many genes within 15q11-q13 are differentially expressed, depending on whether the region is inherited from the father or the mother. In other words, some genes expressed by paternal 15q11-q13 are not expressed by maternal 15q11-q13, and other genes expressed by maternal 15q11-q13 are not expressed by paternal 15q11-q13. This phenomenon of differential expression of a gene according to whether it is inherited from the father or mother is known as imprinting (see Chapters 3 and 6). Maintenance of correct expression of imprinted genes requires switching of the imprint on passage through the germline; all imprints are switched "off" in the gonadal cells, and maternal imprints are then activated in the egg cells, whereas paternal imprints are activated in the sperm cells. Switching of imprinting on passage through the germline is regulated by an imprinting control element and reflected by epigenetic changes in DNA methylation and chromatin that regulate gene expression.

Deletion of 15q11-q13 during male meiosis gives rise to children with PWS because children formed from a sperm carrying the deletion will be missing genes that are active only on the paternally derived 15q11-q13. The mechanism underlying this recurrent deletion is illegitimate recombination between low-copy repeat sequences flanking the deletion interval (see Chapter 6). Less commonly, inheritance of a deletion spanning this region occurs if a patient inherits an unbalanced karyotype from a parent who has a balanced translocation.

Failure to switch the maternal imprints to paternal imprints during male meiosis gives rise to children with PWS because children formed from a sperm with a maternally imprinted 15q11-q13 will not be able to express genes active only on the paternally imprinted 15q11-q13. Imprinting failure arises from mutations within the imprinting control element.

Maternal uniparental disomy also gives rise to PWS because the child has two maternal chromosomes 15 and no paternal chromosome 15. Maternal uniparental disomy is thought to develop secondary to trisomy rescue, that is, loss of the paternal chromosome 15 from a conceptus with chromosome 15 trisomy secondary to maternal nondisjunction.

Despite the observations that loss of a paternally imprinted 15q11-q13 gives rise to PWS, and despite the identification of many imprinted genes within this region, the precise cause of PWS is still unknown. PWS has not yet been shown definitively to result from a mutation in any one specific gene.

Phenotype and Natural History

In early infancy, PWS is characterized by severe hypotonia, feeding difficulties, and hypogonadism, with cryptorchidism in males. The hypotonia improves over time, although adults remain mildly hypotonic. The hypogonadism, which is of hypothalamic origin, does not improve with age and usually causes delayed and incomplete pubertal development as well as infertility. The feeding difficulties usually resolve within the first year of life, and between 1 and 6 years, patients develop extreme hyperphagia and food-seeking behavior (hoarding, foraging, and stealing). This behavior and a low metabolic rate cause marked obesity. The obesity is a major cause of morbidity due largely to cardiopulmonary disease and

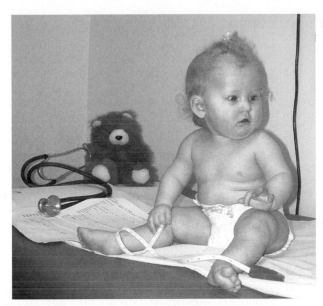

Figure C-38 **A 12-month-old girl with Prader-Willi syndrome.** Note her fair coloring, narrow bifrontal diameter, almond-shaped eyes, and down-turned mouth. The hyperphagia, with resulting central obesity, generally does not begin until the age of 2 to 6 years. *See Sources & Acknowledgments.*

non–insulin-dependent (type 2) diabetes mellitus. Longevity can be nearly normal if obesity is avoided.

Most children with PWS have delayed motor and language development as well as mild intellectual disability (mean IQ, 60 to 80). They also have behavioral problems, including temper tantrums, obsessive-compulsive disorders, and poor adaptation to changes in routine. These behavioral problems continue into adulthood and remain disabling. Approximately 5% to 10% of patients also develop psychoses during early adulthood.

Other anomalies associated with PWS include short stature, scoliosis, osteoporosis, and dysmorphism. Dysmorphic features include a narrow bifrontal diameter, almond-shaped eyes, triangular mouth, and small hands and feet (Fig. C-38). Also, many patients have hypopigmentation of the hair, eyes, and skin.

Management

Although it is often suspected on the basis of history and physical features, a diagnosis of PWS is defined by the absence of a paternally imprinted 15q11-q13. Loss of the paternal imprint is detected by DNA analyses showing that the imprinted genes have only a maternal methylation pattern. If the DNA studies confirm PWS, genetic counseling requires a subsequent karyotype and FISH for 15q11-q13 to determine whether PWS arose from inheritance of a chromosomal translocation.

No medications are currently available to treat the hyperphagia; a very low-calorie and restrictive diet and exercise remain the mainstays for controlling the obesity. Growth hormone replacement can normalize height and improve lean body mass. Sex hormone replacement promotes secondary sexual features but frequently worsens behavioral problems in males and increases the risk for stroke in females. Behavioral management and serotonin reuptake inhibitors are the most effective therapies currently available for the behavioral disorder. Adult patients usually perform best in sheltered living (group homes) and employment environments.

INHERITANCE RISK

The risk for recurrence of PWS in future children of parents is related to the molecular cause. For imprinting defects, the risk can be as high as 50%, whereas for either deletion of 15q11-q13 or maternal uniparental disomy, the recurrence risk is less than 1%. The risk for recurrence if a parent carries a balanced translocation depends on the nature of the translocation but can be as high as 25%; in contrast, all PWS patients reported to date with an unbalanced translocation have had a de novo chromosomal rearrangement.

QUESTIONS FOR SMALL GROUP DISCUSSION

1. Angelman syndrome also arises from imprinting defects of 15q11-q13. Compare and contrast the phenotypes and causative molecular mechanisms of Prader-Willi syndrome and Angelman syndrome.
2. How might imprinting explain the phenotypes associated with triploidy?
3. Beckwith-Wiedemann syndrome and Russell-Silver syndrome also appear to be caused by abnormal expression of imprinted genes. Explain.
4. J.T.'s parents gave her up for adoption. Should the genetic counseling have been done differently? What is nondirective genetic counseling?

REFERENCES

Cassidy SB, Schwartz S, Miller JL, et al: Prader-Willi syndrome, *Genet Med* 14:10–26, 2012.
Driscoll DJ, Miller JL, Schwartz S, et al: Prader-Willi syndrome. Available from: http://www.ncbi.nlm.nih.gov/books/NBK1330/.

RETINOBLASTOMA (*RB1* Mutation, MIM 180200)

Autosomal Dominant

PRINCIPLES

- Tumor suppressor gene
- Two-hit hypothesis
- Somatic mutation
- Tumor predisposition
- Cell-cycle regulation
- Variable expressivity

MAJOR PHENOTYPIC FEATURES

- Age at onset: Childhood
- Leukocoria
- Strabismus
- Visual deterioration
- Conjunctivitis

HISTORY AND PHYSICAL FINDINGS

J.V., a 1-year-old girl, was referred by her pediatrician for evaluation of right strabismus and leukocoria, a reflection from a white mass within the eye giving the appearance of a white pupil (see Fig. 15-7). Her mother reported that she had developed progressive right esotropia in the month before seeing her pediatrician. She had not complained of pain, swelling, or redness of her right eye. She was otherwise healthy. She had healthy parents and a 4-year-old sister; no other family members had had ocular disease. Except for the leukocoria and strabismus, the findings from her physical examination were normal. Her ophthalmological examination defined a solitary retinal tumor of 8 disc diameters arising near the macula. Magnetic resonance imaging of the head did not show extension of the tumor outside the globe and no evidence for an independent primary tumor involving the pineal gland, which is referred to as trilateral disease. She received chemotherapy combined with focal irradiation. DNA analysis showed that she had a germline nonsense mutation (C to T transition) in one allele of her retinoblastoma (*RB1*) gene.

BACKGROUND

Disease Etiology and Incidence

Retinoblastoma (MIM 180200) is a rare embryonic neoplasm of retinal origin (Fig. C-39) that results from germline and/or somatic mutations in both alleles of the *RB1* gene. It occurs worldwide with an incidence of 1 in 18,000 to 30,000.

Pathogenesis

The retinoblastoma protein (Rb) is a tumor suppressor that plays an important role in regulating the progression of proliferating cells through the cell cycle and the exit of differentiating cells from the cell cycle. Rb affects these two functions by sequestration of other transcription factors and by promoting deacetylation of histones, a chromatin modification associated with gene silencing.

Retinoblastoma-associated *RB1* mutations occur throughout the coding region and promoter of the gene. Mutations within the coding region of the gene either destabilize Rb or compromise its association with enzymes necessary for histone deacetylation. Mutations within the promoter reduce expression of normal Rb. Both types of mutations result in a loss of functional Rb.

An *RB1* germline mutation is found in 40% of patients with retinoblastoma, but only 10% to 15% of all patients have a history of other affected family members. *RB1* mutations include cytogenetic abnormalities of chromosome 13q14, single-base substitutions, and small insertions or deletions. Some evidence suggests that the majority of new germline mutations arise on the paternal allele, whereas somatic mutations arise with equal frequency on the maternal and paternal alleles. Nearly half of the mutations occur at CpG dinucleotides. After either the inheritance of a mutated allele or the generation of a somatic mutation on one allele, the other *RB1* allele must also lose function (the second "hit" of the two-hit hypothesis; see Chapter 15) for a cell to proliferate unchecked and develop into a retinoblastoma. Loss of a functional second allele occurs by a novel mutation, loss of heterozygosity, or promoter CpG island hypermethylation; deletion or the development of isodisomy occurs most frequently, and promoter hypermethylation occurs least frequently.

Retinoblastoma usually segregates as an autosomal dominant disorder with full penetrance although a few families have been described with reduced penetrance. The *RB1* mutations identified in these families include missense mutations, in-frame deletions, and promoter mutations. In contrast to the more common null *RB1* alleles, these mutations are believed to represent alleles with some residual function.

Phenotype and Natural History

Patients with bilateral retinoblastoma generally present during the first year of life, whereas those with unilateral disease present somewhat later with a peak between 24 and 30 months. Approximately 70% of patients have unilateral retinoblastoma and 30% bilateral retinoblastoma. All patients with bilateral disease have germline *RB1* mutations, but not all patients with germline mutations develop bilateral disease.

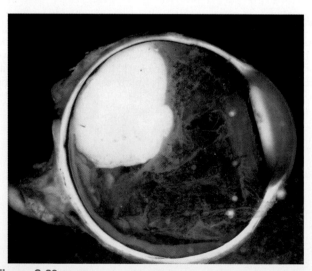

Figure C-39 Midline cross section of an enucleated eye from a patient with retinoblastoma. Note the large primary tumor in the posterior third of the globe and a few white vitreous seeds. (The brown discoloration of the vitreous is a fixation artifact.) *See Sources & Acknowledgments.*

The disease is diagnosed before 5 years of age in 80% to 95% of patients. Retinoblastoma is uniformly fatal if untreated; with appropriate therapy, however, more than 80% to 90% of patients are free of disease 5 years after diagnosis.

As might be expected with mutation of a key cell-cycle regulator, patients with germline *RB1* mutations have a markedly increased risk for secondary neoplasms; this risk is increased by environmental factors, such as treatment of the initial retinoblastoma with radiotherapy. The most common secondary neoplasms are osteosarcomas, soft tissue sarcomas, and melanomas. There is no increase in second malignant neoplasms in patients with nonhereditary retinoblastoma.

Management

Early detection and treatment are essential for optimal outcome. The goals of therapy are to cure the disease and to preserve as much vision as possible. Treatment is tailored to the tumor size and involvement of adjacent tissues. Treatment options for intraocular retinoblastoma include enucleation, various modes of radiotherapy, cryotherapy, light coagulation, and chemotherapy, including direct arterial infusion.

If the disease is unilateral at the time of the patient's presentation, the patient needs frequent examinations to detect any new retinoblastomas in the unaffected eye because 30% of apparently sporadic cases are caused by the inheritance of a new germline mutation. Such frequent examinations are usually continued until at least 7 years of age.

To direct follow-up more efficiently, patients should have molecular testing to identify the mutations in the *RB1* gene. Ideally, a tumor sample is examined first, and then another tissue, such as blood, is analyzed to determine whether one of those mutations is a germline mutation. If neither mutation is a germline mutation, the patient does not require such frequent follow-up.

INHERITANCE RISK

If a parent had bilateral retinoblastoma and thus probably carries a germline mutation, the empirical risk for an affected child is 45%; this risk reflects the high likelihood of a second, somatic mutation (or "hit") in the second *RB1* allele of the child. On the other hand, if the parent had unilateral disease, the empirical risk for an affected child is 7% to 15%; this reflects the relative proportion of germline mutations versus somatic mutations in patients with unilateral disease. Nearly 90% of children who develop retinoblastoma are the first individuals affected within the family. Interestingly, 1% of unaffected parents of an affected child have evidence of a spontaneously resolved retinoblastoma on retinal examination; for these families, therefore, the risk for an affected child is 45%. Except for the rare situation in which one parent is a nonpenetrant carrier of an *RB1* mutation, families in which neither parent had retinoblastoma have a risk for recurrence equivalent to that of the general population.

QUESTIONS FOR SMALL GROUP DISCUSSION

1. What other diseases develop as a result of a high frequency of mutations in CpG dinucleotides? What is the mechanism of mutation at CpG dinucleotides? What could explain the increased frequency of CpG dinucleotide mutations with increasing paternal age?

2. Compare and contrast the type and frequency of tumors observed in Li-Fraumeni syndrome with those observed in retinoblastoma. Both Rb and p53 are tumor suppressors; why are *TP53* mutations associated with a different phenotype than *RB1* mutations?

3. Discuss four diseases that arise as a result of somatic mutations. Examples should illustrate chromosomal recombination, loss of heterozygosity, gene amplification, and accumulation of point mutations.

4. Both SRY (see Chapter 6) and Rb regulate development by modulating gene expression through the modification of chromatin structure. Compare and contrast the two different mechanisms that each uses to modify chromatin structure.

REFERENCES

Lohmann DR, Gallie BL: Retinoblastoma. Available from: http://www.ncbi.nlm .nih.gov/books/NBK1452/.

Villegas VM, Hess DJ, Wildner A, et al: Retinoblastoma, *Curr Opin Ophthalmol* 24:581–588, 2013.

RETT SYNDROME (*Mepc2* Mutations, MIM 312750)

X-Linked Dominant

PRINCIPLES

- Loss-of-function mutation
- Variable expressivity
- Sex-dependent phenotype

MAJOR PHENOTYPIC FEATURES

- Age at onset: Neonatal to early childhood
- Acquired microcephaly
- Neurodevelopmental regression
- Repetitive stereotypic hand movements

HISTORY AND PHYSICAL FINDINGS

P.J. had had normal growth and development until 18 months of age. At 24 months, she was referred because of decelerating head growth and progressive loss of language and motor skills. She had lost purposeful hand movements and developed repetitive hand wringing by 30 months. She also had mild microcephaly, truncal ataxia, gait apraxia, and severely impaired expressive and receptive language. No other family members had any neurological diseases. On the basis of these findings, the neurologist suggested that P.J. might have Rett syndrome. The physician explained that Rett syndrome is a result of mutations in the methyl-CpG–binding protein 2 gene (*MECP2*) in most patients and that testing for such mutations could help confirm the diagnosis. Subsequent testing of P.J.'s DNA identified a heterozygous *MECP2* mutation; she carried the transition 763C>T, which causes Arg255Ter. Neither parent carried the mutation.

BACKGROUND

Disease Etiology and Incidence

Rett syndrome (MIM 312750) is a panethnic X-linked dominant disorder with a female prevalence of 1 in 10,000 to 1 in 15,000. It is caused by loss-of-function mutations of the *MECP2* gene. With the advent of array comparative genomic hybridization (array CGH) technology, males with duplications on the X chromosome in the region of *MECP2* have been demonstrated; these males typically have severe intellectual disabilities. Males with a mutation in *MECP2* and 47,XXY phenotype can also have Rett syndrome with a phenotype similar to females. Two other genes, *CDKL5* and *FOXG1*, can lead to Rett-like phenotypes. *CDKL5* is an X-linked serine/threonine kinase that regulates neuronal proliferation and differentiation, and mutations in this gene cause microcephaly, seizures, and severe intellectual disability. Mutations in *FOXG1* cause an autosomal dominant disorder with similar features, including brain abnormalities, such as oligogyria, and defects in the corpus callosum.

Pathogenesis

MECP2 encodes a nuclear protein that binds methylated DNA and recruits histone deacetylases to regions of methylated DNA. The precise function of MeCP2 has not been fully defined, but it is hypothesized to mediate transcriptional silencing and epigenetic regulation of genes in these regions of methylated DNA. Accordingly, dysfunction or loss of MeCP2, as observed in Rett syndrome, would be predicted to cause inappropriate activation of target genes.

The brains of patients with Rett syndrome are small and have cortical and cerebellar atrophy without neuronal loss; Rett syndrome is therefore not a typical neurodegenerative disease. Within much of the cortex and hippocampus, the neurons from Rett patients are smaller and more densely packed than normal and have a simplified dendritic branching pattern. These observations suggest that MeCP2 is important for establishing and maintaining neuronal interactions rather than for neuronal precursor proliferation or neuronal determination.

Phenotype and Natural History

Classic Rett syndrome is a progressive neurodevelopmental disorder occurring almost exclusively in girls (Fig. C-40). After apparently normal development until 6 to 18 months of age, patients enter a short period of developmental slowing and stagnation with decelerating head growth. Subsequently, they rapidly lose speech and acquired motor skills, particularly purposeful hand use. With continued disease progression, they develop stereotypic hand movements, breathing irregularities, ataxia, and seizures. After a brief period of apparent stabilization, usually during the preschool to early school years, the patients deteriorate further to become severely intellectually disabled and develop progressive spasticity, rigidity, and scoliosis. Patients usually live into adulthood, but their life span is short due to an increased incidence of unexplained sudden death.

Besides Rett syndrome, *MECP2* mutations cause a broad spectrum of diseases affecting both boys and girls. Among girls, the range extends from severely affected patients who never learn to speak, turn, sit, or walk, and develop severe epilepsy, to mildly affected patients who speak and have good gross motor function as well as relatively well-preserved hand function. Among boys, the range of phenotypes encompasses intrauterine death, congenital encephalopathy, intellectual disability with various neurological symptoms, and mild intellectual disability only.

Management

Suspected on the basis of clinical features, the diagnosis of Rett syndrome is usually confirmed by DNA testing; however, current testing detects *MECP2* mutations in only 80% to 90% of patients with typical Rett syndrome. The clinical diagnostic criteria for typical Rett syndrome include normal prenatal and perinatal periods, normal head circumference at birth, relatively normal development through 6 months of age, deceleration of head growth between 6 and 48 months of age, loss of acquired hand skills and purposeful hand movements by 5 to 30 months of age and subsequent development of stereotyped hand movements, impaired expressive and receptive language, severe psychomotor retardation, and development of gait apraxia and truncal ataxia between 12 and 48 months of age.

Currently there are no curative treatments of Rett syndrome and management focuses on supportive and symptomatic therapy. Current medical therapy includes anticonvulsants for seizures, serotonin uptake inhibitors for agitation, carbidopa or levodopa for rigidity, and melatonin to ameliorate sleep disturbances. Families often have problems with social adjustment and coping and should therefore be provided with the opportunity to interact with similarly affected families

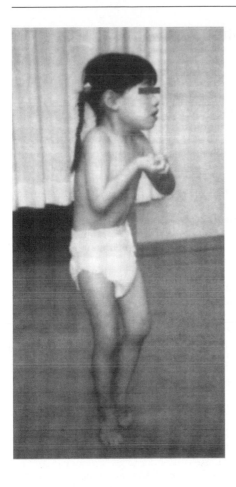
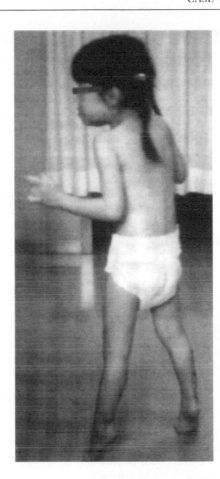

Figure C-40 A 5-year, 3-month-old girl with Rett syndrome demonstrating toe walking. *See Sources & Acknowledgments.*

through support groups and be referred for professional counseling as needed.

INHERITANCE RISK

Approximately 99% of Rett syndrome is sporadic; most *MECP2* mutations are de novo, although in rare cases they can be inherited from an unaffected or mildly affected mother with skewed X chromosome inactivation. At least 70% of de novo mutations arise in the paternal germline.

If a couple has an affected child but a *MECP2, CDKL5,* or *FOXG1* mutation is not identified in either parent, the risk to future siblings is low, although it is higher than among the general population because of the possibility of undetected germline mosaicism. In contrast, if the mother carries a disease-causing mutation, each daughter and son has a 50% risk for inheriting the mutation. However, the poor genotype-phenotype correlation among patients with *MECP2* mutations generally prohibits prediction of whether a female fetus with a *MECP2* mutation will develop classic Rett syndrome or another *MECP2*-associated disease. Similarly, identification of a *MECP2* mutation in a male fetus does not predict intrauterine demise, the development of congenital encephalopathy, or another *MECP2*-associated disease.

QUESTIONS FOR SMALL GROUP DISCUSSION

1. *MECP2* is on the X chromosome. Discuss how this could affect the phenotypic variability observed among females with *MECP2* mutations. Discuss how this might account for the fewer numbers of males with *MECP2* mutations and the differences in disease severity observed generally between males and females.

2. Given that MeCP2 is an epigenetic mediator of gene expression, discuss possible molecular mechanisms by which genetic background, environment, and stochastic factors could cause the phenotypic variability observed among males with *MECP2* mutations.

3. Rett syndrome is a neurodevelopmental disorder without neurodegeneration. Why might the absence of neurodegeneration make this disease more amenable to treatment than Alzheimer disease or Parkinson disease? Why less amenable? In this context, also discuss possible molecular mechanisms for the neurodevelopmental regression observed with Rett syndrome.

4. What defines a disease, the molecular mutation or the clinical phenotype?

REFERENCES

Ausio J, Paz AM, Esteller M: MeCP2: the long trip from a chromatin protein to neurological disorders, *Trends Mol Med* 20(9):487–498, 2014.

Christodoulou J, Ho G: *MECP2*-related disorders. Available from: http://www.ncbi.nlm.nih.gov/books/NBK1497/.

Neul JL: The relationship of Rett syndrome and MeCP2 disorders to autism, *Dialogues Clin Neurosci* 14:253–262, 2012.

SEX DEVELOPMENT DISORDER (46,XX MALE)
(*SRY* Translocation, MIM 400045)
Y-Linked or Chromosomal

PRINCIPLES

- Disorder of sex development
- Developmental regulatory gene
- Pseudoautosomal regions of the X and Y chromosomes
- Illegitimate recombination
- Incomplete penetrance
- Fertility loci

MAJOR PHENOTYPIC FEATURES

- Age at onset: Prenatal
- Sterility
- Reduced secondary sexual features
- Unambiguous genitalia mismatched to chromosomal sex

HISTORY AND PHYSICAL FINDINGS

Ms. R., a 37-year-old executive, was pregnant with her first child. Because of her age-related risk for having a child with a chromosomal abnormality, she elected to have an amniocentesis to assess the fetal karyotype; the karyotype result was normal 46,XX. At 18 weeks' gestation, however, a fetal ultrasound scan revealed a normal male fetus; a subsequent detailed ultrasound scan confirmed a male fetus. Ms. R. had been in good health before and during the pregnancy with no infections or exposures to drugs during the pregnancy. Neither she nor her partner had a family history of a disorder of sexual development, sterility, or congenital anomalies. Reevaluation of the chromosome analysis confirmed a normal 46,XX karyotype, but fluorescence in situ hybridization identified a sex-determining region Y gene (*SRY*) signal on one X chromosome (Fig. C-41). At 38 weeks of gestation, Ms. R. had an uncomplicated spontaneous vaginal delivery of a phenotypically normal male child.

BACKGROUND
Disease Etiology and Incidence

Disorders of sex development (DSDs) are panethnic and genetically heterogeneous. In patients with complete gonadal dysgenesis, point mutations, deletions, or translocations of *SRY* are among the most common causes of such disorders (see Chapter 6). Approximately 80% of 46,XX males with complete gonadal dysgenesis have a translocation of *SRY* onto an X chromosome, and 20% to 30% of 46,XY females with complete gonadal dysgenesis have a mutation or deletion of the *SRY* gene. The incidence of males with 46,XX testicular DSD and females with 46,XY complete gonadal dysgenesis is approximately 1 in 20,000 each.

Pathogenesis

SRY is a DNA-binding protein that alters chromatin structure by bending DNA. These DNA-binding and DNA-bending properties suggest that SRY regulates gene expression. During normal human development, SRY is necessary for the formation of male genitalia, and its absence is permissive for the formation of female genitalia. The precise mechanism through which SRY effects development of male genitalia is undefined,

although some observations suggest that SRY, together with other related transcription factors encoded by autosomal or X-linked genes, is part of a critically balanced network of repressors and activators of the developmental pathways that lead to development of normal testes or ovaries (see Chapter 6).

SRY mutations identified in females with a 46,XY karyotype cause a loss of SRY function. Approximately 10% of XY females have a deletion of *SRY* (*SRY⁻* XY females [MIM 400044]), and an additional 10% have point mutations within *SRY*. The point mutations within *SRY* impair either DNA binding or DNA bending.

The *SRY* alteration observed in males with a 46,XX karyotype is a translocation of *SRY* from Yp to Xp (*SRY⁺* XX males [MIM 400045]; Fig. C-41). During male meiosis, an obligatory crossing over occurs between the pseudoautosomal regions of Xp and Yp; this crossing over ensures proper segregation of the chromosomes and maintains sequence identity between the X and Y pseudoautosomal regions. On occasion, however, recombination occurs centromeric to the pseudoautosomal region and results in the transfer of Yp-specific sequences, including *SRY*, to Xp (see Chapter 6).

In addition to *SRY*, the Y chromosome contains at least three loci (azoospermic factor loci AZFa, AZFb, and AZFc) required for normal sperm development. The absence of these loci at least partially explains the infertility of males with 46,XX testicular DSD.

The X chromosome also contains several loci necessary for ovarian maintenance and female fertility. Oocyte development requires only a single X chromosome, but maintenance of

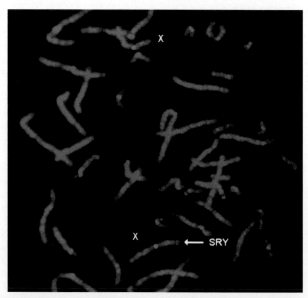

Figure C-41 **Fluorescence in situ hybridization (FISH) analysis for the detection of the t(X;Y)(p22.3;p11.2) translocation in an *SRY⁺* XX male.** The chromosomes are counterstained with DAPI. The probe for *SRY* is a mixture of locus-specific sequences (*red*). X chromosomes are detected with sequences that map to the centromeric DNA (*green*). In normal cells, the red signal is observed only on the Y chromosome. In cells with the t(X;Y)(p22.3;p11.2) translocation, a red signal is observed on the abnormal chromosome X and a green signal on both X chromosomes. *See Sources & Acknowledgments.*

those oocytes requires two X chromosomes. Consistent with these observations, female fetuses with 46,XY complete gonadal dysgenesis develop oocytes, but their ovarian follicles degenerate by birth or shortly thereafter. The absence of a second X chromosome therefore explains the infertility of such females (see Chapter 6).

Phenotype and Natural History

Males with 46,XX testicular DSD have many features of Klinefelter syndrome (47,XXY), including hypogonadism, azoospermia, hyalinization of seminiferous tubules, and gynecomastia. Despite decreased testosterone production, most patients enter puberty spontaneously, although they may require testosterone supplementation to attain full virilization. In contrast to patients with Klinefelter syndrome, most 46,XX male patients have normal to short stature, normal skeletal proportions, normal intelligence, and fewer psychosocial problems. Patients with an extensive portion of Yp on an X chromosome more closely resemble patients with Klinefelter syndrome.

Females with a 46,XY karyotype have complete gonadal dysgenesis and are usually taller than average women. These patients have physical features of Turner syndrome only when the deletion of SRY is associated with an extensive deletion of Yp. Because these patients have only streak gonads, they do not enter puberty spontaneously.

In contrast to the complete penetrance and relatively uniform expressivity observed with translocation or deletion of SRY, point mutations of SRY exhibit both incomplete penetrance and variable expressivity. Patients with SRY point mutations usually have complete gonadal dysgenesis, are taller than average women, and do not spontaneously develop secondary sexual characteristics. A few SRY point mutations, however, have been associated with both an infertile (complete gonadal dysgenesis) female phenotype and a fertile male phenotype within the same family.

Management

In patients with complete gonadal dysgenesis, the diagnosis of a DSD usually arises either because of discordance between the fetal ultrasound scan and fetal karyotype or because of absent or incomplete secondary sexual development and infertility. Confirmation that the DSD is secondary to an abnormality of SRY expression requires demonstration of the relevant SRY alteration.

For 46,XX testicular DSD patients, androgen supplementation is usually effective for virilization, but treatment of the azoospermia is not currently possible. Administration of supplemental androgens does not prevent gynecomastia. Patients need surgical treatment if the gynecomastia becomes sufficiently disconcerting or severe.

For 46,XY complete gonadal dysgenesis females, estrogen therapy is usually initiated at approximately 14 to 15 years of age to promote development of secondary sexual characteristics. Progesterone therapy is added to the regimen to induce menses either at the time of the first vaginal breakthrough bleeding or in the second year of estrogen therapy. In addition, because of the risk for development of gonadoblastoma, it is recommended that dysgenic gonads be removed once skeletal growth is complete.

As with all disorders of genital ambiguity or of discordance between genetic and phenotypic sex, the psychosocial management and counseling of the family and patient are extremely important. Many families and patients have difficulty understanding the medical data and making appropriate psychosocial adjustments.

INHERITANCE RISK

De novo illegitimate recombination is the most common cause of DSDs involving translocation or mutation of SRY; therefore most couples with an affected child have a low risk for recurrence in future children. Rarely, however, some cases arise as a result of inheriting an SRY deletion or translocation from a father with a balanced translocation between Xp and Yp. If the father is a translocation carrier, all children will be either an SRY⁺ XX boy or an SRY⁻ XY girl. Because such patients are invariably sterile, they are at no risk for passing on the disorder.

Most 46,XY complete gonadal dysgenesis females with point mutations in SRY have de novo mutations. Parents of an affected child therefore usually have a low risk for recurrence in future children; however, because some SRY mutations have incomplete penetrance, normal fertile fathers can carry SRY mutations that may or may not cause DSDs among their XY children.

QUESTIONS FOR SMALL GROUP DISCUSSION

1. Mutations of other genes, such as WT1, SOX9, NR5A1, and DAX1, can also result in a DSD. Compare and contrast the phenotypes observed with mutations in these genes with those observed with SRY mutations.
2. The association of SRY point mutations with an infertile female phenotype and a fertile male phenotype within the same family suggests either stochastic variation dependent on the reduced SRY activity or segregation of another locus that interacts with SRY. Why? How could this be resolved?
3. Mutations affecting steroid synthesis or steroid responsiveness are usually associated with ambiguous genitalia, whereas SRY mutations are generally associated with genitalia that, while mismatched with the chromosomal sex, are unambiguously male or female. Discuss the reasons for this generalization.
4. Discuss chromosomal, gonadal, and phenotypic sex, as well as psychological gender, and the importance of each to genetic counseling.

REFERENCES

Ono M, Harley VR: Disorders of sex development: new genes, new concepts, *Nat Rev Endocrinol* 9:79–91, 2013.

Ostrer H: 46,XY disorder of sex development and 46,XY complete gonadal dysgenesis. Available from: http://www.ncbi.nlm.nih.gov/books/NBK1547/.

Ostrer H: Disorders of sex development: an update, *J Clin Endocrin Metab* 99:1503–1509, 2014.

Vilain EJ: 46,XX testicular disorder of sex development. Available from: http://www.ncbi.nlm.nih.gov/books/NBK1416/.

SICKLE CELL DISEASE (β-Globin Glu6Val Mutation, MIM 603903)

Autosomal Recessive

PRINCIPLES

- Heterozygote advantage
- Novel property mutation
- Genetic compound
- Ethnic variation in allele frequencies

MAJOR PHENOTYPIC FEATURES

- Age at onset: Childhood
- Anemia
- Infarction
- Asplenia

HISTORY AND PHYSICAL FINDINGS

For the second time in 6 months, a Caribbean couple brought their 24-month-old daughter, C.W., to the emergency department because she refused to bear weight on her feet. There was no history of fever, infection, or trauma, and her medical history was otherwise unremarkable; findings from the previous visit were normal except for a low hemoglobin level and a mildly enlarged spleen. Findings from the physical examination were normal except for a palpable spleen tip and swollen feet. Her feet were tender to palpation, and she would not stand up. Both parents had siblings who died in childhood of infection and others who may have had sickle cell disease. In view of this history and the recurrent painful swelling of her feet, her physician tested C.W. for sickle cell disease by hemoglobin electrophoresis. This test result documented sickle cell hemoglobin, Hb S, in C.W.

BACKGROUND

Disease Etiology and Incidence

Sickle cell disease (MIM 603903) is an autosomal recessive disorder of hemoglobin in which the β subunit genes have a missense mutation that substitutes valine for glutamic acid at amino acid 6. The disease is most commonly due to homozygosity for the sickle cell mutation, although compound heterozygosity for the sickle allele and a hemoglobin C or a β-thalassemia allele can also cause sickle cell disease (see Chapter 11). The prevalence of sickle cell disease varies widely among populations in proportion to past and present exposure to malaria (see Table). The sickle cell mutation appears to confer some resistance to malaria and thus a survival advantage to individuals heterozygous for the mutation.

Pathogenesis

Hemoglobin is composed of four subunits, two α subunits encoded by *HBA* on chromosome 16 and two β subunits encoded by the *HBB* gene on chromosome 11 (see Chapter 11). The Glu6Val mutation in β-globin decreases the solubility of deoxygenated hemoglobin and causes it to form a gelatinous network of stiff fibrous polymers that distort the red blood cell, giving it a sickle shape (see Fig. 11-5). These sickled erythrocytes occlude capillaries and cause infarctions. Initially, oxygenation causes the hemoglobin polymer to dissolve and

the erythrocyte to regain its normal shape; repeated sickling and unsickling, however, produce irreversibly sickled cells that are removed from the circulation by the spleen. The rate of removal of erythrocytes from the circulation exceeds the production capacity of the marrow and causes a hemolytic anemia.

As discussed in Chapter 11, allelic heterogeneity is common in most mendelian disorders, particularly when the mutant alleles cause loss of function. Sickle cell disease is an important exception to the rule because one specific mutation is responsible for the unique novel properties of Hb S. Hb C is also less soluble than Hb A, and also tends to crystallize in red cells, decreasing their deformability in capillaries and causing mild hemolysis, but Hb C does not form the rod-shaped polymers of Hb S.

Phenotype and Natural History

Patients with sickle cell disease generally present in the first 2 years of life with anemia, failure to thrive, splenomegaly, repeated infections, and dactylitis (the painful swelling of the hands or feet from the occlusion of the capillaries in small bones seen in patient C.W.; Fig. C-42). Vaso-occlusive infarctions occur in many tissues, causing strokes, acute chest syndrome, renal papillary necrosis, autosplenectomy, leg ulcers, priapism, bone aseptic necrosis, and visual loss. Bone vaso-occlusion causes painful "crises," and, if untreated, these painful episodes can persist for days or weeks. The functional asplenia, from infarction and other poorly understood factors, increases susceptibility to bacterial infections such as pneumococcal sepsis and *Salmonella* osteomyelitis. Infection is a major cause of death at all ages, although progressive renal failure and pulmonary failure are also common causes of death in the fourth and fifth decades. Patients also have a high risk for development of life-threatening aplastic anemia after parvovirus infection because parvovirus infection causes a temporary cessation of erythrocyte production.

Heterozygotes for the mutation (who are said to have sickle cell trait) do not have anemia and are usually clinically normal. Under conditions of severe hypoxia, however, such as ascent to high altitudes, erythrocytes of patients with sickle cell trait may sickle and cause symptoms similar to those observed with sickle cell disease. With extreme exertion and dehydration, there is increased risk for rhabdomyolysis in sickle cell heterozygotes.

Frequencies of the Sickle Cell Mutation among California Newborns

Ethnicity	Hb SS (Homozygote)	Hb AS (Heterozygote)
African American	1/700	1/14
Asian Indian	0/1600	1/700
Hispanic	1/46,000	1/180
Middle Eastern	0/22,000	1/360
Native American	1/17,000	1/180
Northern European	1/160,000	1/600
Asian	0/200,000	1/1300

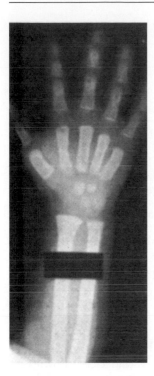
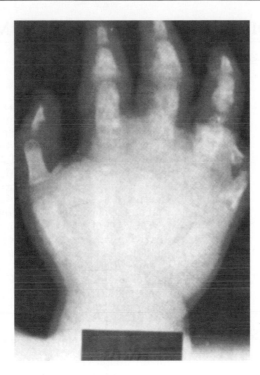

Figure C-42 **Acute dactylitis in a child with sickle cell disease.** Radiographs of a child's hand during (*left*) and 2 weeks following (*right*) an attack of dactylitis. Note the development of destructive bony lesions. *See Sources & Acknowledgments.*

Management

In a given patient with sickle cell disease, there are no accurate predictors for the severity of the disease course. Although the molecular basis of the disease has been known longer than that of any other single-gene defect, current treatment is only supportive. No specific therapy that prevents or reverses the sickling process in vivo has been identified. Persistence of fetal hemoglobin greatly ameliorates disease severity. Several pharmacological interventions aimed at increasing fetal hemoglobin concentrations are under investigation (see Chapter 13), and hydroxyurea has been approved for this indication. Although gene therapy has the potential to ameliorate and cure this disease, effective β-globin gene transfer has been achieved in only a single patient (see Chapter 13). Allogeneic bone marrow transplantation is the only treatment currently available that can cure sickle cell disease.

Because of the 11% mortality from sepsis in the first 6 months of life, most states in the United States offer newborn screening for sickle cell disease to initiate antibiotic prophylaxis that is maintained through 5 years of age (see Chapter 18).

INHERITANCE RISK

Because sickle cell disease is an autosomal recessive disorder, future siblings of an affected child have a 25% risk for sickle cell disease and a 50% risk for sickle cell trait. With use of fetal DNA derived from chorionic villi or amniocytes, prenatal diagnosis is available by molecular analysis for the sickle cell mutation.

QUESTIONS FOR SMALL GROUP DISCUSSION

1. What are the difficulties with gene therapy for this disorder?
2. Name two other diseases that may have become prevalent because of a heterozygote survival advantage. What is the rationale for hypothesizing a heterozygote advantage for those diseases?
3. Although it is always a severe disease, the severity of sickle cell disease is determined partially by the haplotype on which the mutation occurs. How could the haplotype affect disease severity?
4. Using the incidence figures in the Table, what is the risk that an unrelated African American woman and man of Northern European descent will have a child affected with sickle cell disease? with sickle cell trait?

REFERENCES

Bender MA, Hobbs W: Sickle cell disease. Available from: http://www.ncbi.nlm.nih.gov/books/NBK1377/.

Kanter J, Kruse-Jarres R: Management of sickle cell disease from childhood through adulthood, *Blood Rev* 27:279–287, 2013.

McGann PT, Nero AC, Ware RE: Current management of sickle cell anemia, *Cold Spring Harb Perspect Med* 3:a011817, 2013.

Serjeant GR: The natural history of sickle cell disease, *Cold Spring Harb Perspect Med* 3:a011783, 2013.

TAY-SACHS DISEASE (*HEXA* Mutation, MIM 272800)

Autosomal Recessive

PRINCIPLES

- Lysosomal storage disease
- Ethnic variation in allele frequencies
- Genetic drift
- Pseudodeficiency
- Population screening

MAJOR PHENOTYPIC FEATURES

- Age at onset: Infancy through adulthood
- Neurodegeneration
- Retinal cherry-red spot
- Psychosis

HISTORY AND PHYSICAL FINDINGS

R.T. and S.T., an Ashkenazi Jewish couple, were referred to the genetics clinic for evaluation of their risk for having a child with Tay-Sachs disease. S.T. had a sister who died of Tay-Sachs disease as a child. R.T. had a paternal uncle living in a psychiatric home, but he did not know what disease his uncle had. Both R.T. and S.T. had declined screening for Tay-Sachs carrier status as teenagers. Enzymatic carrier testing showed that both R.T. and S.T. had reduced hexosaminidase A activity. Subsequent molecular analysis for *HEXA* mutations predominant in Ashkenazi Jews confirmed that S.T. carried a disease-causing mutation, whereas R.T. had only a pseudodeficiency allele but no disease-causing mutation.

BACKGROUND

Disease Etiology and Incidence

Tay-Sachs disease (MIM 272800), infantile GM_2 gangliosidosis, is a panethnic autosomal recessive disorder of ganglioside catabolism that is caused by a deficiency of hexosaminidase A (see Chapter 12). In addition to severe infantile-onset disease, hexosaminidase A deficiency causes milder disease with juvenile or adult onset.

The incidence of hexosaminidase A deficiency varies widely among different populations; the incidence of Tay-Sachs disease ranges from 1 in 3600 Ashkenazi Jewish births to 1 in 360,000 non–Ashkenazi Jewish North American births. French Canadians, Louisiana Cajuns, and Pennsylvania Amish have an incidence of Tay-Sachs disease comparable to that of Ashkenazi Jews. The increased carrier frequency in these four populations appears to be due to genetic drift, although heterozygote advantage cannot be excluded (see Chapter 9).

Pathogenesis

Gangliosides are ceramide oligosaccharides present in all cell surface membranes but most abundant in the brain. Gangliosides are concentrated in neuronal surface membranes, particularly in dendrites and axon termini. They function as receptors for various glycoprotein hormones and bacterial toxins and are involved in cell differentiation and cell-cell interaction.

Hexosaminidase A is a lysosomal enzyme composed of two subunits. The α subunit is encoded by the *HEXA* gene, and the β subunit is encoded by the *HEXB* gene. In the presence of activator protein, hexosaminidase A removes the terminal

N-acetylgalactosamine from the ganglioside GM_2. Mutations of the α subunit or the activator protein cause the accumulation of GM_2 in the lysosome and thereby Tay-Sachs disease of the infantile, juvenile, or adult type. (Mutation of the β subunit causes Sandhoff disease [MIM 268800].) The mechanism by which the accumulation of GM_2 ganglioside causes neuronal death has not been fully defined, although by analogy with Gaucher disease (see Chapter 12), toxic byproducts of GM_2 ganglioside may cause the neuropathology.

The level of residual hexosaminidase A activity correlates inversely with the severity of the disease. Patients with infantile-onset GM_2 gangliosidosis have two null alleles, that is, no hexosaminidase A enzymatic activity. Patients with juvenile- or adult-onset forms of GM_2 gangliosidosis are usually compound heterozygotes for a null *HEXA* allele and an allele with low residual hexosaminidase A activity.

Phenotype and Natural History

Infantile-onset GM_2 gangliosidosis is characterized by neurological deterioration beginning between the ages of 3 and 6 months and progressing to death by 2 to 4 years. Motor development usually plateaus or begins to regress by 8 to 10 months and progresses to loss of voluntary movement within the second year of life. Visual loss begins within the first year and progresses rapidly; it is almost uniformly associated with a cherry-red spot on funduscopic examination (Fig. C-43). Seizures usually begin near the end of the first year and progressively worsen. Further deterioration in the second year of life results in decerebrate posturing, swallowing difficulties, worse seizures, and finally an unresponsive, vegetative state.

Juvenile-onset GM_2 gangliosidosis manifests between 2 and 4 years and is characterized by neurological deterioration beginning with ataxia and uncoordination. By the end of the first decade, most patients experience spasticity and seizures; by 10 to 15 years, most develop decerebrate rigidity and enter a vegetative state with death generally occurring in the second decade. Loss of vision occurs, but there may not be a cherry-red spot; optic atrophy and retinitis pigmentosa often occur late in the disease course.

Adult-onset GM_2 gangliosidosis exhibits marked clinical variability (progressive dystonia, spinocerebellar degeneration, motor neuron disease, or psychiatric abnormalities). As many as 40% of patients have progressive psychiatric manifestations without dementia. Vision is rarely affected, and results of the ophthalmological examination are generally normal.

Management

The diagnosis of a GM_2 gangliosidosis relies on the demonstration of both absent to nearly absent hexosaminidase A activity in the serum or white blood cells and normal to elevated activity of hexosaminidase B. Mutation analysis of the *HEXA* gene can also be used for diagnosis but is more typically only used to clarify carrier status and for prenatal testing.

Tay-Sachs disease is currently an incurable disorder; therefore, treatment focuses on the management of symptoms and palliative care. Nearly all patients require pharmacological management of their seizures. The psychiatric manifestations of patients with adult-onset GM_2 gangliosidosis are not usually responsive to conventional antipsychotic or antidepressant

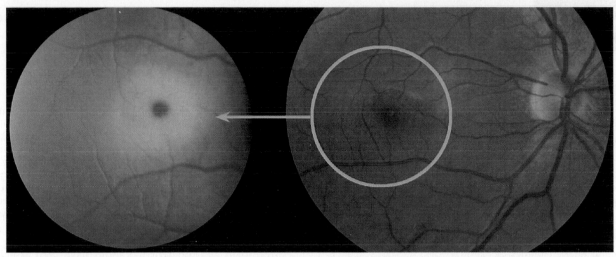

Figure C-43 **Cherry-red spot in Tay-Sachs disease.** *Right,* Normal retina. The *circle* surrounds the macula, lateral to the optic nerve. *Left,* The macula of a child with Tay-Sachs disease. The cherry-red center is the normal retina of the fovea at the center of the macula that is surrounded by a macular retina made white by abnormal storage of GM_2 in retinal neurons. *See Sources & Acknowledgments.*

medications; lithium and electroconvulsive therapy are most effective.

INHERITANCE RISK

For potential parents without a family history of GM_2 gangliosidosis, the empirical risk for having a child affected with GM_2 gangliosidosis depends on the frequency of GM_2 gangliosidosis in their ethnic groups. For most North Americans, the empirical risk for being a carrier is approximately 1 in 250 to 1 in 300, whereas individuals of Ashkenazi Jewish descent have an empirical carrier risk of 1 in 30. For couples who are both carriers, the risk for having a child with GM_2 gangliosidosis is 25%.

Prenatal diagnosis relies on identification of the *HEXA* mutations or hexosaminidase A deficiency in fetal tissue such as chorionic villi or amniocytes. Effective identification of affected fetuses by *HEXA* mutation analysis usually requires that the mutations responsible for GM_2 gangliosidosis in a family have already been identified.

Screening of high-risk populations for carriers and subsequent prevention has reduced the incidence of Tay-Sachs disease among Ashkenazi Jews by nearly 90% (see Chapters 12 and 18). Traditionally, such screening is performed by determining the serum activity of hexosaminidase A with an artificial substrate. This sensitive assay, however, cannot distinguish between pathological mutations and pseudodeficiency (reduced catabolism of the artificial substrate but normal catabolism of the natural substrate); therefore, carrier status is usually confirmed by molecular analysis of *HEXA*. Two pseudodeficiency alleles and more than 70 pathological mutations

have been identified in the *HEXA* gene. Among Ashkenazi Jews who are positive by enzymatic carrier screening, 2% are heterozygous for a pseudodeficiency allele and 95% to 98% are heterozygous for one of three pathological mutations, two causing infantile-onset and one causing adult-onset GM^2 gangliosidosis (see Chapter 12). In contrast, among non-Jewish North Americans who are positive by enzymatic carrier screening, 35% are heterozygous for a pseudodeficiency allele.

QUESTIONS FOR SMALL GROUP DISCUSSION

1. Screening for what other diseases is complicated by pseudodeficiency?
2. What is genetic drift? What are causes of genetic drift? Name two other diseases that exhibit genetic drift.
3. Should population screening be instituted to identify carriers of other diseases?
4. What diseases are genocopies of adult-onset hexosaminidase A deficiency? Consider psychiatric disorders and adult-onset neuronal ceroid-lipofuscinosis. What diseases are genocopies of infantile-onset hexosaminidase A deficiency? Consider GM_2 activator mutations. How would you distinguish between a genocopy and hexosaminidase A deficiency?

REFERENCES

Bley AE, Giannikopoulos OA, Hayden D, et al: Natural history of infantile GM_2 gangliosidosis, *Pediatrics* 128:e1233–e1241, 2011.

Kaback MM, Desnick RJ: Hexosaminidase A deficiency. Available from: http://www.ncbi.nlm.nih.gov/books/NBK1218/.

THALASSEMIA (α- or β-Globin Deficiency, MIM 141800 and MIM 613985)

Autosomal Recessive

PRINCIPLES

- Heterozygote advantage
- Ethnic variation in allele frequencies
- Gene dosage
- Compound heterozygote

MAJOR PHENOTYPIC FEATURES

- Age at onset: Childhood
- Hypochromic microcytic anemia
- Hepatosplenomegaly
- Extramedullary hematopoiesis

HISTORY AND PHYSICAL FINDINGS

J.Z., a 25-year-old healthy Canadian woman, presented to her obstetrician for routine prenatal care. Results of her complete blood count showed a mild microcytic anemia (hemoglobin, 98 g/L; mean corpuscular volume, 75 μm³). She was of Vietnamese origin, and her spouse, T.Z., was of Greek origin. J.Z. was unaware of any blood disorders in her or T.Z's family. Nonetheless, hemoglobin (Hb) electrophoresis showed a mildly elevated Hb A_2 ($\alpha_2\delta_2$) and Hb F ($\alpha_2\gamma_2$), which suggested that J.Z. had β-thalassemia trait; molecular testing detected a nonsense mutation in one β-globin allele and no α-globin deletions. The results of T.Z.'s testing showed that he also had a nonsense mutation of one β-globin allele and no α-globin deletions. After referral to the genetics clinic, the geneticist explained to J.Z. and T.Z. that their risk for a child with β-thalassemia major was 25%. After discussing prenatal diagnosis and postnatal prognosis, J.Z. and T.Z. chose to carry the pregnancy to term without further investigation.

BACKGROUND

Disease Etiology and Incidence

Thalassemias are autosomal recessive anemias caused by deficient synthesis of either α-globin or β-globin. A relative deficiency of α-globin causes α-thalassemia (MIM 141800), and a relative deficiency of β-globin causes β-thalassemia (MIM 613985) (see Chapter 11).

Thalassemia is most common among persons of Mediterranean, African, Middle Eastern, Indian, Chinese, and Southeast Asian descent. Thalassemias appear to have evolved because they confer heterozygote advantage in providing some resistance to malaria (see Chapter 9); the prevalence of thalassemia in an ethnic group therefore reflects past and present exposure of a population to malaria. The prevalence of α-thalassemia trait ranges from less than 0.01% in natives from nonmalarial areas such as the United Kingdom, Iceland, and Japan to approximately 49% among natives of some Southwest Pacific islands; Hb H disease and hydrops fetalis (see Table 11-4) are restricted to the Mediterranean and Southeast Asia. The incidence of β-thalassemia trait ranges from approximately 1% to 2% among Africans and African Americans to 30% in some villages of Sardinia.

Pathogenesis

Thalassemia arises from inadequate hemoglobin production and unbalanced accumulation of globin subunits. Inadequate hemoglobin production causes hypochromia and microcytosis. Unbalanced accumulation of globin causes ineffective erythropoiesis and hemolytic anemia. The severity of thalassemia is proportionate to the severity of the imbalance between α-globin and β-globin production.

More than 200 different mutations have been associated with thalassemia, although only a few mutations account for most thalassemia cases. Deletion of α-globin genes accounts for 80% to 85% of α-thalassemia, and approximately 15 mutations account for more than 90% of β-thalassemia. Molecular studies of both α-globin and β-globin mutations strongly suggest that the various mutations have arisen independently in different populations and then achieved their high frequency by selection.

Phenotype and Natural History

The α-globin mutations are separated into four clinical groups that reflect the impairment of α-globin production (see Table 11-4).

The phenotypes observed in a population reflect the nature of the α-globin mutations in that population. Chromosomes with deletion of both α-globin genes are observed in Southeast Asia and the Mediterranean basin; therefore, Hb H disease and hydrops fetalis usually occur in these populations and not in Africans, who usually have chromosomes with deletion of only one α-globin gene on a chromosome.

The β-globin mutations are also divided into clinical groups reflecting the impairment of β-globin production. β-Thalassemia trait is associated with a mutation in one β-globin allele and β-thalassemia major with mutations in both β-globin alleles. In general, patients with β-thalassemia trait have a mild hypochromic microcytic anemia, mild bone marrow erythroid hyperplasia, and occasionally hepatosplenomegaly; they are usually asymptomatic. Patients with β-thalassemia major present with severe hemolytic anemia when the postnatal production of Hb F decreases. The anemia and ineffective erythropoiesis cause growth retardation, jaundice, hepatosplenomegaly (extramedullary hematopoiesis), and bone marrow expansion (Fig. C-44). Patients usually present within the first 2 years of life, and approximately 80% of untreated patients die by 5 years of age. Patients receiving transfusion therapy alone die before 30 years of infection or hemochromatosis, whereas patients receiving both transfusion therapy and iron chelation therapy usually survive beyond the third decade. Iron overload from repeated transfusions and increased intestinal absorption causes cardiac, hepatic, and endocrine complications.

Management

Initial screening for α- or β-thalassemia trait is usually done by determination of erythrocyte indices. For patients without iron deficiency anemia, the diagnosis of β-thalassemia trait is usually confirmed by finding increased levels of Hb A_2 ($\alpha_2\delta_2$)

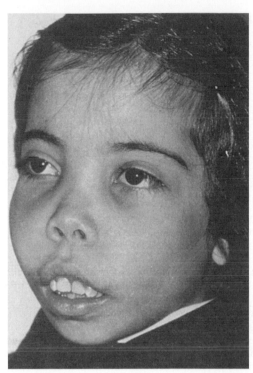

Figure C-44 **The typical facial appearance of a child with untreated β-thalassemia.** Note the prominent cheekbones and the protrusion of the upper jaw that results from the expansion of the marrow cavity in the bones of the skull and face. *See Sources & Acknowledgments.*

and Hb F ($\alpha_2\gamma_2$) (which contain other β-like globin chains from the β-globin cluster), or DNA mutation analysis, or both. In contrast, α-thalassemia trait is not associated with changes in Hb A_2 or Hb F and is confirmed by DNA mutation analysis or demonstration of a high β-globin/α-globin ratio.

Treatment of Hb H disease is primarily supportive. Therapy includes folate supplementation, avoidance of oxidant drugs and iron, prompt treatment of infection, and judicious transfusion. Splenectomy is rarely required.

Treatment of β-thalassemia includes blood transfusions, iron chelation, prompt treatment of infection, and frequently splenectomy. Bone marrow transplantation is the only currently available cure. Clinical trials are currently under way for drugs that will increase the expression of fetal hemoglobin, which would ameliorate β-thalassemia (but not α-thalassemia) (see Chapter 13).

INHERITANCE RISK

If each parent has β-thalassemia trait, the couple has a 25% risk for having a child with β-thalassemia major and a 50%

risk for having a child with β-thalassemia trait. If one parent has β-thalassemia trait and the other parent a triplication of the α-globin gene, this couple could also have a 25% risk for having a child with β-thalassemia major.

For parents with α-thalassemia trait, their risk for a child with Hb H disease or hydrops fetalis depends on the nature of their α-globin mutations. Parents with α-thalassemia trait can have either the $-\alpha/-\alpha$ or $--/\alpha\alpha$ genotype (see Chapter 11); therefore, depending on their genotypes, all their children will have α-thalassemia trait ($-\alpha/-\alpha$), or they may have a 25% risk for having a child with Hb H disease ($-\alpha/--$) or hydrops fetalis ($--/--$).

For both α- and β-thalassemia, prenatal diagnosis is possible by molecular analysis of fetal DNA from either chorionic villi or amniocytes. Molecular prenatal diagnosis of thalassemia is most efficient if the mutations have already been identified in the carrier parents. Preimplantation diagnosis has been achieved but requires knowing which mutations might be expected.

QUESTIONS FOR SMALL GROUP DISCUSSION

1. A father has the genotype $\alpha\alpha\alpha/\alpha-$, β/β and a mother $\alpha\alpha/\alpha\alpha$, β/ . If their child has the genotype $\alpha-/\alpha\alpha$, β/−, what is the most likely phenotype? Why? If the child's genotype is $\alpha\alpha\alpha/\alpha\alpha$, β/−, what is the most likely phenotype? Why?
2. What are the molecular mechanisms of α-globin gene deletion? Of α-globin gene triplication?
3. How does expression of γ-globin protect against β-thalassemia?
4. Describe carrier screening for thalassemia. To what ethnic groups should carrier screening be applied? Should individuals from classically low-risk ethnic groups be screened if their partner has α- or β-thalassemia trait? Consider population admixture.
5. α-Thalassemia is the most common single-gene disorder in the world. Three mechanisms can increase the frequency of a mutation in a population: selection, genetic drift, and founder effects. Describe each mechanism and the reason that selection is likely to account for the high frequency of α-thalassemia.

REFERENCES

Cao A, Galanello R, Origa R: Beta-thalassemia. Available from: http://www.ncbi.nlm.nih.gov/books/NBK1426/.

Cao A, Kan YW: The prevention of thalassemia, *Cold Spring Harbor Perspect Med* 3:a011775, 2013.

Origa R, Moi P, Galanello R, et al: Alpha-thalassemia. Available from: http://www.ncbi.nlm.nih.gov/books/NBK1435/.

THIOPURINE *S*-METHYLTRANSFERASE DEFICIENCY
(*TPMT* Polymorphisms, MIM 610460)

Autosomal Semidominant

PRINCIPLES

- Pharmacogenetics
- Precision medicine
- Cancer and immunosuppression chemotherapy
- Ethnic variation

MAJOR PHENOTYPIC FEATURES

- Age at onset: Deficiency is present at birth, manifestation requires drug exposure
- Myelosuppression
- Increased risk for brain tumor in thiopurine methyltransferase–deficient patients with acute lymphoblastic leukemia receiving brain irradiation

HISTORY AND PHYSICAL FINDINGS

J.B. is a 19-year-old man with long-standing ulcerative colitis. Because he has been refractory to steroid treatment, his physician prescribed azathioprine at a standard dose of 2.5 mg/kg/day. After a few weeks, J.B. developed severe leukopenia. The physician measured thiopurine methyltransferase activity in the red cells and found it to be normal. The physician remembered that J.B. had received a red blood cell transfusion 3 weeks previously and decided to determine his *TPMT* genotype. J.B. was found to be a compound heterozygote for the *TPMT*2 and -*3A alleles. Consequently, he should have been started and maintained on 6% to 10% of the standard dose of azathioprine.

BACKGROUND
Disease Etiology and Incidence

Thiopurine methyltransferase (TPMT) is the enzyme responsible for phase II metabolism of 6-mercaptopurine (6-MP) and 6-thioguanine by catalyzing *S*-methylation and thus inactivating these compounds (see Chapter 18). Azathioprine, a commonly used immunosuppressant, is activated by conversion to 6-MP, and so its metabolism is also affected by TPMT activity. These agents are used as immunosuppressants for various systemic inflammatory diseases, such as inflammatory bowel disease and lupus, and to prevent the rejection of solid tumor transplants. 6-MP is also a component in standard treatment of acute lymphoblastic leukemia. Approximately 10% of whites carry at least one slow metabolizer variant that causes accumulation of high levels of toxic metabolites, which can cause fatal hematopoietic toxicity (Fig. C-45). One in 300 whites is homozygous for an allele that causes complete deficiency of TPMT activity (MIM 610460). Deficiency is much less common in other ethnic groups.

Phenotype and Natural History

Toxicity from thiopurines was first recognized in patients receiving 6-MP for acute lymphoblastic leukemia. Although patients with 6-MP toxicity had a risk for life-threatening leukopenia, those who survived were noted to have longer periods of leukemia-free survival. Among TPMT-deficient patients with acute lymphoblastic leukemia, there was an increased risk for radiation-induced brain tumors and of chemotherapy-induced acute myelogenous leukemia. Fifteen different mutations in the *TPMT* gene have been associated with decreased activity in red cell assay. The wild-type allele is *TPMT*1. *TPMT*2 is a missense mutation that results in an alanine to proline substitution at codon 80 (Ala80Pro), which has only been seen in whites. Approximately 75% of affected whites have the *TPMT*3A allele, in which two mutations are present in cis: Tyr240Cys and Ala154Thr. The *TPMT*3C allele contains only the Tyr240Cys mutation and is found in 14.8% of Ghanaians and 2% of Chinese, Koreans, and Japanese. The Ala154Thr mutation has not been seen in isolation and presumably occurred on a chromosome that already carried the Tyr240Cys allele after the European migration.

Testing for the *TPMT* mutations by polymerase chain reaction is inexpensive and accurate and can prevent azathioprine toxicity by allowing dose adjustment before starting therapy. TPMT testing is the standard of care for acute lymphoblastic leukemia and has a favorable cost-benefit analysis for inflammatory bowel disease. Because TPMT activity is measured in red cells, false negatives are common in patients who have received transfusions as long as 3 months before testing, therefore direct genotyping of a DNA is preferred.

Management

Patients with complete TPMT deficiency should receive 6% to 10% of the standard dose of thiopurine medications. Heterozygous patients may start at the full dose but should have a dose reduction to half within 6 months or as soon as any myelosuppression is observed. The effect of *TPMT* polymorphism is an instructive example of the clinical importance of pharmacogenetics in personalized medicine (see Chapter 18).

INHERITANCE RISK

The *a priori* risk of a white individual carrying a *TPMT* deficiency allele is approximately 10%. In other ethnic groups, it is 2% to 5%. Because this is a simple semi-dominant trait, siblings of heterozygous individuals have a 50% chance of being heterozygous. Siblings of a deficient individual have a 25% chance of being deficient and a 50% chance of being heterozygous. Children of heterozygous carriers have a 25% chance of being deficient, and all children of deficient individuals will be heterozygous carriers if the other parent is *1*1 homozygote.

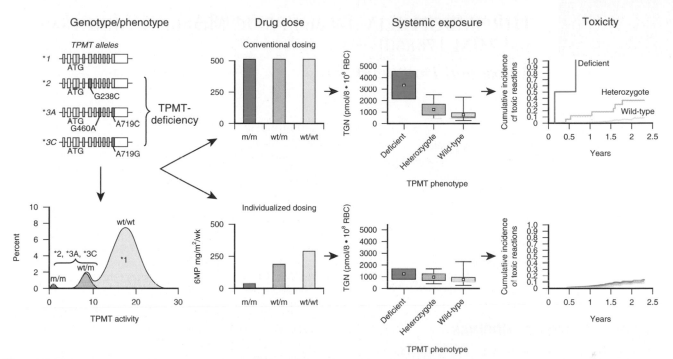

Figure C-45 Genetic polymorphism of thiopurine *S*-methyltransferase (TPMT) and its role in determining response to thiopurine medications (azathioprine, mercaptopurine, and thioguanine). The *upper left* panel depicts the predominant *TPMT* mutant alleles causing autosomal semidominant inheritance of TPMT activity in humans. As depicted in the adjacent *top three* panels, when uniform (conventional) dosages of thiopurine medications are given to all patients, TPMT homozygous mutant patients accumulate 10-fold higher cellular concentrations of the active thioguanine nucleotides (TGNs); heterozygous patients accumulate approximately twofold higher TGN concentrations. These differences translate into a significantly higher frequency of toxicity (*far right* panels). As depicted in the *bottom left three* panels, when genotype-specific dosing is used, similar cellular TGN concentrations are achieved, and all three TPMT phenotypes can be treated without acute toxicity. 6MP, 6-Mercaptopurine; RBC, red blood cell. *See Sources & Acknowledgments.*

QUESTIONS FOR SMALL GROUP DISCUSSION

1. *VKORC1* polymorphisms account for significant variation in warfarin metabolism. Name several conditions in which warfarin therapy is commonly used.
2. The P450 enzymes encoded by the *CYP* genes are important to drug metabolism. Which *CYP* genes metabolize selective serotonin reuptake inhibitors? Does this result in toxicity or decreased effect?
3. Why do humans have genes for drug metabolism?
4. Suggest explanations for ethnic variation in these genes.

REFERENCES

Relling MV, Gardner EE, Sandborn WJ, et al: Clinical pharmacogenetics implementation consortium guidelines for thiopurine methyltransferase genotype and thiopurine dosing, *Clin Pharmacol Ther* 89:387–391, 2011

Scott SA: Personalizing medicine with clinical pharmacogenetics, *Genet Med* 13.987–995, 2011.

THROMBOPHILIA (*FV* and *PROC* Mutations, MIM 188055 and MIM 176860)

Autosomal Dominant

PRINCIPLES

- Gain-of-function mutation (factor V Leiden)
- Loss-of-function mutation (protein C mutations)
- Incomplete penetrance
- Genetic modifiers
- Environmental modifiers
- Heterozygote advantage
- Founder effect

MAJOR PHENOTYPIC FEATURES

- Age at onset: Adulthood
- Deep venous thrombosis

HISTORY AND PHYSICAL FINDINGS

J.J., a 45-year-old businessman of French and Swedish descent, suddenly developed shortness of breath on the day after a trans–Pacific Ocean flight. His right leg was swollen and warm. Subsequent studies identified a thrombus in the popliteal and iliac veins and a pulmonary embolus. Both of his parents had had leg venous thromboses, and his sister had died of a pulmonary embolus during pregnancy. Based on his age and family history, J.J. was believed to have inherited a predisposition to thrombophilia. Screening for inherited causes of thrombophilia identified that J.J. was a carrier of factor V Leiden. Subsequent studies of other family members identified the same heterozygous mutation in J.J.'s father, deceased sister, and unaffected older brother. J.J. and his mother, deceased sister, and unaffected older sister were all found to be heterozygous for a frameshift mutation (3363insC) in *PROC*, the gene encoding protein C. Thus J.J. is a double heterozygote for two variants—in two unlinked genes—that predispose to thrombosis.

BACKGROUND

Disease Etiology and Incidence

Venous thrombosis (MIM 188050) is a panethnic multifactorial disorder (see Chapter 8); its incidence increases with age and varies among races. The incidence is low among Asians and Africans and higher among whites. Major predisposing influences are stasis, endothelial damage, and hypercoagulability. Identifiable genetic factors, present in 25% of unselected patients, include defects in coagulation factor inhibition and impaired clot lysis. Factor V Leiden occurs in 12% to 14%, prothrombin mutations in 6% to 18%, and deficiency of antithrombin III or protein C or S in 5% to 15% of patients with venous thromboses.

Factor V Leiden, an Arg506Gln mutation in the *FV* gene, has a prevalence of 2% to 15% among healthy European populations; it is highest in Swedes and Greeks and rare in Asians and Africans. Factor V Leiden apparently arose from a mutation in a white founder after the divergence from Africans and Asians.

Protein C deficiency (MIM 176860) is a panethnic disorder with a prevalence of 0.2% to 0.4%. Mutations of *PROC* are usually associated with activity levels of less than 55% of normal.

Pathogenesis

The coagulation system maintains a delicate balance of clot formation and inhibition; however, venous thrombi arise if coagulation overwhelms the anticoagulant and fibrinolytic systems. The proteases and protein cofactors of the clotting cascade must be activated at the site of injury to form a fibrin clot and then be inactivated to prevent disseminated coagulation (see Fig. 8-8). Activated factor V, a cofactor of activated factor X, accelerates the conversion of prothrombin to thrombin. Factor V is inactivated by activated protein C, which cleaves activated factor V at three sites (Arg306, Arg506, and Arg679). Cleavage at Arg506 occurs first and accelerates cleavage at the other two sites; cleavage at Arg506 reduces activated factor V function, whereas cleavage at Arg306 abolishes its function. Protein S, a cofactor for protein C, both accelerates the inactivation of activated factor V by protein C and enhances cleavage at Arg306.

The factor V Leiden mutation removes the preferred site for protein C proteolysis of activated factor V, thereby slowing inactivation of activated factor V and predisposing patients to thrombophilia. The risk for thrombophilia is higher for patients homozygous for factor V Leiden; the lifetime risks for venous thrombosis for factor V Leiden heterozygosity and homozygosity are approximately 10% and 80%, respectively.

Inherited deficiency of protein C arises from mutations in the *PROC* coding and regulatory sequences. Many mutations are sporadic, although some, such as the French-Canadian mutation 3363insC, entered populations through a founder. Unlike the gain-of-function factor V Leiden mutation, *PROC* mutations impair protein C function, thereby slowing inactivation of activated clotting factors V and VIII and predisposing to thrombus formation. Inheritance of two mutant *PROC* alleles usually results in purpura fulminans, a form of disseminated intravascular coagulation that is often fatal if it is not treated promptly. Heterozygous protein C mutations predispose to thrombophilia and carry a 20% to 75% lifetime risk for venous thrombosis.

In general, for patients heterozygous for the factor V Leiden polymorphism or a *PROC* mutation, progression from a hypercoagulable state to venous thrombi requires coexisting genetic or environmental factors. Associated nongenetic factors include smoking, pregnancy, oral contraceptive use, surgery, advanced age, neoplasia, immobility, and cardiac disease. Tobacco and oral contraceptive use act synergistically and increase the risk 8.8-fold over those who do not use either. Associated genetic abnormalities include other disorders of coagulation factor inhibition and impaired clot lysis.

Phenotype and Natural History

Although thrombi can develop in any vein, most arise at sites of injury or in the large venous sinuses or valve cusp pockets of the legs. Leg thrombi are usually confined to the veins of the calf, but approximately 20% extend into more proximal vessels. Obstruction of the deep leg veins can cause swelling, warmth, erythema, tenderness, distention of superficial veins, and prominent venous collaterals, although many patients are asymptomatic (Fig. C-46).

Once formed, a venous thrombus can propagate along the vein and eventually obstruct other veins, give rise to an

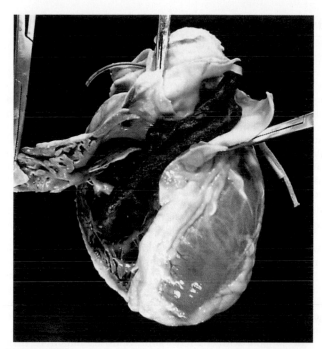

Figure C-46 **Autopsy picture of the cardiac right ventricle from a 58-year-old man who had had a cervical laminectomy and decompression.** He complained of right calf pain 33 days after surgery, and Homans sign was present. Venous ultrasonography detected a thrombus extending from the post-tibial and popliteal veins into the femoral vein. Despite anticoagulation with heparin, the patient was found 2 days later unresponsive and with a low oxygen saturation; he did not respond to cardiopulmonary resuscitation and died. Autopsy showed a thromboembolus in the right ventricle occluding the pulmonary artery. *See Sources & Acknowledgments.*

embolus, be removed by fibrinolysis, or be organized and possibly recanalized. Embolism is serious and can be acutely fatal if it obstructs the pulmonary arterial system; pulmonary embolism occurs in 5% to 20% of patients presenting initially with deep calf vein thrombosis. In contrast, organization of proximal vein thrombi chronically impedes venous return and causes post-thrombotic syndrome, characterized by leg pain, edema, and frequent skin ulceration.

With the possible exception of an increased risk for recurrence, the symptoms, course, and outcomes of patients with *PROC* mutations and factor V Leiden are similar to those of other thrombophilia patients. In general, untreated patients with proximal vein thrombosis have a 40% risk for recurrent venous thrombosis.

Management

The diagnosis of deep venous thrombosis of the calf is difficult because patients are often asymptomatic and most tests are relatively insensitive until the thrombus extends proximal to the deep calf veins. Duplex venous ultrasonography is used most often to diagnose deep venous thrombosis; the thrombus is detected either by direct visualization or by inference when the vein does not collapse on compressive maneuvers. Doppler ultrasonography detects flow abnormalities in the veins.

Factor V Leiden can be diagnosed directly by DNA analysis or can be suspected on the basis of activated protein C resistance. Protein C deficiency is diagnosed by measuring protein C activity; *PROC* mutations are identified by analysis of the *PROC* gene.

Acute treatment focuses on minimizing thrombus propagation and associated complications, especially pulmonary embolism; it usually involves anticoagulation and elevation of the affected extremity. Subsequent therapy focuses on prevention of recurrent venous thrombosis through identification and amelioration of predispositions, and anticoagulant prophylaxis. Treatment recommendations for patients with protein C deficiency and factor V Leiden are still evolving. All patients should receive standard initial therapy followed by at least 3 months of anticoagulant therapy. It is unclear which patients with a single mutant allele should receive prolonged, perhaps lifelong anticoagulation, but long-term anticoagulation is generally prescribed for patients with a second episode of deep venous thrombosis. In contrast, homozygous factor V Leiden patients as well as those who are homozygous for other mutations or are combined carriers (like J.J.) are placed on long-term anticoagulation after their initial episode.

INHERITANCE RISK

Each child of a couple in which one parent is heterozygous for factor V Leiden has a 50% risk for inheriting a mutant allele. Assuming 10% penetrance, each child has an a priori 5% lifetime risk for development of a venous thrombosis.

Each child of a couple in which one parent is heterozygous for a *PROC* mutation also has a 50% risk for inheriting a mutant allele. Estimates for penetrance of protein C deficiency range from 20% to 75%; therefore each child has an a priori 10% to 38% lifetime risk for development of a venous thrombosis.

Because of the incomplete penetrance and availability of effective therapy for factor V Leiden and heterozygous *PROC* mutations, prenatal diagnostic testing is not routinely used except for detection of homozygous or compound heterozygous *PROC* mutations. Prenatal detection of homozygous or compound heterozygous *PROC* mutations is helpful because of the severity of the disease and the need for prompt neonatal treatment.

QUESTIONS FOR SMALL GROUP DISCUSSION

1. Some studies of oral contraceptives suggest that such drugs decrease the blood levels of protein S. How would this predispose to thrombosis? At a molecular level, why would this be expected to enhance the development of venous thromboses in women with the factor V Leiden mutation? Should such women avoid the use of oral contraceptives? Should women be tested for factor V Leiden before using oral contraceptives?
2. Testing of asymptomatic relatives for the factor V Leiden mutation is controversial. For it to be of clear utility, what should presymptomatic testing allow?
3. Synergism is the multiplication of risk with the co-occurrence of risk factors. Illustrate this with factor V Leiden and protein C deficiency (the family of J.J. is an example), factor V Leiden and oral contraceptive use, and factor V Leiden and hyperhomocystinemia.
4. Factor V Leiden is thought to reduce intrapartum bleeding. How would this lead to a heterozygote advantage and maintenance of a high allele frequency in the population?

REFERENCES

Kujovich JL: Factor V Leiden thrombophilia. Available from: http://www.ncbi.nlm.nih.gov/books/NBK1368/.

Varga EA, Kujovich JL: Management of inherited thrombophilia: guide for genetics professionals, *Clin Genet* 81:7–17, 2012.

TURNER SYNDROME (Female Monosomy X)

Chromosomal

PRINCIPLES

- Nondisjunction
- Prenatal selection
- Haploinsufficiency

MAJOR PHENOTYPIC FEATURES

- Age at onset: Prenatal
- Short stature
- Ovarian dysgenesis
- Sexual immaturity

HISTORY AND PHYSICAL FINDINGS

L.W., a 14-year-old girl, was referred to the endocrinology clinic for evaluation of absent secondary sexual characteristics (menses and breast development). Although born small for gestational age, she had been in good health and had normal intellect. No other family members had similar problems. Her examination was normal except for short stature, Tanner stage I sexual development, and broad chest with widely spaced nipples. After briefly discussing causes of short stature and delayed or absent sexual development, her physician requested follicle-stimulating hormone (FSH) level, growth hormone (GH) level, bone age study, and chromosome analysis. These tests showed a normal GH level, an elevated FSH level, and an abnormal karyotype (45,X). The physician explained that L.W. had Turner syndrome. L.W. was treated with GH supplements to maximize her linear growth; 1 year later, she started estrogen and progesterone therapy to induce the development of secondary sexual characteristics.

BACKGROUND

Disease Etiology and Incidence

Turner syndrome (TS) is a panethnic disorder caused by complete or partial absence of a second X chromosome in females. It has an incidence of between 1 in 2000 and 1 in 5000 liveborn girls. Approximately 50% of TS cases are associated with a 45,X karyotype, 25% with a structural abnormality of the second X chromosome, and 25% with 45,X mosaicism (see Chapter 6).

Monosomy for the X chromosome can arise either by the failure to transmit a sex chromosome to one of the gametes or by loss of a sex chromosome from the zygote or early embryo. Failure to transmit a paternal sex chromosome to a gamete is the most common cause of a 45,X karyotype; 70% to 80% of patients with a 45,X karyotype are conceived from a sperm lacking a sex chromosome. Loss of a sex chromosome from a cell in the early embryo is the likely cause of 45,X mosaicism.

Pathogenesis

The mechanism by which X chromosome monosomy causes TS in girls is poorly understood. The X chromosome contains many loci that do not undergo complete X inactivation (see Chapter 6), several of which appear to be necessary for ovarian maintenance and female fertility. Although oocyte development requires only a single X chromosome, oocyte maintenance requires two X chromosomes. In the absence of

a second X chromosome, therefore, oocytes in fetuses and neonates with TS degenerate, and their ovaries atrophy into streaks of fibrous tissue. The genetic bases for the other features of TS, such as the cystic hygroma, lymphedema, broad chest, cardiac anomalies, renal anomalies, and sensorineural hearing deficit, have not been defined but presumably reflect haploinsufficiency for one or more X-linked genes that do not normally undergo inactivation in the female.

Phenotype and Natural History

Although 45,X conceptuses account for between 1% and 2% of all pregnancies, less than 1% of 45,X conceptions result in a liveborn infant. In view of the mild phenotype observed in patients with TS, this high rate of miscarriage and its timing is remarkable and suggests that a second sex chromosome is generally required for intrauterine survival, particularly at the beginning of the second trimester.

All patients with TS have short stature, and more than 90% have ovarian dysgenesis. The ovarian dysgenesis is sufficiently severe that only 10% to 20% of patients have spontaneous pubertal development (breast budding and pubic hair growth), and only 2% to 5% have spontaneous menses. Many individuals also have physical anomalies, such as webbed neck, low nuchal hairline, broad chest, cardiac anomalies, renal anomalies, sensorineural hearing deficit, edema of the hands and feet, and dysplastic nails. Nearly 50% of patients have a bicuspid aortic valve and therefore an increased risk for aortic root dilatation and dissection; nearly 60% have renal anomalies and an increased risk for renal dysfunction.

Most patients have normal intellectual development. Those with intellectual impairment usually have an X chromosome structural abnormality. Socially, individuals with TS tend to be shy and withdrawn (see Chapter 6).

In addition to the complications resulting from their congenital anomalies, women with TS have an increased incidence of osteoporotic fractures, thyroiditis, diabetes mellitus type 1 and type 2, inflammatory bowel disease, and cardiovascular disease. The causes of the diabetes mellitus, thyroid disorders, and inflammatory bowel disease are unclear. Estrogen deficiency is probably largely responsible for the osteoporosis and the increased incidence of atherosclerosis, ischemic heart disease, and stroke, although diabetes mellitus probably accentuates the cardiovascular effects of estrogen deficiency.

Management

When stature in an individual with TS falls below the 5th percentile, she is usually treated with GH supplements until her bone age reaches 15 years (Fig. C-47). On average, this treatment results in a gain of 10 cm in predicted height; the improvement in final height is less, however, the later GH therapy is started. Concurrent estrogen therapy decreases the effectiveness of GH.

Estrogen therapy is usually initiated at approximately 14 to 15 years of age to promote development of secondary sexual characteristics and reduce the risk for osteoporosis. Progesterone therapy is added to the regimen to induce menses either at the time of the first vaginal breakthrough bleeding or in the second year of estrogen therapy. Both are associated with an increased risk for thrombosis, and case reports

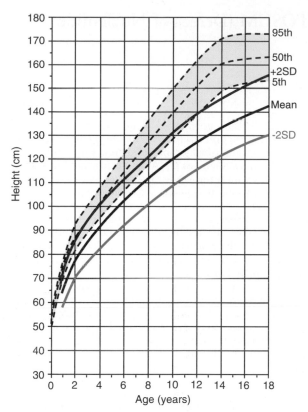

Figure C-47 Growth curves for normal (*shaded dotted lines*) and approximately 350 Turner syndrome girls (*solid lines*). None of the subjects received hormone treatment. *See Sources & Acknowledgments.*

indicate that there may be an increased risk in patients with TS above the general population of hormone therapy users.

In addition, medical management usually includes serial echocardiography to evaluate aortic root dilatation and valvar heart disease, renal ultrasonography to find congenital renal anomalies, and a glucose tolerance test to detect diabetes.

Patients who have complete ovarian dysgenesis do not ovulate spontaneously or conceive children. If they have adequate cardiovascular and renal function, women with TS can have children by in vitro fertilization and ovum donation. They do, however, have a significantly increased risk for aortic dissection and rupture with pregnancy.

INHERITANCE RISK

TS is not associated with advanced maternal or paternal age. Although there have been a few familial recurrences, TS is usually sporadic, and the empirical recurrence risk for future pregnancies is not increased above that of the general population. If TS is suspected on the basis of fetal ultrasound findings, such as a cystic hygroma, the diagnosis should be confirmed by karyotyping of chorionic villi or amniocytes.

Only a few pregnancies have been reported among spontaneously menstruating patients with TS. Among the resulting offspring, one in three has had congenital anomalies, such as congenital heart disease, Down syndrome, and spina bifida. The apparently increased risk for congenital anomalies may be due to ascertainment bias in reporting, because pregnancy is unusual in TS. If the increased risk is a real finding, the cause is unknown.

QUESTIONS FOR SMALL GROUP DISCUSSION

1. Some observations have suggested that patients with Turner syndrome who inherit a paternal X chromosome are more outgoing and have better social adaptation than those who inherit a maternal X chromosome. What molecular mechanisms could explain this?
2. X-chromosome monosomy is the only viable human monosomy (other than the Y in males). Discuss possible reasons.
3. Discuss possible reasons for the high rate of birth defects among the children of women with Turner syndrome.
4. Maternal meiotic nondisjunction gives rise more frequently to Down syndrome and paternal meiotic nondisjunction to Turner syndrome. Discuss possible reasons.
5. Discuss the psychosocial support and counseling that are appropriate and necessary for patients with Turner syndrome.

REFERENCES

Gonzalez L, Witchel SF: The patient with Turner syndrome: puberty and medical management concerns, *Fertil Steril* 98:780–786, 2012.

Hong DS, Reiss AL: Cognitive and neurological aspects of sex chromosome aneuploidies, *Lancet Neurol* 13:306–318, 2014.

Hook EB, Warburton D: Turner syndrome revisited: review of new data supports the hypothesis that all viable 45,X cases are cryptic mosaics with a rescue cell line, implying an origin by mitotic loss, *Hum Genet* 133:417–424, 2014.

Legro RS: Turner syndrome: new insights into an old disorder, *Fertil Steril* 98:773–774, 2012.

XERODERMA PIGMENTOSUM (Defect of Nucleotide Excision Repair)

Autosomal Recessive

PRINCIPLES

- Variable expressivity
- Genetic heterogeneity
- Genetic complementation
- Caretaker tumor-suppressor genes

MAJOR PHENOTYPIC FEATURES

- Age at onset: Childhood
- Ultraviolet light sensitivity
- Skin cancer
- Neurological dysfunction

HISTORY AND PHYSICAL FINDINGS

W.S., a 3-year-old girl, was referred to the dermatology clinic for evaluation of severe sun sensitivity and freckling. On physical examination, she was photophobic and had conjunctivitis and prominent freckled hyperpigmentation in sun-exposed areas; her development and physical examination were otherwise normal. W.S. was the child of nonconsanguineous Japanese parents; no one else in the family was similarly affected. The dermatologist explained that W.S. had classic features of xeroderma pigmentosum, that is, "parchment-like, pigmented skin." To confirm the diagnosis, W.S. had a skin biopsy to evaluate DNA repair and ultraviolet (UV) radiation sensitivity in her skin fibroblasts. The results of this testing confirmed the diagnosis of xeroderma pigmentosum. Despite appropriate preventive measures, W.S. developed metastatic melanoma at 15 years of age and died 2 years later. Her parents had two other children; neither was affected with xeroderma pigmentosum.

BACKGROUND

Disease Etiology and Incidence

Xeroderma pigmentosum (XP) is a genetically heterogeneous, panethnic, autosomal recessive disorder of DNA repair that causes marked sensitivity to UV irradiation (see Table). In the United States and Europe, the prevalence is approximately 1 in 1 million, but in Japan, the prevalence is 1 in 100,000.

Pathogenesis

Repair of DNA damaged by UV irradiation occurs by three mechanisms: excision repair, postreplication repair, and photoreactivation. Excision repair mends DNA damage by nucleotide excision repair or base excision repair. Postreplication repair is a damage tolerance mechanism that allows replication of DNA across a damaged template. Photoreactivation reverts damaged DNA to the normal chemical state without removing or exchanging any genetic material.

Nucleotide excision repair is a complex but versatile process involving at least 30 proteins. The basic principle is the removal of a small single-stranded DNA segment containing a lesion by incision to either side of the damaged segment and subsequent gap-filling repair synthesis with use of the intact complementary strand as a template. Within transcribed genes, DNA damage blocks RNA polymerase II progression.

The stalled RNA polymerase II initiates nucleotide excision repair (transcription-coupled repair). In the rest of the genome and on nontranscribed strands of genes, a nucleotide excision repair complex identifies DNA damage by detection of helical distortions within the DNA (global genome repair).

On occasion, nucleotide excision repair will not have repaired a lesion before DNA replication. Because such lesions inhibit the progression of DNA replication, postreplication repair bypasses the lesion, allowing DNA synthesis to continue. DNA polymerase η mediates translesional DNA synthesis; it efficiently and accurately catalyzes synthesis past dithymidine lesions.

XP is caused by mutations affecting the global genome repair subpathway of nucleotide excision repair or by mutations affecting postreplication repair. In contrast, Cockayne syndrome, a related disorder, is caused by mutations affecting the transcription-coupled repair subpathway of nucleotide excision repair. XP and Cockayne syndrome have been separated into 10 biochemical complementation groups; each group reflects a mutation of a different component of nucleotide excision repair or postreplication repair (see Table).

The reduced or absent capacity for global genome repair or postreplication repair represents a loss of caretaker functions required for maintenance of genome integrity and results in the accumulation of oncogenic mutations (see Chapter 15). Cutaneous neoplasms from patients with XP have a higher level of oncogene and tumor suppressor gene mutations than tumors from the normal population, and those mutations appear to be highly UV specific.

Phenotype and Natural History

Patients with XP develop symptoms at a median age of 1 to 2 years, although onset after 14 years is seen in approximately 5% of patients. Initial symptoms commonly include easy sunburning, acute photosensitivity, freckling, and photophobia. Continued cutaneous damage causes premature skin aging (thinning, wrinkling, solar lentigines, telangiectasias), premalignant actinic keratoses, and benign and malignant neoplasms (Fig. C-48). Nearly 45% of patients develop basal cell or squamous cell carcinomas, or both, and approximately 5% develop melanomas. Approximately 90% of the carcinomas occur at the sites of greatest UV exposure—the face, neck, head, and tip of the tongue. Before the introduction of preventive measures, the median age for development of cutaneous neoplasms was 8 years, 50 years younger than in the general population, and the frequency of such neoplasms was more than 1000-fold greater than that of the general population.

In addition to cutaneous signs, 60% to 90% of patients experience ocular abnormalities, including photophobia, conjunctivitis, blepharitis, ectropion, and neoplasia. Again, the distribution of ocular damage and neoplasms corresponds to the sites of greatest UV exposure.

Approximately 18% of patients experience progressive neuronal degeneration. Features include sensorineural deafness, mental retardation, spasticity, hyporeflexia or areflexia, segmental demyelination, ataxia, choreoathetosis, and supranuclear ophthalmoplegia. The severity of neurological symptoms is usually proportionate to the severity of the nucleotide excision repair deficit. The neurodegeneration may result from

Complementation Groups in Xeroderma Pigmentosum and Related Disorders

Complementation Group	MIM	Gene	Process Affected	Phenotype
XPA	278700	XPA	DNA damage recognition	XP
XPB	133510	ERCC3	DNA unwinding	XP-CS, TTD
XPC	2788720	XPC	DNA damage recognition	XP
XPD	278730	ERCC2	DNA unwinding	XP, TTD, XP-CS
XPE	278740	DDB2	DNA damage recognition	XP
XPF	278760	ERCC4	Endonuclease	XP
XPG	278780	ERCC5	Endonuclease	XP, XP-CS
XPV	278750	POLH	Translesional DNA synthesis	XP
CSA	216400	ERCC8	Transcription-coupled repair	CS
CSB	133540	ERCC6	Transcription-coupled repair	CS

CS, Cockayne syndrome; TTD, trichothiodystrophy; XP, xeroderma pigmentosum; XP-CS, combined XP and Cockayne syndrome phenotype.

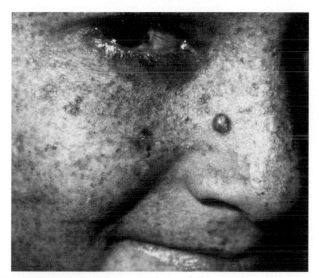

Figure C-48 Cutaneous and ocular findings of xeroderma pigmentosum. Note the freckled hyperpigmentation, the papillomatous and verrucous lesions on the skin, and the conjunctivitis. *See Sources & Acknowledgments.*

an inability to repair DNA damaged by endogenously generated oxygen free radicals.

Nucleotide excision repair also corrects DNA damage from many chemical carcinogens, such as cigarette smoke, charred food, and cisplatin. Consequently, patients have a 10- to 20-fold increase in the incidence of internal neoplasms, such as brain tumors, leukemia, lung tumors, and gastric carcinomas.

Patients with XP have a shortened life span; without preventive protection, their life span is approximately 30 years shorter than that of individuals without XP. Metastatic melanoma and squamous cell skin carcinoma are the most common causes of death.

Two related disorders, Cockayne syndrome and trichothiodystrophy, are also caused by defects in other components of the cellular mechanism for repair of UV-induced DNA damage. Both are characterized by poor postnatal growth, diminished subcutaneous tissue, joint contractures, thin papery skin with photosensitivity, mental retardation, and neurological deterioration. Children with Cockayne syndrome also have retinal degeneration and deafness; children with trichothiodystrophy have ichthyosis and brittle hair and nails. In both syndromes, affected patients rarely live past the second decade. Interestingly, neither syndrome shows an increase in the frequency of skin cancers. However, defects in some repair genes (*ERCC2*, *ERCC3*, and *ERCC5*) produce phenotypes that combine characteristics of XP and either Cockayne syndrome or both Cockayne syndrome and trichothiodystrophy (see Table).

Management

Confirmation of the diagnosis of XP relies on functional tests of DNA repair and UV sensitivity; such tests are usually performed on cultured skin fibroblasts. Diagnostic confirmation by identification of mutations in an XP-associated gene is clinically available for some of the genes associated with the condition. However, failure to identify a causative mutation in one of these genes does not rule out XP as a clinical diagnosis.

The management of patients with XP includes avoidance of exposure to sunlight, protective clothing, physical and chemical sunscreens, and careful surveillance for and excision of cutaneous malignant neoplasms. No curative treatments are currently available.

INHERITANCE RISK

Because XP is an autosomal recessive disease, many patients do not have a family history of the disease. For parents who already have a child affected with XP, the risk for future children to have XP is 25%. Prenatal diagnosis is possible by functional testing of DNA repair and UV sensitivity in cultured amniocytes or chorionic villi or by molecular testing if the mutations have been identified.

QUESTIONS FOR SMALL GROUP DISCUSSION

1. Define complementation groups and explain their use for defining the biochemical basis of disease.
2. Compare and contrast XP and Cockayne syndrome. Why is Cockayne syndrome not associated with an increased risk for neoplasia?
3. Patients with XP have a defect of cutaneous cellular immunity. How could the sensitivity of patients with XP to UV irradiation explain this immunodeficiency? How could this immunodeficiency contribute to cancer susceptibility?
4. Werner syndrome, Bloom syndrome, XP, ataxia-telangiectasia, and Fanconi anemia are inherited diseases of genomic instability. What are the molecular mechanisms underlying each of these disorders? What types of genomic instability are associated with each disorder?

REFERENCES

DiGiovanna JJ, Kraemer KH: Shining a light on xeroderma pigmentosum, *J Invest Dermatol* 132:785–796, 2012.

Kraemer KH, DiGiovanna JJ: Xeroderma pigmentosum. Available from: http://www.ncbi.nlm.nih.gov/books/NBK1397/.

Menck CF, Munford V: DNA repair diseases: what do they tell us about cancer and aging? *Genet Mol Biol* 37:220–233, 2014.

Glossary

Acceptor splice site The boundary between the 3' end of an intron and the 5' end of the following exon. Also called *3' splice site*.

Acrocentric A type of chromosome with the centromere near one end. The human acrocentric chromosomes (13, 14, 15, 21, and 22) have satellited short arms that carry genes for ribosomal RNA.

Active chromatin hub Nuclear domain where the proteins bound to the locus control region and the β-globin locus colocalize to permit globin gene expression.

Adjacent segregation Pattern of chromosome segregation in a cell with a balanced reciprocal translocation following synapsis of a *quadrivalent* in which *unbalanced* gametes are formed that have only one copy of each of the centromeres of the translocated chromosomes (adjacent-1) or have two copies of one or the other of the translocated chromosomes but not both (adjacent-2) (see Fig. 5-12).

Adverse selection A term used in the insurance industry to describe the situation in which individuals with private knowledge of having an increased risk for illness, disability, or death buy disproportionately more coverage than those at a lower risk. As a result, insurance premiums, which are based on averaging risk across the population, are inadequate to cover future claims.

Allele One of the alternative versions of a gene or DNA sequence at a given locus.

Allele-specific oligonucleotide (ASO) An oligonucleotide probe synthesized to match a particular DNA sequence precisely and allow the discrimination of alleles that differ by only a single base.

Allelic heterogeneity In a population, there may be a number of different mutant alleles at a single locus. In an individual, the same or similar phenotypes may be caused by different mutant alleles rather than by identical alleles at the locus.

Allelic imbalance Unequal expression of the two alleles of a gene. The most extreme example is monoallelic expression, which can be random, as in *X-inactivation*, or determined by parent of origin of the allele (*genomic imprinting*).

Allogenic In transplantation, denotes individuals (or tissues) that are of the same species but have different antigens (alternative spelling: allogeneic).

Alpha-fetoprotein (AFP) A fetal glycoprotein excreted into the amniotic fluid that reaches abnormally high concentration in amniotic fluid (and maternal serum) when the fetus has certain abnormalities, especially an open neural tube defect.

Alternate segregation Pattern of chromosome segregation in a cell with a balanced reciprocal translocation following synapsis of a quadrivalent in which balanced gametes are formed that have either a normal chromosome complement or contain the two reciprocal balanced translocation chromosomes.

Alu family of repetitive DNA In the human genome, approximately 10% of the DNA is made up of a set of approximately 1,000,000 dispersed, related sequences, each approximately 300 base pairs long, so named because they are cleaved by the restriction enzyme *Alu*I.

Amniocentesis A procedure used in prenatal diagnosis to obtain amniotic fluid, which contains cells of fetal origin that can be cultured for analysis. Amniotic fluid is withdrawn from the amniotic sac by syringe after insertion of a hollow needle into the amnion through the abdominal wall and uterine wall.

Analytic validity In reference to a clinical laboratory test, the ability of that test to perform correctly, that is, measure what it is designed to measure.

Anaphase Stage in mitosis when chromosomes separate at the centromere and sister chromatids become independent daughter chromosomes, which move to opposite poles of the dividing cell. Immediately follows metaphase.

Ancestry informative markers Loci with alleles that show large differences in frequency among populations originating in different parts of the world.

Aneuploidy Any chromosome number that is not an exact multiple of the haploid number. The common forms of aneuploidy in humans are *trisomy* (the presence of an extra chromosome) and *monosomy* (the absence of a single chromosome).

Anomalies Birth defects resulting from malformations, deformations, or disruptions.

Anticipation The progressively earlier onset and increased severity of certain diseases in successive generations of a family. Anticipation is caused by expansion of the number of repeats that constitute a dynamic mutation within the gene responsible for the disease.

Anticodon A three-base segment of transfer RNA (tRNA) complementary to a codon in messenger RNA (mRNA).

Antisense oligonucleotides (ASOs) Synthetic short (usually 12 to 35 nucleotides) single-stranded molecules that can hybridize to corresponding sequences in a specific target pre-mRNA or microRNA, causing its degradation, inhibiting its translation, or modulating its splicing. Excellent therapeutic potential.

Antisense strand of DNA The noncoding DNA strand, which is complementary to mRNA and serves as the template for RNA synthesis. Also called the *transcribed strand*.

Apoenzyme The protein component of an enzyme that also requires a cofactor to become active. The apoenzyme with the cofactor is termed the *holoenzyme*.

Apoptosis Programmed cell death characterized by a stereotypic pattern of mitochondrial breakdown and chromatin degradation.

Array CGH Comparative genome hybridization performed by hybridizing to a wafer ("chip") made of glass, plastic, or silicon onto which a large number of different nucleic acids have been individually spotted in a matrix pattern. See *microarray*.

Ascertainment The method of selection of individuals for inclusion in a genetic study.

Ascertainment bias A difference in the likelihood that affected relatives of affected individuals will be identified, compared with similarly affected relatives of controls. A possible source of error in family studies.

Association 1. In genetic epidemiology, describes the situation in which a particular allele is found either significantly more or significantly less frequently in a group of affected individuals than would be expected from the frequency of the allele in the general population from which the affected individuals were drawn; not to be confused with *linkage*. 2. In dysmorphology, a group of abnormalities of unknown etiology and pathogenesis that is seen together more often than would be expected by chance.

Assortative mating Selection of a mate with preference for a particular genotype; that is, nonrandom mating. Usually positive (preference for a mate of the same genotype), less frequently negative (preference for a mate of a different genotype).

Assortment The random distribution of different combinations of the parental chromosomes to the gametes. Nonallelic genes assort independently, unless they are linked.

Autologous Refers to grafts in the same animal from one part to another or to malignant cells and the cells of the individual in which they have arisen.

Autosome Any nuclear chromosome other than the sex chromosomes; 22 pairs in the human karyotype. A disease caused by mutation in an autosomal gene or gene pair shows *autosomal inheritance*.

Balanced polymorphism A polymorphism maintained in the population by heterozygote advantage, allowing an allele, even one that is deleterious in the homozygous state, to persist at a relatively high frequency in the population.

Barr body The sex chromatin as seen in female somatic cells, representing an inactive X chromosome.

Base pair (bp) A pair of complementary nucleotide bases, as in double-stranded DNA. Used as the unit of measurement of the length of a DNA sequence.

Bayesian analysis A mathematical method widely used in genetic counseling to calculate recurrence risks. The method combines information from several sources (genetics, pedigree information, and test results) to determine the probability that a specific individual might develop or transmit a certain disorder.

Beneficence The ethical principle of behaving in a way that promotes the well-being of others. See *maleficence*.

Binomial expansion When there are two alternative classes, one with probability p and the other with probability $1 - p = q$, the frequencies of the possible combinations of p and q in a series of n trials is $(p + q)^n$.

Biochemical genetics The study of the genetic basis for phenotype at the level of proteins, biochemical pathways, and metabolism.

Bioinformatics Computational analysis and storage of biological and experimental data, widely applied to genomic and proteomic studies.

Birth defect An abnormality present at birth, not necessarily genetic.

Bivalent A pair of homologous chromosomes in association, as seen at metaphase of the first meiotic division.

Blastocyst A stage in early embryogenesis in which the initial ball of cells derived from the fertilized egg (the morula) secrete fluid and form a fluid-filled internal cavity within which is a separate group of cells, the *inner cell mass*.

Blood group The phenotype produced by genetically determined antigens on a red blood cell. The antigens formed by a set of allelic genes make up a blood group system.

Body mass index (BMI) A measure of weight used for classifying individuals as underweight, appropriate for weight, overweight, or obese that is corrected for height. Expressed as weight divided by square of the height (kg/m^2).

5′-Cap A modified nucleotide added to the 5′ end of a growing mRNA chain, required for normal processing, stability, and translation of mRNA.

Caretaker genes Tumor-suppressor genes that are indirectly involved in controlling cellular proliferation by repairing DNA damage and maintaining genomic integrity, thereby protecting proto-oncogenes and gatekeeper tumor-suppressor genes from mutations that could lead to cancer.

Carrier An individual heterozygous for a particular mutant allele. The term is used for heterozygotes for autosomal recessive alleles, for females heterozygous for X-linked alleles, or, less commonly, for an individual heterozygous for an autosomal dominant allele but not expressing it (e.g., a heterozygote for a Huntington disease allele in the presymptomatic stage).

Case-control study An epidemiological method in which patients with a disease (the cases) are compared with suitably chosen individuals without the disease (the controls) with respect to the relative frequency of various putative risk factors.

cDNA See *complementary DNA*.

Cell cycle The stages between two successive mitotic divisions, described in the text. Consists of the G_1, S, G_2, and M stages.

Cell-free DNA DNA detectable in body fluids that is not packaged in chromatin inside the nucleus of a cell.

Centimorgan (cM) The unit of distance between genes along chromosomes, named for Thomas Hunt Morgan. Two loci are 1 cM apart if recombination is detected between them in 1% of meioses.

Centromere The primary constriction on the chromosome, a region at which the sister chromatids are held together and at which the kinetochore is formed. Required for normal segregation in mitosis and meiosis.

Centrosomes A pair of centers that organize the growth of the microtubules of the mitotic spindle; visible at the poles of the dividing cell in late prophase.

CGH See *comparative genome hybridization*.

Chain termination mutation See *termination codon*.

Checkpoint Positions in the cell cycle, usually at the junction between the G_1 and S or the G_2 and M stages, at which the cell determines whether to proceed to the next stage of the cycle.

Chemical individuality A term coined by Archibald Garrod to describe the naturally occurring differences in the genetic and biochemical makeup of each individual.

Chemical library An annotated collection of hundreds to tens of thousands of small molecules, increasingly used in drug discovery. High-throughput screening against a drug target may identify a compound that interacts with the target, for example to restore activity to a mutant protein. Such chemicals, or derivatives of them, may then be developed into drugs.

Chimera An individual composed of cells derived from two genetically different zygotes. In humans, *blood group chimeras* result from exchange of hematopoietic stem cells by dizygotic twins in utero; *dispermic chimeras*, which are very rare, result from fusion of two zygotes into one individual. Chimerism is also an inevitable result of transplantation.

Chorionic villus sampling (CVS) A procedure used for prenatal diagnosis at 8 to 10 weeks of gestation. Fetal tissue for analysis is withdrawn from the villous area of the chorion either transcervically or transabdominally under ultrasonographic guidance.

Chromatids The two parallel strands of chromatin (referred to as *sister chromatids*), connected at the centromere, that constitute a chromosome after DNA synthesis.

Chromatin The complex of DNA and proteins of which chromosomes are composed. See also *nucleosome*.

Chromatin remodeling DNA packaged in nucleosomes is subject to remodeling between chromatin states through the activity of enzymatic chromatin remodeling complexes. Packaged DNA can thereby be made accessible to facilitate the regulation of transcription, DNA repair, recombination, and replication.

Chromosomal satellite A small mass of chromatin containing genes for ribosomal RNA, at the end of the short arm of each chromatid of an acrocentric chromosome; not to be confused with *satellite DNA*.

Chromosome One of the threadlike structures in the cell nucleus; consists of chromatin. Each contains a single molecule of DNA in the *interphase* nucleus.

Chromosome arm The portion of a chromosome from the centromere to the telomere. Each chromosome has two arms of varying sizes. See *p* and *q*.

Chromosome disorder A clinical condition caused by an abnormal chromosome constitution in which there is duplication, loss, or rearrangement of chromosomal material.

Chromosome instability syndrome Hereditary condition that predisposes to a high frequency of chromosome breakage and rearrangements. Often associated with markedly increased risk for a variety of cancers.

Chromosome mutation Mutation that leaves a chromosome intact but changes the number of chromosomes in a cell.

Chromosome segregation The separation of chromosomes or chromatids in cell division so that each daughter cell gets an equal number of chromosomes.

Chromosome shattering Phenomenon seen in some cancer cells in which novel and complex chromosome rearrangements occur because the chromosomes break into numerous pieces and rejoin. The mechanism is unknown.

Chromosome spread The chromosomes of a dividing cell as seen under the microscope in metaphase or prometaphase.

Cis Refers to the relationship between two sequences that are on the same chromosome, literally meaning "on the near side of." Contrast with *trans*.

Clinical heterogeneity The term describing the occurrence of clinically different phenotypes from mutations in the same gene.

Clinical utility In reference to a clinical laboratory test, the ability of that test to improve the medical care that an individual receives.

Clinical validity In reference to a clinical laboratory test, the ability of that test to detect the disease that the test was designed to detect.

Clonal evolution The multistep process of successive genetic changes that occur in a developing tumor cell population.

Clone 1. A cell line derived by mitosis from a single ancestral diploid cell; in embryology, a cell lineage in which the cells have remained geographically close to each other. 2. In molecular biology, a recombinant DNA molecule containing a gene or other DNA sequence of interest. Also, the act of generating such a cell line or clone.

CNP See *copy number variant*.

CNV See *copy number variant*.

Coding strand In double-stranded DNA, the strand that has the same 5′ to 3′ sense (and sequence, except that in mRNA, U substitutes for T) as mRNA. The coding strand is the strand that is *not* transcribed by RNA polymerase. Also called the *sense strand*.

Codominant If both alleles of a pair are expressed in the heterozygous state, then the alleles (or the traits determined by them, or both) are codominant.

Codon A triplet of three bases in a DNA or RNA molecule, specifying a single amino acid.

Cohort study A random sample of the entire population is analyzed for whether the subjects currently have or, after being followed over time, develop a particular disease.

Colinearity The parallel relationship between the base sequence of the DNA of a gene (or the RNA transcribed from it) and the amino acid sequence of the corresponding polypeptide.

Commitment The transition of an embryonic cell from pluripotency to its particular fate.

Comparative genome hybridization (CGH) A fluorescence hybridization technique used to compare two different DNA samples with respect to their relative content of a particular DNA segment or segments. CGH can be used with fluorescence in situ hybridization (FISH) of metaphase chromosomes or with hybridization to large numbers of DNA fragments fixed to a solid support (*CGH array*).

Complementarity The complementary nature of base pairing in DNA.

Complementary DNA (cDNA) DNA synthesized from a messenger RNA template, through the action of the enzyme reverse transcriptase. See *genomic DNA* for comparison.

Complex inheritance A pattern of inheritance that is not mendelian. A trait with complex inheritance usually results from alleles at more than one locus interacting with environmental factors.

Compound (compound heterozygote) An individual, or a genotype, with two different mutant alleles at the same locus. Not to be confused with *homozygote*, in which the two mutant alleles are identical.

Concordance Describes a pair of relatives in which (1) both members of the pair have a certain qualitative trait or (2) both members have values of a quantitative trait that are similar in magnitude. See *discordance*.

Conditional probability 1. In Bayesian analysis, this is the chance of an observed outcome given that a consultand has a particular genotype. The product of the prior and conditional probabilities is the joint probability. 2. More generally, a synonym for *Bayesian analysis*.

Confined placental mosaicism Mosaicism in a *chorionic villus sampling* (CVS) specimen obtained from the placenta that is not present in the fetus itself.

Congenital Present at birth; not necessarily genetic.

Consanguinity Related by descent from a common ancestor (the adjective is *consanguineous*).

Consensus sequence In genes or proteins, an idealized sequence in which each base or amino acid residue represents the one most frequently found at that position when many actual sequences are compared; for example, the consensus sequence for splice donor or acceptor sites.

Consultand In genetic counseling, anyone who consults a genetic counselor for genetic information.

Contiguous gene syndrome A syndrome resulting from a *microdeletion* of chromosomal DNA extending over two or more contiguous loci. Also called *segmental aneusomy*.

Copy number variant (CNV) A variation in DNA sequence defined by the presence or absence of a segment of DNA, ranging from approximately 200 bp to approximately 2 Mb. Copy number variants may also have alleles that are tandem duplications of two, three, four, or more copies of a DNA segment. If a variant has an allele frequency greater than 1%, it is referred to as a *copy number polymorphism* (CNP).

Cordocentesis A procedure used in prenatal diagnosis to obtain a sample of fetal blood directly from the placenta.

Correlation A statistical tool applied to a set of paired measurements. A positive correlation exists when the larger the first measurement in the pair is, the larger the second measurement of the pair is. A negative correlation is the opposite, that is, the larger the first measurement, the smaller the second. Measured by the correlation coefficient *r*.

Coupling Describes the phase of two alleles at two different but syntenic loci, in which one allele at one of the loci *is* on the same chromosome as the allele at the second locus. See *phase* and *repulsion*.

CpG island Segments of genomic DNA that are particularly rich in the dinucleotide 5′-CG-3′ and are found in the promoters of many housekeeping genes. The "p" in CpG refers to the phosphate of the DNA backbone connecting the cytidine and guanidine nucleosides.

Crossover, crossing over The reciprocal exchange of segments between chromatids of homologous chromosomes, a characteristic of prophase of the first meiotic division. See also *recombination*. Unequal crossing over between misaligned chromatids can lead to duplication of the involved segment on one chromatid and deletion on the other and is a frequent cause of mutation.

Cryptic splice site A DNA sequence similar to the consensus splice site but not normally used. It is used when the normal splice site is altered by mutation or when a mutation in the cryptic site increases its use by the splicing apparatus. May be in a coding or a noncoding sequence.

Cytogenetics The study of chromosomes.

Cytokinesis Cleavage of cytoplasm at the end of mitosis, resulting in two separate cells, each with a full complement of 46 chromosomes.

Cytotrophoblast The fetal cells of the chorionic villi that are sampled for karyotyping and DNA analysis.

Daughter chromosomes The two individual chromosomes formed when a single chromosome composed of paired chromatids separates at the centromere in anaphase of cell division.

Deformation syndrome A recognizable pattern of dysmorphic features caused by extrinsic factors that affect the fetus in utero.

Degeneracy of the code The genetic code is described as degenerate because most of the 20 amino acids are specified by more than 1 of the 64 codons.

Degree of relationship The distance between two individuals in a pedigree. First-degree relatives include parents, siblings, and children. Second-degree relatives are aunts and uncles, nephews and nieces, grandparents and grandchildren.

Deletion The loss of a sequence of DNA from a chromosome. The deleted DNA may be of any length, from a single base to a large part of chromosome. A chromosome deletion may occur at the end of a chromosome (terminal deletion) or within a chromosome arm (interstitial).

Deoxyribonucleic acid See *DNA*.

Determination During development, the second stage of commitment in which a cell follows its developmental program regardless of whether it is transplanted to a different region of the embryo.

Developmental disorder Disorder resulting from disruption of the normal developmental program. Usually prenatal in onset but can first present postnatally.

Developmental program The process by which a cell in an embryo achieves its fate.

Deviation (D) Extent to which haplotype frequencies diverge from the expected based on the allele frequencies. A measure of linkage disequilibrium, usually normalized to allele frequencies using the D' metric.

Dicentric A structurally abnormal chromosome with two centromeres. If a dicentric chromosome segregates as if it has only one centromere, it is referred to as *pseudodicentric*.

Dictyotene The stage of the first meiotic division in which a human oocyte remains from late fetal life until ovulation.

Differentiation The process whereby a cell acquires a tissue-specific pattern of expression of genes and proteins and a characteristic phenotype.

Diploid The number of chromosomes in most somatic cells, which is double the number found in the gametes. In humans, the diploid chromosome number is 46.

Discordance The situation in which (1) one member of the pair has a certain qualitative trait and the other does not or (2) the relatives have values of a quantitative trait that are at opposite ends of the distribution. See *concordance*.

Disjunction See *nondisjunction*.

Disomy See *uniparental disomy*.

Disorder of sexual development (DSD) A phenotype reflecting a mismatch between chromosomal sex and phenotypic sex.

Disruption A birth defect caused by destruction of tissue; may be caused by vascular occlusion, a teratogen, or rupture of the amniotic sac with entrapment.

Dizygotic (DZ) twins Twins produced by two separate ova, separately fertilized. Also called *fraternal twins*.

DNA (deoxyribonucleic acid) The molecule that encodes the genes responsible for the structure and function of living organisms and allows the transmission of genetic information from generation to generation.

DNA fingerprint A set of genotypes at a sufficient number of loci (e.g., at 13 STRP loci in the Combined DNA Index System [CODIS] used by the Federal Bureau of Investigation [FBI] in the United States) to unambiguously and uniquely specify the individual from which the DNA was obtained (except for monozygotic twins).

DNA methylation In eukaryotes, the addition of a methyl residue to the 5-position of the pyrimidine ring of a cytosine base in DNA to form 5-methylcytosine.

DNA polymerase An enzyme that can synthesize a new DNA strand by use of a previously synthesized DNA strand as a template.

DNA proofreading Recognition and removal of a noncomplementary DNA base inserted during replication, followed by its replacement by the correct complementary base.

Dominant A trait is dominant if it is phenotypically expressed in heterozygotes. If heterozygotes and homozygotes for the variant allele have the same phenotype, the disorder is a *pure* dominant (rare in human genetics). If homozygotes have a more severe phenotype than do heterozygotes, the disorder is termed *semidominant* or *incompletely dominant*.

Dominant negative A disease-causing allele, or the effect of such an allele, that disrupts the function of a wild-type allele in the same cell.

Donor splice site The boundary between the 3′ end of an exon and the 5′ end of the next intron. Also called *5′ splice site*.

Dosage compensation As a consequence of X inactivation, the amount of product formed by the two copies of an X-linked gene in females is equivalent to the amount formed by the single gene in males. See *X inactivation*.

Double heterozygote An individual who is heterozygous at each of two different loci. Contrast with *compound heterozygote*.

Double minutes Very small accessory chromosomes, a form of gene amplification.

Driver gene A gene that has been found repeatedly to carry somatic mutations in many samples of the same type of cancer or even in multiple different types of cancers. The mutations are far too frequent to be simply the product of random events. These genes are thus presumed to be involved in the development or progression of the cancer itself. See *passenger gene mutation*.

Dynamic mutation Mutations caused by amplification of a simple nucleotide repeat sequence. They tend to increase in size from one generation to the next, thus the term *dynamic*. Most commonly, the nucleotide unit involved in the expansion contains three nucleotides (*triplet repeat expansion*), as with CAG in Huntington disease or CGG in fragile X syndrome.

Dysmorphic features Morphological developmental abnormalities, as seen in many syndromes of genetic or environmental origin.

Ecogenetic disorder A disorder resulting from the interaction of a genetic predisposition to a specific disease with an environmental factor.

Ectoderm One of the three primary *germ layers* of the early embryo. It begins as the layer farthest from the yolk sac and ultimately gives rise to the nervous system, the skin, and neural crest derivatives such as craniofacial structures and melanocytes.

Ectopic expression Expression of a gene in places where it is not normally expressed.

Embryonic stem cell A cell derived from the inner cell mass that is self-renewing in culture and, when reintroduced into the inner cell mass of a blastocyst, can repopulate all the tissues of the embryo.

Empirical risk In human genetics, the probability that a familial trait will occur in a family member, based on observed numbers of affected and unaffected individuals in family studies rather than on knowledge of the causative mechanism.

ENCODE Project Acronym for *Encyclopedia of DNA Elements*, a large-scale effort to identify and map all of the *regulatory elements* and *epigenetic* regulators of gene expression in a wide spectrum of cell types and tissues.

Endoderm One of the three primary *germ layers* of the early embryo. Ultimately gives rise to the gut, liver, and portions of the urogenital system.

Endophenotype A heritable quantitative biological trait that is a marker of risk for a genetically complex disorder. The concept is commonly used in psychiatric genetics but is used widely in genetic epidemiology.

Enhancer A DNA sequence that acts in *cis* (i.e., on the same chromosome) to increase transcription of a gene. The enhancer may be upstream or downstream to the gene and may be in the same or the reverse orientation. Contrast with *silencer*.

Enzymopathy A metabolic disorder resulting from deficiency or abnormality of a specific enzyme.

Epigenetic The term that refers to any factor that can affect gene function without change in the genotype. Some typical epigenetic factors involve alterations in DNA methylation, chromatin structure, histone modifications, and transcription factor binding that change genome structure and affect gene expression without changing the primary DNA sequence.

Episome A DNA element that either can exist as an autonomously replicating sequence in the cytoplasm or can integrate into chromosomal DNA. Adeno-associated viral vectors, used in gene therapy, are episomes that exist in the cytoplasm for long periods and can, although rarely, be inserted into the nuclear genome.

Euchromatin The major component of chromatin. It stains lightly with G banding, decondensing and becoming light-staining during interphase. Contrast with *heterochromatin*.

Eugenics Refers to increasing the prevalence of desirable traits in a population by decreasing the frequency of deleterious alleles at relevant loci through controlled, selective breeding. The opposite is *dysgenics*.

Eukaryote A unicellular or multicellular organism in which the cells have a nucleus with a nuclear membrane and other specialized characteristics.

Euploid Any chromosome number that is an exact multiple of the number in a haploid gamete (n). Most somatic cells are diploid (2n). Contrast with *aneuploid*.

Exon A transcribed region of a gene that is present in mature messenger RNA.

Exon skipping The use of molecular interventions to exclude an exon from a pre-mRNA that encodes a reading frame–disrupting mutation, thereby rescuing expression of the mutant gene.

Expression profile A quantitative assessment of the mRNAs present in a cell type, tissue, or tumor. Often used to characterize a cell, tissue, or tumor in comparison to the expression profile of another cell, tissue, or tumor.

Expressivity The extent to which a genetic defect is expressed. If there is variable expressivity, the trait may vary in expression from mild to severe but is never completely unexpressed in individuals who have the corresponding genotype. Contrast with *penetrance*.

Familial Any trait that is more common in relatives of an affected individual than in the general population, whether the cause is genetic, environmental, or both.

Fate The ultimate destination for a cell that has traveled down a developmental pathway. The embryonic *fate map* is a complete description of all the fates of all the different parts of the embryo.

Fetal cells Placental cells, obtained by *chorionic villus sampling*; skin, respiratory, and urinary tract cells obtained from amniotic fluid by *amniocentesis*; or fetal blood cells obtained by *cordocentesis*.

Fetal phase Stage of intrauterine development from weeks 9 to 40.

Fibroblast Normal cells derived from subcutaneous tissue that can be easily obtained from a skin biopsy specimen and grown for many generations in culture.

FISH Fluorescence in situ hybridization. See *in situ hybridization*.

Fitness (*f*) The probability of transmitting one's genes to the next generation compared with the average probability for the population.

Flanking sequence A region of DNA preceding or following a transcribed region or, more generally, preceding or following any segment of DNA or chromosome.

Founder effect A high frequency of a mutant allele in a population founded by a small ancestral group when one or more of the founders was a carrier of the mutant allele.

Fragile site Nonstaining gap in the chromatin of a metaphase chromosome, such as the fragile site at Xq27 in fragile X syndrome.

Frameshift mutation A mutation involving a deletion or insertion that is not an exact multiple of three base pairs and thus changes the reading frame of the gene downstream of the mutation.

G banding (Giemsa staining) Method of staining chromosomes to generate reproducible alternating light and dark bands.

Gain-of-function mutation A mutation associated with an increase in one or more of the normal functions of a protein. To be distinguished from *novel property mutation*.

Gamete A mature reproductive cell (ovum or sperm) with the haploid chromosome number.

Gene A hereditary unit; in molecular terms, a sequence of chromosomal DNA that is required for the production of a functional product.

Gene dosage The number of copies of a particular gene in the genome.

Gene family A set of genes containing similarly sized exons and containing highly similar DNA sequences, indicating that the genes have evolved from an ancestral gene by duplication and subsequent divergence.

Gene flow Gradual diffusion of genes from one population to another across a barrier. The barrier may be physical or cultural and may be breached by migration or mixing.

Gene map The characteristic arrangement of the genes on the chromosomes.

Gene mutation Alteration of the sequence of DNA involving as little as one nucleotide up to an arbitrarily set limit of 100 kilobase pairs.

Gene pool All the alleles present at a given locus or, more broadly, at all loci in the population.

Gene transfer therapy (gene therapy) Treatment of a disease by introduction of DNA sequences that will have a therapeutic benefit.

Genetic Determined by genes; not to be confused with *congenital*.

Genetic admixture The merging into the gene pool of an immigrant population with allele frequencies different from the already existing population, which, if there is random mating, results in new allele frequencies reflecting the mixing of the two populations.

Genetic code The 64 triplets of bases that specify the 20 amino acids found in proteins (see Table 3-1).

Genetic counseling The provision of information and assistance to affected individuals or family members at risk for a disorder that may be genetic, concerning the consequences of the disorder, the probability of developing or transmitting it, and the ways in which it may be prevented or ameliorated.

Genetic disorder A defect wholly or partly caused by a gene abnormality.

Genetic drift Random fluctuation of allele frequencies in small populations.

Genetic epidemiology A branch of public health research concerned with characterizing and quantifying the influence of genetic variation in the population on the incidence, prevalence, and causation of disease.

Genetic heterogeneity The production of the same or similar phenotypes by different genetic mechanisms. See *allelic heterogeneity, clinical heterogeneity, locus heterogeneity*.

Genetic lethal A mutant allele or genetically determined trait that leads to failure to reproduce, although not necessarily to death before reproduction.

Genetic load The sum total of death and disease caused by mutant genes.

Genetic map The relative positions of the genes on the chromosomes, as shown by linkage analysis.

Genetic marker A locus that has readily classifiable alleles and can be used in genetic studies. It may be a gene variant or a single nucleotide polymorphism (SNP) or short tandem repeat polymorphism (STRP) or any characteristic of DNA that allows different versions of a locus (or its product) to be distinguished one from another and followed through families. See *polymorphism.*

Genetic screening Testing of family members of an affected proband or of the population as a whole to identify individuals at risk for developing or transmitting a specific disorder.

Genocopy A genotype that determines a phenotype closely similar to that determined by a different genotype.

Genome The complete DNA sequence, containing the entire genetic information, of a gamete, an individual, a population, or a species.

Genome editing A new technology that uses proteins adapted from bacteria or plants (e.g., CRISPR/Cas9) to target a particular site within the genome of a cell with high efficiency and specificity. Once the site is targeted, it can be mutated, undergo repair of a preexisting mutation, or undergo alteration of an epigenetic imprint.

Genome-wide association study (GWAS) A genetic association study using hundreds of thousands to millions of polymorphic variants distributed throughout the genome.

Genomic DNA The chromosomal DNA sequence of a gene or segment of a gene, including the DNA sequence of noncoding as well as coding regions. Also, DNA that has been isolated directly from cells or chromosomes or the cloned copies of all or part of such DNA.

Genomic imprinting Monoallelic expression in which the choice of the allele to be expressed is determined by the parental origin of each allele. See *Prader-Willi syndrome* and *Angelman syndrome* in the text for examples.

Genomic medicine The practice of medicine based on large-scale genomic information, such as sequencing of large gene panels, exomes, or whole genomes; expression profiling to characterize tumors or to define prognosis in cancer; genotyping of variants in genes involved in drug metabolism or action to determine an individual's correct therapeutic dosage; or analysis of multiple protein biomarkers to monitor therapy or to provide predictive information in presymptomatic individuals.

Genomics The field of genetics concerned with structural and functional studies of the genome.

Genotype 1. The genetic constitution of an individual, as distinguished from the phenotype. 2. More specifically, the alleles present at one or more loci.

Germ layer One of three distinct layers of cells that arise in the inner cell mass, the *ectoderm, mesoderm,* and *endoderm,* each of which develops into a distinctly different tissue in the embryo.

Germline The cell line from which gametes are derived.

Germline mosaicism In an individual, the presence of two or more genetically different germline cells, resulting from mutation during the proliferation and differentiation of the germline.

Gonadal dysgenesis Disorder in which chromosomal and phenotypic sex do not match and normal gonads fail to develop. In complete gonadal dysgenesis, external genitalia are normal, whereas in incomplete gonadal dysgenesis, the external genitalia are ambiguous. Mixed gonadal dysgenesis has a range of phenotypes from male to Turner syndrome–like, resulting from 45,X/46,XY mosaicism.

Haploid The chromosome number of a normal gamete, with only one member of each chromosome pair. In humans, the haploid number is 23.

Haploinsufficiency A cause of genetic disease in which the contribution from a normal allele is insufficient to prevent disease because of a loss-of-function mutation at the other allele.

Haplotype A group of alleles in coupling at closely linked loci, usually inherited as a unit.

Hardy-Weinberg law The law that relates allele frequency to genotype frequency, used in population genetics to determine allele frequency and heterozygote frequency when the incidence of a disorder is known.

Hemizygous A term for the genotype of an individual with only one representative of a chromosome or chromosome segment, rather than the usual two; refers especially to X-linked genes in the male but also applies to genes on any chromosome segment that is deleted on the homologous chromosome.

Hemoglobin switching Change in expression of the various globin genes during development of hematopoiesis.

Heritability (h^2) The fraction of total phenotypic variance of a quantitative trait that is due to genotypic differences. May be viewed as a statistical estimate of the hereditary contribution to a quantitative trait.

Heterochromatin Chromatin that stains darkly throughout the cell cycle, even in interphase. Generally thought to be late replicating and genetically inactive. Satellite DNA in regions such as centromeres, acrocentric short arms, and 1qh, 9qh, 16qh, and Yqh constitute *constitutive heterochromatin,* whereas the chromatin of the inactive X chromosome is referred to as *facultative heterochromatin.* Contrast with *euchromatin.*

Heterodisomy See *uniparental disomy.*

Heterogeneity See *allelic heterogeneity, clinical heterogeneity, genetic heterogeneity, locus heterogeneity.*

Heteroplasmy The presence of more than one type of mitochondrial DNA in the mitochondria of a single individual. Contrast with *homoplasmy.*

Heteroploid Any chromosome number other than the normal.

Heterozygote (heterozygous) An individual or genotype with two different alleles, one of which is wild-type, at a given locus on a pair of homologous chromosomes. See *compound heterozygote.*

Heterozygote advantage Situation in which heterozygotes for some mendelian diseases have *increased* fitness not only over homozygotes for the mutant allele but even over homozygotes for the normal allele. See *balanced polymorphism.*

Histocompatibility A host will accept a particular graft only if it is histocompatible—that is, if the graft contains no antigens that the host lacks.

Histones Proteins associated with DNA in the chromosomes that are rich in basic amino acids (lysine or arginine) and virtually invariant throughout eukaryote evolution. Covalent modifications of histones are important *epigenetic* regulators of gene expression. The pattern of histones and their modifications constitute the epigenetic "histone code."

Holoenzyme The functional compound formed by the binding of an apoenzyme and its appropriate coenzyme.

Homeobox gene A gene that contains a conserved 180–base pair sequence termed a *homeobox* in its coding region, encoding a protein motif known as the *homeodomain.* The 60 amino acid residues of the homeodomain are a DNA-binding motif, which is consistent with the role of homeodomain proteins in transcriptional regulation of genes involved in development.

Homogeneously staining regions (HSRs) Chromosome regions that stain uniformly and represent amplified copies of a DNA segment.

Homology A commonly used term in genetics but with different meanings in different contexts. 1. In bioinformatics, homologous sequences are DNA or protein sequences that have similar nucleotide or amino acid sequences, as seen between *orthologous* or *paralogous* genes. 2. In cytogenetics, homologous chromosomes are a pair of chromosomes, one inherited paternally, the other maternally, that are generally of similar size and shape when they are viewed under the microscope and contain the same loci, except for the two sex chromosomes in males (X and Y), which are only partially homologous. Homologous chromosomes pair with each other during meiosis I, undergo crossing over, and separate at anaphase I of meiosis. 3. In evolution, structures in different organisms are termed *homologous* if they evolved from a structure present in a common ancestor.

Homoplasmy The presence of only one type of mitochondrial DNA in the mitochondria of a single individual. Contrast with *heteroplasmy.*

Homozygote (homozygous) An individual or genotype with identical alleles at a given locus on a pair of homologous chromosomes.

Housekeeping genes Genes expressed in most or all cells because their products provide basic functions.

Housekeeping proteins Proteins expressed in virtually every cell that have fundamental roles in the maintenance of cell structure and function (versus *specialty proteins*).

Human Genome Project A major research project, international in scope, that took place in 1990 to 2003 and resulted in the sequencing and assembly of a representative human genome and the genomes of many model organisms.

Human leukocyte antigen (HLA) See *major histocompatibility complex.*

Hybridization In molecular biology, the bonding of two complementary single-stranded nucleic acid molecules according to the rules of base pairing. See *comparative genome hybridization* and *fluorescence in situ hybridization.*

Hydatidiform mole An abnormality of the placenta in which it grows to resemble a hydatid cyst or bunch of grapes, associated with very abnormal fetal development. In a *complete mole,* the karyotype is usually 46,XX, but can be 46,XY, representing duplication of the chromosomes of the sperm with no maternal contribution. A *partial mole* is triploid, usually with an extra paternal chromosome set.

Immunoglobulin gene superfamily A family of evolutionarily related genes composed of human leukocyte antigen (HLA) class I and class II genes, immunoglobulin genes, T-cell receptor genes, and other genes encoding cell surface molecules.

Imprinting See *genomic imprinting.*

Imprinting center Segment of DNA located in and around imprinted genes that regulates genomic imprinting and causes the imprinting of genes on a chromosome inherited from a parent of one sex to switch to the appropriate imprint in the germline of a child of the opposite sex.

In situ hybridization Mapping a gene or segment of DNA by molecular hybridization to a chromosome spread or cell nucleus on a slide by use of a labeled DNA sequence as a probe corresponding to the gene or DNA segment to be mapped. Usually involves fluorescently labeled probes, in which case it is referred to as *fluorescence in situ hybridization (FISH).*

In vitro fertilization A reproductive technology in which sperm are allowed to fertilize an egg in tissue culture and the fertilized eggs are then introduced back into the uterus to allow implantation.

Inborn error of metabolism A genetically determined biochemical disorder in which a specific protein defect disturbs a metabolic pathway that may have pathological consequences.

Inbreeding The mating of closely related individuals. The progeny of close relatives are said to be *inbred.* (Note that some consider the term *inbreeding* to be pejorative when it is applied to human populations.)

Inbreeding, coefficient of (F) The probability that a child of a consanguineous mating will be homozygous at any given locus for an allele inherited through each parent from a common ancestor.

Incompletely dominant A trait that is inherited in a dominant manner but is more severe in a homozygote than in a heterozygote (synonym: *semidominant*).

Indel A polymorphism defined by the presence or absence of a segment of DNA, ranging from one base to a few hundred base pairs. Includes simple indels, microsatellites, and minisatellite polymorphisms. Sometimes abbreviated *in/del*.

Index case The family member affected with a genetic disorder who is the first to draw attention to a pedigree. See *proband*.

Induced pluripotent stem cells (iPS cells) *Pluripotent stem cells* derived not from embryonic cells but from differentiated adult somatic cells that have been induced to lose their differentiated state and revert to pluripotency by artificially expressing a small number of specific transcription factors in these cells.

Induction The determination of the fate of one region of an embryo by extracellular signals from a second, usually neighboring, region.

In-frame deletion A deletion that does not destroy the normal reading frame of the gene.

Initiator codon The codon for methionine (AUG) that signals the start of translation in a mRNA. See *messenger RNA*.

Inner cell mass A small group of cells within the pre-implantation mammalian embryo that will become the primitive ectoderm (or epiblast) after implantation and, ultimately, give rise to the embryo proper and not the placenta.

Insertion A chromosomal abnormality in which a DNA segment from one chromosome, or from an exogenous source such as a retrovirus, is inserted into another chromosome.

Intergenic DNA The mostly untranscribed DNA of unknown function that makes up a large proportion of the total DNA in the genome.

Interphase The stage of the cell cycle between two successive mitoses.

Intervening sequence See *intron*.

Intron A segment of a gene that is initially transcribed but then removed from within the primary RNA transcript by splicing together the sequences (exons) on either side of it.

Inversion A chromosomal rearrangement in which a segment of a chromosome is reversed end to end. If the centromere is included in the inversion, the inversion is *pericentric*; if not, it is *paracentric*.

Isochromosome An abnormal chromosome in which one arm is duplicated (forming two arms of equal length, with the same loci in reverse sequence) and the other arm is missing.

Isodisomy See *uniparental disomy*.

Isolate A subpopulation in which matings take place exclusively or usually with other members of the same subpopulation.

Isolated case An individual who is the only member of his or her kindred affected by a genetic disorder, either by chance or by new mutation. See also *sporadic*.

Karyotype The chromosome constitution of an individual. The term is also used for a photomicrograph of the chromosomes of an individual systematically arranged as well as to describe the process of preparing such a photomicrograph.

kb (kilobase or kilobase pair) A unit of 1000 bases in a DNA or RNA sequence.

Kindred An extended family.

Kinetochore A structure at the centromere to which the spindle fibers are attached.

LINE sequences A class of repetitive DNA made up of long interspersed DNA segments, up to 6 kb in length, occurring in several hundred thousand copies in the genome (also called *L1 family*).

Lineage The progeny of a cell, generally determined by experimentally labeling the cell so that all of its descendants can be identified. See *clone*.

Linkage Genes on the same chromosome are *linked* if they are transmitted together in meiosis more frequently than chance would allow. Compare with *synteny*.

Linkage analysis A statistical method in which the genotypes and phenotypes of parents and offspring in families are studied to determine whether two or more loci are assorting independently or exhibiting linkage during meiosis.

Linkage disequilibrium (LD) The occurrence of specific combinations of alleles in coupling phase at two or more linked loci (haplotypes) more frequently than expected by chance from the frequency of the alleles in the population. Opposite is linkage equilibrium.

Linkage disequilibrium block (LD block) A set of polymorphic markers forming a haplotype whose alleles are in strong linkage disequilibrium with each other. Usually occupies a region of the genome from a few kilobases to a few dozen kilobases in length.

Linkage map A chromosome map showing the relative positions of genes and other DNA markers on the chromosomes, as determined by linkage analysis.

lncRNA See *noncoding RNA*.

Locus The position occupied by a gene on a chromosome. Different forms of the gene (*alleles*) may occupy the locus.

Locus control region (LCR) A DNA domain, situated outside a cluster of structural genes, responsible for the appropriate expression of the genes within the cluster.

Locus heterogeneity The production of identical phenotypes by mutations at two or more different loci.

LOD score A statistical method that tests genetic marker data in families to determine whether two loci are linked. The LOD score is the *logarithm* of the *odds* in favor of linkage. By convention, a LOD score of 3 (odds of 1000:1 in favor) is accepted as proof of linkage, and a LOD score of −2 (100:1 against) as proof that the loci are unlinked.

Loops Three-dimensional arrangement of chromatin, packaged as solenoids, attached to the chromosome scaffold. Thought to be a structural or functional unit of chromosomes.

Loss of heterozygosity (LOH) Loss of a normal allele from a region of one chromosome of a pair, allowing a defective allele on the homologous chromosome to be clinically manifest. A feature of many cases of retinoblastoma, breast cancer, and other tumors due to mutation in a tumor-suppressor gene.

Loss-of-function mutation A mutation associated with a reduction or a complete loss of one or more of the normal functions of a protein.

Lymphoblastoid cells B-lymphocytes immortalized in culture by infection with Epstein-Barr virus.

Lyon's law (or hypothesis) Basic features of the phenomenon of X inactivation, which was first described by the late British geneticist Mary Lyon. Originally called the Lyon hypothesis, but upgraded to a law at the 50th anniversary of her discovery. Silencing of gene expression is sometimes referred to as *lyonization*. See *X inactivation*.

Major histocompatibility complex (MHC) The complex locus on chromosome 6p that includes the highly polymorphic human leukocyte antigen (HLA) genes.

Maleficence Behavior that harms others. Avoidance of maleficence is one of the cardinal principles of ethics. See *beneficence*.

Male to male transmission A pattern of inheritance of a trait from a father to all of his sons and none of his daughters (also referred to as *holandric inheritance*).

Malformation syndrome A recognizable pattern of dysmorphic features having a single cause, either genetic or environmental.

Manhattan plot A graph of all of the *P* values for an association between a trait and all of the single nucleotide polymorphism (SNPs) used in a genome-wide association study (GWAS) study. The SNPs are placed on the x-axis based on their location in the genome, starting with the tip of chromosome 1p on the left and going through each arm of the 22 autosomes. The *P* values are given as $-\log_{10}$ on the y-axis so that the more significant the association, the greater the value. It is called a *Manhattan plot* because the peaks of strong association resemble the tips of skyscrapers seen in the Manhattan skyline (see Fig. 10-11 for example).

Manifesting heterozygote A female heterozygous for an X-linked disorder in whom, because of nonrandom X inactivation, the trait is expressed clinically, although usually not to the same degree of severity as in hemizygous affected males.

Map distance A *theoretical* concept that is based on how often recombination between the loci is observed. Measured in units of centimorgans, defined as the genetic length over which recombination occurs in 1% of meioses.

Marker chromosome A small unidentified chromosome seen in a chromosome preparation. Also referred to as a *supernumerary chromosome* or *extra structurally abnormal chromosome*.

Maternal inheritance The transmission of genetic information only through the mother.

Maternal serum screening Laboratory test that relies on measurement of the levels of particular substances, such as alpha-fetoprotein, human chorionic gonadotropin, and unconjugated estriol, in a pregnant woman's blood to screen for fetuses affected with certain trisomies or with neural tube defects.

Mb (megabase or megabase pair) A unit of 1,000,000 bases or base pairs in genomic DNA.

Meiosis The type of cell division occurring in the germ cells, by which gametes containing the haploid chromosome number are produced from diploid cells. Two meiotic divisions occur: meiosis I and meiosis II. Reduction in chromosome number takes place during meiosis I.

Mendelian Patterns of inheritance that follow the classic laws of Mendel: autosomal dominant, autosomal recessive, and X-linked. See *single-gene disorder*.

Mesoderm The middle *germ layer* in the early embryo; the source of cells that go on to make bones, muscles, connective tissue, heart, hematopoietic system, kidney, and other organs.

Mesonephric duct Structure in the genital ridges of the early embryo that will develop into male internal sexual organs in the male. Also called the *Wolffian duct*.

Messenger RNA (mRNA) An RNA, transcribed from the DNA of a gene, that directs the sequence of amino acids of the encoded polypeptide. Contrast with *noncoding RNA*.

Metacentric A type of chromosome with a central centromere and arms of apparently equal length.

Metaphase The stage of mitosis or meiosis in which the chromosomes have reached their maximal condensation and are lined up on the equatorial plane of the cell, attached to the spindle fibers. This is the stage at which chromosomes are most easily examined. Immediately follows prometaphase.

Metastasis Spread of malignant cells to other sites in the body.

Methemoglobin The oxidized form of hemoglobin, containing iron in the ferric rather than the ferrous state, which is incapable of binding oxygen.

Microarray Miniaturized wafer ("chip") made of glass, plastic, or silicon onto which a large number of different nucleic acids have been individually spotted. See also *comparative genome hybridization, expression profile.*

Microdeletion A chromosomal deletion that is too small to be seen under the microscope. See also *contiguous gene syndrome.*

MicroRNAs (miRNAs) A class of several thousand small single-stranded RNAs of only approximately 22 bases, constituting one of the most abundant classes of gene regulatory molecules. They suppress gene expression post-transcriptionally, by targeting specific mRNAs for cleavage or by suppressing mRNA translation.

Microsatellite See *short tandem repeat polymorphism (STRP).*

Microsatellite instability (MSI) A phenotype of cancer cells in which loss of function of mismatch repair genes causes errors such as slipped mispairing to go unrepaired when microsatellite sequences are replicated. These errors lead to tissue mosaicism so that the cancer appears to contain more than two alleles at many short tandem repeat polymorphic loci.

Minisatellite See *VNTR.*

Missense mutation A mutation that changes a codon specific for one amino acid to specify another amino acid.

Mitochondrial bottleneck A step in oogenesis in which only a small sample of the total number of mitochondria in an oocyte precursor is passed on to daughter cells, thereby allowing significant variation in the proportions of mutant and wild-type mitochondria inherited by the daughter cells.

Mitochondrial DNA (mtDNA) The DNA in the circular chromosome of the mitochondria. Mitochondrial DNA is present in many copies per cell, is maternally inherited, and evolves 5 to 10 times as rapidly as genomic DNA.

Mitochondrial inheritance The inheritance of a trait encoded in the mitochondrial genome. Because the mitochondrial genome is strictly maternally inherited, mitochondrial inheritance occurs solely through the female line.

Mitosis The process of ordinary cell division, resulting in the formation of two cells genetically identical to the parent cell.

Mitotic spindle Microtubular structure inside a mitotic cell to which centromeres attach. Guides the separation of sister chromatids to opposite poles during anaphase of mitosis.

Modifier gene A gene whose alleles alter the phenotype associated with mutations in a nonallelic gene. Often applied to the effect on the expressivity of mendelian disorders caused by variants at other loci.

Monosomy A chromosome constitution in which one member of a chromosome pair is missing, as in 45,X Turner syndrome.

Monozygotic (MZ) twins Twins derived from a single zygote and thus genetically identical. Also termed *identical twins.*

Morphogen A substance produced during development in a localized region of the organism that diffuses out to form a concentration gradient and directs cells into two or more specific developmental pathways, depending on its concentration.

Morphogenesis The process whereby changes in cell shape, adhesion, movement, and number lead to a three-dimensional structure.

Mosaic development Embryological development in which different regions of the embryo develop independently from surrounding regions. See *regulative development.*

Mosaicism Describes an individual or tissue with at least two cell lines, differing in genotype or karyotype, that were derived from a single zygote; not to be confused with *chimera.*

Multifactorial disease Disorders resulting from a combination of multiple factors, genetic and environmental. Demonstrate *complex inheritance* rather than mendelian inheritance patterns.

Multiple hypothesis testing A cause of false-positive statistically significant tests when one hypothesis among many being tested shows statistical significance by chance alone and not because the result is truly significant.

Multiplex ligation-dependent probe amplification A laboratory technique that allows the simultaneous measurement of the copy number of various segments of a gene through repeated rounds of ligation and polymerase chain reaction (PCR). Used to detect deletions and duplications within genes.

Mutagen An agent that increases the spontaneous mutation rate by causing changes in DNA.

Mutant A gene that has been altered by mutation; also used to refer to a nonhuman organism carrying a mutant gene.

Mutation Any permanent heritable change in the sequence of genomic DNA. See *variant.*

Mutation rate (μ) The frequency of mutation at a given locus, expressed as mutations per locus per gamete (or per generation, which is the same).

Negative predictive value With respect to a clinical test for a disease, the extent to which testing negative indicates that one does not have or will not develop the disease.

Neoplasia An abnormal growth produced by imbalance between normal cellular proliferation and normal cellular attrition. May be benign or malignant (cancer).

Noncoding gene See *noncoding RNA.*

Noncoding RNA (ncRNA) After transcription, an RNA product that will not be translated to a protein product. Contrast with *messenger RNA*. To avoid confusion with short ncRNAs such as *miRNAs* or *siRNAs*, sometimes referred to as *long noncoding RNAs* or *lncRNAs*. For example, see XIST RNA, under *X inactivation*.

Noncoding strand See *antisense strand of DNA*.

Nondisjunction The failure of two members of a chromosome pair to disjoin during meiosis I, or of two chromatids of a chromosome to disjoin during meiosis II or mitosis, so that both pass to one daughter cell and the other daughter cell receives neither. Also called *chromosome missegregation*.

Noninvasive prenatal screening (NIPS) Use of cell-free DNA of fetal origin in maternal blood to screen for fetal aneuploidy.

Nonsense-mediated mRNA decay A mechanism for quality control of mRNAs that recognizes and degrades mRNAs carrying mutant premature translation termination (nonsense) codons, thereby preventing the translation of truncated proteins.

Nonsense mutation A single-base substitution in DNA resulting in a *termination codon*.

Nonsynonymous Describes a single nucleotide variant (SNV) that alters a codon and therefore the resulting amino acid sequence of the encoded peptide.

Normal distribution The symmetrical, bell-shaped curve describing the frequency of particular values of a measured quantity in a population.

Novel property mutation A mutation that confers a new property on the protein.

Nuchal translucency An ultrasonographic finding of an echo-free space between the skin line and the soft tissue overlying the cervical spine in the subcutaneous tissue of the fetal neck. Associated with fetal aneuploidy.

Nucleoside Nitrogenous base plus a five-carbon sugar. Adenosine, cytosine, guanosine, and thymidine (in DNA) or uridine (in RNA).

Nucleosome The primary structural unit of chromatin, consisting of 146 base pairs of DNA wrapped twice around a core of eight histone molecules.

Nucleotide A molecule composed of a nitrogenous base and a five-carbon sugar (nucleoside) with a phosphate group attached to the 5' carbon of the sugar molecule. A nucleic acid is a polymer of many nucleotides.

Null allele An allele that results either in the total absence of the gene product or in the total loss of function of the product.

Obligate heterozygote An individual who may be clinically unaffected but on the basis of pedigree analysis must carry a specific mutant allele.

Odds A ratio of probabilities or risks. Often calculated as a ratio of the probability of an event's occurring versus the probability of the event's not occurring, as one way to assess the relative chance of the event. Odds can vary in value from 0 to infinity.

Odds ratio A comparison of the odds that individuals who share a particular feature or factor (e.g., a genotype, an environmental exposure, or a drug) will have a disease or trait versus the odds for individuals who lack the factor.

	Affected	Unaffected	Total
Factor present	a	b	a + b
Factor absent	c	d	c + d
Total	a + c	b + d	a + b + c + d

Among individuals in whom the factor is present, the *odds* of being affected = (a/b). Among individuals in whom the factor is absent, the *odds* of being affected = (c/d), and the odds ratio = (a/b)/(c/d) = ad/bc. [Strictly speaking, this definition of odds ratio is a **disease** odds ratio. A more traditional odds ratio used in epidemiology is an **exposure** odds ratio, which is a comparison of the odds that individuals affected with a particular disease were exposed to a particular factor = (a/c) versus the odds that unaffected individuals were exposed = (b/d), giving an odds ratio of (a/c)/(b/d). Note that both formulations result in the same ratio = ad/bc. Using a disease odds ratio formulation makes it easier to show arithmetically that a disease odds ratio approximates the relative risk ratio when the disease is rare (c ≪ d and a ≪ b)]. See *relative risk*.

Oligonucleotide A short DNA molecule (usually 8 to 50 base pairs), synthesized for hybridization or for use in the polymerase chain reaction.

Oncogene A dominantly acting gene responsible for tumor development. When activated by mutation, overexpression, or amplification, oncogenes in somatic cells may lead to neoplastic transformation. Contrast with *proto-oncogene*, *driver gene*, and *tumor-suppressor gene*.

Oogonia Cells derived from primordial germ cells in females that develop into primary oocytes at the end of the third month of fetal life. Primary oocytes enter prophase of meiosis I and then stop and only complete meiosis and their differentiation into mature ova at the time of ovulation and fertilization.

Open reading frame The interval between the start and stop codons of a nucleotide sequence that encodes a protein.

Origin of replication One of the hundreds of thousands of sites along each chromosome at which DNA replication begins during the S phase of the cell cycle.

Orphan disease A disease is considered to be a rare or orphan disease if it affects less than 200,000 Americans or, in Europe, less than 1 in 2000 persons. The majority of genetic orphan diseases are monogenic.

Orthologous Refers to genes in different species that are similar in DNA sequence and also encode proteins that have the same function—at least at the biochemical level—in each species. Orthologous genes originate from the same gene in a common ancestor. Contrast with *paralogous*.

p 1. In cytogenetics, the short arm of a chromosome (from the French *petit*). 2. In population genetics, the frequency of the more common allele of a pair. 3. In biochemistry, abbreviation of *protein* (e.g., p53 is a 53-kD protein).

Pachytene Stage in meiosis I following synapsis when meiotic recombination takes place.

Paralogous Refers to two or more genes in a single species that are similar in DNA sequence and are likely to encode proteins with similar and perhaps overlapping but not identical functions. Paralogous genes are likely to have originated from a common ancestral gene. Examples are α- and β-globin genes.

Paramesonephric duct Structure in the genital ridges of the early embryo that will develop into female internal sexual organs in the female. Also called the *müllerian duct*.

Parental haplotype A haplotype in a gamete that is also present in the parent (i.e., no crossing over occurred in meiosis during gametogenesis to disrupt the haplotype). Also referred to as a *nonrecombinant haplotype*. The opposite is a nonparental or recombinant haplotype.

Parental transmission bias A phenomenon seen with the inheritance of unstable repeat expansion mutations in which expansions of the repeat occur preferentially when the mutation is transmitted by one parent versus the other.

Partial monosomy Subchromosomal mutation leading to loss of one copy of a segment of a chromosome.

Partial trisomy Subchromosomal mutation leading to gain of a third, extra copy of a segment of a chromosome.

Passenger gene mutation The vast majority of somatic mutations in cancers, which appear to be random, are not recurrent in particular cancer types, and probably occurred as the cancer developed, rather than directly causing the cancer to develop or progress. Compare with *driver gene*.

PAX gene family A family of transcription factors that share a DNA-binding motif originally described in the *Drosophila* "paired gene."

PCR See *polymerase chain reaction*.

Pedigree In medical genetics, a family history of a hereditary condition, or a diagram of a family history indicating the family members, their relationship to the proband, and their status with respect to a particular hereditary condition.

Penetrance The fraction of individuals with a genotype known to cause a disease who have any signs or symptoms of the disease. Contrast with *expressivity*.

Pharmacodynamics The effects of a drug or its metabolites on physiological function and metabolic pathways.

Pharmacogenetics The area of biochemical genetics concerned with the impact of genetic variation on drug response and metabolism.

Pharmacogenomics The application of genomic information or methods to pharmacogenetic problems.

Pharmacokinetics The rate at which the body absorbs, transports, metabolizes, or excretes a drug or its metabolites.

Phase In an individual heterozygous at two syntenic loci, the designation of which allele at the first locus is on the same chromosome as which allele at the second locus. Not to be confused with the phases (stages) of the *cell cycle*. See *coupling* and *repulsion*.

Phenocopy A mimic of a phenotype that is usually determined by a specific genotype, produced instead by the interaction of some environmental factor with a normal genotype.

Phenotype The observed biochemical, physiological, and morphological characteristics of an individual, as determined by his or her genotype and the environment in which it is expressed. Also, in a more limited sense, the abnormalities resulting from a particular mutant gene.

Phenotypic threshold In mitochondrial genetics, the level of heteroplasmy for a given mutant mitochondrial genome at which phenotypic expression or disease occurs.

Philadelphia chromosome (Ph[1]) The structurally abnormal chromosome 22 that typically occurs in a proportion of the bone marrow cells in most patients with chronic myelogenous leukemia. The abnormality is a reciprocal translocation between the distal portion of 22q and the distal portion of 9q that fuses the BCR region of 9 to the *ABL1* oncogene on 22.

Pleiotropy Multiple phenotypic effects of a single allele or pair of alleles. The term is used particularly when the effects are not obviously related.

Pluripotent Describes an embryonic cell that is capable of giving rise to different types of differentiated tissues or structures, depending on its location and environmental influences.

Point mutation See *SNV*.

Polar bodies One of two cells generated during meiosis I and one of two cells generated in meiosis II in oogenesis that receive very little cytoplasm and are not functional ova.

Polyadenylation site In the synthesis of mature mRNA, a site at which a sequence of 20 to 200 adenosine residues (the polyA tail) is added to the 3′ end of an RNA transcript, aiding its transport out of the nucleus and, usually, its stability.

Polygenic Inheritance determined by many genes at different loci, with small additive effects; to be distinguished from the *complex inheritance* seen with *multifactorial disease* in which environmental as well as genetic factors may be involved.

Polymerase chain reaction (PCR) The molecular genetic technique by which a short DNA or RNA sequence is amplified enormously by means of two flanking oligonucleotide primers used in repeated cycles of primer extension and DNA synthesis with DNA polymerase.

Polymorphism The occurrence together in a population of two or more alternative genotypes, each at a frequency greater than that which could be maintained by recurrent mutation alone. A locus is arbitrarily considered to be polymorphic if the rarer allele has a frequency of at least 0.01, so that the heterozygote frequency is at least 0.02. By convention, any allele rarer than this is a *rare variant*.

Population genetics The quantitative study of the frequencies of genetic variants in populations and of how these frequencies change over time both within and between populations.

Positive predictive value With respect to a clinical test for a disease, the extent to which testing positive indicates that one has or will develop the disease.

Preimplantation diagnosis A type of prenatal diagnosis in which one or more cells are removed, either at the blastomere or blastocyst stage, from a multicellular embryo that had been generated by in vitro fertilization and tested for the presence of a disease-causing mutation. An unaffected embryo can then be implanted in the uterus to establish a pregnancy. Avoids the need for considering abortion of an affected fetus that occurs with chorionic villus sampling (CVS) or amniocentesis.

Premature ovarian failure Loss of normal ovarian function before age 40.

Premutation In unstable repeat disorders (e.g., fragile X syndrome), a moderate expansion of the number of repeats that is at increased risk for undergoing further expansion during meiosis and causing the full disorder in the offspring. Premutations can be asymptomatic, as in Huntington disease, or they may be associated with a distinct syndrome, such as the *fragile X–associated tremor/ataxia syndrome* in individuals with triplet repeat expansions in their *FMR1* gene in the premutation range.

Primary constriction See *centromere*.

Primary transcript The initial, unprocessed RNA transcript of a gene that is collinear with the genomic DNA, containing introns as well as exons.

Proband The affected family member through whom the family is ascertained. Also called the *propositus* or *index case*.

Prometaphase The stage of mitosis in which the nuclear membrane dissolves and chromosomes attach to the mitotic spindle. Immediately follows *prophase*.

Promoter A DNA sequence located in the 5′ end of a gene at which transcription is initiated.

Pronuclei The chromosomes of sperm and egg enclosed in separate nuclear membranes immediately following fertilization. After the first mitotic cell division, at which point the two sets of chromosomes join in a single nuclear membrane.

Prophase The first stage of cell division, during which the chromosomes become visible as discrete structures and subsequently thicken and shorten. Prophase of the first meiotic division is further characterized by pairing (*synapsis*) of homologous chromosomes.

Propositus See *proband*.

Proteome The collection of all proteins present in a cell, tissue, or organism at a particular time. Contrast with *transcriptome*, the collection of all RNA transcripts, and *genome*, the collection of all DNA sequences.

Proteomics A field of biochemistry encompassing the comprehensive analysis and cataloguing of the structure and function of all the proteins present in a given cell or tissue (see *proteome*). Parallels *genomics*, a similarly comprehensive approach to the analysis of DNA sequence and mRNA expression.

Proto-oncogene A normal gene involved in cell division or proliferation that may become activated by mutation or other mechanism to become an oncogene.

Pseudoautosomal region Segment of the X and Y chromosome, located at the most distal portion of their respective p and q arms, at which crossing over occurs during male meiosis. Traits due to alleles at pseudoautosomal loci will appear to be inherited as autosomal traits despite the physical location of these loci on the sex chromosomes.

Pseudodeficiency allele A clinically benign allele that has a reduction in functional activity detected by in vitro assays but that has sufficient activity in vivo to prevent haploinsufficiency.

Pseudogene 1. An inactive gene within a gene family, derived by mutation of an ancestral active gene and frequently located in the same region of the chromosome as its functional counterpart (*nonprocessed pseudogene*). 2. A DNA copy of an mRNA, created by retrotransposition and inserted randomly in the genome (*processed pseudogene*). Processed pseudogenes are probably never functional.

Pseudomosaicism The occurrence of a cytogenetically abnormal cell that arose after the tissue was put into culture; generally considered to be artifactual and of no clinical significance.

Purines One of the two types of nitrogen-containing bases (the other being pyrimidine) in DNA and RNA (adenine and guanine).

Pyrimidine One of the two types of nitrogen-containing bases (the other being purines) in DNA and RNA (cytosine and thymine in DNA, cytosine and uracil in RNA).

q 1. In cytogenetics, the long arm of a chromosome. 2. In population genetics, the frequency of the less common allele of a pair. Compare with *p*.

Quadrivalent In a cell with a balanced translocation, the complex of four chromosomes that forms in meiosis I synapsis consisting of the two translocated chromosomes paired with the two normal chromosomes corresponding to the chromosomes involved in the translocation.

Qualitative trait A trait that an individual either has or does not have. Contrast with *quantitative trait*.

Quantitative trait A measurable quantity that differs among different individuals, often following a normal distribution in the population. Contrast with *qualitative trait*.

Random mating Selection of a mate without regard to the genotype of the mate. In a randomly mating population, the frequencies of the various matings are determined solely by the frequencies of the alleles concerned.

Reading frame One of the three possible ways of reading a nucleotide sequence as a series of triplets. An *open reading frame* contains no termination codons and thus is potentially translatable into protein.

Rearrangement Chromosome breakage followed by reconstitution in an abnormal combination. If *unbalanced*, the rearrangement can produce an abnormal phenotype.

Recessive A trait that is expressed only in homozygotes, compound heterozygotes, or hemizygotes.

Reciprocal translocation See *translocation*.

Recombinant An individual who has a new combination of alleles in a haplotype not present in either parent.

Recombinant chromosome A chromosome that results from exchange of reciprocal segments by crossing over between a homologous pair of parental chromosomes during meiosis.

Recombination The formation of new combinations of alleles in coupling by crossing over between their loci.

Recombination fraction (θ) The fraction of offspring of a parent heterozygous at two loci who have inherited a chromosome carrying a recombination between the loci. Usually symbolized by the Greek letter θ.

Recurrence risk The probability that a genetic disorder present in one or more members of a family will recur in another member of the same or a subsequent generation.

Reduction division The first meiotic division, so-called because at this stage the chromosome number per cell is reduced from diploid to haploid.

Redundancy The situation in which genes (often *paralogous*) have overlapping functions.

Regulative development A developmental stage during which removal or destruction of a particular region of the embryo is compensated for by other embryonic regions, thereby allowing normal development.

Regulatory element A DNA segment, such as a promoter, insulator, enhancer, or locus control region, within or near a gene that regulates the expression of the gene.

Regulatory gene A gene that codes for an RNA or protein molecule that regulates the expression of other genes.

Relative risk A comparison of the *risk* for a disease or trait in individuals who share a particular factor (e.g., genotype, an environmental exposure, or a drug) versus the *risk* among individuals who lack the factor.

	Affected	Unaffected	Total
Factor present	a	b	a + b
Factor absent	c	d	c + d
Total	a + c	b + d	a + b + c + d

The *risk* for being affected in individuals who have the factor = (a/a + b), the *risk* for being affected when the factor is absent = (c/c + d), and the relative risk = (a/a + b)/(c/c + d) = a(c + d)/c(a + b). Note that relative risk ≈ ad/bc, the odds ratio, when the disease is relatively rare (c ≪ d and a ≪ b). See *odds ratio*.

Relative risk ratio (λ_r) In complex disorders, the risk that a disease will occur in a relative of an affected person compared with the risk for disease in any random person in the general population.

Repetitive DNA (repeats) DNA sequences that are present in multiple copies in the genome.

Replicative segregation Random distribution of mitochondria into daughter cells.

Repulsion Describes the phase of two alleles at two different but syntenic loci, in which one allele at one of the loci is *not* on the same chromosome as the allele at the second locus. See *phase* and *coupling*.

Retroposition A process by which a molecule of RNA, often derived from transcription of a repetitive element such as *Alu* or *LINE*, is transcribed by *reverse transcriptase* into a molecule of DNA, which is then inserted into another site in the genome.

Retrovirus A virus, with an RNA genome, that propagates by conversion of the RNA into DNA by the enzyme reverse transcriptase.

Reverse transcriptase An enzyme, RNA-dependent DNA polymerase, that catalyzes the synthesis of DNA on an RNA template.

Ribonucleic acid See *RNA*.

Ribosome A cytoplasmic organelle composed of ribosomal RNA and protein, on which polypeptides are synthesized based on the sequences of messenger RNA.

Ring chromosome A structurally abnormal chromosome in which the telomere of each chromosome arm has been deleted and the broken arms have reunited in ring formation.

Risk The probability of an event's occurring. Often calculated as the number of times the event occurs divided by the total number of opportunities there were for the event to occur. As with all probabilities, risk varies from 0 to 1.

RNA (ribonucleic acid) A nucleic acid formed on a DNA template, containing ribose instead of deoxyribose. *Messenger RNA (mRNA)* is the template on which polypeptides are synthesized. *Transfer RNA (tRNA)*, in cooperation with the ribosomes, brings activated amino acids into position along the mRNA template. *Ribosomal RNA (rRNA)*, a component of the ribosomes, functions as a nonspecific site of polypeptide synthesis. *Noncoding RNAs (ncRNAs)* are transcribed RNA molecules that do not encode proteins or directly carry out translation, as tRNA or rRNA do, but instead have a variety of roles in the regulation of gene expression. Some ncRNAs are very long (such as the *XIST* gene product involved in X-inactivation) and referred to as long ncRNA or *lncRNA*.

RNA editing Post-transcriptional modification of RNA transcripts, thereby changing certain codons in the mRNA so they differ from the codon specified in the original DNA template. Most frequently occurs through deamination of adenine to generate an inosine, which is read by the translational machinery as a guanine.

RNA polymerase An enzyme that synthesizes RNA on a DNA template. Different RNA polymerases synthesize different RNA molecules; for example, mRNAs are transcribed by RNA polymerase II.

RNA splicing The excision of introns from primary RNA transcripts and splicing together of exons in the generation of mature mRNA from the primary transcript.

RNAi RNA interference. A system for regulating gene expression in which short RNA segments, approximately 22 bases in length, form double-stranded structures with an mRNA and either target it for destruction or block its translation. (See *microRNA*.) Scientists have taken advantage of this normal, endogenous system of gene regulation to design new and powerful technologies for gene silencing by use of exogenously supplied RNAi sequences.

Robertsonian translocation A translocation between two acrocentric chromosomes by fusion at or near the centromere, with loss of the short arms.

Satellite DNA DNA containing many tandem repeats of a short basic repeating unit; not to be confused with *chromosomal satellite*, the chromatin at the distal end of the short arms of the acrocentric chromosomes.

Scaffold The nonhistone structure observed when histones are experimentally removed from chromosomes.

Believed to represent a structural component of the nucleus and of chromosomes.

Segmental aneusomy Loss or gain of a small segment from one chromosome of a pair, resulting either in hemizygosity or partial trisomy for genes in that segment on the homologous chromosome. See also *contiguous gene syndrome*.

Segmental duplication Blocks of homologous sequences distributed across a region of the genome that mediate duplication and deletion of the segments of DNA located between them.

Segregation In genetics, the distribution of genetic material into daughter cells. For chromosomes, it is the orderly disjunction of the appropriate haploid set of homologous chromosomes at meiosis I or sister chromatids at meiosis II. For mitochondria, refers to distribution of newly formed mitochondria into daughter cells during mitosis. See *nondisjunction* and *replicative segregation*.

Selection In population genetics, the operation of forces that determine the relative fitness of a genotype in the population, thus affecting the frequency of particular alleles. The coefficient of selection, s, is a measure of the proportion of mutant alleles at a locus that are *not* passed on to the next generation and is given by 1-f, where f is the *coefficient of fitness*.

Sense strand See *coding strand*.

Sensitivity In diagnostic tests, the frequency with which the test result is positive when the disorder is present. Not to be confused with *positive predictive value*.

Sequence 1. In genomics and molecular genetics, the order of nucleotides in a segment of DNA or RNA. 2. In clinical genetics, a recognizable pattern of dysmorphic features due to a number of different causes; to be distinguished from *malformation syndrome*.

Sex chromatin See *Barr body*.

Sex chromosomes The X and Y chromosomes.

Sex-influenced A trait that is not X-linked in its pattern of inheritance but is expressed differently, either in degree or in frequency, in males and females.

Sex-limited A trait that is expressed in only one sex, although the gene that determines the trait is not X-linked.

Sex-linked A general term referring to linkage to either of the sex chromosomes. In human and medical genetics, the term is typically not used and is replaced by *X-linkage* or *Y-linkage*.

Short tandem repeat polymorphism (STRP) A polymorphic locus consisting of a variable number of tandemly repeated dinucleotide, trinucleotide, or tetranucleotide units such as $(TG)_n$, $(CAA)_n$, or $(GATA)_n$; different numbers of units constitute the different alleles. Also termed a *microsatellite marker*.

Sib, sibling A brother or sister.

Sibship All the sibs in a family.

Silencer A DNA sequence that acts in *cis* (i.e., on the same chromosome) to decrease transcription of a nearby gene. The silencer may be upstream or downstream to the gene and may be in the same or the reverse orientation (contrast with *enhancer*).

Single nucleotide variant (SNV) A change in DNA sequence at a single base pair in DNA. If the SNV is frequent enough to be a polymorphism, it is referred to as a single nucleotide polymorphism (*SNP*).

Single-copy DNA DNA whose linear order of specific nucleotides is represented only once (or at most a few times) around the entire genome. The type of DNA that makes up most of the genome.

Single-gene disorder A disorder due to one or a pair of mutant alleles at a single locus.

Slipped mispairing A mutational mechanism that occurs during DNA replication of sequences with repeats of one or more nucleotides, in which a repeat on one strand mispairs with a similar repeat on the complementary strand, generating a deletion or expansion of the number of repeats.

Small (or short) interfering RNAs (siRNAs) A class of naturally occurring or synthesized 20- to 25-nucleotide–long, double-stranded RNA molecules that regulate gene expression by inducing the degradation of complementary mRNA, by the process of RNA interference. They have substantial therapeutic potential against targets that otherwise cannot be treated by drugs.

SNP See *single nucleotide variant*.

SNP array A type of microarray that uses *oligonucleotides* corresponding to high-frequency SNPs around the genome to determine if there is a chromosomal or subchromosomal deletion or duplication. Provides an alternative approach to detecting CNVs to *comparative genome hybridization*.

SNV See *single nucleotide variant*.

Solenoid A fiber composed of compacted strings of nucleosomes, that forms the fundamental unit of chromatin organization.

Somatic cell All the cells that contribute to one's body, excluding the cells of the *germline*.

Somatic mutation A mutation occurring in a somatic cell rather than in the germline.

Somatic rearrangement Rearrangement of DNA sequences in the chromosomes of lymphocyte precursor cells, thus generating antibody and T-cell receptor diversity.

Southern blotting A technique, devised by the British biochemist Ed Southern, for preparation of a filter to which DNA has been transferred, following restriction enzyme digestion and gel electrophoresis to separate the DNA molecules by size. The size of specific DNA molecules can then be determined on the filter by their hybridization to labeled DNA sequences complementary to the sequences one wishes to visualize.

Specialty proteins Proteins, expressed in only one or a limited number of cell types, that have unique functions contributing to the individuality of the cells in which they are expressed. Contrast with *housekeeping proteins*.

Specification The first stage of commitment in which a cell will follow its developmental program if it is explanted but can still be reprogrammed to a different fate if it is transplanted to a different part of the embryo.

Specificity In diagnostic tests, the frequency with which a test result is negative when the disease is absent. Not to be confused with *negative predictive value*.

Spermatogonia Diploid cells derived from early germ cells in the male that both divide to replenish their population and, at puberty, undergo a series of developmental steps, including meiosis, leading to terminal differentiation into mature spermatozoa.

Spliceopathy A disorder, typified by myotonic dystrophy, in which mRNAs containing massive expansions of an untranslated unstable repeat mutation sequester splicing factors and deprive the cell of the proteins needed to carry out normal splicing of other mRNAs.

Splicing See *RNA splicing*.

Sporadic In medical genetics, a disease that is not the result of inheritance of a disease-causing allele from a parent. Often the result of a new germline or somatic mutation.

Standardized incidence ratio (SIR) The ratio of the incidence of cases of cancer during a given time period in relatives of a proband with a disorder such as cancer divided by the number expected from the incidence over that same time period in an age-matched group in the general population.

Stem cell A type of cell capable both of self-renewal and of proliferation and differentiation.

Stop codon See *termination codon*.

Stratification The situation in which a population contains a number of subgroups whose members have not freely and randomly mated with the members of other subgroups.

Structural gene A gene coding for any RNA or protein product.

Structural protein A protein that serves a structural role in the body, such as collagen.

Structural rearrangements Rearrangements of one or more chromosomes, which may be *balanced* if there is no change from normal genomic content, or *unbalanced*, if genomic content is abnormal.

Subchromosomal mutation Mutations that change only a portion of a chromosome. Can lead to *partial trisomy* or *partial monosomy*.

Submetacentric A type of chromosome with arms of different lengths.

Synapsis Close pairing of homologous chromosomes in prophase of the first meiotic division.

Synaptonemal complex The protein complex that forms at sites of meiotic recombination and mediates the recombination during synapsis in meiosis I.

Syndrome A characteristic pattern of anomalies, presumed to be causally related.

Synonymous Describes a single nucleotide variant (SNV) that does not alter a codon and therefore does not change the resulting amino acid sequence of the encoded peptide.

Synpolydactyly A birth defect of the hands and feet characterized by extra digits and the fusion of adjoining digits.

Synteny The physical presence together on the same chromosome of two or more gene loci, whether or not they are close enough together for linkage to be demonstrated (the adjective is *syntenic*).

Tandem repeats Two or more copies of the same (or similar) DNA sequence arranged in a direct head to tail succession along a chromosome.

TATA box A consensus sequence in the promoter region of many genes that is located approximately 25 base pairs upstream from the start site of transcription and that determines the start site.

T-cell antigen receptor (TCR) Genetically coded receptor on the surface of T lymphocytes that specifically recognizes antigen molecules.

Telocentric A type of chromosome in which the centromere is at one end of a chromosome and there is only a single arm; telocentric chromosomes do not occur in normal human karyotypes but occasionally occur in chromosomal rearrangements.

Telomerase A ribonucleoprotein reverse transcriptase that contains its own RNA molecule to use as a template for synthesizing species-specific hexamers (e.g., TTAGGG in humans) and adding them to the ends of telomeres.

Telomere The end of each chromosome arm. Human telomeres end with tandem copies of the sequence (TTAGGG)$_n$, which is required for the proper replication of chromosome ends.

Telophase The stage of cell division that begins when the daughter chromosomes reach the poles of the dividing cell and that lasts until the two daughter cells take on the appearance of interphase cells.

Teratogen An agent that produces congenital malformations or increases their incidence.

Termination codon One of the three codons (UAG, UAA, and UGA) that terminate synthesis of a polypeptide. Also called a *stop codon* or *chain-termination codon*.

Tertiary structure Three-dimensional configuration of a molecule.

Tetraploid A cell with four (4n) copies of each chromosome, or an individual made up of such cells.

The Cancer Genome Atlas (TCGA) A vast public database of mutations, epigenomic modifications, and abnormal gene expression profiles found in a wide variety of cancers.

Trans Refers to the relationship between two sequences located across from each other on the two homologous chromosomes, or to interactions between a protein and a chromosome locus. Literally means "across from." Contrast with *cis*.

Transcription The synthesis of a single-stranded RNA molecule from a DNA template in the cell nucleus, catalyzed by RNA polymerase.

Transcription factor One of a large class of proteins that regulate transcription by forming large complexes with other transcription factors and RNA polymerase; these complexes then bind to regulatory regions of genes either to promote or to inhibit transcription.

Transcriptome The collection of all RNA transcripts made in a cell.

Transfer RNA (tRNA) See *RNA*.

Transformation The in vivo process by which a normal cell in a tissue becomes a cancerous cell.

Translation The synthesis of a polypeptide from its mRNA template.

Translocation The transfer of a segment of one chromosome to another chromosome. If two nonhomologous chromosomes exchange pieces, the translocation is *reciprocal*. See also *Robertsonian translocation*.

Triploid A cell with three copies of each chromosome (3n), or an individual made up of such cells.

Trisomy The state of having three representatives of a given chromosome instead of the usual pair, as in trisomy 21 (Down syndrome).

tRNA Transfer RNA; see *RNA*.

Tumor-suppressor gene A normal gene involved in the regulation of cell proliferation. Loss-of-function mutations in both alleles can lead to tumor development, as in the retinoblastoma gene or the *TP53* gene. Contrast with *oncogene*.

Two-hit model The hypothesis that some forms of cancer can be initiated when both alleles of a tumor-suppressor gene become inactivated in the same cell.

Ultrasonography A technique in which high-frequency sound waves are used to examine internal body structures; useful in prenatal diagnosis.

Unbalanced X inactivation The situation in which the two X chromosomes in a female are not inactivated equally, leading to a substantial deviation from the expected 50% chance for each X remaining active. Also called *skewed inactivation*.

Uniparental disomy The presence in the karyotype of two copies of a specific chromosome, both inherited from one parent, with no representative of that chromosome from the other parent. If both homologues of the parental pair are present, the situation is *heterodisomy*; if one parental homologue is present in duplicate, the situation is *isodisomy*. See *Prader-Willi syndrome* and *Angelman syndrome* in the text.

Unstable repeat expansion disorders See *dynamic mutation*.

Untranslated region (UTR) The segments of an mRNA that either precede the initiator codon (5'-UTR) or follow the stop codon (3'-UTR).

Variant An allele that differs from *wild-type*.

Variant of uncertain significance (VUS) A difference between an individual's sequence and that of an arbitrary reference sequence, but whose pathogenetic significance is unknown. Missense mutations are frequent VUSs in whole-exome or whole-genome sequences, although indels may also be of uncertain significance.

Vector In gene transfer therapy, a virus whose genome has been engineered to contain and express a therapeutic DNA sequence of interest. The virus is used to deliver the DNA sequence into a cell.

VNTR (variable number of tandem repeats) A type of DNA polymorphism created by a tandem arrangement of varying numbers of copies of short DNA sequences. Highly polymorphic, used in linkage studies and in DNA "fingerprinting" for paternity testing and forensic medicine.

Whole-exome sequencing (WES) Use of high-throughput sequencing methods to sequence the approximately 2% of the genome containing just the exons of protein-coding genes in an individual. See *whole-genome sequencing*.

Whole-genome sequencing (WGS) Use of high-throughput sequencing methods to sequence an individual's entire genome (minus the few percent that current technologies are not capable of sequencing). See *whole-exome sequencing*.

Wild-type A term used to indicate the normal allele (often symbolized as +) or the normal phenotype.

X;autosome translocation Reciprocal translocation between an X chromosome and an autosome.

X chromosome The larger of the two sex chromosomes, normally present in two copies in females and one copy in males.

X inactivation Inactivation of genes on one X chromosome in somatic cells of female mammals, occurring early in embryonic life, at about the time of implantation. The *X inactivation* center is a segment of the X chromosome that determines which X will be designated the inactive X and contains the *ncRNA XIST*. See *Lyon's law (hypothesis)*.

X linkage The distinctive inheritance pattern of alleles at loci on the X chromosome that do not undergo recombination (crossing over) during male meiosis. Genes on the X chromosome, or traits determined by such genes, are X-linked.

Y chromosome The smaller of the two sex chromosomes, normally present in one copy in males only.

Y linkage Genes on the Y chromosome, or traits (e.g., the male sex) determined by such genes, are Y-linked.

Zone of polarizing activity Region in a developing limb bud that secretes morphogens like sonic hedgehog to establish a gradient that specifies the posterior side of the developing limb bud.

Zygosity The number of zygotes from which a multiple birth is derived. For example, twins may be either monozygotic (MZ) or dizygotic (DZ). To determine whether a certain pair of twins is MZ or DZ is to determine their zygosity.

Zygote A fertilized ovum.

Zygotene Stage in meiosis I when homologous chromosomes align along their entire length to permit *synapsis*.

Sources and Acknowledgments

CHAPTER 2

Figure 2-1 Based on Brown TA: *Genomes*, ed 2, New York, 2002, Wiley-Liss. Inset from Paulson JR, Laemmli UK: The structure of histone-depleted metaphase chromosomes. *Cell* 12:817-828, 1977. Reprinted by permission of the authors and Cell Press.

Figure 2-3 Based on Watson JD, Crick FHC: Molecular structure of nucleic acids: a structure for deoxyribose nucleic acid. *Nature* 171:737-738, 1953.

Figure 2-7 Based on data from European Bioinformatics Institute and Wellcome Trust Sanger Institute: *Ensembl release 70*, January 2013. Available from http://www.ensembl.org, v37.

Figure 2-10 Courtesy Stuart Schwartz, University Hospitals of Cleveland, Ohio.

Figure 2-11 Courtesy Stuart Schwartz, University Hospitals of Cleveland, Ohio.

Figure 2-16 Modified from Moore KL, Persaud TVN: *The developing human: clinically oriented embryology*, ed 6, Philadelphia, 1998, WB Saunders.

CHAPTER 3

Figure 3-2 Data from European Bioinformatics Institute and Wellcome Trust Sanger Institute: *Ensembl release 70*, January 2013. Available from http://www.ensembl.org.

Figure 3-7 Original data from Lawn RM, Efstratiadis A, O'Connell C, et al: The nucleotide sequence of the human β-globin gene. *Cell* 21:647-651, 1980.

CHAPTER 5

Figure 5-2 Redrawn from ISCN 2013.

Figure 5-3 Redrawn from ISCN 2013.

Figure 5-4 Ideograms redrawn from ISCN 2013; photomicrographs courtesy Genetics Department, The Hospital for Sick Children, Toronto, Canada.

Figure 5-5 Images courtesy M. Katharine Rudd, Emory Genetics Laboratory, Atlanta, Georgia.

Figure 5-6 A and **B** reprinted from Lee C: Structural genomic variation in the human genome. In Ginsburg GS, Willard HF, editors: *Genomic and personalized medicine*, ed 2, New York, 2013, Elsevier, pp. 123-132. **C** courtesy M. Katharine Rudd, Emory Genetics Laboratory, Atlanta, Georgia.

Figure 5-8 Data summarized from Hsu LYF: Prenatal diagnosis of chromosomal abnormalities through amniocentesis. In Milunsky A, editor: *Genetic disorders and the fetus*, ed 4, Baltimore, 1998, Johns Hopkins University Press, pp 179-248.

Figure 5-9 A courtesy Center for Human Genetics Laboratory, University Hospitals of Cleveland. **B** courtesy M. Katharine Rudd, Emory Genetics Laboratory. **C** courtesy Daynna J. Wolff, Medical University of South Carolina. **D** original data from Dan S, Chen F, Choy KW, et al: Prenatal detection of aneuploidy and imbalanced chromosomal arrangements by massively parallel sequencing. *PLoS One* 7:e27835, 2012.

CHAPTER 6

Figure 6-1 Data from Hook EB, Cross PK, Schreinemachers DM: Chromosomal abnormality rates at amniocentesis and in live-born infants. *JAMA* 249:2034-2038, 1983.

Figure 6-2 From Jones KL, Jones MC, del Campo M: *Smith's recognizable patterns of human malformation*, ed 7, Philadelphia, 2013, WB Saunders.

Figure 6-5 C fluorescence in situ hybridization image courtesy Hutton Kearney, Duke University Medical Center.

Figure 6-6 B and **C** from Jones KL, Jones MC, del Campo M: *Smith's recognizable patterns of human malformation*, ed 7, Philadelphia, 2013, WB Saunders. **D** based on data from Zhang X, Snijders A, Segraves R, et al: High-resolution mapping of genotype-phenotype relationships in cri du chat syndrome using array comparative genome hybridization. *Am J Hum Genet* 76:312-326, 2005. **E** courtesy M. Katharine Rudd, Emory Genetics Laboratory, Atlanta, Georgia.

Figure 6-7 A from Jones KL: *Smith's recognizable patterns of human malformation*, ed 4, Philadelphia, 1988, WB Saunders, p 173. **B** courtesy Jan Friedman, University of British Columbia. From Magenis RE, Toth-Fejel S, Allen LJ, et al: Comparison of the 15q deletions in Prader-Willi and Angelman syndromes: specific regions, extent of deletions, parental origin, and clinical consequences. *Am J Med Genet* 35:333-349, 1990. Copyright © 1990, Wiley-Liss, Inc. Reprinted by permission of John Wiley and Sons, Inc. **C** courtesy M. Katharine Rudd, Emory Genetics Laboratory, Atlanta, Georgia. **D** modified from *GeneReviews*. Available from www.ncbi.nlm.nih.gov/books/NBK1116/. Copyright © University of Washington.

Figure 6-13 B data from Amos-Landfraf JM, Cottle A, Plenge RM, et al: X chromosome inactivation patterns of 1005 phenotypically unaffected females. *Am J Hum Genet* 79:493-499, 2006.

Figure 6-15 **A** from Jones KL, Jones MC, del Campo M: *Smith's recognizable patterns of human malformation*, ed 7, Philadelphia, 2013, WB Saunders. **B** from Grumbach MM, Hughes IA, Conte FA: Disorders of sex differentiation. In Larsen PR, Kronenberg HM, Melmed S, Polonsky KS, editors: *Williams textbook of endocrinology*, ed 10, Philadelphia, 2003, WB Saunders.

Figure 6-17 From Moore KL, Persaud TVN: *The developing human: clinically oriented embryology*, ed 5, Philadelphia, 1993, WB Saunders.

Figure 6-18 Courtesy L. Pinsky, McGill University, Montreal, Canada.

Figure 6-19 Modified from Moreno-De-Luca A, Myers SM, Challman TD, et al: Developmental brain dysfunction: revival and expansion of old concepts based on new genetic evidence. *Lancet Neurol* 12:406-414, 2013, with permission.

CHAPTER 7

Figure 7-9 From Kelikian H: *Congenital deformities of the hand and forearm*, Philadelphia, 1974, WB Saunders.

Figure 7-11 Images courtesy K. Arahata, National Institute of Neuroscience, Tokyo.

Figure 7-16 From Shears DJ, Vassal HJ, Goodman FR, et al: Mutation and deletion of the pseudoautosomal gene SHOX cause Leri-Weill dyschondrosteosis. *Nat Genet* 19:70-73, 1998.

Figure 7-20 Data courtesy Dr. M. Macdonald, Massachusetts General Hospital, Boston.

Figure 7-21 Data courtesy Dr. Ben Roa, Baylor College of Medicine, Houston, Texas.

Figure 7-22 Courtesy Peter Ray, The Hospital for Sick Children, Toronto, Canada.

Figure 7-23 From Nolin SL: Familial transmission of the *FMR1* CGG repeat. *Am J Hum Genet* 59:1252-1261, 1996. The University of Chicago Press.

CHAPTER 8

Figure 8-1 **B** data from Sive PH, Medalie JH, Kahn HA, et al: Distribution and multiple regression analysis of blood pressure in 10,000 Israeli men. *Am J Epidemiol* 93:317-327, 1971.

Figure 8-3 Data from Johnson BC, Epstein FH, Kjelsberg MO: Distributions and familial studies of blood pressure and serum cholesterol levels in a total community—Tecumseh, Michigan. *J Chronic Dis* 18:147-160, 1965.

Figure 8-4 Courtesy Sir Alec Jeffreys, University of Leicester, United Kingdom.

Figure 8-6 Modified from an original figure courtesy Larry Almonte, with permission.

Figure 8-7 Redrawn from Kajiwara K, Berson EL, Dryja TP: Digenic retinitis pigmentosa due to mutations at the unlinked peripherin/RDS and ROM1 loci. *Science* 264:1604-1608, 1994.

Figure 8-9 Original data provided by A. Chakravarti, Johns Hopkins University, Baltimore, Maryland.

Figure 8-10 Modified from Trowsdale J, Knight JC: Major histocompatibility complex genomics and human disease. *Annu Rev Genomics Hum Genet* 14:301-323, 2013.

Figure 8-11 Modified from Roberts JS, Cupples LA, Relkin NR, et al: *J Geriatr Psychiatry Neurol* 2005 18:250-255.

CHAPTER 9

Figure 9-1 From Novembre J, Galvani AP, Slatkin M: The geographic spread of the CCR5 Δ32 HIV-resistance allele. *PLoS Biol* 3:e339, 2005.

Figure 9-2 From Levran O, Awolesi O, Shen PH, et al: Estimating ancestral proportions in a multi-ethnic US sample: implications for studies of admixed populations. *Hum Genomics* 6:2, 2012.

Figure 9-3 From Paschou P, Ziv E, Burchard EG, et al: PCA-correlated SNPs for structure identification in worldwide human populations. *PLoS Genet* 3:1672-1686, 2007.

CHAPTER 10

Figure 10-8 Modified from original figures of Thomas Hudson, McGill University, Canada.

Figure 10-9 Based on data and diagrams provided by Thomas Hudson, Quebec Genome Center, Montreal, Canada.

Figure 10-11 From Fritsche LG, Chen W, Schu M, et al: Seven new loci associated with age-related macular degeneration. *Nature Genet* 17:1783-1786, 2013.

CHAPTER 11

Figure 11-3 **A** redrawn from Stamatoyannopoulos G, Nienhuis AW: Hemoglobin switching. In Stamatoyannopoulos G, Nienhuis AW, Leder P, Majerus PW, editors: *The molecular basis of blood diseases*, Philadelphia, 1987, WB Saunders. **B** redrawn from Wood WG: Haemoglobin synthesis during fetal development. *Br Med Bull* 32:282-287, 1976.

Figure 11-4 Redrawn from Kazazian HH Jr, Antonarakis S: Molecular genetics of the globin genes. In Singer M, Berg P, editors: *Exploring genetic mechanisms*, Sausalito, CA, 1997, University Science Books.

Figure 11-5 From Kaul DK, Fabry ME, Windisch P, et al: Erythrocytes in sickle cell anemia are heterogeneous in their rheological and hemodynamic characteristics. *J Clin Invest* 72:22, 1983.

Figure 11-6 Redrawn from Ingram V: Sickle cell disease: molecular and cellular pathogenesis. In Bunn HF, Forget BG, editors: *Hemoglobin: molecular, genetic, and clinical aspects*, Philadelphia, 1986, WB Saunders.

Figure 11-7 Redrawn from Orkin SH: Disorders of hemoglobin synthesis: the thalassemias. In Stamatoyannopoulos G, Nienhuis AW, Leder P, Majerus PW, editors: *The molecular basis of blood diseases*, Philadelphia, 1987, WB Saunders, pp 106-126.

Figure 11-8 From Hoffman R, Furie B, McGlave P, et al: *Hematology: basic principles and practice*, ed 5, 2008, Elsevier.

Figure 11-9 Redrawn from Kazazian HH: The thalassemia syndromes: molecular basis and prenatal diagnosis in 1990. *Semin Hematol* 27:209-228, 1990.

Figure 11-11 Modified from Stamatoyannopoulos G, Grosveld F: Hemoglobin switching. In Stamatoyannopoulos G, Majerus PW, Perlmutter RM, Varmus H, editors: *The molecular basis of blood diseases*, ed 3, Philadelphia, 2001, WB Saunders.

CHAPTER 12

Figure 12-4 Derived from Nowacki PM, Byck S, Prevost L, Scriver CR: PAH mutation analysis consortium database: 1997. Prototype for relational locus-specific mutation databases. *Nucl Acids Res* 26:220-225, 1998, by permission of Oxford University Press.

Figure 12-5 Modified from Sandhoff K, Conzelmann E, Neufeld EF, et al: The GM_2 gangliosidoses. In Scriver CR, Beaudet AL, Sly WS, Valle D, editors: *The metabolic bases of inherited disease*, ed 6, New York, 1989, McGraw-Hill, pp 1807-1839.

Figure 12-7 From McIntosh N, Helms P, Smyth R, Logan S: Inborn errors of metabolism. In *Forfar and Arneil's textbook of pediatrics*. Edinburgh, 2008, Churchill Livingstone.

Figure 12-9 Redrawn from Larson C: Natural history and life expectancy in severe α_1-antitrypsin deficiency, Pi Z. *Acta Med Scand* 204:345-351, 1978.

Figure 12-10 From Stoller JK, Aboussouan LS: α_1-Antitrypsin deficiency. *Lancet* 365:2225-2236, 2005.

Figure 12-11 Redrawn from Kappas A, Sassa S, Galbraith RA, Nordmann Y: The porphyrias. In Scriver CR, Beaudet AL, Sly WS, Valle D, editors: *The metabolic bases of inherited disease*, ed 6, New York, 1989, McGraw-Hill, pp 1305-1365.

Figure 12-13 Redrawn from Goldstein JL, Brown MS: Familial hypercholesterolemia. In Scriver CR, Beaudet AL, Sly WS, Valle D, editors: *The metabolic bases of inherited disease*, ed 6, New York, 1989, McGraw-Hill, pp 1215-1250.

Figure 12-14 Modified from Brown MS, Goldstein JL: The LDL receptor and HMG-CoA reductase—two membrane molecules that regulate cholesterol homeostasis. *Curr Top Cell Regul* 26:3-15, 1985.

Figure 12-15 Based on Zielinski J: Genotype and phenotype in cystic fibrosis. *Respiration* 67:117-133, 2000.

Figure 12-16 Courtesy R. H. A. Haslam, The Hospital for Sick Children, Toronto.

Figure 12-17 Courtesy K. Arahata, National Institute of Neuroscience, Tokyo.

Figure 12-20 Courtesy P. N. Ray, The Hospital for Sick Children, Toronto.

Figure 12-21 Courtesy T. Costa, The Hospital for Sick Children, Toronto.

Figure 12-22 Redrawn from Byers PH: Disorders of collagen biosynthesis and structure. In Scriver CR, Beaudet AL, Sly WS, Valle D, editors: *The metabolic bases of inherited disease*, ed 6, New York, 1989, McGraw-Hill, pp 2805-2842.

Figure 12-24 Reproduced with permission from Nussbaum RL, Ellis CE: Alzheimer's disease and Parkinson's disease. *N Engl J Med* 348:1356-1364, 2003.

Figure 12-25 Reproduced with permission from Nussbaum RL, Ellis CE: Alzheimer's disease and Parkinson's disease. *N Engl J Med* 348:1356-1364, 2003.

Figure 12-26 Modified in part from Shoffner JM, Wallace DC: Oxidative phosphorylation disease. In Scriver CR, Beaudet AL, Sly WS, Valle D, editors: *The metabolic and molecular bases of inherited disease*, ed 7, New York, 1995, McGraw-Hill. The location of some of the disorders is taken from DiMauro S, Schon EA: Mitochondrial respiratory-chain diseases. *N Engl J Med* 348:2656-2568, 2003.

Figure 12-27 Modified from Chinnery PF, Turnbull DM: Mitochondrial DNA and disease. *Lancet* 354:SI17-SI21, 1999.

Figure 12-28 Based partly on an unpublished figure courtesy John A. Phillips III, Vanderbilt University Nashville.

CHAPTER 13

Figure 13-1 Modified from Valle D: Genetic disease: an overview of current therapy. *Hosp Pract* 22:167-182, 1987.

Figure 13-2 From Campeau PM, Scriver CR, Mitchell JJ: A 25 year longitudinal analysis of treatment efficacy in inborn errors of metabolism. *Mol Genet Metab* 95:11-16, 2008.

Figure 13-3 From Campeau PM, Scriver CR, Mitchell JJ: A 25-year longitudinal analysis of treatment efficacy in inborn errors of metabolism. *Mol Genet Metab* 95:11-16, 2008.

Figure 13-5 From Brown MS, Goldstein JL: A receptor-mediated pathway for cholesterol homeostasis. *Science* 232:4, 1986. Copyright by the Nobel Foundation.

Figure 13-6 From Goya M, Alvarez M, Teixido-Tura G, et al: Abdominal aortic dilatation during pregnancy in Marfan syndrome. *Rev Esp Cardiol (Engl Ed)* 65:288-289, 2012.

Figure 13-8 From Ramsey BW, Davies J, McElvaney NG, et al: A CFTR potentiator in patients with cystic fibrosis and the G551D mutation. *N Engl J Med* 365:1663-1672, 2011.

Figure 13-9 Redrawn from Valle D: Genetic disease: an overview of current therapy. *Hosp Pract* 22:167-182, 1987.

Figure 13-11 Redrawn from Barton NW, Furbish FS, Murray GJ, et al: Therapeutic response to intravenous infusions of glucocerebrosidase in a patient with Gaucher disease. *Proc Natl Acad Sci U S A* 87:1913-1916, 1990.

Figure 13-12 Modified from Saunthararajah Y, Lavelle D, DeSimone J: DNA hypomethylating reagents and sickle cell disease. *Br J Haematol* 126:629-636, 2004.

Figure 13-13 From van Deutekom JC, Janson AA, Ginjaar IB, et al: Local dystrophin restoration with antisense oligonucleotide PRO051. *N Engl J Med* 357:2677-2686, 2007.

Figure 13-15 From Staba SL, Escolar ML, Poe M, et al: Cord-blood transplantation from unrelated donors in patients with Hurler's syndrome. *N Engl J Med* 350:1960-1969, 2004.

Figure 13-17 From Biffi A, Montini E, Lorioli L, et al: Lentiviral hematopoietic stem cell gene therapy benefits metachromatic leukodystrophy. *Science* 341:1233158, 2013.

CHAPTER 14

Figure 14-1 Images courtesy Dr. Leslie Biesecker, Bethesda, Maryland.

Figure 14-2 Image courtesy Dr. Judith Hall, University of British Columbia, Vancouver, Canada.

Figure 14-3 Image courtesy Dr. Mason Barr, Jr., University of Michigan, Ann Arbor, Michigan.

Figure 14-5 Reprinted with permission from Jones KL, Jones MC, del Campo M: *Smith's recognizable patterns of human malformation*, ed 7, Philadelphia, 2013, WB Saunders.

Figure 14-6 A-C adapted in modified form from Wolpert L: *Principles of development*, New York, 2002, Oxford University Press. **D** from Pooh RK, Kurjak A: Recent advances in 3D assessment of various fetal anomalies. *J Ultrasound Obstet Gynecol* 3:1-23, 2009.

Figure 14-7 Redrawn from Hauk R: *Frequently asked questions about bats*, 2011, Western National Parks Association. Available from http://www.batsrule.info/batsrule-helpsavewildlife/2013/7/7/bat-wings-have-evolved-to-be-different-yet-similar-to-other-species.

Figure 14-8 Reprinted with permission from Ogilvie CM, Braude PR, Scriven PN: Preimplantation diagnosis—an overview. *J Histochem Cytochem* 53:255-260, 2005.

Figure 14-9 Reprinted with permission from Jones KL: *Smith's recognizable patterns of human malformation*, ed 6, Philadelphia, 2005, WB Saunders.

Figure 14-10 Reprinted with permission from Moore KL, Persaud TVN: *The developing human: clinically oriented embryology*, ed 6, Philadelphia, 1998, WB Saunders.

Figure 14-11 Reprinted with permission from Stamatoyannopoulos G, Nienhuis AW, Majerus PW, Varmus H: *The molecular basis of blood diseases*, ed 2, Philadelphia, 1994, WB Saunders.

Figure 14-13 Reprinted with permission from Ogilvie CM, Braude PR, Scriven PN: Preimplantation diagnosis—an overview. *J Histochem Cytochem* 53:255-260, 2005.

Figure 14-15 From Wolpert L, Beddington R, Brockes J, et al: *Principles of development*, New York, 1998, Oxford University Press. Copyright 1998, Oxford University Press.

Figure 14-16 Redrawn from Tjian R: Molecular machines that control genes. *Sci Am* 272:54-61, 1995.

Figure 14-17 Reprinted with permission from Muragaki Y, Mundlos S, Upton J, et al: Altered growth and branching patterns in synpolydactyly caused by mutations in *HOXD13*. *Science* 272:548-551, 1996.

Figure 14-18 A from Lumsden A, Graham A: Neural patterning: a forward role for hedgehog. *Curr Biol* 5:1347-1350, 1995. Copyright 1995, Elsevier Science. **B** from Wolpert L, Beddington R, Brockes J, et al: *Principles of development*, New York, 1998, Oxford University Press.

Figure 14-19 From Roessler E, Belloni E, Gaudenz K, et al: Mutations in the human Sonic Hedgehog gene cause holoprosencephaly. *Nat Genet* 14:357-360, 1996.

Figure 14-20 Modified from Wilson PD: Polycystic kidney disease. *N Engl J Med* 350:151-164, 2004. Copyright 2004, Massachusetts Medical Society.

Figure 14-21 Diagram modified from Gupta A, Tsai L-H, Wynshaw-Boris A: Life is a journey: a genetic look at neocortical development. *Nat Rev Genet* 3:342-355, 2002.

Figure 14-22 A from Partington MW: An English family with Waardenburg's syndrome. *Arch Dis Child* 34:154-157, 1959. **B** from DiGeorge AM, Olmsted RW, Harley RD: Waardenburg's syndrome. A syndrome of heterochromia of the irides, lateral displacement of the medial canthi and lacrimal puncta, congenital deafness, and other characteristic associated defects. *J Pediatr* 57:649-669, 1960. **C** from Jones KL: *Smith's recognizable patterns of human malformation*, ed 6, Philadelphia, 2005, WB Saunders.

Figure 14-23 From Carlson BM: *Human embryology and developmental biology*, ed 3, Philadelphia, 2004, Mosby.

Figure 14-24 Modified from Gilbert SF: *Developmental biology*, ed 7, Sunderland, Massachusetts, 2003, Sinauer Associates.

CHAPTER 15

Figure 15-7 Photograph courtesy B. L. Gallie, The Hospital for Sick Children, Toronto.

Figure 15-10 Adapted from Hemminki K, Sundquist J, Lorenzo Bermejo J: Familial risks for cancer as the basis for evidence-based clinical referral and counseling. *Oncologist* 13:239-247, 2008.

Figure 15-13 Adapted from Reis-Filho J, Pusztai L: Gene expression profiling in breast cancer: classification, prognostication, and prediction. *Lancet* 378:1812-1823, 2011.

CHAPTER 17

Figure 17-2 From Moore KL: *The developing human: clinically oriented embryology*, ed 4, Philadelphia, 1988, WB Saunders.

Figure 17-3 Images courtesy A. Toi, Toronto General Hospital, Toronto, Canada.

Figure 17-4 Images courtesy A. Toi, Toronto General Hospital, Toronto, Canada.

Figure 17-5 Redrawn from Wald NJ, Cuckle HS: Recent advances in screening for neural tube defects and Down syndrome. In Rodeck C, editor: *Prenatal diagnosis*, London, 1987, Bailliére Tindall, pp 649-676.

Figure 17-6 Courtesy Mary Norton, University of California, San Francisco.

Figure 17-9 Modified from Kalousek DK: Current topic: confined placental mosaicism and intrauterine fetal development. *Placenta* 15:219-230, 1994.

CHAPTER 18

Figure 18-1 Data from Fuchs CS, Giovannucci EL, Colditz GA, et al: A prospective study of family history and the risk of colorectal cancer. *N Engl J Med* 331:1669-1674, 1994.

Figure 18-2 Modified with permission from Guengerich F: Cytochrome P450s and other enzymes in drug metabolism and toxicity. *AAPS J* 8:E101-E111, 2006.

CASE STUDIES

Figure C-1 A from French LE, Prins C: Erythema multiforme, Stevens-Johnson syndrome and toxic epidermal necrolysis. In Bolognia JL, Jorizzo JL, Schaffer JV, editors: *Dermatology*, ed 3, Philadelphia, 2012, Elsevier, pp 319-333. © 2012, Elsevier Limited. All rights reserved. **B** from Armstrong AW: Erythema multiforme, Stevens-Johnson syndrome, and toxic epidermal necrolysis. In Schwarzenberger K, Werchniak AE, Ko CJ: *General dermatology*, Philadelphia, 2009, Elsevier, pp 23-28. © 2009, Elsevier Limited. All rights reserved.

Figure C-2 Courtesy S. Unger, R. S. Lachman, and D. L. Rimoin, Cedars-Sinai Medical Center, Los Angeles.

Figure C-3 Courtesy Alan Bird, Moorfields Eye Hospital, London.

Figure C-4 Courtesy D. Armstrong, Baylor College of Medicine and Texas Children's Hospital, Houston.

Figure C-5 Courtesy Christa Lese Martin, Autism and Developmental Medicine Institute, Geisinger Health System, Danville, Pennsylvania.

Figure C-6 Courtesy Rosanna Weksberg and Cheryl Shuman, Hospital for Sick Children, Toronto, Canada.

Figure C-7 Courtesy A. Liede and S. Narod, Women's College Hospital and University of Toronto, Canada.

Figure C-8 Courtesy J. R. Lupski, Department of Molecular and Human Genetics, Baylor College of Medicine, Houston, and C. Garcia, Department of Neurology, Tulane University, New Orleans.

Figure C-9 From Jones K: *Smith's recognizable patterns of human malformation*, ed 6, Philadelphia, 2005, Elsevier.

Figure C-10 Courtesy M. M. LeBeau and H. T. Abelson, University of Chicago.

Figure C-11 Courtesy Harris Yfantis and Raymond Cross, University of Maryland and Veterans Administration Medical Center, Baltimore.

Figure C-12 Courtesy J. Rutledge, University of Washington and Children's Hospital and Medical Center, Seattle.

Figure C-13 Audiogram courtesy Virginia W. Norris, Gallaudet University.

Figure C-14 From Gowers WR: Pseudohypertrophic muscular paralysis. A clinical lecture. London, 1879, J. and A. Churchill.

Figure C-15 Courtesy J. Rutledge, University of Washington and Children's Hospital and Medical Center, Seattle, Washington.

Figure C-16 Courtesy M. L. Levy, Department of Dermatology, Baylor College of Medicine, Houston.

Figure C-17 Courtesy Lori Bean and Katie Rudd, Emory Genetics Laboratory, Emory University, Atlanta, Georgia.

Figure C-18 From Helms CA, Major NM, Anderson MW, et al: *Musculoskeletal MRI*, ed 2, Philadelphia, 2009, WB Saunders, pp. 20-49.

Figure C-19 Redrawn from WHO Working Group: Glucose-6-phosphate dehydrogenase deficiency. *Bull World Health Organ* 67:601, 1989, by permission.

Figure C-20 Courtesy Victor Gordeuk, Howard University, Washington, DC.

Figure C-21 Modified from Stefanini M, Dameshek W: *The hemorrhagic disorders: a clinical and therapeutic approach*, New York, 1962, Grune & Stratton, p 252, by permission. Photographic restoration courtesy B. Moseley-Fernandini.

Figure C-22 A courtesy D. Goodman and S. Sargeant, Dartmouth University, Hanover, New Hampshire. **B** and **C** courtesy Raj Kapur, University of Washington, Seattle.

Figure C-23 Courtesy M. Muenke, National Human Genome Research Institute, National Institutes of Health, Bethesda, Maryland. Modified by permission from Nanni L, Ming JE, Bocian M, et al: The mutational spectrum of the sonic hedgehog gene in holoprosencephaly: SHH mutations cause a significant proportion of autosomal dominant holoprosencephaly. *Hum Mol Genet* 8:2479-2488, 1999.

Figure C-24 Courtesy M. R. Hayden, University of British Columbia, Vancouver, Canada.

Figure C-25 A and **B** from Schoen FJ: The heart. In Kumar V, Abbas AK, Aster JC: *Robbins and Cotran pathologic basis of disease*, Philadelphia, 2015, WB Saunders, pp 523-578. **C** from Issa ZF, Miller JM, Zipes DP: *Clinical arrhythmology and electrophysiology: a companion to Braunwald's heart disease*, Philadelphia, 2012, WB Saunders, pp 618-624.

Figure C-26 Modified from Oakley WG, Pyke DA, Taylor KW: *Clinical diabetes and its biochemical basis*. Oxford, 1968, Blackwell Scientific Publications, p 258, by permission. Photographic restoration courtesy B. Moseley-Fernandini.

Figure C-27 Reproduced with permission from Peleg D, Kennedy CM, Hunter SK: Intrauterine growth restriction: identification and management. *Am Fam Physician* 58:453-460, 466-467, 1998.

Figure C-28 A modified with permission from Liu BA, Juurlink DN: Drugs and the QT interval—caveat doctor. *N Engl J Med* 351:1053-1056, 2004. **B** modified from Chiang C, Roden DM: The long QT syndromes: genetic basis and clinical implications. *J Am Coll Cardiol* 36:1-12, 2000.

Figure C-29 Courtesy T. Pal and S. Narod, Women's College Hospital and University of Toronto, Canada.

Figure C-30 Courtesy A. V. Levin, The Hospital for Sick Children and University of Toronto, Canada.

Figure C-31 Courtesy Tina Cowan, Stanford School of Medicine.

Figure C-32 Courtesy D. Chitayat, The Hospital for Sick Children and University of Toronto, Canada.

Figure C-33 Courtesy Annette Feigenbaum, The Hospital for Sick Children, Toronto, Canada.

Figure C-34 Courtesy K. Yohay, Johns Hopkins School of Medicine, Baltimore, Maryland.

Figure C-35 Courtesy R. A. Lewis, Baylor College of Medicine, Houston.

Figure C-37 Courtesy J. Rutledge, Department of Pathology, University of Washington, Seattle.

Figure C-38 Courtesy S. Heeger, University Hospitals of Cleveland.

Figure C-39 Courtesy R. A. Lewis, Baylor College of Medicine, Houston.

Figure C-40 Courtesy M. Segawa, Segawa Neurological Clinic for Children, Tokyo. Modified from Segawa M: Pathophysiology of Rett syndrome from the stand point of clinical characteristics. *Brain Dev* 23:S94-S98, 2001.

Figure C-41 Courtesy B. Bejjani and L. Shaffer, Baylor College of Medicine, Houston.

Figure C-42 From Nathan DG, Oski FA: *Hematology of infancy and childhood*, Philadelphia, 1981, WB Saunders.

Figure C-43 Courtesy A. V. Levin, The Hospital for Sick Children and University of Toronto, Canada.

Figure C-44 Courtesy N. Olivieri, The Hospital for Sick Children and University of Toronto, Canada.

Figure C-45 From Eichelbaum M, Ingelman-Sundberg M, Evans WE: Pharmacogenomics and individualized drug therapy. *Annu Rev Med* 57:119-137, 2006.

Figure C-46 Courtesy H. Meyerson and Robert Hoffman, Case Western Reserve University, Cleveland, Ohio.

Figure C-47 Modified from Lyon AJ, Preece MA, Grant DB: Growth curve for girls with Turner syndrome. *Arch Dis Child* 60:932, 1985, by permission.

Figure C-48 Courtesy M. L. Levy, Baylor College of Medicine and Texas Children's Hospital, Houston.

Answers to Problems

CHAPTER 2   *Introduction to the Human Genome*

1. (a) *A* or *a*.
   (b) i. At meiosis I. ii. At meiosis II.
2. Meiotic nondisjunction.
3. $(\frac{1}{2})^{23} \times (\frac{1}{2})^{23}$; you would be female.
4. (a) 23; 46.
   (b) 23; 23.
   (c) At fertilization; at S phase of the next cell cycle.
5. Chromosome 1, $\approx$9 genes/Mb; chromosome 13, $\approx$3-4 genes/Mb; chromosome 18, $\approx$4 genes/Mb; chromosome 19, $\approx$19 genes/Mb; chromosome 21, $\approx$5 genes/Mb; chromosome 22, $\approx$10 genes/Mb. Because of the higher density of genes, one would expect that a chromosome abnormality of chromosome 19 would have a greater impact on phenotype than an abnormality of chromosome 18. Similarly, chromosome 22 defects are expected to be more deleterious than those of chromosome 21.

CHAPTER 3   *The Human Genome: Gene Structure and Function*

1. There are several possible sequences because of the degeneracy of the genetic code. One possible sequence of the double-stranded DNA is

   5′ AAA AGA CAT CAT TAT CTA 3′
   3′ TTT TCT GTA GTA ATA GAT 5′

   RNA polymerase "reads" the bottom (3′ to 5′) strand. The sequence of the resulting mRNA would be 5′ AAA AGA CAU CAU UAU CUA 3′.

   The mutants represent the following kinds of mutations:

   Mutant 1: single-nucleotide substitution in fifth codon; for example, UAU → UGU.
   Mutant 2: frameshift mutation, deletion in first nucleotide of third codon.
   Mutant 3: frameshift mutation, insertion of G between first and second codons.
   Mutant 4: in-frame deletion of three codons (nine nucleotides), beginning at the third base.
2. The sequence of the haploid human genome consists of nearly 3 billion nucleotides, organized into 24 types of human chromosome. Chromosomes contain chromatin, consisting of nucleosomes. Chromosomes contain G bands that contain several thousand kilobase pairs of DNA (or several million base pairs) and hundreds of genes, each containing (usually) both introns and exons. The exons are a series of codons, each of which is three base pairs in length. Each gene contains a promoter at its 5′ end that directs transcription of the gene under appropriate conditions.
3. A mutation in a promoter could interfere with or eliminate transcription of the gene. Mutation of the initiator codon would prevent normal translation. Mutations at splice sites can interfere with the normal process of RNA splicing, leading to abnormal mRNAs. A 1-bp deletion in the coding sequence would lead to a frameshift mutation, thus changing the frame in which the genetic code is read; this would alter the encoded amino acids and change the sequence of the protein. (See examples in Chapter 11.) A mutation in the stop codon would allow translation to continue beyond its normal stopping point, thus adding new, incorrect amino acids to the end of the encoded protein.
4. Mutations in introns can influence RNA splicing, thus leading to an abnormally spliced mRNA (see Chapter 11). *Alu* or L1 sequences can be involved in abnormal recombination events between different copies of the repeat, thus deleting or rearranging genes. L1 repeats can also actively transpose around the genome, potentially inserting into a functional gene and disrupting its normal function. Locus control regions influence the proper expression of genes in time and space; deletion of a locus control region can thus disrupt normal expression of a gene (see Chapter 11). Pseudogenes are, generally, nonfunctional copies of genes; thus, in most instances, mutations in a pseudogene would not be expected to contribute to disease, although there are rare exceptions.
5. RNA splicing generates a mature RNA from the primary RNA transcript by combining segments of exonic RNA and eliminating RNA from introns. RNA splicing is a critical step in normal gene expression in all tissues of the body and operates at the level of RNA. Thus the genomic DNA is unchanged. In contrast, in somatic rearrangement, segments of genomic DNA are rearranged to eliminate certain sequences and generate mature genes during lymphocyte precursor cell development as part of the normal process of generating immunoglobulin and T-cell receptor diversity. Somatic rearrangement is a highly specialized process, specific only to these genes and specific cell types.
6. Variation in epigenetic modifications might lead to overexpression or underexpression of a gene or genes. DNA methylation might lead to epigenetic silencing

of a gene. miRNAs can be involved in regulating the expression of other genes, and mutations in such an miRNA might be expected to alter gene expression patterns. The product of lncRNA genes are RNAs that can be involved in epigenetic regulation or other regulatory pathways; deletion or improper expression of such an miRNA might therefore lead to abnormalities of developmental pathways.

7. Genomic imprinting involves epigenetic silencing of an allele (or alleles at a number of closely located genes) based solely on parental origin due to epigenetic marks inherited through the germline. X inactivation involves epigenetic silencing of alleles along much of an entire chromosome based not on parental origin, but rather on a random choice of one or the other X chromosome at the time of initiation of the process in early embryonic development.

CHAPTER 4 *Human Genetic Diversity: Mutation and Polymorphism*

1. (a) CNV.
   (b) Indel.
   (c) A mutation in a splice site.
   (d) An inversion.
   (e) A SNP (or in/del) in a noncoding region or intron, or a SNP that leads to a synonymous substitution.

2. Assuming 20 years is one generation, 41 mutations/9 million alleles/2 generations = $\approx 2.3 \times 10^{-6}$ mutations/generation at the aniridia locus. The estimate is based on assumptions that ascertained cases result from new mutation, that the disease is fully penetrant, that all new mutants are liveborn (and ascertained), and that there is only a single locus at which mutations can lead to aniridia. If there are multiple loci, the estimated rate is too high. If some mutations are not ascertained (because of lack of penetrance or death in utero), the estimated rate might be too low.

3. A microsatellite polymorphism, because microsatellite polymorphisms typically have more alleles, which provides greater capacity to distinguish genomes. A single SNP or indel would only have two alleles.

4. Based on information in this chapter, each cell division leads to less than one new single-bp mutation per genome. Rounding this up a bit to be one mutation per cell division, then one would expect at most 100 single-bp differences between cells at the end points of the two lineages mentioned. The rate of CNV de novo changes is much higher, and thus one would expect many, many such differences between the two lineages. Improved technologies now allow sequencing of single-cell genomes (i.e., rather than sequencing DNA from a collection of millions of cells). Thus it will now be possible to determine the answer to this question experimentally, rather than just theoretically.

5. Different types of mutations are sensitive to maternal or paternal age. Both single-bp mutations and CNVs show an increase in frequency with increasing age of the father. In contrast, meiotic nondisjunction for many chromosomes (including chromosome 21) shows an increase with increasing age of the mother. The rate of mutation (per bp) varies greatly in different locations around the genome; hot spots of mutation show much higher rates, although the basis for this is poorly understood. Intrachromosomal homologous recombination can lead to copy number variation in gene families or to deletion/duplications for regions flanked by homologous sequences (e.g., segmental duplications). Overall, the rate of mutation can be influenced also by genetic variation, both at a population level and in specific parental genomes. In any individual genome, this may influence where one falls in the ranges observed in typical genomes, as summarized in the Box on page 55.

CHAPTER 5 *Principles of Clinical Cytogenetics and Genome Analysis*

1. (a) Forty-six chromosomes, male; one of the chromosome 18s has a shorter long arm than is normal.
   (b) To determine whether the abnormality is de novo or inherited from a balanced carrier parent.
   (c) Forty-six chromosomes, male, only one normal 7 and one normal 18, plus a reciprocal translocation between chromosomes 7 and 18. This is a balanced karyotype. For meiotic pairing and segregation, see text, particularly Figure 5-12.
   (d) The del(18q) chromosome is the der(18) translocation chromosome, 18pter → 18q12::7q35 → 7qter. The boy's karyotype is unbalanced; he is monosomic for the distal long arm of 18 and trisomic for the distal long arm of 7. Given the number of genes on chromosomes 7 and 18 (see Fig. 2-7), one would predict that the boy is monosomic for approximately 100 genes on chromosome 18 and trisomic for approximately 100 genes on chromosome 7.

2. (a) Approximately 95%.
   (b) No increased risk, but prenatal diagnosis may be offered.

3. Postzygotic nondisjunction, in an early mitotic division. Although the clinical course cannot be predicted with complete accuracy, it is likely that she will be somewhat less severely affected than would a nonmosaic trisomy 21 child.

4. (a) Abnormal phenotype, unless the marker is exceptionally small and restricted only to the centromeric sequences themselves. Gametes may be normal or abnormal; prenatal diagnosis indicated.
   (b) Abnormal phenotype (trisomy 13; see Chapter 6); will not reproduce.

(c) Abnormal phenotype in proband and approximately 50% of offspring.

(d) Normal phenotype, but risk for unbalanced offspring (see text).

(e) Normal phenotype, but risk for unbalanced offspring, depending on the size of the inverted segment (see text).

5. (a) Not indicated.

(b) Fetal karyotyping indicated; at risk for trisomy 21, in particular.

(c) Karyotype indicated for the child to determine whether it is trisomy 21 or translocation Down syndrome. If it is translocation, parental karyotypes are indicated.

(d) Not indicated, unless other clinical findings might suggest a contiguous gene syndrome (see Chapter 6).

(e) Karyotype indicated for the boys to rule out deletion or other chromosomal abnormality. If clinical findings indicate possibility of fragile X syndrome, a specific DNA diagnostic test would be indicated.

6. (a) Paracentric inversion of the X chromosome, between bands Xq21 and Xq26, determined by karyotyping.

(b) Terminal deletion of 1p in a female, determined by karyotyping.

(c) Female with deletion within band q11.2 of chromosome 15, determined by in situ hybridization with probes for the *SNRPN* gene and D15S10 locus.

(d) Female with interstitial deletion of chromosome 15, between bands q11 and q13, determined by karyotyping. In situ hybridization analysis confirmed deletion of sequences within 15q11.2, with use of probes for the *SNRPN* gene and D15S10 locus.

(e) Female with deletion of sequences in band 1q36.3, determined by array CGH with the three BAC probes indicated.

(f) Male with an extra marker chromosome, determined by karyotyping. Marker was identified as an r(8) chromosome by in situ hybridization with a probe for D8Z1 at the centromere.

(g) Female with Down syndrome, with a 13q;21q Robertsonian translocation in addition to two normal chromosome 21s, determined by karyotyping.

(h) Presumably normal male carrier of a 13q;21q Robertsonian translocation, in addition to a single normal chromosome 21 (and a single normal chromosome 13), as determined by karyotyping.

7. (a) For Figure 5-6C: 46,XY,dup(X)(q28). The increased ratio of sequences in Xq28 indicates a duplication.

(b) For Figure 5-9C: 47,XX,+21. The individual is a female, because the intensity of sequences on the X are equivalent to all of the autosomes (other than 21). (The scattering of very low-intensity signals from the Y sequences is simply background noise.)

CHAPTER 6 *The Chromosomal and Genomic Basis of Disease: Disorders of the Autosomes and the Sex Chromosomes*

1. Theoretically, X and XX gametes in equal proportions; expected XX, XY, XXX, and XXY offspring (25% each). In actuality, XXX women have virtually all chromosomally normal offspring, XX and XY, implying that XX gametes are at a significant disadvantage or are lost.

2. It is possible that the relevant region of chromosome 9 has very few genes and that the inversion does not interfere with gene structure or function. Carriers are not genetically imbalanced. Their potential risk might be to the offspring, as seen with other pericentric inversions. However, the flanking regions of 9p and 9q are so large (i.e., most of those chromosome arms) that duplication or deletion resulting from meiotic crossing over may be incompatible with life. Alternatively, centromeric regions of chromosomes are relatively poor in recombination; thus there may be very few crossovers in this region, and the inv(9) could pass to the next generation unchanged.

3. No. XYY can result only from meiosis II nondisjunction in the male, whereas XXY can result from nondisjunction at meiosis I in the male or at either division in the female.

4. Translocation of Y chromosome material containing the sex-determining region (and the *SRY* gene) to the X chromosome or to an autosome.

5. The small r(X) may contain genes that normally would undergo X inactivation but fail to do this in this abnormal chromosome lacking an X inactivation center. Such genes would show biallelic expression and be expressed at higher levels than seen in typical males (one X) or typical females (one active X, one inactive X). This abnormal gene expression may underlie the intellectual disability.

In the second family, the larger r(X) contains the X inactivation center. Thus one predicts that X inactivation would proceed normally and the r(X) would be the inactive X in all cells (due to secondary cell selection; see Fig. 6-13B). The phenotype is somewhat uncertain, however, because this individual may be deficient for genes that would typically escape X inactivation and would be expressed biallelically; some features of Turner syndrome may therefore be present.

6. 46,XX; autosomal recessive; prenatal diagnosis possible; need for clinical attention in neonatal period to determine sex and to forestall salt-losing crises.

7. (a) None; the short arms of all acrocentric chromosomes are believed to be identical and contain multiple copies of rRNA genes.

(b) None if the deletion involves only heterochromatin (Yq12). A more proximal deletion might delete genes important in spermatogenesis (see Fig. 6-9).

(c) Cri du chat syndrome, severity depending on the amount of DNA deleted (see Fig. 6-6).

(d) Some features of Turner syndrome, but with normal stature; the Xq⁻ chromosome is preferentially inactivated in all cells (provided that the X inactivation center is not deleted), thus reducing the potential severity of such a deletion.

Different parts of the genome contain different density of genes. Thus deletion of the same amount of DNA on different chromosomes might delete a vastly different number of genes, thus leading to different phenotypic expectations (see Fig. 2-7).

8. Question for discussion. See text for possible explanations.

9. (a) A 1% risk is often quoted, but the risk is probably not greater than the population age-related risk.

(b) Age-related risk is greater than 1%.

(c) No increased risk if the niece with Down syndrome has trisomy 21, but if the niece carries a Robertsonian translocation, the consultand may be a carrier and at high risk.

(d) 10% to 15%; see text.

(e) Only a few percent; see text. The woman's age-related risk may be relevant.

10. 46,XX,rob(21;21)(q10;q10) or 46,XX,der(21;21)(q10;q10). (There is no need to add +21 to the karyotype because the 46 designates that she must have a normal 21 in addition to the translocation.)

11. Crossing over leads to either balanced gametes or nonviable gametes (see Fig. 5-13). Thus liveborn offspring are genomically balanced.

CHAPTER 7 *Patterns of Single-Gene Inheritance*

1. (b) Autosomal recessive; 1 in 4, assuming same paternity as her first child.

(c) Calvin and Cathy are obligate heterozygotes. Given that Calvin and Cathy are first cousins, it is also very likely that they inherited their mutant allele through Betty and Barbara from the same grandparent. Thus Betty and Barbara are very likely to be carriers, but it is not obligatory. It is theoretically possible that Cathy inherited her CF allele from Bob and that Calvin inherited his from his father, Barbara's husband. DNA-based carrier testing will answer the question definitively.

2. (a) Heterozygous at each of two loci; for example, *A/a B/b*.

(b) George and Grace are likely carriers for an autosomal recessive form of deafness; Horace is either a homozygote or compound heterozygote at this same deafness locus. Gilbert and Gisele are both homozygotes or compound heterozygotes for deafness due to mutations at a deafness locus as well. The fact that all of Horace and Hedy's children are deaf suggests that the deafness locus in Gilbert and Gisele's family and the locus in George and Grace's family are the same locus. Isaac and Ingrid, however, although deaf, are deaf due to each being homozygous or compound heterozygote at two different deafness loci, so all of their children are double heterozygotes (as shown in part a of this question).

3. Variable expressivity—d; uniparental disomy—i; consanguinity—j; inbreeding—c; X-linked dominant inheritance—g; new mutation—e; allelic heterogeneity—h; locus heterogeneity—a; homozygosity for an autosomal dominant trait—b; pleiotropy—f.

4. (b) They are homozygous.

(c) 100% for a son of Elise; virtually zero for a daughter unless Elise's partner has hemophilia A.

(d) Enid is an obligatory carrier (heterozygote for hemophilia A mutation) because she has an affected father but is herself not affected, so the chance a son will be affected is 50%. The probability a daughter will be a carrier is 50%, but the daughter's chance of being affected is virtually zero unless Enid's partner is himself affected with hemophilia A, which would give a 50% chance of being affected, or if there is some very unusual circumstance of highly skewed X inactivation or if her daughter has Turner syndrome with a single maternal X carrying the mutant hemophilia A gene.

5. All are possible except (c), which is unlikely if the parents are completely unaffected.

6. (a) New mutation or germline mosaicism in one of the parents.

(b) Mutation rate at the *NF1* locus if truly a new mutation; if one of the parents is a germline mosaic, the risk in the next pregnancy is a function of the fraction of gametes carrying the mutation, which is unknown.

(c) Mutation rate at the *NF1* locus if truly a new mutation; if the father is a germline mosaic for NF1, the risk in the next pregnancy is a function of the fraction of sperm carrying the mutation, which is unknown.

(d) 50%.

7. The consultand and her partner are first cousins once removed. The probability that a child of this mating will be homozygous at any given locus for an allele inherited through each parent from a common

ancestor is known as the **coefficient of inbreeding (F)**. In the accompanying figure, suppose individual I-1 is heterozygous for alleles *1* and *2*, whereas individual I-2 is a *3,4* heterozygote. The chance II-2 inherits the *2* allele is $\frac{1}{2}$, and the chance III-2 inherits it from II-2 is $\frac{1}{2}$, so the chance III-2 inherits allele *2* from I-1 is $\frac{1}{2} \times \frac{1}{2} = \frac{1}{4}$. Similarly, the chance IV-1 is a carrier of allele *2* inherited from I-1 is $\frac{1}{2} \times \frac{1}{2} \times \frac{1}{2} = \frac{1}{8}$, and the chance III-2 and IV-1 are both heterozygotes for *2* inherited from I-1 is $\frac{1}{4} \times \frac{1}{8} = \frac{1}{32}$. The probability their child could be a *22* homozygote inherited from I-1 as a common ancestor is therefore $\frac{1}{4} \times \frac{1}{32} = \frac{1}{128}$. Repeat this calculation for allele *1* in I-1, and for either allele *3* or *4* in II-1, which means the probability the child could be homozygous *1,1*, *2,2*, *3,3*, or *4,4* is equal to $4 \times \frac{1}{128} = \frac{1}{32}$ because there are four ways the child could be a homozygote for an allele inherited from either of the two common ancestors. This is the coefficient of inbreeding.

A simple way to calculate F in simple pedigrees like this is the path method, in which one determines all the paths by which an allele from a common ancestor can be transmitted to the individual whose coefficient of inbreeding one is seeking to calculate.

Form all the paths connecting all the pertinent individuals in this pedigree (see Figure). Each path that gives a closed loop is a consanguineous path. There are two closed loops: A-D-H-K-L-I-E-A and B-D-H-K-L-I-E-B. To calculate F, count all the "nodes" (the dots representing each of the individuals) in each of the closed loops, counting each node only once. Call that n. The coefficient of inbreeding due to that loop is then given by $(\frac{1}{2})^{n-1}$. So, in this example, the loop A-D-H-K-L-I-E-A contains seven unique nodes, n = 7. Add all the coefficients for each loop together to find F. For the pedigree, then:

$$(\tfrac{1}{2})^{n-1} = (\tfrac{1}{2})^{6} = \tfrac{1}{64} \text{ for loop A-D-H-K-L-I-E-A}$$

$$(\tfrac{1}{64})^{n-1} = (\tfrac{1}{64})^{6} = \tfrac{1}{64} \text{ for loop B-D-H-K-L-I-E-B}$$

$$\text{and thus F} = \tfrac{1}{32}$$

8. AD is most likely. Vertical transmission, including male to male, from generation to generation, males and females affected.

AR and XR are possible but unlikely. AR would require that both of the spouses of the two affected individuals in generations I and II be carriers, which is unlikely unless the pedigree comes from a genetic isolate (so-called pseudodominant inheritance of a recessive disorder due to high frequency of carriers in the population). XR would require that the same two women be carriers and, in addition, that there be something unusual in the X inactivation pattern for the female in generation III to be affected while neither of the females in generation II (who are both obligatory carriers) is affected.

Mitochondrial and XD inheritance are incompatible. There is male to male transmission, which eliminates both of these modes of inheritance. In addition, there are female offspring of affected males who are not affected.

9. The probability a child of two carriers of cystic fibrosis will be affected is 0.25, based on autosomal recessive inheritance.

The probability a child will be affected by cystic fibrosis whose mother is a carrier, but whose father is not, is the chance of inheriting the mother's mutant allele and a new mutation in the sperm, which is $0.5 \times 1 \times 10^{-6}$, and therefore the odds both parents are carriers versus only the mother is $0.25/(0.5 \times 10^{-6}) = 5 \times 10^{7}$. The odds in favor of both parents being carriers are overwhelming. Indeed, the probability of misattributed paternity, with the biological father being a carrier, dwarfs the probability of a new mutation.

CHAPTER 8 *Complex Inheritance of Common Multifactorial Disorders*

1. (a) Autosomal dominant with reduced penetrance. If it were truly multifactorial, the risk for more distantly related relatives would drop by more than 50% with each increase in the degree of relatedness.

   (b) In dominant disease, a study of multiple families with the condition would reveal the expected 50% ratio of affected to unaffected in the children of an affected individual (after correcting

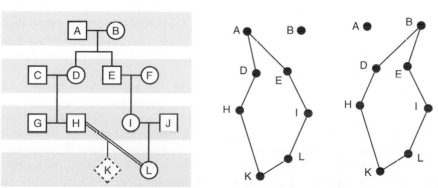

Figure for Chapter 7, question 7.

for bias of ascertainment of the families). In multifactorial inheritance, there would be fewer than the expected 50% affected in the children of an affected individual.

2. Male to male transmission can disprove X-linkage; other criteria of multifactorial inheritance can be examined, as in the text.

3. For autosomal recessive but not for multifactorial inheritance, all of the affected individuals in a family tend to be in the same sibship, with unaffected parents, whereas diseases with multifactorial inheritance can present as affected parents with affected children. It is generally rare for a parent of children with an autosomal recessive disorder to be himself or herself affected because it would require a homozygote or compound heterozygote–affected parent to mate with a carrier of a mutant allele at that same locus. There can be an increased incidence of such rare matings, however, when there *is assortative mating* or if the couple is *consanguineous* or comes from an *inbred population*.

CHAPTER 9   *Genetic Variation in Populations*

1. One way of determining this is to reverse the question and ask instead what proportion of individuals would be *homozygous*. Then the proportion that is heterozygous is 1 minus the proportion that is homozygous. The frequency of homozygotes for the first allele would be $0.40 \times 0.40 = 0.16$, $0.30 \times 0.30 = 0.09$ for the second allele, $0.15 \times 0.15 = 0.0225$ for allele three, etc. Adding these up for the five alleles ($0.16 + 0.09 + 0.0225 + 0.01 + 0.0025 = 0.285$) means 28.5% of individuals would be homozygous for allele 1 *or* allele 2 *or* … allele 5. Therefore 71.5% of individuals would be heterozygous at this locus.

2. $q = 0.26$, $p = \approx 0.74$, $p^2 = \approx 0.55$, $2pq = \approx 0.38$, $q^2 = \approx 0.07$

Frequency of $Rh-/-$ genotype in mother = 0.07. Frequency of $Rh+/+$ in father = 0.55. Frequency of $Rh+/-$ in father = 0.38.

First pregnancy:

Probability of $Rh-/-$ mother $\times$ $Rh+/+$ father mating = $0.07 \times 0.55 = 3.8\%$.

Probability of $Rh-/-$ mother $\times$ $Rh+/-$ father mating = $0.07 \times 0.38 = 2.66\%$.

Second pregnancy:

All second pregnancies of $Rh-/-$ mother $\times$ $Rh+/+$ father will be sensitized by the first pregnancy = 3.8%, and all are at risk for Rh incompatibility in the next pregnancy.

Only half of the first pregnancies of an $Rh-/-$ mother by $Rh+/-$ father will have sensitized the mother ($Rh+/-$), so the risk for a sensitized $Rh-/-$ mother with an $Rh+/-$ partner in the second pregnancy = $\frac{1}{2} \times 2.66\% = 1.33\%$ and the chance a sensitized $Rh-/-$ mother will have

a $Rh+/-$ child when her mate is $Rh+/- = \frac{1}{2} \times 1.33\% = 0.66\%$.

Total at risk for incompatibility $3.8\% + 0.66\% \approx 4.5\%$ in the population at the time of a second pregnancy in the absence of any prophylaxis.

3. (a) Assume there are 100 individuals in the population, carrying 200 alleles at a particular locus. The frequency of A is $(2 \times \frac{81}{200}) + (\frac{18}{200}) = 0.9$ and frequency of $a = 0.1$

(b) The genotype frequencies will be the same as in this generation, assuming Hardy-Weinberg equilibrium.

(c) Frequency of A/a by A/a matings = $0.18 \times 0.18 = \approx 0.0324$.

4. (a) When q is small, $p = \approx 1$, and so $2pq = \approx 2q$. Therefore, if $2pq = 0.04$, then the allele frequency of β-thalassemia $q = \approx 0.02$. (You can also calculate q exactly by letting $2pq = 2(1-q)q = 0.04$, or $q^2 - q + 0.02 = 0$ and solve the quadratic equation.)

(b) Assuming only heterozygotes for β-thalassemia are likely to reproduce (a reasonable assumption because the fitness in β-thalassemia is quite low), then $0.04 \times 0.04 = 0.0016 = 0.16\%$ of matings will be between heterozygotes.

(c) Incidence of affected fetuses or newborns = 0.04% assuming no increased fetal losses in β-thalassemia, which is a reasonable assumption because the disorder has a postnatal onset.

(d) Incidence of carriers among the offspring of couples both found to be heterozygous is 50%.

5. Only (d) is in equilibrium. Possible explanations include selection for or against particular genotypes, nonrandom mating, and recent migration.

6. (a) Abby has a $\frac{2}{3}$ chance of being a carrier. Andrew has approximately a $\frac{1}{150}$ chance of being a carrier. Therefore their risk for having an affected child is $\frac{2}{3} \times \frac{1}{150} \times \frac{1}{4}$, or $\frac{1}{900}$.

(b) $\frac{2}{3} \times \frac{1}{4} \times \frac{1}{4} = \frac{1}{24}$.

(c) $\frac{2}{3} \times \frac{1}{22} \times \frac{1}{4} = \frac{1}{132}$; $\frac{2}{3} \times \frac{1}{4} \times \frac{1}{4} = \frac{1}{24}$.

7. (a) Facioscapulohumeral muscular dystrophy: $q = \frac{1}{50,000}$, $2pq = \frac{1}{25,000}$. Friedreich ataxia: $q = \frac{1}{158}$, $2pq = \frac{1}{79}$. Duchenne muscular dystrophy is X-linked recessive and occurs mostly in males, so we will ignore any of the rare females affected. If it occurs in the population at a frequency of 1 in 25,000, then, assuming half of the population is male, the frequency in males must be 1 in 12,500, so $q = \frac{1}{12,500}$, $2pq = \frac{1}{6,250}$.

(b) The autosomal dominant and X-linked disorders would increase rapidly, within one generation, to reach a new balance. The autosomal recessive disorders would increase also, but only very slowly, because the majority of the mutant alleles are not subject to selection.

8. Mutant allele frequencies are approximately $\frac{1}{26}$ and $\frac{1}{316}$. Two possible explanations for the difference in

allele frequency could be (1) founder effect (or more generally genetic drift) in the early Quebec population when it was very small and inbred, resulting in an increased mutant allele frequency, or (2) environmental conditions of unknown type that provided a heterozygote advantage in Quebec that raised the allele frequency through increased fitness of heterozygote carriers.

CHAPTER 10   *Identifying the Genetic Basis for Human Disease*

1. The HD and MNSs loci map far enough apart on chromosome 4 to be unlinked, even though they are syntenic.

2. The LOD scores indicate that this polymorphism in the α-globin gene locus is closely linked to the polycystic kidney disease gene. The peak LOD score, 25.85, occurs at 5 cM. The odds in favor of linkage at this distance compared with no linkage at all are $10^{25.85}:1$ (i.e., almost $10^{26}:1$). The data in the second study indicate that there is *no* linkage between the disease gene and the polymorphism in this family. Thus there is genetic heterogeneity in this disorder, and linkage information can therefore be used for diagnosis only if there is previous evidence that the disease in that particular family is linked to the polymorphism.

3. Every parent who passed on the cataract was also informative at the γ-crystallin locus, that is, was heterozygous for the polymorphic alleles at this locus. The phase is known by inspecting the pedigree in individuals IV-7 and IV-8 because these two individuals received both the cataract allele and the *A* allele at the γ-crystallin locus from their father (but note, we do not know what the phase was in the father simply by inspection). We do not know the phase in individuals IV-3 or IV-4 because we do not know if they inherited the cataract mutation along with the *A* or the *B* allele at the γ-crystallin locus from their mother. Phase is also known in individuals V-1, V-2, V-6, and V-7. The cataract seems to cosegregate with the "A" allele. There are no crossovers. A complete LOD score analysis should be performed. In addition, one might examine the γ-crystallin gene itself for mutations in affected persons because it would be a reasonable candidate for being the gene in which mutations could cause cataracts.

4. (a) The phase in the mother is probably *B-WAS* (where *WAS* is the disease-causing allele), according to the genotype of the affected boy. This phase can be determined with only 95% certainty because there is a 5% chance that a crossover occurred in the meiosis leading to the affected boy. On the basis of this information, there is a $(0.95 \times 0.95) + (0.05 \times 0.05) = 0.905$ chance that the fetus (who is male) will be *unaffected*.

(b) This surprising result (assuming paternity is as stated) indicates that the mother has inherited the *A* allele (and the *WAS* allele) from her mother and her phase is therefore *A-WAS*, not *B-WAS*. Thus there must have been a crossover in the meiosis leading to the affected boy. To confirm this, one should examine polymorphisms on either side of this one on the X chromosome to make sure that the segregation patterns are consistent with a crossover. On the basis of this new information, there is now a 95% chance that the fetus in the current pregnancy is *affected*.

5. No, because you would not know if II-2 inherited the mutant allele *D* along with the *A* from her father or the *A* from her mother. Phase becomes unknown again, as in Figure 10-10A.

6. Yes, phase is known in the mother of the two affected boys because she must have received the mutant factor VIII allele (*h*) and the *M* allele at the polymorphic locus on the X she received from her father.

7. In 10-7A, D = 0, so D' = 0.
   In 10-7B, D = −0.05, and because D < 0, F = smaller of freq(A)freq(S) versus freq (a)freq(s), so F = (0.5)(0.1) = 0.05 versus (0.5)(0.9) = 0.45, so F = 0.05 and D' = −0.05/0.05 = −1, reflecting the complete linkage disequilibrium.
   In 10-7C, D = −0.04, F = 0.05 again, and D' = −0.8, reflecting a high degree but not perfect LD.

8. Odds ratio for the variant and disease = (a/b)/(c/d) = ad/bc.
   Relative risk = [a/(a + b)]/[c/(c + d)] = a(c + d)/c(a + b).
   With three times as many controls, the odds ratio = a(3b)/c(3d) = 3ad/3bc = ad/bc, which is unchanged from the previous odds ratio.
   Relative risk = [a/(a + 3b)]/[c/(c + 3d)] = a(c + 3d)/c(a + 3b) which is *not* the same as the previously calculated relative risk.

CHAPTER 11   *The Molecular Basis of Genetic Disease*

1. The pedigree should contain the following information: Hydrops fetalis is due to a total absence of α chains. The parents each must have the genotype αα/−−. The α− genotype is common in some populations, including Melanesians. Parents with this genotype cannot transmit a −−/−− genotype to their offspring.

2. Except in isolated populations, patients with β-thalassemia will often be genetic compounds because there are usually many alleles present in a population in which β-thalassemia is common. In isolated populations, the chance that a patient is a true homozygote of a single allele is greater than it

would be in a population in which thalassemia is rare. In the latter group, more "private mutations" might be expected (ones found solely or almost solely in a single pedigree). A patient is more likely to have identical alleles if he or she belongs to a geographical isolate with a high frequency of a single allele or a few alleles, or if his or her parents are consanguineous. See text in Chapter 7.

3. Three bands on the RNA blot could indicate, among other possibilities, that (a) one allele is producing two mRNAs, one normal in size and the other abnormal, and the other allele is producing one mRNA of abnormal size; (b) both alleles are making a normal-sized transcript and an abnormal transcript, but the aberrant ones are of different sizes; or (c) one allele is producing three mRNAs of different sizes, and the other allele is making no transcripts.

Scenario (c) is highly improbable, if possible at all. Two mRNAs from a single allele could result from a splicing defect that allows the normal mRNA to be made, but at reduced efficiency, while leading to the synthesis of another transcript of abnormal size, which results from either the incorporation of intron sequences in the mRNA or the loss of exon sequences from the mRNA. In this case, the other abnormal band comes from the other allele. A larger band from the other allele could result from a splicing defect or an insertion, whereas a smaller band could be due to a splicing defect or a deletion. Hb E is caused by an allele from which both a normal and a shortened transcript are made (see Fig. 11-10); the normal mRNA makes up 40% of the total β-globin mRNA, producing only a mild anemia.

4. These two mutations affect different globin chains. The expected offspring are $\frac{1}{4}$ normal, $\frac{1}{4}$ Hb M Saskatoon heterozygotes with methemoglobinemia, $\frac{1}{4}$ Hb M Boston heterozygotes with methemoglobinemia, and $\frac{1}{4}$ double heterozygotes with four hemoglobin types: normal, both types of Hb M, and a type with abnormalities in both chains. In the double heterozygotes, the clinical consequences are unknown—probably more severe methemoglobinemia.

5. $\frac{2}{3} \times \frac{2}{3} \times \frac{1}{4} = \frac{1}{9}$.

6. $\frac{1}{4}$.

7. 8, 1, 2, 7, 10, 4, 9, 5, 6, and 3.

8. Exceptions to this rule can arise, for example, from splice site mutations that lead to the mis-splicing of an exon. The exon may be excluded from the mRNA, generating an in-frame deletion of the protein sequence or causing a change in the reading frame, leading to the inclusion of different amino acids in the protein sequence.

9. Approximately two thirds of the couples to whom such infants were born did not know about thalassemia or the prevention programs. Approximately 20% refused abortion, and false paternity was identified in 13% of cases.

CHAPTER 12 *The Molecular, Biochemical, and Cellular Basis of Genetic Disease*

1. Three types of mutations could explain a mutant protein that is 50 kD larger than the normal polypeptide:

   A mutation in the normal stop codon that allows translation to continue.

   A splice mutation that results in the inclusion of intron sequences in the coding region. The intron sequences would have to be free of stop codons for sufficient length to allow the extra 50 kD of translation.

   An insertion, with an open reading frame, into the coding sequence.

   For any of these, approximately 500 extra residues would be added to the protein if the average molecular weight of an amino acid is approximately 100. Five hundred amino acids would be encoded by 1500 nucleotides.

2. Autosomal dominant *PCSK9* gain-of-function mutations that cause familial hypercholesterolemia are genocopies of autosomal dominant loss-of-function mutations in the LDL receptor gene (*LDLR*), because a genocopy is a genotype that determines a phenotype closely similar to that determined by a different genotype (for comparison, see *Glossary* for the definition of phenocopy).

3. A nucleotide substitution that changes one amino acid residue to another should be termed a *putative pathogenic mutation*, and possibly a *polymorphism*, unless (1) it has been demonstrated, through a functional assay of the protein, that the change impairs the function to a degree consistent with the phenotype of the patient, or (2) instead of or in addition to a functional assay, it can be demonstrated that the nucleotide change is found *only* on mutant chromosomes, which can be identified by haplotype analysis in the population of patients and their parents and *not* on normal chromosomes in this population.

   The fact that the nucleotide change is only rarely observed in the normal population and found with significantly higher frequency in a mutant population is strong supportive evidence but not proof that the substitution is a pathogenic mutation.

4. If Johnny has CF, the chances are approximately 0.85 × 0.85, or 70%, that he has a previously described mutation that could be readily identified by DNA analysis. His parents are from northern Europe; therefore the probability that he is homozygous for the ΔF508 mutation is 0.7 × 0.7, or 50%, because approximately 70% of CF carriers in northern Europe have this mutation. If he does not have the ΔF508 mutation, he could certainly still have CF, because approximately 30% of the alleles (in the northern European population, at least) are not

ΔF508. Steps to DNA diagnosis for CF include the following: (1) look directly for the ΔF508 mutation; if it is not present, (2) look for other mutations common in the specific population; (3) then look directly for other mutations based on probabilities suggested by the haplotype data; (4) if all efforts to identify a mutation fail (or if time does not allow), perform linkage analysis with polymorphic DNA markers closely linked to CF.

5. James may have a new mutation on the X chromosome because Joe inherited the same X chromosome from his mother, and the deletion was present in neither Joe nor his mother. If this is the case, there is no risk for recurrence. Alternatively, the mother may be a mosaic, and the mosaicism includes her germline. In this case, there is a definite risk that the mutant X could be inherited by another son or passed to a carrier daughter. Approximately 5% to 15% of cases of this type appear to be due to maternal germline mosaicism. Thus the risk is half of this figure for her male offspring because the chance that a son will inherit the mutant X is $\frac{1}{2} \times 5\%$ to $15\% = 2.5\%$ to 7.5%.

6. For DMD, as a classic X-linked recessive disease that is lethal in males, one third of cases are predicted to be new mutations. The large size of the gene is likely to account for the high mutation rate at this locus (i.e., it is a large target for mutation). The ethnic origin of the patient will have no effect on either of these phenomena.

7. A DMD female like T.N. might have the disease because she carries a *DMD* gene mutation on the X chromosome inherited from her mother. T.N. could show clinical symptoms if her paternal X (carrying a normal allele at this locus) was subject to nonrandom inactivation in most or all cells. An alternative explanation would be that she has Turner syndrome and that her only X chromosome (inherited from her mother) carries a DMD gene mutation. A third explanation would be that she has a balanced X;autosome translocation that disrupts the *DMD* gene on the translocated X chromosome. Although her normal X chromosome carries a normal allele at the *DMD* locus, balanced X;autosome translocations show nonrandom inactivation of the structurally normal X due to secondary cell selection (see Chapter 6).

8. The limited number of amino acids that have been observed to substitute for glycine in collagen mutants reflects the nature of the genetic code. Single-nucleotide substitutions at the three positions of the glycine codons allow only a limited number of missense mutations (see Table 3-1).

9. Two bands of *G6PD* on electrophoresis of a red cell lysate indicate that the woman has a different *G6PD* allele on each X chromosome and that each allele is being expressed in her red cell population. However, no single cell expresses both alleles because of X inactivation. Males have only a single X chromosome and thus express only one *G6PD* allele. A female with two bands could have two normal alleles with different electrophoretic mobility, one normal allele and one mutant allele with different electrophoretic mobility, or two mutant alleles with different electrophoretic mobility. Because the two common deficiency alleles (A^- and B^-) migrate to the same position as the common normal-activity alleles (A and B), the woman is unlikely to have a common deficiency allele at both loci. Apart from that, one cannot say much about the possible pathological significance of the two bands without measuring the enzymatic activity. If one of the alleles has low activity, she would be at risk for hemolysis to the extent that the high-activity allele is inactivated as a result of X inactivation.

10. The box in Chapter 12 entitled "Mutant Enzymes and Disease: General Concepts" lists the possible causes of loss of multiple enzyme activities: they may share a cofactor whose synthesis or transport is defective; they may share a subunit encoded by the mutant gene; they may be processed by a common enzyme whose activity is critical to their becoming active; or they may normally be located in the same organelle, and a defect in biological processes of the organelle can affect all four enzymes. For example, they may not be imported normally into the organelle and may be degraded in the cytoplasm. Almost all enzymopathies are recessive (see text), and most genes are autosomal.

11. Haploinsufficiency. Thus, in some situations, the contributions of both alleles are required to provide a sufficient amount of protein to prevent disease. An example of haploinsufficiency is provided by heterozygous carriers of LDL receptor deficiency.

12. This situation is well illustrated by diseases due to mutations in mtDNA or in the nuclear genome that impair the function of the oxidative phosphorylation complex. Nearly all cells have mitochondria, and therefore oxidative phosphorylation occurs in nearly all cells, yet the phenotypes associated with defects in oxidative phosphorylation damage only a subset of organs, particularly the neuromuscular system with its high energy requirements.

13. One example is phenylketonuria, in which intellectual disability is the only significant pathological effect of deficiency of phenylalanine hydroxylase, which is found not in the brain but solely in the liver and kidneys, organs that are unaffected by this biochemical defect. Hypercholesterolemia due to deficiency of the LDL receptor is another example. Although the LDL receptor is found in many cell types, its hepatic deficiency is primarily responsible for the increase in LDL cholesterol levels in blood.

14. There are two defining characteristics of these alleles: the hex A activity that they encode is sufficiently reduced to allow their detection in screening assays (when the other allele is a common Tay-Sachs mutation with virtually no activity), and their hex A activity is nevertheless adequate to prevent the accumulation of the natural substrate (GM_2 ganglioside). There are probably only a few substitutions in the hex A protein that would reduce activity to only a modest degree (i.e., without crippling the protein more substantially). Thus the region of residues 247 to 249 appears to be relatively tolerant of substitutions, at least of Trp for Arg. Substitutions that more dramatically alter the charge or bulk of the residues at these positions may well be disease-causing alleles.

15. A gain-of-function mutation leads to an abnormal increase in the activities performed by the wild-type protein. Consequently, the overall integrity of the protein and each of its functional domains must remain intact despite the gain-of-function mutation. In addition, of course, the mutation must confer the gain of function. Consequently, the mutation must do nothing to disturb the normal properties of the protein and must enhance at least one of them, if not more, to confer the gain of function. Mutations other than missense mutations (e.g., deletions, insertions, premature stops) are almost uniformly highly disruptive to protein structure.

16. The presence of three common alleles for Tay-Sachs disease in the Ashkenazi population seems likely to be due either to a heterozygote advantage or to genetic drift (one form of which is the founder effect, as explained in Chapter 9). The high frequency of these alleles might also be due to gene flow, although the population of origin of the three common mutations is not apparent, making this explanation seem less likely (in contrast, say, to the evidence indicating that the most common PKU alleles in many populations around the world are of Celtic origin).

17. As with any genetically complex disorder, the other sources of genetic variation in Alzheimer disease (AD) may include the following: (1) additional AD loci, with lower effect sizes, that have not yet been identified; (2) synergistic effects between known AD genes (or between known genes and environmental risks) that may have a bigger effect than each of the genes or environments individually; (3) genes that harbor multiple very rare mutations of large effect, but which are undetectable by genome-wide association studies methods because each mutation occurs on a different SNP background.

18. The two forms of myotonic dystrophy are characterized by an expansion of a CUG trinucleotide in the RNA, which is thought to lead to an RNA-mediated pathogenesis. According to this model, the greatly enhanced number of CUG repeats binds an abnormally large fraction of RNA-binding proteins, including, for example, regulators of splicing, thereby depleting the cell of these critical proteins. One approach to therapy might be to prevent this binding. This might be achieved by introducing, perhaps by gene transfer (see Chapter 13), a viral vector that expresses a GAC trinucleotide repeat, which would bind to the CUG repeat sequences in the RNA and block the binding of the RNA-binding proteins to the expanded CUG repeats. Expression of too large a number of GAC repeat–containing molecules might itself have undesirable side effects, however, including binding to CUG codons that encode leucine, blocking their translation.

CHAPTER 13 *The Treatment of Genetic Disease*

1. Unresponsive patients may have mutations that drastically impair the synthesis of a functional gene product. Responsive patients may have mutations in the regulatory region of the gene. The effects of these mutations may be counteracted by the administration of IFN-γ. These mutations could be in the DNA-binding site that responds to the interferon stimulus or in some other regulatory element that participates in the response to IFN-γ. Alternatively, responsive patients may produce a defective cytochrome *b* polypeptide that retains a small degree of residual function. The production of more of this mutant protein, in response to IFN-γ, increases the oxidase activity slightly but significantly.

2. An enzyme that is normally intracellular can function extracellularly if the substrate is in equilibrium between the intracellular and extracellular fluids and if the product is either nonessential inside the cell or in a similar equilibrium state. Thus enzymes with substrates and products that do not fit these criteria would not be suitable for this strategy. This approach may not work for phenylalanine hydroxylase because of its need for tetrahydrobiopterin. However, if tetrahydrobiopterin could diffuse freely across the polyethylene glycol layer around the enzyme, the administration of tetrahydrobiopterin orally may suffice. This strategy would not work for storage diseases because the substrate of the enzyme is trapped inside the lysosome. In Lesch-Nyhan syndrome, the most important pathological process is in the brain, and the enzyme in the extracellular fluid would not be able to cross the blood-brain barrier. Tay-Sachs disease could not be treated in this way because of the nondiffusibility of the substrate from the lysosome.

3. Rhonda's mutations prevent the production of any LDL receptor. Thus the combination of a bile acid–binding resin and a drug (e.g., lovastatin) to inhibit cholesterol synthesis would have no effect on increasing

the synthesis of LDL receptors. The boy must have one or two mutant alleles that produce a receptor with some residual function, and the increased expression of these mutant receptors on the surface of the hepatocyte reduces the plasma LDL-bound cholesterol.

4. Unresponsive patients probably have alleles that do not make any protein, that decrease its cellular abundance in some other way (e.g., make an unstable protein), or that disrupt the conformation of the protein so extensively that its pyridoxal-phosphate binding site has no affinity for the cofactor, even at high concentrations. The answer to the second part of this question is less straightforward. The answer given here is based on the generalization that most patients with a rare autosomal recessive disease are likely to have two different alleles, which assumes that there is no mutational hot spot in the gene and that the patients are not descended from a "founder" and are not members of an ethnic group in which the disease has a high frequency. In this context, Tom is likely to have two alleles that are responsive; first cousins with the same recessive disease are likely to share only one allele, so that Allan is likely to have one responsive allele that he shares with Tom and another allele that is either unresponsive or that responds more poorly to the cofactor than Tom's other allele.

5. (a) You need both a promoter that will allow the synthesis of sufficient levels of the mRNA in the target tissue of choice and the phenylalanine hydroxylase cDNA. In reality, you also need a vector to deliver the "gene" into the cell, but this aspect of the problem has not been dealt with much in the text.

(b) A phenylalanine hydroxylase "gene" will probably be effective in any tissue that has a good blood supply for the delivery of phenylalanine and an adequate source of the cofactor of the enzyme, tetrahydrobiopterin. The promoter would have to be capable of driving transcription in the target tissue chosen for the treatment.

(c) Any mutation that severely reduces the abundance of the protein in the cell but has no effect on transcription. This group includes those mutations that impair translation or that render the protein highly unstable. The thalassemias include examples of all these types.

(d) Liver cells are capable of making tetrahydrobiopterin, whereas other cells may not be. The target cell for the gene transfer should thus be capable of making this cofactor; otherwise, the enzyme will not function unless the cofactor is administered in large amounts.

(e) Human phenylalanine hydroxylase probably exists as a homodimer or homotrimer. In patients whose alleles produce a mutant polypeptide

(versus none at all), these alleles may manifest a dominant negative effect on the product of the transferred gene. This effect could be overcome by making a gene construct that produces more of the normal phenylalanine hydroxylase protein (thus diluting out the effect of the mutant polypeptide) or by transferring the gene into a cell type that does not normally express phenylalanine hydroxylase and that would therefore not be subject to the dominant negative effect.

6. One must consider the kinds of mutations that decrease the abundance of a protein but that are associated with residual function. One class of such mutations are those that decrease the abundance of the mRNA but do not alter the protein sequence (i.e., each protein molecule produced has normal activity, but there are fewer molecules). Mutations of this type might include enhancer or promoter mutations, splice mutations, or others that destabilize the mRNA. In this case, one could consider strategies to increase expression from the normal allele and perhaps also the mutant allele, as is done with hereditary angioedema, in which danazol administration increases the expression of the product from both the wild-type and mutant alleles. A second class of such mutations are those within the coding sequence that destabilize the protein but still allow some residual function. Here, a strategy to increase the stability or the function of the mutant protein should be considered. For example, if the affected protein has a cofactor, one could administer increased amounts of the cofactor, provided such an approach would not have unacceptable side effects.

7. The drug can facilitate the skipping of a premature stop codon, allowing the translational apparatus to misincorporate an amino acid that has a codon comparable to that of the mutant termination codon. This treatment might allow the synthesis of a protein of normal size in both patients. In the responsive patient, the nonsense codon is located in a functionally "neutral" part of the protein, and the amino acid that is substituted in place of the nonsense codon allowed normal folding, processing, and function of the "corrected" protein. In the nonresponsive patient, however, the nonsense mutation is located in codon 117, which in wild-type CFTR is an Arg residue (see Fig. 12-15). This Arg residue contributes to the Cl^- channel of CFTR. In this unresponsive patient, the drug did not lead to incorporation of Arg at this position, and the Cl^- channel had defective conduction as a result.

CHAPTER 14 *Developmental Genetics and Birth Defects*

1. Before determination, an embryo can lose one or more cells, and the remaining cells can undergo

specification and ultimately develop into a complete embryo. Once cells are determined, however, mosaic development takes place—an embryonic tissue will follow its developmental program regardless of what happens elsewhere in the embryo. Regulative development means that an embryonic cell can be removed by blastomere biopsy for the purpose of preimplantation diagnosis without harming the rest of the embryo.

2. a–3, b–2, c–4, d–1.

3. a–4, b–3, c–5, d–2, e–1.

4. Mature T or B cells that have somatically rearranged their T-cell receptor or immunoglobulin loci would not be appropriate. This change is not epigenetic; it is a permanent alteration of the DNA sequence itself. Animals derived from a single nucleus from a mature T or B cell are incapable of mounting an appropriately broad immune response.

5. Consider issues of regulation versus simple capacity to carry out a biochemical reaction. Also, consider dominant negative effects of transcription factors, taking into account the frequent binary nature of such factors (DNA-binding and activation domains).

CHAPTER 15 *Cancer Genetics and Genomics*

1. Approximately 15% of unilateral retinoblastoma is actually the heritable form but affecting only one eye. You need family history; careful examination of both parents' retinas, looking for signs of a scar that could have been a spontaneously regressed retinoblastoma; cytogenetic analysis if the tumor is associated with other malformations and, very importantly, seek to find a mutation in the child in DNA from peripheral blood to see if it is a germline mutation. If the child carries a germline mutation, it is a heritable retinoblastoma, and the child is at risk for tumor in the other eye, in the pineal gland, and for sarcomas later in life, particularly associated with radiation therapy. With the mutation in hand, the parents can be tested to see if one or the other is a nonpenetrant carrier, and prenatal diagnosis could be offered for future pregnancies. Even if no mutation is found in a parent, given that a parent could be a germline mosaic with some increase in recurrence risk over simply the rate of new mutation, prenatal diagnosis can be offered using the mutation found in the affected child. If prenatal diagnosis is not used or if the fetus carries a mutation and the parents choose to allow the pregnancy to go to term, the newborn would need examination under anesthesia as soon as possible after birth and then at frequent intervals after that, with institution of therapy as soon as a tumor is found.

  If the child is not clearly heterozygous for a pathogenic mutation in his or her blood, it is still possible there is somatic mosaicism and the child is still at increased risk for tumor in the other eye or, more

generally, for sarcomas later in life. Sequencing of the tumor itself may show a mutation that could have been easily missed but could be specifically looked for at low levels using next-generation sequencing of the peripheral blood DNA.

2. Colorectal cancer seems to require a number of sequential mutations in several genes, a process that may take longer than one (in hereditary) or two (in sporadic) mutations in the retinoblastoma gene. Age dependence may also reflect the number, timing, and rate of cell divisions in colon cells and in retinoblasts.

3. A cell line with i(17q) is monosomic for 17p and trisomic for 17q. Thus formation of the isochromosome leads to loss of heterozygosity for genes on 17p. This may be particularly important because one or more tumor suppressor genes (such as *TP53*) are present on 17p; a "second hit" on the other copy of *TP53* would lead to complete loss of the p53 protein function. In addition, a number of proto-oncogenes map to 17q. It is possible that increasing their dosage may confer a growth advantage on cells containing the i(17q).

4. The chief concern is the need to reduce radiation exposure to the lowest possible level because of the risk for cancer in children with this genetic defect.

5. Although most (>95%) breast cancer appears to follow multifactorial inheritance, there are two known genes (*BRCA1* and *BRCA2*) in which mutations confer a substantial increased lifetime risk for breast cancer (fivefold to sevenfold) inherited in an autosomal dominant manner. Certain mutations in certain other genes, such as *ATM*, *BARD1*, *BRIP1*, *CDH1*, *CHEK2*, *PALB2*, *PTEN*, and *TP53*, among others, increase the lifetime breast cancer risk significantly over the 12% background risk in the population as well, but generally not to the extent seen with mutations in *BRCA1* or *BRCA2*. In the absence of a gene mutation in a hereditary breast cancer gene, the empirical risk figures are consistent with an overall multifactorial model with admixture of dominant forms of the disease with somewhat reduced lifetime penetrance; thus there is an approximately twofold increased risk for breast cancer in any woman with a first-degree female relative with breast cancer. Direct mutation detection could be performed if desired by the probands in Wanda's and Wilma's families, and if a mutation were found in *BRCA1*, *BRCA2*, or one of the other genes that cause substantial increased risk for breast cancer, a direct test for cancer risk could be offered to their relatives. More recently, one leading breast cancer researcher suggested that population-wide screening for disease-causing mutations in *BRCA1* or *BRCA2* should be initiated independent of family history, either restricted to high-risk ethnic groups or, more widely, to the entire population.

6. It is likely that many activated oncogenes, if inherited in the germline, would disrupt normal development and be incompatible with survival. There are a few exceptions, such as activating *RET* mutations in MEN2 and activating *MET* mutations in hereditary papillary kidney cancer. These activated oncogenes must have tissue-specific oncogenic effects without affecting development. Although it is not known why such specific types of cancers occur in individuals who inherit germline mutations in these oncogenes, one plausible theory is that other genes expressed in most of the tissues of the body counteract the effect of these activating mutations, thereby allowing normal development and suppressing oncogenic effects in most of the tissues in heterozygotes.

CHAPTER 16   *Risk Assessment and Genetic Counseling*

1. (a) Prior risk, $\frac{1}{4}$; posterior risk (two normal brothers), $\frac{1}{10}$.

   (b) Zero, unless the autosomal dominant form can show nonpenetrance, in which case there is a very small probability that Rosemary, Dorothy, and Elsie would all be nonpenetrant carriers. Without knowing the penetrance, we cannot calculate the exact risk that Elsie is heterozygous.

2. (a) Restrict your attention and conditional probability calculations to those women for whom we have conditional probability information that could alter their carrier risk. These individuals are the maternal grandmother (Lucy, see pedigree), who has an affected grandson and two unaffected grandsons, her daughter Molly, who has an affected son, and Martha, who has two unaffected sons. Maud does not contribute any additional information because she has no sons.

Write down an abbreviated pedigree (see illustration), and calculate all the possible prior probabilities. There are four scenarios:

In A, Nathan is a new mutation with probability μ.

In B, Molly is the new mutation—but because Lucy is *not* a carrier, Molly can only carry a new mutation and did not inherit the mutation; her prior probability is 2μ (*not* 4μ) because the new mutation could have occurred on either her paternal or her maternal X chromosome.

In C, Lucy is a carrier. As shown earlier in this chapter in the Box describing the calculation for the probability that any female is a carrier of an X-linked lethal disorder, Lucy's prior probability is 4μ. Molly inherits the mutant gene, but Martha does not, so the probability her two sons would be unaffected is essentially 1.

In D, Lucy is a carrier, as is Molly, but so is Martha, and yet she did not pass the mutant gene to her two sons.

(We do not consider all the other combinations of carrier states; because they are so unlikely, they can be ignored. For example, the possibility that Lucy is a mutation carrier, but Molly does not inherit a mutation from Lucy, and then Nathan is *another* new mutation is vanishingly small because the joint probability of such an event would require *two* new mutations and would contain μ^2 terms in the joint probability that are too small to contribute to the posterior probability.)

The conditional probabilities can then be calculated from these various joint probabilities.

For Molly, she is a carrier in situations B, C, and D, so her probability of being a carrier is $\frac{13}{21}$.

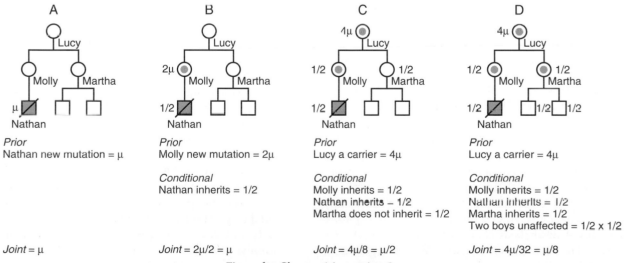

A	B	C	D
Prior	Prior	Prior	Prior
Nathan new mutation = μ	Molly new mutation = 2μ	Lucy a carrier = 4μ	Lucy a carrier = 4μ
	Conditional	Conditional	Conditional
	Nathan inherits = 1/2	Molly inherits = 1/2	Molly inherits = 1/2
		Nathan inherits = 1/2	Nathan inherits = 1/2
		Martha does not inherit = 1/2	Martha inherits = 1/2
			Two boys unaffected = 1/2 x 1/2
Joint = μ	Joint = $2\mu/2 = \mu$	Joint = $4\mu/8 = \mu/2$	Joint = $4\mu/32 = \mu/8$

Figure for Chapter 16, question 2.

Similarly, Molly's mother, Lucy, 5/21; Norma and Nancy, 13/42; Olive and Odette, 13/84; Martha, 1/21; Nora and Nellie, 1/42; Maud, 5/42; Naomi, 5/84.

(b) To have a 2% risk for having an affected son, a woman must have an 8% chance of being a carrier; thus Martha, Nora, and Nellie would not be obvious candidates for prenatal diagnosis by DNA analysis because their carrier risk is less than 8%.

3. $(\frac{1}{2})^{13}$ for 13 successive male births.

$(\frac{1}{2})^{13} \times 2$ for 13 consecutive births of the same sex. (The 2 arises because this is the chance of 13 consecutive male births *or* 13 consecutive female births, before any children are born.)

$\frac{1}{2}$. The probability of a boy is $\frac{1}{2}$ for each pregnancy, regardless of how many previous boys were born (assuming there is straightforward chromosome segregation, no abnormality in sexual development that would alter the underlying 50% to 50% segregation of the X and Y chromosomes during spermatogenesis, and no sex-specific lethal gene carried by a parent).

4. (a) Use the first equation, $I = \mu + \frac{1}{2}H$. To solve for H and substitute it for H in the second equation, $H = 2\mu + \frac{1}{2}H + If$. Solve for I, $I = 3\mu/(1 - f)$. Substituting 0.7 for f gives:
The incidence of affected males $I = 10\mu$.
The incidence of carrier females $H = 18\mu$.
Chance next son will be affected is $\frac{1}{2} \times 0.9 = 0.45$.

(b) Substitute $f = 0$ into the equations and you get $I = 3\mu$ and $H = 4\mu$.

(c) 0.147.

	Both Carriers	Not Both Carriers
Prior	$\frac{4}{9}$	$\frac{5}{9}$
Conditional (3 normal children)	$(\frac{3}{4})^3$	1
Joint	$\frac{4}{9} \times (\frac{3}{4})^3 = 0.19$	$\frac{5}{9} = 0.56$
Posterior	$0.19/(0.19 + 0.56) = \frac{1}{4}$	$0.56/0.75 = \approx \frac{3}{4}$

6. The child's prior probability of carrying a mutant myotonic dystrophy gene is $\frac{1}{2}$. If it is assumed that he has a $\frac{1}{2}$ chance of being asymptomatic, even if he carries the mutant gene, then his chance of carrying it and showing no symptoms is $\frac{1}{3}$. Testing is a complex issue. Many would think that testing an asymptomatic child for an incurable illness with adult onset is improper because the child should be allowed to make that decision for himself or herself (see Chapter 19).

7. (a) Yes; autosomal recessive, autosomal dominant (new mutation), X-linked recessive, and multifactorial inheritance and a chromosome disorder all need to be considered, as would nongenetic factors such as prenatal teratogen exposure and intrauterine infection. A careful physical examination and laboratory testing are required for a proper assessment of risks for this couple.

(b) This increases suspicion that the disorder is autosomal recessive, but the possibility of consanguinity does not prove autosomal recessive inheritance, and all other causes must still be investigated thoroughly.

(c) This fact certainly supports the likelihood that the problem has a genetic explanation. The pedigree pattern would be consistent with autosomal recessive inheritance only if the sister's husband were carrying the same defect (possible if he is from the same village, for example). An X-linked recessive pattern (particularly if the affected children are all boys) or a chromosome defect (e.g., the mothers of the affected children having balanced translocations with unbalanced karyotypes in the affected children) ought to be considered. The mother and her son should receive a genetic evaluation appropriate to the clinical findings, such as karyotyping and fragile X analysis.

8. The woman needs genetic counseling. She has a 50% risk for passing the mutant *NF1* gene to her offspring. The fact that she carries a new mutation only reduces the recurrence risk elsewhere in the family.

9. The seven scenarios are shown in the table.

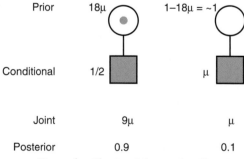

Prior	18μ	$1 - 18\mu = \sim 1$
Conditional	1/2	μ
Joint	9μ	μ
Posterior	0.9	0.1

Figure for Chapter 16, question 4.

5. (a) The prior risk that either Ira or Margie is a cystic fibrosis carrier is $\frac{2}{3}$; therefore the probability that both are carriers is $\frac{2}{3} \times \frac{2}{3} = \frac{4}{9}$.

(b) Their risk for having an affected child in any pregnancy is $\frac{1}{4} \times \frac{4}{9} = \frac{1}{9}$.

(c) Bayesian analysis is carried out.
Thus the chance that Ira's and Margie's next child will be affected is $\frac{1}{4} \times \frac{1}{4} = \frac{1}{16}$.

Conditional Probability Calculation

Situation	Female Carrier Status I-1	II-1	II-3	III-2	Joint Probabilities*
A	No	No	No	No	μ
B1	No	Yes (new mutation)	No	No	$\{2\mu \times \frac{1}{2}\} \times [1] \times [\frac{1}{2}] = \mu/2$
B2	No	Yes (new mutation)	No	Yes	$\{2\mu \times \frac{1}{2}\} \times [1] \times [\frac{1}{2} \times (\frac{1}{2})^2] = \mu/8$
C1	Yes	Yes	No	No	$\{4\mu \times \frac{1}{2} \times \frac{1}{2}\} \times [\frac{1}{2}] \times [\frac{1}{2}] = \mu/4$
C2	Yes	Yes	Yes	No	$\{4\mu \times \frac{1}{2} \times \frac{1}{2}\} \times [\frac{1}{2} \times (\frac{1}{2})^2] \times [\frac{1}{2}] = \mu/16$
C3	Yes	Yes	No	Yes	$\{4\mu \times \frac{1}{2} \times \frac{1}{2}\} \times [\frac{1}{2}] \times [\frac{1}{2} \times (\frac{1}{2})^2] = \mu/16$
C4	Yes	Yes	Yes	Yes	$\{4\mu \times \frac{1}{2} \times \frac{1}{2}\} \times [\frac{1}{2} \times (\frac{1}{2})^2] \times [\frac{1}{2} \times (\frac{1}{2})^2] = \mu/64$

*The joint probabilities for the core individuals in the pedigree (I-1, II-1, and III-1) are enclosed in curly brackets { }, and the probabilities for individuals II-3 and III-2 are shown in square brackets [ ]. See Figure 19-7.

The scenarios in which III-2 is a carrier are B2, C3, and C4. Her posterior probability of being a carrier is therefore

$$\frac{\mu/8 + \mu/16 + \mu/64}{\mu + \mu/2 + \mu/8 + \mu/4 + \mu/16 + \mu/16 + \mu/64}$$

10. Make II-1 the dummy consultand. Proceeding as if III-2 and her two unaffected children were not present, the risk that II-1 is a carrier is covered in situations B, C1, and C2 in the accompanying table, giving a posterior probability of

$$\frac{\mu + \mu/2 + \mu/8}{\mu + \mu + \mu/2 + \mu/8} = 13/21$$

Step One of Dummy Consultand Method

Situation	Female Carrier Status I-1	II-1	II-3	Joint Probabilities
A	No	No	No	$\{1 \times 1 \times \mu\} = \mu$
B	No	Yes (new mutation)	No	$\{1 \times 2\mu \times \frac{1}{2}\} = \mu$
C1	Yes	Yes	No	$\{4\mu \times \frac{1}{2} \times \frac{1}{2}\} \times [\frac{1}{2}] = \mu/2$
C2	Yes	Yes	Yes	$\{4\mu \times \frac{1}{2} \times \frac{1}{2}\} \times [\frac{1}{2} \times (\frac{1}{2})^2] = \mu/8$

One can then use this calculation as a starting point to determine that the prior probability that III-2 is a carrier, ignoring her two unaffected sons, is $\frac{1}{2}$ the probability that her mother, II-1, is a carrier

$= \frac{1}{2} \times \frac{13}{21} = \frac{13}{42}$; the prior probability that she is not a carrier is $1 - (\frac{13}{42}) = \frac{29}{42}$ (see Table for step two). Then we use another round of conditional probability to see what effect the two unaffected sons of III-2 have, to determine the posterior risk that III-2 is a carrier.

Step Two of Dummy Consultand Method

	III-2 Carrier	III-2 Not a Carrier
Prior probability	$\frac{13}{42}$	$\frac{29}{42}$
Conditional (2 unaffected sons)	$(\frac{1}{2})^2$	1
Joint probability	$\frac{13}{168}$	$\frac{29}{42}$
Posterior probability	$\frac{13}{129}$	$\frac{13}{116}$

Thus the posterior probability that III-2 is a carrier using the dummy consultand method, given she has two unaffected sons, is $\frac{13}{129}$, the same as when we used the approach in Table 16-3. So far, so good.

Some consider the dummy consultand method to be faster than the comprehensive approach of drawing out all the scenarios, but it is also easy to misapply, resulting in miscalculation. For example, the dummy consultand method, as outlined here, gives the correct result only for the consultand III-2 herself and not necessarily for other females in the pedigree. For example, the $\frac{13}{21}$ (62%) carrier risk for individual II-1, calculated in the first step of the two-step dummy consultand method, which ignores the information for individual III-2, is actually incorrect. The correct result for II-1 is the posterior probability of all the situations except A in the conditional probability calculation table, which equals $\frac{65}{129}$ (50%). (We thank Susan Hodge from Columbia University for pointing out this problem with the dummy consultand method.)

CHAPTER 17 *Prenatal Diagnosis and Screening*

1. c, e, f, i and j, d, h, g, b, i (and, in part, j), and a.
2. The child can have only Down syndrome or monosomy 21, which is almost always lethal. Thus they should receive counseling and consider other alternatives for having children.
3. No, not necessarily; the problem could be maternal cell contamination.
4. The level of maternal serum alpha-fetoprotein (MSAFP) is typically elevated when the fetus has an open neural tube defect. The levels of MSAFP and unconjugated estriol are usually reduced and the human chorionic gonadotropin level is usually elevated when the fetus has Down syndrome.
5. (a) Approximately 15% (see Table 5-2).
   (b) At least 50% are chromosomally abnormal.

(c) Prenatal diagnosis or karyotyping of the parents is usually not indicated after a single miscarriage; most practitioners would offer parental chromosome analysis and prenatal diagnosis after three spontaneous unexplained miscarriages (although some practitioners suggest offering testing after only two), provided there are no other indications.

6. (a) Yes. Given that her CPK levels indicate she is a carrier of DMD and she had an affected brother, she must have inherited the mutation from her mother because her brother could not have received the mutation from her father. The phase can be determined from analysis of her father, who has to have transmitted a normal X chromosome to his daughter, the consultand.

(b) Yes. A male fetus that receives her father's allele linked to the *DMD* locus will be unaffected. If a male fetus receives her mother's allele linked to *DMD*, it will be affected. This, of course, assumes no recombination between the microsatellite marker and the mutation in the DMD gene in the transmitted chromosome.

(c) First, the consultand should have DMD gene testing. The most common mutations in DMD are deletions (and less commonly duplications), although point mutations are also possible (see Chapter 12). The advent of powerful new sequencing technologies and new methods for determining deletions and duplications, such as the multiplex ligation-dependent polymerase assay (MLPA) or copy number measurements by comparative genome hybridization, has made carrier detection for DMD much more sensitive than in the past, when it was limited by the very large size of the gene and difficulties determining a partial gene deletion in a female with two copies of the gene.

7. Question for discussion. Consider issues of sensitivity and specificity of each of the different forms of testing, the psychosocial issues of prenatal diagnosis and termination at different stages of pregnancy, and risk for complications of the two invasive methods.

8. 600,000 women, 1000 pregnancies affected.

Assume everyone is willing to participate in the sequential screening. Of 1000 true positives, first-trimester screening will identify 840 high-risk "positives" (84%) who undergo CVS; 160 are low risk, and they get a second-trimester screen. Of these 160, 130 (81%) are positive and undergo amniocentesis and are found to have an affected fetus; 30 affected pregnancies are missed.

Of the 599,000 unaffected false positives in the first-trimester screen, 29,950 positives need CVS. The remaining 569,050 are low risk and get a second-trimester screen. You get 28,452 positives in the

second-trimester screen who undergo amniocentesis; the remaining 540,598 unaffected pregnancies are reassured.

In summary, with sequential screening, you will detect 970 of the 1000 (97%) and will miss 30 (3%). You will do 970 invasive tests in affected pregnancies while also doing 29,950 + 28,452 = 58,402 invasive tests in unaffected pregnancies.

Thus you will do 62 invasive tests to detect each affected pregnancy.

This compares to the situation if you just offered invasive testing to everyone. Depending on the uptake, you will miss some fraction of affected fetuses. If the uptake were 97% (very, very unlikely for an invasive test), you would end up doing 582,000 invasive tests to find 970 affected pregnancies. You would miss the same 30 affected pregnancies as with the sequential testing but would have to do a 10 times greater number of invasive tests to achieve the same detection rate.

CHAPTER 18 *Application of Genomics to Medicine and Personalized Health Care*

1.

Idiopathic Cerebral Vein Thrombosis and Factor V Leiden

	iCVT		
Genotype	Affected	Unaffected	Total
Homozygous FVL	1	624	625
Heterozygous FVL	2	48,748	48,750
Wild-type	15	950,610	950,625
Total	18	999,982	1,000,000

FVL, Factor V Leiden; iCVT, idiopathic cerebral vein thrombosis.

You would expect 625 FVL homozygotes and 48,750 heterozygotes.

Relative risk for iCVT in FVL homozygotes = $(1/625)/(15/950,625) = \approx 101$.

Relative risk for iCVT in FVL heterozygotes = $(2/48,750)/(15/950,625) = \approx 3$.

Sensitivity of testing positive for either one or two FVL alleles = $3/18 = 17\%$.

Positive predictive value for homozygotes = $1/625 = 0.16\%$.

Positive predictive value for heterozygotes = $2/48,748 = 0.004\%$.

Although the relative risks are elevated with FVL, particularly when the individual is homozygous for the allele, the disorder itself is very rare and thus the PPV is low.

This example highlights the concept that a relative risk is always a comparison to people who do not carry a particular marker whereas a PPV is the actual (or absolute) risk for someone who carries the marker.

2.

Deep Venous Thrombosis in the Legs, Oral Contraceptive Use, and Factor V Leiden

Genotype	DVT		
	Affected	Unaffected	Total
Homozygous FVL	3	59	62
Heterozygous FVL	58	4,825	4,875
Wild-type	39	95,025	95,063
Total	100	99,000	100,000

DVT, Deep venous thrombosis; FVL, factor V Leiden.

You would expect ≈62 FVL homozygotes and 4875 heterozygotes.

Relative risk for DVT in FVL homozygotes taking oral contraceptives (OCs) = ≈118.

Relative risk for DVT in FVL heterozygotes taking OCs = ≈30.

Sensitivity of testing positive for either one or two FVL alleles = 62%.

Positive predictive value for homozygotes = 3/62 = ≈5%.

Positive predictive value for heterozygotes = 58/4875 = 1.2%.

Note that DVT is more common than the example of idiopathic cerebral vein thrombosis given in question 1, whereas the relative risks for homozygotes are of similar magnitude (101 vs. 118); thus the PPV of testing homozygotes is accordingly much higher but is still only 5%.

3. You should first explain to the parents that the test is a routine one performed on all newborns and that the results, as in many screening tests, are often falsely positive. The parents should also be told that the test result may be a true positive, and if it is, a more accurate and definitive test needs to be done before we will know what the child's condition really is and what treatment will be required. The child should be brought in as soon as possible for examination and the appropriate samples obtained to confirm the elevated phenylalanine level to determine if the child has classic or variant PKU or hyperphenylalaninemia, and to test for abnormalities in tetrabiopterin metabolism. Once a diagnosis is made, dietary phenylalanine restriction is instituted to bring blood phenylalanine levels down below the range considered toxic (>300 μmol/L). The child must then be observed for dietary adjustments to be made to keep the blood phenylalanine levels under control.

4. Questions to consider in formulating your response are as follows:

Consider the benefits of preventing disease by knowing a newborn's genotype at the β-globin locus. Can knowing the genotype help prevent pneumococcal sepsis or other complications of sickle cell anemia?

Distinguish between *SS* homozygotes and *AS* heterozygotes. What harm might accrue from the identification of *AS* individuals by newborn screening? What does identification of a newborn with *SS* or *AS* tell you about the genotypes of the parents and genetic risks for future offspring to the parents?

5.

Carbamazepine-Induced TEN or SJS

HLA-B*1502 allele present	TEN or SJS		
	Affected	Unaffected	Total
+	44	3	47
−	0	98	98
Total	44	101	145

SJS, Stevens-Johnson syndrome; TEN, toxic epidermal necrolysis.

Sensitivity = 44/44 = 100%.

Specificity = 98/101 = 97%.

Positive predictive value = 44/47 = 94%.

6. Terfenadine blocks the HERG cardiac-specific potassium channel encoded by *KCNH2*.

Various alleles in the coding portion of *KCNH2* are associated with prolongation of the QT interval on electrocardiography, which is associated with sudden death.

Terfenadine is metabolized by the cytochrome P450 enzyme CYP3A4, which has numerous alleles associated with reduced metabolism.

Itraconazole is an antifungal that blocks CYP3A4 cytochrome and increases serum levels of drugs metabolized by this cytochrome.

Grapefruit juice contains a series of naturally occurring compounds, furanocoumarins, that are thought to interfere with CYP3A4 metabolism of numerous drugs, including terfenadine.

Caffeine is unlikely to be involved in that caffeine has very little effect on CYP3A4, which has only a minor role in caffeine metabolism. Most caffeine is metabolized by CYP1A2.

CHAPTER 19 *Ethical and Social Issues in Genetics and Genomics*

1. The first consideration is testing the boy for an incurable disease. Because the boy has symptoms and the family is seeking a diagnosis, this is *not* the same situation as if an asymptomatic child is being considered for predictive testing for an adult-onset disorder, such as classic myotonic dystrophy. However, because Huntington disease in a child is overwhelmingly the result of an expansion of an enlarged triplet repeat in one of the parents, usually the father, finding a markedly enlarged expansion in the child will automatically raise the possibility that one of the parents, probably the father, is a carrier of a repeat that is

enlarged enough to cause adult-onset Huntington disease in him. Thus, by testing the child, one might inadvertently discover something about a parent's risk. Testing should therefore be carried out with informed consent from the parents. Other issues: If one of the parents carries the *HD* gene, what do you do about testing the asymptomatic older sib?

2. To justify screening, one must show that the good that comes from screening, the beneficence of the testing, outweighs the harm. Consider the issue of autonomy because implicit in the act of informing families that their child has a chromosomal abnormality is the fact that the child cannot decide whether she or he wants such testing later in life. How predictive is the test? Are we making a diagnosis of a possible chronic disability that may or may not develop or, if it does, may vary in severity and for which there is little the parents can do? One might ask whether there are effective interventions for the abnormalities in learning and behavior that occur in some individuals with sex chromosome anomalies. In fact, there is evidence that informing the parents and providing educational and psychological intervention before major problems arise proves beneficial. There is also, however, the concern about "the self-fulfilling prophecy," that telling parents there might be a problem increases the risk that there will be a problem by altering parental attitudes toward the child. There is a large amount of literature on this subject that is worth investigating and reading. See, for instance:

Bender BG, Harmon RJ, Linden MG, Robinson A: Psychosocial adaptation of 39 adolescents with sex chromosome abnormalities. *Pediatrics* 96(pt 1): 302-308, 1995.

Puck MH: Some considerations bearing on the doctrine of self-fulfilling prophecy in sex chromosome aneuploidy. *Am J Med Genet* 9:129-137, 1981.

Robinson A, Bender BG, Borelli JB, et al: Sex chromosomal aneuploidy: prospective and longitudinal studies. *Birth Defects Orig Artic Ser* 22:23-71, 1986.

3. You must consider the extent to which withholding information constitutes "a serious threat to another person's health or safety." In these different disorders, consider how serious the threat is and whether there is any effective intervention if the relative were informed of his or her risk.

4. Give your rationale for picking the disorders you choose. Consider such factors as how great a threat to is health the disorder, whether the disorder is likely to remain undiscovered and a potential cause of serious illness if not found before symptoms develop by sequencing, how predictive is finding a gene mutation for the disease, and how effective, how invasive, and how risky are any interventions.

An initial (and somewhat controversial) list of 56 such disorders as proposed by a committee of the American College of Medical Genetics and Genomics can be found in:

Green RC, Berg JS, Grody WW, et al: ACMG recommendations for reporting of incidental findings in clinical exome and genome sequencing. *Genet Med* 15:565-574, 2013.

A framework for considering potentially pathogenic sequence variants detected by whole-exome or whole-genome sequencing can be found in:

Richards S, Aziz N, Bale S, et al: Standards and guidelines for the interpretation of sequence variants: a joint consensus recommendation of the American College of Medical Genetics and Genomics and the Association for Molecular Pathology. *Genet Med* doi:10.1038/gim.2015.30, 2015.

Index